Springer Undergraduate Mathematics Series

The Springer Undergraduate Mathematics Series (SUMS) is a series designed for undergraduates in mathematics and the sciences worldwide. From core foundational material to final year topics, SUMS books take a fresh and modern approach. Textual explanations are supported by a wealth of examples, problems and fully-worked solutions, with particular attention paid to universal areas of difficulty. These practical and concise texts are designed for a one- or two-semester course but the self-study approach makes them ideal for independent use.

More information about this series at http://www.springer.com/series/3423

Satish Shirali · Harkrishan Lal Vasudeva

Measure and Integration

Springer

Satish Shirali
Formerly of Panjab University
Chandigarh, India

University of Bahrain
Zallaq, Bahrain

Harkrishan Lal Vasudeva
Formerly of Panjab University
Chandigarh, India

IISER Mohali
Mohali, India

ISSN 1615-2085 ISSN 2197-4144 (electronic)
Springer Undergraduate Mathematics Series
ISBN 978-3-030-18746-0 ISBN 978-3-030-18747-7 (eBook)
https://doi.org/10.1007/978-3-030-18747-7

Mathematics Subject Classification (2010): 28-01, 28A05, 28A10, 28A12, 28A20, 28A25, 28A35

This Springer imprint is published by the registered company Springer Nature Switzerland AG
The registered company address is: Gewerbestrasse 11, 6330 Cham, Switzerland

Preface

A good part of the subject matter of the present book is concerned with Measure Theory and Integration. It provides a substantial part of the necessary background for several branches of modern mathematics, applied mathematics and probability.

The Riemann theory of integration for bounded functions $f:[a, b] \to \mathbb{R}$, where $[a, b]$ is a bounded closed interval of $\mathbb{R}$, serves all the needs of elementary calculus. However, it has some limitations that are inherent in its definition:

(i) It is defined for bounded functions only. The function $x^{-1/2}$ is not Riemann integrable on $[0,1]$ because it is not bounded on the interval. However, it is integrable in an extended sense and the value of the integral is 2.

(ii) It is defined on bounded intervals only. On the interval $[1,\infty)$, which is not bounded, the function $1/x^2$ is not Riemann integrable; however, it is well known that the value of its integral over $[1,\infty)$ is 1 in an extended sense.

On the other hand, the following limitations cannot be overcome by using the same kind of extension:

(iii) Functions having finite range, even if defined on a bounded interval, need not be Riemann integrable. For example, the Dirichlet function $f(x) = 1$ if x is rational and 0 otherwise is not Riemann integrable. Indeed, $\overline{\int_0^1} f(x)dx = 1$ but $\underline{\int_0^1} f(x)dx = 0$.

(iv) The interchange of the notions of limit and integral is not always possible in the Riemann theory of integration. The following example will substantiate the foregoing statement:

Let $\{r_1,r_2,\ldots\}$ be an enumeration of the rationals in $[0,1]$. Define for every $n \in \mathbb{N}$,

$$f_n(x) = \begin{cases} 1 & \text{if } x \in \{r_1, r_2, \cdots, r_n\} \\ 0 & \text{otherwise.} \end{cases}$$

Then $\lim_{n\to\infty} f_n(x) = f(x)$, the Dirichlet function of (iii) above. It is not even Riemann integrable, leave alone the possibility that

$$\lim_{n\to\infty} \int_0^1 f_n(x)dx = \int_0^1 f(x)dx.$$

In modern mathematics the need was felt for a theory of integration which is free of the drawbacks (iii) and (iv) above. Mathematicians of the time made several attempts to enlarge the class of functions which could be brought under a satisfactory theory of integration. É. Borel (1898) had presented a theory of measure for a large class of sets that included unions of countably many disjoint intervals and differences of such sets. The decisive advance was made by H. Lebesgue (1902), who formulated a meaningful theory of integration based on an improvement and generalisation of the work of Borel. He generalised the concept of length of an interval to the measure of an even larger class of sets of real numbers than Borel did. The definition of the Lebesgue integral permits a much more general class of functions, including certain unbounded functions, as integrands and replaces the interval $[a,b]$ by more general kinds of sets called "measurable" sets. It also facilitates the interchange of the notions of limit and integral. More specifically, the equality

$$\lim_{n\to\infty} \int_X f_n dm = \int_X f\, dm,$$

where $\lim_{n\to\infty} f_n(x) = f(x)$, X is a measurable subset of $\mathbb{R}$ and "dm" indicates Lebesgue integration, holds under mild conditions on the functions f_n and f.

While the Riemann theory begins by partitioning the domain of the function sought to be integrated, Lebesgue's approach starts by partitioning the range, or more precisely, partitioning an interval containing the range. It then becomes necessary to have some sort of length for the inverse image of each subinterval of the partition, although the inverse image may not be an interval. In other words, one needs to assign a size, or "measure", to a larger class of sets than just the intervals, and that too, in a manner that is consistent with expectations about the measure of a union of disjoint sets.

The procedure for developing the theory of measure and integration follows a well established procedure: outer measure and measure, measurable functions, integrable functions, convergence theorems for sequences of integrable functions are introduced successively. This is carried out in Chaps. 2 and 3. The Lebesgue theory of integration developed in the latter enables us to define certain spaces such as L^p spaces $1 \le p \le \infty$, which are Banach spaces. These were introduced by F. Riesz. The primary purpose of Sect. 3.3 is the construction of these spaces and the derivation of their properties. The fact that the vital property of completeness can be achieved is a direct consequence of the "congenial" behavior of the integral when limits are taken. The class of square integrable functions is obtained for $p = 2$. In this case $p = q = 2$ and Hölder's Inequality takes the form

$$\int_X |fg| d\mu \leq \left(\int_X f^2 d\mu\right)^{1/2} \left(\int_X g^2 d\mu\right)^{1/2}$$

and is referred to as the Cauchy–Schwarz Inequality, where "$d\mu$" indicates integration with respect to a general measure μ. This inequality leads to the introduction of inner product spaces and orthogonality of two elements of the space L^2. In Sect. 3.4, dense subsets of L^p, $1 \leq p < \infty$, for the case when μ is Lebesgue measure are pointed out.

In Analysis, it is often desirable and fruitful to approximate objects by simpler objects, e.g., irrational numbers are approximated by rational numbers and continuous functions by polynomials. We approximate measurable sets by open sets, measurable functions by simple functions as well as by continuous functions. That every convergent sequence of measurable functions is nearly uniformly convergent is proved. It is shown that each of the classes of simple functions, continuous functions of compact support and step functions is dense in the space of L^p-functions. The class of square integrable functions and the concept of mean square convergence of a sequence of functions of this class are included in Chap. 3.

An arbitrary integrable 2π-periodic function can be expressed formally as

$$f(x) = \frac{1}{2}a_0 + \sum_{n=1}^{\infty} (a_n \cos nx + b_n \sin nx),$$

with coefficients a_n and b_n given by the formulae

$$a_n = \frac{1}{\pi}\int_{-\pi}^{\pi} f(x) \cos nx \, dx, \; n = 0, 1, 2, \ldots$$

and

$$b_n = \frac{1}{\pi}\int_{-\pi}^{\pi} f(x) \sin nx \, dx, \; n = 1, 2, 3, \ldots.$$

The above series is called the "Fourier series" of the function f. In Chap. 4, the representation of a square integrable 2π-periodic function by Fourier series and its convergence in the mean are studied. The important result connecting measure theory and Fourier series, namely the Riesz–Fischer Theorem, constitutes a part of the chapter.

The pointwise convergence (Dirichlet's Theorem), L^2 convergence and Cesàro summability of Fourier series (Fejér's Theorem) are proved under appropriate hypotheses. The Weierstrass Polynomial Approximation Theorem is derived as an application of Fejér's Theorem. The chapter concludes with Fejér's example of a continuous function whose Fourier series diverges at 0.

In Chap. 5, we study monotone functions, functions of bounded variation and absolutely continuous functions. As is well known, a continuous function need not be differentiable. K. Weierstrass (1861) constructed a continuous function having no derivative at any point whatsoever. B. Bolzano (1831) knew the result but was debarred from publishing by an imperial decree. However, if we additionally suppose that the continuous function in question is monotone, it possesses a derivative almost everywhere. This result was proved by H. Lebesgue in 1904. W. H. Young (1911) gave a proof of the Lebesgue Theorem without the continuity assumption. These and other proofs of the seminal result, including the one by L. A. Rubel (1963), are provided. Equally important and related to the class of monotone functions is the class of functions of bounded variation. The important properties of the two classes and the relationship between them are obtained. After proving some of the important properties of the special class of 'absolutely continuous' functions, we provide a characterisation of an absolutely continuous function f on a domain $[a,b]$:

$$f(x) = c + \int_{[a,x]} g,$$

where c is a constant and g is in $L^1[a,b]$. The relationship between rectifiable curves and functions of bounded variation is indicated in the problems. Also included is the change of variables formula in Lebesgue integration. Signed measures, which are a useful generalisation of measures, the Hahn Decomposition Theorem and the Jordan Decomposition Theorem constitute Sect. 5.9. In Sect. 5.10, the Radon–Nikodým Theorem and Lebesgue decomposition of a measure into an absolutely continuous part and singular part are discussed. The connection between bounded increasing left continuous real-valued functions on the real line and finite measures on Borel subsets of the real line are discussed in Sect. 5.11.

Chapter 6 begins with a quick overview of real normed linear spaces and bounded linear functionals on them, the focus being on the L^p spaces. Hölder's Inequality suggests a way of defining bounded linear functionals on these spaces. The Riesz Representation Theorem, which asserts in the σ-finite case that all bounded linear functionals on L^p are indeed as suggested by Hölder's Inequality when $1 \leq p < \infty$, is proved for Lebesgue measure on [0,1]. A proof of the Riesz Representation Theorem for a general finite measure is also provided and is based on the Radon–Nikodým Theorem. Up to this point in the book, various types of convergence of sequences of measurable functions have been encountered, such as pointwise, almost everywhere, uniform and in the L^p norm. Other modes of convergence of sequences of measurable functions that also play a significant role in Analysis are almost uniform convergence and convergence in measure. The interplay between them is thoroughly dwelt upon in Chap. 6.

Let $(X, \mathcal{F}, \mu)$ and $(Y, \mathcal{G}, \nu)$ be two σ-finite measure spaces and $(X \times Y, \mathcal{F} \times \mathcal{G}, \mu \times \nu)$ be the product measure space, where $\mathcal{F} \times \mathcal{G}$ denotes the product σ-algebra generated by the measurable rectangles $A \times B$, $A \in \mathcal{F}$, $B \in \mathcal{G}$ and $\mu \times \nu$ the product measure. If f is an $\mathcal{F} \times \mathcal{G}$-measurable function defined on $X \times Y$, the integral of f on $X \times Y$ is denoted by

$$\int_{X \times Y} f(x,y)d(\mu \times \nu)(x,y),$$

or some variant thereof, and is called a double integral. We may integrate with respect to y and obtain a function of x, namely,

$$\int_Y f(x,y)d\nu(y).$$

Integrating it with respect to x, we obtain

$$\int_X d\mu(x) \int_Y f(x,y)d\nu(y).$$

In Chap. 7, we seek conditions which ensure the following equality:

$$\int_X d\mu(x) \int_Y f(x,y)d\nu(y) = \int_{X \times Y} f(x,y)d(\mu \times \nu)(x,y) = \int_Y d\nu(y) \int_X f(x,y)d\mu(x).$$

The application of the above equality to convergence of Fourier series in L^p spaces is indicated.

A reader with a modest background in Mathematical Analysis in one variable, including rudiments of Metric Spaces, which are included in Chap. 1, will find the material covered in the book well within reach. The treatment of the Cantor set and the Cantor function, though not new, is well worth the reader's attention.

The authors are grateful to the referees for useful suggestions.

Gurugram, India
Chandigarh, India
March 2019

Satish Shirali
Harkrishan Lal Vasudeva

Contents

Chapter 1
Preliminaries

We shall find it convenient to use logical symbols such as $\forall$, $\exists$, $\ni$, $\Rightarrow$ and $\Leftrightarrow$. These are listed below with their meanings. A brief summary of Set Algebra, Functions, Elementary Real Analysis, which will be used throughout this book, is included in this chapter. Our purpose is descriptive and no attempt has been made to give proofs of the results stated, except in Sect. 1.8. The reader is expected to be familiar with the material up to Sect. 1.7.

The words 'set', 'class', 'collection' and 'family' are regarded as synonymous and we do not define these terms.

1.1 Sets and Functions

The following commonly used symbols will be employed in this book:

$\forall$ means "for all" or "for every"
$\exists$ means "there exists"
$\ni$ means "such that"
$\Rightarrow$ means "implies that" or simply "implies"
$\Leftrightarrow$ means "if and only if".

The concept of *set* plays an important role in every branch of modern mathematics. Although it is easy and natural to define a set as a collection of objects, it has been shown that this definition leads to a contradiction. The notion of set is, therefore, left undefined and a set is described by simply listing its elements or by its properties or properties of its elements. Thus $\{x_1, x_2, \ldots, x_n\}$ is the set whose elements are $x_1, x_2, \ldots, x_n$; and $\{x\}$ is the set whose only element is x. If X is the set of all elements x such that some property $P(x)$ is true, we shall write

$$X = \{x : P(x)\}.$$

S. Shirali and H. L. Vasudeva, *Measure and Integration*,
Springer Undergraduate Mathematics Series,
https://doi.org/10.1007/978-3-030-18747-7_1

The symbol $\varnothing$ denotes the empty set.

We write $x \in X$ if x is a member of the set X; otherwise $x \notin X$. If Y is a *subset* of X, that is, if $x \in Y$ implies $x \in X$, we write $Y \subseteq X$. If $Y \subseteq X$ and $X \subseteq Y$, then $X = Y$. If $Y \subseteq X$ and $Y \neq X$, then Y is *proper subset* of X. Observe that $\varnothing \subseteq X$ for every set X.

Given a set X, the collection of all its subsets is called the *power set* of X and will be denoted by $\mathcal{P}(X)$.

We list below the standard notations for the most important sets of numbers:

$\mathbb{N}$ the set of all natural numbers
$\mathbb{Z}$ the set of all integers
$\mathbb{Q}$ the set of all rational numbers
$\mathbb{R}$ the set of all real numbers
$\mathbb{C}$ the set of all complex numbers.

Given two sets X and Y, we can form the following new sets from them:

$$X \cup Y = \{x : x \in X \text{ or } x \in Y\},$$
$$X \cap Y = \{x : x \in X \text{ and } x \in Y\}.$$

$X \cup Y$ and $X \cap Y$ are the *union* and *intersection* respectively of X and Y. More generally, given a nonempty family $\mathcal{F}$ of sets, we define their union to be

$$\bigcup_{F \in \mathcal{F}} F = \{x : x \in F \text{ for at least one } F \in \mathcal{F}\}.$$

It is also denoted by $\bigcup \mathcal{F}$ or $\cup \mathcal{F}$. Given a nonempty family $\mathcal{F}$ of sets, we define their intersection to be

$$\bigcap_{F \in \mathcal{F}} F = \{x : x \in F \text{ for every } F \in \mathcal{F}\}.$$

It is also denoted by $\bigcap \mathcal{F}$ or $\cap \mathcal{F}$.

The symbol

$$f : X \to Y$$

means that f is a *function* (or *mapping* or *map*) from the nonempty set X into the set Y; that is, f assigns to each $x \in X$ an element $f(x) \in Y$. The elements assigned to members of X by f are often called values of f. If $A \subseteq X$ and $B \subseteq Y$, the *image* of A and *inverse image* of B are respectively

$$f(A) = \{f(x) : x \in A\},$$
$$f^{-1}(B) = \{x : f(x) \in B\}.$$

Note that $f^{-1}(B)$ may be empty even when $B \neq \varnothing$.

The *domain* of f is X and the *range* is $f(X)$; the *range space* is Y. If $f(X) = Y$, the function f is said to map X *onto* Y (or the function is said to be *surjective*). We write

$f^{-1}(y)$ instead of $f^{-1}(\{y\})$ for every $y \in Y$. When $f^{-1}(y)$ consists of at most one element for each $y \in Y$, f is said to be *one-to-one* (or *injective*). If f is one-to-one, then f^{-1}, which is actually a map from $\mathcal{P}(Y)$ to $\mathcal{P}(X)$, gives rise to a function with domain $f(X)$ and range X. The latter is also denoted by f^{-1}, as it does not cause any confusion. A function which is both injective and surjective is said to be *bijective* and is often spoken of as a *bijection* or *one-to-one correspondence*. If f is bijective, then f^{-1} is a function with domain Y and range X, and is called the *inverse of f*. A map is said to be *invertible* if it has an inverse. It is obvious that being invertible is equivalent to being bijective and that the inverse is unique if it exists.

There can never exist a bijection between any set X and its power set $\mathcal{P}(X)$.

It is sometimes necessary to consider a function f only on a nonempty subset S of its domain X. Technically that makes it a different function and it is called the *restriction* of f to S. Introducing a new symbol to denote a restriction can clutter the notation and we shall avoid it as far as possible.

Let $g: U \to V$ and $f: X \to Y$ be maps, where X has a nonempty intersection with the range $g(U)$. Then the inverse image $Z = g^{-1}(X \cap g(U)) \subseteq U$ is nonempty and the function $f \circ g : Z \to Y$ such that $(f \circ g)(z) = f(g(z))$ is called the *composition* of f and g. For most theoretical purposes, it is sufficient to work with the case $X \supseteq g(U)$, because this ensures that $X \cap g(U) = g(U)$ and hence that $Z = U$.

Among the important examples of functions are those defined on subsets of the set $\mathbb{N}$ of natural numbers.

By a *finite sequence* of n terms we understand a function f whose domain is the set $\{1, 2, \ldots, n\}$. The range of f is the set $\{f(1), \ldots, f(n)\}$. The function value $a_k = f(k)$ is called the kth *term* of the sequence. By an *infinite sequence* we shall mean a function f whose domain is $\mathbb{N}$. The range of f is the set $\{f(1), f(2), \ldots\}$, which is also written as $\{f(k): k \geq 1\}$. Here again, the function value $a_k = f(k)$ is called the kth *term* of the sequence. For brevity, we shall use the notation $\{f(k)\}_{k \geq 1}$ or $\{f(k)\}_{k=1}^{\infty}$, or simply $\{f(k)\}$, to denote the infinite sequence whose kth term is $f(k)$. It is sometimes more convenient to denote the sequence by $\{a_k\}_{k \geq 1}$ or $\{a_k\}_{k=1}^{\infty}$, or simply $\{a_k\}$, especially when a letter such as f may not have been introduced to denote the function.

Sometimes the range space X of a sequence is specified by speaking of a sequence *in X*.

Ordinarily, a sequence will be understood to be an infinite sequence unless the context indicates the contrary. A sequence of n terms may be denoted by $\{f(k)\}_{k=1}^{n}$ or by $\{a_k\}_{k=1}^{n}$.

Given a sequence $\{a_k\}_{k=1}^{\infty}$, consider a sequence $\{k_n\}$ of positive integers such that $k_1 < k_2 < k_3 < \cdots$. Then the sequence $\{a_{k_n}\}$ is called a *subsequence* of $\{a_k\}_{k=1}^{\infty}$.

When we have a map from a set Λ into the power set of some set X, we speak of an *indexed family* or *indexed collection* of subsets of X. Here Λ is often referred to as the *indexing set*. One can create an indexed family from any nonempty family $\mathcal{F}$ by taking $\mathcal{F}$ itself as the indexing set and selecting the map to be the identity map. This often makes it possible to carry over results and definitions about an indexed family of sets with a *general type* of indexing set to an arbitrary nonempty family of subsets.

If $\{X_\alpha\}$ is an indexed collection of sets, where α runs through some indexing set Λ, we write

$$\bigcup_{\alpha\in\Lambda} X_\alpha \text{ and } \bigcap_{\alpha\in\Lambda} X_\alpha$$

for the union and intersection respectively of X_α:

$$\bigcup_{\alpha\in\Lambda} X_\alpha = \{x : x \in X_\alpha \text{ for at least one } \alpha \in \Lambda\},$$
$$\bigcap_{\alpha\in\Lambda} X_\alpha = \{x : x \in X_\alpha \text{ for every } \alpha \in \Lambda\}.$$

If $\Lambda = \mathbb{N}$, the set of all natural numbers, the customary notations are

$$\bigcup_{n=1}^{\infty} X_n \text{ and } \bigcap_{n=1}^{\infty} X_n.$$

Two sets A and B are said to be *disjoint* if they have no element in common, i.e., $A \cap B = \varnothing$. If no two members of a collection of sets, indexed or otherwise, have any element in common, then the sets or the collection is said to be *pairwise disjoint*. We shall often omit the word "pairwise", because no confusion will arise.

A union of disjoint sets is often called a *disjoint union.*

If $Y \subseteq X$, the *complement* of Y in X is the set of elements that are in X but not in Y, that is

$$X\backslash Y = \{x : x \in X, x \notin Y\}.$$

The complement of Y is denoted by Y^c whenever it is clear from the context with respect to which larger set the complement is taken.

If $\{X_\alpha\}$ is a collection of subsets of X, then the following *De Morgan's laws* hold:

$$\Big(\bigcup_{\alpha\in\Lambda} X_\alpha\Big)^c = \bigcap_{\alpha\in\Lambda} (X_\alpha)^c \quad \text{and} \quad \Big(\bigcap_{\alpha\in\Lambda} X_\alpha\Big)^c = \bigcup_{\alpha\in\Lambda} (X_\alpha)^c.$$

Given finitely many sets $X_1, X_2, \ldots, X_n$, by an ordered *n-tuple* $(x_1, x_2, \ldots, x_n)$, we understand a finite sequence $\{x_k\}_{k=1}^n$, where $x_k \in X_k$ for $k = 1, 2, \ldots, n$. The *Cartesian product* $X_1 \times X_2 \times \cdots \times X_n$ of the sets $X_1, X_2, \ldots, X_n$ is the set of all ordered n-tuples $(x_1, x_2, \ldots, x_n)$. The element x_k is called the kth *coordinate* of $(x_1, x_2, \ldots, x_n)$ and is the same as the kth term of the sequence $\{x_k\}_{k=1}^n$.

If $\{X_\alpha : \alpha \in \Lambda\}$ is any family of subsets of X, then

$$f\Big(\bigcup_{\alpha\in\Lambda} X_\alpha\Big) = \bigcup_{\alpha\in\Lambda} f(X_\alpha)$$

and

$$f\Big(\bigcap_{\alpha\in\Lambda} X_\alpha\Big) \subseteq \bigcap_{\alpha\in\Lambda} f(X_\alpha).$$

Also, if $\{Y_\alpha: \alpha \in \Lambda\}$ is a family of subsets of Y, then

$$f^{-1}\Big(\bigcup_{\alpha\in\Lambda} Y_\alpha\Big) = \bigcup_{\alpha\in\Lambda} f^{-1}(Y_\alpha)$$

and

$$f^{-1}\Big(\bigcap_{\alpha\in\Lambda} Y_\alpha\Big) = \bigcap_{\alpha\in\Lambda} f^{-1}(Y_\alpha).$$

If Y_1 and Y_2 are subsets of Y, then

$$f^{-1}(Y_1 \backslash Y_2) = f^{-1}(Y_1) \backslash f^{-1}(Y_2).$$

One of the methods of producing subsets from a given family of sets is via the Axiom of Choice.

Axiom of Choice 1.1.1 Given a nonempty family $\{A_\alpha: \alpha \in \Lambda\}$ of nonempty disjoint subsets of a set X, there exists an $S \subseteq X$ containing exactly one element from each A_α.

A *binary relation* R on a set X is a subset of $X \times X$. The statement that $(x, y) \in R$ is written as xRy.

Let X be a set with a binary relation R. The relation R is said to be an *equivalence relation* if, for all $x, y, z \in X$,

(i) xRx,
(ii) $xRy \Rightarrow yRx$,
(iii) $xRy, yRz \Rightarrow xRz$.

The properties (i), (ii) and (iii) are, respectively, known as *reflexivity*, *symmetry* and *transitivity*.

Let R be an equivalence relation on X. For each $x \in X$, the subset $Rx = \{y \in X: yRx\}$ is called the *equivalence class* of x and often denoted by $[\![x]\!]$. The fundamental result in this direction is the following:

(i) $\cup\{Rx: x \in X\} = X$, (ii) if xRy then $Rx = Ry$, (iii) if $x \not{R} y$, i.e., x is not related to y, then $Rx \cap Ry = \varnothing$.

Thus equivalence classes constitute a *partition* of X, by which we mean a family of disjoint subsets whose union is X. Conversely, a partition of X yields an equivalence relation on X, whose equivalence classes are precisely the disjoint sets of the partition. An element of an equivalence class is called a *representative* of that class.

Let $\mathbb{R}^*$ be the set which is the union of $\mathbb{R}$ with two elements, written ∞ and $-\infty$ (points at infinity). The enlarged set, when endowed with the structure described in (a)–(i) below, in addition to that already available on its subset $\mathbb{R}$, is called the **extended real numbers**:

(a) $-\infty < x < \infty$ for every $x \in \mathbb{R}$;
(b) $x + \infty = \infty + x = \infty$ and $x + (-\infty) = (-\infty) + x = -\infty$ for every $x \in \mathbb{R}$;
(c) $x \cdot \infty = \infty \cdot x = \infty$ and $x \cdot (-\infty) = (-\infty) \cdot x = -\infty$ for every positive $x \in \mathbb{R}$;
(d) $x \cdot (\infty) = (\infty) \cdot x = -\infty$ and $x \cdot (-\infty) = (-\infty) \cdot x = \infty$ for every negative $x \in \mathbb{R}$;
(e) $\pm\infty \cdot 0 = 0 \cdot (\pm\infty) = 0$;
(f) $\infty + \infty = \infty$ and $(-\infty) + (-\infty) = -\infty$;
(g) $\infty \cdot (\pm\infty) = (\pm\infty) \cdot \infty = \pm\infty$ and $(\pm\infty) \cdot (-\infty) = (-\infty) \cdot (\pm\infty) = \mp\infty$;
(h) $-(-\infty) = \infty$ and $-(\infty) = -\infty$;
(i) $(\infty)^p = \infty$ for real $p > 0$.

It can be verified on a case by case basis that the associative property of addition and the distributive property continue to be valid as long as ∞ and $-\infty$ do not both appear in a sum. This fact is essential for working with infinite series in $\mathbb{R}^*$. The question of whether multiplication is associative will not arise anywhere. In view of the restricted validity of the associative and distributive properties, the extended real numbers do not form a number "system" in the sense usually understood.

The reader is cautioned that there is no unanimity among authors about the properties (a)–(i). For instance [11, p. 1] and [19, p. 12] include a property to the effect that $(x/\pm\infty) = 0$ for real x but [12, pp. 54–55] does not; also, [11] includes our (g) but [12] specifically excludes all parts of it except for $\infty \cdot \infty = \infty$. The work [20, pp. 18–19] does not define any addition or multiplication involving $-\infty$. While [20] defines $0 \cdot \infty$ to be 0, the same author leaves it undefined in [19, p. 12].

The absolute value is defined in the obvious manner and the Triangle Inequality can be verified, whenever the smaller side is well defined.

Let $A \subseteq \mathbb{R}^*$. The smallest extended real number which is greater than or equal to each $x \in A$ is called the *supremum* of A and is denoted by $\sup A$ or by $\sup_{x \in A} x$, or by some variant thereof, depending on how the set A has been described. If $A = \varnothing$, then every extended real number is vacuously greater than or equal to each $x \in A$ and hence $\sup \varnothing$ is $-\infty$. Analogously for the *infimum*, $\inf A$ and $\inf \varnothing$. It may be noted in passing that $\varnothing$ is the only set for which the infimum is greater than the supremum.

Let $\{x_n\}_{n \geq 1}$ be a sequence in $\mathbb{R}^*$. Then $\sup_{n \geq 1} x_n$ and $\inf_{n \geq 1} x_n$ are respectively the supremum and infimum of the range of the sequence.

One of the advantages of the extended real numbers is that every subset has a supremum and an infimum. Another is that results about the integral of a sum or product of two functions can be stated without recourse to separate cases.

1.2 Countable and Uncountable Sets

Two sets A and B are said to be *equivalent*, and we write $A \sim B$, if there exists a one-to-one mapping, i.e., a bijection of A onto B. (Two finite sets are equivalent if and only if they have the same number of elements.) This relation clearly has the following properties:

It is *reflexive*: $A \sim A$ for all A,
it is *symmetric*: if $A \sim B$, then $B \sim A$,
it is *transitive*: if $A \sim B$ and $B \sim C$, then $A \sim C$.

The transitivity is a consequence of the fact that the composition of two bijections is a bijection.

Here are some examples of equivalent sets:

1. The set $X = \{2n : n \in \mathbb{N}\}$ and the set $\mathbb{N}$ are equivalent. Indeed, $n \to 2n$ is a bijection of $\mathbb{N}$ onto X.
2. It is easy to see that if I is an interval (open, closed, semi-open) with more than one point, then there is a bijection between I and $\mathbb{R}$. Indeed, the function $f : (0, 1) \to \mathbb{R}$ given by

$$f(x) = \begin{cases} \frac{2x-1}{x}, & 0<x<\frac{1}{2} \\ \frac{2x-1}{1-x} & \frac{1}{2} \leq x<1 \end{cases}$$

defines a bijection between (0, 1) and $\mathbb{R}$. Another such function f is given by $f(x) = \ln\frac{x}{1-x}$. Each of these functions is continuous along with its inverse. If we extend the domain to [0, 1] by setting $f(0) = -\infty$ and $f(1) = \infty$, we obtain a bijection between [0, 1] and $\mathbb{R}^*$. On the other hand, the function g: [0, 1] $\to$ (0, 1) given by

$$g(x) = \begin{cases} \frac{1}{2} & x = 0 \\ \frac{x}{2x+1} & x = \frac{1}{n}, \quad n \in \mathbb{N} \\ x & x \neq 0, \frac{1}{n}, \quad n \in \mathbb{N} \end{cases}$$

defines a bijection between [0, 1] and (0, 1). The latter function, suitably modified, shows that $[0, 1) \sim (0, 1)$.

Thus $\mathbb{R} \sim (0, 1) \sim [0, 1) \sim [0, 1] \sim \mathbb{R}^*$. However, we emphasise that this equivalence is only a set-theoretic one in the sense defined at the beginning of this section. For example, the elements $\frac{n}{n+2}$ and 1 of [0, 1], where $n \in \mathbb{N}$, would be considered as being "arbitrarily" close but not $g(\frac{n}{n+2}) = \frac{n}{n+2}$ and $g(1) = \frac{1}{3}$. This means some of the mathematical features of the set [0, 1] are not "preserved" by g. In more precise terminology, g is not continuous. The knowledgeable reader will observe that the bijection g is far from being what is called a "homeomorphism".

Theorem 1.2.1 (Schröder–Bernstein) *Let A and B be two sets. If each of them is equivalent to a subset of the other, then* $A \sim B$.

For any positive integer n, let $\mathbb{N}_n$ be the set whose elements are 1, 2, ..., n. A set A is finite if $A \sim \mathbb{N}_n$ for some n. A set is infinite if it is not finite. A set A is said to be *countable* if $A \sim \mathbb{N}$. It is said to be *uncountable* if it is neither finite nor countable and is said to be *at most countable* if it is either finite or countable. An infinite subset of a countable set is countable. When a set A is countable, a bijection that maps $\mathbb{N}$ onto A is often called an *enumeration* of the elements of A and we speak of elements $a_1, a_2, \ldots$ of A.

An important result that will often be used in the sequel is that the Cartesian product of $\mathbb{N}$ with itself finitely many times is countable.

The set $\mathbb{Q}$ of rational numbers is countable, whereas the set $\mathbb{R}$ is uncountable.

Another important result that will often be used in the sequel is that if $\{A_n\}_{n\geq 1}$ is a sequence of countable sets, then $\bigcup_{n\geq 1} A_n$ is countable. Also, if each A_n is at most countable, then so is $\bigcup_{n\geq 1} A_n$.

Let X be any nonempty set. The set of all maps $X \to \{0, 1\}$, where $\{0, 1\}$ is the set consisting of the two integers 0 and 1, is denoted by 2^X.

For any set $A \subseteq X \neq \emptyset$, the function $\chi_A : X \to \mathbb{R}$ such that $\chi_A(x)$ is 0 if $x \notin A$ and 1 if $x \in A$ is called the *characteristic function* of A. Observe that χ_A, where $A \subseteq X$, is a member of the set 2^X. The following simple result has important consequences: $2^X \sim \mathcal{P}(X)$. Indeed the mapping $f : \mathcal{P}(X) \to 2^X$ defined by $f(A) = \chi_A, A \subseteq X$, is clearly a bijection between $\mathcal{P}(X)$ and 2^X.

In subsequent chapters, we shall need to compare the size of two sets. It turns out that with every set A is associated a unique set called its cardinal number, denoted by card(A), such that A and card(A) are equivalent. If card(A) and card(B) are two distinct cardinal numbers, then A and B are not equivalent but one of the sets, say A, and a proper subset of B are equivalent. In this case, card(A) is said to be a smaller cardinal number and we write card(A) < card(B).

We shall not go into the rigorous treatment of cardinal numbers. For a rigorous treatment of this topic, the reader may refer to [10].

The cardinal number of $\mathbb{R}$ occurs frequently and we shall denote it by c when necessary. The cardinal number of $\mathbb{N}$ is denoted by $\aleph_0$.

The following results, which we state without proof, will be needed in the sequel:

1. For any set A, $\text{card}(A) < \text{card}(\mathcal{P}(A))$;
2. $\text{card}(\mathcal{P}(\mathbb{N})) = \text{card}(2^{\mathbb{N}}) = \text{card}(\mathbb{R}) = \text{card}([0, 1]) = c$.

1.3 Metric Spaces

The ideas of convergence and open sets in a variety of spaces encountered in the present text can be best described in a metric space. The general reference for Sects. 1.3 to 1.5 is [26].

Definition 1.3.1 Let X be a nonempty set. A nonnegative function d defined on the set of ordered pairs $\{(x, y): x \in X, y \in X\}$ is a **metric** on X if it satisfies the following properties:

(i) $d(x, y) = 0$ if and only if $x = y$,
(ii) $d(x, y) = d(y, x)$,
(iii) $d(x, z) \leq d(x, y) + d(y, z)$,

for all $x, y, z \in X$.

The pair (X, d) is called a **metric space**. Customarily, one calls X the metric space when d is understood from the context.

$(\mathbb{R}, d)$, where $d(x, y) = |x - y|$, $x, y \in \mathbb{R}$, is the real line with the *usual metric*. The n-fold Cartesian product $\mathbb{R} \times \mathbb{R} \times \cdots \times \mathbb{R}$ is denoted by $\mathbb{R}^n$ and is called **Euclidean n-space**. There are three useful metrics on it, given by

$$d_1(x, y) = \sum_{i=1}^{n} |x_i - y_i|,$$

$$d_2(x, y) = \sqrt{\sum_{i=1}^{n} (x_i - y_i)^2}$$

and

$$d_\infty(x, y) = \max_{1 \leq i \leq n} |x_i - y_i|,$$

where $x = (x_1, x_2, \ldots, x_n) \in \mathbb{R}^n$ and $y = (y_1, y_2, \ldots, y_n) \in \mathbb{R}^n$.
The inequality

$$d_\infty(x, y) \leq d_1(x, y) \leq \sqrt{n} d_2(x, y) \leq n d_\infty(x, y) \quad (1.1)$$

holds.

Definition 1.3.2 Any nonempty subset $Y \subseteq X$ together with the **induced metric** defined by $d_Y(x, y) = d(x, y)$, $x, y \in Y$, is a **metric subspace** of (X, d). It is denoted by (Y, d_Y).

The subscript Y of d_Y is often dropped when there is no scope for confusion. In fact, the subset Y itself is referred to as a metric subspace when d is understood.

The subset $(0, \infty) = \{x \in \mathbb{R} : x > 0\}$ of $\mathbb{R}$, with the induced metric, denoted yet again by d, is a metric subspace of $\mathbb{R}$.

If we relax the condition (i) above to read as $d(x, y) = 0$ if $x = y$, we obtain a *pseudometric space*. More precisely, we have the following definition.

Definition 1.3.3 Let X be a nonempty set. A nonnegative function d defined on the set of ordered pairs $\{(x, y): x \in X, y \in X\}$ is a **pseudometric** on X if it satisfies the following properties:

(i) $d(x, y) = 0$ if $x = y$,
(ii) $d(x, y) = d(y, x)$,
(iii) $d(x, z) \leq d(x, y) + d(y, z)$,

for all $x, y, z \in X$.

The pair (X, d) is called a **pseudometric space**. Customarily, one calls X the pseudometric space when d is understood from the context.

The notion of pseudometric space will prove useful in Chaps. 3 and 6.

In a pseudometric space, it is standard procedure to generate a related metric space by "identifying" x and y whenever $d(x, y) = 0$, i.e. by forming equivalence classes corresponding to the binary relation defined by $d(x, y) = 0$—quite obviously an equivalence relation. One can verify for equivalence classes $[\![x]\!], [\![y]\!]$ with respective representatives x, y that $d([\![x]\!], [\![y]\!]) = d(x, y)$ is unambiguous (it is independent of the choice of representatives x, y) and that it gives rise to a metric on the set of equivalence classes.

We next consider the notion of convergence of sequences.

Definition 1.3.4 Let (X, d) be a metric space and $\{x_n\}_{n\geq 1}$ a sequence in X. An element $x \in X$ is said to be a **limit** of the sequence $\{x_n\}_{n\geq 1}$ if, for every $\varepsilon > 0$, there exists a natural number n_0 such that $d(x, x_n) < \varepsilon$ whenever $n \geq n_0$. If this is the case, we say that the sequence **converges** to x and write it in symbols as $\boldsymbol{x_n \to x}$ **as** $\boldsymbol{n \to \infty}$, or alternatively, as $\mathbf{lim}_n \boldsymbol{x_n = x}$. If there is no such x, we say that the sequence $\{x_n\}_{n\geq 1}$ diverges.

The symbol $\lim_n$ is often written as $\mathbf{\lim\limits_{n\to\infty}}$ or simply as **lim**.

When a limit exists in a metric space, it is unique.

Let $x_n = \frac{1}{n}$, $n = 1, 2, \ldots$. Then $\{x_n\}_{n\geq 1}$ is a sequence in $\mathbb{R}$ and $\lim_n x_n = 0$. The sequence $\{x_n\}_{n\geq 1}$, where $x_n = (-1)^n$, $n = 1, 2, \ldots$, is divergent.

Suppose a sequence in $\mathbb{R}^n$ converges to a limit in the sense of any one of the three metrics d_1, d_2, d_∞ described above. In view of the inequality (1.1) between the three metrics, it follows that it converges to the same limit in the sense of the other two metrics.

To enable us to operate in the extended real number system, we shall enlarge the scope of the above definition.

We say that a sequence $\{x_n\}_{n\geq 1}$ of extended real numbers diverges to ∞, and write $x_n \to \infty$ as $n \to \infty$ if for any $M > 0$, we can find an n_0 such that $x_n > M$ for any $n \geq n_0$. In this case, $\lim x_n$ is the extended real number ∞.

Similarly, a sequence $\{x_n\}_{n\geq 1}$ diverges to $-\infty$ and we write $x_n \to -\infty$ as $n \to \infty$ if for any $M > 0$, we can find an n_0 such that $x_n < -M$ for any $n \geq n_0$. In this case, $\lim x_n$ is the extended real number $-\infty$.

Whether $\lim x_n$ is an extended real number or a real number, it is unique whenever it exists.

Definition 1.3.5 A sequence $\{x_n\}_{n\geq 1}$ in a metric space X is said to be **Cauchy** if, for every $\varepsilon > 0$, there exists a natural number $n_0 = n_0(\varepsilon)$ such that $d(x_n, x_m) < \varepsilon$ whenever $n \geq n_0$ and $m \geq n_0$. A metric space is said to be **complete** if every Cauchy sequence in it converges.

It is trivial that every sequence that converges is a Cauchy sequence.

The space $(\mathbb{R}, d)$ is a complete metric space, where $d(x, y) = |x - y|$, $x, y \in \mathbb{R}$, is the usual metric. This is a consequence of the following well-known result in Analysis:

Theorem 1.3.6 (Cauchy's Principle of Convergence) *Let $\{x_n\}_{n \geq 1}$ be a sequence in $\mathbb{R}$ satisfying the condition that, for every $\varepsilon > 0$, there exists a natural number $n_0 = n_0(\varepsilon)$ such that $d(x_n, x_m) < \varepsilon$ whenever $n \geq n_0$ and $m \geq n_0$. Then the sequence converges to a limit in $\mathbb{R}$.*

The subspace $(\mathbb{Q}, d)$ is incomplete. It is well known that there is a sequence of rational numbers that converges to $\sqrt{2}$. It therefore converges to a point that does not belong to the space.

Suppose a sequence in $\mathbb{R}^n$ is Cauchy in the sense of any one of the three metrics d_1, d_2, d_∞ described above. In view of the inequality (1.1) between the three metrics, it follows that it is Cauchy in the sense of the other two metrics. Moreover, $\mathbb{R}^n$ is complete in the sense of each of the metrics.

Proposition 1.3.7 *If a subsequence of a Cauchy sequence in a metric space converges, then the Cauchy sequence converges to the same limit as the subsequence.*

The open and closed sets, defined below, will subsequently play a distinguished role.

Definition 1.3.8 Let (X, d) be a metric space. The set $S(x_0, r) = \{x \in X: d(x, x_0) < r\}$, where $r > 0$ and $x_0 \in X$, is called the **open ball** with centre x_0 and radius r. The set $\overline{S}(x_0, r) = \{x \in X : d(x, x_0) \leq r\}$, where $r \geq 0$ and $x_0 \in X$, is called the **closed ball** with centre x_0 and radius r.

The open ball $S(x_0, r)$ on the real line is the bounded open interval $(x_0 - r, x_0 + r)$ with midpoint x_0 and total length $2r$. Conversely, any nonempty bounded open interval on the real line is an open ball. So the open balls on the real line are precisely the nonempty bounded open intervals. The closed balls on the real line are precisely the nonempty bounded closed intervals.

In $\mathbb{R}^n$, $n > 1$, the open ball of radius r with centre $(x_1, x_2, \ldots, x_n)$ depends on which metric is being used. When the metric is d_∞, it is the Cartesian product of the n open intervals of length $2r$ with midpoints at $x_1, x_2, \ldots, x_n$, respectively. It is an example of what will later be called a "cuboid".

Definition 1.3.9 A set U in a metric space (X, d) is an **open set** if, given any $x_0 \in U$, there exists an $r > 0$ such that $S(x_0, r) \subseteq U$. That is, U contains an open ball around each of its points.

So, X, $\varnothing$ are open and it also follows that any union of open sets is open and that any finite intersection of open sets is open.

An arbitrary intersection of open sets need not be open. Indeed, the collection $\left\{\left(-\frac{1}{n}, \frac{1}{n}\right) : n \in \mathbb{N}\right\}$ has intersection $\{0\}$, which is certainly not open.

The inequalities (1.1) have the consequence that a subset of $\mathbb{R}^n$ which is open in the sense of any one of the three metrics d_1, d_2, d_∞ is open in the sense of the other two. Therefore we need not specify the metric when we mention an open subset.

Let Y be a nonempty subset of the metric space (X, d) together with the induced metric. A subset in Y is open if and only if it is the intersection of an open set in X with Y.

By a **neighbourhood of a point** x we mean an open set containing x. Thus a sequence $\{x_n\}_{n \geq 1}$ converges to x if and only if, for every neighbourhood of x, there exists a natural number n_0 such that x_n lies in the neighbourhood whenever $n \geq n_0$.

Definition 1.3.10 A set F in a metric space (X, d) is a **closed set** if its complement F^c is open. A set in a metric space (X, d) is **bounded** if it is contained in some ball.

For example, $[0, \infty) = (-\infty, 0)^c$ is closed in $(\mathbb{R}, d)$, where d is the usual metric. A subset of a metric space may be neither open nor closed: the semi-open interval $(0, 1]$ is neither open nor closed in $\mathbb{R}$.

Definition 1.3.11 The **closure** $\overline{F}$ of a set $F \subseteq X$, where (X, d) is a metric space, is the intersection of all closed subsets of X that contain F.

Definition 1.3.12 A point a is said to be a **limit point** of F if, given $\varepsilon > 0$, there exists an $x \in F$, $x \neq a$, such that $d(x, a) < \varepsilon$.

In other words, a limit point of a set is characterised by the property that every neighbourhood of it contains a point of the set other than the point in question.

The limit points of an interval are those numbers that belong to it and also any endpoints it may have, regardless of whether the endpoints belong to the interval or not.

A point of a set F is said to be *isolated* if it is not a limit point of F, which is to say, some neighbourhood of it contains no other point of the set.

A set is closed if and only if it is the same as its closure. The closure is the union of the set with the set of its limits points (if any). A set is closed if and only if its limit points, if any, belong to it; in particular, a set having no limit points is always closed.

Let f be an extended real-valued function defined on a subset F of a metric space X and a be a limit point of F. We say that $f(x)$ *tends to* $l \in \mathbb{R}$ *as* x *tends to* a if, for every $\varepsilon > 0$, there exists some $\delta > 0$ such that

$$|f(x) - l| < \varepsilon \quad \forall x \in F \quad \text{satisfying} \quad 0 < d(x, a) < \delta.$$

Also, we say that $f(x)$ *tends to* $l = \infty \in \mathbb{R}^*$ *as* x *tends to* a if, for every real M, there exists some $\delta > 0$ such that

$$f(x) > M \quad \forall x \in F \quad \text{satisfying} \quad 0 < d(x, a) < \delta.$$

Similarly for $l = -\infty$.

The value $l \in \mathbb{R}^*$ is uniquely determined when it exists.

The extended real number l is said to be *the limit of* $f(x)$ *as* x *tends to* a and we write

$$\lim_{x \to a} f(x) = l \quad \text{or} \quad \lim_{x \to a} f(x) = l \quad \text{or} \quad f(x) \to l \text{ as } x \to a.$$

Note that one among $f(a)$ and $\lim_{x \to a} f(x)$ may make sense but not the other. More precisely, if a is a limit point not belonging to the domain F, only $\lim_{x \to a} f(x)$ makes sense, while if a belongs to the domain F but is not a limit point (is an isolated point), only $f(a)$ makes sense.

If $\lim_{x \to a} f(x)$ and $\lim_{x \to a} g(x)$ both exist and are real, then so do the limits $\lim_{x \to a}(f(x) + g(x))$ and $\lim_{x \to a}(f(x)g(x))$; moreover,

$$\lim_{x \to a}(f(x) + g(x)) = \lim_{x \to a} f(x) + \lim_{x \to a} g(x)$$

and

$$\lim_{x \to a}(f(x)g(x)) = (\lim_{x \to a} f(x))(\lim_{x \to a} g(x)).$$

If $\lim_{x \to a} f(x)$ and $\lim_{x \to a} g(x)$ both exist and one or both are $\pm\infty$, the above assertion about the sum holds unless the sum $\lim_{x \to a} f(x) + \lim_{x \to a} g(x)$ is undefined, i.e., unless one of the limits is ∞ and the other is $-\infty$. The assertion regarding the product holds unless one limit is 0 and the other is $\pm\infty$.

If α is any real number, then $\lim_{x \to a}(\alpha f(x)) = \alpha(\lim_{x \to a} f(x))$. This holds even if $\alpha = \pm\infty$ but with the proviso that $\lim_{x \to a} f(x) \neq 0$.

If there exists an $\eta > 0$ such that $f(x) \leq h(x) \leq g(x)$ whenever $0 < d(x, a) < \eta$ and if $\lim_{x \to a} f(x)$ and $\lim_{x \to a} g(x)$ both exist and are equal, then $\lim_{x \to a} h(x)$ also exists and $\lim_{x \to a} h(x) = \lim_{x \to a} f(x) = \lim_{x \to a} g(x)$.

The following special aspect of limits when the domain is an interval is crucial in Chap. 5.

When we speak of the limit of a function defined on a subset F of $\mathbb{R}$, it is understood that $\mathbb{R}$ has the usual metric, which is given by $d(x, y) = |x - y|$. So the meaning of "$\lim_{x \to a} f(x) = l \in \mathbb{R}$", is that, for every $\varepsilon > 0$, there exists some $\delta > 0$ such that

$$|f(x) - l| < \varepsilon \ \forall x \in F \text{ satisfying } 0 < |x - a| < \delta.$$

By definition of absolute value, this is equivalent to the conjunction of the two statements:

$$|f(x) - l| < \varepsilon \ \forall x \in F \text{ satisfying } 0 < x - a < \delta,$$
$$|f(x) - l| < \varepsilon \ \forall x \in F \text{ satisfying } 0 < a - x < \delta.$$

Suppose the domain F is an interval. (Since a is understood to be a limit point of F, it is either a point of F or an endpoint.) If for every $\varepsilon > 0$, there exists some $\delta > 0$ such that the first of these statements holds and a is not the right endpoint of F, the number l must be unique and we speak of l as the *right-hand the limit of $f(x)$ as x tends to a* or as *the limit of $f(x)$ as x tends to $a+$*. Similarly, if for every $\varepsilon > 0$, there exists some $\delta > 0$ such that the second of these statements holds and a is not the left endpoint of F, the number l must be unique and we speak of l as the *left-hand limit of $f(x)$ as x tends to a* or as *the limit of $f(x)$ as x tends to $a-$*. There is no left- [resp. right-] hand limit at the left [resp. right] endpoint of an interval. The modifications required when $l = \pm\infty$ are obvious.

The left-hand and right-hand limits are denoted by

$$f(a-) \quad \text{or} \quad \lim_{x\to a-} f(x)$$

and

$$f(a+) \quad \text{or} \quad \lim_{x\to a+} f(x)$$

respectively. Together, they are referred to as *one-sided* limits. At the left [resp. right] endpoint of F, the right- [resp. left-] hand limit is the same as the limit. At an interior point, a limit exists if and only if both one-sided limits exist and are equal, in which case, their common value is the limit, sometimes called the *two-sided* limit for emphasis.

It will be proved in Proposition 5.2.2 that for an increasing function, the left [resp. right] hand limit always exists except at the left- [resp. right-] endpoint of the domain.

It is trivial that $f(a+) = \lim_{h\to 0+} f(a+h)$ and $f(a-) = \lim_{h\to 0-} f(a+h)$.

Suppose f is an extended real-valued function defined on a subset F of $\mathbb{R}$, and $\sup F = \infty$. We say that *$f(x)$ tends to $l \in \mathbb{R}$ as x tends to* ∞ if, for every $\varepsilon > 0$, there exists some $K \in \mathbb{R}$ such that

$$|f(x) - l| < \varepsilon \; \forall x \in F \text{ satisfying } x > K.$$

Also, we say that *$f(x)$ tends to $l = \infty \in \mathbb{R}^*$ as x tends to* ∞ if, for every real M, there exists some $K \in \mathbb{R}$ such that

$$f(x) > M \; \forall x \in F \text{ satisfying } x > K.$$

Similarly for $l = -\infty$.

The value $l \in \mathbb{R}^*$ is uniquely determined when it exists.

The extended real number l is said to be *the limit of $f(x)$ as x tends to* ∞ and we write

$$\lim_{x\to\infty} f(x) = l \quad \text{or} \quad \lim\nolimits_{x\to\infty} f(x) = l \quad \text{or} \quad f(x) \to l \text{ as } x \to \infty.$$

The meaning of $\lim_{x\to-\infty} f(x)$ when the domain F satisfies $\inf F = -\infty$ is clear.

The observations made earlier about sums and products of limits are of course valid for one-sided limits and also for $\lim_{x\to\infty} f(x)$ and $\lim_{x\to-\infty} f(x)$.

The following important formulation of the limit of a function is phrased in terms of limits of sequences.

Proposition 1.3.13 *Let $f : F \to \mathbb{R}$, where F is a subset of some metric space and let a be a limit point of F. Then $\lim_{x\to a} f(x) = l \in \mathbb{R}^*$ if and only if, for every sequence $\{x_n\}_{n\geq 1}$ in F that converges to a, and which satisfies $x_n \neq a$ for every n, the sequence $\{f(x_n)\}_{n\geq 1}$ converges/diverges to l.*

There is an analogous formulation of one-sided limits of functions in terms of limits of sequences. Suppose the domain F is an interval and let a be a limit point of F but not be its right endpoint. Then $f(a+) = l \in \mathbb{R}^*$ if and only if, for every sequence $\{x_n\}_{n\geq 1}$ in F that converges to a and satisfies $x_n > a$ for every n, the sequence $\{f(x_n)\}_{n\geq 1}$ converges/diverges to l. Analogously for $f(a-)$, where a is not the left endpoint.

Let f be a real-valued function whose domain of definition is a subset F of a metric space. We say that f is *continuous at the point* $a \in F$ if, given $\varepsilon > 0$, there exists a $\delta > 0$ such that for all $x \in X$ with $d(x, a) < \delta$, we have $|f(x) - f(a)| < \varepsilon$. The function is said to be *continuous on* F if it is continuous at every point of F. If we merely say that a function is continuous, we mean that it is continuous on its domain.

It may be checked that f is continuous at a limit point $a \in F$ if and only if $\lim\limits_{x\to a} f(x) = f(a)$. The following criterion of continuity follows from the preceding paragraph and Proposition 1.3.13.

Proposition 1.3.14 *Let f be a real-valued function defined on a subset F of a metric space and $a \in F$. Then f is continuous at a if and only if, for every sequence $\{x_n\}_{n\geq 1}$ in F that converges to a, we have* $\lim f(x_n) = f(\lim x_n) = f(a)$.

This result shows that continuous functions are precisely those which send every convergent sequence and its limit into a convergent sequence and its limit, in other words, 'preserve' limits of sequences. As an aside, we note that if a function 'preserves' convergence of sequences, it necessarily preserves their limits as well.

Suppose once again that the domain F is an interval, and let $a \in F$ but not be its right endpoint. Then f is said to be *right continuous at the point* a if, given $\varepsilon > 0$, there exists a $\delta > 0$ such that for all $x \in F$ with $0 < x - a < \delta$, we have $|f(x) - f(a)| < \varepsilon$. It is easy to verify that right continuity at a means the same thing as $f(a+) = f(a)$. The formulation in terms of sequences is that f is right continuous at a if and only if, for every sequence $\{x_n\}_{n\geq 1}$ in F that converges to a and satisfies $x_n > a$ for every n, the sequence $\{f(x_n)\}_{n\geq 1}$ converges to $f(a)$. Analogous observations apply to *left continuity*, where $a \in F$ is not the left endpoint.

If $a \in F$ is neither the left endpoint nor the right endpoint, then f is continuous at a if and only if it is left as well as right continuous at a. On the other hand, if $a \in F$ is the left [resp. right] endpoint, then f is continuous at a if and only if it is right [resp. left] continuous at a.

Let f be an extended real-valued function whose domain of definition is a metric space X and $Y \subseteq X$. We define $\sup(f, Y)$ and $\inf(f, Y)$ as follows:

$$\sup(f, Y) = \sup\{f(y) : y \in Y\} \quad \text{and} \quad \inf(f, Y) = \inf\{f(y) : y \in Y\}.$$

A set consisting of only isolated points is said to be **discrete**; it is always closed.

A set F is said to be **dense** in X, where (X, d) is a metric space, if $\overline{F} = X$. The set $\mathbb{Q}$ of rational numbers in $\mathbb{R}$ is dense in $(\mathbb{R}, d)$, where d is the usual metric. The set $\mathbb{Q}^n$ is dense in $\mathbb{R}^n$ (regardless of which metric).

An *open cover* (*covering*) of a subset A of a metric space is a collection $\mathcal{C} = \{G_\alpha : \alpha \in \Lambda\}$ of open sets in the metric space whose union contains A, that is,

$$A \subseteq \bigcup_{\alpha} G_\alpha.$$

If $\mathcal{C}'$ is a subcollection of $\mathcal{C}$ such that the union of sets in $\mathcal{C}'$ also contains A, then $\mathcal{C}'$ is called a *subcover* (or *subcovering*) from $\mathcal{C}$ of A. If $\mathcal{C}'$ consists of finitely many sets, then we say that $\mathcal{C}'$ is a *finite subcover* (or *finite subcovering*).

Definition 1.3.15 A nonempty subset K of a metric space is said to be **compact** if every open cover of K contains a finite subcover.

Obviously any nonempty finite subset of a metric space is compact.

A nonempty subset K is compact if and only if every sequence in it has a subsequence which converges to a limit belonging to K. This in turn is equivalent to every infinite subset of K having a limit point belonging to K.

Nontrivial examples in $\mathbb{R}^n$ are given by the following result.

Theorem 1.3.16 (Heine–Borel) *A nonempty subset of* $\mathbb{R}^n$ *is compact if and only if it is closed and bounded.*

An intersection of a sequence of sets belonging to a class $\mathcal{A}$ of sets is called a **countable intersection of** sets of $\mathcal{A}$. The intersection of a finite sequence of sets belonging to $\mathcal{A}$ can obviously be represented as the intersection of an infinite sequence of sets belonging to $\mathcal{A}$ and is therefore spoken of as a countable intersection. Similarly, one speaks of a **countable union**.

Sometimes the sets in a countable union are required to be (pairwise) disjoint and it is often convenient to use the abbreviation *countable disjoint union*. If $\varnothing \in \mathcal{A}$, as happens often, the union of a finite sequence of disjoint sets in $\mathcal{A}$ is also a countable disjoint union of sets in $\mathcal{A}$.

We shall also use the following results.

Theorem 1.3.17 *Let* U *be an open subset of* $\mathbb{R}$. *Then* U *can be written as a countable disjoint union of open intervals.*

Theorem 1.3.18 *An open set* V *contained in* $\mathbb{R}^n$ *can be written as a countable union of Cartesian products of open intervals.*

Definition 1.3.19 Let (X, d_X) and (Y, d_Y) be metric spaces and $A \subseteq X$. A function $f: A \to Y$ is said to be **continuous at** $a \in A$ if, for $\varepsilon > 0$, there exists some $\delta > 0$ such that

$$d_Y(f(x), f(a)) < \varepsilon \text{ whenever } x \in A \text{ and } d_X(x, a) < \delta.$$

If f is continuous at every point of A, then f is said to be **continuous on** A.

The analogue of Proposition 1.3.14 is valid for a function from a subset of a metric space to a metric space.

Theorem 1.3.20 (Extreme Value Theorem) *A continuous real-valued function on a compact set is bounded and has a maximum value as well as a minimum value.*

Suppose a function defined on a subset of $\mathbb{R}^n$ is continuous at a point of its domain in the sense of any one of the three metrics d_1, d_2, d_∞ described above. In view of the inequality (1.1) between the three metrics, it follows that it is continuous at that point in the sense of the other two metrics. This allows us to speak of continuity without specifying which metric we have in mind.

Theorem 1.3.21 *Let* (X, d_X) *and* (Y, d_Y) *be metric spaces and* $A \subseteq X$. *A function* $f: A \to Y$ *is continuous on A if and only if for every open set* $V \subseteq Y$, *there exists an open set* $U \subseteq X$ *such that* $f^{-1}(V) = U \cap A$.

Definition 1.3.22 Let (X, d_X) and (Y, d_Y) be metric spaces and $A \subseteq X$. A function $f: A \to Y$ is said to be **uniformly continuous on** A if, for $\varepsilon > 0$, there exists some $\delta > 0$ such that

$$d_Y(f(x), f(a)) < \varepsilon \text{ whenever } x, a \in A \text{ and } d_X(x, a) < \delta.$$

Theorem 1.3.23 *Let* (X, d_X) *and* (Y, d_Y) *be metric spaces and* $A \subseteq X$ *be compact. If the function* $f: A \to Y$ *is continuous on A, then f is uniformly continuous on A.*

Proposition 1.3.24 *Let A be a nonempty subset of a set M with metric* ρ. *The distance of a point* $x \in M$ *from the subset A is defined as* $d(x, A) = \inf\{\rho(x, a): a \in A\}$. *The map* $\phi : M \to \mathbb{R}$ *defined by* $\phi(x) = d(x, A)$ *is continuous.*

Proof Let $x_1, x_2 \in M$ and $p \in A$ be arbitrary. Then

$$\rho(x_1, p) \leq \rho(x_1, x_2) + \rho(x_2, p);$$

this implies $d(x_1, A) \leq \rho(x_1, x_2) + \rho(x_2, p)$. Choosing p appropriately, we have $\rho(x_2, p) \leq d(x_2, A) + \varepsilon$, where $\varepsilon > 0$ is arbitrary. Then $d(x_1, A) \leq \rho(x_1, x_2) + d(x_2, A) + \varepsilon$. Therefore $d(x_1, A) - d(x_2, A) \leq \rho(x_1, x_2)$, since $\varepsilon > 0$ is arbitrary. Similarly, $d(x_2, A) - d(x_1, A) \leq \rho(x_1, x_2)$. Consequently, $|d(x_1, A) - d(x_2, A)| \leq \rho(x_1, x_2)$, which, in fact, implies uniform continuity. □

Proposition 1.3.25 (needed in Example 5.5.2) *Let* (X, d_X) *and* (Y, d_Y) *be metric spaces. If* $f: X \to Y$ *and* $g: X \to Y$ *are both continuous and agree on a dense subset of X, then they agree everywhere on X.*

The next result is relevant only in the special metric space $\mathbb{R}$. Known as the Bolzano Intermediate Value Theorem, it guarantees that a continuous function on an interval assumes (at least once) every value that lies between any two of its values.

Theorem 1.3.26 (Intermediate Value Theorem) *Let I be an interval and* $f : I \to \mathbb{R}$ *be a continuous mapping on I. If* $a, b \in I$ *and* $\alpha \in \mathbb{R}$ *satisfy* $f(a) < \alpha < f(b)$ or $f(a) > \alpha > f(b)$, *then there exists a point* $c \in I$ *between a and b such that* $f(c) = \alpha$.

1.4 Sequences and Series of Extended Real Numbers

In Sect. 1.1, sequences with values in an arbitrary set were defined, and in Definition 1.3.4, convergence of sequences with values in a metric space was introduced. Sequences in $\mathbb{R}^*$, often called *extended real sequences*, enjoy additional properties, which will be needed in the sequel. We record them below for convenience.

It is important to distinguish between the extended real sequence $\{x_n\}_{n\geq 1}$ and its range $\{x_n : n \in \mathbb{N}\}$, which is a subset of $\mathbb{R}^*$.

In accordance with Definition 1.3.4, a real number $l \in \mathbb{R}$ is a limit of the sequence $\{x_n\}_{n\geq 1}$ if for each $\varepsilon > 0$, there is a positive integer n_0 such that for all $n \geq n_0$, we have $|x_n - l| < \varepsilon$. As in any metric space, a sequence has at most one limit. When $\{x_n\}_{n\geq 1}$ does have a limit, we denote it by $\lim x_n$. In symbols, $l = \lim x_n$ if

$$\forall\, \varepsilon > 0, \exists\, n_0 \ni n \geq n_0 \Rightarrow |x_n - l| < \varepsilon.$$

A real sequence that has a real limit is said to *converge* (or to be *convergent*).

Recall that the concept of limit of a sequence was enlarged to include the case of extended real numbers.

If $\lim x_n$ and $\lim y_n$ both exist and both are real, then so do $\lim(x_n + y_n)$ and $\lim(x_n y_n)$; moreover,

$$\lim(x_n + y_n) = \lim x_n + \lim y_n \quad \text{and} \quad \lim(x_n y_n) = (\lim x_n)(\lim y_n).$$

If α is any real number, then $\lim (\alpha x_n) = \alpha(\lim x_n)$.

Suppose one or both limits are $\pm\infty$. Then the assertion about the sum holds provided that the sum $\lim x_n + \lim y_n$ is defined, and the assertion regarding the product holds unless one limit is 0 and the other is $\pm\infty$. The assertion regarding αx_n holds even if $\alpha = \pm\infty$ but with the proviso that $\lim x_n \neq 0$.

The sequence $x_n = (1 + \frac{1}{n})^n$ has a limit denoted by e; this number is irrational and lies between 2 and 3. Also, $\lim n^{1/n} = 1$.

A sequence $\{x_n\}_{n\geq 1}$ of extended real numbers is said to be *increasing* if it satisfies the inequalities $x_n \leq x_{n+1}$, $n = 1, 2, \ldots$, and *decreasing* if it satisfies the inequalities $x_n \geq x_{n+1}$, $n = 1, 2, \ldots$. We say that the sequence is *monotone* if it is either increasing or it is decreasing.

A sequence $\{x_n\}_{n\geq 1}$ of real numbers is said to be *bounded* if there exists a real number $M > 0$ such that $|x_n| \leq M$ for all $n \in \mathbb{N}$. The following simple criterion for the convergence of a monotone sequence is very useful.

Proposition 1.4.1 *A monotone sequence of real numbers is convergent if and only if it is bounded.*

Bolzano–Weierstrass Theorem 1.4.2 *A bounded sequence of real numbers has a convergent subsequence.*

The convergence criterion described in Proposition 1.4.1 is restricted to monotone sequences. It is important to have a condition implying the convergence of a sequence of real numbers that is applicable to a larger class and preferably does not require knowledge of the value of the limit. The Cauchy Principle of Convergence stated in Theorem 1.3.6 provides such a condition.

Definition 1.4.3 A sequence of intervals is said to be **nested** if each interval contains the next.

Nested Interval Theorem 1.4.4 *Given a nested sequence of closed intervals whose lengths tend to* 0, *there exists a unique real number that belongs to all of them.*

Definition 1.4.5 Let $\{x_n\}_{n\geq 1}$ be a sequence in $\mathbb{R}^*$. We define the **limit superior** of $\{x_n\}_{n\geq 1}$ to be

$$\textbf{lim sup}\, \boldsymbol{x_n} = \inf_{n\geq 1}(\sup_{k\geq n} x_k) = \lim_n(\sup\{x_n, x_{n+1}, \ldots\})$$

and the **limit inferior** to be

$$\textbf{lim inf}\, \boldsymbol{x_n} = \sup_{n\geq 1}(\inf_{k\geq n} x_k) = \lim_n(\inf\{x_n, x_{n+1}, \ldots\}).$$

It can be checked that the limit exists if and only if the limits superior and inferior both exist, in which case they agree with the limit.

For the sequence $\{x_n\}_{n\geq 1}$, where $x_n = (-1)^n$, $n = 1, 2, \ldots$, $\limsup x_n = 1$ and $\liminf x_n = -1$. In particular, $\lim_n x_n$ does not exist.

Definition 1.4.6 Suppose $\{f_n\}_{n\geq 1}$ is a sequence of extended real-valued functions defined on a nonempty set X. Then for $x \in X$, we define

$$\begin{aligned}
(\sup_n f_n)(x) &= \sup_n f_n(x)\\
(\inf_n f_n)(x) &= \inf_n f_n(x)\\
(\limsup_n f_n)(x) &= \limsup_n f_n(x)\\
(\liminf_n f_n)(x) &= \liminf_n f_n(x).
\end{aligned}$$

For the sequence $\{f_n\}_{n\geq 1}$, where $f_n(x) = x^n$, $n = 1, 2, \ldots$ and $x \in [0, 1)$, $\sup_n f_n = f_1$ and $\inf_n f_n = 0$.

Let $\{a_k\}_{k\geq 1}$ be a sequence of extended real numbers, in which ∞ and $-\infty$ do not both appear as terms. Let $\{s_n\}_{n\geq 1}$, where $s_n = \sum_{k=1}^{n} a_k$, be the sequence of *partial sums* associated with $\{a_k\}_{k\geq 1}$. A sequence of partial sums is called a *series*. The limit $\lim s_n$, if it exists, is called the *sum* of the series and the series is said to *converge* or be *convergent* if the sum is a real number. If either the sum does not exist or is $\pm\infty$, the series is said to be *divergent*. The symbol $\sum_{k=1}^{\infty} a_k$, or its abbreviated form $\sum a_k$, denotes the series as well as the sum, if any. The context determines which of the two is intended. By the kth *term* of the series we mean a_k.

The series $\sum_{k=1}^{\infty} k^{-p}$ is convergent if and only if $p > 1$.

Proposition 1.4.7 (Comparison Test for Series) *If* $0 \leq a_k \leq b_k$ *for each k and the series* $\sum_{k=1}^{\infty} b_k$ *is convergent, then the series* $\sum_{k=1}^{\infty} a_k$ *is also convergent and* $\sum_{k=1}^{\infty} a_k \leq \sum_{k=1}^{\infty} b_k$. *Suppose further that there exists a j such that* $a_j < b_j$. *Then* $\sum_{k=1}^{\infty} a_k < \sum_{k=1}^{\infty} b_k$.

A series $\sum_{k=1}^{\infty} a_k$ of real numbers is said to *converge absolutely* if the related series $\sum_{k=1}^{\infty} |a_k|$ converges.

Proposition 1.4.8 *If a series converges absolutely then it converges.*

Proposition 1.4.9 *If the terms of an absolutely convergent series are rearranged in any manner whatsoever, the resulting series also converges and has the same sum.*

1.5 Sequences and Series of Functions

As elsewhere in Mathematical Analysis, the distinction between pointwise and uniform convergence of a sequence or series of functions is crucial in Measure Theory. Here we summarise some of the basic facts concerning the two kinds of convergence.

Definition 1.5.1 Let $\{f_n\}_{n\geq 1}$ and f be extended real-valued functions defined on some set X. We say that $\{f_n\}_{n\geq 1}$ **converges pointwise** to f if and only if $\lim_n f_n(x) = f(x)$ for each $x \in X$.

If f_n converges to f pointwise, we write (according to convenience)

$$\lim_n f_n = f \text{ pointwise} \quad \text{or} \quad f_n \to f$$
$$\text{or} \quad f_n(x) \to f(x), x \in X.$$

The sequence $\{f_n\}_{n\geq 1}$, where $f_n(x) = x^n$, $x \in [0, 1]$, converges pointwise to the function f defined by $f(x) = 0$, $0 \leq x < 1$, $f(1) = 1$.

Definition 1.5.2 The sequence $\{f_n\}_{n\geq 1}$ of real-valued functions **converges uniformly** to a real-valued function f if and only if for each $\varepsilon > 0$, there exists an integer $n_0(\varepsilon)$ such that

$$|f(x) - f_n(x)| < \varepsilon \quad \text{for all } n \geq n_0(\varepsilon) \text{ and all } x \in X.$$

If f_n converges to f uniformly, we write (according to convenience)

$$\lim_n f_n = f(\text{unif}) \text{ or } f_n \overset{unif}{\rightarrow} f.$$

The sequence $\{f_n\}_{n\geq 1}$, where $f_n(x) = x^n$, $x \in [0, \alpha]$, and $0 < \alpha < 1$, converges uniformly to the identically zero function.

It is immediate that uniform convergence implies pointwise convergence. The reverse implication is, however, not true. The sequence $\{f_n\}_{n\geq 1}$, where $f_n(x) = x^n$, $x \in [0, 1]$, is not uniformly convergent. That it is pointwise convergent has been observed above.

Proposition 1.5.3 *Let $\{f_n\}_{n\geq 1}$ be sequence of real-valued functions defined on a metric space X and $x_0 \in X$. Suppose that $f_n \to f$ uniformly on X. If each f_n is continuous at x_0, then f is continuous at x_0. Briefly put, a uniform limit of functions continuous at a point is continuous at that point.*

Remark It follows from Proposition 1.5.3 that the sequence $\{f_n\}_{n\geq 1}$, where $f_n(x) = x^n$, $x \in [0, 1]$, does not converge uniformly to the function f defined by $f(x) = 0$, $0 \leq x < 1$, $f(1) = 1$, as the limit function is discontinuous at $x = 1$.

The Cauchy criterion of uniform convergence of sequences of functions is as follows.

Theorem 1.5.4 (Cauchy Criterion of Uniform Convergence) *The sequence of real-valued functions $\{f_n\}_{n\geq 1}$ defined on X converges uniformly on X if and only if, given $\varepsilon > 0$, there exists an integer n_0 such that, for all $x \in X$ and all $n \geq n_0$, $m \geq n_0$, we have $|f_n(x) - f_m(x)| < \varepsilon$.*

The above necessary and sufficient condition for uniform convergence is called the **uniform Cauchy property** and a sequence satisfying it is said to be **uniformly Cauchy**.

A finite sum

$$\alpha_0 + \sum_{k=1}^{n} (\alpha_k \cos kx + \beta_k \sin kx)$$

is called a *trigonometric polynomial*. For a proof of the result below, the reader may consult Problem 6.3.P8 of [28]; it will be used in Proposition 3.4.4 and an independent proof based on Fourier series will later be given in Sect. 4.3.

Theorem 1.5.5 (Weierstrass Approximation Theorem) *Let f be a continuous function on $[-\pi, \pi]$ satisfying $f(-\pi) = f(\pi)$. Then for any $\varepsilon > 0$, there exists a trigonometric polynomial*

$$P(x) = a_0 + \sum_{k=1}^{n} (a_k \cos kx + b_k \sin kx)$$

such that

$$|f(x) - P(x)| < \varepsilon \quad \textit{for all } x \in [-\pi, \pi].$$

Recall that a series $\sum_{k=1}^{\infty} x_k$ of real numbers converges to $x \in \mathbb{R}$ if the sequence $\{s_n\}_{n\geq 1}$, where $s_n = \sum_{k=1}^{n} x_k$ (the nth partial sum), converges to x. We write

$x = \lim s_n = \sum_{k=1}^{\infty} x_k$ and x is called the sum of the series. If $\sum_{n=1}^{\infty} f_n(x)$ converges for every $x \in X$, and if we define $f(x) = \sum_{n=1}^{\infty} f_n(x), x \in X$, the function f is called the *sum* of the series $\sum_{n=1}^{\infty} f_n$. We say that the *series* $\sum_{n=1}^{\infty} f_n$ *converges pointwise.*

We say that the *series* $\sum_{n=1}^{\infty} f_n$ *converges uniformly* on X if the sequence $\{s_n\}_{n \geq 1}$ of functions, where $s_n(x) = \sum_{k=1}^{n} f_k(x),\ x \in X$, converges uniformly on X. As with series of numbers, we speak of the series $\sum_{n=1}^{\infty} f_n$ converging *absolutely* when the sequence $\sum_{k=1}^{n} |f_k|$ of partial sums converges pointwise. It is easy to check that when $\sum_{n=1}^{\infty} |f_n|$ converges uniformly, so does $\sum_{n=1}^{\infty} f_n$. For an example when $\sum_{n=1}^{\infty} f_n$ converges uniformly as well as absolutely but $\sum_{n=1}^{\infty} |f_n|$ does not converge uniformly, see Problem 5.2.P9 on p. 160 of [28].

The following test for uniform convergence of series of functions is due to Weierstrass.

Proposition 1.5.6 (Weierstrass M-test) *Suppose* $\{f_n\}_{n \geq 1}$ *is a sequence of functions defined on X and suppose that* $|f_n(x)| \leq M_n$, $x \in X$ *and* $n = 1, 2, \ldots$. *Then* $\sum_{n=1}^{\infty} f_n$ *converges uniformly on X if* $\sum_{n=1}^{\infty} M_n$ *converges.*

Proposition 1.5.7 (needed in Proposition 5.2.11) *Suppose* $\{f_n\}_{n \geq 1}$ *is a sequence of functions defined on an interval, converging uniformly to a limit function f and c is a point in the interval such that* $f_n(c-)$ *exists for every n. Then* $\lim_n f_n(c-)$ *exists and equals* $f(c-)$. *Similarly for right-hand limits.*

The gist of the proof is the same as that of Proposition 6.2.6 on p. 178 of [28].

Uniform convergence cannot be dropped. Take $f_n(x) = x^n$ on $[0, 1]$ and $c = 1$. Then $f_n(c-) = 1$ for every n but $f = \lim f_n$ satisfies $f(c-) = 0$.

1.6 Derivatives

Let $S \subseteq \mathbb{R}$ and $x \in S$ be a limit point of the set. A function $f : S \to \mathbb{R}$ is *differentiable at x* if

$$\lim_{h\to 0}\frac{f(x+h)-f(x)}{h}$$

exists, in which case, the limit is called the *derivative of f at x* and is denoted by $f'(x)$. It is often more convenient to write $\frac{d}{dx}f(x)$ for $f'(x)$. The *derivative function f′* is the one that maps each point of differentiability into the derivative at that point and is called simply the *derivative of f.*

If $f'(x)$ and $g'(x)$ both exist, then so do $(f+g)'(x)$ and $(fg)'$; moreover, $(f+g)'(x)=f'(x)+g'(x)$ and $(fg)'(x)=f'(x)g(x)+f(x)g'(x)$.

If α is any real number, then $(\alpha f)'(x)=\alpha(f'(x))$. If x is a limit point of the set on which $g\neq 0$ and also belongs to the set, then

$$\left(\frac{f}{g}\right)'(x)=\frac{f'(x)g(x)-f(x)g'(x)}{g(x)^2}.$$

We assume that the reader is aware of trigonometric functions, the exponential and natural logarithm functions, and also of their limit and differentiation properties, such as

$$\frac{d}{dx}\sin x=\cos x,\quad \frac{d}{dx}\tan^{-1}x=\frac{1}{1+x^2},\quad \frac{d}{dx}\ln x=\frac{1}{x}\quad \text{and so forth.}$$

The functions can be defined variously via limit processes and all their properties learned in calculus can be derived from there. The manner in which this is done will be of no consequence for the material in this book.

Proposition 1.6.1 (Chain Rule) *Suppose $f : S\to\mathbb{R}$ is differentiable at $x\in S$ and g maps a set containing $f(S)$ into $\mathbb{R}$. If g is differentiable at $f(x)\in f(S)$, then the composition $g\circ f$ is differentiable at x and*

$$(g\circ f)'(x)=g'(f(x))\cdot f'(x).$$

Let I denote an interval. A function $f : I\to\mathbb{R}$ is said to have a *local maximum* at $c\in I$ if there exists a $\delta>0$ such that $x\in I$, $|x-c|<\delta\Rightarrow f(x)\leq f(c)$. Similarly for a *local minimum.*

A function $f : D\to\mathbb{R}$, where D is a subset of $\mathbb{R}$, is said to be *increasing* if, for all $x_1, x_2\in D$,

$$x_1<x_2\Rightarrow f(x_1)\leq f(x_2)$$

and *strictly increasing* if

$$x_1<x_2\Rightarrow f(x_1)<f(x_2).$$

Correspondingly for *decreasing* and *strictly decreasing*. A *monotone* function on an interval is one which is either increasing or decreasing.

Proposition 1.6.2 *If* $f : [a, b] \to \mathbb{R}$ *satisfies* $f'(x) \geq 0$ *for every* $x \in [a, b]$, *then* f *is increasing. Similarly, if* $f : [a, b] \to \mathbb{R}$ *satisfies* $f'(x) \leq 0$ *for every* $x \in [a, b]$, *then* f *is decreasing.*

Proposition 1.6.3 *If* $f : [a, b] \to \mathbb{R}$ *has a local maximum or a local minimum at* $c \in (a, b)$ *and is differentiable at* c, *then* $f'(c) = 0$.

Proposition 1.6.4 *Suppose* $f : [a, b] \to \mathbb{R}$ *satisfies* $f'(c) = 0$, *where* $c \in (a, b)$. *If* $f''(c) < 0$ *then* f *has a local maximum at* c *and if* $f''(c) > 0$ *then* f *has a local minimum at* c.

Mean Value Theorem 1.6.5 *Suppose the continuous function* $f : [a, b] \to \mathbb{R}$ *is differentiable on* (a, b). *Then there exists some* $\xi \in (a, b)$ *such that*

$$f(b) - f(a) = f'(\xi)(b - a).$$

Proposition 1.6.6 *Suppose* $f : [a, b] \to \mathbb{R}$ *has derivative zero at every point of its domain. Then the function is a constant.*

Proposition 1.6.7 (needed in Proposition 5.2.11) *Suppose* $\{f_n\}_{n \geq 1}$ *is a pointwise convergent sequence of functions defined on an interval. If each* f_n *is increasing, then the limit function is also increasing.*

1.7 Riemann Integration

By a partition P of an interval $[a, b]$ we mean a finite sequence of points x_k in the interval such that

$$P : a = x_0 < x_1 < \cdots < x_n = b.$$

For a bounded function $f : [a, b] \to \mathbb{R}$ and any partition, the nonempty set $\{f(x): x_{k-1} \leq x \leq x_k\}$ is bounded above as well as below for each k. Consequently it has supremum M_k and an infimum m_k. The *upper* and *lower sums of* f *over the partition* P are respectively

$$\sum_{k=1}^{n} M_k(x_k - x_{k-1}) \quad \text{and} \quad \sum_{k=1}^{n} m_k(x_k - x_{k-1}).$$

Their respective infimum and supremum are called the *upper* and *lower integrals* respectively of f and are denoted by $\overline{\int_a^b} f$ and $\underline{\int_a^b} f$.

Thus

$$\overline{\int_a^b} f = \inf\left\{\sum_{k=1}^{n} M_k(x_k - x_{k-1}) : \text{all partitions } P\right\}$$

and

$$\underline{\int_a^b} f = \sup\left\{\sum_{k=1}^{n} m_k(x_k - x_{k-1}) : \text{all partitions } P\right\}.$$

It turns out that $\overline{\int_a^b} f \geq \underline{\int_a^b} f$ for every bounded function f. If equality holds, then the function f is said to be *Riemann integrable*, and the *Riemann integral* of f from a to b is the common value of the upper and lower integrals, denoted by

$$\int_a^b f \quad \text{or} \quad \int_a^b f(x)\,dx.$$

Sometimes it is convenient to speak of f being *Riemann integrable on* $[a, b]$.

The Riemann integral exists, for instance, if f is continuous or monotone.

In the present section, we shall abbreviate "Riemann integrable" to simply "integrable", although in the context of Measure Theory later, the word will have a technically different meaning.

If the restriction of f to $[\alpha, \beta] \subseteq [a, b]$ is integrable, we say that f is integrable on $[\alpha, \beta]$.

If $f : [a, b] \to \mathbb{R}$ and $g : [a, b] \to \mathbb{R}$ are both integrable on $[a, b]$, then so are $f + g$, fg and αf (α a real number); moreover,

$$\int_a^b (f+g) = \int_a^b f + \int_a^b g \quad \text{and} \quad \int_a^b (\alpha f) = \alpha \int_a^b f.$$

Suppose $f : [a, b] \to \mathbb{R}$ is bounded and $a < c < b$. If f is integrable on $[a, b]$, then it is integrable on $[a, c]$ as well as $[c, b]$, and the equality

$$\int_a^b f = \int_a^c f + \int_c^b f$$

holds. Conversely, if f is integrable on $[a, c]$ as well as $[c, b]$, then it is integrable on $[a, b]$ and the foregoing equality holds. It holds without the proviso that $a < c < b$ if we agree that

$$\int_a^b f = -\int_b^a f.$$

If $f : [a,b] \to \mathbb{R}$ and $g : [a,b] \to \mathbb{R}$ are both integrable and $f(x) \leq g(x)$ for each $x \in [a, b]$, then $\int_a^b f \leq \int_a^b g$.

If f is integrable on $[a, b]$, and $[\alpha, \beta] \subseteq [a, b]$, then f is integrable on $[\alpha, \beta]$. If also $f \geq 0$ on $[a, b]$ then $\int_\alpha^\beta f \leq \int_a^b f$.

Proposition 1.7.1 *Let $f : [a,b] \to \mathbb{R}$ be integrable. Then $|f| : [a,b] \to \mathbb{R}$ is also integrable and*

$$\left| \int_a^b f \right| \leq \int_a^b |f|.$$

Theorem 1.7.2 (a) First Fundamental Theorem of Calculus *Let $f : [a,b] \to \mathbb{R}$ be integrable and let*

$$F(x) = \int_a^x f, \quad a \leq x \leq b.$$

Then F is continuous on $[a, b]$. Moreover, if f is continuous at a point $c \in [a, b]$, then F is differentiable at c and

$$F'(c) = f(c).$$

(b) Second Fundamental Theorem of Calculus *Let f be Riemann integrable on $[a, b]$ and suppose that there exists a function F on the same interval such that $F'(x) = f(x)$ everywhere. Then*

$$\int_a^b f(x)\,dx = F(b) - F(a).$$

There are two versions of the *Substitution Rule* or *Change of Variables Formula.*

Proposition 1.7.3 (Version 1) *Suppose $\varphi : [\alpha, \beta] \to [a, b]$ is a bijection having a continuous derivative that vanishes nowhere. If $(f \circ \varphi)|\varphi'|$ is integrable on $[\alpha, \beta]$ and f is integrable on the image $\varphi([\alpha, \beta]) = [a, b]$, then*

$$\int_a^b f = \int_\alpha^\beta (f \circ \varphi)|\varphi'|.$$

The reason for the absolute value on the right-hand side is that if $\varphi' < 0$ everywhere, then we have $\varphi(\alpha) = b$ and $\varphi(\beta) = a$. It is possible to deduce the integrability of $(f \circ \varphi)|\varphi'|$ from the remaining hypotheses. See [15, pp. 170–171].

Proposition 1.7.3 (Version 2) *Let $F : [a,b] \to \mathbb{R}$ and $\varphi : [\alpha, \beta] \to [a,b]$ both be differentiable. If F' and $(F' \circ \varphi)\varphi'$ are both Riemann integrable, then*

$$\int_{\varphi(\alpha)}^{\varphi(\beta)} F' = \int_{\alpha}^{\beta} (F' \circ \varphi)\varphi'.$$

It is not presumed in either version of the above proposition that $\varphi(\alpha) < \varphi(\beta)$. The next result is the *Formula of Integration by Parts.*

Proposition 1.7.4 *Let f and g be differentiable functions on $[a, b]$ having integrable derivatives f' and g'. Then the products fg' and $f'g$ are integrable, and*

$$\int_a^b fg' = f(b)g(b) - f(a)g(a) - \int_a^b f'g.$$

Proposition 1.7.5 *Suppose $\{f_n\}_{n \geq 1}$ is a sequence of Riemann integrable functions on $[a, b]$ with uniform limit f. Then f is Riemann integrable on $[a, b]$ and*

$$\lim \int_a^b f_n = \int_a^b f.$$

If a function $f : [a, \infty) \to \mathbb{R}$ is integrable on $[a, b]$ whenever $a < b$, the symbol $\int_a^\infty f$ means $\lim\limits_{b \to \infty} \int_a^b f$, even if the limit does not exist. If it does not, we say that $\int_a^\infty f$ is *divergent*; otherwise *convergent*. In either case, $\int_a^\infty f$ is called the *improper integral of f over* $[a, \infty)$. If it is convergent, we speak of f being improper Riemann integrable over $[a, \infty)$. If $\int_a^\infty |f|$ is convergent, we say that $\int_a^\infty f$ is *absolutely convergent*. If $\int_a^\infty |f|$ is divergent but $\int_a^\infty f$ is convergent, then we say that the latter is conditionally convergent. If $\int_a^\infty f$ is absolutely convergent, then it is convergent. Corresponding remarks apply to a function $f : (-\infty, b] \to \mathbb{R}$ and the improper integral $\int_{-\infty}^b f$.

As an illustration,

$$\int_1^\infty \frac{1}{x} dx = \lim_{b \to \infty} \int_1^b \frac{1}{x} dx = \lim_{b \to \infty} (\ln b - \ln 1) = \infty$$

and the improper integral $\int_1^\infty \frac{1}{x}\, dx$ is therefore divergent.

If a function $f : (a, b] \to \mathbb{R}$ is integrable on $[c, b]$ whenever $a < c < b$, but is unbounded on $(a, b]$, the symbol $\int_a^b f$ means $\lim\limits_{c \to a+} \int_c^b f$, even if the limit does not exist. If it does not, we say that $\int_a^b f$ is *divergent*; otherwise *convergent*. In either case, $\int_a^b f$ is called the *improper integral of f over* $[a, b]$. If it is convergent, we speak of f being improper Riemann integrable over $[a, b]$. Corresponding remarks apply to a function $f : [a, b) \to \mathbb{R}$ and the improper integral $\int_a^b f$. The meaning of absolute convergence in this context is clear.

As an illustration,

$$\int_0^1 x^{-\frac{1}{2}}\,dx = \lim_{c\to 0+}\int_c^1 x^{-\frac{1}{2}}\,dx = \lim_{c\to 0+}(2-2c^{\frac{1}{2}}) = 2$$

and the improper integral $\int_0^1 x^{-\frac{1}{2}}\,dx$ is therefore convergent.

Proposition 1.7.6 (Comparison Test for Improper Integrals) *Suppose the functions* $f:[a,\infty)\to\mathbb{R}$ *and* $g:[a,\infty)\to\mathbb{R}$ *are integrable on* $[a, b]$ *for every* $b > a$. *If* $0 \le f(x) \le g(x)$ *for every* $x \in [a, \infty)$ *and if* $\int_a^\infty g$ *is convergent, then* $\int_a^\infty f$ *is convergent; moreover,* $\int_a^\infty f \le \int_a^\infty g$. *Similar assertions apply to improper integrals over bounded intervals.*

1.8 Decimals and the Cantor Set

Let $b \in \mathbb{N}$ and $n \in \mathbb{N}$, where $b > 1$. Every integer j such that $0 \le j < b^{n-1}$ then has a representation as $j = \sum_{k=1}^{n-1} \alpha_k b^{n-k-1}$, where $\alpha_1, \alpha_2, \ldots, \alpha_{n-1}$ are integers satisfying $0 \le \alpha_k \le b-1$. For $n = 1$, this is trivially true, because $j = 0$ in this case and the sum is empty, which is understood to have value 0. For $n > 1$, the assertion is essentially Theorem 1.11.7 of [28, p. 51].

In the context of such representations, the integer b is called the *base* and is tacitly taken to be greater than 1. In much of our discussion later, there will be particular emphasis on the bases $b = 2$ and $b = 3$.

Consider an arbitrary $x \in [0, 1]$. In what follows, we shall first exhibit an infinite sequence $\{\alpha_k : k \in \mathbb{N}\}$ of integers such that

$$x = \alpha_1 b^{-1} + \cdots + \alpha_k b^{-k} + \cdots = \sum_{k=1}^{\infty} \alpha_k b^{-k},$$

$$0 \le \alpha_k \le b-1.$$

Divide the interval [0, 1] into b equal parts. Then x lies in some interval $[\frac{\alpha_1}{b}, \frac{\alpha_1+1}{b}]$, where α_1 is an integer in $\{0, 1, \ldots, b-1\}$. If $x\,(x \ne 0, x \ne 1)$ is one of the points of subdivision, then two values of α_1 are possible and either may be chosen. Regardless of the choice, we have the inequality

$$\frac{\alpha_1}{b} \le x \le \frac{\alpha_1+1}{b},$$

where $\alpha_1 \in \{0, 1, \ldots, b-1\}$. The chosen subinterval is again divided into b equal subintervals and the process is continued indefinitely. In this way, we obtain a sequence of nested intervals

$$\frac{\alpha_1}{b}+\frac{\alpha_2}{b^2}+\cdots+\frac{\alpha_n}{b^n}\leq x\leq\frac{\alpha_1}{b}+\frac{\alpha_2}{b^2}+\cdots+\frac{\alpha_n}{b^n}+\frac{1}{b^n}, \tag{1.2}$$

where

$$\alpha_1, \alpha_2, \ldots, \alpha_k, \ldots$$

is a sequence of integers such that $0 \leq \alpha_k \leq b-1$, $k \in \mathbb{N}$. By virtue of (1.2), we have $x=\sum_{k=1}^{\infty}\alpha_k b^{-k}$ and we have thus exhibited the kind of sequence that was promised.

We call any representation of x by a series $\sum_{k=1}^{\infty}\alpha_k b^{-k}$ a *decimal representation* of x in the base b and write $x = .\alpha_1\alpha_2\ldots\alpha_k\ldots$. The matter of its uniqueness or otherwise will be taken up further below, culminating in Theorem 1.8.3. The expression $.\alpha_1\alpha_2\ldots\alpha_k\ldots$ represents the series $\sum_{k=1}^{\infty}\alpha_k b^{-k}$ in an abbreviated form that does not mention b explicitly. Here the integers α_1, α_2, …, α_k, … are called the 1st digit, 2nd digit and so on, of the decimal representation. They may be spoken of as digits "of x" when a representation is understood to be given.

We shall speak of the cases $b = 2$ and 3 as *binary* and *ternary* respectively.

Conversely, given a sequence α_1, α_2, …, α_k, … of integers such that $0 \leq \alpha_k \leq b-1$, $k \in \mathbb{N}$, we claim that there exists a unique $x \in [0, 1]$ satisfying (1.2) for all n and hence also satisfying $x=\sum_{k=1}^{\infty}\alpha_k b^{-k}$. To obtain such an x, first construct intervals defined by (1.2). They have the following properties: (i) each succeeding interval is a subinterval of the preceding one (keeping in mind that $\frac{\alpha_{n+1}}{b^{n+1}}+\frac{1}{b^{n+1}}\leq\frac{1}{b^n}$); (ii) the lengths of the intervals tend to 0 as $n\to\infty$. It follows from the Nested Interval Theorem 1.4.4 that there exists a unique $x \in [0, 1]$ as claimed.

The reader will recall from previous knowledge that the number of sequences of b terms, with each term belonging to a given set consisting of b elements (such as $\{0, 1, \ldots, b-1\}$), is b^n.

Remark 1.8.1 The endpoints of the interval described in (1.2) have a useful alternative description that we proceed to derive. Given α_n, the b^{n-1} possibilities for the left-hand side in (1.2) can be described without explicit reference to α_1, α_2, …, α_{n-1}. Indeed, the left endpoint of the interval can be recast as follows (including when $n = 1$): $\frac{1}{b^n}(\sum_{k=1}^{n}\alpha_k b^{n-k})=\frac{1}{b^n}(\sum_{k=1}^{n-1}\alpha_k b^{n-k}+\alpha_n)=\frac{1}{b^n}(b\sum_{k=1}^{n-1}\alpha_k b^{n-k-1}+\alpha_n)$ $=\frac{1}{b^n}(bj+\alpha_n)$, where $j=\sum_{k=1}^{n-1}\alpha_k b^{n-k-1}$ is an integer satisfying $0\leq j\leq(b-1)\sum_{k=1}^{n-1}b^{n-k-1}=b^{n-1}-1$. Conversely, if j is an integer satisfying $0 \leq j \leq b^{n-1}-1$, then it follows from the representation of j mentioned in the opening paragraph of this section that $\frac{1}{b^n}(bj+\alpha_n)=\frac{1}{b^n}(b\sum_{k=1}^{n-1}\alpha_k b^{n-k-1}+\alpha_n)=\frac{1}{b^n}(\sum_{k=1}^{n-1}\alpha_k b^{n-k}+\alpha_n)=\frac{1}{b^n}(\sum_{k=1}^{n}\alpha_k b^{n-k})$. Thus, for a given α_n, the possible left-hand sides in (1.2) can be alternatively described as precisely the b^{n-1} numbers of the form $\frac{1}{b^n}(bj+\alpha_n)$ with $0 \le j \le b^{n-1}-1$.

This alternative description yields the consequence that, for a given α_n, a number lies in an interval of the form (1.2) if and only if it lies in some interval $[\frac{bj+\alpha_n}{b^n},\frac{bj+\alpha_n+1}{b^n}]$, where $0 \le j \le b^{n-1}-1$. The case when $\alpha_n = 1$ is of special interest: a number $x \in [0, 1]$ has a representation $x=\sum_{k=1}^{\infty}\alpha_k b^{-k}$ with nth digit $\alpha_n = 1$ if and only if it lies in some interval $[\frac{bj+1}{b^n},\frac{bj+2}{b^n}]$ with $0 \le j \le b^{n-1}-1$. For $b = 3$, we thus find that a number $x \in [0, 1]$ has a ternary representation $x=\sum_{k=1}^{\infty}\alpha_k 3^{-k}$ with nth digit $\alpha_n = 1$ if and only if it lies in an interval of the form $[\frac{3j+1}{3^n},\frac{3j+2}{3^n}]$, where $0 \le j \le 3^{n-1}-1$.

Since the “if” part and “only if” part will be used separately, we record them separately for ease of reference when needed later.

(A) If $x \in [\frac{3j+1}{3^n},\frac{3j+2}{3^n}]$ for some j satisfying $0 \le j \le 3^{n-1}-1$, then x has a ternary representation with 1 as the nth digit.

(B) If $x \in [0, 1]$ does not belong to an interval $[\frac{3j+1}{3^n},\frac{3j+2}{3^n}]$ for any j satisfying $0 \le j \le 3^{n-1}-1$, then x cannot have a ternary representation with 1 as the nth digit.

The interval can also be described as $[\frac{3j}{3^n}+\frac{1}{3^n},\frac{3j}{3^n}+\frac{2}{3^n}]$. The endpoints of any such interval will be seen to have more than one ternary representation, the details of which are discussed below in the third paragraph of Remark 1.8.6.

Before proceeding, we remind the reader of the elementary equalities

$$(b-1)\sum_{k=1}^{n} b^{n-k}=b^n-1 \quad \text{and} \quad (b-1)\sum_{k=n+1}^{\infty} b^{-k}=b^{-n}.$$

The latter of these has the consequence that

$$(b-1)\sum_{k=n+1}^{m} b^{-k}<b^{-n} \text{ for all } m>n.$$

Consider $x=\sum_{k=1}^{\infty}\alpha_k b^{-k}$ and $y=\sum_{k=1}^{\infty}\beta_k b^{-k}$, where $0 \le \alpha_k \le b-1$ and $0 \le \beta_k \le b-1$ for all $k \in \mathbb{N}$. Then $\alpha_k - \beta_k \ge -(b-1)$ for all $k \in \mathbb{N}$ and equality obtains if and only if $\alpha_k = 0$, $\beta_k = b-1$, a fact that will be used presently.

Suppose the sequences $\{\alpha_k\}$ and $\{\beta_k\}$ are distinct, without prejudice to whether x and y are distinct or not. Then there must exist a smallest k such that $\alpha_k \neq \beta_k$; denote it by n. This means $\alpha_n \neq \beta_n$ and $k < n$ implies $\alpha_k = \beta_k$ (without ruling out the possibility that $n = 1$). We have

$$\begin{aligned} x - y &= \sum_{k=1}^{\infty} (\alpha_k - \beta_k) b^{-k} = \sum_{k=n}^{\infty} (\alpha_k - \beta_k) b^{-k} \\ &= (\alpha_n - \beta_n) b^{-n} + \sum_{k=n+1}^{\infty} (\alpha_k - \beta_k) b^{-k}. \end{aligned} \tag{1.3}$$

Using the fact that $\alpha_k - \beta_k \geq -(b-1)$ for all $k \in \mathbb{N}$, we deduce from (1.3) that

$$\begin{aligned} x - y &\geq (\alpha_n - \beta_n) b^{-n} - (b-1) \sum_{k=n+1}^{\infty} b^{-k} = (\alpha_n - \beta_n) b^{-n} - b^{-n} \\ &= b^{-n}((\alpha_n - \beta_n) - 1). \end{aligned} \tag{1.4}$$

This shows that $\alpha_n > \beta_n$ implies $x \geq y$. Remembering that $\alpha_n \neq \beta_n$, the contrapositive is that $x < y$ implies $\alpha_n < \beta_n$, or equivalently, $x > y$ implies $\alpha_n > \beta_n$.

Remark 1.8.2 In the event that there exists a positive integer $m > n$ such that $\alpha_k = \beta_k$ for all $k > m$, i.e., the digits agree beyond a certain stage, we deduce from (1.3) the stronger inequality that

$$\begin{aligned} x - y &\geq (\alpha_n - \beta_n) b^{-n} - (b-1) \sum_{k=n+1}^{m} b^{-k} > (\alpha_n - \beta_n) b^{-n} - b^{-n} \\ &= b^{-n}((\alpha_n - \beta_n) - 1). \end{aligned}$$

Therefore $\alpha_n > \beta_n$ implies $x > y$. Combining this with the conclusion of the paragraph preceding this remark, we infer that, if the digits agree beyond a certain stage, then the inequality $x > y$ is equivalent to the condition that $\alpha_n > \beta_n$, where n denotes the smallest positive integer k for which $\alpha_k \neq \beta_k$. It will be convenient to rephrase the condition without explicit mention of the symbols α_k, β_k and n as

the earliest digit of x that differs from the corresponding digit of y is the greater one.

The equivalence of the condition with the inequality $x > y$ when digits agree beyond a certain stage will be invoked in Remark 1.8.14(e) below.

To consider the matter of uniqueness of a decimal representation, we specialise to $x = y$ in the foregoing discussion up to and including (1.4). Since $\alpha_n \neq \beta_n$, we may assume without loss of generality that $\alpha_n > \beta_n$. Then it follows from (1.4) that

$$0 \geq b^{-n}((\alpha_n - \beta_n) - 1) \geq 0.$$

Consequently, $b^{-n}((\alpha_n - \beta_n) - 1) = 0$ and hence $\alpha_n - \beta_n = 1$.

Using this in (1.3), we find that

$$0 = b^{-n} + \sum_{k=n+1}^{\infty} (\alpha_k - \beta_k)b^{-k}.$$

We can deduce from here that $k > n$ implies $\alpha_k = 0$, $\beta_k = b - 1$. Indeed, if this were not so, there would exist some $k > n$ such that $\alpha_k - \beta_k > -(b-1)$ and the preceding equality would lead to $0 > b^{-n} - (b-1)\sum_{k=n+1}^{\infty} b^{-k} = b^{-n} - b^{-n} = 0$, a contradiction.

Since $k > n$ implies $\alpha_k = 0$, we can write

$$x = \sum_{k=1}^{n} \alpha_k b^{-k} = b^{-n}\sum_{k=1}^{n} \alpha_k b^{n-k} = \frac{m}{b^n}, \text{ where } m = \sum_{k=1}^{n} \alpha_k b^{n-k}.$$

Here,

$$0 < b^{-n} \leq (\beta_n + 1)b^{-n} = \alpha_n b^{-n} \leq \alpha_n b^{n-n} \leq \sum_{k=1}^{n} \alpha_k b^{n-k} = m \leq (b-1)\sum_{k=1}^{n} b^{n-k}$$
$$= b^n - 1,$$

which is to say, $0 < m < b^n$. As $b - 1 \geq \alpha_n = \beta_n + 1 > 0$, the integer α_n is not divisible by b, and hence the integer

$$m = \sum_{k=1}^{n} \alpha_k b^{n-k} = \alpha_1 b^{n-1} + \alpha_2 b^{n-2} + \cdots + \alpha_n$$

is also not divisible by b.

Let us summarise what has been proved about uniqueness in the preceding three paragraphs.

Theorem 1.8.3 *Suppose $x \in [0, 1]$ has two distinct representations*

$$x = \sum_{k=1}^{\infty} \alpha_k b^{-k} = \sum_{k=1}^{\infty} \beta_k b^{-k},$$

where $0 \leq \alpha_k \leq b-1$, $0 \leq \beta_k \leq b-1$, $k \in \mathbb{N}$, and n is the smallest value of k such that $\alpha_k \neq \beta_k$. If $\alpha_n > \beta_n$, then

(a) $\alpha_n - \beta_n = 1$,
(b) *$k > n$ implies $\alpha_k = 0$, $\beta_k = b - 1$,*
(c) *$x = \frac{m}{b^n}$, where $0 < m < b^n$ (in particular, $x \neq 0$ and $x \neq 1$) and m is not divisible by b.*

It follows from (b) that there cannot be a third representation distinct from the two described above. So, there can be at most two distinct representations.

Although (c) tells us that numbers in [0, 1] having at least two distinct representations must be of a certain form, it does not tell us that numbers of that form must conversely have at least two distinct representations. We prove the converse by proceeding as below.

Let $x = \frac{m}{b^n}$, where $0 < m < b^n$ and m is not divisible by b. By Theorem 1.11.7 of [28, p. 51], there exists a finite sequence $\gamma_1, \ldots, \gamma_n$ of integers such that $0 \leq \gamma_k \leq b - 1$ and $m = \sum_{k=1}^{n} \gamma_k b^{n-k}$. Therefore

$$\frac{m}{b^n} = \sum_{k=1}^{n} \gamma_k b^{-k}.$$

This can be rewritten as $\frac{m}{b^n} = \sum_{k=1}^{\infty} \alpha_k b^{-k}$, where $\alpha_k = \gamma_k$ for $k \leq n$ and $\alpha_k = 0$ for $k > n$. In other words,

$$\frac{m}{b^n} = .\gamma_1\gamma_2\ldots\gamma_n 00.\ldots$$

Since m is not divisible by b, the representation $m = \sum_{k=1}^{n} \gamma_k b^{n-k}$ shows that $\gamma_n \neq 0$ and therefore $\gamma_n \geq 1$. But $\alpha_n = \gamma_n$ and therefore $\alpha_n \geq 1$. If we define the sequence $\{\beta_k : k \in \mathbb{N}\}$ by setting

$$\beta_k = \alpha_k = \gamma_k \quad \text{for} \quad k<n, \quad \beta_n = \alpha_n - 1 = \gamma_n - 1, \quad \beta_k = b - 1 \quad \text{for} \quad k > n,$$

obviously the smallest value of k such that $\alpha_k \neq \beta_k$ is n, $\alpha_n > \beta_n$, and $0 \leq \beta_k \leq b - 1$ for $k \neq n$. Since $\alpha_n \geq 1$, the inequality $0 \leq \beta_k \leq b - 1$ holds for $k = n$ as well and hence for all k. Furthermore,

$$\begin{aligned}
\sum_{k=1}^{\infty} \beta_k b^{-k} &= \sum_{k=1}^{n-1} \beta_k b^{-k} + \beta_n b^{-n} + \sum_{k=n+1}^{\infty} \beta_k b^{-k} \\
&= \sum_{k=1}^{n-1} \alpha_k b^{-k} + (\alpha_n - 1) b^{-n} + (b - 1) \sum_{k=n+1}^{\infty} b^{-k} \\
&= \sum_{k=1}^{n} \alpha_k b^{-k} - b^{-n} + (b - 1) \sum_{k=n+1}^{\infty} b^{-k} \\
&= \sum_{k=1}^{n} \alpha_k b^{-k} - b^{-n} + b^{-n} \\
&= \sum_{k=1}^{n} \alpha_k b^{-k} = \sum_{k=1}^{n} \gamma_k b^{-k} = \frac{m}{b^n}.
\end{aligned}$$

In other words,

$$\frac{m}{b^n} = .\beta_1\beta_2...\beta_k... = .\gamma_1\gamma_2...\gamma_{n-1}(\gamma_n - 1)(b-1)(b-1)....$$

This completes the converse proof that numbers of the form mentioned in (c) have at least two distinct representations of the form $\sum_{k=1}^{\infty} \alpha_k b^{-k}$, where $\alpha_1, \alpha_2, \ldots$ are integers satisfying $0 \leq \alpha_k \leq b-1$, $k \in \mathbb{N}$.

Taking into account Theorem 1.8.3, we have actually proved something more, which we now formulate:

Corollary 1.8.4 *Any $x \in [0, 1]$ either has unique representation as $\sum_{k=1}^{\infty} \alpha_k b^{-k}$, i.e., as $.\alpha_1\alpha_2...\alpha_k...$, or has precisely two such representations; in the latter case, there exists an integer $n \geq 1$ such that one of the representations is*

$$.\gamma_1\gamma_2...\gamma_n 00... \text{ with } \gamma_n \geq 1$$

and the other one is

$$.\gamma_1\gamma_2...\gamma_{n-1}(\gamma_n - 1)(b-1)(b-1)....$$

(It is of course understood that when $n = 1$, there are no $\gamma_1, \gamma_2, \ldots, \gamma_{n-1}$.)

For instance,

$$\frac{1}{2} = .0111\cdots = .1000\cdots \quad \text{(base 2)},$$

$$\frac{1}{3} = .0222\cdots = .1000\cdots \quad \text{(base 3)},$$

$$\frac{2}{3} = .1222\cdots = .2000\cdots \quad \text{(base 3)}.$$

If m is divisible by b, then the number $x = \frac{m}{b^n}$ still has two distinct representations $x = \sum_{k=1}^{\infty} \alpha_k b^{-k} = \sum_{k=1}^{\infty} \beta_k b^{-k}$ but the smallest k such that $\alpha_k \neq \beta_k$ is less than n.

The hypothesis of the next corollary may seem a bit contrived, but we shall encounter such a situation later.

Corollary 1.8.5 *Let $x \in [0, 1]$ and $b > 2$. Suppose that, for every $n \in \mathbb{N}$, x has a representation $.\alpha_1\alpha_2...\alpha_k...$ in which the* n*th digit satisfies $\alpha_n \neq 1$. Then it has a representation $.\alpha_1\alpha_2...\alpha_k...$ in which, for every $n \in \mathbb{N}$, the* n*th digit satisfies $\alpha_n \neq 1$, which is to say, none of the digits is* 1.

Proof If x has a unique representation as $.\alpha_1\alpha_2...\alpha_k...$, there is nothing to prove. So, consider the case when the representation $x = .\alpha_1\alpha_2...\alpha_k...$ is not unique. By Corollary 1.8.4, there exists an integer $k \geq 1$ such that x has precisely two representations

$$x = .\gamma_1\gamma_2 \cdots \gamma_k 00 \ldots \quad \text{and} \quad x = .\gamma_1\gamma_2 \cdots \gamma_{k-1}(\gamma_k - 1)(b-1)(b-1)\ldots,$$

where $0 \leq \gamma_j \leq b-1$ for $1 \leq j \leq k$ and $\gamma_k \geq 1$. We have to show that in at least one of these representations, none of the digits is 1. Assume that one of the digits in the first representation is 1. We need only argue that none of the digits in the second representation is 1. Let the rth digit in the first representation be 1. Then $1 \leq r \leq k$ and $\gamma_r = 1$. If $r < k$, then the rth digit in the second representation is also 1, whereby the hypothesis is violated for $n = r$. So we must have $r = k$ and $\gamma_j \neq 1$ for $1 \leq j \leq k-1$. What this implies regarding the second representation is that none of the first $k-1$ digits is 1, the kth digit is 0 and the remaining ones are $b-1$. As $b-1 > 1$, none of the digits in the second representation can be 1. □

Remark 1.8.6 Let $n \in \mathbb{N}$ be given. When the numerator $m = \sum_{k=1}^{n} \gamma_k b^{n-k}$ of the number $x = \frac{m}{b^n}$ exceeds a multiple of b by 1, we have $\gamma_n = 1$ and the two representations are $x = .\gamma_1\gamma_2 \cdots \gamma_{n-1}100\ldots$ and $x = .\gamma_1\gamma_2 \cdots \gamma_{n-1}0(b-1)(b-1)\ldots$.

In the rest of this remark, we consider ternary representations, i.e., $b = 3$.

Let j be an integer such that $0 \leq j \leq 3^{n-1} - 1$. The integer $3j + 1$ satisfies $3j + 1 < 3^n$ and exceeds a multiple of 3 by 1, and therefore we have $\frac{3j+1}{3^n} = .\gamma_1\gamma_2 \cdots \gamma_{n-1}1000\ldots = .\gamma_1\gamma_2 \cdots \gamma_{n-1}0222\ldots$. It follows that $\frac{3j+2}{3^n} = .\gamma_1\gamma_2 \cdots \gamma_{n-1}2000\ldots = .\gamma_1\gamma_2 \cdots \gamma_{n-1}1222\ldots$. This shows that the endpoints of the interval $(\frac{3j+1}{3^n}, \frac{3j+2}{3^n})$ have ternary representations without 1 as the nth digit. We shall argue that interior points of the interval do not have this property. We have recorded in (A) of Remark 1.8.1 that any x in the interval has a ternary representation with 1 as the nth digit. If it also has a representation without 1 as the nth digit, then the smallest k for which the two representations differ is some integer $q \leq n$. It follows by (c) of Theorem 1.8.3 that the number must be of the form $\frac{p}{3^q}$, where p is some integer. If it is in the interior of the interval in question, we must have $3j + 1 < 3^{n-q}p < 3j + 2$, which means $3^{n-q}p$ is not an integer even though $q \leq n$. This contradiction shows that the number in the interior can have no ternary representation without 1 as the nth digit. What we have seen about the intervals

$$\left(\frac{3j+1}{3^n}, \frac{3j+2}{3^n}\right), \quad 0 \leq j \leq 3^{n-1} - 1$$

so far is that

(A) Any number in the interior of such an interval must have 1 as the nth digit in each of its ternary representations.

(B) Any endpoint of such an interval has at least one ternary representation without 1 as the nth digit.

(C) A number that is neither of the above two kinds cannot have a ternary representation with 1 as the nth digit (see (B) of Remark 1.8.1); *a fortiori*, such a number has at least one ternary representation without 1 as the nth digit.

In other words, the points of [0, 1] lying in the complement of the union of all these intervals, $0 \le j \le 3^{n-1} - 1$, are precisely those that have at least one ternary representation without 1 as the nth digit. What we have said in this paragraph up to this point applies to any given $n \in \mathbb{N}$. We finally consider the consequence for all $n \in \mathbb{N}$: The intersection of the complements in [0, 1] of *all* these intervals, $n = 1, 2, \ldots$, consists of all those numbers in [0, 1] which, for any given n, have at least one ternary representation without 1 as the nth digit. By Corollary 1.8.5 and its converse (which is trivial), these are precisely all those numbers in [0, 1] which have a ternary representation in which none of the digits is 1.

What is known as the *Cantor set* will be obtained by removing open intervals from the closed interval $C_0 = [0, 1]$ in stages, finitely many at a time. At the first stage, we remove the open middle third $I_{1,1} = \left(\frac{1}{3}, \frac{2}{3}\right)$ of C_0 to obtain $C_1 = \left[0, \frac{1}{3}\right] \cup \left[\frac{2}{3}, 1\right]$, a union of 2 disjoint closed intervals of length $\frac{1}{3}$ each. At the second stage, we remove the open middle thirds $I_{2,1} = \left(\frac{1}{9}, \frac{2}{9}\right)$, $I_{2,2} = \left(\frac{7}{9}, \frac{8}{9}\right)$ of the closed intervals $\left[0, \frac{1}{3}\right]$ and $\left[\frac{2}{3}, 1\right]$ to obtain the set

$$C_2 = \left[0, \frac{1}{9}\right] \cup \left[\frac{2}{9}, \frac{1}{3}\right] \cup \left[\frac{2}{3}, \frac{7}{9}\right] \cup \left[\frac{8}{9}, 1\right].$$

Observe that C_2 consists of $2^2 = 4$ disjoint closed intervals, each of the form $\left[\frac{i}{3^2}, \frac{i+1}{3^2}\right]$, of length $\frac{1}{3^2}$ each. At the third stage, we remove the open middle thirds of each of the closed intervals in C_2 to obtain $2^3 = 8$ disjoint closed intervals of length $\frac{1}{3^3}$ each. We continue in this way. In general, if C_n has been constructed and is a union of 2^n disjoint closed intervals of length $\frac{1}{3^n}$ each, then we obtain C_{n+1} by removing the open middle thirds of each of the closed intervals in C_n. Note that C_{n+1} is a union of 2^{n+1} disjoint closed intervals of length $\frac{1}{3^{n+1}}$ each. We shall denote the 2^n disjoint closed intervals, of which C_n is the union, by $J_{n,k}$ and call them the *component intervals* of C_n. Here, k ranges from 1 to 2^n and the intervals $J_{n,k}$ will be numbered from left to right. The middle third of any $J_{n,k}$ is flanked on either side by component intervals of C_{n+1} and all three have length $\frac{1}{3^{n+1}}$. The Cantor set C is what remains after the process of removing the open middle third of each component interval of C_n has been carried out for every $n \in \mathbb{N}$.

Definition 1.8.7 The **Cantor set** C is the intersection of the sets C_n, $n \in \mathbb{N}$, obtained by removing the middle thirds, starting with $C_0 = [0, 1]$:

$$C = \bigcap_{n=1}^{\infty} C_n.$$

For each n, the Cantor set C is a subset of C_n, which is a union of intervals of total length $\left(\frac{2}{3}\right)^n$.

The interval $I_{1,1} = \left(\frac{1}{3}, \frac{2}{3}\right)$, which is removed from $C_0 = [0, 1]$ at the first stage to obtain C_1, will be called the *removed* (or *complementary*) *interval* "of C_1", notwithstanding the fact that it is disjoint from C_1. It is of the same length as the 2 component intervals of C_1, which is $\frac{1}{3}$. In general, the 2^{n-1} intervals that are removed from C_{n-1} at the nth stage in obtaining C_n will be called the *removed* (or *complementary*) *intervals* "of C_n" notwithstanding the fact that they are disjoint from C_n. They are of the same length as the component intervals of C_n, which is $\frac{1}{3^n}$. They will be numbered from left to right as $I_{n,k}$, where k ranges from 1 to 2^{n-1}.

If we denote by E_n the union of the open intervals that are removed at the nth stage $\left(E_n = \bigcup_{k=1}^{2^{n-1}} I_{n,k}\right)$, then

$$C = [0, 1]\backslash \bigcup_{n=1}^{\infty} E_n = [0, 1]\backslash \left(\bigcup_{n=1}^{\infty} \bigcup_{k=1}^{2^{n-1}} I_{n,k} \right).$$

The first removed interval $I_{n,1}$ has only one component interval of C_n to its left, namely, $[0, \frac{1}{3^n}]$. Between any pair of consecutive removed intervals of C_n, there are two component intervals of C_n. Thus, the kth removed interval $I_{n,k}$ has $2k - 1$ component intervals of C_n to its left.

Remark 1.8.8 One can show by induction that $I_{n,k} = (\frac{3j+1}{3^n}, \frac{3j+2}{3^n})$ for some j, $0 \leq j \leq 3^{n-1} - 1$. This is obvious for $n = 1$, because in this case, j can only be 0 and $(\frac{3j+1}{3^n}, \frac{3j+2}{3^n})$ works out to be $\left(\frac{1}{3}, \frac{2}{3}\right)$, which is indeed the interval removed at the first stage. Assume for some n that every interval removed at the nth stage is of the form $(\frac{3j+1}{3^n}, \frac{3j+2}{3^n})$ for some j, $0 \leq j \leq 3^{n-1} - 1$. Then the intervals that remain at the nth stage are $[\frac{3j+0}{3^n}, \frac{3j+1}{3^n}]$ and $[\frac{3j+2}{3^n}, \frac{3j+3}{3^n}]$. It is these intervals whose middle thirds are removed at the $(n + 1)$th stage. However, the middle thirds of these intervals are $(\frac{3(3j)+1}{3^{n+1}}, \frac{3(3j)+2}{3^{n+1}})$ and $(\frac{3(3j+2)+1}{3^{n+1}}, \frac{3(3j+2)+2}{3^{n+1}})$, both of which have the required form because $0 \leq 3j \leq 3^n - 3 < 3^n - 1$ and $0 < 3j + 2 \leq 3^n - 3 + 2 = 3^n - 1$. This completes the induction argument that $I_{n,k} = (\frac{3j+1}{3^n}, \frac{3j+2}{3^n})$ for some j, $0 \leq j \leq 3^{n-1} - 1$.

In the light of Remark 1.8.6, it now follows that the right endpoint of the removed interval $I_{n,k}$ has a ternary representation of the form $.\gamma_1\gamma_2 \cdots \gamma_{n-1}2000\ldots$ and the other ternary representation contains the digit 1. (We shall see in Remark 1.8.14(d) later that none of the digits in $.\gamma_1\gamma_2 \cdots \gamma_{n-1}2000\ldots$ can be 1.) Another way to express this representation is $\sum\limits_{j=1}^{n-1} \frac{a_j}{3^j} + \frac{2}{3^n}$, with each a_j even. Remark 1.8.6 also shows that the left endpoint of $I_{n,k}$ has the representation $\sum\limits_{j=1}^{n-1} \frac{a_j}{3^j} + \sum\limits_{j=n+1}^{\infty} \frac{2}{3^j}$.

Having shown that an interval removed at the nth stage is indeed of the form $(\frac{3j+1}{3^n}, \frac{3j+2}{3^n})$, we can now assert that the intervals that remain at the nth stage are of the form $[\frac{3j+0}{3^n}, \frac{3j+1}{3^n}]$ and $[\frac{3j+2}{3^n}, \frac{3j+3}{3^n}]$.

It may be noted that not every interval $(\frac{3j+1}{3^n}, \frac{3j+2}{3^n})$ with $0 \leq j \leq 3^{n-1} - 1$ is removed at the nth stage; in fact, these intervals are 3^{n-1} in number while only 2^{n-1}

intervals are removed at the nth stage. However, it is important to take into account that all the 3^{n-1} intervals of this form, whether removed at the nth stage or not, are disjoint from C_n, as we shall now demonstrate. In the light of what was recorded in the preceding paragraph, it is sufficient to show that an interval of the form $(\frac{3j+1}{3^n}, \frac{3j+2}{3^n})$ is disjoint from all intervals of the form either $[\frac{3j+0}{3^n}, \frac{3j+1}{3^n}]$ or $[\frac{3j+2}{3^n}, \frac{3j+3}{3^n}]$, even with a different j. To show this, it turns out that we need not restrict ourselves only to $0 \le j \le 3^{n-1} - 1$. Suppose $(\frac{3j+1}{3^n}, \frac{3j+2}{3^n}) \cap [\frac{3p+0}{3^n}, \frac{3p+1}{3^n}] \neq \varnothing$, where $j, p \in \mathbb{Z}$. Then there exists some x such that

$$3j+1<x<3j+2 \quad \text{as well as} \quad 3p+0\le x\le 3p+1.$$

If either $3j + 2 \le 3p + 0$ or $3p + 1 \le 3j + 1$, we immediately obtain a contradiction. Therefore we have $3j + 2 > 3p + 0$ as well as $3p + 1 > 3j + 1$. The second of these inequalities implies $p > j$, which leads to $p \ge j + 1$ and hence to $3p \ge 3j + 3$, contradicting the first inequality. This contradiction establishes that no such x can exist, which implies that $(\frac{3j+1}{3^n}, \frac{3j+2}{3^n}) \cap [\frac{3p+0}{3^n}, \frac{3p+1}{3^n}] = \varnothing$. Next consider the possibility that $(\frac{3j+1}{3^n}, \frac{3j+2}{3^n}) \cap [\frac{3p+2}{3^n}, \frac{3p+3}{3^n}] \neq \varnothing$, where $j, p \in \mathbb{Z}$. Then there exists some x such that

$$3j+1<x<3j+2 \quad \text{as well as} \quad 3p+2\le x\le 3p+3.$$

If either $3j + 2 \le 3p + 2$ or $3p + 3 \le 3j + 1$, we immediately obtain a contradiction. Therefore we have $3j + 2 > 3p + 2$ as well as $3p + 3 > 3j + 1$. The first of these inequalities implies $j > p$, which leads to $j \ge p + 1$ and hence to $3j + 1 \ge 3p + 4$, contradicting the second inequality. This contradiction establishes that no such x can exist, which implies that $(\frac{3j+1}{3^n}, \frac{3j+2}{3^n}) \cap [\frac{3p+2}{3^n}, \frac{3p+3}{3^n}] = \varnothing$.

Thus all the 3^{n-1} intervals of the form $(\frac{3j+1}{3^n}, \frac{3j+2}{3^n})$ are disjoint from C_n. It follows from (B) and (C) of Remark 1.8.6 that a number in C_n has one ternary representation without 1 as its nth digit, and may or may not have another ternary representation. Since a number in the Cantor set is in C_n for every n, we conclude from Corollary 1.8.5 that it has a ternary representation with no digit equal to 1. We claim that the converse also holds:

Suppose $x \in [0, 1]$ has a ternary representation with no digit equal to 1. By (A) of Remark 1.8.6, it cannot be in the interval $(\frac{3j+1}{3^n}, \frac{3j+2}{3^n})$ for any j and for any n. Since all the removed intervals are of this form, x cannot be in any of the removed intervals at any stage. This means it is in the Cantor set.

We summarise the discussion as the following theorem.

Theorem 1.8.9 *A point $x \in [0, 1]$ is in the Cantor set if and only if it has a ternary representation which consists of the digits* 0 *and* 2 *only* (*i.e., every digit is even*).

Remark 1.8.10 Moreover, this representation of a point in C is unique, as is easily deduced from (a) of Theorem 1.8.3. However, we can give an independent proof as below.

If $\sum_{n=1}^{\infty} \frac{\alpha_n}{3^n} = \sum_{n=1}^{\infty} \frac{\beta_n}{3^n}$, where each of α_n [resp. β_n] is either 0 or 2, we shall conclude that $\alpha_n = \beta_n$ for every n. Suppose that there exists an n such that $\alpha_n \neq \beta_n$. Let m be the smallest integer such that $\alpha_m \neq \beta_m$. Then $|\alpha_m - \beta_m| = 2$ and $|\alpha_n - \beta_n| \leq 2$ for every n, so that

$$\begin{aligned}
0 &= \left| \sum_{n=m}^{\infty} \frac{\alpha_n}{3^n} - \sum_{n=m}^{\infty} \frac{\beta_n}{3^n} \right| = \left| \sum_{n=m}^{\infty} \frac{\alpha_n - \beta_n}{3^n} \right| \\
&= \left| \frac{\alpha_m - \beta_m}{3^m} + \sum_{n=m+1}^{\infty} \frac{\alpha_n - \beta_n}{3^n} \right| \\
&\geq \frac{|\alpha_m - \beta_m|}{3^m} - \left| \sum_{n=m+1}^{\infty} \frac{\alpha_n - \beta_n}{3^n} \right| \\
&\geq \frac{|\alpha_m - \beta_m|}{3^m} - \sum_{n=m+1}^{\infty} \frac{|\alpha_n - \beta_n|}{3^n} \\
&\geq \frac{1}{3^m} |\alpha_m - \beta_m| - \sum_{n=m+1}^{\infty} \frac{2}{3^n} \\
&= \frac{1}{3^m}, \quad \text{a contradiction.}
\end{aligned}$$

Hence $\alpha_n = \beta_n$ for every n.

Terminology If in a decimal representation in any arbitrary base b, every digit from a certain point onward is 0, we shall call it a representation with '0 recurring'. The term '$b - 1$ recurring' has the corresponding meaning. In the rest of this section, 'two representations' will be understood to mean at least two *distinct* representations.

Remark 1.8.11 According to Theorem 1.8.3 and the discussion thereafter, the following are equivalent for any number $x \in (0, 1)$, the base b being arbitrary:

(1) x has two decimal representations;
(2) x has a decimal representation with 0 recurring;
(3) x has a decimal representation with $b - 1$ recurring;
(4) x has no decimal representation with neither 0 recurring nor $b - 1$ recurring;
(5) x is of the form $\frac{m}{b^n}$, where $0 < m < b^n$.

Proposition 1.8.12 $\text{card}(C) = 2^{\mathbb{N}} = c$.

Proof Define $g : C \to \mathbb{R}$ as follows: By Theorem 1.8.9 and Remark 1.8.10, any $x \in C$ has a unique representation with every digit even:

$$x = \sum_{k=1}^{\infty} \frac{a_k}{3^k}, \quad \text{each } a_k \text{ being either 0 or 2.}$$

Put $g(x) = \sum_{k=1}^{\infty} \frac{b_k}{2^k}$, where $b_k = \frac{1}{2} a_k$.

Observe that $g(0) = 0$, $g(1) = 1$. Clearly, the range of g is contained in $[0, 1]$. In fact, the range of g can be shown to be $[0, 1]$. Consider $y \in [0, 1]$ with binary representation $y = .d_1 d_2 \ldots$, where $d_n \in \{0, 1\}$ for all $n \in \mathbb{N}$. Let $x = .(2d_1)(2d_2)\ldots$ in base 3. Then by Theorem 1.8.9, we have $x \in C$ and the definition of g yields $g(x) = y$. [Thus, e.g., $g\left(\frac{1}{3}\right) = g\left(\frac{2}{3}\right) = \frac{1}{2}$, illustrating that g is not one-to-one.]

Suppose $g(x) = 0$. Since 0 has a unique representation in any base (see Theorem 1.8.3), it follows that $b_k = 0$ for each k. Hence $a_k = 0$ for each 0 and $x = 0$. Consequently, $x > 0$ implies $g(x) > 0$. Since 1 also has unique representation in any base (see Theorem 1.8.3), a similar argument shows that $x < 1$ implies $g(x) < 1$. Thus, $x \in (0, 1)$ if and only if $g(x) \in (0, 1)$.

Let $A \subseteq C$ consist of numbers having two ternary representations and $B \subseteq [0, 1]$ consist of numbers having two binary representations. Since 0 and 1 have unique representations in any base, we have $A, B \subseteq (0, 1)$. We shall prove that $x \in A$ if and only if $g(x) \in B$.

First suppose $x \in A$. Then $x \in (0, 1)$ and hence $g(x) \in (0, 1)$. By Remark 1.8.11, any ternary representation of x has either 0 recurring or 2 recurring. It follows from the definition of g that one binary representation of $g(x)$ has either 0 recurring or 1 recurring. Since $g(x) \in (0, 1)$, we conclude on the basis of Remark 1.8.11 that $g(x)$ has two binary representations, i.e., $g(x) \in B$.

For the converse, suppose $g(x) \in B$. Then $g(x) \in (0, 1)$ and hence $x \in (0, 1)$. By Remark 1.8.11, any binary representation of $g(x)$ has either 0 recurring or 1 recurring. It follows from the definition of g that one ternary representation of x has either 0 recurring or 2 recurring. Since $x \in (0, 1)$, we conclude on the basis of Remark 1.8.11 that x has two ternary representations, i.e., $x \in A$. This proves the converse and hence completes the proof that $x \in A$ if and only if $g(x) \in B$.

The fact established earlier that g maps C onto $[0, 1]$, when combined with the fact that $x \in A$ only if $g(x) \in B$, implies that g maps $C \backslash A$ onto $[0, 1] \backslash B$. We shall now argue that it does so in a one-to-one manner.

Let $x, x' \in C \backslash A$ and $x \neq x'$. Since it has been proved that $x \in A$ if $g(x) \in B$, we know that $g(x), g(x') \notin B$. We write $x = \sum_{k=1}^{\infty} \frac{a_k}{3^k}$, and $x' = \sum_{k=1}^{\infty} \frac{a'_k}{3^k}$ (each a_k as well as each a'_k being even). Then there exists some k such that $\alpha_k \neq \alpha'_k$ and hence $\frac{1}{2} a_k \neq \frac{1}{2} a'_k$. Therefore the binary representations by means of which $g(x)$ and $g(x')$ are defined are distinct from each other. However, since $g(x), g(x') \notin B$, neither can have two binary representations. This implies $g(x) \neq g(x')$. Thus g has been shown to be one-to-one on $C \backslash A$.

Consequently, the mapping g sets up a one-to-one correspondence between $C\backslash A$ and $[0, 1]\backslash B$. By Remark 1.8.11, the set A consists of all numbers of the form $\frac{m}{3^n}$, $0<m<3^n$ and the set B consist of all numbers of the form $\frac{m}{2^n}$, $0<m<2^n$. Therefore both A and B are countable. By making the countable subsets A and B correspond to each other, we obtain a one-to-one correspondence between the Cantor set and $[0, 1]$. Consequently, $\text{card}(C) = 2^{\mathbb{N}} = c$. □

Remark 1.8.13 We have noted in paragraph 3 of the above proof that $x > 0$ implies $g(x) > 0$. Combining this with the consequence of Theorem 1.8.9 that C contains arbitrarily small positive numbers, we infer that there exist arbitrarily small numbers at which g takes positive values.

Another property of the function g that will be useful in Sect. 5.5 is that the value of $2^n g$ at the right endpoint of any removed interval equals the number of component intervals to the left of that removed interval. To arrive at this conclusion, we need to avail of some facts about removed and component intervals that were noted earlier and also some fresh ones. We present the deduction in a sequence of remarks.

Remarks 1.8.14

(a) The set C_n has 2^n component intervals, each being closed and having length 3^{-n}. There are half as many removed intervals $I_{n,k}$, each being open with length 3^{-n} and flanked on either side by a component interval. These 2^{n-1} open intervals are the ones that were removed from C_{n-1} to obtain C_n, and they are disjoint from C_n.
(b) The kth removed interval $I_{n,k}$ (numbered from left to right) of C_n has $2k - 1$ component intervals of C_n to its left. Therefore, if we arrange the removed intervals from left to right and count the number of component intervals to the left of the respective intervals, we get the arithmetic progression $1, 3, \ldots, 2^n - 1$.
(c) The endpoints of a removed interval of C_n do not belong to the removed interval but do belong to the component intervals of C_n as well as to the component intervals of C_m with $m > n$, and hence to C.
(d) As recorded in Remark 1.8.8, the right endpoint of a removed interval has a ternary representation of the form $.\gamma_1\gamma_2\cdots\gamma_{n-1}2000\ldots$ and its other ternary representation contains the digit 1. Since it belongs to C (as observed in (c)), it must have a ternary representation with every digit even (Theorem 1.8.9), which must then be the one having the form $.\gamma_1\gamma_2\cdots\gamma_{n-1}2000\ldots$. Consequently, every digit in this representation of the right endpoint of a removed interval must be even.
(e) Another consequence is that for right endpoints x and y of any two removed intervals, the ternary representations with every digit even have the property that the digits agree beyond a certain stage. Suppose $x > y$. Then by Remark 1.8.2, the earliest digit of x that differs from the corresponding digit of y is the greater one. It is transparent that the same must also be true of the binary representations of $g(x)$ and $g(y)$ in the definition of $g : C \to \mathbb{R}$ in

Proposition 1.8.12. Moreover, they also have the property that the digits agree beyond a certain stage. It follows by Remark 1.8.2 again that $g(x) > g(y)$. Thus we have shown that when x and y are right endpoints of removed intervals, $x > y$ implies $g(x) > g(y)$.

(f) Consider again the 2^{n-1} removed intervals $I_{n,k}$ of C_n arranged from left to right as in (b); their right endpoints are in strictly increasing order and have the form noted in (d). Therefore, upon applying g to them, we get numbers in strictly increasing order in view of (e) and having binary representations $\cdot\bar{\gamma}_1\bar{\gamma}_2\cdots\bar{\gamma}_{n-1}1000\ldots$ with every $\bar{\gamma}_p$ either 0 or 1. There are precisely 2^{n-1} such binary representations. If we multiply each one by 2^n, we get 2^{n-1} odd positive integers in strictly increasing order, none exceeding 2^n. But there are only 2^{n-1} odd positive integers not exceeding 2^n. It follows that we get all of them and hence the odd positive integers in strictly increasing order that we get are $1, 3, \ldots, 2^n - 1$. This is the same as the arithmetic progression in (b) above. Thus, upon multiplying the value of g at the right endpoint of a removed interval of C_n by 2^n, we obtain the number of component intervals of C_n to the left of that removed interval.

Chapter 2
Measure in Euclidean Space

2.1 Introduction

The theory of integration has its roots in the method of "exhaustion" which was developed by Archimedes for the purpose of calculating the areas and volumes of certain geometric figures. The work of Newton and Leibniz made this method into a systematic tool for such calculations but it was Riemann who gave a formal definition of the integral.

The definition due to Riemann may be summarised as follows: The Riemann integral of a bounded real-valued function f defined on $[a, b]$ is the limit (when it exists) of the Riemann sums

$$\sum_{j=1}^{n} f(t_j)(x_j - x_{j-1}) = \sum_{j=1}^{n} f(t_j)\ell(I_j),$$

where $I_j = [x_{j-1}, x_j]$ are contiguous intervals with union $[a, b]$, $\ell(I_j)$ denotes the length $x_j - x_{j-1}$ of I_j and t_j belongs to I_j for $1 \le j \le n$. The integral exists for a large class of functions, which includes, among others, continuous functions and monotone functions.

The reader will discern that this definition and the one given in Sect. 1.7 seem to be at variance. However, they are equivalent (see Darboux's Theorem 10.1.7, [28]).

It is well known that if a sequence f_n of integrable functions converges uniformly to a limit function f, then the latter has an integral, which is equal to the limit of the integrals of f_n. In other words, the Riemann integral "behaves well" with respect to uniform convergence. Unfortunately, it does not behave well with respect to what is called "monotone convergence", because a sequence f_n of integrable functions such that $f_n \le f_{n+1}$ can converge pointwise to a limit function f that is not even integrable. We give an illustration.

S. Shirali and H. L. Vasudeva, *Measure and Integration*,
Springer Undergraduate Mathematics Series,
https://doi.org/10.1007/978-3-030-18747-7_2

Example 2.1.1 The reader is undoubtedly familiar with the standard example of the function f: $[0, 1] \to \mathbb{R}$ defined by

$$f(x) = \begin{cases} 1 & \text{if } x \text{ is rational} \\ 0 & \text{if } x \text{ is irrational,} \end{cases}$$

which is not integrable in the Riemann theory. Now take any enumeration of the rationals in [0, 1] and let E_n denote the set of the first n rationals in the enumeration. Then $E_n \subseteq E_{n+1}$ and hence the characteristic functions f_n of E_n satisfy $0 \leq f_n \leq f_{n+1}$ and clearly converge pointwise to f. Each f_n differs from the identically zero function at only a finite number of points and therefore has Riemann integral 0, but the limit function f is not Riemann integrable. (The fact that each set $E_{n+1} \backslash E_n$ contains only one point, and is therefore an interval of length 0, will be needed shortly.) Thus problems which involve integration with a limiting process are often awkward with this integral.

However, if we extend the concept of the length of an interval to a notion of size for more general sets, technically called the "measure" of the set, then the integral of the function f in Example 2.1.1 may be taken as representing the area of a figure of height 1 and base $A = \{x \in [0, 1]: f(x) = 1\}$ and should therefore be just the measure of the set A. On the other hand, since each f_n has integral 0, the limit of the sequence of integrals is also 0, and hence we would like the integral of f to be 0 as well. Thus the measure of A should be taken as 0. Note that A is the countable disjoint union of singleton sets $E_{n+1} \backslash E_n$, which are intervals of length 0. This observation already provides an inkling of what is involved in extending the concept of length of an interval to that of a measure of more general sets. More precisely, for a set which is a countable disjoint union of intervals, the measure should be the sum of the lengths of those intervals. Unfortunately, there are two reasons why this is unsatisfactory as a definition. One is that the representation of the set as a union of disjoint intervals may not be unique; another is that we may need to include sets more general than such unions.

Towards the end of the nineteenth century, many mathematicians considered it desirable to replace the Riemann integral by some other type of integral which is more general and better suited for dealing with limiting processes. Amongst the attempts made in this direction, it was Lebesgue's contribution, dating back to the beginning of the last century, that turned out to be the most successful. His definition enables us to integrate some, but not all, functions for which the Riemann integral does not exist, in particular the one mentioned in Example 2.1.1. Although the enlargement is useful in itself, its main virtue is that theorems relating to the interchange of limit and integral are valid under less stringent assumptions than are required for the Riemann integral. In order to achieve the objective outlined above, Lebesgue approached the matter by partitioning an interval containing the range rather than partitioning the domain.

For a controversy about whether it makes any difference which integration theory (Riemann or Lebesgue) is used in aircraft design, see the article [7].

The length $\ell(I)$ of an interval I is defined, as usual, to be the difference of its endpoints. If $I = (a, b)$ $(a, b]$, $[a, b)$ or $[a, b]$, then $\ell(I) = b - a$. This also makes sense for the empty set $\varnothing = (a, a)$ and for unbounded intervals; recall that that $\infty - a = \infty$ and so on (see Chap. 1, Sect. 1.1). Thus $\ell(I)$ is a nonnegative extended real number. It will be convenient to include the empty set as a special case of an interval.

Length is an example of a "set function" in the following sense.

Definition 2.1.2 A **set function** is a function whose domain is some class of subsets of a given set.

In the case of length, the given set is $\mathbb{R}$ and the class of subsets that constitute the domain is the class of intervals; the range is the set of nonnegative extended reals.

We remind the reader that some of the properties satisfied by ℓ are as follows:

(i) $\ell(I) \geq 0$ for all intervals I;
(ii) if $\{I_j\}_{j\geq 1}$ is a sequence of disjoint intervals such that $\bigcup_{j=1}^{\infty} I_j$ is an interval, then $\ell(\bigcup_{j=1}^{\infty} I_j) = \sum_{j=1}^{\infty} \ell(I_j)$;
(iii) if x is any fixed real number and I is any interval (so that $I + x = \{y + x: y \in I\}$ is also an interval), then $\ell(I+x) = \ell(I)$.

When we extend the concept of the length ℓ of an interval to a notion of size or measure m for more general subsets of $\mathbb{R}$, we would like the analogues of the above properties to continue to hold for the extension. To recap, we now list some of the desired properties in terms of m.

(i) m is an nonnegative extended real-valued set function;
(ii) $m(E) = \ell(E)$ if E is an interval;
(iii) m is "countably additive", meaning thereby that if $\{E_j\}_{j\geq 1}$ is a sequence of disjoint subsets in the domain of m and $\cup_{j\geq 1} E_j$ is also in the domain of m, then

$$m(\bigcup_{j=1}^{\infty} E_j) = \sum_{j=1}^{\infty} m(E_j);$$

(iv) for E in the domain of m and x a fixed real number, $m(E + x) = m(E)$, provided $E + x$ is in the domain of m, where $E + x = \{y + x: y \in E\}$; this is called "translation invariance" of m.

It can be shown that it is impossible to have such a measure m with domain consisting of *all* subsets of $\mathbb{R}$ (see [30]). Later we shall construct one which is defined on a much larger class of subsets of $\mathbb{R}$ than just intervals; in particular, the set A of Example 2.1.1 will satisfy $m(A) = 0$. It turns out that in order to obtain satisfactory theorems relating to the interchange of limit and integral, property (iii) of m needs to be strengthened to read as:

(iii) if $\{E_j\}_{j\geq 1}$ is a sequence of disjoint subsets in the domain of m, then first, $\cup_{j\geq 1} E_j$ is also in the domain of m, and second,

$$m\left(\bigcup_{j=1}^{\infty} E_j\right) = \sum_{j=1}^{\infty} m(E_j).$$

Lebesgue was the first to establish the existence of such a measure, but by a construction that differs from the one we shall present.

Problem Set 2.1

2.1.P1. From properties (i) and (iii) of m, show that $m(\varnothing) = 0$, provided that at least one set C in the domain of m satisfies $m(C) < \infty$.
2.1.P2. From properties (i) and (iii) of m, show that, if $A \cap B = \varnothing$, where A, B, $A \cup B$ are in the domain of m, then $m(A \cup B) = m(A) + m(B)$.
2.1.P3. From properties (i) and (iii) of m, show that, if $A \backslash B$, $B \backslash A$, A, B, $A \cup B$ and $A \cap B$ are in the domain of m, then $m(A \cup B) + m(A \cap B) = m(A) + m(B)$.

2.2 Lebesgue Outer Measure

As a first attempt at extending the concept of the length ℓ of an interval (the empty set being regarded as an open interval of length 0) to a notion of size or measure m for more general subsets of $\mathbb{R}$, we introduce the following concept. Let A be any subset of $\mathbb{R}$. Consider a sequence $\{I_n\}_{n\geq 1}$ of open intervals that covers A, that is, $A \subseteq \cup_{n\geq 1} I_n$. For each such sequence, the sum $\sum_{n=1}^{\infty} \ell(I_n)$ is independent of the order of the terms, since $\ell(I_n) \geq 0$ for each n.

Definition 2.2.1 For any subset A of $\mathbb{R}$, we define the **Lebesgue outer measure** $m^*(A)$ to be the infimum of all numbers

$$\sum_{n=1}^{\infty} \ell(I_n),$$

where $\{I_n\}_{n\geq 1}$ is a sequence of open intervals that covers A and the infimum is taken over all such sequences.

For most of this chapter, we shall refer to Lebesgue outer measure as simply "outer measure", as there will be no other outer measure under discussion, except in some problems, where the matter is clarified.

Remarks 2.2.2

(a) Some or all of the I_n may be empty. It follows in particular that $m^*(\varnothing) = 0$.
(b) The outer measure m^* is a set function defined on *all* subsets of $\mathbb{R}$ and has values in the set of nonnegative extended real numbers.

The following properties of the outer measure are almost immediate from the definition. The property stated in (c) is called **monotonicity**.

Proposition 2.2.3

(a) $0 \leq m^*(A) \leq \infty$ *for all subsets of* $\mathbb{R}$;
(b) $m^*(\varnothing) = 0$;
(c) *If* $A \subseteq B \subseteq \mathbb{R}$, *then* $m^*(A) \leq m^*(B)$;
(d) *For* $x \in \mathbb{R}$, *we have* $m^*(\{x\}) = 0$.

Proof (a) and (b) are obvious. If $\{I_n\}_{n\geq 1}$ is a sequence of open intervals that covers B, then it also covers A and hence $m^*(A)$ is the infimum taken over a larger or as large a set of numbers as compared to $m^*(B)$. So, $m^*(A) \leq m^*(B)$. This proves (c). For the proof of (d), note that, for every $\varepsilon > 0$, the sequence $\{I_n\}_{n\geq 1}$ of intervals, where $I_1 = \left(x - \frac{\varepsilon}{2}, x + \frac{\varepsilon}{2}\right)$ and $I_n = \varnothing$ for $n \geq 2$, covers $\{x\}$, while $\sum_{n=1}^{\infty} \ell(I_n) = \varepsilon$. □

It is immediate from Proposition 2.2.3(a) and (c) that if $B \subseteq A$ and $m^*(A) = 0$, then $m^*(B) = 0$.

Suppose that A is any subset of $\mathbb{R}$ and x is any real number. The **translate** of A by x is the set $\boldsymbol{A + x} = \{y + x\colon y \in A\}$. The assertion of the next proposition is called **translation invariance** of the Lebesgue outer measure m^*.

Proposition 2.2.4 *For any* $A \subseteq \mathbb{R}$ *and* $x \in \mathbb{R}$, *we have* $m^*(A + x) = m^*(A)$.

Proof Given any open interval I, bounded or unbounded, its translate $I + x$ is also an open interval and $\ell(I + x) = \ell(I)$. Now let any $\varepsilon > 0$ be given. Suppose $m^*(A) < \infty$. Then there is a sequence $\{I_n\}_{n\geq 1}$ of open intervals such that $A \subseteq \bigcup_{n\geq 1} I_n$ and satisfies

$$\sum_{n=1}^{\infty} \ell(I_n) < m^*(A) + \varepsilon.$$

Clearly, $A + x \subseteq \cup_{n\geq 1}(I_n + x)$. Therefore

$$m^*(A + x) \leq \sum_{n=1}^{\infty} \ell(I_n + x) = \sum_{n=1}^{\infty} \ell(I_n) < m^*(A) + \varepsilon.$$

Since $\varepsilon > 0$ is arbitrary, we have $m^*(A + x) \leq m^*(A)$. As A is the translate of $A + x$ by $-x$, it follows (on replacing A by $A + x$ and x by $-x$ in the above inequality) that $m^*(A) \leq m^*(A + x)$. Since the reverse inequality has already been proved, it follows that $m^*(A + x) = m^*(A)$ when $m^*(A) < \infty$. The same argument with obvious modifications works when $m^*(A) = \infty$. □

Example 2.2.5 Suppose that A has at most countably many points. Then $m^*(A) = 0$. In view of the monotonicity of m^*, we need consider only countable A. So, let

$\{x_k\}_{k\geq 1}$ be an enumeration of the points of A. Given $\varepsilon > 0$, let I_k be an open interval of length $\varepsilon/2^k$ that contains the point x_k. The sequence $\{I_k\}_{k\geq 1}$ then covers A and satisfies $\sum_{k=1}^{\infty} \ell(I_k) = \sum_{k=1}^{\infty} \varepsilon/2^k = \varepsilon$. So, $m^*(A) \leq \varepsilon$. In particular $m^*(\mathbb{Q}) = m^*(\mathbb{N}) = m^*(\mathbb{Z}) = 0$.

It follows from what has just been noted that a set of positive outer measure must be uncountable.

Proposition 2.2.6 *If A is any interval, then its outer measure is the same as its length*: $m^*(A) = \ell(A)$.

Proof The case when A is empty is trivial. So we shall consider only the case of nonempty A.

First consider the case when A is a closed bounded interval $[a, b]$. Given $\varepsilon > 0$, we can choose $I_1 = \left(a - \frac{\varepsilon}{2}, b + \frac{\varepsilon}{2}\right)$ and $I_k = \varnothing$ for $k \geq 2$ to see that $m^*(A) \leq b - a + \varepsilon$. This being true for each $\varepsilon > 0$, we have $m^*(A) \leq b - a = \ell(A)$. We shall show that $m^*(A) \geq b - a$.

Let $\{I_k\}_{k\geq 1}$ be a sequence of open intervals that covers A. Then the family of intervals $\{I_k\colon k \geq 1\}$ is an open cover of A and, by the Heine–Borel Theorem 1.3.16, contains a finite subcover, which we shall name as $\mathfrak{F}$. We may assume that each interval in $\mathfrak{F}$ has a nonempty intersection with A, so that the latter is an interval with well-defined endpoints. It will follow that $\sum_{k=1}^{\infty} \ell(I_k) > b - a$ if we show that the total length of the intervals in $\mathfrak{F}$ is greater than or equal to $b - a$. To show the latter, we proceed as follows.

By renumbering the intervals if necessary, we may assume that the finite subcover $\mathfrak{F}$ consists of intervals I_k, $k = 1, \ldots, p$. Let the intervals $A \cap I_k$ have endpoints $c^{(k)}$, $d^{(k)}$. Arrange all the numbers $c^{(k)}$ and $d^{(k)}$, $k = 1, \ldots, p$, as a single increasing sequence $a^{(0)} < a^{(1)} < \cdots < a^{(m)}$. Note that we must necessarily have $a = a^{(0)}$ and $b = a^{(m)}$ because the union of the intervals $A \cap I_k$ is precisely equal to A. Consequently, the single increasing sequence provides a partition T of the interval $[a, b]$ and also gives rise to a partition of each interval $A \cap I_k$. Let T and T_k denote respectively the systems of subintervals generated by the partitions.

Since the length of an interval equals the total length of all the subintervals of any partition of it, we have on the one hand

$$\ell(A) = \sum_{J\in T} \ell(J) \tag{2.1}$$

and on the other hand

$$\ell(I_k) \geq \ell(A \cap I_k) = \sum_{J\in T_k} \ell(J),$$

so that

$$\sum_{k=1}^{p} \ell(I_k) \geq \sum_{k=1}^{p} \sum_{J \in T_k} \ell(J). \tag{2.2}$$

Note that each interval belonging to T is a constituent of at least one collection T_k but could belong to several T_k. Therefore

$$\sum_{k=1}^{p} \sum_{J \in T_k} \ell(J) \geq \sum_{J \in T} \ell(J).$$

Together with (2.1) and (2.2), this implies the required conclusion.

Next, consider the case when A is any bounded nonempty interval. Then $\bar{A}$ is a closed bounded interval and not only do we have $\ell(\bar{A}) = \ell(A)$, but also, according to the case already proved, $m^*(\bar{A}) = \ell(\bar{A})$, so that

$$m^*(\overline{A}) = \ell(A).$$

Now, given $\varepsilon > 0$, there exists a closed bounded interval I such that $I \subseteq A$ and

$$\ell(I) > \ell(A) - \varepsilon.$$

According to what has already been proved,

$$\ell(I) = m^*(I).$$

Since $I \subseteq A \subseteq \bar{A}$, we know from Proposition 2.2.3(c) that

$$m^*(I) \leq m^*(A) \leq m^*(\overline{A}).$$

It follows from the four statements displayed above that

$$\ell(A) - \varepsilon < m^*(A) \leq \ell(A).$$

Since this is true for an arbitrary $\varepsilon > 0$, we have $m^*(A) = \ell(A)$.

Finally, consider the case when A is an unbounded interval. Then for any $M > 0$, there exists a closed bounded interval $I \subseteq A$ such that $\ell(I) \geq M$. Hence

$$m^*(A) \geq m^*(I) = \ell(I) \geq M.$$

Thus $m^*(A) \geq M$ for every positive real number M. It follows that $m^*(A) = \infty = \ell(A)$. □

A part of the proof above consisted in showing that the total length of a finite family $\mathfrak{F}$ of open intervals covering a closed bounded interval $[a, b]$ is no less than that of the latter. An alternative argument, which actually proves that the total length *exceeds* the latter, is as follows.

Clearly, we may assume that the intervals of $\mathfrak{F}$ are bounded. Since $\mathfrak{F}$ covers $[a, b]$, the number a belongs to some interval $(u_1, v_1) \in \mathfrak{F}$ Then

$$u_1 < a < v_1. \tag{2.3}$$

Suppose the n intervals $(u_j, v_j) \in \mathfrak{F}$, $1 \leq j \leq n$, have the property that

$$u_1 < a < v_1 \quad \text{and} \quad u_{j+1} < v_j < v_{j+1} \text{ whenever } 1 \leq j \leq n-1. \tag{2.4}$$

When $n = 1$, (2.4) asserts the same thing as (2.3), because there can be no j satisfying $1 \leq j \leq n-1$. If $v_n < b$, then we have $a < v_n < b$, so that $v_n \in [a, b]$ and, in view of the hypothesis that $\mathfrak{F}$ covers $[a, b]$, there must exist $(u_{n+1}, v_{n+1}) \in \mathfrak{F}$ for which $u_{n+1} < v_n < v_{n+1}$. This means we now have $n + 1$ intervals $(u_j, v_j) \in \mathfrak{F}$ having the property that

$$u_1 < a < v_1 \quad \text{and} \quad u_{j+1} < v_j < v_{j+1} \quad \text{whenever} \quad 1 \leq j \leq n.$$

This has been deduced from the assumption that $v_n < b$, and says that (2.4) holds with $n + 1$ in place of n. We have already noted that (2.4) holds when $n = 1$. It follows that, if $v_n < b$ for every n, then there is a sequence of intervals $\{(u_n, v_n)\}$ of $\mathfrak{F}$ satisfying (2.4). However, (2.4) also forces the intervals to be different from each other, thus we get an infinite subset of $\mathfrak{F}$, which contradicts the finiteness of the latter. Therefore $v_n \geq b$ for some n. So, we get only finitely many intervals $(u_j, v_j) \in \mathfrak{F}$, $1 \leq j \leq n$, and they satisfy

$$u_1 < a < v_1, u_{j+1} < v_j < v_{j+1} \quad \text{for} \quad 1 \leq j \leq n-1 \quad \text{and} \quad b \leq v_n. \tag{2.5}$$

For any pair of finite sequences $\{u_j\}$ and $\{v_j\}$ of real numbers, we shall prove by induction that (2.5) implies

$$\sum_{j=1}^{n} (v_j - u_j) > b - a. \tag{2.6}$$

If $n = 1$, then the last inequality in (2.5) implies $b \leq v_1$ and the first implies $\sum\limits_{j=1}^{n} (v_j - u_j) = v_1 - u_1 \geq b - u_1 > b - a$. Thus (2.5) implies (2.6) when $n = 1$. Assume (induction hypothesis) for some n that (2.5) implies (2.6), and consider a pair of finite sequences $\{u_j\}$ and $\{v_j\}$, $1 \leq j \leq n + 1$, which satisfy (2.5) with $n + 1$ in place of n:

$$u_1 < a < v_1, \quad u_{j+1} < v_j < v_{j+1} \quad \text{for} \quad 1 \leq j \leq n \quad \text{and} \quad b \leq v_{n+1}.$$

Taking $j = n$, we see that $u_{n+1} < v_n$, and so the finite sequences $\{u_j\}$ and $\{v_j\}$, $1 \leq j \leq n$, consisting of the first n terms, satisfy (2.5) with b replaced by u_{n+1}. The induction hypothesis yields

$$\sum_{j=1}^{n} (v_j - u_j) > u_{n+1} - a,$$

which leads to $\sum_{j=1}^{n+1} \left(v_j - u_j\right) > v_{n+1} - u_{n+1} + u_{n+1} - a = v_{n+1} - a \geq b - a.$

This completes the induction argument that (2.5) always implies (2.6). Since we have shown above that $\mathfrak{F}$ contains a subfamily $\{(u_j,v_j)\colon 1 \leq j \leq n\}$ satisfying (2.5), it follows that the total length of the intervals in $\mathfrak{F}$ is greater than $b - a$.

For the next important property of outer measure, the following definition will turn out to be useful.

Definition 2.2.7 Let $\mathcal{A}$ be a nonempty collection of subsets of a set X and ν be an extended real-valued set function with domain $\mathcal{A}$. The set function ν is said to be **finitely additive** if, for every finite sequence of sets $\{A_j\}_{1 \leq j \leq n}$ such that each $A_j \in \mathcal{A}$, $\cup_{1 \leq j \leq n} A_j \in \mathcal{A}$, and $A_j \cap A_k = \varnothing$ whenever $j \neq k$ (i.e. disjoint), we have

$$\nu\left(\bigcup_{j=1}^{n} A_j\right) = \sum_{j=1}^{n} \nu(A_j).$$

It is said to be **finitely subadditive** if, for every finite sequence of sets $\{A_j\}_{1 \leq j \leq n}$ such that each $A_j \in \mathcal{A}$ and $\cup_{1 \leq j \leq n} A_j \in \mathcal{A}$, we have

$$\nu\left(\bigcup_{j=1}^{n} A_j\right) \leq \sum_{j=1}^{n} \nu(A_j),$$

and **countably subadditive** *if* for every sequence of sets $\{A_j\}_{j \geq 1}$ such that each $A_j \in \mathcal{A}$ and $\cup_{j \geq 1} A_j \in \mathcal{A}$, we have

$$\nu\left(\bigcup_{j=1}^{\infty} A_j\right) \leq \sum_{j=1}^{\infty} \nu(A_j).$$

It is said to be **countably additive** if for every sequence of sets $\{A_j\}_{j \geq 1}$ such that each $A_j \in \mathcal{A}$ and $\cup_{j \geq 1} A_j \in \mathcal{A}$, and $A_j \cap A_k = \varnothing$ whenever $j \neq k$ (i.e. disjoint), we have

$$\nu(\bigcup_{j=1}^{\infty} A_j) = \sum_{j=1}^{\infty} \nu(A_j).$$

Before proving that the Lebesgue outer measure is countably subadditive, we give some simple instances of finitely and countably subadditive set functions in a general set X.

Examples 2.2.8

(a) Let X be any infinite set and suppose $\mathcal{A}$ consists of all nonempty finite subsets of X. Take $\nu(A)$ to be the number of elements in A. Then ν is finitely as well as countably subadditive. In fact, $\sum_{j=1}^{\infty} \nu(A_j)$ will always be ∞, because each term will be 1 or greater; this will cease to be so if we enlarge $\mathcal{A}$ to include $\varnothing$. But ν will nevertheless be finitely as well as countably subadditive. Indeed, ν is finitely as well as countably additive regardless of whether we enlarge $\mathcal{A}$ to include $\varnothing$. This example works even when X is finite but nonempty, although for slightly different reasons.

(b) Let X be any set containing 4 or more elements and suppose $\mathcal{A}$ consists of all finite subsets of X containing 2 or more elements. Take $\nu(A)$ to be 1 less than the number of elements in A. Then $\sum_{j=1}^{\infty} \nu(A_j)$ will always be ∞, and therefore ν is countably subadditive. However, it is not finitely subadditive, as can be seen by considering two disjoint subsets containing 2 elements each.

Proposition 2.2.9 *The outer measure m^* is countably subadditive; that is, if $\{A_j\}_{j\geq 1}$ is a sequence of subsets of $\mathbb{R}$, then*

$$m^*(\bigcup_{j=1}^{\infty} A_j) \leq \sum_{j=1}^{\infty} m^*(A_j). \tag{2.7}$$

Proof Assume that $\varepsilon > 0$ is given. For each $j \in \mathbb{N}$, choose a sequence $\{I_{j,k}\}_{k\geq 1}$ of open intervals that covers A_j and satisfies

$$\sum_{k=1}^{\infty} \ell(I_{j,k}) \leq m^*(A_j) + \varepsilon/2^j.$$

Then the sequence $\{I_{j,k}\}_{j,k\geq 1}$ covers $\cup_{j\geq 1} A_j$ and has total length less than or equal to $\sum_{j=1}^{\infty} m^*(A_j) + \varepsilon$. Hence

$$m^*(\bigcup_{j=1}^{\infty} A_j) \leq \sum_{j=1}^{\infty} m^*(A_j) + \varepsilon.$$

Since $\varepsilon > 0$ is arbitrary, we conclude that (2.7) holds. □

Corollary 2.2.10 *The outer measure m^* is finitely subadditive; that is, if $\{A_j\}_{1 \le j \le n}$ is a finite sequence of subsets of $\mathbb{R}$, then*

$$m^*\left(\bigcup_{j=1}^{n} A_j\right) \le \sum_{j=1}^{n} m^*(A_j).$$

Proof Set $A_j = \varnothing$ for $j > n$ and use Propositions 2.2.9 and 2.2.3(b). □

Remarks 2.2.11

(a) If $m^*(A) = 0$, then $m^*(A \cup B) = m^*(B)$. In fact, by Proposition 2.2.3(c) and Corollary 2.2.10,

$$m^*(B) \le m^*(A \cup B) \le m^*(A) + m^*(B) = m^*(B).$$

(b) Let A be any subset of $\mathbb{R}$ and A_I denote the set of irrationals in A, then $m^*(A) = m^*(A_I)$. This follows from (a) above and the result noted in Example 2.2.5.

(c) The set C, known as the Cantor set (see Sect. 1.8 for details), has outer measure 0. Recall that for each natural number $n \in \mathbb{N}$, the set C is a subset of a union C_n of intervals whose total length is $(2/3)^n$. It follows by Propositions 2.2.6 and 2.2.9 that $m^*(C_n) \le (2/3)^n$. Since $C \subseteq C_n$, it further follows that $m^*(C) \le (2/3)^n \to 0$ as $n \to \infty$.

In the rest of this section, we explore more closely the role of open sets in determining the Lebesgue outer measure of an arbitrary subset of $\mathbb{R}$. The material can be omitted for now without loss of continuity and the reader may wish to take it up only when it is needed later.

Proposition 2.2.12 *Let $A \subseteq \mathbb{R}$ be arbitrary and $\varepsilon > 0$ be given. Then there exists an open set $O \supseteq A$ such that $m^*(O) \le m^*(A) + \varepsilon$. Moreover, in the case when $m^*(A) < \infty$, the set O can be so chosen that the inequality is strict.*

Proof If $m^*(A) = \infty$, the assertion is trivial, because we can choose $O = \mathbb{R}$. So, suppose $m^*(A) < \infty$. By definition, $m^*(A)$ is the infimum of all possible sums $\sum_{n=1}^{\infty} \ell(I_n)$, where $\{I_n\}_{n \ge 1}$ is a sequence of open intervals that covers A. Therefore, there exists such a covering sequence $\{I_n\}_{n \ge 1}$ such that

$$\sum_{n=1}^{\infty} \ell(I_n) < m^*(A) + \varepsilon.$$

Now $O = \cup_{n \ge 1} I_n$ is a union of open intervals (hence open sets) and is therefore an open set. As $\{I_n\}_{n \ge 1}$ is a sequence of open intervals that covers A, we have $A \subseteq O$. Also, by countable subadditivity of m^* (Proposition 2.2.9) and the fact that the outer measure of an interval is its length (Proposition 2.2.6),

$$m^*(O) = m^*(\bigcup_{n=1}^{\infty} I_n) \leq \sum_{n=1}^{\infty} m^*(I_n) = \sum_{n=1}^{\infty} \ell(I_n) < m^*(A) + \varepsilon. \qquad \square$$

It is natural to ask whether one can sharpen the preceding result to eliminate the ε. In other words, can one always find an open set $O \supseteq A$ such that $m^*(O) = m^*(A)$? Of course, one cannot: if A contains a single point, then its outer measure is 0 but any open set containing A has to be a nonempty open set and must therefore contain an open interval of positive length, which makes its outer measure greater than that of A. The next natural question is what happens if we take a sequence of open sets O_n such that $m^*(O_n) < m^*(A) + \frac{1}{n}$ and then set $G = \cap_{n \geq 1} O_n$. Of course, we get $G \supseteq A$ and $m^*(G) = m^*(A)$, but an intersection of a sequence of open sets need not be open (example: $\bigcap_{n \geq 1} (-\frac{1}{n}, \frac{1}{n}) = \{0\}$, which is not open). Here is the best we can do:

Proposition 2.2.13 *Let $A \subseteq \mathbb{R}$ be arbitrary. Then there exists a set $G \supseteq A$ such that $m^*(G) = m^*(A)$ and G is an intersection of a sequence of open sets.*

Proof If $m^*(A) = \infty$, we may take $G = \mathbb{R}$. So, suppose $m^*(A) < \infty$. Then by Proposition 2.2.12, for each $n \in \mathbb{N}$, there exists an open set $O_n \supseteq A$ such that $m^*(O_n) < m^*(A) + \frac{1}{n}$. Let $G = \cap_{n \geq 1} O_n$. Then G is an intersection of a sequence of open sets, $G \supseteq A$, and

$$m^*(A) \leq m^*(G) \leq m^*(O_n) \leq m^*(A) + \frac{1}{n} \quad \text{for each } n \in \mathbb{N}.$$

Hence $m^*(G) = m^*(A)$. $\square$

The above proposition can be rephrased as saying that a given subset of $\mathbb{R}$ is a subset of a set which is a countable intersection of open sets and has the same outer measure as the given set. This can be shortened further by introducing the following terminology.

Definition 2.2.14 The class of those subsets of a metric space that are countable intersections of open sets is called $\mathcal{G}_\delta$. Sets in this class are called **$\mathcal{G}_\delta$-sets** or said to be **of type $\mathcal{G}_\delta$**. The class of those subsets of a metric space that are countable unions of closed sets is called $\mathcal{F}_\sigma$. Sets in this class are called **$\mathcal{F}_\sigma$-sets** or said to be **of type $\mathcal{F}_\sigma$**.

Thus every subset of $\mathbb{R}$ is contained in a $\mathcal{G}_\delta$-set of the same outer measure as itself.

Recall that if we take the union of a countable number of families, each of which is itself countable, the resulting family is again countable. It is a consequence that a countable intersection [resp. union] of $\mathcal{G}_\delta$-sets [resp. $\mathcal{F}_\sigma$-sets] is a $\mathcal{G}_\delta$-set [$\mathcal{F}_\sigma$-set].

Examples 2.2.15

(a) Consider the subset $\mathbb{Q}$ of all rational numbers in $\mathbb{R}$. Let $x_1, x_2, \ldots$ be an enumeration of $\mathbb{Q}$. We can now write $\mathbb{Q} = \bigcup_{i \geq 1} \{x_i\}$. Since each of the single point sets $\{x_i\}$ is closed, it follows that $\mathbb{Q}$ is an $\mathcal{F}_\sigma$-set.

(b) If $E \in \mathcal{F}_\sigma$, then by using the definition of $\mathcal{F}_\sigma$ and taking complements, we conclude via De Morgan's Laws that $E^c \in \mathcal{G}_\delta$. In particular, it follows that the set of irrationals in $\mathbb{R}$ is a set of type $\mathcal{G}_\delta$. Similar considerations show that, if $E \in \mathcal{G}_\delta$ then $E^c \in \mathcal{F}_\sigma$.

(c) Since $(a, b) = \bigcup_{n \geq 1} [a + \frac{1}{n}, b - \frac{1}{n}]$ and $\{0\} = \bigcap_{n \geq 1} (-\frac{1}{n}, \frac{1}{n})$, it follows that (a, b) is a set of type $\mathcal{F}_\sigma$ and that $\{0\}$ is of type $\mathcal{G}_\delta$.

(d) It is clear that every open set is of type $\mathcal{G}_\delta$ and every closed set is of type $\mathcal{F}_\sigma$. It is easily seen that every open set is also of type $\mathcal{F}_\sigma$, because it is a countable union of open intervals, each of which is an $\mathcal{F}_\sigma$-set by (c). On taking complements, we see that every closed set is of type $\mathcal{G}_\delta$.

Problem Set 2.2

2.2.P1. Use the results of this section to show that [0, 1] is uncountable.

2.2.P2. If the domain of a countably subadditive set function ν includes $\varnothing$ and if $\nu(\varnothing) = 0$, show that ν is finitely subadditive.

2.2.P3. Show that for any two sets A and B with union [0, 1], the outer measure satisfies

$$m^*(A) \geq 1 - m^*(B).$$

2.2.P4. Let $\{I_j\}_{1 \leq j \leq n}$ be a finite sequence of open intervals covering the rationals in [0, 1]. Show that $\sum_{j=1}^{n} \ell(I_j) \geq 1$. (Note: With a little extra effort, it can be shown that the sum of lengths is strictly greater than 1; however, we shall not need this fact.)

2.2.P5. Show that if we were to define outer measure as approximation by finitely many open intervals, i.e. if $m^*(A)$ were to be

$$\inf\{\sum_{i=1}^{n} \ell(I_i) : A \subseteq \bigcup_{i=1}^{n} I_i, \quad \text{each } I_i \text{ an open interval}\},$$

then it would not be countably subadditive.

2.2.P6. Show that if we were to define outer measure as approximation from within, i.e. if $m^*(A)$ were to be

$$\sup\{\sum_{i=1}^{\infty} \ell(I_i) : A \supseteq \bigcup_{i=1}^{\infty} I_i, \text{ each } I_i \text{ an open interval and } i \neq j \Rightarrow I_i \cap I_j = \varnothing\},$$

then it would not be finitely subadditive.

2.2.P7. Prove that, if the open set A is the union of a sequence $\{I_n\}_{n\geq 1}$ of disjoint open intervals, then

$$m^*(A) = \sum_{n=1}^{\infty} \ell(I_n).$$

2.2.P8. Show that using closed intervals instead of open intervals in the definition of outer measure does not change the evaluation of $m^*(A)$.

2.2.P9. Show that using intervals closed only on the left (or on the right, or a mixture of various types of intervals) rather than open intervals does not change the evaluation of $m^*(A)$.

2.2.P10. (a) For $k > 0$ and $A \subseteq \mathbb{R}$, let kA denote $\{x\colon k^{-1}x \in A\}$. Show that $m^*(kA) = k \cdot m^*(A)$.

(b) For $A \subseteq \mathbb{R}$ let $-A$ denote $\{x\colon -x \in A\}$. Show that $m^*(-A) = m^*(A)$.

2.2.P11. Let $A = \{x \in [0, 1]\colon x$ has a decimal expansion not containing the digit $5\}$. Note that $0.5 \in A$, because $0.5 = 0.4999\ldots$, but $0.51 \notin A$. Then show that $m^*(A) = 0$.

2.2.P12. Suppose $m^*(A \cap I) \leq \frac{1}{2} m^*(I)$ for every interval I. Prove that $m^*(A) = 0$.

2.2.P13. For nonempty $E \subseteq \mathbb{R}$, the diameter is $\operatorname{diam} E = \sup\{|x - y|\colon x, y \in E\}$. Show that $m^*(E) \leq \operatorname{diam} E$.

2.3 Measurable Sets and Lebesgue Measure

The Lebesgue outer measure m^* of Definition 2.2.1 is an example of a nonnegative, monotone, countably subadditive set function, which is defined on all subsets of $\mathbb{R}$ and satisfies $m^*(\varnothing) = 0$. (A set function μ is said to be **monotone** if $A \subseteq B$ implies $\mu(A) \leq \mu(B)$.) There is one further essential property that one would like to have, namely, countable additivity. In fact, m^* is not even finitely additive. For an example of this phenomenon, see Problem 2.3.P16. So to speak, in our attempt to extend the concept of length of an interval to a notion of measure to more general sets, we have gone too far, and the problem now is to restrict m^* to a subclass on which m^* is countably additive, as will unfold in this section. Such a subclass exists and is indeed a "σ-algebra" (formally described in Definition 2.3.10 below) containing open as well as closed sets. It turns out to be useful to select precisely those subsets $E \subseteq \mathbb{R}$ for which

$$m^*(A) = m^*(A \cap E) + m^*(A \cap E^c) \quad \text{for every } A \subseteq \mathbb{R},$$

where E^c denotes the complement of E in $\mathbb{R}$. This property of E can be regarded as saying informally that E "splits" every subset A "additively".

Definition 2.3.1 A subset $E \subseteq \mathbb{R}$ is said to be **measurable** if

$$m^*(A) = m^*(A \cap E) + m^*(A \cap E^c) \quad \text{for every } A \subseteq \mathbb{R}.$$

Remarks 2.3.2

(a) Observe that the finite subadditivity property (see Corollary 2.2.10) of m^* implies

$$m^*(A) \leq m^*(A \cap E) + m^*(A \cap E^c) \quad \text{for every } A \subseteq \mathbb{R}.$$

Thus in testing the measurability of E, it is enough to show the reverse of this inequality. This condition in Definition 2.3.1 is known as the **Carathéodory condition**.

(b) If E is measurable, so is its complement E^c.
(c) For any subset A of $\mathbb{R}$, we have $m^*(A) = m^*(A) + 0 = m^*(A \cap \mathbb{R}) + m^*(A \cap \varnothing)$. Thus $\mathbb{R}$ is measurable, and hence by (b), the same is true of $\varnothing$.
(d) If $m^*(E) = 0$, where $E \subseteq \mathbb{R}$, then E is measurable; in fact, every subset of E is also measurable. Indeed, for $A \subseteq \mathbb{R}$, we have $m^*(A \cap E) \leq m^*(E) = 0$ and $m^*(A \cap E^c) \leq m^*(A)$. So, $m^*(A) \geq m^*(A \cap E^c) + 0 \geq m^*(A \cap E) + m^*(A \cap E^c)$. In particular, the Cantor set (see Remark 2.2.11(c)) is measurable and so is every subset of it.
(e) A countable subset of $\mathbb{R}$ is measurable. This follows from Example 2.2.5 and (d) above. Consequently, the set $\mathbb{Q}$ of rationals in $\mathbb{R}$ is measurable. The set of irrationals in $\mathbb{R}$, being the complement of $\mathbb{Q}$, is measurable in view of (b) above.

Notation 2.3.3 We denote the collection of all measurable subsets of $\mathbb{R}$ by $\mathfrak{M}$. In what follows, we prove some of the properties of the collection $\mathfrak{M}$.

Lemma 2.3.4 *If E_1 and E_2 are in $\mathfrak{M}$, then so is $E_1 \cup E_2$.*

Proof Let A be any subset of $\mathbb{R}$. From the measurability of E_1, we have

$$m^*(A) = m^*(A \cap E_1) + m^*(A \cap E_1^c). \tag{2.8}$$

Since $A \cap E_1$ and $A \cap E_1^c$ are subsets of $\mathbb{R}$, it follows from the measurability of E_2 that

$$m^*(A \cap E_1) = m^*(A \cap E_1 \cap E_2) + m^*(A \cap E_1 \cap E_2^c) \tag{2.9}$$

and

$$m^*(A \cap E_1^c) = m^*(A \cap E_1^c \cap E_2) + m^*(A \cap E_1^c \cap E_2^c). \tag{2.10}$$

Now,

$$(E_1 \cap E_2) \cup (E_1 \cap E_2^c) \cup (E_1^c \cap E_2) = E_1 \cup E_2$$

and

$$(A \cap E_1 \cap E_2) \cup (A \cap E_1 \cap E_2^c) \cup (A \cap E_1^c \cap E_2) = A \cap (E_1 \cup E_2). \tag{2.11}$$

Also,

$$(A \cap E_1^c \cap E_2^c) = A \cap (E_1 \cup E_2)^c. \tag{2.12}$$

On using the finite subadditivity of outer measure (Corollary 2.2.10), it follows from (2.11) and (2.12) that

$$\begin{aligned} &m^*(A \cap (E_1 \cup E_2)) + m^*(A \cap (E_1 \cup E_2)^c) \\ &\quad \leq m^*(A \cap E_1 \cap E_2) + m^*(A \cap E_1 \cap E_2^c) + m^*(A \cap E_1^c \cap E_2) + m^*(A \cap E_1^c \cap E_2^c). \end{aligned}$$

Using (2.9) and (2.10), and then (2.8), we further obtain from here that

$$m^*(A \cap (E_1 \cup E_2)) + m^*(A \cap (E_1 \cup E_2)^c) \leq m^*(A \cap E_1) + m^*(A \cap E_1^c) = m^*(A).$$

By Remark 2.3.2(a), this completes the proof. □

We introduce some terminology that will make it possible to summarise the important properties of $\mathfrak{M}$ conveniently.

Definition 2.3.5 Let X be a nonempty set. A family $\mathcal{F}$ of subsets of X is called an **algebra** if

(a) $X \in \mathcal{F}$ (b) $A \in \mathcal{F}, B \in \mathcal{F} \Rightarrow A \cup B \in \mathcal{F}$ (c) $A \in \mathcal{F} \Rightarrow A^c \in \mathcal{F}$.

Remarks 2.3.6

(a) $\emptyset \in \mathcal{F}$.
(b) $A_1, A_2, \ldots, A_n \in \mathcal{F} \Rightarrow \bigcup_{j=1}^n A_j \in \mathcal{F}$.
(c) If $A \in \mathcal{F}, B \in \mathcal{F} \Rightarrow A \cap B \in \mathcal{F}$.
(d) $A_1, A_2, \ldots, A_n \in \mathcal{F} \Rightarrow \bigcap_{j=1}^n A_j \in \mathcal{F}$.
(e) $A \in \mathcal{F}, B \in \mathcal{F} \Rightarrow A \backslash B \in \mathcal{F}$.

Proposition 2.3.7 *The collection $\mathfrak{M}$ of measurable subsets of $\mathbb{R}$ is an algebra. In particular, if E_1, E_2, …, E_n are measurable subsets, then so are $\cup_{1 \leq k \leq n} E_k$ and $\cap_{1 \leq k \leq n} E_k$.*

Proof This is merely a summary of Remarks 2.3.2(b) and 2.3.2(c), Lemma 2.3.4 and Remarks 2.3.6(b) and 2.3.6(d). □

We now show that the outer measure restricted to $\mathfrak{M}$ is finitely additive.

Proposition 2.3.8 *If E_1, E_2, ..., E_n is a finite sequence of disjoint measurable subsets of $\mathbb{R}$ and $A \subseteq \mathbb{R}$ is arbitrary, then*

$$m^*(A \cap \bigcup_{j=1}^{n} E_j) = \sum_{j=1}^{n} m^*(A \cap E_j). \tag{2.13}$$

Taking $A = \mathbb{R}$ in particular, we have

$$m^*(\bigcup_{j=1}^{n} E_j) = \sum_{j=1}^{n} m^*(E_j).$$

Proof We shall prove this by induction on n. For $n = 1$, the equality (2.13) is obvious. Suppose it holds for $n = k$, that is (induction hypothesis),

$$m^*(A \cap \bigcup_{j=1}^{k} E_j) = \sum_{j=1}^{k} m^*(A \cap E_j) \tag{2.14}$$

whenever E_1, E_2, ..., E_k is a finite sequence of disjoint measurable subsets of $\mathbb{R}$ and $A \subseteq \mathbb{R}$ is arbitrary. In order to deduce the validity of (2.13) for $n = k + 1$, observe that, in view of the measurability of E_{k+1},

$$m^*(A \cap \bigcup_{j=1}^{k+1} E_j) = m^*((A \cap \bigcup_{j=1}^{k+1} E_j) \cap E_{k+1}) + m^*((A \cap \bigcup_{j=1}^{k+1} E_j) \cap E_{k+1}^c). \tag{2.15}$$

Using the disjointness, we have

$$(A \cap \bigcup_{j=1}^{k+1} E_j) \cap E_{k+1} = A \cap E_{k+1} \text{ and } (A \cap \bigcup_{j=1}^{k+1} E_j) \cap E_{k+1}^c = A \cap \bigcup_{j=1}^{k} E_j.$$

Therefore (2.15) becomes

$$m^*(A \cap \bigcup_{j=1}^{k+1} E_j) = m^*(A \cap E_{k+1}) + m^*(A \cap \bigcup_{j=1}^{k} E_j).$$

The induction hypothesis (2.14) now yields the desired equality (2.13). □

The outer measure m^* restricted to $\mathfrak{M}$ is in fact countably additive, as we shall show below in Theorem 2.3.13. The first step in this direction is to show that $\mathfrak{M}$ is closed under countable unions. We begin with a lemma.

Lemma 2.3.9 *Let $\mathcal{F}$ be an algebra of subsets of X. If $\{A_j\}_{j \geq 1}$ is a sequence of sets in $\mathcal{F}$, then there exists a sequence $\{B_j\}_{j \geq 1}$ of disjoint sets in $\mathcal{F}$ such that*

$$\bigcup_{j=1}^{\infty} B_j = \bigcup_{j=1}^{\infty} A_j.$$

Proof Set

$$B_1 = A_1,\ B_2 = A_2 \cap A_1^c,\ B_3 = A_3 \cap (A_1 \cup A_2)^c, \ldots,$$
$$B_j = A_j \cap (A_1 \cup A_2 \cup \cdots \cup A_{j-1})^c, \ldots.$$

Clearly, each $B_j \in \mathcal{F}$ because the latter is an algebra. Moreover, each $B_j \subseteq A_j$. Suppose $m < n$; then

$$B_m \cap B_n \subseteq A_m \cap A_n \cap (A_1^c \cap A_2^c \cap \cdots \cap A_{n-1}^c) \subseteq A_m \cap A_m^c = \varnothing.$$

Since each $B_j \subseteq A_j$, the inclusion

$$\bigcup_{j=1}^{\infty} B_j \subseteq \bigcup_{j=1}^{\infty} A_j$$

holds. On the other hand, suppose that $x \in \cup_{j \geq 1} A_j$ and let j be the smallest positive integer such that $x \in A_j$. Then $x \in A_j$ and $x \in \left(A_1 \cup A_2 \cup \cdots \cup A_{j-1}\right)^c$. So $x \in B_j$ and hence $x \in \cup_{j \geq 1} B_j$. Thus the reverse inclusion also holds. □

Definition 2.3.10 An algebra $\mathcal{F}$ of subsets of X is called a **σ-algebra** (or a **σ-field**) if every union of a sequence of sets in $\mathcal{F}$ is again in $\mathcal{F}$. That is, if $\{A_j\}_{j \geq 1}$ is a sequence of sets in $\mathcal{F}$, then $\cup_{j \geq 1} A_j$ must be in $\mathcal{F}$.

Remarks 2.3.11

(a) If $\{A_j\}_{j \geq 1}$ is a sequence of sets in a σ-algebra $\mathcal{F}$, then $\cap_{j \geq 1} A_j \in \mathcal{F}$. In fact, if $A_j \in \mathcal{F}$, then so is A_j^c; since $\mathcal{F}$ is a σ-algebra, $\cup_{j \geq 1} A_j^c \in \mathcal{F}$ and hence $\cap_{j \geq 1} A_j = \left(\cup_{j \geq 1} A_j^c\right)^c \in \mathcal{F}$.

(b) A class $\mathcal{F}$ of subsets of X is a σ-algebra if and only if

$$\text{(i) } X \in \mathcal{F} \quad \text{(ii) } A_j \in \mathcal{F} \quad \forall j \in \mathbb{N} \Rightarrow \bigcup\nolimits_{j \geq 1} A_j \in \mathcal{F} \quad \text{(iii) } A \in \mathcal{F} \Rightarrow A^c \in \mathcal{F}.$$

(c) Trivially, a finite union or intersection of sets in a σ-algebra is again in the σ-algebra.

(d) A knowlegable reader will discern the difference between the postulates of a σ-algebra and those of a "topology" on a set X.

Theorem 2.3.12 *The collection $\mathfrak{M}$ of measurable subsets of $\mathbb{R}$ is a σ-algebra.*

Proof We have shown in Proposition 2.3.7 that the collection $\mathfrak{M}$ of measurable sets is an algebra. All we need to show now is that if $E_j \subseteq \mathbb{R}(j \in \mathbb{N})$ are measurable, then so is their union $\cup_{j\geq 1}E_j$.

Since $\mathfrak{M}$ is an algebra, by Lemma 2.3.9, there exists a sequence $\{F_j\}_{j\geq 1}$ of disjoint measurable sets such that $\cup_{j\geq 1}E_j = \cup_{j\geq 1}F_j$. To complete the proof, we show

$$m^*(A) \geq m^*(A \cap (\bigcup_{j=1}^{\infty} F_j)) + m^*(A \cap (\bigcup_{j=1}^{\infty} F_j)^c)$$

for any set $A \subseteq \mathbb{R}$.

Observe that, for any positive integer p, $\cup_{1\leq j\leq p}F_j$ is measurable, because $\mathfrak{M}$ is an algebra. Therefore

$$\begin{aligned} m^*(A) &= m^*(A \cap (\bigcup_{j=1}^{p} F_j)) + m^*(A \cap (\bigcup_{j=1}^{p} F_j)^c) \\ &\geq \sum_{j=1}^{p} m^*(A \cap F_j) + m^*(A \cap (\bigcup_{j=1}^{\infty} F_j)^c), \end{aligned}$$

where we have used Proposition 2.3.8, the fact that $\left(\cup_{1\leq j\leq p}F_j\right)^c \supseteq \left(\cup_{j\geq 1}F_j\right)^c$ and Proposition 2.2.3(c). Since this inequality is valid for all p and the sequence $\left\{\sum_{j=1}^{p} m^*(A \cap F_j)\right\}_{p\geq 1}$ has supremum $\sum_{j=1}^{\infty} m^*(A \cap F_j)$, we have

$$\begin{aligned} m^*(A) &\geq \sum_{j=1}^{\infty} m^*(A \cap F_j) + m^*(A \cap (\bigcup_{j=1}^{\infty} F_j)^c) \\ &\geq m^*(\bigcup_{j=1}^{\infty} (A \cap F_j)) + m^*(A \cap (\bigcup_{j=1}^{\infty} F_j)^c) \quad \text{by Propostion 2.2.9} \\ &\geq m^*(A \cap (\bigcup_{j=1}^{\infty} F_j)) + m^*(A \cap (\bigcup_{j=1}^{\infty} F_j)^c). \end{aligned}$$

□

Theorem 2.3.13 *The outer measure m^* is countably additive on the σ-algebra $\mathfrak{M}$ of measurable subsets of $\mathbb{R}$.*

Proof Let $\{E_j\}_{j\geq 1}$ be a sequence of disjoint measurable subsets. From Proposition 2.2.3(c) and Proposition 2.3.8, it follows for an arbitrary positive integer p that

$$m^*(\bigcup_{j=1}^{\infty} E_j) \geq m^*(\bigcup_{j=1}^{p} E_j) = \sum_{j=1}^{p} m^*(E_j).$$

Therefore

$$m^*(\bigcup_{j=1}^{\infty} E_j) \geq \sum_{j=1}^{\infty} m^*(E_j).$$

The reverse inequality is a consequence of countable subadditivity (see Proposition 2.2.9). □

Definition 2.3.14 The restriction of m^* to $\mathfrak{M}$ is called **(Lebesgue) measure** and will be denoted by ***m***. For any $E \in \mathfrak{M}$, the (extended) real number $m(E) = m^*(E)$ is called the (Lebesgue) measure of the set E.

Since m^* is countably subadditive according to Proposition 2.2.9, it follows trivially that Lebesgue measure is also countably subadditive. We shall have occasion to use this fact in Chap. 5.

Proposition 2.3.15 *The interval* (a, ∞) *is measurable.*

Proof Let A be any subset of $\mathbb{R}$ and let $A_1 = A \cap (a, \infty)$ and $A_2 = A \cap (a, \infty)^c = A \cap (-\infty, a]$. We must show that $m^*(A) \geq m^*(A_1) + m^*(A_2)$. If $m^*(A) = \infty$, then there is nothing to prove. Suppose that $m^*(A) < \infty$.

By the definition of outer measure, for every $\varepsilon > 0$, there exists a sequence $\{I_n\}_{n \geq 1}$ of open intervals such that $A \subseteq \cup_{n \geq 1} I_n$ and

$$\sum_{n=1}^{\infty} \ell(I_n) \leq m^*(A) + \varepsilon.$$

Let $I_n' = I_n \cap (a, \infty)$ and $I_n'' = I_n \cap (-\infty, a]$. Then I_n' and I_n'' are intervals (some of them possibly empty) and

$$\ell(I_n) = \ell(I_n') + \ell(I_n'') = m^*(I_n') + m^*(I_n'').$$

Since $A_1 \subseteq \cup_{n \geq 1} I_n'$ and $A_2 \subseteq \cup_{n \geq 1} I_n''$, we have

$$m^*(A_1) \leq m^*(\bigcup_{n=1}^{\infty} I_n') \leq \sum_{n=1}^{\infty} m^*(I_n')$$

and

$$m^*(A_2) \leq m^*(\bigcup_{n=1}^{\infty} I_n'') \leq \sum_{n=1}^{\infty} m^*(I_n'').$$

Hence

$$m^*(A_1)+m^*(A_2)\leq \sum_{n=1}^{\infty} m^*(I'_n)+\sum_{n=1}^{\infty} m^*(I''_n)$$
$$=\sum_{n=1}^{\infty}[m^*(I'_n)+m^*(I''_n)]=\sum_{n=1}^{\infty}\ell(I_n)\leq m^*(A)+\varepsilon.$$

Since $\varepsilon > 0$ is arbitrary, we have $m^*(A) \geq m^*(A_1) + m^*(A_2)$, as required. □

Proposition 2.3.16 *A subset of* $\mathbb{R}$ *that is either open or closed must be measurable.*

Proof Since the collection $\mathfrak{M}$ of measurable sets is a σ-algebra, it follows from Proposition 2.3.15 that $(-\infty, a] = (a,\infty)^c$ is measurable for every real a. Since $(-\infty,b)=\bigcup_{n\geq 1}(-\infty,b-\frac{1}{n}]$, it further follows that $(-\infty, b)$ is measurable and hence each open interval $(a,b)=(a,\infty)\cap(-\infty,b)$ is measurable. We can now conclude that each open set, being a countable union of open intervals (see Theorem 1.3.17), must be measurable. Since $\mathfrak{M}$ is a σ-algebra, the complement of an open set, i.e. a closed set, must be measurable. □

Proposition 2.3.17 *Let* $\mathcal{A}$ *be a class of subsets of* X. *Then there is a smallest σ-algebra* $\mathcal{S}$ *of subsets of* X *that contains* $\mathcal{A}$.

Proof Let $\{\mathcal{S}_\alpha\}$ be the collection of all those σ-algebras of subsets of X that contain $\mathcal{A}$. Since the collection of all subsets of X is a σ-algebra, the collection $\{\mathcal{S}_\alpha\}$ is nonempty. It is easily verified that $\cap_\alpha \mathcal{S}_\alpha$ is a σ-algebra and contains $\mathcal{A}$. It is necessarily the smallest σ-algebra of subsets of X that contains $\mathcal{A}$. □

The σ-algebra in the above proposition is called the σ-algebra **generated by** $\mathcal{A}$ (or **by the sets of** $\mathcal{A}$).

Definition 2.3.18 The σ-algebra generated by all the open subsets of $\mathbb{R}$ is called the **Borel algebra** and will be denoted by $\mathfrak{B}$. A set in the Borel algebra is called a **Borel measurable set** or simply a **Borel set**. A set in the algebra $\mathfrak{M}$ (hitherto called simply "measurable") will be called a **Lebesgue measurable set** when a distinction needs to be made.

Remarks 2.3.19

(a) Since every open set is Lebesgue measurable (Proposition 2.3.16) and Lebesgue measurable sets constitute a σ-algebra, it follows by Proposition 2.3.17 and Definition 2.3.18 that $\mathfrak{B}\subseteq\mathfrak{M}$. In other words, every Borel set is Lebesgue measurable. The restriction of m^* to $\mathfrak{B}$ is called **Borel measure** and will again be denoted by m. Thus, for $E \in \mathfrak{B}$, $m(E) = m^*(E)$ is called the Borel measure of E and is the same as its Lebesgue measure.

(b) The cardinality of the Borel algebra $\mathfrak{B}$ is c and that of $\mathfrak{M}$ is strictly larger. The proof of the former may be found in [12, p. 134]. For the latter, see Problem 2.3.P20.

The triples ($\mathbb{R}$, $\mathfrak{B}$, m) and ($\mathbb{R}$, $\mathfrak{M}$, m) are called the **Borel measure space** and **Lebesgue measure space** respectively.

We have verified the following properties of Lebesgue measurable sets and Lebesgue measure:

(i) Complements, countable unions and countable intersections of measurable sets are measurable.
(ii) Any interval is measurable and its measure is its length.
(iii) If $\{E_n\}_{n\geq 1}$ is a sequence of disjoint measurable sets, then

$$m\Big(\bigcup_{n=1}^{\infty} E_n\Big) = \sum_{n=1}^{\infty} m(E_n).$$

The above properties are also valid for Borel measurable sets and Borel measure: (i) is a direct consequence of the definition of $\mathfrak{B}$ as being a σ-algebra, while (ii) and (iii) follow from their Lebesgue counterparts taken with Remark 2.3.19.

Some additional properties of measure will now be verified. We begin with what is known as "continuity" of measure. The following definition will be needed.

Definition 2.3.20 For any sequence $\{E_n\}_{n\geq 1}$ of sets, the **limit superior** and **limit inferior** are respectively

$$\limsup E_n = \bigcap_{i=1}^{\infty} \bigcup_{n\geq i} E_n \quad \text{and} \quad \liminf E_n = \bigcup_{i=1}^{\infty} \bigcap_{n\geq i} E_n.$$

If $\limsup E_n = \liminf E_n$, then we call this set the **limit** of the sequence $\{E_n\}_{n\geq 1}$ of sets and denote it by $\lim E_n$.

It is easily seen from the definition that $\limsup E_n$ is the set of all points belonging to infinitely many of the sets E_n and $\liminf E_n$ is the set of all points belonging to all but finitely many. Consequently, $\liminf E_n \subseteq \limsup E_n$. If $E_1 \subseteq E_2 \subseteq \cdots$, then we have $\lim E_n = \cup_{n\geq 1} E_n$; if $E_1 \supseteq E_2 \supseteq \cdots$, then we have $\lim E_n = \cap_{n\geq 1} E_n$.

Proposition 2.3.21 (Continuity) *Let $\{E_n\}_{n\geq 1}$ be a sequence of Lebesgue [or Borel] measurable sets. Then*

(a) $E_1 \subseteq E_2 \subseteq \cdots$ *implies* $m(\lim E_n) = \lim_{n\to\infty} m(E_n)$ (**inner continuity**),
(b) $E_1 \supseteq E_2 \supseteq \cdots$ *with* $m(E_1) < \infty$ *implies* $m(\lim E_n) = \lim_{n\to\infty} m(E_n)$ (**outer continuity**),
(c) $m(\liminf E_n) \leq \liminf m(E_n)$,
(d) $m(\limsup E_n) \geq \limsup m(E_n)$ *provided* $m\Big(\bigcup_{k=i}^{\infty} E_k\Big) < \infty$ *for some* i.

Proof It is sufficient to argue the case of Lebesgue measurable sets only, because the result will then follow for Borel measurable sets by Remark 2.3.19.

(a) Write $F_1 = E_1$, and $F_k = E_k \cap E_{k-1}^c$ for $k > 1$. Then $E_n = F_1 \cup F_2 \cup \cdots \cup F_n$ and also, $\cup_{k\geq 1} E_k = \cup_{k\geq 1} F_k$; besides, the sets F_k are measurable and disjoint. Hence, Theorem 2.3.13 yields

$$m(\lim E_n) = m(\bigcup_{k=1}^{\infty} E_k) = m(\bigcup_{k=1}^{\infty} F_k) = \sum_{k=1}^{\infty} m(F_k) = \lim_{n\to\infty} \sum_{k=1}^{n} m(F_k)$$
$$= \lim_{n\to\infty} m(\bigcup_{k=1}^{n} F_k) = \lim_{n\to\infty} m(E_n).$$

(b) Write $F_k = E_1 \cap E_k^c$. Since $E_k \subseteq E_1$, therefore E_1 is the disjoint union of E_k and F_k, where F_k is measurable. Before proceeding, note that $F_1 \subseteq F_2 \subseteq \cdots$ and consequently, $\lim F_n = \cup_{k\geq 1} F_k$. Now, by Theorem 2.3.13, we know that $m(E_1) = m(F_k) + m(E_k)$. Since $m(E_1) < \infty$, we can rewrite this as

$$m(F_k) = m(E_1) - m(E_k). \tag{2.16}$$

Moreover,

$$\lim F_n = \bigcup_{k=1}^{\infty} F_k = \bigcup_{k=1}^{\infty} (E_1 \cap E_k^c) = E_1 \cap \bigcup_{k=1}^{\infty} E_k^c = E_1 \cap (\bigcap_{k=1}^{\infty} E_k)^c$$
$$= E_1 \cap (\lim E_n)^c, \quad \text{because } \lim E_n = \bigcap\nolimits_{k\geq 1} E_k.$$

Since $\lim E_n \subseteq E_1$, a similar argument as for (2.16) now leads to

$$m(\lim F_n) = m(E_1) - m(\lim E_n). \tag{2.17}$$

Also, $F_1 \subseteq F_2 \subseteq \cdots$ and hence by (a) and (2.16), we have

$$m(\lim F_n) = \lim_{n\to\infty} m(F_n) = m(E_1) - \lim_{n\to\infty} m(E_n). \tag{2.18}$$

On comparing (2.17) and (2.18), and once again using the fact that $m(E_1) < \infty$, we obtain the required equality.

(c) Set $F_n = \cap_{k\geq n} E_k$. Then by definition, we have

$$\liminf E_n = \bigcup_{n=1}^{\infty} F_n.$$

Observe that $\{F_n\}$ is an increasing sequence of measurable sets. Therefore by part (a),

$$m\left(\bigcup_{n=1}^{\infty} F_n\right) = \lim_{n\to\infty} m(F_n),$$

so that

$$m(\liminf E_n) = \lim_{n\to\infty} m(F_n). \tag{2.19}$$

Let $n \in \mathbb{N}$ be fixed. Then $F_n \subseteq E_{n+k}$ and so

$$m(F_n) \leq m(E_{n+k}) \quad \text{for all } k \in \mathbb{N}.$$

Thus we have

$$m(F_n) \leq \liminf_k m(E_{n+k}) = \liminf\, m(E_k).$$

This is true for all $n \in \mathbb{N}$. Hence

$$\lim_{n\to\infty} m(F_n) \leq \liminf m(E_k). \tag{2.20}$$

The result now follows from (2.19) and (2.20).

(d) Similar to (c). □

Remark 2.3.22 Part (b) of the above proposition does not hold without the hypothesis that $m(E_1) < \infty$. Indeed, if $E_n = (n, \infty)$ for each $n \in \mathbb{N}$, then $\bigcup_{n\geq 1} E_n = \varnothing$; however, $\lim_{n\to\infty} m(E_n) = \infty$ and $m(\lim E_n) = m(\varnothing) = 0$.

Proposition 2.3.23 (Translation Invariance of Lebesgue measure on $\mathbb{R}$) *Let $E \in \mathfrak{M}$ [resp. $\mathfrak{B}$], $x \in \mathbb{R}$. Then*

(a) $E + x \in \mathfrak{M}$ [resp. $\mathfrak{B}$],
(b) $m(E + x) = m(E)$.

That is, the Lebesgue [resp. Borel] measure on $\mathfrak{M}$ [resp. $\mathfrak{B}$] is translation invariant.

Proof

(a) First we prove this for $\mathfrak{M}$. Let $A \subseteq \mathbb{R}$ be arbitrary and $E \in \mathfrak{M}$. Note that

$$A \cap (E + x) = [(A - x) \cap E] + x \quad \text{and} \quad A \cap (E + x)^c = [(A - x) \cap E^c] + x.$$

It follows immediately from the measurability of E and Proposition 2.2.4 (translation invariance of m^*) that

$$\begin{aligned} & m^*(A \cap (E+x)) + m^*(A \cap (E+x)^c) \\ & \quad = m^*([(A-x) \cap E] + x) + m^*([(A-x) \cap E^c] + x) \\ & \quad = m^*([(A-x) \cap E]) + m^*([(A-x) \cap E^c]) \\ & \quad = m^*(A-x) = m^*(A). \end{aligned}$$

This proves that $E+x \in \mathfrak{M}$, thus establishing (a) for $\mathfrak{M}$.

To prove the same for $\mathfrak{B}$, consider the class $\mathcal{F} = \{F \subseteq \mathbb{R} : F+x \in \mathfrak{B}\}$. First, since $\mathbb{R}+x = \mathbb{R} \in \mathfrak{B}$, we have $\mathbb{R} \in \mathcal{F}$. Second, $A_j \in \mathcal{F} \forall j \in \mathbb{N} \Rightarrow \cup_{j\geq 1} A_j \in \mathcal{F}$. To see why, suppose $A_j \in \mathcal{F} \forall j \in \mathbb{N}$. Then every $A_j + x \in \mathfrak{B}$. Now, $y \in A_j + x \Leftrightarrow y - x \in A_j$. Therefore $y \in \cup_{j\geq 1}(A_j + x) \Leftrightarrow y - x \in \cup_{j\geq 1} A_j \Leftrightarrow y \in (\cup_{j\geq 1} A_j) + x$. In other words, $\cup_{j\geq 1}(A_j + x) = (\cup_{j\geq 1} A_j) + x$. But since $\mathfrak{B}$ is a σ-algebra, we have $\cup_{j\geq 1}(A_j + x) \in \mathfrak{B}$. Therefore $(\cup_{j\geq 1} A_j) + x \in \mathfrak{B}$, so that $\cup_{j\geq 1} A_j \in \mathcal{F}$. And third, $A \in \mathcal{F} \Rightarrow A^c \in \mathcal{F}$. This is because $(A + x)^c$ can be shown to be the same as $A^c + x$ by arguing that $y \in (A + x)^c \Leftrightarrow y \notin A + x \Leftrightarrow y - x \notin A \Leftrightarrow y - x \in A^c$. By Remark 2.3.11(b), it follows that $\mathcal{F}$ is a σ-algebra. Surely, $O + x$ is open, and hence belongs to $\mathfrak{B}$ whenever $O \subseteq \mathbb{R}$ is open. Therefore the σ-algebra $\mathcal{F}$ contains all open sets and hence contains $\mathfrak{B}$. Thus $F \in \mathfrak{B} \Rightarrow F \in \mathcal{F} \Rightarrow F + x \in \mathfrak{B}$.

(b) Since $E+x \in \mathfrak{M}$ (or $\mathfrak{B}$) and m is the restriction of m^* to $\mathfrak{M}$ (or $\mathfrak{B}$), by Proposition 2.2.4 (translation invariance of m^*), we have

$$m(E+x) = m^*(E+x) = m^*(E) = m(E).$$ □

Finally, we prove the following characterisation of Lebesgue measurable subsets of $\mathbb{R}$.

Proposition 2.3.24 *Let E be a given subset of* $\mathbb{R}$. *The following five statements are equivalent*:

(α) *E is Lebesgue measurable*;
(β) *For every* $\varepsilon > 0$, *there exists an open set* $O \supseteq E$ *such that* $m^*(O \backslash E) < \varepsilon$;
(γ) *There exists a* $\mathcal{G}_\delta$*-set* $G \supseteq E$ *such that* $m^*(G \backslash E) = 0$;
(δ) *For every* $\varepsilon > 0$, *there exists a closed set* $F \subseteq E$ *such that* $m^*(E \backslash F) < \varepsilon$;
(η) *There exists an* $\mathcal{F}_\sigma$*-set* $F \subseteq E$ *such that* $m^*(E \backslash F) = 0$.

Proof We shall prove (α) ⇒ (β) ⇒ (γ) ⇒ (α) and then (δ) ⇔ (α) ⇔ (η).

(α) ⇒ (β). For any set E and $\varepsilon > 0$, there exists (see Proposition 2.2.12) an open set $O \supseteq E$ such that

$$m^*(O) \leq m^*(E) + \frac{\varepsilon}{2}.$$

Since $O \supseteq E$, we have $O \cap E = E$. From the Lebesgue measurability of E, we have

$$m^*(O) = m^*(O \cap E) + m^*(O \cap E^c) = m^*(E) + m^*(O \backslash E).$$

Thus

$$m^*(E) + \frac{\varepsilon}{2} \geq m^*(O) = m^*(E) + m^*(O \backslash E).$$

In the case when $m^*(E) < \infty$, this shows that $m^*(O \backslash E) \leq \frac{\varepsilon}{2} < \varepsilon$.

Now let $m^*(E) = \infty$ and set $E_n = E \cap [-n, n]$ for $n \in \mathbb{N}$. Then $E_n \in \mathfrak{M}$ and $m^*(E_n) \leq m^*([-n, n]) < \infty$. Moreover, $E = \cup_{n \in \mathbb{N}} E_n$.

Since $m^*(E_n) < \infty$, it follows on applying the case of finite measure proved above that, for any $\varepsilon > 0$, there exists an open set $O_n \supseteq E_n$ such that

$$m^*(O_n \backslash E_n) < \frac{\varepsilon}{2^n}.$$

Note that

$$O = \bigcup_{n \in \mathbb{N}} O_n \supseteq \bigcup_{n \in \mathbb{N}} E_n = E$$

and

$$O \backslash E = \bigcup_{n \in \mathbb{N}} O_n \backslash \bigcup_{n \in \mathbb{N}} E_n \subseteq \bigcup_{n \in \mathbb{N}} (O_n \backslash E_n).$$

So,

$$m^*(O \backslash E) \leq \sum_{n=1}^{\infty} m^*(O_n \backslash E_n) < \sum_{n=1}^{\infty} \frac{\varepsilon}{2^n} = \varepsilon.$$

(β) $\Rightarrow$ (γ). It follows from (β) that, for any $n \in \mathbb{N}$, there exists an open set $O_n \supseteq E$ such that

$$m^*(O_n \backslash E) < \frac{1}{n}.$$

Let $G = \cap_{n \geq 1} O_n$; then G is a $\mathcal{G}_\delta$-set containing E. Moreover,

$$m^*(G \backslash E) = m^*(\bigcap_{n=1}^{\infty} O_n \backslash E) \leq m^*(O_n \backslash E) < \frac{1}{n} \quad \text{for every } n \in \mathbb{N}.$$

Hence $m^*(G \backslash E) = 0$.

(γ) $\Rightarrow$ (α). Note that $E = G \backslash (G \backslash E)$ and that the set G, being a countable intersection of Lebesgue measurable sets, is Lebesgue measurable by Theorem 2.3.12; also, $G \backslash E$ is Lebesgue measurable by Remark 2.3.2(d). Use Theorem 2.3.12 once again.

We have shown so far that $(\alpha) \Rightarrow (\beta) \Rightarrow (\gamma) \Rightarrow (\alpha)$. The consequence that $(\beta) \Leftrightarrow (\alpha)$ will shortly be used in proving that $(\alpha) \Leftrightarrow (\delta)$.

$(\alpha) \Leftrightarrow (\delta)$. Assume (α), i.e. $E \in \mathfrak{M}$. Then $E^c \in \mathfrak{M}$. Since $(\alpha) \Rightarrow (\beta)$, it follows that, for any $\varepsilon > 0$, there exists an open set $O \supseteq E^c$ such that $m^*(O \backslash E^c) < \varepsilon$. So the closed set $F = O^c$ satisfies $F \subseteq E$ and also $m^*(E \backslash F) = m^*(E \backslash O^c) = m^*(E \cap O) = m^*(O \backslash E^c) < \varepsilon$. Thus (δ) holds.

Conversely, assume (δ). Then the open set $O = F^c$ satisfies $O \supseteq E^c$ and $m^*(O \backslash E^c) < \varepsilon$. In other words, (β) holds with E^c in place of E. Since $(\beta) \Rightarrow (\alpha)$, it follows that $E^c \in \mathfrak{M}$ and hence $E \in \mathfrak{M}$. Thus (α) holds.

$(\alpha) \Leftrightarrow (\eta)$. This is analogous to the argument that $(\alpha) \Leftrightarrow (\delta)$. □

Remark 2.3.25 Let $E \subseteq \mathbb{R}$ be Lebesgue measurable, i.e. $E \in \mathfrak{M}$. Then it follows from the preceding proposition that there exist $F, G \in \mathfrak{B}$ such that $F \subseteq E \subseteq G$ with $m(G \backslash F) = 0$. Conversely, the existence of such F and G can be proved to imply that $E \in \mathfrak{M}$ by arguing in the following manner: Suppose such F, G exist. Then $F \in \mathfrak{M}$ because $\mathfrak{B} \subseteq \mathfrak{M}$; besides, $E \backslash F \subseteq G \backslash F$, which implies $m^*(E \backslash F) \leq m^*(G \backslash F) = 0$ and hence $E \backslash F \in \mathfrak{M}$ by Remark 2.3.2(d). Therefore $E = (E \backslash F) \cup F \in \mathfrak{M}$. Note that in this situation, $m(E) = m(F) = m(G)$. The foregoing equivalence is related to the fact that Lebesgue measure is the "completion" of Borel measure (see Definition 7.2.3 and the paragraph preceding Theorem 7.2.2).

Problem Set 2.3

2.3.P1. Let A, B be subsets of $\mathbb{R}$ and $A \subseteq B$. Show that

(a) If $m^*(B \backslash A) = 0$, then $m^*(B) = m^*(A)$.
(b) If $m^*(B) = m^*(A) < \infty$ and $A \in \mathfrak{M}$, then $m^*(B \backslash A) = 0$.

2.3.P2. Suppose that A is a subset of $\mathbb{R}$ with the property that, for $\varepsilon > 0$, there exist measurable sets B and C such that $B \subseteq A \subseteq C$ and $m(C \cap B^c) < \varepsilon$. Show that A is measurable.

2.3.P3. Let $\{E_n\}_{n \geq 1}$ be a sequence of sets such that $E_1 \subseteq E_2 \subseteq \cdots$. Show that $m^*(\lim E_n) = \lim_{n \to \infty} m^*(E_n)$.

2.3.P4. Given a subset A of $\mathbb{R}$, let $A_n = A \cap [-n, n]$ for $n \in \mathbb{N}$. Show that $m^*(A) = \lim_{n \to \infty} m^*(A_n)$.

2.3.P5. The **symmetric difference** of sets A, B is defined to be $A \Delta B = (A \backslash B) \cup (B \backslash A)$, consisting of points belonging to one of the two sets but not to both. Show that if $A \in \mathfrak{M}$ and $m^*(A \Delta B) = 0$, then $B \in \mathfrak{M}$.

2.3.P6. Show that every nonempty open subset O has positive measure.

2.3.P7. Let $O = \cup_{n \geq 1}(x_n - \frac{1}{n^2}, x_n + \frac{1}{n^2})$, where $x_1, x_2, \ldots$ is an enumeration of all the rationals. Prove that $m(O \Delta F) > 0$ for any closed set F.

2.3.P8. The number of elements in a σ-algebra generated by n given sets cannot exceed 2^{2^n}.

2.3.P9. Show that a σ-algebra consisting of infinitely many distinct sets contains an uncountable number of sets.

2.3.P10. Show that, if E_1 and E_2 are measurable, then

$$m(E_1 \cup E_2) + m(E_1 \cap E_2) = m(E_1) + m(E_2).$$

2.3.P11. If $E_1 \in \mathfrak{M}$ and $E_2 \in \mathfrak{M}$ are such that $E_1 \supseteq E_2$ and $m(E_2) < \infty$, then $m(E_1 \backslash E_2) = m(E_1) - m(E_2)$. (Since $\mathfrak{B} \subseteq \mathfrak{M}$, it follows trivially that this is true also when $E_1 \in \mathfrak{B}$ and $E_2 \in \mathfrak{B}$.)
2.3.P12. Let $0 < \alpha < 1$. Construct a measurable set $E \subseteq [0, 1]$ of measure $1 - \alpha$ and containing no interval of positive length.
2.3.P13. Show that, if E is such that $0 < m^*(E) < \infty$ and $0 < \alpha < 1$, then there exists an open interval I such that $m^*(I \cap E) > \alpha\ell(I)$.
2.3.P14. (a) Suppose $m^*(E) < \infty$ and, for every $\varepsilon > 0$, there exists an open set U such that $m^*(E\Delta U) < \varepsilon$. Show that $E \in \mathfrak{M}$.
(b) If E is measurable and $m(E) < \infty$, then show that, for every $\varepsilon > 0$, there exists a finite union U of disjoint open intervals such that $m^*(E\Delta U) < \varepsilon$.
2.3.P15. Give an example of a nonmeasurable set.
2.3.P16. Show that the outer measure m^* is not finitely additive.
2.3.P17. Let f be defined on $[0, 1]$ by $f(0) = 0$ and $f(x) = x \sin \frac{1}{x}$ for $x > 0$. Show that the measure of the set $\{x: f(x) > 0\}$ is $1 - (\ln 2)/\pi$.
2.3.P18. Let E be Lebesgue measurable with $0 < m(E) < \infty$ and let $\varepsilon > 0$ be given. Then there exists a compact set $K \subseteq E$ such that $m(E \backslash K) = m(E) - m(K) < \varepsilon$.

In what follows, we provide a characterisation of the σ-algebra generated by an algebra $\mathcal{A}$ in terms of what are called "monotone classes".

A family $\mathfrak{M}$ of subsets of a nonempty set X is called a **monotone class** if it satisfies the following two conditions:

(i) if $A_1 \subseteq A_2 \subseteq \cdots$ and each $A_j \in \mathfrak{M}$, then $\cup_{j \geq 1} A_j \in \mathfrak{M}$;
(ii) if $B_1 \supseteq B_2 \supseteq \cdots$ and each $B_j \in \mathfrak{M}$, then $\cap_{j \geq 1} B_j \in \mathfrak{M}$.

Any σ-algebra is a monotone class. Let $A \subseteq X$, where X is any nonempty set and $\mathfrak{M} = \{A\}$. Then $\mathfrak{M}$ is a monotone class which is not a σ-algebra.
2.3.P19. (a) If $\mathcal{Y}$ is any class of subsets of a nonempty set X, then show that there exists a smallest monotone class containing $\mathcal{Y}$. We shall denote it by $\mathfrak{M}_0(\mathcal{Y})$.
(b) If $\mathcal{A}$ is an algebra, show that $\mathcal{S}(\mathcal{A}) = \mathfrak{M}_0(\mathcal{A})$; that is, the σ-algebra generated by an algebra is also the smallest monotone class containing the algebra.
2.3.P20. Show that (a) the cardinality of $\mathfrak{M}$ is greater than c and (b) the Cantor set has a subset which is not a Borel set.
2.3.P21. Let $\mathcal{F}$ be an algebra of subsets of a set X. If A and B are subsets of X such that B, $B \cap A^c$, $A \cap B^c$ all belong to $\mathcal{F}$, show that A also belongs to $\mathcal{F}$.
2.3.P22. Let f be a real-valued function on $[a, b]$. Suppose $E \subseteq [a, b]$, f' exists and satisfies $|f'(x)| \leq M$ for all $x \in [a, b]$. Prove that $m^*(f(E)) \leq M\, m^*(E)$.
2.3.P23. (a) Let F and Y be subsets of a set X. Then show that $F^c \cap Y$ is the complement in Y of $F \cap Y$.

(b) Let $Y \subseteq X$ and $\mathcal{G}$ be a σ-algebra of subsets of Y. Show that the family

$$\mathcal{F}_0 = \{F \subseteq X: F \cap Y \in \mathcal{G}\}$$

of subsets of X is a σ-algebra.

2.3.P24. Let $\mathcal{A}$ be a family of subsets of a set X and $\mathcal{F}$ the σ-algebra generated by it. Suppose $Y \subseteq X$ and

$$\mathcal{A}_Y = \{A \cap Y : A \in \mathcal{A}\}, \quad \mathcal{F}_Y = \{F \cap Y: F \in \mathcal{F}\}.$$

Show that $\mathcal{F}_Y$ is the σ-algebra of subsets of Y generated by $\mathcal{A}_Y$.

2.4 Measurable Functions

The class of what are called *measurable functions* includes continuous and monotone functions among others, and plays a vital role in the theory of Lebesgue integration, which will be discussed in the sequel. The concept of measurability for functions depends only on a prescribed σ-algebra of subsets of the domain set (in our case $\mathfrak{M}_X = \{X \cap Y: Y \in \mathfrak{M}\}$ or $\mathfrak{B}_X = \{X \cap Y: Y \in \mathfrak{B}\}$, where $X \subseteq \mathbb{R}$ is measurable and $\mathfrak{M}$ is the σ-algebra of Lebesgue measurable subsets of $\mathbb{R}$ and $\mathfrak{B}$ is the σ-algebra of Borel measurable subsets of $\mathbb{R}$). It is analogous to the concept of continuous functions, with which the reader is undoubtedly familiar. We recall the appropriate characterisation of continuous functions to emphasise the analogy between the definition of measurable functions and that of continuous functions: A real-valued function f with domain X is continuous if and only if for every open set $V \subseteq \mathbb{R}$, the inverse image $f^{-1}(V) = \{x \in X: f(x) \in V\}$ is open in X, that is, there is an open set $U \subseteq \mathbb{R}$ such that $f^{-1}(V) = X \cap U$ (see Theorem 1.3.21).

Let X be a subset of $\mathbb{R}$ and α be an arbitrary real number. Suppose that f is a real-valued function defined on X. The following notations will be employed:

$$\begin{aligned} X(f \geq \alpha) &= \{x \in X : f(x) \geq \alpha\}, \\ X(f = \alpha) &= \{x \in X : f(x) = \alpha\}, \\ X(f \leq \alpha) &= \{x \in X : f(x) \leq \alpha\}, \\ X(f > \alpha) &= \{x \in X : f(x) > \alpha\}, \\ X(f < \alpha) &= \{x \in X : f(x) < \alpha\}. \end{aligned}$$

Before stating the next definition, we note the simple fact that, if $X \in \mathfrak{M}$ [resp. $\mathfrak{B}$], then the class $\{A \subseteq X : A \in \mathfrak{M}\ [\text{resp.}\ \mathfrak{B}]\}$ of those subsets of X that are Lebesgue [resp. Borel] measurable is the same as the class of intersections $\mathfrak{M}_X\ [\text{resp.}\ \mathfrak{B}_X] = \{X \cap Y : Y \in \mathfrak{M}\ [\text{resp.}\ \mathfrak{B}]\}$ and is a σ-algebra. $\mathfrak{M}_X$ [resp. $\mathfrak{B}_X$] is called the σ-algebra of **Lebesgue** [resp. **Borel**] **measurable subsets** of X.

Definition 2.4.1 Let $\mathcal{F}$ be a σ-algebra of subsets of $X \subseteq \mathbb{R}$. A real-valued function f defined on X is said to be **$\mathcal{F}$-measurable** if, for every real number α, the set $X(f > \alpha)$ is in $\mathcal{F}$. In particular, if $X \in \mathfrak{M}$ and if $\mathcal{F}$ is the σ-algebra $\mathfrak{M}_X = \{X \cap Y : Y \in \mathfrak{M}\}$ of Lebesgue measurable subsets of X, then f is said to be **Lebesgue measurable**. Similarly, if $X \in \mathfrak{B}$ and if $\mathcal{F}$ is the σ-algebra $\mathfrak{B}_X = \{X \cap Y : Y \in \mathfrak{B}\}$ of Borel measurable subsets of X, then f is said to be **Borel measurable**. From here onwards, both will be referred to as simply "measurable" when it is clear from the context which σ-algebra is intended.

It may be noted that every Borel measurable function is Lebesgue measurable. The converse does not hold. Indeed, let A be a Lebesgue measurable set which is not Borel measurable (see 2.3.P20). Then it follows from Example 2.4.4(b) below that χ_A is Lebesgue measurable but not Borel measurable.

We now prove a proposition that shows that we could have modified the form of the set in defining measurability.

Proposition 2.4.2 *Let f be a real-valued function defined on a measurable subset $X \subseteq \mathbb{R}$. Then the following statements about the function f are equivalent*:

(α) *For every $\alpha \in \mathbb{R}$, the set $X(f > \alpha)$ is measurable*;
(β) *For every $\alpha \in \mathbb{R}$ the set $X(f \leq \alpha)$ is measurable*;
(γ) *For every $\alpha \in \mathbb{R}$, the set $X(f \geq \alpha)$ is measurable*;
(δ) *For every $\alpha \in \mathbb{R}$, the set $X(f < \alpha)$ is measurable.*

Proof Since the sets $X(f > \alpha)$ and $X(f \leq \alpha)$ are complements of each other, and the collection $\mathfrak{M}_X$ [resp. $\mathfrak{B}_X$] is closed under complementation (being a σ-algebra), it follows that (α) and (β) are equivalent. Similarly, the statements (γ) and (δ) are equivalent.

If (α) holds, then $X(f > \alpha - \frac{1}{n})$ is measurable for each n and

$$X(f \geq \alpha) = \bigcap_{n=1}^{\infty} X(f > \alpha - \frac{1}{n});$$

so, the set $X(f \geq \alpha)$, being the intersection of a sequence of measurable sets, is measurable. Thus (α) implies (γ). Again, since

$$X(f > \alpha) = \bigcup_{n=1}^{\infty} X(f \geq \alpha + \frac{1}{n}),$$

it follows that $X(f > \alpha)$, being the union of a sequence of measurable sets, is measurable. Thus (γ) implies (α). □

Corollary 2.4.3 *A real-valued function f defined on a measurable subset $X \subseteq \mathbb{R}$ is measurable if and only if any one of the statements* (α), (β), (γ) *or* (δ) *in* Proposition 2.4.2 *holds.*

Proof An immediate consequence of the preceding definition and proposition. □

We give below examples of measurable functions.

Examples 2.4.4

(a) Any constant function defined on a measurable set $X \subseteq \mathbb{R}$ is measurable. For, if $f(x) = k$ for all $x \in X$ and $\alpha \geq k$, then $X(f > \alpha) = \emptyset$, whereas, if $\alpha < k$, then $X(f > \alpha) = X$.
(b) Suppose $X \subseteq \mathbb{R}$ is measurable, $A \subseteq X$ and χ_A is defined on X, that is, $\chi_A(x) = 1$ if $x \in A$ and 0 otherwise. Then χ_A is a measurable function if and only if A is measurable. In fact, for an arbitrary real number α, the set $\{x \in X: \chi_A(x) > \alpha\}$ is X, A or $\emptyset$, depending on whether $\alpha < 0$, $0 \leq \alpha < 1$ or $\alpha \geq 1$ respectively.
(c) If $X \subseteq \mathbb{R}$ is measurable and $f: X \rightarrow \mathbb{R}$ is continuous, then $X(f > \alpha)$ is open and is therefore a Borel set (see Definitions 2.3.18 and 2.4.1); this means f is Borel measurable and hence also Lebesgue measurable.
(d) Let I be an interval and $f: I \rightarrow \mathbb{R}$ be monotonically increasing, that is, $x < x'$ $\Rightarrow f(x) \leq f(x')$. Then f is Borel measurable, because $I(f > \alpha)$ is an interval; indeed, if x_1 and x_2 are any two points in it, then so is any y lying between x_1 and x_2. Similarly, a monotonically decreasing function is always Borel measurable.
(e) Suppose $X \subseteq \mathbb{R}$ is a countable union $\cup_{k \geq 1} X_k$ of sets X_k, not necessarily disjoint. Then for any function $f: X \rightarrow \mathbb{R}$, we obviously have

$$X(f > \alpha) = \bigcup_{k=1}^{\infty} X_k(f > \alpha).$$

Therefore if each X_k is measurable and the restriction of f to each X_k is measurable, then f is measurable. As this applies when the sets X_k are not disjoint, it follows that the result remains true when they are finite in number.
(f) Suppose $X \subseteq \mathbb{R}$ is measurable and let $f: X \rightarrow \mathbb{R}$ be measurable. Then the truncated function

$$f_A(x) = \begin{cases} f(x) & |f(x)| \leq A \\ A & f(x) > A \\ -A & f(x) < -A \end{cases}$$

can be seen to be measurable. Since its restriction to each of the three measurable sets (whichever ones may be nonempty)

$$\{x \in X : f(x) > A\}, \quad \{x \in X : -A \leq f(x) \leq A\} \quad \text{and} \quad \{x \in X : f(x) < -A\},$$

having union X, is measurable, it follows from part (e) above that f_A is measurable. Alternatively, $f_A = \min\{A, \max\{-A, f\}\}$ and Theorem 2.4.6(e) below implies that f_A is measurable.

Proposition 2.4.5 *If f is a measurable function defined on a measurable subset $X \subseteq \mathbb{R}$, then the set $X(f = \alpha)$ is measurable for each real number α.*

Proof For a real number α, the sets $X(f \geq \alpha)$ and $X(f \leq \alpha)$ are both measurable, as f is measurable. Therefore, so is their intersection $X(f \geq \alpha) \cap X(f \leq \alpha)$. But the intersection is precisely $X(f = \alpha)$. □

Remark The converse of Proposition 2.4.5 is, however, not true: Let $A \subseteq [0, 1)$ be a nonmeasurable set described in the solution to Problem 2.3.P15. Consider the function f defined on $[0, 1)$ by

$$f(x) = \begin{cases} x^2 & x \in A \\ -x^2 & x \in [0, 1) \backslash A \end{cases}$$

Then for each real number α, the set $\{x \in [0, 1): f(x) = \alpha\}$ consists of at most two elements and is therefore measurable. However, the set $\{x \in [0, 1): f(x) \geq 0\}$ is A, which is not measurable. Consequently, the function f is not measurable.

We shall next show that certain simple algebraic combinations of measurable functions are measurable ("algebra of measurable functions"). Their limits are deferred to a later section.

Theorem 2.4.6 *Let X be a measurable subset of* $\mathbb{R}$ *and suppose that* $f, g: X \to \mathbb{R}$ *are measurable. Then the following are also measurable*:

(a) af $(a \in \mathbb{R})$;
(b) $|f|$;
(c) $f + g$;
(d) fg;
(e) $\min\{f, g\}$ *and* $\max\{f, g\}$.

Proof

(a) If $a = 0$, the statement is trivial. If $a > 0$, then

$$\{x \in X : af(x) > \alpha\} = \{x \in X : f(x) > \frac{\alpha}{a}\}$$

and the set on the right-hand side is measurable. The case when $a < 0$ is similar.

(b) If $\alpha < 0$, then the set $\{x \in X : |f(x)| > \alpha\} = X$, whereas, if $\alpha \geq 0$, then

$$\{x \in X : |f(x)| > \alpha\} = \{x \in X : f(x) > \alpha\} \cup \{x \in X : f(x) < -\alpha\}$$

and since each of the sets on the right is measurable, so is their union.

(c) By hypothesis, if r is a rational number, then

$$A_r = \{x \in X : f(x) > r\} \cap \{x \in X : g(x) > \alpha - r\}$$

is measurable. Since

$$\{x \in X : (f+g)(x) > \alpha\} = \bigcup_{r\in\mathbb{Q}} A_r$$

and a countable union of measurable sets is measurable, it follows that the set on the left-hand side is measurable.

(d) We shall first show that, if f is measurable, then so is f^2. If $\alpha < 0$, then

$$\{x \in X : f(x)^2 > \alpha\} = X,$$

which is a measurable set; if $\alpha \geq 0$, then

$$\{x \in X : f(x)^2 > \alpha\} = \{x \in X : f(x) > \sqrt{\alpha}\} \cup \{x \in X : f(x) < -\sqrt{\alpha}\}$$

and since each of the sets on the right is measurable, the one on the left is measurable. It follows that f^2 is measurable. This fact, together with (a), (c) and the equality

$$fg = \frac{1}{2}[(f+g)^2 - (f-g)^2],$$

implies that fg is measurable.

(e) For an arbitrary real number α,

$$\begin{aligned}\{x \in X : \max\{f, g\}(x) > \alpha\} &= \{x \in X : f(x) > \alpha \text{ or } g(x) > \alpha\}\\ &= \{x \in X : f(x) > \alpha\} \cup \{x \in X : g(x) > \alpha\}.\end{aligned}$$

Since each of the sets on the right is measurable, the one on the left is measurable. It follows that $\max\{f, g\}$ is measurable. The argument for $\min\{f, g\}$ is similar. □

Remarks 2.4.7

(a) If f is any real-valued function defined on X, let f^+ and f^- be the nonnegative functions on X defined as

$$f^+(x) = \max\{f(x), 0\} \quad \text{and} \quad f^-(x) = \max\{-f(x), 0\}.$$

The function f^+ is called the **positive part of** f and f^- is called the **negative part of** f. The following identities are obvious:

$$f = f^+ - f^- \quad \text{and} \quad |f| = f^+ + f^-.$$

It is a consequence of these identities that

$$f^+ = \frac{1}{2}(|f| + f) \quad \text{and} \quad f^- = \frac{1}{2}(|f| - f).$$

From Theorem 2.4.6, we infer that f is measurable if and only if f^+ and f^- are.

(b) The converse of (b) in Theorem 2.4.6 is not true. If $A \subseteq [0, 1)$ is a nonmeasurable subset, define

$$f(x) = \begin{cases} 1 & x \in A \\ -1 & x \in [0,1)\backslash A. \end{cases}$$

Then f is not measurable, because the set $\{x \in [0, 1): f(x) > 0\} = A$ is nonmeasurable. But $|f|$ is the constant function 1 and is therefore measurable.

(c) The above example also shows that f^2 can be measurable when f is not.

Problem Set 2.4

2.4.P1. Let the function f: $[0, 1] \to \mathbb{R}$ be defined by $f(x) = \frac{1}{x}$ if $0 < x \leq 1$, $f(0) = 0$. Show that f is measurable.

2.4.P2. Show that, if f is a real-valued function on a measurable subset $X \subseteq \mathbb{R}$ such that $X(f \geq r)$ is measurable for every rational number r, then f is measurable.

2.4.P3. Let $X \subseteq \mathbb{R}$ be measurable. Without using Theorem 2.4.6, show that if f and g are measurable functions defined on X, then the set $\{x \in X: f(x) > g(x)\}$ is measurable.

2.4.P4. Let f be a real-valued function defined on a measurable subset $X \subseteq \mathbb{R}$. Then f is measurable if and only if, for every open set $V \subseteq \mathbb{R}$,

$$f^{-1}(V) = \{x \in X : f(x) \in V\}$$

is measurable.

2.4.P5. Let f be a measurable function defined on a measurable subset $X \subseteq \mathbb{R}$ and ϕ be defined and continuous on the range of f. Then $\phi \circ f$ is a measurable function on X.

2.4.P6. Show that any function f defined on a set X of Lebesgue measure zero is Lebesgue measurable.

2.4.P7. Let $X \subseteq \mathbb{R}$ and f: $X \to \mathbb{R}$ be any function. Show that the family of subsets of X given by $\mathcal{F} = \{f^{-1}(V)\text{: } V \in \mathfrak{B}\}$ is a σ-algebra. (The same is true if $\mathfrak{B}$ is replaced by $\mathfrak{M}$.)

2.4.P8. Let $X \subseteq \mathbb{R}$ and $f : X \to \mathbb{R}$ be any function. If $\mathcal{F}$ is a σ-algebra of subsets of X, show that the family of subsets of $\mathbb{R}$ given by $\mathcal{G} = \{V \subseteq \mathbb{R} : f^{-1}(V) \in \mathcal{F}\}$ is a σ-algebra. Hence show that, if f is Borel measurable, then $V \in \mathfrak{B} \Rightarrow f^{-1}(V) \in \{X \cap U : U \in \mathfrak{B}\}$. Is it true that, if f is Lebesgue measurable, then $V \in \mathfrak{B} \Rightarrow f^{-1}(V) \in \{X \cap U : U \in \mathfrak{M}\}$?

2.4.P9. Let f be a real-valued measurable function defined on a measurable set $X \subseteq \mathbb{R}$. Prove that

(a) if $\alpha > 0$, then the function $|f|^\alpha$ is measurable;
(b) if $f(x) \neq 0$ on X and $\alpha < 0$, then $|f|^\alpha$ is measurable.

2.4.P10. A complex-valued function f with domain a measurable set $X \subseteq \mathbb{R}$ is said to be measurable if its real and imaginary parts, which are real-valued functions on X, are measurable. Prove that a complex-valued function is measurable if and only if $f^{-1}(V)$ is measurable for every open set $V \subseteq \mathbb{C}$ (the complex plane).

2.5 Extended Real-Valued Functions

The preceding discussion of measurable functions pertained to real-valued functions. However, in dealing with a sequence of measurable functions, we need to form its supremum, infimum, limit superior, limit inferior or limit. It is then convenient to work with the extended real number system $\mathbb{R} \cup \{-\infty\} \cup \{\infty\}$ as the range space. With this in mind, we proceed to define measurability for extended real valued functions. Much of what we say is intended for Borel as well as Lebesgue measurability, except when we explicitly mention only one of them.

Definition 2.5.1 Let f be an extended real-valued function defined on a measurable subset $X \subseteq \mathbb{R}$. Then f is said to be **measurable** if, for each real number α, the set $X(f > \alpha)$ is measurable.

Proposition 2.5.2 *An extended real-valued function f on a measurable set $X \subseteq \mathbb{R}$ is measurable if and only if* (a) *the sets $A = X(f = \infty)$, $B = X(f = -\infty)$ are measurable and* (b) *the restriction of f to $X \backslash (A \cup B)$, which is real-valued, is measurable.*

Proof We take up the "only if" part first. Observe that, for each positive integer n, the set $X(f > n)$ is measurable by definition; hence, so is their intersection $\bigcap_{n=1}^{\infty} X(f > n)$, which is $A = X(f = \infty)$. This shows that A is measurable. The set $B = X(f = -\infty)$ is measurable because it is the complement of $\bigcup_{n \in \mathbb{Z}} X(f > n)$. Moreover, for each real number α, the set $\{x \in X \backslash (A \cup B) : f(x) > \alpha\}$ is measurable, because it equals the intersection $\{x \in X : f(x) > \alpha\} \cap \{X \backslash (A \cup B)\}$. We have thus proved that if an extended real-valued function f is measurable, then the sets $A = X(f = \infty)$, $B = X(f = -\infty)$ and $\{x \in X \backslash (A \cup B) : f(x) > \alpha\}$, where α is an arbitrary real number, are measurable. This proves the "only if" part.

The converse is immediate from the fact that, for any real number α, we have

$$X(f > \alpha) = \{x \in X \backslash (A \cup B) : f(x) > \alpha\} \cup A. \qquad \square$$

Remarks 2.5.3

(a) It is a consequence of Theorem 2.4.6 and Proposition 2.5.2 that, if f and g are extended real-valued measurable functions, then so are

$$cf\ (c\,\text{a real number}),\quad |f|,\quad \min\{f,g\}\quad \text{and}\quad \max\{f,g\},$$

keeping in mind the conventions (see Chap. 1) that $0 \cdot (\pm\infty) = 0$.

(b) If f and g are extended real-valued measurable functions, then so are $f + g$ and fg. In connection with the measurability of $f + g$, we proceed as follows.

Consider the sets $A = X(f = \infty)$, $B = X(f = -\infty)$, $C = X(g = \infty)$ and $D = X(g = -\infty)$. The domain of definition of $f + g$ is $Y = X\backslash((A \cap D) \cup (B \cap C))$. By Proposition 2.5.2, the sets A, B, C and D are all measurable and hence the domain of definition of $f + g$ is also measurable. To prove the measurability of $f + g$ by using Proposition 2.5.2, we first note that the sets,

$$Y(f + g = \infty) = (A\backslash D) \cup (C\backslash B) = E, \quad \text{say},$$

and

$$Y(f + g = -\infty) = (B\backslash C) \cup (D\backslash A) = F, \quad \text{say},$$

are both measurable. On the complement of their union, i.e., $Y\backslash(E \cup F)$, the function $f + g$ is real-valued and therefore f and g are also real-valued, so that Theorem 2.4.6 is applicable to them on the domain $Y\backslash(E \cup F)$. We thereby obtain the measurability of the restriction of $f + g$ to $Y\backslash(E \cup F)$. Since E and F have already been seen to be measurable, an application of Proposition 2.5.2 to $f + g$ on the measurable set Y leads to the measurability of the function.

We next consider the measurability of the product. For any real number $\alpha < 0$,

$$\{x : f(x)g(x) > \alpha\} = \begin{cases} \{x : f(x), g(x) \text{ are finite and } f(x)g(x) > \alpha\} \\ \cup \{x : f(x),\ g(x) \text{ are of the same sign and at least one is infinite}\} \\ \cup \{x : f(x) = 0 \text{ or } g(x) = 0\}. \end{cases}$$

By Proposition 2.5.2 and Theorem 2.4.6, all the sets on the right-hand side are measurable. If $\alpha \geq 0$, then

$$\{x : f(x)g(x) > \alpha\} = \begin{cases} \{x : f(x), g(x) \text{ are finite and } f(x)g(x) > \alpha\} \\ \cup \{x : f(x), g(x) \text{ are of the same sign and at least one is infinite}\}. \end{cases}$$

Again by Proposition 2.5.2 and Theorem 2.4.6, all the sets on the right-hand side are measurable.

If f is any extended real-valued function defined on X, the positive and negative parts f^+ and f^- respectively are defined exactly as for real-valued functions:

$$f^+(x) = \max\{f(x), 0\} \quad \text{and} \quad f^-(x) = \max\{-f(x), 0\}.$$

The following identities continue to be valid:

$$f = f^+ - f^-, \quad |f| = f^+ + f^-, \quad f^+ = \frac{1}{2}(|f| + f) \quad \text{and} \quad f^- = \frac{1}{2}(|f| - f).$$

As before, when f is measurable, the associated functions f^+ and f^- are also measurable.

We recall the definitions of the limit superior and limit inferior of a sequence $\{a_n\}_{n\geq 1}$ in the extended real number system $\mathbb{R}^* = \mathbb{R} \cup \{-\infty\} \cup \{\infty\}$:

$$\begin{aligned} \limsup a_n &= \inf_{n\geq 1}(\sup_{k\geq n} a_k) = \lim_n(\sup\{a_n, a_{n+1}, \ldots\}), \\ \liminf a_n &= \sup_{n\geq 1}(\inf_{k\geq n} a_k) = \lim_n(\inf\{a_n, a_{n+1}, \ldots\}). \end{aligned}$$

If $\{a_n\}_{n\geq 1}$ converges or diverges to $\pm\infty$, then evidently,

$$\lim_n a_n = \limsup a_n = \liminf a_n.$$

From the definition we easily get

$$\limsup a_n = -\liminf(-a_n)$$

and

$$\begin{aligned} a_n \leq b_n \, \forall n \in \mathbb{N} &\Rightarrow \liminf a_n \leq \liminf b_n \\ &\Rightarrow \limsup a_n \leq \limsup b_n. \end{aligned}$$

Suppose $\{f_n\}_{n\geq 1}$ is a sequence of extended real-valued functions on a domain $X \subseteq \mathbb{R}$. Then **lim sup f_n** and **lim inf f_n** are functions defined on the domain X as

$$(\limsup f_n)(x) = \limsup(f_n(x)) \quad \text{and} \quad (\liminf f_n)(x) = \liminf(f_n(x)).$$

If $f(x) = \lim_n f_n(x)$, the limit being assumed to exist at every $x \in X$, then we call f the **(pointwise) limit** of the sequence $\{f_n\}_{n\geq 1}$ and denote it by **$\lim_n f_n$** or **$\lim f_n$**. We also define **$\sup_{1\leq i\leq n} f_i$** and **$\inf_{1\leq i\leq n} f_i$** respectively as

$$(\sup_{1\leq i\leq n} f_i)(x) = \sup_{1\leq i\leq n}(f_i(x)) \quad \text{and} \quad (\inf_{1\leq i\leq n} f_i)(x) = \inf_{1\leq i\leq n}(f_i(x)),$$

and **sup** f_n and **inf** f_n respectively as

$$(\sup f_n)(x) = \sup(f_n(x)) \quad \text{and} \quad (\inf f_n)(x) = \inf(f_n(x)).$$

Theorem 2.5.4 *If* $\{f_n\}_{n\geq 1}$ *is a sequence of extended real-valued measurable functions on the same measurable set* $X \subseteq \mathbb{R}$, *then the following hold*:

(a) $\sup_{1\leq i\leq n} f_i$ *is measurable for each* n;
(b) $\inf_{1\leq i\leq n} f_i$ *is measurable for each* n;
(c) $\sup f_n$ *is measurable*;
(d) $\inf f_n$ *is measurable*;
(e) $\lim\sup f_n$ *is measurable*;
(f) $\lim\inf f_n$ *is measurable*.

Proof

(a) For each real number α,

$$X(\sup_{1\leq i\leq n} f_i > \alpha) = \bigcup_{i=1}^{n} X(f_i > \alpha);$$

the right-hand side, being a countable (actually finite) union of measurable sets, is measurable. It follows that the left-hand side is measurable. Since this is true for each real number α, the function $\sup_{1\leq i\leq n} f_i$ is measurable.

(b) By Remark 2.5.3, each $-f_n$ is measurable. Moreover, $\inf_{1\leq i\leq n} f_i = -\sup_{1\leq i\leq n}(-f_i)$. Therefore it follows from (a) that $\inf_{1\leq i\leq n} f_i$ is also measurable.

(c) For each real number α,

$$X(\sup f_n > \alpha) = \bigcup_{n=1}^{\infty} X(f_n > \alpha);$$

the right-hand side, being a countable union of measurable sets, is measurable. It follows that the left-hand side is measurable. Since this is true for each real number α, the function $\sup f_n$ is measurable.

(d) Follows from (c) and Remark 2.5.3, because $\inf f_n = -\sup(-f_n)$.

(e) $\lim\sup f_n = \inf_{n\geq 1}(\sup_{k\geq n} f_k)$ and is therefore measurable by (c) and (d).

(f) $\lim\inf f_n = -\lim\sup(-f_n)$ and is therefore measurable by (e) and Remark 2.5.3. □

Corollary 2.5.5 *Let* $\{f_n\}_{n\geq 1}$ *be a sequence of extended real-valued measurable functions on a measurable set* $X \subseteq \mathbb{R}$. *If* $\lim f_n$ *exists, then it is measurable.*

Proof An immediate consequence of the theorem above. □

The definition of the Lebesgue integral is based on what are called "simple" functions. They are a generalisation of step functions, which take only finitely many values and each value is taken on an interval; thus step functions are linear combinations of characteristic functions of intervals. For example, if $s:[1, 3]\to\mathbb{R}$ is the step function which equals 5 on [1, 2), 8 on (2, 3] and $s(2) = 6$, then $s = 5\chi_{[1,2)} + 6\chi_{[2,2]} + 8\chi_{(2,3]}$. The generalisation is achieved by replacing intervals by measurable sets, so that a simple function is a linear combination of characteristic functions of measurable sets. However, we require the values to be nonnegative and finite. The formal description is as follows.

Definition 2.5.6 A measurable function defined on a measurable subset of $\mathbb{R}$ is called a **simple function** if its range consists of finitely many points of $[0, \infty)$.

Examples 2.5.7

(a) The characteristic function χ_A of a nonempty measurable set $A \subset \mathbb{R}$ is a simple function with range consisting of the two numbers 0 and 1. If A is empty or equals $\mathbb{R}$, then the range consists of a single number, namely, 0 or 1.
(b) Suppose A and B are nonempty measurable subsets of $\mathbb{R}$, both distinct from $\mathbb{R}$. Then each of χ_A and χ_B has range consisting of two numbers; but the range of the sum $\chi_A + \chi_B$ consists of the

$$\begin{array}{ll} \text{single number 1} & \text{if } A\cap B=\varnothing \text{ and } A\cup B=\mathbb{R} \\ \text{two numbers 0 and 1} & \text{if } A\cap B=\varnothing \text{ and } A\cup B\neq\mathbb{R} \\ \text{two numbers 0 and 2} & \text{if } A=B (\text{in which case } A\cup B\neq\mathbb{R}) \\ \text{two numbers 1 and 2} & \text{if } A\cup B=\mathbb{R} \text{ and } A\cap B\neq\varnothing \\ \text{three numbers } 0,1,2 & \text{if } A\cup B\neq\mathbb{R}, A\neq B \text{ and } A\cap B\neq\varnothing. \end{array}$$

(c) The sum of two functions with finite range has a finite range. Since it is also true that the sum of two measurable functions is measurable, we conclude that the sum of two simple functions is simple. Therefore a finite sum $\Sigma_{1\le j\le n}\alpha_j\chi_{Aj}$, where $A_1, \ldots, A_n$ are measurable and $\alpha_1, \alpha_2, \ldots, \alpha_n$ are nonnegative real numbers, is always a simple function. As illustrated in the preceding example, the values of the simple function need not be among the numbers α_j.

Remarks 2.5.8

(a) An extended real-valued measurable function on a measurable set with finite range is also sometimes called a simple function. The above situation (with real and nonnegative values) is the one that will mainly interest us in the sequel.
(b) The value ∞ is explicitly excluded from the range of a simple function.
(c) If $\alpha_1, \alpha_2, \ldots, \alpha_n$ are the distinct values assumed by a simple function s and A_j denotes the set $X(s = \alpha_j)$, then the sets A_j are nonempty, disjoint with $\cup_{1\le j\le n}A_j = X$, and are measurable (by Proposition 2.4.5); moreover,

$$s(x) = \sum_{j=1}^{n} \alpha_j \chi_{A_j}(x).$$

Conversely, if the sets A_j are nonempty, disjoint with $\cup_{1 \leq j \leq n} A_j = X$ and are measurable, and if $\alpha_1, \alpha_2, \ldots, \alpha_n$ are distinct, then the simple function $s = \sum_{1 \leq j \leq n} \alpha_j \chi_{A_j}$ has the coefficients $\alpha_1, \alpha_2, \ldots, \alpha_n$ as its distinct values and satisfies $X(s = \alpha_j) = A_j$. When these conditions are fulfilled, we shall refer to the sum as the **canonical representation** of s. The range $s(X)$ of the function is $\{\alpha_j: 1 \leq j \leq n\}$.

(d) In Example 2.5.7(b), the canonical representations of $\chi_A + \chi_B$ in the five cases displayed there are respectively

$$1 \cdot \chi_{\mathbb{R}}, \quad 0 \cdot \chi_{(A \cup B)^c} + 1 \cdot \chi_{A \cup B}, \quad 2 \cdot \chi_A + 0 \cdot \chi_{A^c}, \quad 1 \cdot \chi_{(A \cap B)^c} + 2 \cdot \chi_{A \cap B},$$
$$0 \cdot \chi_{(A \cup B)^c} + 1 \cdot \chi_{A \Delta B} + 2 \cdot \chi_{A \cap B}.$$

This shows that simple functions have simple representations as linear combinations of characteristic functions, and yet they are not so simple to add!

We have seen in Corollary 2.5.5 that the limit of a sequence of measurable functions is measurable. We shall next show that a measurable nonnegative extended real-valued function is the limit of an increasing sequence of nonnegative *simple* (measurable) functions.

Theorem 2.5.9 *Let f be a measurable extended nonnegative real-valued function on a measurable subset X of $\mathbb{R}$. Then there exists a sequence $\{s_n\}_{n \geq 1}$ of simple functions such that for every $x \in X$,*

(a) $0 \leq s_1(x) \leq s_2(x) \leq \cdots \leq f(x)$,
(b) $s_n(x) \to f(x)$ *as* $n \to \infty$.

Moreover, if f is bounded, then $\{s_n\}_{n \geq 1}$ converges to f uniformly on X.

Proof Since f is an extended nonnegative real-valued function, its range is contained in $[0, \infty]$. For each positive integer n, split $[0, \infty]$ by the sequence of points

$$0, \frac{1}{2^n}, \frac{2}{2^n}, \ldots, \frac{n2^n - 1}{n}, n, \infty,$$

so that

$$X = f^{-1}([0, \infty]) = \bigcup_{k=1}^{n2^n} f^{-1}([\frac{k-1}{2^n}, \frac{k}{2^n})) \cup f^{-1}([n, \infty]).$$

For $1 \leq k \leq n2^n$, set

$$E_{n,k} = f^{-1}([\frac{k-1}{2^n}, \frac{k}{2^n})) \quad \text{and} \quad F_n = f^{-1}([n, \infty]).$$

Note that, for $n \in \mathbb{N}$ and $1 \leq k \leq n2^n$, the sets $E_{n,k}$ and F_n are measurable by hypothesis and are also disjoint. Put

$$s_n = \sum_{k=1}^{n2^n} \frac{k-1}{2^n} \chi_{E_{n,k}} + n\chi_{F_n} \quad \text{for } n \in \mathbb{N}.$$

It is clear that $0 \leq s_n(x) \leq f(x)$ everywhere. We shall now show for any $x \in X$ that

$$s_n(x) \leq s_{n+1}(x), \tag{2.21}$$

which will prove (a).

To this end, first consider any $x \in X$ such that $f(x) \geq n + 1$. Then $x \in F_{n+1} \subseteq F_n$, whence $s_n(x) = n$ and $s_{n+1}(x) = n + 1$, so that (2.21) holds.

Next, consider an $x \in X$ such that $n \leq f(x) < n + 1$. Then $x \in F_n$ and therefore $s_n(x) = n$. Also, $n2^{n+1} \leq f(x)2^{n+1} < (n + 1)2^{n+1}$, so that $k - 1 \leq f(x)2^{n+1} < k$ for some integer k satisfying $n2^{n+1} < k \leq (n + 1)2^{n+1}$. (Take k to be 1 plus the integer part of $f(x)2^{n+1}$.)This implies that $x \in E_{n+1,k}$, where $n2^{n+1} < k \leq (n + 1)2^{n+1}$. Consequently, $s_{n+1}(x) = (k - 1)/2^{n+1} \geq n$. Since $s_n(x) = n$ as already noted, we conclude that (2.21) holds if $n \leq f(x) < n + 1$.

Now consider the remaining case, i.e., $f(x) < n$. Here, $0 \leq f(x)2^n < n2^n$ so that $k - 1 \leq f(x)2^n < k$ for some integer k (again, 1 plus the integer part of $f(x)2^n$) satisfying $0 < k \leq n2^n$. This implies that $x \in E_{n,k}$, where $0 < k \leq n2^n$. Consequently, $s_n(x) = (k - 1)/2^n$. Note that the inequality $k - 1 \leq f(x)2^n < k$ holds if and only if

$$\text{either} \quad 2(k-1) \leq f(x)2^{n+1} < 2k-1 \quad \text{or} \quad 2k-1 \leq f(x)2^{n+1} < 2k.$$

This is equivalent to

$$\text{either} \quad x \in E_{n+1,2k-1} \quad \text{or} \quad x \in E_{n+1,2k},$$

bearing in mind that $1 \leq 2k - 1 < 2k \leq (n + 1)2^{n+1}$. If $x \in E_{n+1,2k-1}$, then $s_{n+1}(x) = (2k - 2)/2^{n+1} = (k - 1)/2^n$. Since $s_n(x) = (k - 1)/2^n$ as already noted, we find that (2.21) holds if $x \in E_{n+1,2k-1}$. On the other hand, if $x \in E_{n+1,2k}$, then $s_{n+1}(x) = (2k - 1)/2^{n+1} > (k - 1)/2^n = s_n(x)$. We can therefore conclude that (2.21) always holds even in this remaining case.

If x is such that $f(x) < \infty$, then $x \in F_n^c$ for all sufficiently large n and therefore $s_n(x) \geq f(x) - 2^{-n}$. (Since $x \in E_{n,k}$ for some k, it follows that $s_n(x) = (k - 1)/2^n$. As $f(x) < k/2^n$, we have $f(x) - 2^{-n} < (k - 1)/2^n = s_n(x)$.) So, $s_n(x)$ increases monotonically to $f(x)$. If $f(x) = \infty$, then $x \in \cap_{n \geq 1} F_n$, so that $s_n(x) = n$ for all n and again $s_n(x)$ increases monotonically to $f(x)$.

Finally, if $0 \leq f(x) \leq M$ for every $x \in X$, then $s_n(x) \geq f(x) - 2^{-n}$ for every $n \geq n_0$, where $n_0 \geq M$, that is, $0 \leq f(x) - s_n(x) \leq 2^{-n}$ for every $x \in X$ and for every $n \geq n_0$, where $n_0 \geq M$. This implies that $\{s_n\}_{n \geq 1}$ converges to f uniformly on X. □

We now prepare to prove that, if two functions differ on a set of Lebesgue measure zero and one of them is Lebesgue measurable, then so is the other. This is best phrased in terminology (formally introduced later) that is useful in other contexts as well. We shall begin by illustrating the contexts.

The function f defined on $\mathbb{R}$ by $f(x) = 1$ for rational x and 0 for irrational x has the property that $f(x) = 0$ except on the set of rationals, which has Lebesgue measure zero. The phenomenon is described by saying that f is zero "almost everywhere". The greatest integer function $f(x) = [x]$ is continuous except on the set $\mathbb{Z}$ of integers, which has Lebesgue measure zero. We say that this function is continuous "almost everywhere".

The two properties considered in the preceding paragraph, namely vanishing at x and being continuous at x, are "hereditary" properties in the sense that: if either of them holds on some set, then it holds on every subset thereof. The same is true of uniform continuity.

More generally, consider a hereditary property that holds everywhere on a measurable set $X \subseteq \mathbb{R}$ except on a subset $E \in \mathfrak{B}_X$ with $m(E) = 0$. Since $\mathfrak{B} \subseteq \mathfrak{M}$, it is trivial that the property holds everywhere on a measurable set $X \subseteq \mathbb{R}$ except on a set $E_1 \in \mathfrak{M}_X$ with $m(E_1) = 0$. The converse is also true: suppose a hereditary property holds everywhere on a measurable set $X \subseteq \mathbb{R}$ except on a subset $E_1 \in \mathfrak{M}_X$ with $m(E_1) = 0$. Then $E_1 \in \mathfrak{M}_X$ and by the equivalence $(\alpha) \Leftrightarrow (\gamma)$ in Proposition 2.3.24, there exists a $\mathcal{G}_\delta$-set G, which has to be in $\mathfrak{B}$ by virtue of being $\mathcal{G}_\delta$, such that $G \supseteq E_1$ and $m(G) = m(E_1) = 0$. Now, $E_1 \subseteq X \cap G \in \mathfrak{B}_X$ and $m(X \cap G) = 0$; by heredity, the property holds everywhere on X except $X \cap G$. Thus $E = X \cap G$ serves as the subset of X such that the hereditary property holds everywhere on the measurable set X except on the subset $E \in \mathfrak{B}_X$ with $m(E) = 0$. The upshot of this discussion is that, in the forthcoming definition of "almost everywhere", it makes no difference whether the exceptional set is in $\mathfrak{M}_X$ or in $\mathfrak{B}_X$.

Definition 2.5.10 If a hereditary property holds everywhere on a measurable set X except on a subset E in either $\mathfrak{M}_X$ or $\mathfrak{B}_X$ having $m(E) = 0$, we say that it holds **almost everywhere**. The phrase is abbreviated as **a.e.** When the property is described in terms of an explicit $x \in X$, we can say that it holds for **almost all** x.

We note that convergence of a sequence of functions, whether pointwise or uniform, is also a hereditary property. Therefore it makes sense to speak of pointwise or uniform convergence almost everywhere of a sequence of functions.

The exceptional set referred to in the definition is not required to be nonempty; in other words, a property that holds everywhere also holds a.e.

Examples 2.5.11

(a) Let $X = [0, 1]$ and suppose $f: X \to \mathbb{R}$ is 0 everywhere, while $g : X \to \mathbb{R}$ is 1 on $\mathbb{Q} \cap X$ but 0 on the rest of X. Then the property that $f(x) = g(x)$ holds everywhere on X except on the subset $\mathbb{Q} \cap X$. Since this subset has measure zero, we have $f = g$ a.e.

(b) Let $X = [0, 1]$ and $\{f_n\}_{n \geq 1}$ be the sequence of functions on X given by $f_n(x) = x^n$. This sequence converges to 0 except at 1. Since the subset $\{1\}$ consisting of the exceptional point 1 has measure 0, we have $\lim_{n\to\infty} f_n = 0$ a.e. on X.

(c) Let $X = [-1, 1]$ and $\{f_n\}_{n \geq 1}$ be the sequence of functions on X given by $f_n(x) = 1/(1 + x^{2n})$. Then $\lim_{n\to\infty} f_n(x) = 1$ if $x \in (-1, 1)$. Since the complement of $(-1, 1)$ in X, which consists of the two numbers ± 1, has measure zero, we have $\lim_{n\to\infty} f_n = 1$ a.e. on X. It is a fact that 1 $\lim_{n\to\infty} f_n(x) = \frac{1}{2}$ when $x = \pm 1$, but this fact has no bearing on the assertion that $\lim_{n\to\infty} f_n = 1$ a.e. on X.

Proposition 2.5.12 *Let f and g be continuous functions on an interval $I \subseteq \mathbb{R}$ of positive length. Suppose $f = g$ almost everywhere on X. Then $f = g$ everywhere on I.*

Proof Suppose $x_0 \in X$ is such that $f(x_0) \neq g(x_0)$. Then $\{x \in X: f(x) \neq g(x)\}$ is nonempty and the continuity of $f - g$ at x_0 implies that it is an open subset of I. Now, a nonempty open subset of any interval of positive length must contain an interval of positive length. So,

$$m(\{x \in X : f(x) \neq g(x)\}) > 0,$$

which contradicts the hypothesis that $f = g$ a.e. □

Proposition 2.5.13 *Let f and g be two functions on a Lebesgue measurable subset $X \subseteq \mathbb{R}$ such that $f = g$ a.e. on X and f is Lebesgue measurable. Then g is Lebesgue measurable.*

Proof Let $A = X(f > \alpha)$ and $B = X(g > \alpha)$. Note that

$$A\Delta B \subseteq X(f \neq g).$$

Since f is measurable, it follows that A is Lebesgue measurable. Also,

$$m^*(A\Delta B) \leq m^*(X(f \neq g)) = 0.$$

In view of Problem 2.3.P5, this implies that the set B is Lebesgue measurable, i.e. the function g is Lebesgue measurable. □

Proposition 2.5.14 *If a real-valued function f defined on a Lebesgue measurable subset $X \subseteq \mathbb{R}$ is continuous a.e., then f is Lebesgue measurable on X.*

Proof Let $Y = \{x \in X: f$ is not continuous at $x\}$. Note that $m^*(Y) = 0$ and so Y is measurable. Let α be an arbitrary real number. Since f is continuous on the measurable set $X\backslash Y$, it follows that $Z = \{x \in X\backslash Y: f(x) > \alpha\}$ is measurable [analogously to Example 2.4.4(c)]. Also, the set $W = \{x \in Y: f(x) > \alpha\}$, being a subset of the set Y of measure zero, is measurable. Consequently, the union $Z \cup W$ is measurable; but this union is the same as $\{x \in X: f(x) > \alpha\}$. □

Problem Set 2.5

2.5.P1. Let $g(x) = f'(x)$ exist for every $x \in [a, b]$. Prove that g is Lebesgue measurable.

2.5.P2. Prove that the set of points at which a sequence of measurable functions converges or diverges to $\pm\infty$ is a measurable set.

2.5.P3. Show that $\sup\{f_\alpha: \alpha \in \Lambda\}$ is not necessarily measurable even if each f_α is.

2.5.P4. If f is a real-valued Lebesgue measurable function defined on a Borel set $X \subseteq \mathbb{R}$, then there exists a Borel measurable function g on X such that $f(x) = g(x)$ a.e.

2.5.P5. Let f and g be extended real-valued measurable functions defined on a measurable subset $X \subseteq \mathbb{R}$. Then their product fg is also measurable.

2.5.P6. Let s be a simple function on a measurable set X such that $s = \sum_{1 \le i \le p} \alpha_i \chi_{A_i}$, where

$$i \neq i' \Rightarrow A_i \cap A_{i'} = \varnothing \quad \text{and} \quad \bigcup_{i=1}^{p} A_i = X.$$

(a) If $\gamma_1, \gamma_2, \ldots, \gamma_n$ are the distinct elements of the range of s, show that

$$\{\gamma_j : 1 \le j \le n\} = \{\alpha_i : 1 \le i \le p, A_i \neq \varnothing\}.$$

(b) Show that, for any j, $1 \le j \le n$, there must exist some index i ($1 \le i \le p$) for which the coefficient α_i equals γ_j and, at the same time, $A_i \neq \varnothing$. If for each j ($1 \le j \le n$), N_j is the set of all such indices i corresponding to a given j, namely,

$$N_j = \{i : 1 \le i \le p, \alpha_i = \gamma_j \quad \text{and} \quad A_i \neq \varnothing\},$$

show that $\cup_{i \in N_i} A_i = X(s = \gamma_j)$.

(c) Finally, show that the canonical representation of s is $\sum_{1 \le j \le n} \gamma_j \chi_{Bj}$, where $B_j = \cup_{i \in Nj} A_i$.

2.5.P7. Let $\sum_{1 \le i \le p} \alpha_i \chi_{A_i}$ be the canonical representation of a simple function s. If $0 < \alpha < \infty$, show that $\sum_{1 \le i \le p} (\alpha\alpha_i)\chi_{A_j}$ is the canonical representation of αs.

2.5.P8. Give an example when $(f + g)^+$ is not the same as $f^+ + g^+$.

2.5.P9. If f and g are extended real-valued functions and $g \ge 0$, show that $(fg)^+ = f^+g$ and $(fg)^- = f^-g$.

2.5.P10. Let f be a bounded measurable function defined on a bounded closed interval $[a, b]$ and let $\varepsilon > 0$ be arbitrary. Show that there exists a step function g on $[a, b]$ such that

$$m(\{x \in [a, b] : |f(x) - g(x)| \ge \varepsilon\}) < \varepsilon.$$

2.6 Egorov's and Luzin's Theorems

The theorems mentioned in the title of this section are important in their own right. Egorov's Theorem will be used subsequently.

We begin with a preliminary result which is of independent interest.

Proposition 2.6.1 *Suppose $\{f_n\}_{n\geq 1}$ is a sequence of functions on a measurable set X and $\lim_{n\to\infty} f_n = f$ a.e.*

(a) *If each f_n is Lebesgue measurable, then so is f.*
(b) *If X is a Borel set and each f_n is Borel measurable, then there exists a Borel measurable function g on X such that $f = g$ a.e.*

Proof

(a) Indeed, $f = \limsup f_n$ a.e. and the result follows from Theorem 2.5.4 and Proposition 2.5.13.
(b) By part (a), f is Lebesgue measurable. Now use Problem 2.5.P4. □

In the rest of this section, "measure" and "measurable" will be understood exclusively in the sense of Lebesgue, as there will be no occasion to work with Borel measure or measurability.

The first main result here relates the notions of convergence almost everywhere and uniform convergence.

Theorem 2.6.2 (Egorov). *Suppose that a sequence $\{f_n\}_{n\geq 1}$ of measurable functions converges a.e. to f on E, where $m(E) < \infty$. Then, for any $\varepsilon > 0$, there exists a measurable set $E_\varepsilon \subseteq E$ such that $m(E\backslash E_\varepsilon) < \varepsilon$ and the sequence $\{f_n\}_{n\geq 1}$ converges to f uniformly on E_ε.*

Proof According to Proposition 2.6.1, f is measurable. Set

$$E_{n,p} = \bigcap_{i\geq n} \{x \in E : |f_i(x) - f(x)| < \frac{1}{p}\},$$

that is, $E_{n,p}$ is the set of all x for which

$$|f_i(x) - f(x)| < \frac{1}{p} \quad \text{for all} \quad i \geq n.$$

Let $E_p = \cup_{n\geq 1} E_{n,p}$. It is clear from the definition of $E_{n,p}$ that

$$E_{1,p} \subseteq E_{2,p} \subseteq \cdots$$

for fixed p. Therefore, by Proposition 2.3.21 and Problem 2.3.P11, for arbitrary p and $\varepsilon > 0$, there exists an integer $n(p)$ depending on p such that

$$m(E_p \setminus E_{n(p),p}) < \frac{\varepsilon}{2^p}.$$

We set

$$E_\varepsilon = \bigcap_{p=1}^{\infty} E_{n(p),p}$$

and show that E_ε is the required set. For an arbitrary $\eta > 0$, choose p so large that $\frac{1}{p} < \eta$. For $x \in E_\varepsilon$, we have

$$|f_i(x) - f(x)| < \frac{1}{p} < \eta$$

for $i > n(p)$. This proves the uniform convergence of $\{f_n\}_{n\geq 1}$ to f on E_ε. It remains to show that $m(E \setminus E_\varepsilon) < \varepsilon$.

In order to show this, we begin by noting that $m(E \setminus E_p) = 0$ for every p. Indeed, if $x_0 \in E \setminus E_p$, then

$$|f_i(x_0) - f(x_0)| \geq \frac{1}{p}$$

for infinitely many values of i, that is, the sequence $\{f_n\}_{n\geq 1}$ does not converge to f at x_0. Since $f_n \to f$ a.e. by hypothesis, it follows that $m(E \setminus E_p) = 0$. Consequently,

$$m(E \setminus E_{n(p),p}) = m(E_p \setminus E_{n(p),p}) < \frac{\varepsilon}{2^p}.$$

Therefore

$$m(E \setminus E_\varepsilon) = m(E \setminus \bigcap_{p=1}^{\infty} E_{n(p),p}) = m(\bigcup_{p=1}^{\infty} (E \setminus E_{n(p),p}))$$
$$\leq \sum_{p=1}^{\infty} m(E \setminus E_{n(p),p}) < \sum_{p=1}^{\infty} \frac{\varepsilon}{2^p} = \varepsilon.$$

This completes the proof. □

The requirement that $m(E) < \infty$ cannot be dropped in Theorem 2.6.2.

In other words, there is a measurable set E with $m(E) = \infty$ and a sequence $\{f_n\}_{n\geq 1}$ of measurable functions converging a.e. to f on E such that for some $\eta > 0$ and every measurable set $E_\eta \subseteq E$ satisfying $m(E \setminus E_\eta) < \eta$, the sequence $\{f_n\}_{n\geq 1}$ fails to converge uniformly on E_η.

To demonstrate this, we take $E = \mathbb{R}^+ = \{x \in \mathbb{R} : x \geq 0\}$, $f_n = \chi_{[n-1,n]}$ and $\eta = \frac{1}{2}$. Then $\{f_n\}_{n\geq 1}$ converges everywhere to the identically zero function. Note that when $n > m + 1$, the functions f_n and f_m are characteristic functions of *disjoint* sets; this disjointness has the consequence that

$$\text{if } x \in [n-1, n] \text{ and } n > m+1, \text{ then } |f_n(x) - f_m(x)| = 1. \tag{2.22}$$

Now consider an arbitrary measurable set $E_\eta \subseteq E$ satisfying $m(E \backslash E_\eta) < \eta$. We have to show that the sequence $\{f_n\}_{n \geq 1}$ fails to converge uniformly on E_η.

In order to prove this, we first note that $E_\eta \cap [n-1, n] \neq \varnothing$ for every n (for otherwise, $[n-1, n]$ would be contained in $E \backslash E_\eta$, which has measure less than $\frac{1}{n}$). Let n_0 be an arbitrary positive integer. Choose $n > m + 1 > n_0$ and $x \in E_\eta \cap [n-1, n]$. Then by (2.22), we have $|f_n(x) - f_m(x)| = 1$, ruling out uniform convergence on E_η.

The definition of a measurable function is fashioned after a characterisation of a continuous function. In what follows, we shall prove a theorem due to Luzin, which shows that a measurable function can in a certain sense be approximated by a continuous function. The reader who chooses to skip the rest of this section will experience no loss of continuity, because the results will not be used later in this book.

We begin with a lemma which says that every measurable function that is finite a.e. becomes bounded when we disregard a set of arbitrarily small measure.

Lemma 2.6.3 *Let a measurable function f defined on a set E of finite measure be finite a.e. Then for any $\varepsilon > 0$, there exists a bounded measurable function g such that $m(E(f \neq g)) < \varepsilon$.*

Proof For each n, let $E_n = E(|f| > n)$ and $E_\infty = E(|f| = \infty)$. By hypothesis, $m(E_\infty) = 0$. In view of the obvious relations

$$E_1 \supseteq E_2 \supseteq E_3 \supseteq \cdots \quad \text{and} \quad E_\infty = \bigcap_{n=1}^{\infty} E_n$$

and the fact that $m(E) < \infty$, Proposition 2.3.21(b) shows that

$$m(E_n) \to m(E_\infty) \quad \text{as} \quad n \to \infty.$$

Hence there exists an n_0 such that $m(E_{n_0}) < \varepsilon$. Define a function g on E by

$$g(x) = \begin{cases} f(x) & \text{if } x \in E \backslash E_{n_0} \\ 0 & \text{if } x \in E_{n_0}. \end{cases}$$

The function g is then measurable and $E(f \neq g) \subseteq E_{n_0}$. So, $m(E(f \neq g)) \leq m(E_{n_0}) < \varepsilon$. Also, the definition of E_{n_0} ensures that g is bounded above in absolute value by n_0. □

Lemma 2.6.4 *Let the sets $\{F_i\}_{1 \leq i \leq n}$ be closed and disjoint. If the function f defined on their union F is constant on each of the sets F_i, then it is continuous.*

Proof Let $\{x_k\}_{k\geq 1}$ be a sequence of points in F such that $x_k \to x$. Since F is closed, $x \in F$ and hence $x \in F_m$ for some m, $1 \leq m \leq n$. Since $F_i \cap F_j = \emptyset$, $i \neq j$, it follows that $x \notin F_{m'}$ for $m' \neq m$. Since $\cup_{i\neq m} F_i$ is closed and does not contain x, no subsequence of $\{x_k\}_{k\geq 1}$ can lie in it. Thus all terms of the sequence from some index k_0 onwards lie in F_m. Since f is constant on F_m, we have $f(x) = f(x_k)$ for $k \geq k_0$, whence $f(x) = \lim f(x_k)$. □

Lemma 2.6.5 *Let F be a closed set contained in the closed interval* $[a, b]$. *If f is a function defined and continuous on the set F, then there exists a continuous function g on* $[a, b]$ *such that* $g(x) = f(x)$ *for all* $x \in F$ *and* $\max |g(x)| = \max |f(x)|$.

Proof Let $[\alpha, \beta]$ be the smallest closed interval containing F. If a function g of the required kind has been obtained on $[\alpha, \beta]$, then define g on $[a, b]\backslash[\alpha, \beta]$ by

$$g(x) = \begin{cases} f(\alpha) & \text{if } x \in [a, \alpha) \\ f(\beta) & \text{if } x \in (\beta, b]. \end{cases}$$

Then g is the required function on $[a, b]$. So, without loss of generality, assume $[a, b]$ is the smallest closed interval containing F; in particular, $a, b \in F$. If $F = [a, b]$, then there is nothing to prove. Suppose that $F \subset [a, b]$. Then, $[a, b]\backslash F = (a, b)\backslash F$, being open in $\mathbb{R}$, is a countable union of disjoint open intervals (γ, δ) with end-points in F. For the rest of this proof, we shall call them component intervals. On F, define $g(x) = f(x)$. Then for each component interval (γ, δ), we have $g(\gamma) = f(\gamma)$ and $g(\delta) = f(\delta)$; define g on (γ, δ) by a straight line graph. Note that f is bounded because F is compact (see Theorem 1.3.20) and hence that g is bounded (with the same bounds as f); moreover, it is defined on the entire interval $[a, b]$.

We shall show that g is continuous from the right on $[a, b)$. Continuity from the left on $(a, b]$ is proved similarly.

Within each component interval (γ, δ), the function, being linear, is continuous. If $x_0 \in F$ is the left endpoint γ of one of the intervals (γ, δ), then the function g, being linear on (γ, δ), is continuous there. So, let $x_0 \in F$ be different from the left endpoint of any of the component intervals, and $x_0 \neq b$. Let $\{x_m\}_{m\geq 1}$ be a sequence of points in $[a, b]$ converging from the right to x_0, i.e. $x_m \to x_0^+$.

First consider the case that $x_m \in F$ for all m. Then $g(x_m) = f(x_m) \to f(x_0) = g(x_0)$, since f is continuous on F.

Next, consider the case that $x_m \notin F$ for all m. Suppose, if possible, that the sequence $\{g(x_m)\}_{m\geq 1}$ does not converge to $g(x_0)$. It is nonetheless bounded and therefore $\{x_m\}_{m\geq 1}$ has a subsequence, which we shall denote by $\{y_k\}_{k\geq 1}$, such that the corresponding subsequence $\{g(y_k)\}_{k\geq 1}$ converges to a limit other than $g(x_0)$. Denote the component interval of $[a, b]\backslash F$ that contains y_k by (γ_k, δ_k). Since $y_k > x_0 \in F$ and $F \cap (\gamma_k, \delta_k) = \emptyset$, we have $x_0 < \gamma_k < y_k < \delta_k$, so that it is possible to extract a subsequence $\{y_{k(j)}\}_{j\geq 1}$ of $\{y_k\}_{k\geq 1}$ such that $x_0 < y_{k(j+1)} < \gamma_{k(j)} < y_{k(j)}$. This property of the subsequence ensures that the intervals $(\gamma_{k(j)}, \delta_{k(j)})$ are distinct and hence also disjoint. Since the sequence $\{x_m\}_{m\geq 1}$ converges to x_0, its subsequence $\{y_{k(j)}\}_{j\geq 1}$ does the same. It follows that $\gamma_{k(j)} \to x_0$. Now the sums of the lengths of the disjoint intervals $(\gamma_{k(j)}, \delta_{k(j)})$ cannot exceed that of $[a, b]$ and therefore,

$\delta_{k(j)} - \gamma_{k(j)} \to 0$, which implies $\delta_{k(j)} \to x_0$. But $\gamma_{k(j)}$, $\delta_{k(j)}$ and x_0 all lie in F and hence $f(\gamma_{k(j)}) \to f(x_0)$, $f(\delta_{k(j)}) \to f(x_0)$. Since g is linear on the interval $[\gamma_{k(j)}, \delta_{k(j)}]$, each $g(y_{k(j)})$ lies between $f(\gamma_{k(j)})$ and $f(\delta_{k(j)})$ and therefore also converges to $f(x_0) = g(x_0)$. But this is a contradiction, because $\{g(y_{k(j)})\}_{j\geq 1}$ is a subsequence of $\{g(y_k)\}_{k\geq 1}$, which converges to a limit other than $g(x_0)$. The contradiction shows that $\{g(x_m)\}_{m\geq 1}$ indeed converges to $g(x_0)$.

Finally, consider the remaining case that $x_m \in F$ for some m but not others. Again suppose, if possible, that the sequence $\{g(x_m)\}_{m\geq 1}$ does not converge to $g(x_0)$. As before, $\{x_m\}_{m\geq 1}$ has a subsequence, which we shall denote by $\{y_k\}_{k\geq 1}$, such that the corresponding subsequence $\{g(y_k)\}_{k\geq 1}$ converges to a limit other than $g(x_0)$. Now $\{y_k\}_{k\geq 1}$ must either have a subsequence $\{y_{k(j)}\}_{j\geq 1}$ that lies in F or have a subsequence $\{y_{k(j)}\}_{j\geq 1}$ that lies outside F. In either event, one of the above cases applies to $\{y_{k(j)}\}_{j\geq 1}$, whereby $\{g(y_{k(j)})\}_{j\geq 1}$ is seen to converge to $g(x_0)$. Like before, this is a contradiction, showing that $\{g(x_m)\}_{m\geq 1}$ indeed converges to $g(x_0)$. □

Theorem 2.6.6 (Luzin). *If f is a Lebesgue measurable function finite a.e. on $E = [a, b]$, then, given any $\varepsilon > 0$, there is a continuous function g on $[a, b]$ such that $m(E(f \neq g)) < \varepsilon$.*

Proof To begin with, we shall show that, for an arbitrary $\varepsilon > 0$, there exists a closed set $F \subseteq [a, b]$ such that $m([a, b]\backslash F) < \varepsilon$ and the restriction of the function f to the set F is continuous.

We shall assume that f is bounded on $[a, b]$, i.e. $|f(x)| \leq M$. Let $\varepsilon > 0$ be arbitrary. For any $k \in \mathbb{N}$, divide the interval $[-M, M]$ into $2k$ parts by means of the points

$$y_j = -M + \frac{jM}{k}, \quad j = 0, 1, 2, \cdots, 2k$$

and construct the sets

$$E_{k,1} = E(y_0 \leq f \leq y_1)$$

and

$$E_{k,j} = E(y_{j-1} < f \leq y_j), \quad j = 2, 3, \ldots, 2k.$$

These sets are measurable and disjoint, and

$$[a, b] = \bigcup_{j=1}^{2k} E_{k,j}.$$

By Proposition 2.3.24, there exists a closed set $F_{k,j} \subseteq E_{k,j}$ such that

$$m(E_{k,j}\backslash F_{k,j}) < \frac{\varepsilon}{2^{k+1}k}.$$

Let $F_k = \cup_{1\le j\le 2k} F_{k,j}$. Then $[a, b]\backslash F_k = \cup_{1\le j\le 2k}(E_{k,j}\backslash F_{k,j})$ and therefore $m([a, b]\backslash F_k) < \varepsilon/2^k$. Now we define on F_k the function f_k by $f_k(x) = y_i$ for $x \in F_{k,i}$, $i = 1, 2, \ldots, 2k$. From Lemma 2.6.4, it follows that f_k is continuous on F_k. The definition of $E_{k,j}$ implies

$$0 \le f_k(x) - f(x) \le \frac{M}{k} \tag{2.23}$$

for all $x \in F_k$. Put $F = \cap_{k\ge 1} F_k$. Then

$$[a,b]\backslash F = \bigcup_{k=1}^{\infty}([a,b]\backslash F_k)$$

and

$$m([a,b]\backslash F) \le \sum_{k=1}^{\infty} m([a,b]\backslash F_k) < \sum_{k=1}^{\infty}\frac{\varepsilon}{2^k} = \varepsilon.$$

Moreover, for $x \in F$, the inequality (2.23) holds for any k. Therefore $f_n \to f$ uniformly on F. Since all f_k are continuous, the aforementioned uniform convergence implies that f is also continuous on F.

Suppose now that the function is finite a.e. on $[a, b]$ but not bounded. Using Lemma 2.6.3, we get a bounded measurable function $\tilde{g}$ such that

$$m(\{x \in [a,b] : f(x) \ne \tilde{g}(x)\}) < \varepsilon.$$

By applying the result proved in the preceding paragraph to $\tilde{g}$, it follows that this function is continuous on some closed set F satisfying $m([a, b]\backslash F) < \varepsilon$. Using Lemma 2.6.5, we obtain a continuous function g defined on $[a, b]$ such that $g(x) = \tilde{g}(x)$ for all $x \in F$. Since

$$\{x \in [a,b] : f(x) \ne g(x)\} \subseteq \{x \in [a,b] : f(x) \ne \tilde{g}(x)\} \cup \{x \in [a,b] : \tilde{g}(x) \ne g(x)\},$$

it follows that

$$\begin{aligned} m(\{x \in [a,b] : f(x) \ne g(x)\}) \le\; & m(\{x \in [a,b] : f(x) \ne \tilde{g}(x)\}) \\ & + m(\{x \in [a,b] : \tilde{g}(x) \ne g(x)\}) < 2\varepsilon. \end{aligned}$$ □

2.7 Lebesgue Outer Measure in $\mathbb{R}^n$

In defining a measure on $\mathbb{R}$, we extended the concept of length of an interval to the notion of outer measure of an arbitrary subset of $\mathbb{R}$. Although outer measure m^* was defined on all subsets of $\mathbb{R}$, it failed to satisfy the desirable properties of a measure described following Definition 2.1.2. (see Problem 2.3.P16) The collection $\mathfrak{M}$ of Lebesgue measurable subsets of $\mathbb{R}$ was defined to consist of those subsets $E \subseteq \mathbb{R}$ that satisfy the Carathéodory condition of measurability, that is,

$$m^*(A) = m^*(A \cap E) + m^*(A \cap E^c)$$

for all $A \subseteq \mathbb{R}$. It turned out that $\mathfrak{M}$ was a σ-algebra containing all intervals and that the set function $m = m^*|_{\mathfrak{M}}$ satisfied all the desirable properties.

In this section, we propose to replicate the first part of the above procedure in $\mathbb{R}^n$, $n > 1$. We shall use the concept of the volume of a "cuboid" to define a nonnegative set function $m_n{}^*$ on $\mathcal{P}(\mathbb{R}^n)$, called *Lebesgue outer measure* in $\mathbb{R}^n$.

In the next section, we shall select those subsets of $\mathbb{R}^n$ that satisfy the analogue of the above Carathéodory condition. These selected subsets will be shown to constitute a σ-algebra $\mathfrak{M}_n$, called the σ-algebra of measurable subsets of $\mathbb{R}^n$. The Lebesgue outer measure m_n^* restricted to $\mathfrak{M}_n$ will turn out to have the desirable properties of a measure. It is known as *Lebesgue measure in* $\mathbb{R}^n$ and is denoted by m_n (Definition 2.8.10 below).

Several proofs are exactly analogous to those of corresponding results in Sect. 2.2 and will therefore be omitted.

A straightforward analogue of a bounded interval in higher dimensions is a cartesian product of bounded intervals. Although many authors call them intervals, we shall refer to them as "cuboids". They are best visualised as rectangles in $\mathbb{R}^2$ and as "boxes" in $\mathbb{R}^3$. It is clear from our formal definition below that a nonempty cuboid is a Cartesian product of bounded intervals of positive length.

Definition 2.7.1 By a **cuboid** in $\mathbb{R}^n$ we mean either the empty set or a set of points

$$I = \{(x_1, x_2, \ldots, x_n) \in \mathbb{R}^n : a_i \leq x_i \leq b_i \text{ for } 1 \leq i \leq n\}$$

where $a_i, b_i \in \mathbb{R}$ and $a_i < b_i$. A set of points characterised in a similar manner but with any or all of the signs $\leq$ replaced by strict inequalities $<$ will also be called a cuboid, and will be denoted by

$$\langle a_1, b_1; a_2, b_2; \ldots; a_n, b_n \rangle.$$

By the **volume** of any such cuboid I we mean the product

$$v(I) = \prod_{1 \leq i \leq n} (b_i - a_i)$$

and by the volume $v(\emptyset)$ of the empty cuboid, we mean 0.

We consider the empty set as a cuboid with volume zero for the same reason that we consider the empty set as an interval with length zero in the one-dimensional case. In view of the requirement that $a_i < b_i$, the volume of a nonempty cuboid is always positive.

We note that the symbol $\langle a_1, b_1; a_2, b_2; \ldots; a_n, b_n \rangle$ can denote any of 2^{2n} cuboids.

The interior of a set A will be denoted by A°. Two sets A and B are said to *overlap* if $A^\circ \cap B^\circ$ is nonempty. It is clear that two intervals of positive length are nonoverlapping if they are disjoint or have only an endpoint in common.

Suppose that each of the intervals $\langle a_i, b_i \rangle$, which may be open, closed or neither but nonempty, is split into a finite number of nonoverlapping intervals with the help of a partition

$$a_i = a_i^{(0)} < a_i^{(1)} < \cdots < a_i^{(k_i)} = b_i$$

and form the cuboids

$$I_{j_1, j_2, \ldots, j_n} = \left\langle a_1^{(j_1)}, a_1^{(j_1+1)}; \ldots; a_n^{(j_n)}, a_n^{(j_n+1)} \right\rangle.$$

Here the indices satisfy $0 \le j_i \le k_i - 1$. The total number of these cuboids must be $k_1 k_1 \cdots k_n$. The cuboids $I_{j_1, j_2 \ldots, j_n}$ form a **paving** of the (nonempty) cuboid

$$I = \langle a_1, b_1; a_2, b_2; \ldots; a_n, b_n \rangle.$$

The fact that the union of all cuboids in a paving may be a proper subset of I is of no consequence. It should be noted that the cuboids in a paving are nonoverlapping, as this will play a role in the proof of Lemma 2.7.2(i) below.

It follows from the above definition of the volume of a cuboid that the total volume of all the cuboids in the paving agrees with the volume of the cuboid. This is clear in $\mathbb{R}$, $\mathbb{R}^2$ and $\mathbb{R}^3$. It can be verified by induction that the result holds in $\mathbb{R}^n$ when $n > 3$. In the notation of the preceding paragraph, this means:

$$v(I) = \sum_{j_1=0}^{k_1-1} \sum_{j_2=0}^{k_2-1} \cdots \sum_{j_n=0}^{k_n-1} v(I_{j_1, j_2, \ldots, j_n}).$$

The following lemma will be used in the sequel.

Lemma 2.7.2

(i) *If a cuboid I is represented as a union of pairwise nonoverlapping cuboids I_k, $k = 1, \ldots, p$, that is, $I = \bigcup_{1 \le k \le p} I_k$,* where $I_k^\circ \cap I_{k'}^\circ = \emptyset$ *whenever $k \neq k'$, then*

$$v(I) = \sum_{k=1}^{p} v(I_k). \tag{2.24}$$

(ii) *If the cuboids I and I_k, $k = 1, \ldots, p$, are such that $I \subseteq \cup_{1 \le k \le p} I_k$, where the cuboids are allowed to overlap, then*

$$v(I) \le \sum_{k=1}^{p} v(I_k). \tag{2.25}$$

Proof

(i) If $I = \varnothing$, then each $I_k = \varnothing$ and there is nothing to prove.
Let $I = \langle a_1, b_1; a_2, b_2; \ldots; a_n, b_n \rangle$. We may assume that each I_k is nonempty because the empty ones, if any, can be discarded without affecting (2.24). So, let

$$I_k = \left\langle c_1^{(k)}, d_1^{(k)}; c_2^{(k)}, d_2^{(k)}; \ldots; c_n^{(k)}, d_n^{(k)} \right\rangle, \quad k = 1, \ldots, p.$$

For each i ($1 \le i \le n$), arrange all the numbers $c_i^{(k)}$ and $d_i^{(k)}$, $k = 1, \cdots, p$, as a single increasing sequence $a_i^{(0)} < a_i^{(1)} < \cdots < a_1^{k_i}$. Note that we must necessarily have $a_i = a_i^{(0)}$ and $b_i = a_1^{k_i}$ because the union of these cuboids is precisely equal to I. Consequently, the single increasing sequence partitions the interval $[a_i, b_i]$ into a finite number of nonoverlapping subintervals. The cuboids

$$I_{j_1, j_2, \ldots, j_n} = \left\langle a_1^{(j_1)}, a_1^{(j_1+1)}; \ldots; a_n^{(j_n)}, a_n^{(j_n+1)} \right\rangle$$

constructed with the help of these partitions constitute a paving of I and also give rise to a paving of each cuboid I_k. Let T_k denote the system of those cuboids $I_{j_1, J_2, \ldots, j_n}$ which form a paving of a cuboid I_k. Since the volume of a cuboid equals the total volume of all the cuboids of any paving of it, we have on the one handbelongs to one and only

$$v(I) = \sum_{j_1, \ldots, j_n} v(I_{j_1, j_2, \ldots, j_n}) \tag{2.26}$$

and on the other hand

$$v(I_k) = \sum_{I_{j_1, \ldots, j_n} \in T_k} v(I_{j_1, \ldots, j_n}),$$

so that

$$\sum_{k=1}^{p} v(I_k) = \sum_{k=1}^{p} \left(\sum_{I_{j_1,\ldots,j_n} \in T_k} v(I_{j_1,\ldots,j_n}) \right). \tag{2.27}$$

However, since the cuboids I_k are nonoverlapping, so are the cuboids $I_{j_1,J_2,\ldots,j_n}$, and hence we know that each $I_{j_1,J_2,\ldots,j_n}$ belongs to one and only one T_k and therefore, the sum on the right-hand side of (2.27) is equal to the sum on the right-hand side of (2.26). Thus (2.24) is seen to hold.

(ii) Once again, we may assume that I and I_k are all nonempty. Since $v(I) = v(I^\circ)$, the left-hand side of (2.25) remains unaffected if I is replaced by I°. Also, $I \subseteq \cup_{1 \le k \le p} I_k$ implies $I^\circ \subseteq \cup_{1 \le k \le p} I_k$. We may therefore assume without loss of generality that I is open. We can also suppose that the I_k's intersect I because the insertion of some additional summands in the right-hand side of (2.25) only strengthens the inequality.
Let $I_k' = I_k \cap I$, $k = 1, \ldots, p$. We have $I = \cup_{1 \le k \le p} I_k'$. As in the proof of (i), we construct a paving of the cuboid I which generates a paving of all the cuboids I_k'. It may be noted that the construction of such a paving is valid even when the cuboids I_k' are allowed to overlap. As before, we denote the cuboids of the paving by $I_{j_1,J_2,\ldots,j_n}$ and let T_k denote the system of those cuboids which form a paving of I_k'. Since the total volume of all the cuboids in a paving of any cuboid agrees with the volume of that cuboid, we have

$$\sum_{k=1}^{p} \left(\sum_{I_{j_1,\ldots,j_n} \in T_k} v(I_{j_1,\ldots,j_n}) \right) = \sum_{k=1}^{p} v(I_k') \le \sum_{k=1}^{p} v(I_k)$$

and

$$v(I) = \sum_{j_1,\ldots,j_n} v(I_{j_1,\ldots,j_n}).$$

Note that each cuboid $I_{j_1,J_2,\ldots,j_n}$ is a constituent of at least one collection T_k but could belong to several T_k. Therefore

$$\sum_{k=1}^{p} \left(\sum_{I_{j_1,\ldots,j_n} \in T_k} v(I_{j_1,\ldots,j_n}) \right) \ge \sum_{j_1,\ldots,j_n} v(I_{j_1,\ldots,j_n}).$$

Together with the preceding two displayed statements, this implies the required conclusion. □

Since cuboids are the only sets we know how to measure at this stage, it is natural that the definition of outer measure involves coverings with cuboids.

Definition.2.7.3 For any subset A of $\mathbb{R}^n$, we define the **Lebesgue outer measure** $\boldsymbol{m_n{}^*(A)}$ to be the infimum of all numbers

$$\sum_{k=1}^{\infty} v(I_k),$$

where $\{I_k\}_{k\geq 1}$ is a sequence of open cuboids that covers A and the infimum is taken over all such sequences.

As in the case when $n = 1$, we shall refer to Lebesgue outer measure as simply "outer measure", as there will be no other outer measure under discussion except in Problems 2.7.P2 and 2.8.P4.

Remarks 2.7.4

(a) Some or all of the I_k may be empty. It follows in particular that $m_n{}^*(\emptyset) = 0$.
(b) The outer measure $m_n{}^*$ is a set function defined on *all* subsets of $A \subseteq \mathbb{R}$ and has values in the set of nonnegative extended real numbers.

The following properties of the outer measure are almost immediate from the definition. The property stated in (c) is called **monotonicity**.

Proposition 2.7.5

(a) $0 \leq m_n{}^*(A) \leq \infty$ *for all subsets of* $\mathbb{R}^n$;
(b) $m_n{}^*(\emptyset) = 0$;
(c) If $A \subseteq B \subseteq \mathbb{R}^n$, then $m_n{}^*(A) \leq m_n{}^*(B)$;
(d) For $x \in \mathbb{R}^n$, we have $m_n{}^*(\{x\}) = 0$.

Suppose that A is any subset of $\mathbb{R}^n$ and x is any element of $E \subseteq \mathbb{R}$. As in the case of $\mathbb{R}$, the **translate** of A by x is the set $\boldsymbol{A + x} = \{y + x: y \in A\}$. The assertion of the next proposition is called **translation invariance** of the Lebesgue outer measure $m_n{}^*$.

Proposition 2.7.6 *For any* $A \subseteq \mathbb{R}^n$ *and* $x \in \mathbb{R}^n$, *we have* $m_n{}^*(A + x) = m_n{}^*(A)$.

Example 2.7.7 Suppose that A has at most countably many points. Then $m_n{}^*(A) = 0$. In view of the monotonicity of $m_n{}^*$, we need consider only countable A. So, let $\{x_k\}_{k\geq 1}$ be an enumeration of the points of A. Given $\varepsilon > 0$, let I_k be an open cuboid of volume $\varepsilon/2^k$ that contains the point x_k. The sequence $\{I_k\}_{k\geq 1}$ then covers A and satisfies $\sum_{k=1}^{\infty} v(I_k) = \sum_{k=1}^{\infty} \varepsilon/2^k = \varepsilon$. So, $m_n{}^*(A) \leq \varepsilon$. In particular,

$$m_n^*(\mathbb{Q} \times \mathbb{Q} \times \cdots \times \mathbb{Q}) = m_n^*(\mathbb{N} \times \mathbb{N} \times \cdots \times \mathbb{N}) = m_n^*(\mathbb{Z} \times \mathbb{Z} \times \cdots \times \mathbb{Z}) = 0.$$

It follows from what has just been noted that a set of positive outer measure must be uncountable.

Proposition 2.7.8 *If I is any cuboid, then its outer measure is the same as its volume*: $m_n^*(I) = v(I)$.

Proof If I is empty, there is nothing to prove; so assume I nonempty. It is easily seen that, given any $\eta > 0$, there exists an open cuboid $I' \supset I$ such that $v(I') \leq (1 + \eta)v(I)$. Then $\{I', \varnothing, \varnothing, \ldots\}$ is a sequence of open cuboids that covers I and therefore $m_n^*(I) \leq v(I') \leq (1 + \eta)v(I)$. But η is an arbitrary positive number; therefore $m_n^*(I) \leq v(I)$. To prove the reverse inequality, we proceed as follows:

First suppose I is closed and let $\eta > 0$ be given. There is a sequence of open cuboids $\{I_k\}_{k\geq 1}$ such that $I \subseteq \cup_{k\geq 1} I_k$ and $\sum_{k=1}^{\infty} v(I_k) \leq m_n^*(I) + \eta$. By the Heine–Borel Theorem, any collection of open cuboids covering I contains a finite subcollection, which we again denote by $\{I_k\}_{1\leq k\leq p}$, such that $I \subseteq \cup_{1\leq k\leq p} I_k$. By Lemma 2.7.2(ii), we have $v(I) \leq \sum_{k=1}^{p} v(I_k)$. Thus

$$v(I) \leq \sum_{k=1}^{p} v(I_k) \leq \sum_{k=1}^{\infty} v(I_k) \leq m_n^*(I) + \eta.$$

Since $\eta > 0$ is arbitrary, it follows that $v(I) \leq m_n^*(I)$. This completes the proof in the case when I is a closed cuboid.

If I is any nonempty cuboid, then, given $\eta > 0$, there is a closed cuboid $I'' \subseteq I$ such that

$$v(I'') > (1 - \eta)v(I).$$

Hence

$$(1 - \eta)v(I) < v(I'') = m_n^*(I'') \leq m_n^*(I) \leq m_n^*(\bar{I}) = v(\bar{I}) = v(I).$$

Thus for each $\eta > 0$,

$$(1 - \eta)v(I) < m_n^*(I) \leq v(I),$$

and so,

$$v(I) = m_n^*(I).$$ □

The reader is reminded of the concepts of *finitely additive*, *countably subadditive* and *countably additive* as laid down in Definition 2.2.7. Illustrations of these concepts in a general context were given in Examples 2.2.8.

Proposition 2.7.9 *The outer measure $m_n{}^*$ is countably subadditive; that is, if $\{A_j\}_{j\geq 1}$ is a sequence of subsets of $\mathbb{R}^n$, then*

$$m_n{}^*\Big(\bigcup_{j=1}^{\infty} A_j\Big) \leq \sum_{j=1}^{\infty} m_n{}^*(A_j).$$

Corollary 2.7.10 *The outer measure $m_n{}^*$ is finitely subadditive; that is, if $\{A_j\}_{1\leq j\leq p}$ is a finite sequence of subsets of $\mathbb{R}^n$, then*

$$m_n{}^*\Big(\bigcup_{j=1}^{p} A_j\Big) \leq \sum_{j=1}^{p} m_n{}^*(A_j).$$

Remarks 2.7.11

(a) If $m_n{}^*(A) = 0$, then $m_n{}^*(A \cup B) = m_n{}^*(B)$. In fact, by Proposition 2.7.5(c) and Corollary 2.7.10,

$$m_n{}^*(B) \leq m_n{}^*(A \cup B) \leq m_n{}^*(A) + m_n{}^*(B) = m_n{}^*(B).$$

(b) Let A be any subset of $\mathbb{R}^n$ and A_I denote the set of points in A with at least one irrational coordinate. Then $m_n{}^*(A) = m_n{}^*(A_I)$. This follows from (a) above and the result noted in Example 2.7.7.

In the rest of this section, we explore more closely the role of open sets in determining the Lebesgue outer measure of an arbitrary subset of $\mathbb{R}^n$. The material can be omitted for now without loss of continuity and the reader may wish to take it up only when it is needed later.

Proposition 2.7.12 *Let $A \subseteq \mathbb{R}^n$ be arbitrary and $\varepsilon > 0$ be given. Then there exists an open set $O \supseteq A$ such that $m_n{}^*(O) \leq m_n{}^*(A) + \varepsilon$.*

Proposition 2.7.13 *Let $A \subseteq \mathbb{R}^n$ be arbitrary. Then there exists a set $G \supseteq A$ such that $m_n{}^*(G) = m_n{}^*(A)$ and G is an intersection of a sequence of open sets.*

Thus every subset of $\mathbb{R}^n$ is contained in a $\mathcal{G}_\delta$-set of the same outer measure as itself.

Proposition 2.7.14 *Any nonempty open set O is a countable union of open cuboids and also a countable union of closed cuboids.*

Proof Since O is nonempty, each point of it must belong to an open cuboid I contained in O. It is obvious how to "shrink" the open cuboid I slightly in such a manner that the point continues to belong to it while the corresponding closed cuboid remains within I. Moreover, one can choose the smaller cuboid to have rational "corners". Then O is the union of all these smaller cuboids as well as of the corresponding closed cuboids, both of which must necessarily be countable in number. □

Examples 2.7.15

(a) Since the subset $\mathbb{Q} \times \mathbb{Q} \times \cdots \times \mathbb{Q}$ of $\mathbb{R}^n$ is countable and a single point set is closed, it follows that the aforementioned set is an $\mathcal{F}_\sigma$-set.

(b) If $E \in \mathcal{F}_\sigma$, then by using the definition of $\mathcal{F}_\sigma$ and taking complements, we conclude that $E^c \in \mathcal{G}_\delta$. In particular, it follows that the set of points in $\mathbb{R}^n$ with at least one coordinate irrational is a set of type $\mathcal{G}_\delta$.

(c) Since $(a, b) = \cup_{p\geq 1}[a+\frac{1}{p}, b-\frac{1}{p}]$ and $\{a\} = \cap_{p\geq 1}(a-\frac{1}{p}, a+\frac{1}{p})$, it follows that an open cuboid

$$(a_1, b_1; a_2, b_2; \ldots; a_n, b_n) = \bigcup_{p=1}^{\infty}[a_1 + \frac{1}{p}, b_1 - \frac{1}{p}; \ldots; a_n + \frac{1}{p}, b_n - \frac{1}{p}]$$

is a set of type $\mathcal{F}_\sigma$ and that a single point set is of type $\mathcal{G}_\delta$. Here the left-hand side is a Cartesian product of open intervals $\prod_{i=1}^n (a_i, b_i)$ and the right-hand side is a Cartesian product of closed intervals $\prod_{i=1}^n [a_i + \frac{1}{p}, b_i - \frac{1}{p}]$.

(d) It is clear that every open set including $\varnothing$ is of type $\mathcal{G}_\delta$ and every closed set including $\varnothing$ is of type $\mathcal{F}_\sigma$. It is easily seen that every nonempty open set is also of type $\mathcal{F}_\sigma$, because it is a countable union of closed cuboids by Proposition 2.7.14. On taking complements, we see that every closed set is of type $\mathcal{G}_\delta$.

Problem Set 2.7

2.7.P1. Prove that $|m_n{}^*(A) - m_n{}^*(B)| \leq \max\{m_n{}^*(A\backslash B), m_n{}^*(B\backslash A)\} \leq m_n{}^*(A\Delta B)$ provided $m_n{}^*(A)$ and $m_n{}^*(B)$ are finite.

2.7.P2. An *outer measure* on $\mathbb{R}^n$ is an extended real-valued, nonnegative, monotone and countably subadditive set function μ^* defined on all subsets of $\mathbb{R}^n$ and satisfying $\mu^*(\varnothing) = 0$. If $\{\mu_k{}^*\}_{k\geq 1}$ is a sequence of outer measures and $\{a_k\}_{k\geq 1}$ a sequence of positive real numbers, then show that the set function defined by $\mu^*(E) = \sum_{k=1}^{\infty} a_k\mu_k^*(E)$ is an outer measure.

2.8 Measurable Sets and Lebesgue Measure in $\mathbb{R}^n$

The outer measure has the advantage that it is defined on $\mathcal{P}(\mathbb{R}^n)$; however, it is not countably additive (see Definition 2.2.7), which is a desirable property for a measure. The outer measure becomes countably additive if the family of sets on which it is defined is suitably restricted. We pursue this matter in this section. Proofs that are similar to their analogues in Sect. 2.3 will be omitted.

Our first definition is wholly analogous to the one in $\mathbb{R}$.

Definition 2.8.1 A subset $E \subseteq \mathbb{R}^n$ is said to be **measurable** if

$$m_n{}^*(A) = m_n{}^*(A \cap E) + m_n{}^*(A \cap E^c) \quad \text{for every } A \subseteq \mathbb{R}^n.$$

Remarks 2.8.2

(a) Observe that the finite subadditivity property (see **Corollary** 2.7.10) of m_n^* implies

$$m_n{}^*(A) \le m_n{}^*(A \cap E) + m_n{}^*(A \cap E^c) \quad \text{for every } A \subseteq \mathbb{R}^n.$$

Thus in testing the measurability of E, it is enough to show the reverse of this inequality. This condition of measurability in Definition 2.8.1 is known as the **Carathéodory condition.**

(b) If E is measurable, so is its complement E^c.

(c) For $A \subseteq \mathbb{R}^n$, we have $m_n{}^*(A) = m_n{}^*(A) + 0 = m_n{}^*(A \cap \mathbb{R}^n) + m_n{}^*(A \cap \varnothing)$. Thus $\mathbb{R}^n$ is measurable, and hence by (b), so is $\varnothing$.

(d) If $m_n{}^*(E) = 0$, where $E \subseteq \mathbb{R}^n$, then E is measurable; in fact, every subset of E is also measurable. This is because, for $A \subseteq \mathbb{R}^n$, we have $m_n{}^*(A \cap E) \le m_n{}^*(E) = 0$ and $m_n{}^*(A \cap E^c) \le m_n{}^*(A)$. So, $m_n{}^*(A) \ge m_n{}^*(A \cap E^c) + 0 \ge m_n{}^*(A \cap E) + m_n{}^*(A \cap E^c)$.

(e) A countable subset of $\mathbb{R}^n$ is measurable. This follows from Example 2.7.7 and (d) above. Consequently, the set $\mathbb{Q} \times \mathbb{Q} \times \cdots \times \mathbb{Q}$ in $\mathbb{R}^n$ is measurable. The set of points in $\mathbb{R}^n$ with at least one coordinate irrational, being the complement of $\mathbb{Q} \times \mathbb{Q} \times \cdots \times \mathbb{Q}$, is measurable in view of (b) above.

Notation 2.8.3 *We denote the collection of all measurable subsets of* $\mathbb{R}^n$ *by* $\mathfrak{M}_n$.

In what follows, we record some of the properties of the collection $\mathfrak{M}_n$.

Lemma 2.8.4 *If* E_1 *and* E_2 *are in* $\mathfrak{M}_n$, *then so is* $E_1 \cup E_2$.

Remarks 2.8.5

(a) $\varnothing \in \mathfrak{M}_n$.

(b) $A_1, A_2, \ldots, A_p \in \mathcal{F} \Rightarrow \bigcup_{j=1}^{p} A_j \in \mathfrak{M}_n$.

(c) If $A \in \mathcal{F}, B \in \mathcal{F} \Rightarrow A \cap B \in \mathfrak{M}_n$.

(d) $A_1, A_2, \ldots, A_p \in \mathcal{F} \Rightarrow \bigcap_{j=1}^{p} A_j \in \mathfrak{M}_n$.

(e) $A \in \mathcal{F}, B \in \mathcal{F} \Rightarrow A \backslash B \in \mathfrak{M}_n$.

Proposition 2.8.6 *The collection* $\mathfrak{M}_n$ *of measurable subsets of* $\mathbb{R}^n$ *is an algebra. In particular, if* E_1, E_2, …, E_p *are measurable subsets, then so are* $\cup_{1 \le k \le p} E_k$ *and* $\cap_{1 \le k \le p} E_k$.

Proof This is merely a summary of Remarks 2.8.2(b) and 2.8.2(c), Lemma 2.8.4 and Remarks 2.8.5(b) and 2.8.5(d). □

We now note that the outer measure restricted to $\mathfrak{M}_n$ is finitely additive (Cf. Proposition 2.3.8).

Proposition 2.8.7 *If E_1, E_2, …, E_p is a finite sequence of disjoint measurable subsets of $\mathbb{R}^n$ and $A \subseteq \mathbb{R}^n$ is arbitrary, then*

$$m_n^*(A \cap \bigcup_{j=1}^{p} E_j) = \sum_{j=1}^{p} m_n^*(A \cap E_j).$$

In particular, if $A = \mathbb{R}^n$, then

$$m_n^*\Big(\bigcup_{j=1}^{p} E_j\Big) = \sum_{j=1}^{p} m_n^*(E_j).$$

The outer measure m_n^* restricted to $\mathfrak{M}_n$ is in fact countably additive (Theorem 2.8.9).

Theorem 2.8.8 (Cf. Theorem 2.3.12) *The collection $\mathfrak{M}_n$ of measurable subsets of $\mathbb{R}^n$ constitute a σ-algebra.*

Theorem 2.8.9 (Cf. Theorem 2.3.13) *The outer measure m_n^* is countably additive on the σ-algebra $\mathfrak{M}_n$ of measurable subsets of $\mathbb{R}^n$.*

Definition 2.8.10 The restriction of m_n^* to $\mathfrak{M}_n$ is called **(Lebesgue) measure** and will be denoted by $\boldsymbol{m_n}$. For any $E \in \mathfrak{M}_n$, the (extended) real number $m_n(E) = m_n^*(E)$ is called the (Lebesgue) measure of the set E.

Since m_n^* is countably subadditive according to Proposition 2.7.9, it follows trivially that Lebesgue measure is also countably subadditive.

Definition 2.8.11 By a **closed upper half-space** we mean a subset H of $\mathbb{R}^n$ for which there exists j $(1 \leq j \leq n)$ and a real number c_j such that

$$H = \{(x_1, x_2, \ldots, x_n) \in \mathbb{R}^n : c_j \leq x_j\}.$$

If the inequality is replaced by $<$, we speak of an **open upper half-space**. If the inequalities are reversed, we have what are called **closed** and **open lower half-spaces**.

A half-space is the n-dimensional analogue of an interval with one finite endpoint.

Remarks 2.8.12

(a) A cuboid is an intersection of $2n$ half-spaces.
(b) An open [resp. closed] half-space is an open [resp. closed] set.
(c) The complement of an open [resp. closed] upper half space is a closed [resp. open] lower half-space, and vice versa.
(d) If H is an open upper half-space (so that its complement H^c is also a half-space) and I is any open cuboid, then $I \cap H$ and $I \cap H^c$ are both cuboids, possibly empty, and $v(I) = v(I \cap H) + v(I \cap H^c)$. This is obvious in the one-dimensional case. Suppose

$$I = (a_1, b_1; a_2, b_2; \ldots; a_n, b_n) \quad \text{and} \quad H = \{(x_1, x_2, \ldots, x_n) \in \mathbb{R}^n : c_j < x_j\}.$$

Then $(x_1, x_2, \ldots, x_n) \in I \cap H$ must satisfy $a_j < x_j < b_j$ as well as $c_j < x_j$. This is not possible if $c_j \geq b_j$, and therefore $I \cap H = \varnothing$, so that $I \cap H^c = I$ and there is nothing left to prove in this case. Also, $(x_1, x_2, \ldots, x_n) \in I \cap H^c$ must satisfy $a_j < x_j < b_j$ as well as $c_j \geq x_j$. This is not possible if $c_j \leq a_j$, and therefore $I \cap H^c = \varnothing$, so that $I \cap H = I$ and there is nothing left to prove in this case either. So, suppose $a_j < c_j < b_j$. Then $I \cap H$ is given by the inequalities

$$a_i < x_i < b_i \text{ for } i \neq j \quad \text{and} \quad c_j < x_j < b_j.$$

Thus $I \cap H$ is a cuboid with volume

$$v(I \cap H) = (b_j - c_j) \cdot \prod\nolimits_{i \neq j} (b_i - a_i).$$

At the same time, $I \cap H^c$ is given by the inequalities

$$a_i < x_i < b_i \text{ for } i \neq j \quad \text{and} \quad a_j < x_j \leq c_j.$$

Thus $I \cap H^c$ is a cuboid with volume

$$v(I \cap H^c) = (c_j - a_j) \cdot \prod\nolimits_{i \neq j} (b_i - a_i).$$

It is now immediate that $v(I) = v(I \cap H) + v(I \cap H^c)$.

Proposition 2.8.13 *An open upper half-space H is measurable.*

Proof Let A be any subset of $\mathbb{R}^n$ and let $A_1 = A \cap H$ and $A_2 = A \cap H^c$. We must show that $m_n^*(A) \geq m_n^*(A_1) + m_n^*(A_2)$. If $m_n^*(A) = \infty$, then there is nothing to prove. Suppose that $m_n^*(A) < \infty$.

By the definition of outer measure, for every $\varepsilon > 0$, there exists a sequence $\{I_n\}_{n \geq 1}$ of open cuboids such that $A \subseteq \cup_{k \geq 1} I_k$ and

$$\sum_{k=1}^{\infty} v(I_k) \leq m_n^*(A) + \varepsilon.$$

Let $I'_k = I_k \cap H$ and $I''_k = I_k \cap H^c$. By Remark 2.8.12(d), I'_k and I''_k are cuboids (some of them possibly empty) and

$$\begin{aligned} v(I_k) &= v(I'_k) + v(I''_k) \\ &= m_n^*(I'_k) + m_n^*(I''_k) \end{aligned}$$

by Proposition 2.7.8.

Since $A_1 \subseteq \cup_{k \geq 1} I'_k$ and $A_2 \subseteq \cup_{k \geq 1} I''_k$, we have

$$m_n{}^*(A_1) \le m_n{}^*(\bigcup_{k=1}^{\infty} I'_k) \le \sum_{k=1}^{\infty} m_n{}^*(I'_k)$$

and

$$m_n{}^*(A_2) \le m_n{}^*(\bigcup_{k=1}^{\infty} I''_k) \le \sum_{k=1}^{\infty} m_n{}^*(I''_k).$$

Hence

$$\begin{aligned} m_n{}^*(A_1) + m_n{}^*(A_2) &\le \sum_{k=1}^{\infty} m_n{}^*(I'_k) + \sum_{k=1}^{\infty} m_n{}^*(I''_k) \\ &= \sum_{k=1}^{\infty} [m_n{}^*(I'_k) + m_n{}^*(I''_k)] = \sum_{k=1}^{\infty} v(I_k) \le m_n{}^*(A) + \varepsilon. \end{aligned}$$

Since $\varepsilon > 0$ is arbitrary, we have $m_n{}^*(A) \ge m_n{}^*(A_1) + m_n{}^*(A_2)$, as required. □

Proposition 2.8.14 *A subset of $\mathbb{R}^n$ that is either open or closed must be measurable. Every cuboid is measurable. Moreover, $\mathcal{G}_\delta$-sets and $\mathcal{F}_\sigma$-sets are always measurable.*

Proof Since the collection $\mathfrak{M}_n$ of measurable sets is a σ-algebra, it follows from Proposition 2.8.13 that a closed lower half-space is measurable. Since an open lower half-space given by $x_j < c_j$ is the countable union of the closed lower half-spaces given by $x_j \le c_j - \frac{1}{n}$, it further follows that an open lower half-space is measurable. Hence a closed upper half-space is also always measurable. By Remark 2.8.12(a), any cuboid is measurable.

Now any nonempty open set O is a countable union of open cuboids by Proposition 2.7.14. Since $\mathfrak{M}_n$ is a σ-algebra, it follows that O is measurable. Thus, every open set is measurable. Once again, since $\mathfrak{M}_n$ is a σ-algebra, the complement of an open set, i.e. a closed set, must be measurable. The final assertion about $\mathcal{G}_\delta$- and $\mathcal{F}_\sigma$-sets now follows immediately from the fact that $\mathfrak{M}_n$ is a σ-algebra. □

Recall from Proposition 2.3.17 that there exists a smallest σ-algebra generated by any collection of subsets.

Definition 2.8.15 The σ-algebra generated by all the open subsets of $\mathbb{R}^n$ is called the **Borel algebra (in $\mathbb{R}^n$)** and will be denoted by $\mathfrak{B}_n$. A set in the Borel algebra is called a **Borel measurable set (in $\mathbb{R}^n$)** or simply a **Borel set (in $\mathbb{R}^n$)**. A set in the algebra $\mathfrak{M}_n$ (hitherto called simply "measurable") will be called a Lebesgue measurable set when a distinction needs to be made.

Remark 2.8.16 Since every open set is Lebesgue measurable by Proposition 2.8.14 and Lebesgue measurable sets constitute a σ-algebra, it follows by Definition 2.8.15 that $\mathfrak{B} \subseteq \mathfrak{M}$. In other words, every Borel set is Lebesgue measurable. The restriction of $m_n{}^*$ to $\mathfrak{B}_n$ is called **Borel measure (in $\mathbb{R}^n$)** and will again be denoted by m_n. Thus, for $E \in \mathfrak{B}_n$, the extended real number $m_n(E) = m_n{}^*(E)$ is called the Borel measure of E and is the same as its Lebesgue measure.

The triples $(\mathbb{R}^n, \mathfrak{B}_n, m_n)$ and $(\mathbb{R}^n, \mathfrak{M}_n, m_n)$ are called the **Borel measure space** and **Lebesgue measure space** respectively.

We have verified the following properties of Lebesgue measurable sets and Lebesgue measure:

(i) Complements, countable unions and countable intersections of measurable sets are measurable.

(ii) Any cuboid is measurable and its measure is its volume (see Proposition 2.7.8).

(iii) If $\{E_k\}_{k\geq 1}$ is a sequence of disjoint measurable sets, then

$$m_n(\bigcup_{k=1}^{\infty} E_k) = \sum_{k=1}^{\infty} m_n(E_k).$$

The above properties are also valid for Borel measurable sets and Borel measure: (i) is a direct consequence of the definition of $\mathfrak{B}_n$ as being a σ-algebra, while (iii) follows from its Lebesgue counterpart taken with Remark 2.8.16. Assertion (ii) about Borel sets follows from Remark 2.8.12(a) and the fact that every open half space, being an open set, is a Borel set and hence so is every half-space.

Proposition 2.8.17 (Translation Invariance of Lebesgue measure on $\mathbb{R}^n$) *Let $E \in \mathfrak{M}_n[resp.\ \mathfrak{B}_n], x \in \mathbb{R}^n$. Then*

(a) $E + x \in \mathfrak{M}_n[resp. \mathfrak{B}_n]$,

(b) $m_n(E + x) = m_n(E)$.

That is, the Lebesgue [resp. Borel] measure on $\mathfrak{M}_n[resp.\mathfrak{B}_n]$ is translation invariant.

Proof Analogous to the proof of Proposition 2.3.23. □

Finally, we prove the following characterisation of Lebesgue measurable subsets of $\mathbb{R}^n$.

Proposition 2.8.18 *Let E be a given subset of $\mathbb{R}^n$. The following five statements are equivalent:*

(α) *E is Lebesgue measurable;*

(β) *For every $\varepsilon > 0$, there exists an open set $O \supseteq E$ such that $m_n^*(O \backslash E) < \varepsilon$;*

(γ) *There exists a $\mathcal{G}_\delta$-set $G \supseteq E$ such that $m_n^*(G \backslash E) = 0$;*

(δ) *For every $\varepsilon > 0$, there exists a closed set $F \subseteq E$ such that $m_n^*(E \backslash F) < \varepsilon$;*

(η) *There exists an $\mathcal{F}_\sigma$-set $F \subseteq E$ such that $m_n^*(E \backslash F) = 0$.*

Proof We shall prove (α) ⇒ (β) ⇒ (γ) ⇒ (α) and then (δ) ⇔ (α) ⇔ (η).

(α) ⇒ (β). For any set E and $\varepsilon > 0$, there exists [see Proposition 2.7.12] an open set $O \supseteq E$ such that

$$m_n^*(O) \le m_n^*(E) + \frac{\varepsilon}{2}.$$

Since $O \supseteq E$, we have $O \cap E = E$. From the Lebesgue measurability of E, we have

$$m_n^*(O) = m_n^*(O \cap E) + m_n^*(O \cap E^c) = m_n^*(E) + m_n^*(O \backslash E).$$

Thus

$$m_n^*(E) + \frac{\varepsilon}{2} \ge m_n^*(O) = m_n^*(E) + m_n^*(O \backslash E).$$

In the case when $m_n^*(E) < \infty$, this shows that $m_n^*(O \backslash E) \le \frac{\varepsilon}{2} < \varepsilon$.

Now let $m_n^*(E) = \infty$. Observe that for each $k \in \mathbb{N}$, the set $I_k = \{(x_1, x_2, \ldots, x_n) \in \mathbb{R}^n : |x_j| < k, 1 \le j \le n\}$ is a cuboid of volume $(2k)^n$, which is finite. For each $k \in \mathbb{N}$, the set $E_k = E \cap I_k \in \mathfrak{M}_n$ and $m_n^*(E_k) \le m_n^*(I_k) < \infty$. Moreover $E = \cup_{k \ge 1} E_k$. Since $m_n^*(E_k) < \infty$, it follows on applying the case of finite measure proved above that, for any $\varepsilon > 0$, there exists an open set $O_k \supseteq E_k$ such that

$$m_n^*(O_k \backslash E_k) < \frac{\varepsilon}{2^k}.$$

Note that

$$O = \bigcup_{k=1}^{\infty} O_k \supseteq \bigcup_{k=1}^{\infty} E_k = E$$

and

$$O \backslash E = \bigcup_{k=1}^{\infty} O_k \backslash \bigcup_{k=1}^{\infty} E_k \subseteq \bigcup_{k=1}^{\infty} (O_k \backslash E_k).$$

So,

$$m_n^*(O \backslash E) \le \sum_{k=1}^{\infty} m_n^*(O_k \backslash E_k) < \sum_{k=1}^{\infty} \frac{\varepsilon}{2^{k+1}} = \varepsilon.$$

(β) ⇒ (γ). It follows from (β) that, for any $k \in \mathbb{N}$, there exists an open set $O_k \supseteq E$ such that

$$m_n^*(O_k \backslash E) < \frac{1}{k}.$$

Let $G = \cap_{k \ge 1} O_k$; then G is a $\mathcal{G}_\delta$-set containing E. Moreover,

$$m_n{}^*(G\backslash E) = m_n{}^*(\bigcap_{k=1}^{\infty} O_k\backslash E) \leq m_n{}^*(O_k\backslash E) < \frac{1}{k} \quad \text{for every } k \in \mathbb{N}.$$

Hence $m_n{}^*(G\backslash E) = 0$.

(γ) $\Rightarrow$ (α). Note that $E = G\backslash(G\backslash E)$ and that the set G, being a countable intersection of Lebesgue measurable sets, is Lebesgue measurable by Theorem 2.8.8; also, $G\backslash E$ is Lebesgue measurable by Remark 2.8.2(d). Use Theorem 2.8.8 once again.

We have shown so far that (α) $\Rightarrow$ (β) $\Rightarrow$ (γ) $\Rightarrow$ (α). The consequence that (β) $\Leftrightarrow$ (α) will shortly be used in proving that (α) $\Leftrightarrow$ (δ).

(α) $\Leftrightarrow$ (δ). Assume (α), i.e. $E \in \mathfrak{M}_n$. Then $E^c \in \mathfrak{M}_n$. Since (α) $\Rightarrow$ (β), it follows that, for any $\varepsilon > 0$, there exists an open set $O \supseteq E^c$ such that $m_n{}^*(O\backslash E^c) < \varepsilon$. So the closed set $F = O^c$ satisfies $F \subseteq E$ and also $m_n{}^*(E\backslash F) = m_n{}^*(E\backslash O^c) = m_n{}^*(E \cap O) = m_n{}^*(O\backslash E^c) < \varepsilon$. Thus ($\delta$) holds.

Conversely, assume (δ). Then the open set $O = F^c$ satisfies $O \supseteq E^c$ and $m_n{}^*(O\backslash E^c) < \varepsilon$. In other words, ($\beta$) holds with E^c in place of E. Since (β) $\Rightarrow$ (α), it follows that $E^c \in \mathfrak{M}_n$ and hence $E \in \mathfrak{M}_n$. Thus (α) holds.

(α) $\Leftrightarrow$ (η). This is analogous to the argument that (α) $\Leftrightarrow$ (δ). □

Remark 2.8.19 Let $E \subseteq \mathbb{R}^n$ be Lebesgue measurable, i.e. $E \in \mathfrak{M}_n$. Then it follows from the preceding proposition that there exist $F, G \in \mathfrak{B}_n$ such that $F \subseteq E \subseteq G$ with $m_n(G\backslash F) = 0$. Conversely, the existence of such F and G can be proved to imply $E \in \mathfrak{M}$ in the following manner: Suppose such F, G exist. Then $F \in \mathfrak{M}_n$ because $\mathfrak{B}_n \subseteq \mathfrak{M}_n$; besides, $E\backslash F \subseteq G\backslash F$, which implies $m_n{}^*(E\backslash F) \leq m_n{}^*(G\backslash F) = 0$ and hence $E \subseteq \mathbb{R}$ by Remark 2.8.2(d). Therefore $E = (E\backslash F) \cup F \in \mathfrak{M}_n$. Note that in this situation, $m_n(E) = m_n(F) = m_n(G)$. The foregoing equivalence is related to the fact that Lebesgue measure is the "completion" of Borel measure (see Definition 7.2.3 and Theorem 7.2.5).

Problem Set 2.8

2.8.P1. Let $\{E_i\}_{i\geq 1}$ be a sequence of measurable sets. Then

(a) if $E_i \subseteq E_{i+1}$, then $m_n(\lim E_k) = \lim_{k\to\infty} m_n(E_k)$;

(b) if $E_i \supseteq E_{i+1}$ and $m_n(E_1) < \infty$, then $m_n(\lim E_k) = \lim_{k\to\infty} m_n(E_k)$.

2.8.P2. Prove that for subsets A, B, C of any nonempty set X,

$$A\Delta C \subseteq (A\Delta B) \cup (B\Delta C).$$

2.8.P3. Let A, B be subsets of $\mathbb{R}^n$. Show that

$$m_n{}^*(A\Delta B) \leq m_n{}^*(A) + m_n{}^*(B).$$

2.8.P4. If μ^* is an outer measure on $\mathbb{R}^n$ and if A and B are subsets of $\mathbb{R}^n$, of which at least one is μ^*-measurable, then show that

$$\mu^*(A) + \mu^*(B) = \mu^*(A \cup B) + \mu^*(A \cap B).$$

Remark: Under the additional hypothesis that at least one among A and B has finite outer measure, the above equality can be written as

$$\mu^*(A \cup B) = \mu^*(A) + \mu^*(B) - \mu^*(A \cap B).$$

2.8.P5. Let $A \subseteq E$, where E is measurable and $m_n(E) < \infty$. Show that A is measurable provided

$$m_n{}^*(E) = m_n{}^*(A) + m_n{}^*(E \backslash A).$$

2.8.P6. If $F \in \mathfrak{M}_n$ and $m_n{}^*(F \Delta G) = 0$, then show that G is measurable.
2.8.P7. Show that every nonempty open set has positive measure.
2.8.P8. Let $q_1, q_2, \ldots$ be an enumeration of points in $\mathbb{R}^n$ with rational coordinates and let $G = \cup_{k \geq 1} I_k$, where I_k is an open cuboid centred at q_k with volume $1/k^2$. Prove that for any closed set F, $m_n(G \Delta F) > 0$.
2.8.P9. Let E be Lebesgue measurable with $0 < m_n(E) < \infty$ and let $\varepsilon > 0$ be given. Then there exists a compact set $K \subseteq E$ such that $m_n(E \backslash K) = m_n(E) - m_n(K) < \varepsilon$.

Chapter 3
Measure Spaces and Integration

3.1 Integrals of Simple Functions

The Riemann integral of a function over a real interval is defined via upper and lower sums or via Riemann sums, all of which have finitely many terms. What is often not emphasised is that they are in fact integrals of certain step functions. Thus, step functions are used as building blocks for Riemann integration, which is possible because their integrals are finite sums.

In the development of the Lebesgue theory, we prefer to use as building blocks those functions that are constant on measurable sets and have only finitely many nonnegative real values, in other words, simple functions as in Definition 2.5.6. The reason for not using step functions will become clear soon.

All intervals are instances of measurable sets and therefore nonnegative-valued step functions are instances of simple functions. However, they are far from being the only ones, because there exist measurable sets other than intervals. For example, the set of irrational numbers in [0,1] is measurable and therefore its characteristic function is a simple function, although it is not a step function.

A simple function can be expressed as a linear combination of characteristic functions in several ways. We shall define the Lebesgue integral of a simple function in terms of its canonical representation so as to avoid ambiguity. The awkwardness of expressing the canonical representation of the sum of two simple functions in terms of their respective canonical representations [see Remark 2.5.8 (d) and Problem 2.5.P6] creates a slight difficulty, which will be surmounted in Proposition 3.1.5.

The context in which we shall be working is that we have a set function μ having a σ-algebra $\mathcal{F}$ as its domain and satisfying the following properties:

(M1) $\mu(\varnothing) = 0$ and $\mu(E) \geq 0$ for every $E \in \mathcal{F}$;

(M2) μ is countably additive [as in Definition 2.2.7], that is, for every disjoint sequence of sets $\{A_j\}_{j\geq 1}$ such that each $A_j \in \mathcal{F}$, we have

S. Shirali and H. L. Vasudeva, *Measure and Integration*,
Springer Undergraduate Mathematics Series,
https://doi.org/10.1007/978-3-030-18747-7_3

$$\mu\left(\bigcup_{j=1}^{\infty} A_j\right) = \sum_{j=1}^{\infty} \mu(A_j).$$

There will be no provision for special elements of $\mathcal{F}$ called "intervals", thereby taking away all scope for step functions; in particular, μ will not be presumed to have been built up from the concept of length or volume of simpler kinds of sets. The nature of the set X of which the elements of $\mathcal{F}$ are subsets will not matter, as it will not enter into our considerations. In such a general set up, the set function involved is sometimes called a *general* or an *abstract* measure for emphasis.

Definition 3.1.1 *A* **measure space** $(X, \mathcal{F}, \mu)$ is a nonempty set X with a σ-algebra $\mathcal{F}$ of subsets of X and an extended real-valued function μ on $\mathcal{F}$ such that (M1) and (M2) hold. The set function μ is called a **measure** on $\mathcal{F}$.

Sometimes it will be convenient to speak of a "measure on X (or in X)", meaning a measure on some σ-algebra of subsets of X. A subset that belongs to the σ-algebra is called a *measurable* subset without presuming any outer measure.

To recap, a measure on a set X is a countably additive nonnegative extended real-valued function on a σ-algebra of subsets of X that vanishes on the empty set.

The reader may note that the continuity property in Proposition 2.3.21 was proved by using only properties (M1) and (M2) of Lebesgue and Borel measures while the manner in which these measures were constructed from lengths of intervals did not play any role. The property is therefore valid for abstract measures too.

Examples 3.1.2 The only examples we have seen so far are the Borel and Lebesgue measures on measurable subsets of $\mathbb{R}^n$. Here are some more.

(a) For a set $E \subseteq X$, where X is any nonempty set, define $\mu(E) = \infty$ if E is an infinite set and let $\mu(E)$ be the number of points in E if E is finite. Then μ is called the **counting measure** on $\mathcal{P}(X)$, the collection of all subsets of X and $(X, \mathcal{P}(X), \mu)$ is a measure space. In the particular case when $X = \mathbb{N}$, the counting measure on $\mathcal{P}(\mathbb{N})$ is often denoted by γ.
(b) Fix $x_0 \in X$, where X is any nonempty set whatsoever. Define $\mu(E) = 1$ if $x_0 \in E$ and $\mu(E) = 0$ otherwise. This μ may be called the unit mass concentrated at x_0. It is easy to verify that μ is a measure defined on the σ-algebra $\mathcal{P}(X)$.
(c) Let $(X, \mathcal{F}, \mu)$ be a measure space and Y be a measurable set (which means simply that $Y \in \mathcal{F}$). The triple $(Y, \mathcal{F}_Y, \mu|_{\mathcal{F}_Y})$, where $\mathcal{F}_Y = \{Y \cap A : A \in \mathcal{F}\}$ and $\mu|_{\mathcal{F}_Y}$ is the restriction of μ to the σ-algebra $\mathcal{F}_Y$, is a measure space. In particular, if $X = \mathbb{R}$, $\mathcal{F} = \mathfrak{M}$, the σ-algebra of Lebesgue measurable subsets of $\mathbb{R}$ and μ the Lebesgue measure defined on $\mathfrak{M}$, then for any $Y \in \mathcal{F} = \mathfrak{M}$, $\mathcal{F}_Y$ is the collection of Lebesgue measurable subsets of Y and $\mu|_{\mathcal{F}_Y}$ is the Lebesgue measure restricted to the measurable subsets of Y. The corresponding statement with $\mathfrak{B}$ in place of $\mathfrak{M}$ is also true.

The definitions and results of Sect. 2.5 up to Theorem 2.5.9 carry over when the set X is understood to be any set with a given σ-algebra and measure μ.

Sometimes a property of a function or of a sequence of functions, such as vanishing at x or converging to a limit function, is a "hereditary" property in the sense that if it holds on some set, then it holds on every subset thereof. The same is true of uniform continuity. As in the case of Lebesgue and Borel measure, we have a concept of "almost everywhere" as below [cf. Definition 2.5.10].

Definition If a hereditary property holds everywhere on X except on a subset E in $\mathcal{F}$ having $\mu(E) = 0$, we say that it holds **almost everywhere**. The phrase is abbreviated as **a.e.** When the property is described in terms of an explicit $x \in X$, we can say that it holds for **almost all** x.

In the rest of this chapter, X will denote a set with a σ-algebra $\mathcal{F}$ and a measure μ on it. However, when X is specified as a (Lebesgue) measurable subset of $\mathbb{R}^n$ ($n \geq 1$), it will be understood that μ is Lebesgue measure m_n on the σ-algebra of Lebesgue or Borel measurable subsets of X, unless the context calls for some other σ-algebra or measure.

We note in passing that if we were not to work with an abstract measure but were instead to stay with Lebesgue or Borel measure on an interval, then we could have used step functions as building blocks and avoided simple functions altogether. For treatments that proceed in this manner, see [1] or [17]. Also, we would have the option of proceeding via the Henstock–Kurzweil "gauge integral", which is rather more complicated to define than the Riemann integral, but leads directly to the Lebesgue as well as Riemann integrals (including improper) without depending on the concept of measure. For an introduction to the gauge integral, see [6] or the Internet article [22].

Before proceeding, we remind the reader [see Remark 2.5.8c] that $\sum_{1 \leq j \leq n} \alpha_j \chi_{Aj}$ is a canonical representation of a simple function on X if and only if

(a) the measurable sets A_j are disjoint with union X, while
(b) the nonnegative real numbers α_j are distinct and the sets A_j nonempty.

We shall say that the sets A_j **form a partition** of X when they satisfy (a); in particular, the sets must be measurable.

Definition 3.1.3 Let s be a simple function on X, having the canonical representation $s = \sum_{1 \leq j \leq n} \alpha_j \chi_{Aj}$. The **integral of** s is defined to be

$$\int s = \sum_{j=1}^{n} \alpha_j \mu(A_j).$$

It is also denoted by $\int_X s$, $\int_X s\, d\mu$ or $\int s\, d\mu$ according to convenience.

Thus the integral of a simple function is a sum of products, each of which corresponds to a term in the sum that represents the function canonically.

A simple function s with canonical representation $s = \sum_{1 \le j \le n} \alpha_j \chi_{A_j}$ takes the value 0 (vanishes) a.e. if and only if $\mu(A_j) = 0$ for every j such that $\alpha_j \neq 0$. When this is the case, clearly, $\int_X s = 0$.

When μ is Lebesgue measure m on $\mathbb{R}^n$ ($n \ge 1$) or on a measurable subset thereof, we speak of the *Lebesgue integral*. If it is necessary to emphasise that we are working with an abstract measure, we speak of the *measure space integral*.

Examples 3.1.4

(a) Suppose s is zero everywhere on X. Its canonical representation is $0{\cdot}\chi_X$ and therefore $\int_X s\,d\mu = \int_X 0\,d\mu = 0 \cdot \mu(X) = 0$, even if $\mu(X) = \infty$. The function has another representation as $1{\cdot}\chi_\varnothing$ (not canonical) and the corresponding sum of products is $1{\cdot}\mu(\varnothing)$, which also turns out to be 0.

(b) The rationals in [0,1] form a subset of Lebesgue measure zero. Therefore the characteristic function of this set is a simple function with Lebesgue integral 0. This is in contrast to the Riemann theory, in which the function has no Riemann integral at all on [0,1].

(c) Consider $s = \alpha\chi_A$, where α is a nonnegative real number and A is a measurable subset of X, that is, $A \in \mathcal{F}$. If $\alpha > 0$, then the canonical representation of s is $\alpha{\cdot}\chi_A + 0{\cdot}\chi_A^c$ provided A and A^c are both nonempty, and therefore $\int_X s\,d\mu = \alpha{\cdot}\mu(A) + 0{\cdot}\mu(A^c) = \alpha{\cdot}\mu(A)$ even if $\mu(A^c) = \infty$. If $\alpha = 0$, then the canonical representation is $0{\cdot}\chi_X$ and therefore $\int_X s\,d\mu = 0$ as in (a) above; but now $\alpha{\cdot}\mu(A)$ is also 0 and so it is still true that $\int_X s\,d\mu = \alpha \cdot \mu(A)$. Thus $\int_X (\alpha\chi_A) d\mu = \alpha \cdot \mu(A)$ when $\alpha \ge 0$. Note that this holds also when $A = \varnothing$ or X, because $\alpha\chi_\varnothing$ has canonical representation $0{\cdot}\chi_X$ and $\alpha\chi_X$ is already in canonical form.

(d) Let $X = [0,3)$ and s have the values 1,4,5 on [0,1), [1,1], (1,3) respectively. Then $s = \sum_{1 \le j \le 3} \alpha_j \chi_{A_j}$, where $\alpha_1 = 1$, $\alpha_2 = 4$, $\alpha_3 = 5$ and $A_1 = [0,1)$, $A_2 = [1,1]$, $A_3 = (1,3)$. According to the above definition,

$$\int_X s\,d\mu = 1 \cdot (1-0) + 4 \cdot (1-1) + 5 \cdot (3-1) = 11.$$

If we split A_3 as the disjoint union $(1,2] \cup (2,3)$, then we can represent s as $\sum_{1 \le i \le 4} \beta_i \chi_{B_i}$, where $\beta_1 = 1$, $\beta_2 = 4$, $\beta_3 = \beta_4 = 5$ and $B_1 = [0,1)$, $B_2 = [1,1]$, $B_3 = (1,2]$, $B_4 = (2,3)$. This representation is not canonical, but we nevertheless have

$$\sum_{1 \le i \le 4} \beta_i \mu(\chi_{B_i}) = 1 \cdot (1-0) + 4 \cdot (1-1) + 5 \cdot (2-1) + 5 \cdot (3-2) = 11.$$

In this sum, the last two terms can be "grouped" together to form the last term in the previous sum. The purpose of the next proposition is to encapsulate this phenomenon in general.

Proposition 3.1.5 *Let s be a simple function on X such that*

$$s = \sum_{i=1}^{p} \alpha_i \chi_{A_i} = \sum_{j=1}^{n} \beta_j \chi_{B_j},$$

where the sets A_i $(1 \leq i \leq p)$ *as well as the sets* B_j $(1 \leq j \leq n)$ *both form partitions of X, i.e.*

$$i \neq i' \Rightarrow A_i \cap A_{i'} = \varnothing, \quad j \neq j' \Rightarrow B_j \cap B_{j'} = \varnothing \tag{3.1}$$

and

$$\bigcup_{i=1}^{p} A_i = \bigcup_{j=1}^{n} B_j = X. \tag{3.2}$$

Then

$$\sum_{i=1}^{p} \alpha_i \mu(A_i) = \sum_{j=1}^{n} \beta_j \mu(B_j) = \int_X s \, d\mu. \tag{3.3}$$

Proof By (3.1), the sets $A_i \cap B_j$, $1 \leq j \leq n$, are disjoint for each fixed i and, in view of (3.2), their union is

$$\bigcup_{j=1}^{n} (A_i \cap B_j) = A_i \cap \bigcup_{j=1}^{n} B_j = A_i \cap X = A_i.$$

It follows that

$$\mu(A_i) = \sum_{j=1}^{n} \mu(A_i \cap B_j) \quad \text{for each } i,$$

and hence

$$\sum_{i=1}^{p} \alpha_i \mu(A_i) = \sum_{i=1}^{p} \sum_{j=1}^{n} \alpha_i \mu(A_i \cap B_j). \tag{3.4}$$

Similarly,

$$\mu(B_j) = \sum_{i=1}^{p} \mu(A_i \cap B_j) \quad \text{for each } j$$

and hence

$$\sum_{j=1}^{n} \beta_j \mu(B_j) = \sum_{j=1}^{n} \sum_{i=1}^{p} \beta_j \mu(A_i \cap B_j) = \sum_{i=1}^{p} \sum_{j=1}^{n} \beta_j \mu(A_i \cap B_j). \tag{3.5}$$

Consider a fixed i and any j ($1 \leq j \leq n$). If $A_i \cap B_j \neq \varnothing$, then s takes the value α_i as well as β_j on this nonempty set and therefore $\alpha_i = \beta_j$, so that

$$\alpha_i \mu(A_i \cap B_j) = \beta_j \mu(A_i \cap B_j).$$

On the other hand, if $A_i \cap B_j = \varnothing$, then the above equality still holds, because both sides are zero. Thus it holds for all i and j, which implies the first equality in (3.3) in view of (3.4) and (3.5). Since (3.1) and (3.2) are certainly fulfilled when $\sum_{1 \leq i \leq p} \alpha_i \chi_{A_i}$ is a canonical representation of s, it is immediate from Definition 3.1.3 that the second equality in (3.3) also holds. □

Suppose as in Proposition 3.1.5 that the sets A_i ($1 \leq i \leq p$) as well as the sets B_j ($1 \leq j \leq n$) form partitions of X. Then $s = \sum_{i=1}^{p} \alpha_i \chi_{A_i}$, $t = \sum_{j=1}^{n} \beta_j \chi_{B_j}$ are simple functions (not necessarily equal to each other of course). Now, $s + t$ takes the value $\alpha_i + \beta_j$ on $A_i \cap B_j$ and the sets $A_i \cap B_j$ form a partition of X, so that $s+t = \sum_{i=1}^{p} \left(\sum_{j=1}^{n} ((\alpha_i + \beta_j) \chi_{A_i \cap B_j}) \right)$. This means that the proposition above covers the foregoing representations of all the three functions s, t and $s + t$. This is how it helps overcome the difficulty arising from the lack of a conveniently expressible canonical representation of $s + t$ in terms of the respective canonical representations of s and t. [But see Problem 3.1.P8.]

We shall now prove what is called the *additivity* of the integral for simple functions.

Proposition 3.1.6 *Let s and t be simple functions on X. Then*

$$\int_X (s+t) d\mu = \int_X s \, d\mu + \int_X t \, d\mu.$$

Proof Let $s = \sum_{i=1}^{p} \alpha_i \chi_{A_i}$ and $t = \sum_{j=1}^{p} \beta_j \chi_{B_j}$, where the sets A_i ($1 \leq i \leq p$) as well as the sets B_j ($1 \leq j \leq n$) form partitions of X. Then

$$s + t = \sum_{i=1}^{p} \left(\sum_{j=1}^{n} ((\alpha_i + \beta_j) \chi_{A_i \cap B_j}) \right)$$

and the sets $A_i \cap B_j$ also form a partition of X. Therefore by Proposition 3.1.5,

$$\int_X s\,d\mu = \sum_{i=1}^{p} \alpha_i \mu(A_i) \quad \text{and} \quad \int_X t\,d\mu = \sum_{j=1}^{n} \beta_j \mu(B_j) \tag{3.6}$$

and

$$\begin{aligned}\int_X (s+t)d\mu &= \sum_{i=1}^{p}\left(\sum_{j=1}^{n}((\alpha_i+\beta_j)\mu(A_i\cap B_j))\right)\\ &= \sum_{i=1}^{p}\left(\sum_{j=1}^{n}(\alpha_i\mu(A_i\cap B_j))\right)+\sum_{j=1}^{n}\left(\sum_{i=1}^{p}(\beta_j\mu(A_i\cap B_j))\right)\\ &= \sum_{i=1}^{p}\alpha_i\left(\sum_{j=1}^{n}\mu(A_i\cap B_j)\right)+\sum_{j=1}^{n}\beta_j\left(\sum_{i=1}^{p}\mu(A_i\cap B_j)\right).\end{aligned} \tag{3.7}$$

For each i, the set A_i is the disjoint union $\cup_{1\le j\le n}(A_i\cap B_j)$ and therefore

$$\sum_{j=1}^{n}\mu(A_i\cap B_j) = \mu(A_i).$$

Similarly, for each j,

$$\sum_{i=1}^{p}\mu(A_i\cap B_j) = \mu(B_j).$$

Using these equalities in (3.7) and then appealing to (3.6), we get

$$\int_X (s+t)d\mu = \sum_{i=1}^{p}\alpha_i\mu(A_i)+\sum_{j=1}^{n}\beta_j\mu(B_j) = \int_X s\,d\mu + \int_X t\,d\mu. \qquad \square$$

The two properties proved in the next proposition will sometimes be referred to as *linearity* and *monotonicity* respectively of the integral for simple functions.

Proposition 3.1.7 *Let s and t be simple functions on X and $\alpha,\beta \geq 0$ be real numbers. Then*

$$\int_X (\alpha s+\beta t)d\mu = \alpha\int_X s\,d\mu + \beta\int_X t\,d\mu.$$

If $s \geq t$ everywhere, then $\int_X s\,d\mu \geq \int_X t\,d\mu$.

Proof The equality will follow from Proposition 3.1.6 as soon as we show that $\int_X \alpha s\,d\mu = \alpha\int_X s\,d\mu$. [*Homogeneity* of the integral for simple functions.] This equality holds when $\alpha = 0$ because then $\alpha s = 0\cdot\chi_X$ (canonically) and therefore

$\int_X \alpha s\,d\mu = 0 \cdot \mu(X) = 0$, even if $\mu(X) = \infty$, while at the same time, $\alpha \int_X s\,d\mu = 0 \cdot \int_X s\,d\mu = 0$, even if $\int_X s\,d\mu = \infty$. Now consider the case when $0 < \alpha < \infty$. Let $s = \sum_{1 \le i \le p} \alpha_i \chi_{A_i}$, where the sets A_i $(1 \le i \le p)$ form a partition of X. Then $\alpha s = \sum_{1 \le i \le p} (\alpha \alpha_i)\chi_{A_i}$ and, by Proposition 3.1.5,

$$\int_X s\,d\mu = \sum_{i=1}^{p} \alpha_i \mu(A_i), \quad \int_X \alpha s\,d\mu = \sum_{i=1}^{p} (\alpha \alpha_i)\mu(A_i).$$

Hence $\int_X \alpha s\,d\mu = \alpha \int_X s\,d\mu$, as we wished to show.

Let $s = \sum_{1 \le i \le p} \alpha_i \chi_{A_i}$ and $t = \sum_{1 \le j \le n} \beta_j \chi_{B_j}$, where the sets A_i $(1 \le i \le p)$ as well as the sets B_j $(1 \le j \le n)$ form partitions of X. Then the sets $A_i \cap B_j$ $(1 \le i \le p, 1 \le j \le n)$ form a partition of X. Moreover, s takes the value α_i on the disjoint sets $A_i \cap B_j$ $(1 \le j \le n)$, so that [see Problem 3.1.P4(b) for details]

$$s = \sum_{i=1}^{p} \left(\sum_{j=1}^{n} \alpha_i \chi_{A_i \cap B_j} \right).$$

Therefore by Proposition 3.1.5,

$$\int_X s\,d\mu = \sum_{i=1}^{p} \left(\sum_{j=1}^{n} \alpha_i \mu(A_i \cap B_j) \right). \tag{3.8}$$

A similar argument shows that

$$\int_X t\,d\mu = \sum_{j=1}^{n} \left(\sum_{i=1}^{p} \beta_j \mu(A_i \cap B_j) \right). \tag{3.9}$$

Next, we claim that

$$\alpha_i \mu(A_i \cap B_j) \ge \beta_j \mu(A_i \cap B_j) \quad \text{for all } i, j. \tag{3.10}$$

If $A_i \cap B_j = \varnothing$, then there is nothing to prove because $\mu(A_i \cap B_j) = 0$. Consider the case when $A_i \cap B_j \ne \varnothing$. Then some x belongs to the set. For this x, we have $s(x) = \alpha_i$ because $x \in A_i$ and also $t(x) = \beta_j$ because $x \in B_j$. Since $s \ge t$ everywhere, we must have $\alpha_i \ge \beta_j$. This inequality and the fact that $\mu(A_i \cap B_j) \ge 0$ together show that (3.10) holds in the present case as well, thereby completing the proof of the claim.

The inequality resulting from taking the summation in (3.10) over all i and j, when seen in the light of (3.8) and (3.9), is precisely what we sought to prove. □

Proposition 2.3.21 carries over easily to general measures [cf. Problem 2.8.P1].

Proposition 3.1.8 (Continuity) *Let $\{E_n\}_{n\geq 1}$ be a sequence of sets belonging to $\mathcal{F}$. Then*

(a) $E_1 \subseteq E_2 \subseteq \cdots$ *implies* $\mu(\lim E_n) = \lim_{n\to\infty} \mu(E_n)$ **[inner continuity]**,

(b) $E_1 \supseteq E_2 \supseteq \cdots$ *with* $\mu(E_1) < \infty$ *implies* $\mu(\lim E_n) = \lim_{n\to\infty} \mu(E_n)$ **[outer continuity]**.

Proof The argument proceeds along the same lines as in Proposition 2.3.21, with (M2) serving the purpose that Proposition 2.3.13 did. □

The following simple consequence of inner continuity will be superseded by the Monotone Convergence Theorem 3.2.4 later, but will be useful for establishing it.

Proposition 3.1.9 *Let $\{A_n\}_{n\geq 1}$ be a sequence of measurable subsets of X such that $A_1 \subseteq A_2 \subseteq \cdots$, and $\cup_{n\geq 1} A_n = X$. Then, for any simple function s,*

$$\lim_{n\to\infty} \int_X (s\chi_{A_n}) d\mu = \int_X s\, d\mu.$$

Proof Any simple function is a finite sum of functions of the form $\alpha\chi_A$. Therefore by the linearity of the integral for simple functions [Proposition 3.1.7], it is sufficient to prove for any nonnegative real number α that

$$\lim_{n\to\infty} \int_X (\alpha\, \chi_A \chi_{A_n}) d\mu = \int_X (\alpha\, \chi_A) d\mu.$$

Since $\chi_A \chi_{A_n} = \chi_{A\cap A_n}$, the integrals on the left- and right-hand sides here are $\alpha\mu(A\cap A_n)$ and $\alpha\mu(A)$ respectively [see Example 3.1.4c]. Therefore we need only argue that $\lim_{n\to\infty} \mu(A_n \cap A) = \mu(A)$. This equality follows from the inner continuity property of measures. □

Problem Set 3.1

3.1.P1. Let $X = [0,4]$, $s = 2\chi_{[0,2]} + 3\chi_{(2,4]}$ and $t = 6\chi_{(1,3]}$. Find $\int_X s\, dm$ and $\int_X t\, dm$. Also, find the canonical form of $s + t$ and use it to compute $\int_X (s+t)\, dm$ from the definition.

3.1.P2. Let $X = (0,4]$ and $A_n = \left[\frac{1}{n}, 4\right]$, so that $A_1 \subseteq A_2 \subseteq \cdots$ and $\cup_{n\geq 1} A_n = X$. For $s = \chi_{(0,2]}$, find $\int_X (s\chi_{A_n}) dm$, $\int_X s\, dm$. and $\lim_{n\to\infty} \int_X (s\chi_{A_n}) dm$.

3.1.P3. Let X be a measurable subset of $\mathbb{R}$ and $\{A_n\}_{n\geq 1}$ be a sequence of measurable subsets such that $A_1 \subseteq A_2 \subseteq \cdots$. If $X \backslash \cup_{n\geq 1} A_n$ has positive measure, show that there exists a simple function s such that $\lim_{n\to\infty} \int_X (s\chi_{A_n}) d\mu \neq \int_X s\, d\mu$.

3.1.P4. Let X be a measurable set, A_i $(1 \leq i \leq p)$ disjoint measurable subsets of it and α_i $(1 \leq i \leq p)$ nonnegative real numbers.

(a) If the sets A_i do *not* form a partition of X, show that there are infinitely many simple functions taking the respective values α_i on A_i.

(b) If the sets A_i form a partition of X, show that there is a unique simple function s taking the respective values α_i on A_i and that it is given by $s = \sum_{1 \le i \le p} \alpha_i \chi_{A_i}$.

3.1.P5. Is Proposition 3.1.5 valid without the hypothesis that the A_i and the B_i form partitions of X?

3.1.P6. [Needed in 3.2.P13] Let s be a simple function on X. Define ϕ_s: $[0,\infty) \to [0,\infty]$ as $\phi_s(u) = \mu(X(s > u))$. If $\mu(X) < \infty$, then ϕ_s takes values in $[0,\infty)$. Show that

(a) $\phi_s(u) = 0$ if $u \ge M = \max\{s(x) : x \in X\}$.
(b) ϕ_s is a step function.
(c) The Riemann integral $\int_0^M \phi_s(u)du$, which is equal to the improper integral $\int_0^\infty \phi_s(u)du$ in view of (a), is also equal to the measure space integral $\int_X s\,d\mu$.
(d) What happens if $\mu(X) = \infty$?

3.1.P7. Let $\{s_n\}_{n \ge 1}$ be the sequence of simple functions given by $s_n(x) = 1/n$ for $|x| \le n$ and 0 for $|x| > n$. Show that $s_n \to 0$ uniformly on $X = \mathbb{R}$, but $\int_X s_n\,dm = 2$ for every $n \in \mathbb{N}$. [Uniform convergence on a set X of finite measure to a bounded limit function does imply convergence of integrals; this will be seen to be true in the next section even for more general functions in the light of the Dominated Convergence Theorem 3.2.16.]

3.1.P8. *Optional.* [The principle that permits grouping of terms in a sum according to any scheme whatsoever can be expressed in this manner: Suppose we have a sum $\sum_{1 \le i \le p} b_i$ to evaluate and that the set $\{i : 1 \le i \le p\}$ of indices has been partitioned into n subsets N_j, $1 \le j \le n$, meaning thereby that these subsets are nonempty and disjoint with union equal to $\{i : 1 \le i \le p\}$. Then the required sum can be evaluated as the grand total of the n subtotals $\sum_{i \in N_j} b_i$. In other words,

$$\sum_{1 \le i \le p} b_i = \sum_{1 \le j \le n} \left(\sum_{i \in N_j} b_i \right)$$

provided $\bigcup_{j=1}^{n} N_j = \{1, 2, \ldots, p\}$, every $N_j \ne \varnothing$ and $j \ne j' \Rightarrow N_j \cap N_{j'} = \varnothing$.

This is valid even if one or more of the b_i are ∞ as long as each b_i is nonnegative. We shall refer to it as the *grouping principle*.]

Let s be a simple function on a measurable set X such that

$$s = \sum_{i=1}^{p} \alpha_i \chi_{A_i},$$

where

$$i \ne i' \Rightarrow A_i \cap A_{i'} = \varnothing \quad \text{and each } A_i \text{ is measurable}$$

and

$$\bigcup_{i=1}^{p} A_i = X.$$

Using 2.5.P6 and the *grouping principle*, but not Proposition 3.1.5, show that

$$\sum_{i=1}^{p} \alpha_i \mu(A_i) = \int_X s\, d\mu. \tag{3.11}$$

3.1.P9. If in a measure space, $\{A_n\}_{n\geq 1}$ is a sequence of sets of measure 0 and $\{B_n\}_{n\geq 1}$ is a descending sequence of sets of finite measure such that their measure tends to 0, show that $\cap_{n\geq 1}(A_n \cup B_n)$ has measure 0.

3.1.P10. Let $\mathcal{F}$ be a σ-algebra of subsets of a set X and $f : X \to \mathbb{R}$ be any function whatsoever. Show that the family $\mathcal{G} = \{A \subseteq \mathbb{R} : f^{-1}(A) \in \mathcal{F}\}$ of subsets of $\mathbb{R}$ is a σ-algebra. Hence show that if f is $\mathcal{F}$-measurable, then $f^{-1}(A) \in \mathcal{F}$ whenever $A \subseteq \mathbb{R}$ is any Borel subset of $\mathbb{R}$.

3.1.P11. Suppose $f : \mathbb{R} \to \mathbb{R}$ and $g : \mathbb{R} \times \mathbb{R} \to \mathbb{R}$ are both Borel measurable. Show that the composition $f \circ g : \mathbb{R} \times \mathbb{R} \to \mathbb{R}$ is Borel measurable.

3.1.P12. Let μ be a measure on the Borel σ-algebra $\mathfrak{B}$ of subsets of $\mathbb{R}$, satisfying the following properties:

(i) $\mu(0,1) > 0$;
(ii) $\mu(0,1) < \infty$;
(iii) for $E \in \mathfrak{B}$ and $\varepsilon > 0$, there exists an open set $O \supseteq E$ such that $\mu(O \backslash E) < \varepsilon$;
(iv) $\mu(E + x) = \mu(E)$ for an arbitrary $E \in \mathfrak{B}$ and arbitrary $x \in \mathbb{R}$ (translation invariance).

Show that there exists a positive $\alpha \in \mathbb{R}$ such that $\mu(E) = \alpha \cdot m(E)$ for all $E \in \mathfrak{B}$.

3.2 Integrals of Measurable Functions

It is an immediate consequence of the monotonicity of the integral for simple functions [Proposition 3.1.7] that

$$\int_X s\, d\mu = \sup\left\{ \int_X t\, d\mu : 0 \leq t \leq s \quad \text{and} \quad t : X \to \mathbb{R} \text{ a simple function} \right\}.$$

The right-hand side here remains meaningful if we replace s by an arbitrary nonnegative extended real-valued function f. We take advantage of this to extend the concept of integral to a broader class of nonnegative functions, but restrict ourselves to measurable ones (which means that $X(f > \alpha) \in \mathcal{F}$ for every real α); this is because we shall need the restriction for most of our proofs in the sequel:

Definition 3.2.1 For any measurable extended nonnegative real-valued function f on X, we define the **integral of** f to be

$$\int f = \sup\left\{ \int_X t\,d\mu : 0 \le t \le f \quad \text{and} \quad t : X \to \mathbb{R} \text{ a simple function} \right\}.$$

It is also denoted by $\int_X f$, $\int_X f\,d\mu$ or $\int f\,d\mu$ according to convenience.

The above notation for the integral can become inconvenient if we have an expression, such as $f(x) = x^2, 0 \le x < 1$, available for the function but do not wish to introduce a letter f, or the like, to represent it. When this is so, we can use the notation $\int x^2 dm(x)$ or $\int_{[0,1]} x^2 dx$. The symbol $\int_0^1 x^2 dx$ will, as far as possible, be reserved for the Riemann integral.

Remark 3.2.2 The set over which the sup is taken in the definition is nonempty, because the function t which has value 0 everywhere is simple and satisfies $0 \le t \le f$. Therefore the sup is always nonnegative. Note that existence of the sup poses no difficulties, because we allow ∞ as a value. In this connection, we note that, if $\int_X f\,d\mu < \infty$, then $X(f = \infty)$ has measure zero, i.e. f has finite values almost everywhere. Indeed, if the aforementioned set, call it A, were to have positive measure, then for every $n \in \mathbb{N}$, the simple function $t = n\chi_A$ would satisfy $0 \le t \le f$, and hence by the above definition, we would have $\infty > \int_X f\,d\mu \ge \int_X t\,d\mu = n \cdot \mu(A)$ for every $n \in \mathbb{N}$, a contradiction.

It is easy to prove *homogeneity* and *monotonicity* of the integral for nonnegative extended real-valued functions.

Proposition 3.2.3 *Let f and g be measurable extended nonnegative real-valued functions on X and $\alpha \ge 0$ be a real number (not ∞). Then*

$$\int_X (\alpha f)d\mu = \alpha \int_X f\,d\mu.$$

If $f \ge g$ everywhere, then $\int_X f\,d\mu \ge \int_X g\,d\mu$.

Proof If $\alpha = 0$, both sides of the equality to be proved are 0. Suppose $0 < \alpha < \infty$. Then t is a simple function satisfying $0 \le t \le f$ if and only if αt is a simple function satisfying $0 \le \alpha t \le \alpha f$. Also, $\int_X (\alpha t)\,d\mu = \alpha \int_X t\,d\mu$ by Proposition 3.1.7. Together with the Definition 3.2.1, this implies the required equality when $0 < \alpha < \infty$.

The second part (i.e. monotonicity) follows from the observations that firstly, when $f \ge g$, any simple function t satisfying $0 \le t \le g$ also satisfies $0 \le t \le f$, and that secondly, $A \subseteq B \Rightarrow \sup A \le \sup B$. □

We are about to prove one of the fundamental results about Lebesgue integration. One straightforward consequence of it is that the infirmity in handling limiting processes with Riemann integration that was pointed out in Example 2.1.1 disappears in the present theory.

Before stating the result, we point out that it has an alternative version in which the inequalities and convergence in (a) and (b) of the hypothesis are assumed to hold only almost everywhere; however, that complicates the precise statement of the conclusion because the limit function need not be measurable. These matters are relegated to Problem 3.2.P30.

Theorem 3.2.4 (Monotone Convergence Theorem). *Let $\{f_n\}_{n\geq 1}$ be a sequence of measurable nonnegative extended real-valued functions on X such that, for every x belonging to the set,*

(a) $0 \leq f_1(x) \leq f_2(x) \leq \cdots$,
(b) $f_n(x) \to f(x)$ *as* $n \to \infty$.

Then f is measurable and

$$\lim_{n\to\infty} \int_X f_n d\mu = \int_X f\, d\mu.$$

Proof From (a) and (b), we have $f_n(x) \leq f(x)$ everywhere on X. The measurability of f is a consequence of Corollary 2.5.5. In view of monotonicity [Proposition 3.2.3], we have $\int_X f_n d\mu \leq \int_X f_{n+1} d\mu \leq \int_X f\, d\mu$ for all n. Therefore $\lim_{n\to\infty} \int_X f_n\, d\mu$ exists as an extended real number and

$$\lim_{n\to\infty} \int_X f_n d\mu \leq \int_X f\, d\mu. \tag{3.12}$$

To arrive at the reverse inequality, consider any simple function s such that $0 \leq s \leq f$. For arbitrary $\beta \in (0,1)$, let

$$A_n = X(f_n \geq \beta s) \quad \text{for all}\, n \in \mathbb{N}. \tag{3.13}$$

Then each A_n is measurable. Moreover, our hypothesis (a) implies that

$$A_1 \subseteq A_2 \subseteq \cdots. \tag{3.14}$$

It is also true that $x \in X \Rightarrow x \in \cup_{n\geq 1} A_n$; indeed, for those $x \in X$ for which $s(x) = 0$, the implication follows from (a) and for those x for which $s(x) > 0$, it follows from (b) and the fact that $\beta < 1$ (because $\beta s(x) < s(x) \leq f(x)$). Thus

$$\bigcup_{n=1}^{\infty} A_n = X. \tag{3.15}$$

From the definition (3.13) of A_n, it follows that $f_n\chi_{A_n} \geq (\beta s)\chi_{A_n}$ everywhere on X. But $f_n \geq f_n\chi_{A_n}$ and therefore $f_n \geq (\beta s)\chi_{A_n}$. By monotonicity, we have

$$\int_X f_n d\mu \geq \int_X ((\beta s)\chi_{A_n})d\mu. \tag{3.16}$$

Since βs is a simple function, we conclude from (3.14), (3.15) and Proposition 3.1.9 that the right-hand side of (3.16) has limit $\int_X (\beta s)d\mu$ as $n \to \infty$. We have already noted that the left-hand side has a limit as $n \to \infty$. Therefore, we have $\lim_{n\to\infty} \int_X f_n\, d\mu \geq \int_X (\beta s)d\mu = \beta \cdot \int_X s\, d\mu$. Since this is true for all $\beta \in (0,1)$, it follows that

$$\lim_{n\to\infty} \int_X f_n d\mu \geq \int_X s\, d\mu.$$

But this has been established for an arbitrary simple function s such that $0 \leq s \leq f$. So,

$$\lim_{n\to\infty} \int_X f_n d\mu \geq \int_X f\, d\mu.$$

In conjunction with (3.12), this proves the required equality. □

Remark If the hypothesis (a) above is replaced by $f_1 \geq f_2 \geq \cdots \geq 0$, then the conclusion need not hold. To see why, take $f_n = 1/n$ on $X = \mathbb{R}$. Then $f = 0$ on $\mathbb{R}$ and $\int_X f\, d\mu = 0$, but $\int_X f_n\, d\mu = \infty$ for every n. If we add the hypothesis that $\int_X f_1 < \infty$, then the conclusion does indeed hold but the proof requires some results yet to come. A full discussion is therefore postponed to Remark 3.2.11(d).

Theorem 3.2.5 *Let f and g be measurable nonnegative extended real-valued functions on X and $\alpha, \beta \geq 0$ be real. Then*

$$\int_X (\alpha f + \beta g)d\mu = \alpha \int_X f\, d\mu + \beta \int_X g\, d\mu.$$

Proof As homogeneity has already been proved in Proposition 3.2.3, it remains to prove only additivity, i.e.,

$$\int_X (f+g)d\mu = \int_X f\, d\mu + \int_X g\, d\mu.$$

By Proposition 2.5.9, there exist increasing sequences $\{s_n\}_{n\geq 1}$ and $\{t_n\}_{n\geq 1}$ of simple functions with respective limits f and g. It follows that the sequence

$\{s_n + t_n\}_{n\geq 1}$, which has limit $f + g$, is also an increasing sequence of measurable functions. Now,

$$\int_X (s_n + t_n)d\mu = \int_X s_n d\mu + \int_X t_n d\mu.$$

Upon taking limits in this equality and appealing to the Monotone Convergence Theorem 3.2.4, we get additivity. □

Corollary 3.2.6 *Let $\{E_k\}_{k\geq 1}$ be a sequence of disjoint measurable subsets of X such that $\cup_{k\geq 1}E_k = X$, and f be a measurable nonnegative extended real-valued function on X. Then*

$$\int_X f\, d\mu = \sum_{k=1}^{\infty} \int_X (f\chi_{E_k})d\mu.$$

Proof Let $f_n = \sum_{k=1}^{n} (f\chi_{E_k})$. Since the measurable sets of the sequence $\{E_k\}_{k\geq 1}$ are disjoint and $\cup_{k\geq 1}E_k = X$, the sequence $\{f_n\}_{n\geq 1}$ consists of measurable functions satisfying (a) and (b) of the Monotone Convergence Theorem 3.2.4. Therefore $\int_X f\, d\mu = \lim_{n\to\infty} \int_X f_n\, d\mu$. On the other hand, Theorem 3.2.5 implies that $\int_X f_n\, d\mu = \sum_{k=1}^{n} \int_X (f\chi_{E_k})d\mu$. The required conclusion is now immediate. □

Example 3.2.7 We illustrate the use of this Corollary. Let $f : (0, 1] \to \mathbb{R}$ be defined by $f(x) = [-\log x]$, where [ ] denotes the integer part function and log means logarithm to base 10, and $f(0) = 0$. Then f is a step function on $[\varepsilon,1]$ whenever $0 < \varepsilon < 1$. In fact,

$$f(x) = k \quad \text{for } 10^{-k-1} < x \leq 10^{-k}, \quad \text{where } k+1 \in \mathbb{N}.$$

It follows that $\int_{(0,1]} f\chi_{E_k} = k(10^{-k} - 10^{-k-1}) = 9\ k/10^{\ k+1}$, where $E_k = (10^{-k-1},10^{-k}]$. Now $f \geq 0$, the measurable sets E_k are disjoint and $\cup_{k\geq 0}E_k = (0,1]$. Therefore Corollary 3.2.6 yields $\int_{(0,1]} f = \sum_{k=1}^{\infty} 9k/10^{k+1}$, a convergent series by comparison [see Proposition 1.4.7] with the geometric series $\sum_{k=1}^{\infty} 1/(\sqrt{10})^{k+2}$, considering that $k/10^{\ k+1} = (k/(\sqrt{10})^k)(1/(\sqrt{10})^{k+2})$.

Theorem 3.2.8 (Fatou's Lemma) *Let $\{f_n\}_{n\geq 1}$ be a sequence of measurable nonnegative extended real-valued functions on X. Then*

$$\int_X (\liminf f_n)d\mu \leq \liminf \int_X f_n\, d\mu.$$

Proof Let $\{g_n\}_{n\geq 1}$ be the sequence of functions on X defined by

$$g_n(x) = \inf_{k \geq n} f_k(x) \text{ for all } x \in X \text{ and all } n \in \mathbb{N}.$$

Then $g_n \leq f_n$ everywhere on X, so that

$$\int_X g_n \, d\mu \leq \int_X f_n \, d\mu. \tag{3.17}$$

Also, $\{g_n\}_{n \geq 1}$ is an increasing sequence of measurable nonnegative extended real-valued functions such that, by definition of liminf [see Remark 2.5.3], $\lim_{n\to\infty} g_n \to \liminf f_n$ everywhere. Therefore by the Monotone Convergence Theorem 3.2.4, we get

$$\int_X (\liminf f_n) \, d\mu = \lim_{n\to\infty} \int_X g_n \, d\mu = \liminf \left(\int_X g_n \, d\mu \right)$$
$$\leq \liminf \int_X f_n \, d\mu \quad \text{by (3.17) and Remark 2.5.3.} \qquad \square$$

For an example when strict inequality holds in the above Theorem, see Problem 3.2.P2.

So far we have dealt with nonnegative functions only. We now extend the concept of Lebesgue integral to measurable functions that take positive as well as negative values.

Definition 3.2.9 For any measurable extended real-valued function f on X, the **integral of** f is defined to be

$$\int f = \int f^+ - \int f^-,$$

provided that at least one among $\int f^+$ and $\int f^-$ is finite. If both happen to be ∞, then $\int f$ is taken as undefined. *If* $\int f < \infty$, the function is said to be **integrable**. The integral is also denoted by $\int_X f$, $\int_X f \, d\mu$ or $\int f \, d\mu$ according to convenience.

Proposition 3.2.10

(a) *Let A be a measurable subset of X such that* $\mu(A^c) = 0$. *If* $\int_X f \, d\mu$ *exists, then* $\int_X f \chi_A \, d\mu$ *exists and equals* $\int_X f \, d\mu$.
(b) *Let f and g be measurable extended real-valued functions on X satisfying* $f = g$ *a.e. Then one among f and g is integrable if and only if the other is, in which case* $\int_X f \, d\mu = \int_X g \, d\mu$.

Proof

(a) If B is any measurable subset, then so is $B \cap A$; moreover, $\mu(B \cap A^c) \leq \mu(A^c) = 0$. Therefore $\mu(B \cap A^c) = 0$ and hence

$$\mu(B) = \mu(B \cap (A \cup A^c)) = \mu(B \cap A) + \mu(B \cap A^c) = \mu(B \cap A).$$

For any simple function $s = \sum_{j=1}^{n} \alpha_j \chi_{A_j}$, we have $s\chi_A = \sum_{j=1}^{n} \alpha_j \chi_{A_j \cap A}$ and therefore, by the equality proved above,

$$\int_X (s\chi_A) d\mu = \sum_{j=1}^{n} \alpha_j \mu(A_j \cap A) = \sum_{j=1}^{n} \alpha_j \mu(A_j) = \int_X s\, d\mu.$$

Thus the result holds for simple functions. Consider any measurable nonnegative function f. For any simple function such that $0 \leq s \leq f$, the function $s\chi_A$ is simple and satisfies $0 \leq s\chi_A \leq f\chi_A$. Therefore $\int_X s\, d\mu = \int_X (s\chi_A) d\mu \leq \int_X f\chi_A d\mu$. This holds for every simple function s such that $0 \leq s \leq f$. Consequently, $\int_X f\, d\mu \leq \int_X f\chi_A\, d\mu$. But the reverse inequality also holds, because $f\chi_A \leq f$. This proves the required equality for measurable nonnegative extended real-valued functions.

The equality now follows for measurable extended real-valued functions from the fact that $(f\chi_A)^+ = f^+\chi_A \leq f^+$ and $(f\chi_A)^- = f^-\chi_A \leq f^-$ [see 2.5.P9].

(b) Consider measurable extended real-valued functions f and g on X satisfying $f = g$ a.e. Accordingly, let A be a measurable subset of X such that $\mu(A^c) = 0$ and $f = g$ on A. It is trivial that $f^+ = g^+$ and $f^- = g^-$ on A, so that $f^+\chi_A = g^+\chi_A$ and $f^-\chi_A = g^-\chi_A$ on X. By part (a), we have $\int_X f^+ d\mu = \int_X f^+ \chi_A d\mu = \int_X g^+ \chi_A d\mu = \int_X g^+ d\mu$ and correspondingly for f^- and g^-. Together with Definition 3.2.9, this leads to the conclusion that one among f and g is integrable if and only if the other is, in which case $\int_X f\, d\mu = \int_X g\, d\mu$. □

Remarks 3.2.11

(a) If f is nonnegative-valued, then $f = f^+$ and $\int f$ as defined now is the same as what it would be under Definition. 3.2.1. In particular, it is still true that $f \geq 0 \Rightarrow \int_X f\, d\mu \geq 0$.

(b) If an extended real-valued function f defined on X is integrable, then each of the sets $X(f = \infty)$ and $X(f = -\infty)$ has measure zero [see Remark 3.2.2]. In other words, an integrable function is finite-valued almost everywhere.

(c) Suppose f is a measurable extended real-valued function that is defined a.e. This means there exists some measurable $Y \subseteq X$ such that f is defined and measurable on Y, while $\mu(X \backslash Y) = 0$. Then there are several measurable extended real-valued functions F on X that agree with f on Y, for example, the one obtained by extending f outside Y to be ∞ on $X \backslash Y$. Clearly, the positive and negative parts of all such functions F agree on Y and thus agree a.e. It follows from Proposition 3.2.10(b) that one such F is integrable if and only if all of them are, in which case, they all have the same integral. This integral is what we shall understand by $\int_X f\, d\mu$ when f is defined a.e. on X.

(d) We resume the discussion about replacing the hypothesis (a) of the Monotone Convergence Theorem 3.2.4 by $f_1 \geq f_2 \geq \cdots \geq 0$ and adding the hypothesis that $\int_X f_1 < \infty$. We aim to show that the conclusion of the theorem still holds. The additional hypothesis and Remark 3.2.2 together show that the function f_1 is finite a.e. and accordingly, there exists a measurable subset A of X such that $\mu(A^c) = 0$ and f_1 is finite on A. From the hypothesis that $f_1 \geq f_2 \geq \cdots \geq 0$, it follows that all the functions f_n and the limit f are finite on A. Upon multiplying all functions concerned by χ_A, we obtain functions that not only satisfy the same conditions but are also finite everywhere on X. Moreover, we find from Proposition 3.2.10(a) that their respective integrals are the same as before. Therefore we may assume that the functions f_n and the limit f are all finite-valued on X, which permits subtractions like $f_1 - f_n$ and $f_1 - f$.
Now, the sequence of functions $f_1 - f_n$ satisfies the hypotheses of the Monotone Convergence Theorem 3.2.4 and has limit $f_1 - f$. Hence $\lim_{n\to\infty} \int_X (f_1 - f_n)d\mu = \int_X (f_1 - f)d\mu \leq \int_X f_1\, d\mu < \infty$. Since $(f_1 - f_n) + f_n = f_1 = (f_1 - f) + f$, it follows by Theorem 3.2.5 that

$$\int_X (f_1 - f_n)\, d\mu + \int_X f_n\, d\mu = \int_X f_1\, d\mu = \int_X (f_1 - f)\, d\mu + \int_X f\, d\mu.$$

Since $\lim_{n\to\infty} \int_X f_n\, d\mu$ exists and, as already noted,

$$\lim_{n\to\infty} \int_X (f_1 - f_n)d\mu = \int_X (f_1 - f)\, d\mu,$$

it follows from the above equality that

$$\int_X (f_1 - f)\, d\mu + \lim_{n\to\infty} \int_X f_n\, d\mu = \int_X (f_1 - f)\, d\mu + \int_X f\, d\mu.$$

All four terms involved in this equality are finite and hence the required conclusion follows.

Proposition 3.2.12
(a) *The function f is integrable if and only if its absolute value $|f|$ is integrable.*
(b) *If $\int_X f\, d\mu$ exists (perhaps $\pm\infty$), then $\left|\int_X f\, d\mu\right| \leq \int_X |f|\, d\mu$.*

Proof
(a) In accordance with the above definition, for f to be integrable, it is necessary and sufficient that $\int_X f^+\, d\mu$ and $\int_X f^-\, d\mu$ both be finite. This is the case if and

only if $\int_X |f|\,d\mu < \infty$, because $|f| = f^+ + f^-$, which implies $\int_X |f|\,d\mu = \int_X f^+\,d\mu + \int_X f^-\,d\mu$ by Theorem 3.2.5.

(b) We need consider only the case when $\int_X |f|\,d\mu < \infty$. Since $|f| = f^+ + f^-$, we have $\int_X |f|\,d\mu = \int_X f^+\,d\mu + \int_X f^-\,d\mu$ by Theorem 3.2.5. Hence the finiteness of $\int_X |f|\,d\mu$ implies the finiteness of both $\int_X f^+\,d\mu$ and $\int_X f^-\,d\mu$. Applying the above definition and using the finiteness just proved, we have

$$\left|\int_X f\,d\mu\right| = \left|\int_X f^+\,d\mu - \int_X f^-\,d\mu\right| \le \left|\int_X f^+\,d\mu\right| + \left|\int_X f^-\,d\mu\right| = \int_X f^+\,d\mu + \int_X f^-\,d\mu$$
$$= \int_X |f|\,d\mu.$$ □

As the reader can easily verify, $(f + g)^+$ is not the same as $f^+ + g^+$. Therefore we make a slight detour to establish the additivity of the integral on the basis the above definition.

Proposition 3.2.13 *Suppose $f = g - h$, where f, g, h are extended real-valued functions on X, and g, h are nonnegative. Then $f^+ \le g$ and $f^- \le h$.* [No measurability involved.]

Proof It is sufficient to prove the first inequality, because $f = (-f)^+$ and $-f = -g + h$. Let $x \in X$. If either $g(x) = \infty$ or $f^+(x) = 0$, there is nothing to prove. So, let $g(x) < \infty$ and $f^+(x) > 0$; since $f^+(x) = \max\{f(x),0\}$, we have $f(x) > 0$, and since $f^-(x) = \max\{-f(x),0\}$, we further have $f^-(x) = 0$. Therefore $f^+(x) = f^+(x) - f^-(x) = f(x) = g(x) - h(x)$. Since $h(x) \ge 0$, it follows from here that $f^+(x) \le g(x)$. □

Proposition 3.2.14 *Let $f = g - h$, where f, g, h are measurable extended real-valued functions on X and g, h are nonnegative. Then $\int_X g\,d\mu < \infty \Rightarrow \int_X f^+\,d\mu < \infty$ and $\int_X h\,d\mu < \infty \Rightarrow \int_X f^-\,d\mu < \infty$. If at least one among $\int_X g\,d\mu$ and $\int_X h\,d\mu$ is finite, then the same is true of $\int_X f^+\,d\mu$ and $\int_X f^-\,d\mu$, and furthermore, $\int_X f\,d\mu = \int_X g\,d\mu - \int_X h\,d\mu$.*

Proof The two implications are immediate from Proposition 3.2.13 and they lead directly to the first part of the last statement. Only the equality needs an argument.

To begin with, observe that the equality $f^+ - f^- = f = g - h$ leads to $f^+ + h = f^- + g$ even if not all the numbers involved are finite. Suppose $\int_X g\,d\mu < \infty$. Then $\int_X f^+\,d\mu < \infty$ and

$$\int_X f^+\,d\mu + \int_X h\,d\mu = \int_X f^-\,d\mu + \int_X g\,d\mu.$$

Since $\int_X g\,d\mu$ and $\int_X f^+\,d\mu$ are both finite, we deduce from here that

$$\int_X f^+ d\mu - \int_X f^- d\mu = \int_X g\, d\mu - \int_X h\, d\mu.$$

In view of Definition 3.2.9, this is precisely the equality we wished to prove. Similar considerations apply if $\int_X g\, d\mu < \infty$ instead. □

Theorem 3.2.15 *Let f and g be integrable extended real-valued functions on X and* α, β *be real. Then* $\alpha f + \beta g$ *is integrable and*

$$\int_X (\alpha f + \beta g) d\mu = \alpha \int_X f\, d\mu + \beta \int_X g\, d\mu.$$

If $f \geq g$ *everywhere, then* $\int_X f\, d\mu \geq \int_X g\, d\mu$.

Proof We first prove the result only for finite-valued functions.

As observed in Remark 2.5.3(b), $\alpha f + \beta g$ is measurable. By Proposition 3.2.3 and Theorem 3.2.5,

$$\int_X |\alpha f + \beta g|\, d\mu \leq \int_X (|\alpha||f| + |\beta||g|)\, d\mu = |\alpha| \int_X |f|\, d\mu + |\beta| \int_X |g|\, d\mu < \infty.$$

Thus $\alpha f + \beta g$ is integrable by Proposition 3.2.12.

To establish the equality (called *linearity of the Lebesgue integral*) it is sufficient to prove *additivity*:

$$\int_X (f + g) d\mu = \int_X f\, d\mu + \int_X g\, d\mu$$

and *homogeneity*:

$$\int_X (\alpha f) d\mu = \alpha \int_X f\, d\mu.$$

As f and g are both integrable and real-valued, so are f^+, f^-, g^+, g^- as well as $f^+ + g^+$ and $f^- + g^-$. Since all functions involved are real-valued, we have $f + g = (f^+ + g^+) - (f^- + g^-)$; besides, both the functions $f^+ + g^+$ and $f^- + g^-$ are nonnegative. Therefore it follows by Proposition 3.2.14 that

$$\int_X (f + g) d\mu = \int_X (f^+ + g^+) d\mu - \int_X (f^- + g^-) d\mu.$$

Since all integrals here are finite, we can rearrange terms on the right-hand side to get

$$\int_X (f+g)d\mu = \int_X (f^+ - f^-)d\mu + \int_X (g^+ - g^-)d\mu = \int_X f\,d\mu + \int_X g\,d\mu.$$

This proves additivity for real-valued Lebesgue integrable functions f and g. We next consider the case when the functions involved are extended real-valued.

Let $Y \subseteq X$ be the set of points where neither of the functions is $\pm\infty$, i.e., both are real-valued. By Remark 3.2.11(b), $\mu(X\backslash Y) = 0$ since the functions are integrable. Observe that $f + g$ is defined and real-valued on Y. Thus $f + g$ is defined a.e. By Remark 3.2.11(c), the desired conclusion about $f + g$ will follow if we can prove the existence of some integrable function F on X that agrees with $f + g$ on Y and satisfies $\int_X F\,d\mu = \int_X f\,d\mu + \int_X g\,d\mu$.

Now, $f\chi_Y$ and $g\chi_Y$ are real-valued on all of X and are measurable by the second assertion of Remark 2.5.3(b). Also, $f = f\chi_Y$ and $g = g\chi_Y$ on Y, and hence $F = (f + g)\chi_Y$ agrees with $f + g$ on Y. Now, it follows from Proposition 3.2.10(a) that $\int_X f\,d\mu = \int_X f\chi_Y\,d\mu$ and $\int_X g\,d\mu = \int_X g\chi_Y\,d\mu$. Hence, from the additivity proved above in the real-valued case, we find that $(f + g)\chi_Y$ is integrable and $\int_X (f+g)\chi_Y\,d\mu = \int_X f\,d\mu + \int_X g\,d\mu$. Therefore $F = (f + g)\chi_Y$ is a function of the kind we sought to prove the existence of.

We next prove homogeneity.

For $\alpha \geq 0$, the required equality is an immediate consequence of Definition 3.2.9 and the homogeneity of the integral for nonnegative functions [Proposition 3.2.3]. The case of negative α now follows by using the relations $f^- = (-f)^+$ and $f^+ = (-f)^-$.

For monotonicity, we use the consequence $\int_X (f-g)\,d\mu = \int_X f\,d\mu - \int_X g\,d\mu$ of the linearity that has just been proved and argue that

$$f \geq g \Rightarrow f - g \geq 0 \Rightarrow \int_X (f-g)d\mu \geq 0 \Rightarrow \int_X f\,d\mu \geq \int_X g\,d\mu. \qquad \square$$

We close this section with the second of the two fundamental convergence theorems that highlight the advantages of the measure space integral over the Riemann integral.

Before stating the result, we point out that it has an alternative version in which the inequality and convergence in the hypothesis are assumed to hold only almost everywhere; however, that complicates the precise statement of the conclusion because the limit function need not be measurable. These matters are relegated to Problem 3.2.P30.

Theorem 3.2.16 (Dominated Convergence Theorem) *Let $\{f_n\}_{n\geq 1}$ be a sequence of measurable extended real-valued functions on X, converging everywhere to a function f and let g be an integrable function on X such that*

$$|f_n(x)| \leq g(x) \quad \textit{for every } x \in X.$$

Then f_n as well as f are integrable,

$$\lim_{n\to\infty}\int_X |f_n - f|d\mu = 0 \quad \textit{and} \quad \lim_{n\to\infty}\int_X f_n\, d\mu = \int_X f\, d\mu.$$

Proof Since $|f_n| \le g$ everywhere, it follows by Theorem 3.2.15 and Proposition 3.2.12(a) that all the f_n are integrable. Since the hypotheses imply that $|f| \le g$ everywhere it also follows that f is integrable.

The hypotheses also imply that $\liminf\,(2g - |f_n - f|) = 2g$ and that $|f_n - f| \le 2g$ so that the functions $2g - |f_n - f|$ are nonnegative. Fatou's Lemma 3.2.8 can therefore be applied to obtain via Theorem 3.2.15 that

$$\begin{aligned}\int_X (2g)d\mu &\le \liminf \int_X (2g - |f_n - f|)\, d\mu \\ &\le \int_X (2g)\, d\mu + \liminf\left(-\int_X |f_n - f| d\mu\right) \\ &= \int_X (2g)d\mu - \limsup \int_X |f_n - f|\, d\mu.\end{aligned}$$

Since $\int_X (2g)\, d\mu < \infty$, we further obtain from here that

$$\limsup \int_X |f_n - f| d\mu \le 0.$$

This yields the first of the two equalities in question. We get the second equality from the first upon applying Proposition 3.2.12(a) to $f_n - f$ and using Theorem 3.2.15 yet again. □

Remark The preceding theorem and the Monotone Convergence Theorem 3.2.4 both say that if $f_n \to f$, then under an additional hypothesis, $\int_X f_n\, d\mu \to \int_X f\, d\mu$. The need for an additional hypothesis is underscored by the following example: Let $X = (0,1]$ and $f_n = n\chi_{(0,1/n]}$. Then each f_n is a simple function with

$$\int_X f_n\, dm = n\left(\frac{1}{n} - 0\right) = 1 \quad \text{for every } n \in \mathbb{N}.$$

Also, $f_n \to f$, where f is zero everywhere. Indeed, for $0 < x \le 1$, there exists an $n_0 \in \mathbb{N}$ such that $f_n(x) = 0$ for $n \ge n_0$. Now, $\int_X f\, dm = 0$ but $\lim_{n\to\infty} \int_X f_n\, dm = 1$.

Example 3.2.17 Let $g:(0,1] \to \mathbb{R}$ be defined by

$$g(x) = \begin{cases} 0 & \text{if } x \text{ is rational} \\ [-\log x] & \text{otherwise,} \end{cases}$$

where [ ] denotes the integer part function and log means logarithm to base 10. Let B be the complement in (0,1] of the rationals. Then B^c has measure 0. Since $g = f\chi_B$, where f is as in Example 3.2.7, and the latter has been shown there to be integrable, it follows by Proposition 3.2.10(a) that g has the same integral.

Remark 3.2.18 It will be seen in Problem 3.2.P13 that the Lebesgue integral of a nonnegative measurable function is actually the improper Riemann integral of a related function. In principle, the Lebesgue integral could therefore have been directly defined as being the latter, without bringing in simple functions. However, such an approach would have made it difficult to prove the additivity of the integral. In fact, the improper integral of the related function continues to make sense when (M2) of Sect. 3.1 is replaced by the weaker requirement of monotonicity: $A \subseteq B \Rightarrow \mu(A) \leq \mu(B)$. It is then known as the *Choquet integral.* If we add the further requirement that μ be inner continuous, then we can even prove the monotone convergence property for the Choquet integral, but not additivity. This fact may be regarded as a partial explanation of why we needed to prove the monotone convergence property before establishing additivity: the latter is actually a deeper property, despite being more elementary in appearance!

When (M2) is replaced by monotonicity, μ is often called a *fuzzy measure* and the concept has found applications in the areas of multicriteria decision making and pattern recognition. In such a context, $\mathcal{F}$ is usually taken to consist of *all* subsets of a *finite* set X and μ is assumed to take only finite nonnegative values. The interested reader may consult [9, 13, 14, 31, 32 or 33].

Proposition 3.2.19 *For a bounded function $f : [a,b] \to \mathbb{R}$, there exists an increasing sequence of step functions ϕ_n and a decreasing sequence of step functions ψ_n such that*

(i) $\phi_n \leq f \leq \psi_n$,
(ii) $\phi_n \to \phi$, $\psi_n \to \psi$,
(iii) $\int_{[a,b]} \phi \, dm = \underline{\int_a^b} f \, dx$, $\int_{[a,b]} \psi \, dm = \overline{\int_a^b} f \, dx$,
(iv) $\phi(x) = \psi(x)$ *iff f is continuous at x*,
(v) *f is continuous at x if x is not a point of any partition associated with any* ϕ_n *or* ψ_n *and* $\phi(x) = \psi(x)$.

Proof By definition of lower and upper integrals of any bounded function, there exist sequences $\{P'_n\}_{n\geq 1}$ and $\{P''_n\}_{n\geq 1}$ of partitions of $[a, b]$ satisfying $\lim L(f, P'_n) = \underline{\int_a^b} f(x)dx$ and $\lim U(f, P''_n) = \overline{\int_a^b} f(x)dx$. Let P^*_n be the partition consisting of the points

$$x_k = a + k\frac{b-a}{n}, 0 \leq k \leq n.$$

Now let $P_1 = P_1^*$ and $P_{n+1} = P_n \cup P_{n+1}^* \cup (\cup_{1 \le k \le n+1}(P_k' \cup P_k''))\ \forall\, n \ge 1$. Then each partition is a refinement of the preceding one. Therefore for $n \ge 2$, $L(f, P_n') \le L(f, P_n) \le \underline{\int_a^b} f(x)dx$ and $\overline{\int_a^b} f(x)dx \le U(f, P_n) \le U(f, P_n'')$. Since $\lim L(f, P_n') = \underline{\int_a^b} f(x)dx$ and $\lim U(f, P_n'') = \overline{\int_a^b} f(x)dx$, it follows that $\lim L(f, P_n) = \underline{\int_a^b} f(x)dx$ and $\lim U(f, P_n) = \overline{\int_a^b} f(x)dx$. Note that the length of the longest subinterval of P_n approaches zero. For each n, let ϕ_n be the step function that is constant on each open subinterval of P_n, the constant value being the infimum of f on the corresponding closed subinterval; we set $\phi_n(b)$ equal to its value on the subinterval extending to the left of b and at every other point of P_n, we set ϕ_n equal to its value on the subinterval extending to the right of that point. Then $\int_a^b \phi_n\, dx = L(f, P_n) \le \underline{\int_a^b} f(x)dx$, $\int_a^b \phi_n(x)\, dx \to \underline{\int_a^b} f(x)dx$ and we also have $\phi_n \le f$ everywhere on $[a, b]$. Since P_{n+1} is a refinement of P_n, the bounded sequence $\{\phi_n\}_{n \ge 1}$ of step functions is increasing. By a similar argument with suprema and upper sums, we get a bounded decreasing sequence of step functions $\psi_n \ge f$ such that $\lim \int_a^b \psi_n\, dx = \overline{\int_a^b} f(x)dx$. Thus there exists a pair of bounded increasing and decreasing sequences of step functions ϕ_n and ψ_n respectively, which satisfy the requirement (i) and also satisfy

$$\lim \int_a^b \phi_n(x)dx = \underline{\int_a^b} f(x)dx \quad \text{and} \quad \lim \int_a^b \psi_n(x)dx = \overline{\int_a^b} f(x)dx. \tag{3.18}$$

Note that $\int_a^b \phi_n(x)dx$ and $\int_a^b \psi_n(x)dx$ are equal to the corresponding Lebesgue integrals, because ϕ_n and ψ_n are step functions. Since both sequences of step functions are (pointwise) monotone, they have pointwise limits ϕ and ψ, say. Then (ii) holds, as required. All functions concerned are bounded and therefore the Dominated Convergence Theorem 3.2.16 can be applied to deduce from (3.18) that the requirement (iii) is satisfied.

To prove (iv), suppose f is continuous at x. Then for any $\varepsilon > 0$, there exists a $\delta > 0$ such that $\sup f - \inf f < \varepsilon$, where the sup and inf are taken over $(x - \delta, x + \delta)$. Since the length of the longest subinterval of P_n approaches zero, for sufficiently large n, any interval of P_n containing x will lie in $(x - \delta, x + \delta)$, and hence $0 \le \psi_n(x) - \phi_n(x) < \varepsilon$. Since $\varepsilon > 0$ is arbitrary, it follows that $\phi(x) = \psi(x)$.

Finally, to prove (v), let x not be a point of any P_n and suppose $\phi(x) = \psi(x)$. Then it follows from (i) and (ii) that

$$\phi(x) = \psi(x) = f(x). \tag{3.19}$$

Let $\varepsilon > 0$. Since $\{\phi_n\}$ is increasing and $\{\psi_n\}$ is decreasing, (ii) yields an $n \in \mathbb{N}$ such that

$$\psi(x) \le \psi_n(x) < \psi(x) + \varepsilon \quad \text{and} \quad \phi(x) - \varepsilon < \phi_n(x) \le \phi(x). \tag{3.20}$$

In view of (i), we have $f(x) - \varepsilon < \psi_n(x)$ and $\phi_n(x) < f(x) + \varepsilon$. If x' belongs to the subinterval of P_n containing x, then by definition of ϕ_n and ψ_n, the value of the function f at x' satisfies $\phi_n(x) \le f(x') \le \psi_n(x)$, so that (3.20) leads to

$$\phi(x) - \varepsilon < \phi_n(x) \le f(x') \le \psi_n(x) < \psi(x) + \varepsilon,$$

whereupon (3.19) leads to

$$f(x) - \varepsilon < f(x') < f(x) + \varepsilon.$$

Since x must belong to the interior of the subinterval of P_n that contains it, an interval $(x - \delta, x + \delta)$ is contained in the subinterval; this δ then has the property that $x' \in (x - \delta, x + \delta) \Rightarrow f(x) - \varepsilon < f(x') < f(x) + \varepsilon$. □

Theorem 3.2.20 [Needed in Problems 3.2.P7-11 & 3.2.P23-24] *If a bounded function $f: [a, b] \to \mathbb{R}$ is Riemann integrable, then it is Lebesgue measurable with Lebesgue integral equal to its Riemann integral.*

Proof Let ϕ and ψ be as in Proposition 3.2.19. Then by (i) and (ii) therein, we have $\phi \le f \le \psi$. In view of (iii), the Riemann integrability of f leads to $\int_{[a,b]} (\psi - \phi)dx = 0$. Upon using Problem 3.2.P1(b) it follows that $\phi = \psi$ a.e., which then implies $\phi = f = \psi$ a.e. Therefore by Proposition 2.5.13, f must be Lebesgue measurable. Since $\phi \le f \le \psi$ and all three are Lebesgue measurable, their Lebesgue integrals are in the same order. Together with (iii), this leads to $\int_{[a,b]} f(x)dx = \int_a^b f(x)dx$. □

Problem Set 3.2

3.2.P1.

(a) If $\int_X f\, d\mu > 0$, where f is a measurable extended real-valued function on X, show that $X(f > 0)$ has positive measure.

(b) [Needed in Theorem 3.2.20, Problems 3.2.P6 & 3.2.P14 and Theorem 3.3.5] When f is nonnegative, show that if $X(f > 0)$ has positive measure, then $\int_X f\, d\mu > 0$.

(c) When f and g are measurable, $f \ge g$ and $\int_X f\, d\mu = \int_X g\, d\mu$, does it follow that $f = g$ a.e.?

3.2.P2. Let $\{f_n\}_{n \ge 1}$ be the sequence of functions on $X = [0, 2]$ defined by $f_n = \chi_{[0,1]}$ if n is even and $\chi_{(1,\, 2]}$ if n is odd. Find $\int_X (\liminf f_n)\, d\mu$ and $\liminf \int_X f_n\, d\mu$.

3.2.P3. Let L be the class of all Lebesgue integrable functions on X. Define $\rho(f, g)$ to be $\int_X |f - g|\, d\mu$. Show that $\rho(f, g)$ can be 0 when $f \ne g$ but that ρ has all the other properties of a metric (i.e. ρ is a "pseudometric"). In what way is this different if $X = [a, b]$ and we set $\rho(f, g) = \int_a^b |f(x) - g(x)|\, dx$ in the class of Riemann integrable functions?

3.2.P4. Give an example of a nonnegative real-valued function on [0,1] which is unbounded on every subinterval of positive length (so that it cannot have an improper Riemann integral) but has Lebesgue integral 0.

3.2.P5.

(a) [Needed in Problems 3.2.P8(a) & 3.2.P9] Let Y be a measurable subset of X. If f is a measurable function on X, then its restriction to Y, denoted by $f|_Y$, is a measurable function on Y. Show that the product $f\chi_Y$ (defined on X) has an integral if and only if the restriction $f|_Y$ has an integral, and that when this is so, $\int_X (f\chi_Y) = \int_Y (f|_Y)$. [Note: It is customary to denote this common value by $\int_Y f$ or $\int_Y f\, d\mu$.]

(b) Suppose f is a measurable extended nonnegative real-valued function on $\mathbb{R}$. Let $E \in \mathfrak{M}$ or $\mathfrak{B}$, and let $\int_E f = 0$. Show that f vanishes a.e. on E, i.e. $\{x \in E : f(x) > 0\}$ has measure zero.

3.2.P6. [Theorem due to Lebesgue] Show that a bounded function $f{:}[a, b] \to \mathbb{R}$ is Riemann integrable if and only if it is continuous a.e.

3.2.P7. Show that there does not exist a nonnegative Lebesgue integrable function g on [0,1] such that $g(x) \geq n^2x^n(1 - x)$ everywhere.

3.2.P8.

(a) Let I denote the interval $\left[\frac{1}{(2k+1)\pi}, \frac{1}{2k\pi}\right]$, where $k \in \mathbb{N}$. Using Problem 3.2.P5 and Theorem 3.2.20, show that the Lebesgue integral

$$\int_{[0,1]} \left(\frac{1}{x}\sin\frac{1}{x}\right)\chi_I(x)\, dm(x)$$

is no less than $\frac{2}{(2k+1)\pi}$.

(b) Hence show that, for $f(x) = \frac{1}{x}\sin\frac{1}{x}$, we have $\int_{[0,1]} f^+\, dm = \infty$.

(c) Does $\int_{[0,1]} f\, dm$ exist?

3.2.P9. Suppose that the function $f{:}(0,1] \to \mathbb{R}$ is Riemann integrable on $[c,1]$ whenever $0 < c \leq 1$, and that $\lim_{c\to 0} \int_c^1 |f(x)|dx$ and $\lim_{c\to 0} \int_c^1 f(x)dx$ exist. Then by definition, these limits are respectively equal to $\int_0^1 |f(x)|dx$ and $\int_0^1 f(x)dx$. Show that f is measurable and that its Lebesgue integral $\int_{[0,1]} f$ is the same as $\int_0^1 f(x)dx$. [Remark: The corresponding assertion is true when $f{:}[A,\infty) \to \mathbb{R}$ is Riemann integrable over $[A, B]$ for any finite $B > A$ and the improper integral $\int_A^\infty f(x)dx$ converges absolutely.]

3.2.P10. Suppose $f{:}[0,1] \to \mathbb{R}$ is Lebesgue integrable and is also Riemann integrable on $[\varepsilon,1]$ for every positive $\varepsilon < 1$. Show that $\lim_{\varepsilon\to 0} \int_\varepsilon^1 f(x)dx$ exists and equals $\int_{[0,1]} f\, dm$. In other words, the Riemann integral $\int_0^1 f(x)dx$ exists, possibly as an improper integral, and whether improper or not, equals the Lebesgue integral of f over [0,1]. [Remark: The corresponding assertion is true when $f{:}[A,\infty) \to \mathbb{R}$ is Lebesgue integrable over $[A,\infty)$.]

3.2.P11. Let f_n:[0,1] $\rightarrow$ $\mathbb{R}$ be defined by $f_n(x) = nx/(1+n^{10}x^{10})$. Find the pointwise limit f of $\{f_n\}_{n\geq 1}$ and show that the convergence is neither uniform nor monotone. Also, establish the convergence of Riemann integrals $\int_0^1 f_n(x)dx \rightarrow \int_0^1 f(x)dx$.

3.2.P12. Let f:[0,1] $\times$ [0,1] $\rightarrow$ $\mathbb{R}$ satisfy $|f(x, y)| \leq 1$ everywhere. Suppose that for each fixed $x \in [0,1]$, $f(x, y)$ is a measurable function of y, and also that for each fixed $y \in [0,1]$, $f(x, y)$ is a continuous function of x. If g:[0,1] $\rightarrow$ $\mathbb{R}$ is defined as $g(x) = \int_{[0,1]} f(x,y)dm(y)$, show that g is continuous.

Note: The improper Riemann integral $\int_0^\infty g(u)du$ is the sum of limits of Riemann integrals

$$\lim_{a\to 0}\int_a^c g(u)\,du + \lim_{b\to\infty}\int_c^b g(u)\,du,$$

where c can be taken as any positive real number without affecting the value of the sum. The improper integral is finite if and only if each of the limits is finite. This will be needed in the next problem.

3.2.P13. Let f be a measurable extended nonnegative real-valued function on X. Define ϕ_f: $[0,\infty)$ $\rightarrow$ $[0,\infty]$ as $\phi_f(u) = \mu(X(f > u))$. Show that

(a) ϕ_f is a decreasing function;
(b) $f \leq g \Rightarrow \phi_f \leq \phi_g$;
(c) For an increasing sequence $\{h_n\}$ of measurable functions converging to h, we have $\phi_{h_n} \rightarrow \phi_h$.
(d) $\int_0^\infty \phi_f(u)du$ is equal to the integral $\int_X f\,d\mu$.
(e) $\int_0^\infty \mu(X(f > u))pu^{p-1}du = \int_X f^p\,d\mu$, where $0 < p < \infty$.

3.2.P14.

(a) Let f be an integrable function on a set X such that, for every measurable $E \subseteq X$, we have $\int_E f\,d\mu = 0$ (which means $\int_X f\chi_E\,d\mu = 0$, as explained in Problem 3.2.P5). Show that $f = 0$ a.e.
(b) Let f be a Lebesgue integrable function on $[a, b]$. If $\int_{[a,X]} f\,dm = 0$ for all $x \in [a, b]$, show that $f = 0$ a.e. on $[a, b]$.

3.2.P15. Let f be an integrable function on a set E. Show that for every $\varepsilon > 0$, there exists a $\delta > 0$ such that whenever $A \subseteq E$ with $\mu(A) < \delta$, we have $\left|\int_A f\,d\mu\right| < \varepsilon$ (which means $\left|\int_E (f\chi_A)\,d\mu\right| < \varepsilon$, as explained in Problem 3.2.P5).

3.2.P16. Let f be an integrable function on $[a, b]$. If $F(x) = \int_{[a,x]} f\,dm$, prove that F is continuous on $[a, b]$. If f is an integrable function on $[a,\infty)$, is it true that F is uniformly continuous on $[a,\infty)$?

3.2.P17. Let $\{f_n\}_{n\geq 1}$ be a sequence of measurable nonnegative functions on $\mathbb{R}$ such that f_n $\rightarrow$ f a.e. and suppose that $\int_{\mathbb{R}} f_n\,dm \rightarrow \int_{\mathbb{R}} f\,dm < \infty$. Show that for each measurable set $E \subseteq \mathbb{R}$, $\int_E f_n\,dm \rightarrow \int_E f\,dm$. [The result also holds on a general measure space.]

3.2.P18. Let f be a measurable nonnegative function on $\mathbb{R}$ and E be a measurable subset of finite measure. Show that

(a) $\int_E f = \lim_{N\to\infty} \int_E f_N$, where $f_N(x) = \min\{f(x), N\}$;
(b) $\int_{\mathbb{R}} f = \lim_{N\to\infty} \int_{[-N,N]} f$.

3.2.P19. Let $E \subseteq \mathbb{R}$ be measurable and g an integrable function on E. Suppose $\{f_n\}_{n\geq 1}$ is a sequence of measurable functions on E such that $|f_n| \leq g$ a.e. Show that

$$\int_E \liminf f_n \leq \liminf \int_E f_n \leq \limsup \int_E f_n \leq \int_E \limsup f_n.$$

3.2.P20

(a) Let $\{f_n\}_{n\geq 1}$ and $\{g_n\}_{n\geq 1}$ be sequences of measurable functions on X such that $|f_n| \leq g_n$ for every n. Let f and g be measurable functions such that $\lim_n f_n = f$ a.e. and $\lim_n g_n = g$ a.e. If $\lim_n \int_X g_n = \int_X g < \infty$, show that

$$\lim_n \int_X f_n = \int_X f.$$

(b) Let $\{f_n\}_{n\geq 1}$ be a sequence of integrable functions on X such that $f_n \to f$ a.e., where f is also integrable. Show that $\int_X |f_n - f| d\mu \to 0$ if and only if $\int_X |f_n| d\mu \to \int_X |f| d\mu$.

3.2.P21. Let the function $f: I_1 \times I_2 \to \mathbb{R}$, where I_1 and I_2 are intervals, satisfy the following conditions:

(i) The function $f(x, y)$ is measurable for each fixed $y \in I_2$ and $f(x, y_0)$ is Lebesgue integrable on I_1 for some $y_0 \in I_2$;
(ii) $\frac{\partial}{\partial y} f(x, y)$ exists for each interior point $(x, y) \in I_1 \times I_2$;
(iii) There exists a nonnegative integrable function g on I_1 such that $\left|\frac{\partial}{\partial y} f(x, y)\right| \leq g(x)$ for each interior point $(x, y) \in I_1 \times I_2$.

Show that the Lebesgue integral $\int_{I_1} f(x, y)dx$ exists for every $y \in I_2$ and that the function F on I_2 given by

$$F(y) = \int_{I_1} f(x, y)dx$$

is differentiable at each interior point of I_2, the derivative being

$$F'(y) = \int_{I_1} \frac{\partial}{\partial y} f(x, y)dx.$$

3.2.P22. Use the result of Problem 3.2.P21 to prove that $\int_0^\infty x^n \exp(-x)dx = n!$

3.2.P23. Prove that the improper integral $\int_0^\infty e^{-xy}\frac{\sin x}{x}dx$, where it is understood that $\frac{\sin x}{x}$ is to be replaced by 1 when $x = 0$, converges absolutely for $y > 0$, and evaluate it.

3.2.P24. Assuming that $\int_0^\infty e^{-xy}\frac{\sin x}{x}dx$ converges for $y = 0$, i.e. $\int_0^\infty \frac{\sin x}{x}dx$ converges, find its value by using the Dominated Convergence Theorem 3.2.16. It is understood that $\frac{\sin x}{x}$ is to be replaced by 1 when $x = 0$. [In Problem 4.2.P10, we shall obtain the value without employing Lebesgue integration.] Show that $\frac{\sin x}{x}$ is nevertheless not Lebesgue integrable over $[1,\infty)$.

3.2.P25. Let f be a nonnegative integrable function on $[a, b]$. For each $n \in \mathbb{N}$, let $E_n = X(n-1 \le f < n)$. Prove that $\sum\limits_{k=1}^{\infty}(k-1)\cdot m(E_k) < \infty$.

3.2.P26. Let $x^\gamma f(x)$ be integrable over $(0,\infty)$ for $\gamma = \alpha$ and $\gamma = \beta$, where $0 < \alpha < \beta$. Show that for each $\gamma \in (\alpha,\beta)$, the integral $\int_{(0,\infty)} x^\gamma f(x)dm(x)$ exists and is a continuous function of γ.

3.2.P27. [Complement to Fatou's Lemma] Let $\{f_n\}_{n\ge1}$ be a sequence of measurable functions on X such that $|f_n| \le g$, where g is an integrable function on X. Prove that $\int_X \limsup f_n\, d\mu \ge \limsup \int_X f_n\, d\mu$ and give an example to show that the condition about the function g cannot be dropped.

3.2.P28. Show that the improper Riemann integral $\int_0^\infty \left(1+\frac{x}{n}\right)^{-n}\sin\frac{x}{n}dx$ has limit 0 as $n \to \infty$.

3.2.P29. Evaluate the limit as $n \to \infty$ of the Riemann integral $\int_0^n \left(1-\frac{x}{n}\right)^n e^{x/2}dx$.

3.2.P30. Let $(X, \mathcal{F}, \mu)$ be a measure space and $\{f_k\}_{k\ge1}$ be a sequence of functions on X which converges almost everywhere to a limit function f. In other words, there exists some $N \in \mathcal{F}$ such that $\mu(N) = 0$ and $f_k(x) \to f(x)$ for every $x \notin N$. In symbols, $f_k \xrightarrow{ae} f$.

(a) Give an example to show that f need not be measurable even if each f_k is.
(b) Show that $f_k \xrightarrow{ae} g$ if and only if $f = g$ a.e.
(c) If each f_k is measurable, show that there always exists a measurable g such that $f_k \xrightarrow{ae} g$.
(d) State the Monotone and Dominated Convergence Theorems for the case when the inequalities and the convergence in the hypotheses are assumed to hold only almost everywhere.
(e) If $f_k \xrightarrow{ae} f$ and each f_k is measurable, then under the additional hypothesis that $(X, \mathcal{F}, \mu)$ is *complete*, which is to say,

$$F \subseteq E, \mu(E) = 0 \text{ implies } F \in \mathcal{F},$$

show that f has to be measurable.

3.2.P31. Let γ be the counting measure [see Example 3.1.2(a)] on either $\mathbb{N}$ or $\{1,2,\ldots, n\}$. Show for any nonnegative extended real-valued function f that the integral is given by the sum $\sum_k f(k)$.

3.3 L^p Spaces

It is well known that the uniform limit of a sequence of Riemann integrable functions (a) is Riemann integrable and (b) has integral equal to the limit of the integrals of the functions. It may be noted that the functions as well as the interval in the preceding statement are assumed to be bounded. This does not apply to improper Riemann integrals. For bounded functions, uniform convergence is equivalent to convergence with respect to the *uniform* (also called Chebychev) metric

$$d(f,g) = \sup\{|f(s) - g(s)| : s \in S\},$$

where S is the common domain of the functions in question. In terms of this metric, the above assertions (a) and (b) about uniform convergence leads to completeness of the space of Riemann integrable functions and continuity of the Riemann integral as a real-valued function on it.

Once the monotonicity and linearity of the Riemann integral are established, one can quickly derive the part about continuity, i.e.

$$f_n \to f \text{ uniformly} \quad \Rightarrow \quad \int_a^b f_n(x)\,dx \to \int_a^b f(x)\,dx.$$

One can also prove the analogous statement that:

$$\int_a^b |f_n(x) - f(x)|\,dx \to 0 \quad \Rightarrow \quad \int_a^b f_n(x)\,dx \to \int_a^b f(x)\,dx.$$

This second assertion is broader than the previous one; indeed, if $[a, b] = [0,1]$, $f_n(x) = x^n$ and $f(x) = 0$ for $x < 1$ and $f(1) = 1$, we have $\int_a^b |f_n(x) - f(x)|dx \to 0$ but f_n converges to f on $[0,1]$ only pointwise, not uniformly. It is natural to ask whether the broader assertion can also be formulated as continuity of the integral with reference to some complete metric. The obvious candidate for the metric is

$$d(f,g) = \int_a^b |f(x) - g(x)|dx. \tag{3.21}$$

However, this d will not quite do because it can happen that $d(f, g) = 0$ when $f \neq g$ (e.g. if f and g differ only at finitely many points), although it does have all the other properties of a metric [see Problem 3.2.P3]. In other words, d is only a pseudometric on the space of Riemann integrable functions in the sense of Definition. 1.3.3. As explained there, one can generate a metric space of equivalence classes using the equivalence relation $d(f, g) = 0$ and it turns out for equivalence classes $[\![f]\!], [\![g]\!]$ with respective representatives f, g that $d([\![f]\!], [\![g]\!]) = d(f, g)$ is unambiguous (it is independent of the choice of representatives f, g) and it gives rise to a metric on the set of equivalence classes.

The convergence $[\![f_n]\!] \to [\![f]\!]$ in the metric space generated in the above manner translates in terms of representatives as $\lim_{n\to\infty} d(f_n, f) = 0$ and the Cauchy condition on $\{[\![f_n]\!]\}_{n\geq 1}$ as

for every $\varepsilon > 0$, there exists an $N \in \mathbb{N}$ such that $d(f_n, f_k) < \varepsilon$ whenever $n, k \geq N$.

Regarding the metric space so obtained from (3.21) above, one can ask whether it is complete. The answer is NO if we are working with the Riemann integral but YES if we switch to the Lebesgue integral or measure space integral, even if the domain is of infinite measure! The metric space obtained by switching to the measure space integral in (3.21), where f and g are now assumed to be real-valued and integrable in the sense of Sect. 3.2, is denoted by $\boldsymbol{L^1(X, \mathcal{F}, \mu)}$, or simply $\boldsymbol{L^1(X)}$ or $\boldsymbol{L^1}$ if it is clear from the context what the intended measure space $(X, \mathcal{F}, \mu)$ is. If we are working in the general context of X, $\mathcal{F}$ and μ as in the preceding two sections, then the metric space is denoted by any convenient abbreviation as above.

Below we deal with a family of spaces of the $L^1(X, \mathcal{F}, \mu)$ type, called L^p spaces, $1 \leq p \leq \infty$. We begin with $p < \infty$.

Definition 3.3.1 Let $1 \leq p < \infty$. By $\boldsymbol{L^p(X)}$ *or* $\boldsymbol{L^p(X, \mathcal{F}, \mu)}$ or simply $\boldsymbol{L^p}$ we mean the class of real-valued measurable functions f defined on X for which $\int_X |f|^p d\mu < \infty$, *i.e.*,

$$L^p(X) = \left\{ f : \int_X |f|^p d\mu < \infty \right\}.$$

We associate with each $f \in L^p(X)$ the real number

$$\|f\|_p = \left(\int_X |f|^p d\mu \right)^{1/p}$$

and call $\|f\|_p$ the $\boldsymbol{L^p}$ **norm** of f.

Let f be a nonnegative extended real-valued measurable function defined on X and S be the set of all real $M > 0$ such that

$$\mu(\{x \in X : f(x) > M\}) = 0.$$

If $S = \varnothing$, set $\beta = \infty$. If $S \neq \varnothing$, put $\beta = \inf S$. When this is so,

$$\mu(\{x \in X : f(x) > \beta\}) \leq \sum_{n=1}^{\infty} \mu\left(\left\{ x \in X : f(x) > \beta + \frac{1}{n} \right\} \right) = 0, \qquad (3.22)$$

and hence $\beta \in S$. We call β the **essential supremum** of f and denote it by $\textbf{esssup}\, f$.

When the only set of measure 0 is the empty set, as happens in the case of the counting measure, it is easily verified that the essential supremum is nothing but the ordinary supremum.

Definition 3.3.2 If f is a real-valued measurable function, we define $\|f\|_\infty$ to be the essential supremum of $|f|$ and write

$$\|f\|_\infty = \text{ess sup}|f|,$$

and let $\boldsymbol{L^\infty(X)}$ or $\boldsymbol{L^\infty(X, \mathcal{F}, \mu)}$ consist of all f for which $\|f\|_\infty < \infty$, *i.e.*

$$L^\infty(X) = \{f : \|f\|_\infty < \infty\}.$$

The real number $\|f\|_\infty$ is called the $\boldsymbol{L^\infty}$ **norm** of f.

Remark It is a consequence of (3.22) that $|f(x)| \leq \beta$ for almost all x if and only if $\beta \geq \|f\|_\infty$.

The space $L^\infty(X)$ is also denoted by L^∞ when convenient.

Many of the important classes of spaces employed in Analysis consist of measurable functions. A large number of these spaces have norms defined on them via integrals. The L^p spaces, known as **Lebesgue spaces**, are among such important spaces of functions. These spaces were first introduced by F. Riesz. (D. Hilbert, in connection with his work on integral equations, introduced the L^2 space. It is the natural infinite dimensional analogue of Euclidean spaces.) Aside from their intrinsic value in Analysis, they play a significant role in Fourier Analysis and Mathematical Physics, in particular, in Quantum Mechanics. The main tools in the study of these spaces are the inequalities attributed to Hölder and to Minkowski.

Throughout this chapter, $(X, \mathcal{F}, \mu)$ will denote a measure space. However, sometimes X will be a Lebesgue measurable subset of $\mathbb{R}$ with Lebesgue measure m.

Before proving the Hölder and Minkowski inequalities, we shall establish a lemma about real numbers, which is a generalisation of the inequality between arithmetic and geometric means.

Lemma 3.3.3 *Let a and b be nonnegative real numbers, and suppose* $0 < \lambda \neq 1$. *If* $\lambda < 1$, *then*

$$a^\lambda b^{1-\lambda} \leq \lambda a + (1-\lambda)b$$

with equality if and only if $a = b$.

Proof Consider the continuous function φ defined for nonnegative real numbers x by

$$\varphi(x) = (1-\lambda) + \lambda x - x^\lambda.$$

Then $\varphi'(x) = \lambda(1 - x^{\lambda-1})$ for $x > 0$, and so $x = 1$ is the only possible point for the extrema of φ for $x > 0$. Since $\varphi''(x) = -\lambda(\lambda-1)x^{\lambda-2}$ for all $x > 0$, it follows that

$\varphi''(1) > 0$ because $0 < \lambda < 1$. Thus the function has a local minimum at $x = 1$. As $\varphi(0) > 0$ and $\varphi(1) = 0$, we infer that $x = 1$ is the point of absolute minimum of φ. Consequently, for all $x \geq 0$, we have

$$(1-\lambda)+\lambda x - x^{\lambda} = \varphi(x) \geq \varphi(1) = 0,$$

with equality if and only if $x = 1$. If $b \neq 0$, the lemma follows on substituting a/b for x, while if $b = 0$, the lemma is trivial. □

Definition 3.3.4 If p and q are positive real numbers such that $p + q = pq$, or equivalently,

$$\frac{1}{p}+\frac{1}{q}=1,$$

then p and q are called a **pair of conjugate exponents**.

It is clear that, if p and q are a pair of conjugate exponents, then $1 < p < \infty$ and $1 < q < \infty$. An important special case is $p = q = 2$. Observe that $q \to \infty$ as $p \to 1$. Consequently, 1 and ∞ are also regarded as a conjugate pair.

Theorem 3.3.5 (Hölder's Inequality) *Let p and q be a pair of conjugate exponents, $1 < p < \infty$. Then for nonnegative extended real-valued measurable functions f and g defined on X,*

$$\int_X fg\, d\mu \leq \left(\int_X f^p d\mu\right)^{1/p} \left(\int_X g^q d\mu\right)^{1/q}. \tag{3.23}$$

Furthermore,

(a) *if equality holds in* (3.23) *with both sides finite, then $\alpha f^p = \beta g^q$ a.e. for some nonnegative constants α and β, not both zero (another way to say this is that either $f^p = cg^q$ a.e. or $g^q = cf^p$ a.e. for some nonnegative constant c)*;

(b) *if $\alpha f^p = \beta g^q$ a.e. for some nonnegative constants α and β, not both zero, then equality holds in* (3.23).

Proof Let the factors on the right-hand side of (3.23) be denoted by A and B respectively.

First suppose $A = 0$. Then the right-hand side of (3.23) is zero. Besides, $f = 0$ a.e. by Problem 3.2.P1(b) and hence the same is true of fg. Therefore the left-hand side of (3.23) is also zero. Hence (3.23) holds with equality. A similar argument works when $B = 0$. So, suppose $A > 0$ and $B > 0$. If either A or B is ∞, there is nothing to prove.

It remains to consider the case when $0 < A < \infty$ and $0 < B < \infty$. So, suppose both are finite. In this remaining case, the functions f and g are finite a.e. and we may replace them by finite-valued functions without affecting any of the integrals involved. The finiteness allows us to apply the lemma with

$$a = \left(\frac{f(x)}{A}\right)^p, b = \left(\frac{g(x)}{B}\right)^q, \lambda = \frac{1}{p} \text{ and } 1 - \lambda = \frac{1}{q}.$$

Thus,

$$\frac{f(x)}{A}\frac{g(x)}{B} \leq \frac{1}{p}\left(\frac{f(x)}{A}\right)^p + \frac{1}{q}\left(\frac{g(x)}{B}\right)^q, \tag{3.24}$$

and integrating both sides, we get

$$\frac{1}{AB}\int_X fg\, d\mu \leq \frac{1}{pA^p}\int_X f^p d\mu + \frac{1}{qB^q}\int_X g^q d\mu = 1, \tag{3.25}$$

which immediately implies (3.23) in the remaining case.

For the proof of (a), suppose equality holds in (3.23) and both sides are finite. Again, we first dispose of the cases $A = 0$ and $B = 0$. If $A = 0$, then $f = 0$ a.e. by Problem 3.2.P1(b) and hence the desired equality holds with $\alpha = 1$ and $\beta = 0$. Similarly, if $B = 0$, then the desired equality holds with $\alpha = 0$ and $\beta = 1$. In the remaining case that $A \neq 0 \neq B$, we must have $0 < A < \infty$ and $0 < B < \infty$. As before, (3.24) must hold and (3.25) follows from it upon integrating both sides. Now equality in (3.23) implies equality in (3.25) and hence equality a.e. in (3.24). By Lemma 3.3.3, this further implies

$$\left(\frac{f(x)}{A}\right)^p = \left(\frac{g(x)}{B}\right)^q \quad \text{a.e.,}$$

which leads to the desired equality with $\alpha = B^q$ and $\beta = A^p$, both of which are nonzero. This establishes (a).

To prove (b), suppose that there exist α and β, not both zero, such that $\alpha f^p = \beta g^q$ a.e. Plainly, we may assume that $\alpha \neq 0$. This permits us to write f as $\gamma g^{q/p}$ a.e., where $\gamma = (\beta/\alpha)^{1/p}$. It follows that $fg = \gamma g^{1+(q/p)} = \gamma g^q$ a.e. It is now easy to verify that the two sides of (3.23) are both equal to $\gamma \int_X g^q$. □

Remark In (a) of the above theorem, the hypothesis that both sides are finite cannot be omitted. Indeed, we may have $f(x) = 1$ and $g(x) = x$ everywhere on $[0,\infty)$ with Lebesgue measure, in which case, equality holds in (3.23) with both sides equal to ∞, but the kind of constants α and β claimed in (a) do not exist.

In (b) of the above theorem, the conclusion that both sides are finite cannot be drawn. To wit, we may have f and g both equal to 1 everywhere on $[0,\infty)$ with Lebesgue measure, in which case both sides of (3.23) are ∞.

If $p = q = 2$, Hölder's Inequality is known as the Cauchy–Schwarz Inequality [Theorem 3.3.16].

Theorem 3.3.6 (Minkowski's Inequality) *Let $1 < p < \infty$. Then for any nonnegative extended real-valued measurable functions f and g on X, the following inequality holds*:

$$\left[\int_X (f+g)^p d\mu\right]^{1/p} \leq \left(\int_X f^p d\mu\right)^{1/p} + \left(\int_X g^p d\mu\right)^{1/p}. \tag{3.26}$$

Furthermore,

(a) *if equality holds in* (3.26) *with both sides finite, then $\alpha f = \beta g$ a.e. for some nonnegative constants α and β, not both zero (another way to say this is that either $f = cg$ a.e. or $g = cf$ a.e. for some constant c)*;
(b) *if $\alpha f = \beta g$ a.e. for some nonnegative constants α and β, not both zero, then equality holds in* (3.26).

Proof To prove (3.26), we write

$$(f+g)^p = f(f+g)^{p-1} + g(f+g)^{p-1}.$$

On applying Hölder's Inequality to each of the products on the right, we get

$$\int_X f(f+g)^{p-1} d\mu \leq \left(\int_X f^p d\mu\right)^{\frac{1}{p}} \left(\int_X (f+g)^{(p-1)q} d\mu\right)^{\frac{1}{q}} \tag{3.27}$$

and

$$\int_X g(f+g)^{p-1} d\mu \leq \left(\int_X g^p d\mu\right)^{\frac{1}{p}} \left(\int_X (f+g)^{(p-1)q} d\mu\right)^{\frac{1}{q}}, \tag{3.28}$$

where $1/p + 1/q = 1$. Since $(p - 1)q = p$, adding (3.27) and (3.28) gives

$$\int_X (f+g)^p\, d\mu \leq \left[\int_X (f+g)^p\, d\mu\right]^{\frac{1}{q}} \left[\left[\int_X f^p\, d\mu\right)^{\frac{1}{p}} + \left(\int_X g^p\, d\mu\right)^{\frac{1}{p}}\right]. \tag{3.29}$$

Clearly, it is enough to prove (3.26) in the case when the left-hand side is greater than zero and the right-hand side is less than infinity. Using the well-known inequality [found in [26, Theorem 1.1.7, p. 27]]

$$(f+g)^p \leq 2^{p-1}(f^p + g^p),$$

we deduce that the left-hand side of (3.26) is finite. On dividing both sides of (3.29) by $\left[\int_X (f+g)^p d\mu\right]^{1/q}$, and keeping in view that $1/p + 1/q = 1$, we get (3.26).

For the proof of (a), suppose equality holds in (3.26) and both sides are finite. We first dispose of the case when both sides are 0. In this case, $f = g = 0$ a.e. and we take $\alpha = \beta = 1$.

So, suppose both sides of (3.26) are positive. Since they have been assumed finite, the right-hand sides of (3.27) and (3.28) are finite.

We claim that equality holds in (3.27) as well as (3.28). If not, then adding them and using the finiteness of their right-hand sides, we would obtain (3.29) with strict inequality; upon dividing by $\left[\int_X (f+g)^p d\mu\right]^{1/q}$, which is finite and positive in the present case, we would arrive at (3.26) with strict inequality. This justifies our claim that (3.27) and (3.28) both hold with equality.

Since (3.27) and (3.28) both hold with equality, it follows upon using Theorem 3.3.5 that

$$\rho f^p = \sigma(f+g)^{(p-1)q} \text{ a.e.} \quad \text{and} \quad \gamma g^p = \delta(f+g)^{(p-1)q} \text{ a.e.}$$

for some nonnegative constants ρ and σ, not both zero, and some nonnegative constants γ and δ, not both zero. It is easy to deduce from this that $\alpha f = \beta g$ a.e., where α and β are some nonnegative constants, not both zero. This establishes (a).

To prove (b), suppose that there exist α and β, not both zero, such that $\alpha f = \beta g$ a.e. Plainly, we may assume that $\alpha \neq 0$. This permits us to write f as γg a.e., where $\gamma = \beta/\alpha$. It follows that $f + g = (1 + \gamma)g$ a.e. and it is easy to verify that the two sides of (3.26) are both equal to $(1+\gamma)\left(\int_X g^p\right)^{1/p}$. □

Remark In (a) of the above theorem, the hypothesis that both sides are finite cannot be omitted. Indeed, we may have $f(x) = 1$ and $g(x) = x$ everywhere on $[0,\infty)$ with Lebesgue measure, in which case, equality holds in (3.26) with both sides equal to ∞, but the kind of constants α and β claimed in (a) do not exist.

In (b) of the above theorem, the conclusion that both sides are finite cannot be drawn. To wit, we may have f and g both equal 1 everywhere on $[0,\infty)$ with Lebesgue measure, in which case both sides of (3.26) are ∞.

Theorem 3.3.7 *Let p and q be a pair of conjugate exponents and let* $1 \leq p \leq \infty$. *If* $f \in L^p(X)$ *and* $g \in L^q(X)$, *then* $fg \in L^1$ *and*

$$\|fg\|_1 \leq \|f\|_p \|g\|_q. \tag{3.30}$$

Proof For $1 < p < \infty$, this is just Hölder's Inequality applied to $|f|$ and $|g|$. If $p = \infty$, observe that

$$|f(x)g(x)| \leq \|f\|_\infty |g(x)| \tag{3.31}$$

for almost all x. On integrating (3.31), we obtain (3.30). If $p = 1$, then $q = \infty$ and the same argument applies. □

Theorem 3.3.8 *Suppose* $1 \leq p \leq \infty$, *and* $f, g \in L^p(X)$. *Then* $f + g \in L^p(X)$ *and the inequality*

$$\|f+g\|_p \leq \|f\|_p + \|g\|_p \tag{3.32}$$

holds. Also, $\alpha f \in L^p(X)$ *when* α *is a real number, and*

$$\|\alpha f\|_p = |\alpha| \|f\|_p. \tag{3.33}$$

Proof For $1 < p < \infty$, (3.32) follows from Minkowski's Inequality applied to $|f|$ and $|g|$, since

$$\int_X |f+g|^p d\mu \leq \int_X (|f| + |g|)^p d\mu.$$

For $p = 1$, it is a trivial consequence of the inequality $|f+g| \leq |f| + |g|$. For $p = \infty$, the argument does not involve any integrals but is nevertheless simple enough to be left to the reader.

As regards (3.33), when $1 \leq p < \infty$,

$$\|\alpha f\|_p = \left(\int_X |\alpha f|^p d\mu\right)^{1/p} = |\alpha| \left(\int_X |f|^p d\mu\right)^{1/p} = |\alpha| \|f\|_p.$$

When $p = \infty$, the argument does not involve any integrals but, as before, is simple enough to be left to the reader. □

Remarks 3.3.9

(a) When $X = \{1,2,\ldots, n\}$, a function on X is nothing but an n-tuple $x = (x_1, x_2,\ldots, x_n)$. It is real-valued if and only if $x \in \mathbb{R}^n$. With the counting measure on X, sums can be interpreted as integrals [see Problem 3.2.P31] and

$$\|x\|_p = \left(\sum_{i=1}^n |x_i|^p\right)^{1/p}, \text{where } 1 \leq p < \infty \quad \text{and} \quad \|x\|_\infty = \sup\{|x_i| : 1 \leq i \leq n\}.$$

The sum and sup are both taken over finitely many numbers and are therefore finite if and only if each x_i is finite, i.e., if and only if $x \in \mathbb{R}^n$. Thus $\mathbb{R}^n$ and $L^p(X)$ are the same set, $1 \leq p \leq \infty$. Also, Theorem 3.3.8 leads to

$$\left(\sum_{i=1}^n |x_i + y_i|^p\right)^{1/p} \leq \left(\sum_{i=1}^n |x_i|^p\right)^{1/p} + \left(\sum_{i=1}^n |y_i|^p\right)^{1/p},$$
$$\left(\sum_{i=1}^n |\alpha x_i|^p\right)^{1/p} = |\alpha| \left(\sum_{i=1}^n |x_i|^p\right)^{1/p}$$

for $1 \leq p < \infty$ and to

$$\sup\{|x_i+y_i| : 1\le i\le n\}\le \sup\{|x_i| : 1\le i\le n\}+\sup\{|y_i| : 1\le i\le n\},$$
$$\sup\{|\alpha x_i| : 1\le i\le n\}=|\alpha|\sup\{|x_i| : 1\le i\le n\}$$

for $p = \infty$.
It is left to the reader to formulate what Theorem 3.3.8 leads to when $X = \mathbb{N}$ with the counting measure.

(b) Set

$$d(f,g)=\left[\int_X |f-g|^p d\mu\right]^{1/p},\quad f,g\in L^p(X), 1\le p<\infty$$
$$\text{and } d(f,g)=\|f-g\|_\infty,\quad f,g\in L^\infty(X).$$

Then $0 \le d(f, g) < \infty$, $d(f, f) = 0$, $d(f, g) = d(g, f)$ and it follows from Theorem 3.3.8 that the Triangle Inequality, namely,

$$d(f,g)\le d(f,h)+d(g,h),\quad f,g,h\in L^p(X)$$

is satisfied. The only property of a metric that d fails to satisfy in general is that "$d(f, g) = 0$ implies $f = g$". In other words, d is only a pseudometric on the space $L^p(X)$ in the sense of Definition 1.3.3. As explained there, one can generate a metric space of equivalence classes using the equivalence relation $d(f, g) = 0$. It turns out for equivalence classes $[\![f]\!], [\![g]\!]$ with respective representatives f, g that $d([\![f]\!], [\![g]\!]) = d(f, g)$ is unambiguous (it is independent of the choice of representatives f, g) and it gives rise to a metric on the set of equivalence classes. The resulting metric space whose elements are equivalence classes is again denoted by $L^p(X)$, etc. In the space L^p of equivalence classes $[\![f]\!]$, one can introduce operations by setting

$$[\![f]\!]+[\![g]\!]=[\![f+g]\!]\quad\text{and}\quad \alpha[\![f]\!]=[\![\alpha f]\!].$$

It is left to the reader to verify that these equalities determine the sum of two equivalence classes and multiplication of an equivalence class by a scalar in an unambiguous manner, which is to say, independently of the choice of representatives f and g. After that, it is quite straightforward to check that the operations have the necessary properties to make L^p what is called a linear space [see Sect. 6.1].
In our present situation, $d(f, g) = 0$ precisely when $f = g$ a.e. In the case of the counting measure, the only set of measure 0 is the empty set and equality a.e. amounts to equality everywhere. In particular, d is a metric. This applies to $\mathbb{R}^n$ discussed in part (a) above.
The reader is reminded that the number $\|f\|_p$ is actually associated with the equivalence class $[\![f]\!]$ of the function f. The resulting real-valued function on $L^p(X)$, $1 \le p \le \infty$, has the following properties:

(i) $\|f\|_p \geq 0 \; \forall f \in L^p(X)$;
(ii) $\|f\|_p = 0 \Leftrightarrow f = 0$ (meaning $[\![f]\!] = 0$);
(iii) $\|\alpha f\|_p = |\alpha|\|f\|_p \; \forall f \in L^p(X), \forall \alpha \in \mathbb{R}$;
(iv) $\|f + g\|_p \leq \|f\|_p + \|g\|_p \; \forall f, g \in L^p(X)$.

Here (i), (ii) and (iv) are essentially restatements of what we have already said; (iii) is immediate from the definition of $\|f\|_p$.

Since much of the work is done in terms of the representatives of the equivalence classes, we shall follow the customary practice by writing $[\![f]\!]$ as simply f and speaking of the elements of $L^p(X)$ as though they are functions. Thus for example, we may speak of $[\![f]\!]$ as being continuous or nonnegative, meaning that at least one function in the equivalence class has the property.

The notions of convergence and Cauchy sequence are defined with reference to the metric mentioned in the remark above. More specifically, we have the following:

If $\{f_n\}_{n\geq 1}$ is a sequence in $L^p(X)$, if $f \in L^p(X)$ and $\lim_{n\to\infty} \|f_n - f\|_p = 0$, we say that f_n converges to f in L^p norm or f_n converges to f *in the sense of the mean of order p*. If for every $\varepsilon > 0$, there exists a positive n_0 such that $\|f_n - f_k\|_p < \varepsilon$ whenever n, $k \geq n_0$, we say that the sequence $\{f_n\}_{n\geq 1}$ is a **Cauchy sequence** in $L^p(X)$.

The next result shows that $L^p(X)$, $1 \leq p < \infty$, is a complete metric space in the metric

$$d(f, g) = \|f - g\|_p, \quad f, g \in L^p(X),$$

that is, every Cauchy sequence $\{f_n\}_{n\geq 1}$ in $L^p(X)$ converges to an element of $L^p(X)$.

We state and prove the result in terms of representatives rather than equivalence classes, that is, for the pseudometric space from which $L^p(X)$ is constructed by taking equivalence classes.

Theorem 3.3.10 *If $\{f_n\}_{n\geq 1}$ is a Cauchy sequence in $L^p(X)$, where $1 \leq p < \infty$, then there exists an $f \in L^p(X)$ such that $\lim_{n\to\infty} \|f_n - f\|_p = 0$. In other words, $L^p(X)$ is a complete metric space when $1 \leq p < \infty$.*

Proof Since the sequence $\{f_n\}_{n\geq 1}$ is Cauchy in $L^p(X)$, there exists a subsequence $\{f_{n_i}\}_{i\geq 1}$, $n_1 < n_2 < n_3 < \cdots$, such that

$$\|f_{n_{i+1}} - f_{n_i}\|_p < 2^{-i}, \quad i \geq 1. \tag{3.34}$$

Put

$$g_k = \sum_{i=1}^{k} |f_{n_{i+1}} - f_{n_i}|, \quad g = \sum_{i=1}^{\infty} |f_{n_{i+1}} - f_{n_i}|.$$

Since (3.34) holds, the Minkowski Inequality shows that $\|g_k\|_p < 1$, $k = 1,2,\ldots$. By Fatou's Lemma applied to $\{g_k^p\}$, we have

$$\|g\|_p^p = \int_X \lim_{k\to\infty} g_k^p d\mu \le \liminf_{k\to\infty} \int_X g_k^p d\mu \le 1.$$

Thus $g \in L^p(X)$ and hence is finite a.e. Consequently,

$$f_{n_1} + \sum_{i=1}^{\infty} (f_{n_{i+1}} - f_{n_i}) \tag{3.35}$$

converges absolutely for almost every $x \in X$. Denote the sum of (3.35) by $f(x)$ for those x where (3.35) converges and set $f(x) = 0$ on the remaining set of measure zero. Since

$$f_{n_1} + \sum_{i=1}^{k-1} (f_{n_{i+1}} - f_{n_i}) = f_{n_k},$$

we see that

$$|f_{n_k}| \le |f_{n_1}| + \sum_{i=1}^{k-1} |f_{n_{i+1}} - f_{n_i}| \le |f_{n_1}| + \sum_{i=1}^{\infty} |f_{n_{i+1}} - f_{n_i}| = |f_{n_1}| + g = h, \quad \text{say.}$$

Thus $|f_{n_k}| \le h$ for $k = 1,2,\ldots$, where $h \in L^p(X)$, being the sum of L^p functions.

We have proved that $\lim_{i\to\infty} f_{n_i} = f$ a.e. Consider an $\varepsilon > 0$ and choose N as in the Cauchy condition. Since $\{f_{n_i}\}$ is a subsequence of $\{f_n\}$, we have

$$\int_X |f_{n_i} - f_{n_j}|^p d\mu < \varepsilon \quad \text{whenever } i,j > N.$$

Then

$$\int_X |f_{n_j} - f|^p d\mu \le \liminf_{i\to\infty} \int_X |f_{n_i} - f_{n_j}|^p d\mu \le \varepsilon \quad \text{whenever } j > N,$$

using Fatou's Lemma. Thus, the subsequence $\{f_{n_i}\}_{i\ge1}$ converges in $L^p(X)$ to the limit f. But $\{f_n\}_{n\ge1}$ is Cauchy and it therefore follows that $\lim_{n\to\infty} \|f - f_n\|_p = 0$. □

The argument for the following corollary is contained in the preceding proof.

Corollary 3.3.11 *If $\{f_n\}_{n\ge1}$ is a sequence in $L^p(X)$ converging to $f \in L^p(X)$, where $1 \le p < \infty$, then there exists a subsequence $\{f_{n_i}\}_{i\ge1}$ such that*

$$\lim_{n_i\to\infty} f_{n_i} = f \quad a.e.$$

Any subsequence that converges a.e. must have f as its limit a.e.

In the case $p = \infty$, any Cauchy sequence $\{f_n\}_{n \geq 1}$ converges uniformly to some limit function $f \in L^\infty(X)$ a.e., as will be seen from the proof below.

Theorem 3.3.12 *$L^\infty(X)$ is a complete metric space.*

Proof Suppose $\{f_n\}_{n \geq 1}$ is a Cauchy sequence in $L^\infty(X)$. Let

$$B_{\nu,n}\{x \in X : |f_n(x) - f_\nu(x)| > \|f_n - f_\nu\|_\infty\} \quad \text{and} \quad E = \bigcup_{\nu,n} B_{\nu,n}.$$

Since the absolute value of a function is greater than its essential supremum only on a set of measure zero, each of the sets $B_{\nu,n}$, ν, $n = 1,2,3,\ldots$ has measure zero; it therefore follows that $\mu(E) = 0$.

We assert that $\{f_n\}_{n \geq 1}$ has the uniform Cauchy property of Theorem 1.5.4 on E^c. To prove this, consider any $\varepsilon > 0$. We must show that there exists an integer n_0 such that n, $\nu \geq n_0$ implies $|f_n(x) - f_\nu(x)| < \varepsilon$ for all $x \in E^c$. By the given Cauchy property in $L^\infty(X)$, we know that there exists an integer n_0 such that n, $\nu \geq n_0$ implies $\|f_n - f_\nu\|_\infty < \varepsilon$. Now let $x \in E^c$ be arbitrary. By definition of E, we have $x \in B_{\nu,n}^c$ for all n, ν, i.e., $|f_n(x) - f_\nu(x)| \leq \|f_n - f_\nu\|_\infty$ for all n, ν. Therefore n, $\nu \geq n_0$ implies $|f_n(x) - f_\nu(x)| < \varepsilon$, which shows that n_0 is indeed the kind of integer whose existence was to be shown.

It now follows by Theorem 1.5.4 that $\{f_n\}_{n \geq 1}$ converges uniformly on E^c to some real-valued limit function f. Upon defining $f(x) = 0$ for each $x \in E$, we get a measurable function on X. By virtue of the uniform convergence, for every $\eta > 0$, there exists an n_0 such that $n \geq n_0$ implies $|f_n(x) - f(x)| < \eta$ for all $x \in E^c$, which further implies $\|f_n - f\|_\infty \leq \eta$ because $\mu(E) = 0$. Thus $f_n - f \in L^\infty(X)$ and $\|f_n - f\|_\infty \to 0$. Moreover, $f = (f - f_n) + f_n \in L^\infty(X)$. □

Example 3.3.13 The distinction between the Cauchy condition of Theorem 3.3.10 and the condition that the sequence $\{f_n\}_{n \geq 1}$ is Cauchy almost everywhere can be illustrated by taking $f_n(x) = \frac{n}{1+nx}$ on $[0,1]$. Since $f_n(x) \to \frac{1}{x}$ when $x \neq 0$, this sequence $\{f_n\}_{n \geq 1}$ is Cauchy almost everywhere on $[0,1]$. Each f_n is in $L^1[0,1]$ because $\int_{[0,1]} |f_n| = \int_{[0,1]} \frac{n}{1+nx} dx = \ln(1 + n) < \infty$. However, $\ln(1 + n) \to \infty$ and therefore even a subsequence of $\{f_n\}_{n \geq 1}$ is unbounded in the metric space $L^1[0,1]$ and hence cannot satisfy the Cauchy condition. To see this without using a metric space argument, we begin by noting that, for $n > k$, we have

$$\int_{[0,1]} |f_n - f_k| = \int_{[0,1]} \left|\frac{n}{1+nx} - \frac{k}{1+kx}\right| dm(x) = \int_{[0,1]} \left[\frac{n}{1+nx} - \frac{k}{1+kx}\right] dm(x)$$
$$= \int_0^1 \left[\frac{n}{1+nx} - \frac{k}{1+kx}\right] dx = \ln\frac{1+n}{1+k}.$$

When $n > 2\,k + 1$, this exceeds $\ln 2$, which shows that even a subsequence of $\{f_n\}_{n \geq 1}$ cannot satisfy the Cauchy condition of Theorem 3.3.10.

The following proposition shows that Theorem 3.3.10 breaks down if we interpret the integral as the Riemann integral.

Proposition 3.3.14 *There exists a sequence* $\{f_n\}_{n\geq 1}$ *of continuous functions on* [0,1] *which fulfils the Cauchy condition that*

$$\textit{for every } \varepsilon > 0, \textit{there exists an } N \in \mathbb{N} \textit{ such that}$$
$$\int_0^1 |f_n(x) - f_k(x)|dx < \varepsilon \textit{ whenever } n, k > N$$

but for which there exists no Riemann integrable function f satisfying

$$\lim_{n\to\infty} \int_0^1 |f_n(x) - f(x)|dx = 0.$$

Proof Let $r_1, r_2, \ldots$ be an enumeration of the rationals in [0,1]. For each $i \in \mathbb{N}$, let g_i be a continuous nonnegative function on [0,1] such that

$$g_i(r_i) > i \quad \text{and} \quad \int_0^1 g_i(x)dx = 1/2^i. \tag{3.36}$$

Define $f_k = \sum_{1\leq i\leq k} g_i$. Then each f_k is continuous and satisfies

$$f_k \geq g_k \quad \text{on } [0, 1]. \tag{3.37}$$

Also, $n > k \Rightarrow f_n \geq f_k$ and

$$\begin{aligned}\int_0^1 |f_n(x) - f_k(x)|dx &= \int_0^1 (f_n(x) - f_k(x))dx \\ &= \int_0^1 \sum_{k<i\leq n} g_i(x)dx = \sum_{k<i\leq n} \frac{1}{2^i} < 12^k.\end{aligned}$$

It follows that $\{f_n\}_{n\geq 1}$ is an increasing sequence of continuous functions that satisfies the Cauchy condition.

Suppose, if possible, that f is a Riemann integrable function on [0,1] such that

$$\lim_{n\to\infty} \int_0^1 |f_n(x) - f(x)|dx = 0. \tag{3.38}$$

Consider an arbitrary subinterval $[a, b] \subseteq [0,1]$, where $a < b$; being Riemann integrable, $|f|$ must have an upper bound M on $[a, b]$. Since $a < b$, the interval $[a, b]$ must contain infinitely many rational numbers and hence must contain some r_k with $k > M + 1$. By (3.36) and (3.37), $f_k(r_k) \geq g_k(r_k) > k > M + 1$. By continuity, $[a, b]$ has some subinterval $[\alpha,\beta]$ of positive length on which $f_k > M + 1$.

Now, $|f| \le M$ on $[a, b] \supseteq [\alpha,\beta]$ and therefore $f_k - f > (M + 1) - M = 1$ on $[\alpha,\beta]$. Since $\{f_n\}_{n\ge 1}$ is an increasing sequence, we have

$$f_n - f > 1 \quad \text{on } [\alpha, \beta] \quad \text{for all } n \ge k.$$

Hence for any $n \ge k$, we have

$$\int_0^1 |f_n(x) - f(x)|dx \ge \int_\alpha^\beta |f_n(x) - f(x)|dx \ge \beta - \alpha > 0,$$

in contradiction with (3.38). So, there can be no such Riemann integrable function on [0,1]. □

Remark 3.3.15 It is worth noting that the above proof delivers something slightly more than promised. It shows that in order for f to be a limit of the sequence constructed, it has to be unbounded on *every* subinterval of [0,1] having positive length, which means that it cannot even be eligible to have an improper integral. An informal way to put it is that, in order to reach the completeness target, one has to overshoot improper integrals. Theorem 3.3.10 may be interpreted as saying that this, i.e., completeness is precisely what the Lebesgue integral achieves; besides, it does so without "sacrificing" any Riemann integrable functions in the process, as seen in Theorem 3.2.20. That it does so without "overkill" will be the main point of Proposition 3.4.2. Before embarking on the preliminaries for it, we remind the reader that a conditionally convergent improper (Riemann) integral is not a Lebesgue integral, as illustrated in Problem 3.2.P24.

We have seen that for $1 \le p \le \infty$, the spaces $L^p(X, \mathcal{F}, \mu)$ equipped with a norm

$$\|f\|_p = \left(\int_X |f|^p d\mu\right)^{1/p}, \quad f \in L^p(X), \quad 1 \le p < \infty$$
$$\|f\|_\infty = \text{ess sup}|f|, \quad f \in L^\infty(X)$$

and metric

$$d(f,g) = \left[\int_X |f - g|^p d\mu\right]^{1/p}, \quad f,g \in L^p(X),\ 1 \le p < \infty$$
$$\text{and} \quad d(f,g) = \|f - g\|_\infty, \quad f,g \in L^\infty(X),$$

are complete with respect to the metric.

The case when $p = 2$ plays an important role in the study of mathematical physics, probability theory and several other areas of mathematics. In the study of analytic geometry, we recall that orthogonality of two vectors is determined analytically by considering their inner (or dot) product. The space $L^2(X, \mathcal{F}, \mu)$ admits an inner product that gives rise to a norm coinciding with the one described in the paragraph above. This facilitates the study of the space. These are some of the reasons why we discuss $L^2(X, \mathcal{F}, \mu)$ separately.

Theorem 3.3.16 (Cauchy–Schwarz Inequality) *Let f and g be measurable functions on X. If $f \in L^2(X)$ and $g \in L^2(X)$, then $fg \in L^1(X)$ and*

$$\int_X fg\, d\mu \leq \left(\int_X f^2 d\mu\right)^{\frac{1}{2}} \left(\int_X g^2 d\mu\right)^{\frac{1}{2}}.$$

Moreover, equality holds if and only if either $|f| = c|g|$ or $|g| = c|f|$ a.e. for some constant c.

Proof This follows from Hölder's Inequality of Theorem 3.3.5. □

Theorem 3.3.17 (Minkowski's Inequality) *Let f and g be measurable functions on X. If $f \in L^2(X)$ and $g \in L^2(X)$, then $f + g \in L^2(X)$ and*

$$\left[\int_X |f+g|^2 d\mu\right]^{1/2} \leq \left(\int_X f^2 d\mu\right)^{1/2} + \left(\int_X g^2 d\mu\right)^{1/2}.$$

Moreover, equality holds if and only if either $|f| = c|g|$ or $|g| = c|f|$ a.e. for some constant c.

Proof This follows from Minkowski's Inequality of Theorem 3.3.6. □

If $f \in L^2(X)$ and $g \in L^2(X)$, then the **inner product** of f and g is defined to be

$$\langle f, g\rangle = \int_X fg\, d\mu.$$

By the Cauchy–Schwarz Inequality [Theorem 3.3.16], $\langle f, g\rangle$ is always a (finite) real number. The inner product is also known as the **scalar product** or as the **dot product** and is often denoted by $f \cdot g$. The number $\langle f, g\rangle$ is associated with the pair of equivalence classes $[\![f]\!], [\![g]\!]$, as can be checked by using the Cauchy–Schwarz Inequality, and Theorem 3.3.8 with $p = 1$. The following properties of the inner product are used constantly without reference:

Proposition 3.3.18 *Let $f \in L^2(X)$, $g \in L^2(X)$ and $h \in L^2(X)$. Then*

(a) $\langle f, g\rangle = \langle g, f\rangle$;
(b) $\langle \lambda f, g\rangle = \lambda\langle f, g\rangle\ \forall \lambda \in \mathbb{R}$;
(c) $\langle f + g, h\rangle = \langle f, h\rangle + \langle g, h\rangle$;
(d) $\langle f, f\rangle = \|f\|^2 = \|f\|_2^2$.

Proof The proof is straightforward and is left to the reader. □

Elements f, g of $L^2(X)$ are said to be **orthogonal** if $\langle f, g\rangle = 0$. As the reader can verify by evaluating the relevant integrals, that when k and n are distinct nonnegative integers, the functions in $L^2[-\pi,\pi]$ given by $f(x) = \sin kx$ and $g(x) = \sin nx$ are orthogonal, and so are those given by $f(x) = \cos kx$ and $g(x) = \cos nx$. The functions given by $f(x) = \cos kx$ and $g(x) = \sin nx$, where k and n need not be distinct, can also be verified to be orthogonal.

An element $f \in L^2(X)$ is said to be **normalised** if $\|f\| = 1$, i.e. if $\langle f, f\rangle = 1$. It is left to the reader to verify independently that

(a) when n is a positive integer, the function in $L^2[-\pi,\pi]$ given by

$$f(x) = \frac{\cos nx}{\sqrt{\pi}} \quad \text{or} \quad f(x) = \frac{\sin nx}{\sqrt{\pi}}$$

is normalised;

(b) the functions in $L^2[-1,1]$ given by $f(x) = \sqrt{\frac{1}{2}}$ and $g(x) = \sqrt{\frac{3}{2}}x$ are orthogonal and each one is normalised. The reader may compute the relevant integrals to confirm this.

We have already encountered convergence pointwise, convergence pointwise a.e., uniform convergence and convergence in the L^p metric of a sequence of functions. The following implications are trivial:

uniform convergence $\Rightarrow$ pointwise convergence $\Rightarrow$ pointwise convergence a.e.

It is known from elementary analysis that the reverse of the first implication is not true. The reverse of the second implication is also not true, as demonstrated in Examples 2.5.11(b) and (c).

In the case when $\mu(X) < \infty$, it is easy to see that $\|f\|_p \leq \|f\|_\infty \cdot \mu(X)^{1/p}$. It follows that $L^\infty(X) \subseteq L^p(X)$ and that convergence in the L^∞ metric implies convergence in the L^p metric.

Convergence in the L^2 metric is also called **mean square convergence**: $\lim_{n\to\infty} \int_X [f_n(x) - f(x)]^2 d\mu(x) = 0$, or $\lim_{n\to\infty} \|f_n - f\|_2 = 0$. The notion applies to functions g for which $\int_X g(x)^2 d\mu(x)$ is finite, the so-called **square integrable** functions.

In the next chapter, we shall study mean square convergence of Fourier series of square integrable functions in the context of the Riemann integral as well as the Lebesgue integral. The use of the Lebesgue integral is preferred because of the availability of a completeness property [Theorem 3.3.10].

Problem Set 3.3

3.3.P1. Give an example of a nonnegative extended real-valued function on [0,1] which has Lebesgue integral 1 and is (a) the limit of an increasing sequence of continuous functions (b) unbounded on every subinterval of positive length (so that it cannot have an improper Riemann integral).

3.3.P2. Give an example when $\int_X |f_n - f| d\mu \to 0$ but f_n does not converge pointwise to f anywhere.

3.3.P3. Let $\{f_k\}_{k\geq 1}$ be a sequence of integrable functions on a measurable set X such that $\sum_{k=1}^{\infty} \int_X |f_k| < \infty$. Show that the series $\sum_{k=1}^{\infty} f_k$ converges a.e. to an integrable sum f and that $\int_X f = \sum_{k=1}^{\infty} \int_X f_k$.

3.3.P4. For $f \in L^1(\mathbb{R})$, show that

$$\lim_{h\to 0} \int_{\mathbb{R}} |f(x+h) - f(x)| dx = 0,$$

i.e., the mapping $\varphi : \mathbb{R} \to L^1(\mathbb{R})$ defined by $\varphi(h) = f_h$, where $f_h(x) = f(x + h)$, $x \in \mathbb{R}$, is continuous.

3.3.P5. Suppose that $f \in L^1(\mathbb{R})$. Show that

$$\lim_{|h|\to\infty} \int_{\mathbb{R}} |f(x+h) + f(x)| dx = 2\int_{\mathbb{R}} |f(x)| dx.$$

3.3.P6. If $f \in L^2[0,1]$, show that $f \in L^1[0,1]$. However, if $f \in L^2[0,\infty)$ then f need not belong to $L^1[0,\infty)$.

3.3.P7. The functions in $L^2[-\pi,\pi]$ given by $f_n(x) = \sin nx$, $n \in \mathbb{N}$, form a bounded and closed subset which is not compact.

3.3.P8. Let f and g be positive measurable functions on $[0,1]$ such that $fg \geq 1$. Prove that

$$\left(\int_{[0,1]} f\, dm\right)\left(\int_{[0,1]} g\, dm\right) \geq 1.$$

3.3.P9. Show that the following inequalities are inconsistent for a function $f \in L^2[0,\pi]$:

$$\|f - \cos\| \leq \frac{2}{3}, \quad \|f - \sin\| \leq \frac{1}{3}.$$

3.3.P10. Let $f_n(x) = \frac{n}{1+n\sqrt{x}}$ for $x \in [0,1]$ and $n \in \mathbb{N}$.

(a) Show that the sequence $\{f_n\}_{n\geq 1}$ has a pointwise limit a.e.;
(b) Does f_n belong to $L^2[0,1]$ for each (or some) $n \in \mathbb{N}$?
(c) Does the pointwise limit (a.e.) of f_n belong to $L^2[0,1]$?

3.3.P11. [Cf. Problem 3.3.P14] Let $(X, \mathcal{F}, \mu)$ be a measure space with $\mu(X) < \infty$. Show that $L^\infty(X) \subseteq L^p(X)$ for all p, $0 < p < \infty$. Also, show that, if f is measurable, then

$$\lim_{p\to\infty} \|f\|_p = \|f\|_\infty.$$

3.3.P12.

(a) Let $0 < p < 1$ and let q be such that $\frac{1}{p} + \frac{1}{q} = 1$. Then $q < 0$. If $f \in L^p(X)$ and the function g on X is such that $g \neq 0$ a.e. and $\int_X |g|^q d\mu < \infty$, then $\int_X |g|^q d\mu \neq 0$ unless $\mu(X) = 0$, and

$$\int_X |fg| d\mu \geq \left(\int_X |f|^p d\mu\right)^{1/p} \left(\int_X |g|^q d\mu\right)^{\frac{1}{q}}.$$

(This is Hölder's Inequality for $0 < p < 1$.) If the hypothesis that $\int_X |g|^q d\mu < \infty$ is omitted, then the inequality is valid trivially, provided we interpret a negative power of ∞ to mean 0.

(b) Let $0 < p < 1$ and f, g be measurable functions on X such that $f \geq 0$, $g \geq 0$. Then

$$\|f + g\|_p \geq \|f\|_p + \|g\|_p.$$

(This is Minkowski's Inequality for $0 < p < 1$.)

(c) For a complex-valued measurable function ϕ on X, show that $\left|\int_X \phi \, d\mu\right| \leq \int_X |\phi| \, d\mu$.

3.3.P13. Let $0 < p < 1$ and $X = [0,1]$. Then there exist f, $g \in L^p(X)$ such that

$$\|f + g\|_p > \|f\|_p + \|g\|_p.$$

3.3.P14. [Cf. Problem 3.3.P11] Show that, if $\mu(X) < \infty$ and $0 < p < q \leq \infty$, then $L^q(X) \subseteq L^p(X)$. Also, show that when $\mu(X) = \infty$, the inclusion does not hold.

3.3.P15. Let X be a measurable subset of $\mathbb{R}$ and $0 < p < q < \infty$. If $f \in L^p(X) \cap L^q(X)$, then $f \in L^r(X)$ for all r such that $p < r < q$.

3.3.P16.

(a) Let $p \geq 1$ and let $\|f_n - f\|_p \to 0$ as $n \to \infty$. Show that $\|f_n\|_p \to \|f\|_p$ as $n \to \infty$.

(b) Suppose $\{f_n\}_{n\geq 1}$ is a sequence in $L^p(X)$, $f \in L^p(X)$, $f_n \to f$ a.e. and $\|f_n\|_p \to \|f\|_p$ as $n \to \infty$. Prove that $\|f_n - f\|_p \to 0$ as $n \to \infty$.

3.3.P17. Show that $\int_{[0,\pi]} x^{-1/4} \sin x \, dx \leq \pi^{3/4}$.

3.3.P18. For each n, consider the functions $f_n : \mathbb{R} \to \mathbb{R}$ given by $f_n = \chi_{[n,n+1]}$, $n = 1,2,\ldots$. Show that $f_n(x) \to 0$ as $n \to \infty$ for each $x \in \mathbb{R}$. Also, show that for all p such that $1 \leq p \leq \infty$, we have $f_n \in L^p(\mathbb{R})$, $\|f_n\|_p = 1$ for all n, so that $\|f_n\|_p \not\to 0$

as $n \to \infty$. (This example shows that pointwise convergence does not imply convergence in any L^p norm, $1 \le p \le \infty$.)

3.3.P19. Give an example to show that convergence in the pth norm does not imply convergence a.e.

3.3.P20. Let p and q be conjugate indices and $\lim_{n\to\infty} \|f_n - f\|_p = 0 = \lim_{n\to\infty} \|g_n - g\|_q$, where $f_n, f \in L^p(X)$ and $g_n, g \in L^q(X)$, $n = 1,2,\ldots$. Show that $\lim_{n\to\infty} \|f_n g_n - fg\|_1 = 0$.

3.3.P21. For $1 \le p < \infty$, we denote by ℓ^p the space of all sequences $\{x_\nu\}_{\nu \ge 1}$ such that $\sum_{\nu=1}^{\infty} |x_\nu|^p < \infty$.

(a) Without interpreting sums as integrals with respect to the counting measure, show that, if $\{x_\nu\}_{\nu\ge1} \in \ell^p$ and $\{y_\nu\}_{\nu\ge1} \in \ell^q$, where $1 < p,\ q < \infty$ and $\frac{1}{p} + \frac{1}{q} = 1$, then

$$\sum_{\nu=1}^{\infty} |x_\nu y_\nu| \le \left(\sum_{\nu=1}^{\infty} |x_\nu|^p\right)^{1/p} \left(\sum_{\nu=1}^{\infty} |y_\nu|^q\right)^{1/q}.$$

For $p = 1$, $q = \infty$, show that

$$\sum_{\nu=1}^{\infty} |x_\nu y_\nu| \le \left(\sum_{\nu=1}^{\infty} |x_\nu|\right) \|\{y_\nu\}_{\nu\ge1}\|_\infty.$$

(b) Using the inequality of part (a), show also that, if $\{x_\nu\}_{\nu\ge1} \in \ell^p$ and $\{y_\nu\}_{\nu\ge1} \in \ell^p$, then

$$\left(\sum_{\nu=1}^{\infty} |x_\nu + y_\nu|^p\right)^{1/p} \le \left(\sum_{\nu=1}^{\infty} |x_\nu|^p\right)^{1/p} + \left(\sum_{\nu=1}^{\infty} |y_\nu|^p\right)^{1/p}.$$

[The inequalities in (a) and (b) are known as Hölder and Minkowski Inequalities respectively for sequences.]

3.3.P22. For the space ℓ^p $(1 \le p < \infty)$ of Problem 3.3.P21, show that

$$d_p(x, y) = \left(\sum_{\nu=1}^{\infty} |x_\nu - y_\nu|^p\right)^{1/p}$$

defines a complete metric, without using Theorem 3.3.10.

3.3.P23. Show that, if $k_1, k_2, \ldots, k_n > 1$, $\sum_{i=1}^{n} \frac{1}{k_i} = 1$ and $f_i \in L^{k_i}(X)$ for each i, then

$$\int_X \left| \prod_{i=1}^{n} f_i \right| d\mu \leq \prod_{i=1}^{n} \left(\int_X |f_i|^{k_i} d\mu \right)^{\frac{1}{k_i}}.$$

Equality occurs if and only if either $f_i = 0$ a.e. for some i or there exist positive constants c_i, $1 \leq i \leq n$, such that

$$c_i |f_i|^{k_i} = c_j |f_j|^{k_j} \text{ a.e.} 1 \leq i, j \leq n.$$

3.3.P24.

(a) Let $\infty \geq p \geq 1$ and $f_i \in L^p(X)$ for $i = 1,2,\ldots, n$. Show that

$$\left\| \sum_{i=1}^{n} f_i \right\|_p \leq \sum_{i=1}^{n} \|f_i\|_p.$$

(b) Let $\infty > p > 1$. If equality holds in (a), show that there exist nonnegative constants c_i, $1 \leq i \leq n$, not all 0, such that $c_i f_i = c_j f_j$ a.e. for $1 \leq i, j \leq n$.
(c) If either $p = 1$ or $p = \infty$, show that the analogue of (b) does not hold.

3.3.P25. Show that, if for some p, where $0 < p < \infty, f \in L^p(X) \cap L^\infty(X)$, then for all q such that $p < q < \infty$, we have $f \in L^q(X)$ and

$$\|f\|_q \leq \|f\|_p^{p/q} \|f\|_\infty^{(1-p/q)}.$$

3.4 Dense Subsets of L^p

According to Definition 2.5.6, a simple function is required to be nonnegative. If we drop this requirement, the resulting function f will be a difference of simple functions, because f^+, f^- will both be simple. The next result says essentially that the integrable functions among such differences of simple functions form a dense subset of the space of integrable functions.

Proposition 3.4.1 *Given any $f \in L^p(X)$, $1 \leq p < \infty$, there exists a sequence $\{s_n\}_{n\geq 1}$ of differences of simple functions such that $\lim_{n\to\infty} \int_X |s_n - f|^p d\mu = 0$. If f is nonnegative, then we can choose the sequence $\{s_n\}_{n\geq 1}$ to consist of nonnegative functions. In particular, differences of simple functions are dense in $L^p(X)$.*

Proof Suppose f is nonnegative. By Theorem 2.5.9, there exists a sequence $\{s_n\}_{n\geq 1}$ of simple functions such that

$$\text{(a)}\quad 0 \le s_1(x) \le s_2(x) \le \cdots \le f(x),$$
$$\text{(b)}\quad s_n(x) \to f(x) \text{ as } n \to \infty.$$

Observe that $|s_n(x) - f(x)|^p \le |f(x)|^p$. Since $f \in L^p(X)$, i.e., $\int_X |f|^p d\mu < \infty$, using the Dominated Convergence Theorem 3.2.16, we have $\lim\limits_{n\to\infty} \int_X |s_n - f|^p d\mu = 0$. Observe that the sequence $\{s_n\}_{n\ge 1}$ consists of nonnegative functions.

We get the result for a general $f \in L^p(X)$ by applying what has been proved to f^+, f^- separately. □

Most of our considerations in this chapter have been valid for a general measure μ. In the rest of this section, we shall be considering only Lebesgue measure m.

We begin by proving that continuous functions on a compact set X are dense among integrable functions on X, which is the precise version of what we called "without overkill" in Remark 3.3.15.

Proposition 3.4.2 *Given any function $f \in L^p(X)$, $1 \le p < \infty$, where $X \subseteq \mathbb{R}$ is a compact set, there exists a sequence $\{f_n\}_{n\ge 1}$ of continuous functions with domain X such that $\lim\limits_{n\to\infty} |f_n - f|^p dm = 0$. If f is nonnegative, then we can choose the sequence $\{f_n\}_{n\ge 1}$ to consist of nonnegative functions. In particular, continuous functions are dense in $L^p(X)$.*

Proof It is sufficient to prove this when f is a characteristic function χ_E on X with $E \subseteq X$. For, it will first extend immediately to simple integrable functions and then, by Proposition 3.4.1, to all functions in $L^p(X)$. The case when $m(E) = 0$ needs no argument, because we can take every f_n to be zero everywhere. So we work with the hypothesis that $0 < m(E) < \infty$.

The first step is to reduce to the case when E is compact. From the hypothesis $0 < m(E) < \infty$, it follows by Problem 2.3.P18 that for each $\varepsilon > 0$, one can find a compact $K \subseteq E$ such that $0 \le m(E) - m(K) < \varepsilon$. Then

$$0 \le \int_X |\chi_E - \chi_K|^p dm = \int_X (\chi_E - \chi_K) dm = \int_X \chi_E dm - \int_X \chi_K dm$$
$$= m(E) - m(K) < \varepsilon.$$

This means that if the required kind of sequence can be found for every χ_K with K compact, then the same can be done for every χ_E with $E \subseteq X$. Therefore we need only prove the result for $f = \chi_K$ with nonempty compact K (the case when $K = \emptyset$ being trivial).

Accordingly, consider any nonempty compact $K \subseteq X$. Define the sequence of functions g_n on $\mathbb{R}$ as

$$g_n(x) = \frac{1}{1 + n \cdot d(x, K)}, \tag{3.39}$$

where $d(x, K)$ denotes the distance of x from K as in Proposition 1.3.24. As shown there, it is a continuous function of x and therefore every g_n is continuous. (K has to be assumed nonempty for Proposition 1.3.24 to be applicable.) It follows that the restriction of g_n to X is continuous, as is obviously the case with a restriction of any continuous function. In the rest of this proof, g_n will mean the restriction to X. Then each g_n satisfies

$$0 \leq g_n \leq 1 \text{ everywhere on } X. \tag{3.40}$$

The compactness of K guarantees that $\text{dist}(x, K) = 0 \Leftrightarrow x \in K$. Together with (3.39), this implies

$$g_n \to \chi_K \text{ everywhere on } X.$$

Now (3.40) yields

$$|g_n - \chi_K| \leq 1 \text{ everywhere on } X.$$

Since X is compact, it is bounded and hence its Lebesgue measure must be finite and hence the constant function 1 is integrable on it. Therefore the Dominated Convergence Theorem 3.2.16 can be used to conclude that

$$\lim_{n\to\infty} \int_X |g_n - \chi_K|^p dm = 0,$$

which proves the result for $f = \chi_K$. As already noted, this is all that we needed to prove.

The fact that the g_n are nonnegative justifies the assertion of the theorem that the functions f_n can be chosen to be nonnegative if f is nonnegative. □

When X is an interval $[a, b]$, the space $L^p(X, \mathcal{F}, m)$ with Lebesgue measure m and $\mathcal{F}$ the σ-algebra of either Lebesgue measurable sets or Borel sets will be denoted by $L^p[a, b]$.

Proposition 3.4.3 *Suppose $f \in L^p[a, b]$, $1 \leq p < \infty$ and let $\varepsilon > 0$. Then there exists a step function h such that $\int_{[a,b]} |f - h|^p dm < \varepsilon$. If f is nonnegative, then we can choose the step function h to be nonnegative. In particular, step functions are dense in $L^p(X)$.*

Proof As $f = f^+ - f^-$, we may assume throughout that $f \geq 0$. By Proposition 3.4.1, there exists a simple function $s \geq 0$ such that $\int_{[a,b]} |f - s|^p dm < \varepsilon/2$. The simple function s may be expressed as

$$s = \sum_{i=1}^{n} a_i \chi_{E_i}$$

with $\cup_{1\le i\le n}E_i = [a, b]$. Let $\varepsilon' = \frac{\varepsilon}{2nM^p}$, where $M = \sup s$ on $[a, b]$, and M is supposed to be positive. For each of the measurable sets E_i, there exist a union G_i of finitely many disjoint open intervals such that $m(E_i\Delta G_i) < \varepsilon'$ [see Problem 2.3.P14]. This inequality continues to hold if we replace each open interval by its intersection with $[a, b]$. Then χ_{Gi} is a step function on $[a, b]$ such that $\int_{[a,b]} |\chi_{E_i} - \chi_{G_i}|^p dm = m(E_i\Delta G) < \varepsilon'$. Therefore

$$\int_{[a,b]} \left| s - \sum_{i=1}^{n} a_i\chi_{G_i} \right|^p dm < \sum_{i=1}^{n} a_i^p\varepsilon' \le nM^p\varepsilon' = \frac{\varepsilon}{2}.$$

By Minkowski's Inequality of Theorem 3.3.6,

$$\left[\int_{[a,b]} \left| f - \sum_{i=1}^{n} a_i\chi_{G_i} \right|^p dm\right]^{1/p} \le \left[\int_{[a,b]} |f - s|^p dm\right]^{1/p} + \left[\int_{[a,b]} \left| s - \sum_{i=1}^{n} a_i\chi_{G_i} \right|^p dm\right]^{1/p}$$
$$< 2\left(\frac{\varepsilon}{2}\right)^{1/p},$$

and $\sum_{i=1}^{n} a_i\chi_{Gi}$ is step function. □

We note that a bounded measurable function on a closed bounded interval is integrable and therefore the above proposition is applicable.

Proposition 3.4.4 *Trigonometric polynomials form a dense subset of $L^2[-\pi,\pi]$ and of $L^1[-\pi,\pi]$.*

Proof Consider any $f \in L^2[-\pi,\pi]$ and an arbitrary $\varepsilon > 0$. In view of Proposition 3.4.2, there exists a continuous ϕ such that

$$\|f - \phi\| < \varepsilon/2. \tag{3.41}$$

Define ψ by $\psi(x) = \phi(x)$ in $[-\pi, \pi-\delta]$, $\psi(\pi) = \phi(-\pi)$ and ψ linear in $[\pi-\delta, \pi]$, where $\delta \in (0,2\pi)$ will be chosen subsequently but is arbitrary for now. If $K > 0$ is such that $|\phi(x)| \le K$, then $|\psi(x)| \le K$ as well. Therefore

$$\|\phi - \psi\|^2 = \int_{[\pi-\delta,\pi]} (\phi - \psi)^2 dm \le 4K^2\delta < \frac{\varepsilon^2}{16}$$

for sufficiently small δ. Then by (3.41),

$$\|f - \psi\| < \frac{3\varepsilon}{4}. \tag{3.42}$$

Observe that ψ is continuous and $\psi(\pi) = \psi(-\pi)$. From the Weierstrass Approximation Theorem 1.5.5, it follows that there exists a trigonometric polynomial T which uniformly approximates ψ arbitrarily closely:

$$\sup\{|T(x) - \psi(x)| : x \in [-\pi, \pi]\} < \varepsilon/4(2\pi)^{\frac{1}{2}}.$$

It follows that

$$\|T - \psi\| = \sqrt{\int_{[-\pi,\pi]} (T(x) - \psi(x))^2 dm(x)} \leq \sqrt{\left(\frac{\varepsilon^2}{32\pi} 2\pi\right)} = \frac{\varepsilon}{4}. \tag{3.43}$$

Now (3.42) and (3.43) together imply that $\|f - T\| < \varepsilon$.

A similar argument works in $L^1[-\pi,\pi]$. □

In Proposition 3.4.2, the set X is assumed to be compact. Therefore, not only does it have finite Lebesgue measure, but also every continuous function on it is bounded. This ensures that every continuous function on it is in L^p. If X is to be replaced by $\mathbb{R}$, it will no longer be the case that every continuous function is in L^p. However, every continuous function that vanishes outside some compact set will be. Such functions are said to be of "compact support" in accordance with the definition below.

Definition 3.4.5 The **support** of a function f defined on a metric space X is the closure of the subset $\{x \in X : f(x) \neq 0\}$.

For example, if $X = (0, 2]$ and $f = X_{(0,1)}$, then the support of f is the closure of (0,1) in (0,2], which is (0, 1]. This support is not compact. However, if we understand f to be the same characteristic function $X_{(0,1)}$ but on $\mathbb{R}$, then the support is [0,1], which is compact. The support of any function $f = X_{[1,\alpha)}$ on (0,2], where $1 < \alpha \leq 2$, is $[1,\alpha]$, which is a compact set. An example of a continuous function on $\mathbb{R}$ with compact support is the product $f(x) = \chi_{[0,\pi]}(x) \sin x$. It is trivial that sums and constant multiples of continuous functions of compact support are again continuous functions of compact support.

Proposition 3.4.6 *Given any function $f \in L^p(\mathbb{R})$, $1 \leq p < \infty$, there exists a sequence $\{f_n\}_{n\geq 1}$ of continuous functions with compact support such that $\lim_{n\to\infty} \int_{\mathbb{R}} |f_n - f|^p dm = 0$. If f is nonnegative, then we can choose the sequence $\{f_n\}_{n\geq 1}$ to consist of nonnegative functions. In particular, continuous functions with compact support are dense in $L^p(\mathbb{R})$.*

Proof It is sufficient to prove the result for a nonnegative f.

There exists an increasing sequence $\{s_n\}_{n\geq 1}$ of nonnegative simple functions converging to f on $\mathbb{R}$ [see Theorem 2.5.9]. The sequence of products $s_n\chi_{[-n,\,n]}$ then also converges to f and satisfies $\left|s_n\chi_{[-n,n]} - f\right|^p \leq |f|^p$ and $\left|s_n\chi_{[-n,n]} - f\right|^p \to 0$. Besides, these products are all in $L^p(\mathbb{R})$. By the Dominated Convergence

Theorem 3.2.16, we get $\lim_{n\to\infty} \int_{\mathbb{R}} \left| s_n\chi_{[-n,n]} - f \right|^p dm = 0$. Note that $s_n\chi_{[-n,n]}$ vanishes outside a bounded interval.

Let $\varepsilon > 0$ be arbitrary. It follows from what has been established in the preceding paragraph that there exists a simple nonnegative function s such that $\int_{\mathbb{R}} |s - f|^p dm < \varepsilon$ and s vanishes outside a bounded interval. Since s is a finite linear combination of characteristic functions of bounded sets, we have only to prove the theorem in the case $f = \chi_E$, where $E \subseteq \mathbb{R}$ is a bounded measurable set.

If $m(E) = 0$, then $f = 0$ a.e. and we may take $f_n = 0$ for all n. So, assume $m(E) > 0$.

By Proposition 2.3.24, there exists a closed set F and an open set O such that $F \subseteq E \subseteq O$ and $m(O\backslash F) < \min\{\varepsilon, m(E)\}$. Since E is bounded, it is contained in some bounded open interval and, by intersecting O with that interval if necessary, we may assume that O is bounded. In particular, $O^c \neq \varnothing$. Also, $m(E) - m(F) \leq m(O\backslash F) < m(E) < \infty$, so that $m(F) > 0$ and hence $F \neq \varnothing$. Since $O^c \neq \varnothing \neq F$, the distances $d(x, O^c)$ and $d(x, F)$ are well defined on $\mathbb{R}$. It is clear from the definition of dist(x, A) as being $\inf\{ |x - a| : a \in A\}$ that, when A is closed, we have $d(x, A) = 0 \Leftrightarrow x \in A$. Since O^c and F are both closed, it follows that $d(x, O^c) + d(x, F) = 0 \Leftrightarrow x \in O^c \cap F$; however $O^c \cap F = \varnothing$ because $F \subseteq O$ and hence $d(x, O^c) + d(x, F) > 0$ for all $x \in \mathbb{R}$. So, there exists a real-valued function $g : \mathbb{R} \to \mathbb{R}$ given by

$$g(x) = \frac{d(x, O^c)}{d(x, F) + d(x, O^c)}.$$

Clearly, we have (i) $0 \leq g \leq 1$ everywhere (ii) $g = 1$ on F and (iii) $g = 0$ on O^c, so that $\{x \in X : g(x) \neq 0\} \subseteq O$, a bounded set, which has the consequence that the support of g, which is closed by definition, is bounded and therefore compact. We also know from Proposition 1.3.24 that g is continuous. Thus, g is a nonnegative continuous function of compact support and satisfies $0 \leq g \leq 1$ everywhere.

From the fact that $g = 1$ on F and $g = 0$ on O^c, we conclude that $g - \chi_E$ can be nonzero only on $O\backslash F$. Also, we have $|g - \chi_E| \leq 1$ everywhere on $\mathbb{R}$. It follows that $\int_{\mathbb{R}} |g - \chi_E|^p dm \leq m(O\backslash F) < \varepsilon$.

The fact that g is nonnegative justifies the assertion of the theorem that the functions f_n can be chosen to be nonnegative if f is nonnegative. □

Chapter 4
Fourier Series

4.1 Introduction

Fourier series arose in connection with the study of two physical problems, namely, the problem of the vibrating string and that of heat conduction in solids. The new series became one of the most important tools in mathematical physics and had a far-reaching influence in mathematical research. In 1807, Fourier announced that an 'arbitrary' function f could be represented in the form

$$f(x) = \frac{1}{2}a_0 + \sum_{n=1}^{\infty} (a_n \cos nx + b_n \sin nx),$$

with coefficients a_n and b_n given by the formulae

$$a_n = \frac{1}{\pi}\int_{-\pi}^{\pi} f(x)\cos nx\, dx, \qquad n = 0, 1, 2, \ldots$$

and

$$b_n = \frac{1}{\pi}\int_{-\pi}^{\pi} f(x)\sin nx\, dx, \qquad n = 1, 2, 3, \ldots.$$

The above series is called the "Fourier series" of the function f. The reason for using $\frac{1}{2}a_0$ in the Fourier series is that for the constant function 1, it turns out that $a_0 = 2$ while $a_n = b_n = 0$ for $n > 0$.

This announcement led to the discovery of mathematical theories such as Cantor's theory of infinite sets, the Riemann and Lebesgue integrals and summability of series.

The purpose of this chapter is to acquaint the reader with some of the most important aspects of Fourier series and the role of $L^2[-\pi, \pi]$. Research in the last

S. Shirali and H. L. Vasudeva, *Measure and Integration*,
Springer Undergraduate Mathematics Series,
https://doi.org/10.1007/978-3-030-18747-7_4

fifty years or so has added enormously to our knowledge of Fourier series and has done much to improve the earlier treatments of the subject. The origins of Fourier series in mathematical physics and all its generalisations and ramifications have been scrupulously avoided here, and we shall be content to derive only some basic theorems.

Definition 4.1.1 A function f with domain $\mathbb{R}$ is called **periodic** if there is some positive number l such that

$$f(x+l) = f(x) \qquad \text{for all real } x.$$

Any such number l is called a **period** of f. The graph of such a function is obtained by repetition of its graph over any interval of length l on both sides of that interval.

The trigonometric functions sine and cosine are the simplest examples of nonconstant differentiable periodic functions; they have 2π as a period.

For a function f with period l, we have (from the definition)

$$f(x) = f(x+l) = f((x+l)+l) = f(x+2l) = f(x+3l) = \cdots = f(x+nl)$$

for any positive integer n. Hence $2l$, $3l$, … are also periods of f. Furthermore, if f and g both have l as a period, then the function h defined by

$$h(x) = af(x) + bg(x),$$

where a and b are constants, also has l as a period. If a periodic function f has a smallest positive period, then this period is often called the *fundamental period* of f. Thus sine and cosine have 2π as the fundamental period while a constant function, though periodic, has no fundamental period. An example of a nonconstant periodic function f having no fundamental period is

$$f(x) = \begin{cases} 0 & \text{if } x \text{ is rational} \\ 1 & \text{if } x \text{ is irrational.} \end{cases}$$

The pointwise limit of a sequence of functions with period l again has period l.

We shall be discussing the possibility of representing real-valued integrable periodic functions as the sum of a series, in which the terms are constant multiples of the trigonometric functions 1, $\cos x$, $\sin x$, $\cos 2x$, $\sin 2x$, …, i.e., as

$$\alpha_0 + \alpha_1 \cos x + \beta_1 \sin x + \alpha_2 \cos 2x + \beta_2 \sin 2x + \cdots,$$

where α_0, α_1, α_2, … and β_1, β_2, … are real constants. Such a series is called a *trigonometric series* and the α_n and β_n are called its *coefficients*. It may be compactly written as

$$\alpha_0 + \sum_{n=1}^{\infty} (\alpha_n \cos nx + \beta_n \sin nx).$$

In what follows, we shall need to evaluate some important integrals involving the trigonometric functions 1, cos x, sin x, cos $2x$, sin $2x$, ...

Lemma 4.1.2 *The following equalities hold*:

(a) $\int_{-\pi}^{\pi} \cos kx \cos nx\, dx = 0 \quad \text{for } k, n = 0, 1, 2, \ldots \text{ and } k \neq n;$

(b) $\int_{-\pi}^{\pi} \cos^2 nx\, dx = \begin{cases} \pi & \text{for } n = 1, 2, \ldots \\ 2\pi & \text{for } n = 0; \end{cases}$

(c) $\int_{-\pi}^{\pi} \sin kx \sin nx\, dx = 0 \quad \text{for } k, n = 1, 2, \ldots \text{ and } k \neq n;$

(d) $\int_{-\pi}^{\pi} \sin^2 nx\, dx = \pi \quad \text{for } n = 1, 2, \ldots;$

(e) $\int_{-\pi}^{\pi} \cos kx \sin nx\, dx = 0 \quad \text{for } k, n = 0, 1, 2, \ldots.$

Proof Each of these results follows easily from one or the other of the identities

$$\cos(kx \pm nx) = \cos kx \cos nx \mp \sin kx \sin nx$$
$$\sin(kx \pm nx) = \sin kx \cos nx \pm \cos kx \sin nx,$$

and elementary methods of integration. □

Proposition 4.1.3 *If*

$$f(x) = a_0 + \sum_{n=1}^{\infty} (a_n \cos nx + b_n \sin nx), \tag{4.1}$$

where the trigonometric series converges uniformly on $[-\pi, \pi]$, *then its coefficients are given by the formulae*

$$\left.\begin{array}{c} a_0 = \frac{1}{2\pi}\int_{-\pi}^{\pi} f(x)\, dx \\ a_n = \frac{1}{\pi}\int_{-\pi}^{\pi} f(x) \cos nx\, dx, \; b_n = \frac{1}{\pi}\int_{-\pi}^{\pi} f(x) \sin nx\, dx \quad \text{for } n = 1, 2, \ldots. \end{array}\right\} \tag{4.2}$$

Proof In view of the given uniform convergence in (4.1) the limit function f is continuous, so that all integrals in question exist. Moreover, we have

$$\begin{aligned} \int_{-\pi}^{\pi} f(x) dx &= \int_{-\pi}^{\pi} [a_0 + \sum_{n=1}^{\infty} (a_n \cos nx + b_n \sin nx)] dx \\ &= \int_{-\pi}^{\pi} a_0 dx + \sum_{n=1}^{\infty} [a_n \int_{-\pi}^{\pi} \cos nx\, dx + b_n \int_{-\pi}^{\pi} \sin nx\, dx] \\ &= 2\pi\, a_0, \end{aligned}$$

using Lemma 4.1.2. Thus $a_0 = \frac{1}{2\pi}\int_{-\pi}^{\pi} f(x) dx$.

From (4.1), it follows that

$$f(x)\cos nx = a_0 \cos nx + \sum_{k=1}^{\infty} (a_k \cos kx \cos nx + b_k \sin kx \cos nx). \tag{4.3}$$

What is more, the convergence here is once again uniform. This is because the difference between any two partial sums of (4.3) is obtained by multiplying cos nx by the corresponding difference in (4.1), whereby the uniform Cauchy property is preserved because $|\cos nx| \leq 1$. In view of the uniform convergence in (4.3), we have by Proposition 1.7.5

$$\begin{aligned}\int_{-\pi}^{\pi} f(x)\cos nx\, dx &= \int_{-\pi}^{\pi} a_0 \cos nx\, dx \\ &\quad + \sum_{k=1}^{\infty} [a_n \int_{-\pi}^{\pi} \cos nx \cos kx\, dx + b_n \int_{-\pi}^{\pi} \cos nx \sin kx\, dx] \\ &= a_n \int_{-\pi}^{\pi} \cos^2 nx\, dx = \pi a_n \quad \text{when } n \neq 0,\end{aligned}$$

by Lemma 4.1.2. Thus $a_n = \frac{1}{\pi}\int_{-\pi}^{\pi} f(x) \cos nx\, dx$ for n = 1,2, …. A similar argument establishes the equality claimed for b_n. □

Formulae (4.2) proved above are known as *Euler–Fourier formulae* for the coefficients.

It is easy to write down trigonometric series that diverge somewhere. For example, let $a_0 = a_n = b_n = 1$ for n = 1, 2, …; then the series becomes 1 + 1 + … when $x = 0$ and therefore diverges. Indeed, the series diverges when x is any rational multiple of π, because the terms do not approach 0. However, Proposition 4.1.3 shows that, if a trigonometric series converges to a sum function f uniformly, then the coefficients of the series are given by the Euler–Fourier Formulae (4.2). Therefore the approach we adopt is to start with a function that has period 2π and is integrable over a closed interval of length 2π (it must then be integrable over any closed bounded interval), define the coefficients as the right-hand sides of (4.2) and then study the relationship of the series so formed to the function.

Definition 4.1.4 Let $f : \mathbb{R} \to \mathbb{R}$ have period 2π, i.e.

$$f(x) = f(x + 2\pi) \quad \text{for all } x \in \mathbb{R}.$$

Assume that f is Lebesgue integrable over an interval of length 2π. Then the **Fourier series** of f is the series

$$\frac{1}{2}a_0 + \sum_{n=1}^{\infty} (a_n \cos nx + b_n \sin nx),$$

where

$$a_n = \frac{1}{\pi}\int_{-\pi}^{\pi} f(x)\cos nx\,dx, \quad n = 0, 1, 2, \ldots$$

and

$$b_n = \frac{1}{\pi}\int_{-\pi}^{\pi} f(x)\sin nx\,dx, \quad n = 1, 2, 3, \ldots.$$

The a_n and b_n are called the **Fourier coefficients** of f and we write

$$f(x) \sim \frac{1}{2}a_0 + \sum_{n=1}^{\infty}(a_n \cos\ nx + b_n \sin\ nx).$$

The Fourier series on the right-hand side here is said to 'correspond to' the function. Note that we have used the symbol '$\sim$' and not '=', because the sum of the series may not be equal to $f(x)$ for some values of x, i.e. the Fourier series may not represent the function, as the first of the following examples shows.

Examples 4.1.5

(a) The Fourier series corresponding to the function

$$f(x) = \begin{pmatrix} 0 & -\pi \le x < 0 \\ 1 & 0 \le x \le \pi \end{pmatrix}$$

extended to $\mathbb{R}$ so as to be periodic, fails to converge to the function at $x = 0$. Here

$$a_n = \frac{1}{\pi}\int_{-\pi}^{\pi} f(x)\cos nx\,dx = \frac{1}{\pi}\int_{0}^{\pi}\cos nx\,dx = \begin{cases} 1 & \text{when } n = 0 \\ 0 & \text{when } n \neq 0 \end{cases}$$

and

$$b_n = \frac{1}{\pi}\int_{-\pi}^{\pi} f(x)\sin nx\,dx = \frac{1}{\pi}\int_{0}^{\pi}\sin nx\,dx = \frac{1-\cos n\pi}{n\pi}.$$

Thus

$$b_n = \frac{2}{n\pi} \text{ for } n = 1, 3, 5, \ldots \text{ and } b_n = 0 \text{ for } n = 2, 4, 6, \ldots.$$

Consequently,

$$f(x) \sim \frac{1}{2} + \frac{2}{\pi}\sum_{n=0}^{\infty}\frac{\sin(2n+1)x}{2n+1}.$$

When $x = 0$, the sum of this series is $\frac{1}{2}$ whereas $f(0) = 1$.

(b) Let $f(x) = \cos px$, where $p > 0$ is an integer. Then by Lemma 4.1.2, the Fourier coefficients of f are all 0 except for a_p, which is 1. Thus for example, when $p = 3$,

$$f(x) \sim 0 + (0+0) + (0+0) + (\cos 3x + 0) + (0+0) + \cdots.$$

A corresponding statement is true when $f(x) = \sin px$.

Note the linear dependence of the Fourier coefficients on the function; this has the consequence, for instance, that the Fourier series of a sum of two functions is obtained by simply adding corresponding terms; similarly for constant multiples. Moreover, the Fourier series of a constant function consists of a single term equal to that constant, all terms in the summation being zero.

The following lemma will be useful in a later section.

Lemma 4.1.6 *For any real x,*

$$\frac{1}{2} + \cos x + \cos 2x + \cdots + \cos nx = \begin{cases} \frac{\sin(\frac{2n+1}{2})x}{2\sin\frac{x}{2}} & \text{if } x \neq 2\pi k, k \in \mathbb{Z} \\ \frac{1}{2} + n & \text{if } x = 2\pi k, k \in \mathbb{Z}. \end{cases}$$

Proof Using the identity

$$\sin A - \sin B = 2\sin\frac{A-B}{2}\cos\frac{A+B}{2},$$

we have

$$\begin{aligned} \sin\frac{2n+1}{2}x &= \sin\frac{1}{2}x + \left(\sin\frac{3}{2}x - \sin\frac{1}{2}x\right) + \left(\sin\frac{5}{2}x - \sin\frac{3}{2}x\right) \\ &\quad + \cdots + \left(\sin\frac{2n+1}{2}x - \sin\frac{2n-1}{2}x\right) \\ &= \sin\frac{1}{2}x + 2\sin\frac{1}{2}x\cos x + 2\sin\frac{1}{2}x\cos 2x \\ &\quad + \cdots + 2\sin\frac{1}{2}x\cos nx \\ &= 2\left(\sin\frac{1}{2}x\right)\left(\frac{1}{2} + \cos x + \cos 2x + \cdots + \cos nx\right). \end{aligned}$$

The identity in question follows immediately from here for $x \neq 2k\pi$. The case when $x = 2k\pi$ is straightforward. □

Problem Set 4.1

4.1.P1. (a) [Needed in (b)] Prove Abel's Lemma: If $b_1 \geq b_2 \geq \cdots \geq b_n \geq 0$ and $k \leq \sum_{r=1}^{p} u_r \leq K$ for $1 \leq p \leq n$, where u_1, u_2, …, u_n are any n real numbers, then

$$b_1 k \leq \sum_{r=1}^{n} b_r u_r \leq b_1 K.$$

The next two parts together constitute what is called *Bonnet's form of the Second Mean Value Theorem for Integrals* and will be needed in Theorem 4.2.8 below.

(b) If $\int_a^b f$ and $\int_a^b g$ both exist and f is decreasing on $[a, b]$ with $f(b) \geq 0$, then there exists a $\xi \in [a, b]$ such that

$$\int_a^b fg = f(a) \int_a^\xi g.$$

(c) If $\int_a^b f$ and $\int_a^b g$ both exist and f is increasing on $[a, b]$ with $f(a) \geq 0$, then there exists a $\xi \in [a, b]$ such that

$$\int_a^b fg = f(b) \int_\xi^b g.$$

4.1.P2. Find the Fourier coefficients of the periodic function f for which

$$f(x) = \begin{cases} -\alpha & -\pi \leq x < 0 \\ \alpha & 0 \leq x < \pi. \end{cases}$$

4.1.P3. Using Bonnet's form of the Second Mean Value Theorem for Integrals of 4.1.P1(c), prove that for a monotone function f on $[-\pi, \pi]$, the Fourier coefficients a_n and b_n $(n \neq 0)$ satisfy the inequalities

$$|a_n| \leq \frac{1}{n\pi}|f(\pi) - f(-\pi)|, \quad |b_n| \leq \frac{1}{n\pi}|f(\pi) - f(-\pi)|.$$

4.1.P4. Suppose f is Lebesgue integrable with period 2π and "modulus of continuity" ω, which means

$$\omega(\delta) = \sup\{|f(x) - f(y)| : |x - y| \leq \delta\}.$$

Show that its Fourier coefficients satisfy $|a_n| \leq \omega(\pi/n)$, $|b_n| \leq \omega(\pi/n)$ $(n \neq 0)$.

4.1.P5. For the trigonometric series $\frac{1}{2}a_0 + \sum_{n=1}^{\infty}(a_n \cos nx + b_n \sin nx)$, suppose $\sum_{k=1}^{\infty}(|a_k| + |b_k|)$ converges. Show that the trigonometric series converges uniformly and that it is the Fourier series of its own sum.

4.1.P6. A finite set of continuous functions $g_1, \ldots, g_n$ on $[a, b]$ is said to be an **orthonormal** system with weight function w (Lebesgue integrable and nonnegative) if

$$\langle g_i, g_j \rangle = \int_a^b g_i(x) g_j(x) w(x) dx = \begin{cases} 0 & i \neq j \\ 1 & i = j. \end{cases}$$

(Example: By Lemma 4.1.2, any finite number of functions among $\frac{1}{2}$, cos kx, sin kx form an orthonormal set with domain $[-\pi, \pi]$ and weight $w(x) = \frac{1}{\pi}$.) For the function $\sum_{k=1}^{n} a_k g_k(x)$, show that

$$\sum_{k=1}^{n} |a_k| \leq \sqrt{n} \cdot \left(\int_a^b w(x) dx \right)^{\frac{1}{2}} \max \left\{ \left| \sum_{k=1}^{n} a_k g_k(x) \right| : x \in [a, b] \right\}.$$

4.1.P7. Suppose the sequence of partial sums s_n of the trigonometric series $a_0 + \sum_{n=1}^{\infty} (a_n \cos nx + b_n \sin nx)$ has a subsequence $S_{n(k)}$ that converges uniformly to a function f. Show that the trigonometric series is the Fourier series corresponding to f.

4.2 The Convergence Problem

Let us assume that $f \in L^1[-\pi, \pi]$. For later use, we define f on the entire real line $\mathbb{R}$ by periodicity: $f(x + 2k\pi) = f(x)$ for $x \in [-\pi, \pi)$ and $k \in \mathbb{Z}$ (the number $f(\pi)$ has no importance for $f \in L^1[-\pi, \pi]$). The fundamental questions to be investigated are whether the Fourier series of f [see Definition 4.1.4] converges, and if it does, whether the sum is $f(x)$. Briefly speaking, does the Fourier series of a 2π-periodic Lebesgue integrable function represent that function? Associated with the Fourier series is the sequence of its partial sums

$$s_n(x) = \frac{1}{2} a_0 + \sum_{k=1}^{n} (a_k \cos kx + b_k \sin kx), \quad n = 1, 2, \ldots.$$

We have seen in Example 4.1.5(a) that there are functions whose Fourier series fail to converge to $f(x)$ at some point. In 1876, du Bois-Reymond gave an example of a continuous function whose Fourier series diverges at some point. It surprised many mathematicians of the era, as it was believed until then that the Fourier series of a continuous function should converge at every point. In the appendix to this chapter, we present Fejér's construction of a continuous function whose Fourier series diverges at a point.

If, however, the function has a continuous derivative everywhere and has period 2π, then its Fourier series can be shown to converge to the function everywhere. This is a consequence of Dirichlet's Theorem, which will be proved later in this section. We shall also investigate to what extent the Fourier series determines the function.

Actually, the class of functions that can be represented by their Fourier series is much larger.

Example 4.2.1 For the Fourier series of the function f given by $f(x) = \cos px$, where $p > 0$ is an integer, recall from Example 4.1.5(b) that every Fourier coefficient is 0 except for a_p, which is 1. In other words, when $p > 0$,

$$\cos px \sim 0 + (0+0) + (0+0) + \cdots + (0+0) + (\cos px + 0) + (0+0) + \cdots,$$

where the term $(\cos px + 0)$ corresponds to $n = p$ in the Fourier series. It is understood of course that when $p = 0$, this becomes

$$\cos 0x \sim 1 + (0+0) + (0+0) + \cdots.$$

This shows that the above mentioned partial sums of the Fourier series are

$$s_n(x) = \begin{cases} 0 & n < p \\ \cos px & n \geq p. \end{cases}$$

Similarly for $\sin px$.

Remark 4.2.2 To express the partial sum $s_n(x)$ of the Fourier series of f in a manageable form, we use the Euler–Fourier formulae to obtain

$$\begin{aligned} s_n(x) &= \frac{1}{2}a_0 + \sum_{k=1}^{n}(a_k \cos kx + b_k \sin kx) \\ &= \frac{1}{2\pi}\int_{-\pi}^{\pi} f(x)dx + \frac{1}{\pi}\sum_{k=1}^{n}\left(\cos kx \int_{-\pi}^{\pi} f(t)\cos kt\, dt + \sin kx \int_{-\pi}^{\pi} f(t)\sin kt\, dt\right) \\ &= \frac{1}{\pi}\int_{-\pi}^{\pi} f(t)\Big[\frac{1}{2} + \sum_{k=1}^{n}(\cos kx \cos kt + \sin kx \sin kt)\Big]dt \\ &= \frac{1}{\pi}\int_{-\pi}^{\pi} f(t)\Big[\frac{1}{2} + \sum_{k=1}^{n}\cos k(x-t)\Big]dt. \end{aligned}$$

Using Lemma 4.1.6, we may rewrite this as

$$s_n(x) = \frac{1}{\pi}\int_{-\pi}^{\pi} f(t)D_n(x-t)dt,$$

where $D_n : \mathbb{R} \to \mathbb{R}$ is the function for which

$$D_n(u) = \begin{cases} \frac{\sin(\frac{2n+1}{2})u}{2\sin\frac{u}{2}} & \text{if } u \neq 2k\pi, k \in \mathbb{Z} \\ \frac{1}{2} + n & \text{if } u = 2k\pi, k \in \mathbb{Z} \end{cases}$$

for all $u \in \mathbb{R}$. This achieves the manageable form we set out to obtain; since D_n enters into it in a particularly simple manner, there is reason to expect that a study of this function may yield information about how f is related to its Fourier series. Accordingly, it deserves a special name: D_n is called the **Dirichlet kernel**.

Proposition 4.2.3 *The Dirichlet kernel has the following properties*:

(a) D_n *is continuous*;
(b) $D_n(-t) = D_n(t)$ *for all real* t;
(c) $\frac{2}{\pi}\int_0^\pi D_n(t)dt = 1$.

Proof This is immediate from Lemma 4.1.6, according to which

$$D_n(t) = \frac{1}{2} + \cos t + \cos 2t + \cdots + \cos nt \quad \text{for all } t \in \mathbb{R}.$$

□

Remark 4.2.4 An alternative form of the integral representation for $s_n(x)$ that was obtained above turns out to be useful. Substituting $t = x - u$ in the integral, we get

$$\begin{aligned} s_n(x) &= \frac{1}{\pi}\int_{x-\pi}^{x+\pi} f(x-u)D_n(u)du \\ &= \frac{1}{\pi}[\int_{x-\pi}^{-\pi} f(x-u)D_n(u)du + \int_{-\pi}^{\pi} f(x-u)D_n(u)du + \int_{\pi}^{x+\pi} f(x-u)D_n(u)du]. \end{aligned}$$

Observe that

$$\frac{1}{\pi}\int_{x-\pi}^{-\pi} f(x-u)D_n(u)\,du = \frac{1}{\pi}\int_{x+\pi}^{\pi} f(x-u)D_n(u)\,du,$$

using the periodicity of the integrand. Thus

$$s_n(x) = \frac{1}{\pi}\int_{-\pi}^{\pi} f(x-u)D_n(u)\,du.$$

Consequently,

$$\begin{aligned} s_n(x) &= \frac{1}{\pi}\int_{-\pi}^{0} f(x-u)D_n(u)\,du + \frac{1}{\pi}\int_{0}^{\pi} f(x-u)D_n(u)\,du \\ &= \frac{1}{\pi}\int_{0}^{\pi} f(x+u)D_n(-u)\,du + \frac{1}{\pi}\int_{0}^{\pi} f(x-u)D_n(u)\,du \\ &= \frac{1}{\pi}\int_{0}^{\pi} [f(x+u)+f(x-u)]D_n(-u)\,du, \end{aligned}$$

since $D_n(-u) = D_n(u)$ for all u. This is the alternative form which we shall be using below.

In the proof of Dirichlet's Theorem, we need a well-known and important result. It states that the Fourier coefficients a_k, b_k of a Lebesgue integrable function with period 2π must approach 0 as $k \to \infty$. It is often called the *Riemann–Lebesgue Lemma.*

Theorem 4.2.5 (Riemann–Lebesgue) *Let f be a Lebesgue integrable function on* $\mathbb{R}$. *Then*

$$\lim_{\lambda\to\infty}\int_{\mathbb{R}} f(x)\sin\,\lambda x\,dx = 0 = \lim_{\lambda\to\infty}\int_{\mathbb{R}} f(x)\cos\,\lambda x\,dx.$$

Proof Let $[\alpha, \beta]$ be any closed interval. If f is the characteristic function of $[\alpha, \beta]$, then

$$\left|\int_{\mathbb{R}} f(x)\sin\lambda x\,dx\right| = \left|\frac{\cos\lambda\alpha - \cos\lambda\beta}{\lambda}\right| \le \frac{2}{|\lambda|}.$$

The right-hand side tends to 0 as $\lambda \to \infty$. Thus the result holds for any step function, which is constant on a finite number of bounded intervals and vanishes outside their union.

Let $\varepsilon > 0$ be chosen arbitrarily. Then corresponding to ε, there is a positive number A [see 3.2.P18(b)] such that

$$\int_{|x|>A} |f(x)|dx < \frac{1}{2}\varepsilon. \tag{4.4}$$

By Proposition 3.4.3, there is a step function ϕ vanishing outside $[-A, A]$ such that

$$\int_{-A}^{A} |f(x) - \phi(x)|dx < \frac{1}{2}\varepsilon. \tag{4.5}$$

It follows from (4.4) and (4.5) that

$$\begin{aligned}\int_{\mathbb{R}} |f(x) - \phi(x)|dx &= \int_{-A}^{A} |f(x) - \phi(x)|dx + \int_{|x|>A} |f(x) - \phi(x)|dx \\ &= \int_{-A}^{A} |f(x) - \phi(x)|dx + \int_{|x|>A} |f(x)|dx \\ &< \varepsilon.\end{aligned}$$

Therefore

$$\begin{aligned}\left|\int_{\mathbb{R}} f(x) \sin \lambda x\, dx\right| &\leq \left|\int_{\mathbb{R}} (f(x) - \phi(x)) \sin \lambda x\, dx\right| + \left|\int_{\mathbb{R}} \phi(x) \sin \lambda x\, dx\right| \\ &\leq \int_{\mathbb{R}} |f(x) - \phi(x)|dx + \left|\int_{\mathbb{R}} \phi(x) \sin \lambda x\, dx\right| \\ &< \varepsilon + \left|\int_{\mathbb{R}} \phi(x) \sin \lambda x\, dx\right|.\end{aligned}$$

Since the result has already been shown to be true for step functions, it now follows that $\lim_{\lambda\to\infty} \int_{\mathbb{R}} f(x) \sin \lambda x\, dx = 0$. The argument that $\lim_{\lambda\to\infty} \int_{\mathbb{R}} f(x) \cos \lambda x\, dx = 0$ is similar. □

Corollary 4.2.6 *For any fixed positive $\delta < \pi$, we have*

$$\lim_{n\to\infty} \int_{\delta}^{\pi} D_n(t)dt = 0.$$

Proof In fact,

$$\int_{\delta}^{\pi} D_n(t)dt = \int_{\delta}^{\pi} \left(2 \sin\frac{1}{2}t\right)^{-1} \sin\left(n + \frac{1}{2}\right)t\, dt \to 0 \qquad \text{as} \qquad n \to \infty,$$

by Theorem 4.2.5 applied to the Riemann integrable function $(2 \sin\frac{1}{2}t)^{-1}$ on the interval $[\delta, \pi]$, where $\delta > 0$. □

The next result shows that the behaviour of the sequence $\{s_n(x)\}$ for large n depends only on the behaviour of f on the interval $[x - \delta, x + \delta]$, where $\delta > 0$ is arbitrary but less than π.

Proposition 4.2.7 (Localisation Theorem) *Let $f : \mathbb{R} \to \mathbb{R}$ have period 2π and be Lebesgue integrable on $[-\pi, \pi]$ and $0 < \delta < \pi$. Then for each $x \in \mathbb{R}$,*

$$\lim_{n\to\infty} s_n(x) = \lim_{n\to\infty} \int_{\delta}^{\pi} [f(x+t) + f(x-t)]D_n(t)dt = 0.$$

Proof By definition of D_n,

$$\int_{\delta}^{\pi} [f(x+t)+f(x-t)]D_n(t)\,dt = \int_{\delta}^{\pi} \frac{f(x+t)+f(x-t)}{2\sin\frac{1}{2}t} \sin(n+\frac{1}{2})t\,dt. \quad (4.6)$$

Since $\frac{f(x+t)+f(x-t)}{2\sin\frac{t}{2}}$ is Lebesgue integrable over $[\delta, \pi]$, it follows from Theorem 4.2.5 that the integral on the right-hand side of (4.6) approaches 0 as $n \to \infty$. □

Theorem 4.2.8 *For every Riemann integrable function f of period 2π and for every fixed number δ, $0 < \delta \leq \pi$, the integral*

$$\int_{\delta}^{\pi} [f(x+t)+f(x-t)]D_n(t)\,dt$$

tends to 0 as $n \to \infty$, uniformly for all real values of x.

Proof By definition of D_n,

$$\int_{\delta}^{\pi} [f(x+t)+f(x-t)]D_n(t)\,dt = \int_{\delta}^{\pi} \frac{f(x+t)+f(x-t)}{2\sin\frac{1}{2}t} \sin(n+\frac{1}{2})t\,dt. \quad (4.7)$$

Applying Bonnet's form of the Mean Value Theorem for Integrals [see Problem 4.1.P1], the right-hand side of (4.7) becomes

$$\frac{1}{2\sin\frac{\delta}{2}} \int_{\delta}^{\xi} [f(x+t)+f(x-t)] \sin(n+\frac{1}{2})t\,dt,$$

where $\delta \leq \xi \leq \pi$. We shall show that the last integral tends to 0 uniformly for all real values of x, as $n \to \infty$. It suffices to consider the integral

$$\int_{\delta}^{\pi} f(x+t) \sin(n+\frac{1}{2})t\,dt$$

as the other integral with $f(x-t)$ in place of $f(x+t)$ behaves analogously. The last integral equals

$$\int_{\delta+x}^{\xi+x} f(u)[\sin(n+\frac{1}{2})u\ \cos(n+\frac{1}{2})x - \cos(n+\frac{1}{2})u\ \sin(n+\frac{1}{2})x]du. \quad (4.8)$$

Since the function f has period 2π, it is enough to consider values of x lying in the interval $[0, 2\pi]$. Since

$$0<\delta\leq\delta+x\leq\delta+2\pi, \quad \xi\leq\xi+x\leq\xi+2\pi \quad \text{and} \quad 0<\delta\leq\xi\leq\pi,$$

it follows that the limits of the integral lie in the fixed interval $[0, 3\pi]$. Observe that $\left|\sin\left(n+\frac{1}{2}\right)x\right|\leq 1$ and $\left|\cos\left(n+\frac{1}{2}\right)x\right|\leq 1$ for all x. Using the Riemann–Lebesgue Lemma [see Theorem 4.2.5], the integral (4.8) above is seen to tend to 0 as $n\to\infty$, uniformly for all real x. □

The next result, which is of independent interest, will prove useful in establishing the main theorem.

Proposition 4.2.9 *The partial sums of the series* $\sum_{k=1}^{\infty}\frac{\sin kx}{k}$ *are uniformly bounded in absolute value by* $1+\pi$ *on* $\mathbb{R}$.

Proof We have

$$\begin{aligned}\left|\sum_{k=m}^{n}\frac{\sin kx}{k}\sin\frac{1}{2}x\right| &= \frac{1}{2}\left|\sum_{k=m}^{n}\frac{1}{k}[\cos(k-\frac{1}{2})x-\cos(k+\frac{1}{2})x\right| \\ &= \frac{1}{2}\left|\frac{1}{m}\cos(m-\frac{1}{2})x+\sum_{k=m}^{n-1}\left(\frac{1}{k+1}-\frac{1}{k}\right)\cos(k+\frac{1}{2})x-\frac{1}{n}\cos(n+\frac{1}{2})x\right| \\ &\leq \frac{1}{2}[\frac{1}{m}-\sum_{k=m}^{n-1}\left(\frac{1}{k+1}-\frac{1}{k}\right)+\frac{1}{n}]=\frac{1}{m}.\end{aligned} \tag{4.9}$$

Let $S_n(x)$ denote the nth partial sum of the series, i.e.

$$S_n(x)=\sum_{k=1}^{n}\frac{\sin kx}{k}, \qquad x\in(0,\pi].$$

Assuming that $x\neq 0$, denote the integer part of $\frac{1}{x}$ by ν, so that $\nu\leq\frac{1}{x}<\nu+1$; in particular, $\nu x\leq 1$. Since $|\sin t|\leq t$ for $t\geq 0$ [reason: the functions $t\pm\sin t$ are increasing for $t\geq 0$ and are 0 at $t=0$], we get

$$|S_n(x)|\leq\sum_{k=1}^{n}\frac{kx}{k}=nx.$$

For $n\leq\nu$, we obtain from here that $|S_n(x)|\leq\nu x\leq 1$. On the other hand, for $\nu<n$ and $x\in(0,\pi]$ we have

$$\begin{aligned}|S_n(x)| &\le |S_\nu(x)| + \left|\sum_{k=\nu+1}^{n} \frac{\sin kx}{k}\right| \\ &\le 1 + \frac{1}{\nu+1}, \quad \text{upon using (4.9)} \\ &\le 1 + 2\frac{\frac{1}{2}x}{\sin\frac{1}{2}x} \\ &\le 1 + \pi,\end{aligned}$$

using the fact that $\sin t \ge \frac{2t}{\pi}$ for $0 \le t \le \frac{\pi}{2}$. In fact, the function $\frac{\sin t}{t}$ is decreasing for $0 \le t \le \frac{\pi}{2}$ and its value at $t = \frac{\pi}{2}$ is $\frac{2}{\pi}$. This shows that for $x \in (0, \pi]$ and *all* natural numbers n, we have $|S_n(x)| \le 1 + \pi$. For $x = 0$, this inequality holds trivially, because $S_n(0) = 0$. Thus the upper bound $1 + \pi$ has been established for all partial sums and all $x \in [0, \pi]$.

Since $S_n(-x) = -S_n(x)$, the bound also applies when $x \in [-\pi, 0]$. In view of the periodicity, it applies on all of $\mathbb{R}$. □

A set of sufficient conditions for the convergence of a Fourier series, which is due to Dirichlet, possesses a degree of generality that makes it useful for most purposes. Before stating the theorem, we recall that $f(x+)$ and $f(x-)$ are respectively the right limit and left limit of f at x, and that when the function is monotonic on each of (a, x) and (x, b) and is also bounded, these limits necessarily exist. For, when f is increasing on (a, x), say, given any $\varepsilon > 0$, there exists an $x_0 \in (a, x)$ such that $f(x_0) > \sup_{(a,x)} f - \varepsilon$; then $x_0 < t < x$ implies $\sup_{(a,x)} f \ge f(t) \ge f(x_0)$, which, together with the preceding inequality, implies $\left|f(t) - \sup_{(a,x)} f\right| < \varepsilon$. The function of Example 4.1.5(a) satisfies $f(0-) = 0$, $f(0+) = 1$ and it was noted that, when $x = 0$, its Fourier series converges to $\frac{1}{2}$, a value that agrees with $\frac{1}{2}(f(0-) + f(0+))$. Dirichlet's Theorem states that this happens under broad conditions.

Theorem 4.2.10 (Dirichlet) *Let $f : \mathbb{R} \to \mathbb{R}$ be Riemann integrable on $[-\pi, \pi]$ with period 2π and let $x \in \mathbb{R}$ be arbitrary but fixed. If (a, b) is an interval such that $x \in (a, b)$ and f is increasing on each of (a, x) and (x, b), then*

$$s_n(x) \to \frac{1}{2}[f(x^+) + f(x^-)].$$

Here $s_n(x)$ denotes the nth partial sum of the Fourier series of f. If, moreover, f is continuous on (a, b), then

$$s_n(x) \to f(x) \quad \text{as } n \to \infty$$

uniformly on every closed subinterval $[a + \mu, b - \mu]$.

Proof Since f is increasing on each of (a, x) and (x, b) and is also bounded (because it is Riemann integrable), the limits $f(x+)$ and $f(x-)$ both exist. By Remark 4.2.4 and Proposition 4.2.3(c), we may write

$$\begin{aligned} s_n(x) - \frac{1}{2}[f(x^+) + f(x^-)] &= \frac{1}{\pi}\int_0^\pi [f(x+t) + f(x-t)]D_n(t)dt \\ &\quad - \frac{1}{\pi}\int_0^\pi [f(x^+) + f(x^-)]D_n(t)dt \\ &= \frac{1}{\pi}\int_0^\pi [f(x+t) + f(x-t) - f(x^+) - f(x^-)]D_n(t)dt \\ &= \left(\frac{1}{\pi}\int_0^\delta + \frac{1}{\pi}\int_\delta^\pi\right)[f(x+t) + f(x-t) - f(x^+) - f(x^-)]D_n(t)dt, \\ &\quad \text{for all } \delta \in (0,\pi) \\ &= I + J, \quad \text{say.} \end{aligned} \tag{4.10}$$

Now I can be written in the form

$$I = \frac{1}{\pi}\int_0^\delta [f(x+t) - f(x^+)]D_n(t)dt - \frac{1}{\pi}\int_0^\delta [f(x^-) - f(x-t)]D_n(t)dt. \tag{4.11}$$

Since $x \in (a, b)$, we can choose the number δ so small that $x \pm \delta$ both belong to (a, b). Since f is increasing on $(x, x + \delta)$, the difference $f(x + t) - f(x+)$ is an increasing function of t on the interval $(0, \delta)$ and its limit at 0 is 0. On applying Bonnet's form of the Second Mean Value Theorem for Integrals [see Problem 4.1. P1(c)], we obtain

$$\int_0^\delta [f(x+t) - f(x^+)]D_n(t)dt = [f(x+\delta) - f(x^+)]\int_\eta^\delta D_n(t)dt \quad \text{for some } \eta \in [0,\delta].$$

By Lemma 4.1.6 and Proposition 4.2.9, we have

$$\left|\int_\eta^\delta D_n(t)dt\right| = \left|\frac{\delta-\eta}{2} + \sum_{k=1}^n \frac{\sin k\delta}{k} - \sum_{k=1}^n \frac{\sin k\eta}{k}\right| \le \frac{\delta-\eta}{2} + 2(1+\pi).$$

Denoting $\frac{\delta-\eta}{2} + 2(1+\pi)$ by M_1, we therefore have

$$\left|\int_0^\delta [f(x+t) - f(x^+)]D_n(t)dt\right| \le [f(x+\delta) - f(x^+)]M_1. \tag{4.12}$$

Similarly,

$$\left|\int_0^{\delta} [f(x^-) - f(x-t)]D_n(t)dt\right| \leq f(f(x^-) - f(x-\delta)]M_1. \tag{4.13}$$

The right-hand sides of the above two inequalities approach 0 as $\delta \to 0$. It follows that the left-hand sides do so *uniformly* in n. Therefore by (4.11), for any $\varepsilon > 0$, we can choose δ so small that $x \pm \delta$ both belong to (a, b) and

$$|I| \leq \frac{1}{2}\varepsilon \text{ for every natural number } n. \tag{4.14}$$

Fixing such a δ, we next consider the integral denoted above by J. This integral can be written as

$$J = \frac{1}{\pi}\int_{\delta}^{\pi} [f(x+t) + f(x-t)]D_n(t)dt - \frac{1}{\pi}\int_{\delta}^{\pi} [f(x^+) + f(x^-)]D_n(t)dt.$$

From the Localisation Theorem 4.2.7 applied to f and to the constant function $f(x+) + f(x-)$, we deduce that $J \to 0$ as $n \to \infty$. Therefore, we have

$$|J| < \frac{1}{2}\varepsilon \text{ for sufficiently large } n. \tag{4.15}$$

Since $\varepsilon > 0$ is arbitrary, the first part of the required conclusion now follows from (4.10), (4.14) and (4.15).

Now suppose f is also continuous on (a, b). Then on any closed subinterval $[a + \mu, b - \mu]$, f is uniformly continuous and bounded. Consequently, the right-hand sides of the inequalities (4.12) and (4.13), and hence also $|I|$, are uniformly small for values of x lying in the interval $[a + \mu, b - \mu]$, provided μ is sufficiently small.

For fixed δ, $\frac{1}{\pi}\int_{\delta}^{\pi} [f(x+t) + f(x-t)]D_n(t)dt$ is uniformly small in absolute value in view of Theorem 4.2.8. Also, $\left|\frac{1}{\pi}\int_{\delta}^{\pi} [f(x^+) + f(x^-)]D_n(t)dt\right|$ is uniformly small since f is bounded and $\int_{\delta}^{\pi} D_n(t)dt$ tends to 0 as $n \to \infty$. So J is uniformly small for all x in $[a + \mu, b - \mu]$ as $n \to \infty$.

Since I and J have both been shown to be uniformly small when n is sufficiently large, it follows from (4.10) that $s_n(x) \to f(x)$ as $n \to \infty$ uniformly on every closed subinterval $[a + \mu, b - \mu]$. □

Corollary 4.2.11 *Let $f : \mathbb{R} \to \mathbb{R}$ be Riemann integrable with period 2π and monotonic on $[a, b]$. If f is continuous at a point $x \in (a, b)$, then*

$$s_n(x) \to f(x).$$

Here $s_n(x)$ denotes the nth *partial sum of the Fourier series of f.*

Proof Since f is continuous at x, we have $f(x+) = f(x) = f(x-)$. The result now follows from Theorem 4.2.10. □

The problem of uniform convergence on an interval of a Fourier series of a function satisfying what is called a "Lipschitz condition" is considered in the next theorem.

Definition 4.2.12 A real-valued function f satisfies a **Lipschitz condition of order** α, where $0 < \alpha \leq 1$, on an interval $[a, b]$ contained in its domain if there exist $M > 0$ and $\delta > 0$ such that

$$|f(x+t) - f(x)| \leq M|t|^{\alpha}, \quad a \leq x \leq b, |t| < \delta.$$

A Lipschitz condition is understood to be of order 1 unless specified otherwise. The constant M in the condition is not unique and is called a *Lipschitz constant* for the function.

Theorem 4.2.13 *Suppose* $f : \mathbb{R} \to \mathbb{R}$ *is Riemann integrable with period* 2π *and satisfies a Lipschitz condition of order* α *on the interval* $[a, b]$. *Then the Fourier series of f converges to f uniformly on* $[a, b]$.

Proof Let δ and M be as in Definition 4.2.12; we may suppose that $\delta \leq \pi$. Denote the partial sum of the Fourier series by $s_n(x)$, as usual. From Remark 4.2.4 and Proposition 4.2.3(c), we have

$$\begin{aligned} s_n(x) - f(x) &= \frac{2}{\pi}\int_0^{\pi} [\frac{f(x+t) + f(x-t)}{2} - f(x)]D_n(t)\,dt \\ &= (\frac{2}{\pi}\int_0^{\eta} + \frac{2}{\pi}\int_{\eta}^{\pi})[\frac{f(x+t) + f(x-t)}{2} - f(x)]D_n(t)\,dt \qquad (4.16) \\ &= I + J, \quad \text{say,} \end{aligned}$$

where η will be chosen appropriately later. For the moment suppose that $0 < \eta \leq \delta \leq \pi$. Let $x \in [a, b]$. Then

$$\begin{aligned} |I| &= \left|\frac{2}{\pi}\int_0^{\eta} [\frac{f(x+t) + f(x-t)}{2} - f(x)]D_n(t)\,dt\right| \\ &\leq \frac{1}{\pi}\int_0^{\eta} |f(x+t) + f(x-t) - 2f(x)|\left|\frac{\sin(n+\frac{1}{2})t}{2\sin\frac{1}{2}t}\right| dt \\ &\leq \frac{1}{\pi}\int_0^{\eta} 2Mt^{\alpha}\frac{1}{2\sin\frac{1}{2}t}\,dt, \quad \text{an improper integral.} \end{aligned}$$

As the function $t^{\alpha}\frac{1}{2\sin\frac{1}{2}t} = t^{\alpha-1}\frac{\frac{1}{2}t}{\sin\frac{1}{2}t}$ is such that $\frac{\frac{1}{2}t}{\sin\frac{1}{2}t} \to 1$ as $t \to 0$ while $\alpha - 1 > -1$, it follows that the improper integral converges and therefore approaches 0 as $\eta \to 0$, by Problem 3.2.P15 and the fact that the integrand is nonnegative. Hence, for a given $\varepsilon > 0$, there exists an $\eta > 0$ such that, for any natural number n,

$$|I| \leq \frac{1}{2}\varepsilon, \tag{4.17}$$

for all $x \in [a, b]$.

Now fix such an η and write J as

$$J = \frac{1}{\pi}\int_{\eta}^{\pi} [f(x+t) + f(x-t)]D_n(t)dt - \frac{2}{\pi}f(x)\int_{\eta}^{\pi} D_n(t)dt.$$

By Theorem 4.2.8, the first integral on the right-hand side here tends to 0 uniformly for all real values of x as $n \to \infty$. Since f is Riemann integrable and hence bounded, it follows that the second integral also tends to 0 uniformly for all real values of x by Corollary 4.2.6. Thus, for large enough n, we have

$$|J| < \frac{1}{2}\varepsilon \tag{4.18}$$

for all real values of x. From (4.16), (4.17) and (4.18), it follows that, for large enough n,

$$|s_n(x) - f(x)| < \varepsilon \qquad \text{for all } x \in [a, b].$$

□

Examples 4.2.14

(a) Let f be the function defined by

$$f(x) = \cos ax \qquad \text{for } -\pi \leq x \leq \pi,$$

where a is a real number that is not an integer. Since f is even and $f(-\pi) = f(\pi)$, it can be extended to $\mathbb{R}$ so as to have period 2π and also be even. Since f satisfies a Lipschitz condition on $[-\pi, \pi]$, it follows by Theorem 4.2.13 that its Fourier series converges uniformly on $[-\pi, \pi]$. Observe that the Fourier coefficients b_n are all 0 while

$$\frac{1}{2}a_0 = \frac{1}{\pi}\int_0^{\pi} \cos ax\, dx = \frac{\sin a\pi}{a\pi},$$

and, for $n > 0$,

$$a_n = \frac{1}{\pi}\int_0^{\pi} [\cos(a+n)x + \cos(a-n)x]dx = (-1)^n \frac{2a}{a^2 - n^2}\frac{\sin a\pi}{a\pi}.$$

Therefore, for $-\pi \leq x \leq \pi$, we have

$$\cos ax = \frac{\sin a\pi}{a\pi} + 2\frac{\sin a\pi}{\pi}\sum_{n=1}^{\infty}(-1)^n \frac{a \cos nx}{a^2 - n^2},$$

or equivalently

$$\frac{\pi \cos ax}{2 \sin a\pi} = \frac{1}{2a} + \sum_{n=1}^{\infty} (-1)^n \frac{a \cos nx}{a^2 - n^2}.$$

Setting $x = 0$ in the latter, we obtain

$$\frac{\pi}{2 \sin a\pi} = \frac{1}{2a} + \sum_{n=1}^{\infty} (-1)^n \frac{a}{a^2 - n^2},$$

which may be written as

$$\operatorname{cosec} a\pi = \frac{1}{a\pi} + \frac{2a}{\pi} \sum_{n=1}^{\infty} (-1)^n \frac{1}{a^2 - n^2}.$$

This formula is valid whenever a is not an integer.
Setting $x = \pi$ instead, we obtain

$$\cot a\pi = \frac{1}{a\pi} + \frac{2a}{\pi} \sum_{n=1}^{\infty} \frac{1}{a^2 - n^2},$$

which is also valid whenever a is not an integer.

(b) Let $f(x) = \sin ax$, for $-\pi \leq x \leq \pi$, where a is not an integer. Extend it to all of $\mathbb{R}$ so as to be periodic. We shall find its Fourier series by calculating its coefficients. Since the restriction to $(-\pi, \pi)$ is odd, the Fourier coefficients a_n are all 0 and the remaining Fourier coefficients are given by

$$\begin{aligned} b_n &= \frac{2}{\pi} \int_0^{\pi} \sin ax \sin nx \, dx = \frac{1}{\pi} \int_0^{\pi} [\cos(a-n)x - \cos(a+n)x] dx \\ &= 2 \frac{\sin a\pi}{\pi} (-1)^n \frac{n}{a^2 - n^2}. \end{aligned}$$

The function satisfies a Lipschitz condition on any closed subinterval of the open interval $(-\pi, \pi)$. Therefore by Theorem 4.2.13, its Fourier series converges uniformly on any closed subinterval of $(-\pi, \pi)$. So

$$\sin ax = 2 \frac{\sin a\pi}{\pi} \sum_{n=1}^{\infty} (-1)^n \frac{n}{a^2 - n^2} \sin nx.$$

This series can also be obtained from the one for $\cos ax$ above by term differentiation provided we can prove the uniform convergence of the differentiated series [19, Theorem 7.17] independently. Such a proof can be based on Dirichlet's Test [28, Theorem 4.4.2] and we present it for the sake of those readers who may know the latter:

Choose $a_n = \frac{n}{n^2-a^2}$, $b_n = (-1)^{n+1} \sin nx$. Observe that for $n \geq a$, the sequence $\{a_n\}$ decreases to 0 as $n \to \infty$. Also,

$$
\begin{aligned}
2\cos\frac{1}{2}x[\sin x - \sin 2x + \cdots + (-1)^{n+1}\sin nx] &= (\sin\frac{3}{2}x + \sin\frac{1}{2}x) - (\sin\frac{5}{2}x + \sin\frac{3}{2}x) \\
&\quad + \cdots + (-1)^{n+1}(\sin(n+\frac{1}{2})x + \sin(n-\frac{1}{2})x) \\
&= (-1)^{n+1}\sin(n+\frac{1}{2})x + \sin\frac{1}{2}x.
\end{aligned}
$$

So,

$$
\left|\sin x - \sin 2x + \cdots + (-1)^{n+1}\sin nx\right| \leq \frac{1}{\left|\cos\frac{1}{2}x\right|} \leq \frac{1}{\min\left|\cos\frac{1}{2}x\right|},
$$

where the min is taken over $x \in [\alpha, \beta] \subset (-\pi, \pi)$. Here $\min\left|\cos\frac{1}{2}x\right|$ is guaranteed to be positive because $\cos\frac{1}{2}x$ is continuous and positive on $(-\pi, \pi)$ and hence also on $[\alpha, \beta]$. Thus the partial sums of Σb_n are uniformly bounded on $[\alpha, \beta]$. It follows by Dirichlet's Test that $\Sigma_{n\geq a} a_n b_n$ converges uniformly on $[\alpha, \beta]$.

(c) Consider the function $f(t) = t^2/4$, $t \in [-\pi, \pi]$. It satisfies $f(-\pi) = f(\pi)$ and therefore can be extended to be periodic on $\mathbb{R}$. Moreover, it will satisfy a Lipschitz condition on $[-\pi, \pi]$; in fact,

$$
|f(x) - f(y)| = \frac{1}{4}\left|x^2 - y^2\right| \leq \frac{\pi}{2}|x - y|.
$$

Thus the Fourier series of the function converges uniformly to the function. Since the function is even, its Fourier coefficients b_n all vanish, while

$$
\begin{aligned}
a_n &= \frac{2}{\pi}\int_0^\pi \frac{x^2}{4}\cos nx\,dx = \frac{1}{2\pi}\int_0^\pi x^2\cos nx\,dx \\
&= \frac{1}{2\pi}[x^2\frac{\sin nx}{n} - (2x)\left(-\frac{\cos nx}{n^2}\right) + 2\left(-\frac{\sin nx}{n^3}\right)]_0^\pi \\
&= \frac{1}{2\pi}\frac{2\pi}{n^2}(-1)^n
\end{aligned}
$$

and

$$a_0 = \frac{2}{\pi}\int_0^\pi \frac{x^2}{4}dx = \frac{\pi^2}{6}.$$

Thus the Fourier series is

$$\frac{x^2}{4} = \frac{\pi^2}{12} - \sum_{n=1}^{\infty}(-1)^{n+1}\frac{\cos nx}{n^2}.$$

Setting $x = 0$ here yields

$$\frac{\pi^2}{12} = \sum_{n=1}^{\infty}\frac{(-1)^{n+1}}{n^2}.$$

Problem Set 4.2

4.2.P1. Suppose that f is a 2π-periodic function that is Lebesgue integrable on $[-\pi, \pi]$ and is differentiable at a point x_0. Prove that

$$\lim_{n\to\infty} s_n(x_0) = f(x_0),$$

where s_n denotes the partial sum of the Fourier series of f.

4.2.P2. (**Bessel's Inequality**) If f is any Riemann integrable periodic function, its Fourier coefficients satisfy the inequality

$$\frac{1}{2}a_0^2 + \sum_{n=1}^{\infty}(a_n^2 + b_n^2) \le \frac{1}{\pi}\int_{-\pi}^{\pi} f(x)^2 dx.$$

Note that it is part of the assertion that the series on the left is convergent. [It follows from this result that the Fourier coefficients a_n and b_n of a Riemann integrable function tend to zero as $n \to \infty$.]

4.2.P3. The series $\sum_{n=1}^{\infty}\frac{\cos nx}{\sqrt{n}}$ is uniformly convergent on every closed subinterval $[\alpha, 2\pi-\alpha]$, $0 < \alpha < \pi$ of $(0, 2\pi)$. Show that it is not the Fourier series of a Riemann integrable function.

4.2.P4. Is the series $\sum_{n=1}^{\infty}\frac{\sin nx}{\ln(n+1)}$ the Fourier series of a continuous function?

4.2.P5. Show that $\frac{2}{\pi}\int_0^\pi \left|\frac{\sin(n+\frac{1}{2}t)}{2\sin\frac{1}{2}t}\right|dt \ge \frac{4}{\pi^2}\sum_{k=1}^{n}\frac{1}{k} \ge \frac{4}{\pi^2}\left(\frac{1}{n} + \ln n\right)$.

4.2.P6. Show that $\frac{2}{\pi}\int_0^\pi \left|\frac{\sin(n+\frac{1}{2}t)}{2\sin\frac{1}{2}t}\right|dt \le \ln n + \frac{1}{n\pi} + \ln\pi + \frac{2}{\pi}$.

4.2.P7. Let the Fourier series corresponding to a continuous periodic function $f(x)$ be given by $\frac{1}{2}a_0 + \sum_{n=1}^{\infty}(a_n\cos nx + b_n\sin nx)$. Show that the series obtained by integrating this Fourier series term by term converges to $\int_0^x f(t)dt$. [The remarkable thing about this result is that the Fourier series of f is not assumed to converge to it. Note however that the integrated series is not a trigonometric series.]

4.2.P8. If f is continuously differentiable on $[a, b]$, use integration by parts (but *not* the Riemann–Lebesgue Theorem 4.2.5) to show that

$$\lim_{\lambda\to\infty}\int_a^b f(t)\sin\lambda t\,dt = 0 = \lim_{\lambda\to\infty}\int_a^b f(t)\cos\lambda t\,dt.$$

4.2.P9. Let f be continuous with period 2π and Fourier coefficients a_k, b_k. Show that

$$\sum_{k=0}^{n}(|a_k|+|b_k|)\le 2(2n+1)^{\frac{1}{2}}\sup\{|f(x)| : x\in[-\pi,\pi]\}.$$

4.2.P10. [Cf. Problem 3.2.P24] Show that the improper integral $\int_0^\infty \frac{\sin x}{x}dx$ converges and evaluate it by applying Dirichlet's Theorem 4.2.10. It is understood that $\frac{\sin x}{x}$ is to be replaced by 1 when $x = 0$.

4.3 Cesàro Summability of Fourier Series

As observed in Sect. 4.2, although the Fourier series of even a continuous function need not converge to it, strengthening the hypothesis on the function ensures that the corresponding Fourier series does converge to it. In the other direction, we may weaken the notion of convergence and seek an answer to the question of whether the Fourier series corresponding to a Lebesgue integrable periodic function converges in the weakened sense, and if it does, whether it converges to the function.

It is known [see the example following Proposition 3.1.8 in [28] or Ex. 10 on p. 39 of [4]] that the arithmetic means of the first n terms of a convergent sequence converge to the same limit as the latter. Even for a nonconvergent sequence, such as $\{(-1)^n : n\in\mathbb{N}\}$, the sequence of arithmetic means can be convergent. This leads us to consider the arithmetic means of the first n partial sums s_n of the Fourier series of a Lebesgue integrable function f with period 2π:

$$\sigma_n(x) = \frac{1}{n}[s_0(x)+s_1(x)+\cdots+s_{n-1}(x)],$$

where

$$s_n(x) = \frac{1}{2}a_0+\sum_{k=1}^{n}(a_k\cos kx+b_k\sin kx)$$

and a_n, b_n are the Fourier coefficients of f. In what follows, we shall show that the $\sigma_n(x)$, known as the *Fejér sums*, converge to $f(x)$ for all those x at which f is continuous; the Fourier series is then said to be **Cesàro summable** to $f(x)$.

Example 4.3.1 For the Fourier series of the function f given by $f(x) = \cos px$, where $p > 0$ is an integer, recall from Example 4.2.1 that the partial sums of the Fourier series are

$$s_n(x) = \begin{cases} 0 & \text{if } n<p \\ \cos px & \text{if } n \geq p. \end{cases}$$

It follows that the Fejér sums are

$$\sigma_n(x) = \begin{cases} 0 & \text{if } n \leq p \\ \frac{(n-p)\cos px}{n} & \text{if } n > p. \end{cases}$$

Remark 4.3.2 To express $\sigma_n(x)$ in a more manageable form, we use the definition of $s_n(x)$ to obtain

$$\sigma_n(x) = \frac{1}{n}[s_0(x) + s_1(x) + \cdots + s_{n-1}(x)] = \frac{1}{\pi}\int_0^\pi [f(x+t) + f(x-t)]F_n(t)dt,$$

where $F_n : \mathbb{R} \to \mathbb{R}$ is the function defined as

$$F_n(t) = \frac{1}{n}[D_0(t) + D_1(t) + \cdots + D_{n-1}(t)].$$

In view of the definition of D_n, this becomes

$$F_n(t) = \begin{cases} \frac{1}{n}\sum_{k=0}^{n-1} \frac{\sin(k+\frac{1}{2})t}{2\sin\frac{1}{2}t}, & t \neq 2k\pi, k \in \mathbb{Z} \\ \frac{n}{2}, & t = 2k\pi, k \in \mathbb{Z}. \end{cases}$$

F_n is known as the **Fejér kernel**. When t is not an integral multiple of 2π, it can also be written in the "closed form"

$$F_n(t) = \frac{\sin^2 \frac{nt}{2}}{2n \sin^2 \frac{t}{2}}.$$

This is a consequence of the trigonometric relation

$$\cos A - \cos B = -2\sin\frac{A+B}{2}\sin\frac{A-B}{2},$$

whence one can obtain

$$\sum_{k=0}^{n-1} 2\sin(k+\frac{1}{2})t \sin\frac{1}{2}t = \sum_{k=0}^{n-1} [\cos kt - \cos(k+1)t] = 1 - \cos nt = 2\sin^2\frac{nt}{2}.$$

The following properties of the Fejér kernel are important.

Proposition 4.3.3 *The Fejér kernel F_n has the following properties*:

(a) *F_n is continuous*;
(b) *$F_n(-t) = F_n(t)$ for all real t*;
(c) *$F_n(t) \geq 0$ for all real t*;
(d) *$\frac{2}{\pi}\int_0^\pi F_n(t)dt = 1$ for all n*;
(e) *$F_n(t) \leq \frac{1}{2n\sin^2\frac{\delta}{2}}$ for $\delta < t < \pi$*;
(f) *$\lim_{n\to\infty} \int_\delta^\pi F_n(t)dt = 0$, where $0 < \delta < \pi$.*

Proof Parts (a) and (b) follow from the corresponding properties of D_n. Part (c) follows from the closed form of F_n derived above. Since $F_n(t) = \frac{1}{n}\sum_{k=0}^{n-1} D_k(t)$, we have by Proposition 4.2.3(c)

$$\frac{2}{\pi}\int_0^\pi F_n(t)dt = \frac{2}{n\pi}\int_0^\pi \sum_{k=0}^{n-1} D_k(t)dt = \sum_{k=1}^{n-1}\frac{2}{n\pi}\int_0^\pi D_k(t)dt = \sum_{k=0}^{n-1}\frac{1}{n} = 1.$$

This proves part (d). To prove (e), note that $\sin^2\frac{x}{2}$ increases as x increases from 0 to π. Finally, it follows from (e) that

$$\int_\delta^\pi F_n(t)dt \leq \frac{\pi - \delta}{2n\sin^2\frac{\delta}{2}} \to 0 \quad \text{as} \quad n \to \infty, \quad \text{provided } 0 < \delta < \pi.$$

This completes the proof of (f). □

Theorem 4.3.4 (Fejér) *If $f : \mathbb{R} \to \mathbb{R}$ is Lebesgue integrable with period 2π and continuous at $x \in [-\pi, \pi]$, then its Fourier series at x is Cesàro summable to $f(x)$, i.e. the arithmetic means σ_n of the partial sums of the Fourier series of f converge to f at x. Moreover, if f is continuous on $[-\pi, \pi]$, then the convergence is uniform on $\mathbb{R}$.*

Proof We multiply both sides of the equality in Proposition 4.3.3(d) by $f(x)$ and subtract from the expression for $\sigma_n(x)$ derived in Remark 4.3.2 to obtain

$$\sigma_n(x) - f(x) = \frac{2}{\pi}\int_0^\pi [\frac{f(x+t)+f(x-t)}{2} - f(x)]F_n(t)dt.$$

Fix $\varepsilon > 0$. Because of the above equality, in order to show that $\sigma_n(x) \to f(x)$, it suffices to prove that, for sufficiently large n,

$$\left|\frac{2}{\pi}\int_0^{\pi}\left[\frac{f(x+t)+f(x-t)}{2}-f(x)\right]F_n(t)dt\right|<\varepsilon. \tag{4.19}$$

For uniform convergence, it suffices to prove (4.19) for sufficiently large n and all real x.

Since f is continuous at x, we can find δ satisfying $0 < \delta < \pi$ such that

$$|f(x+t)-f(x)|<\frac{1}{2}\varepsilon \quad \text{for } |t|<\delta.$$

Thus, if $0 \le t < \delta$, then

$$\begin{aligned}\left|\frac{f(x+t)+f(x-t)}{2}-f(x)\right| &\le \frac{1}{2}[|f(x+t)-f(x)|+|f(x-t)-f(x)|]\\ &< \frac{1}{2}\left(\frac{1}{2}\varepsilon+\frac{1}{2}\varepsilon\right)=\frac{1}{2}\varepsilon.\end{aligned}$$

Consequently,

$$\begin{aligned}\left|\frac{2}{\pi}\int_0^{\delta}\left[\frac{f(x+t)+f(x-t)}{2}-f(x)\right]F_n(t)dt\right| &\le \frac{1}{2}\varepsilon\frac{2}{\pi}\int_0^{\delta}F_n(t)dt\\ &\le \frac{1}{2}\varepsilon\frac{2}{\pi}\int_0^{\pi}F_n(t)dt=\frac{1}{2}\varepsilon,\end{aligned} \tag{4.20}$$

using Proposition 4.3.3(c) & (d). Observe that, if f is continuous on $[-\pi, \pi]$, then it is uniformly continuous on $\mathbb{R}$ by virtue of its periodicity, and this makes it possible to choose the number δ independently of x, thereby ensuring that (4.20) holds for all real x.

If $t \ge \delta$, then $F_n(t) \le \frac{1}{2n\sin^2\frac{\delta}{2}}$ by Proposition 4.3.3(e), and thus

$$\begin{aligned}&\left|\frac{2}{\pi}\int_{\delta}^{\pi}\left[\frac{f(x+t)+f(x-t)}{2}-f(x)\right]F_n(t)dt\right|\\ &\quad\le \frac{2}{4\pi n\sin^2\frac{\delta}{2}}\int_{\delta}^{\pi}|f(x+t)+f(x-t)-2f(x)|dt\\ &\quad\le \frac{2}{4\pi n\sin^2\frac{\delta}{2}}\left[2\int_{-\pi}^{\pi}|f(t)|dt+2\pi|f(x)|\right]\\ &\quad<\frac{1}{2}\varepsilon \quad \text{for sufficiently large } n \text{ (depending on } x).\end{aligned} \tag{4.21}$$

Inequalities (4.20) and (4.21) prove (4.19). As already noted, this completes the proof that $\sigma_n(x) \to f(x)$ when f is continuous at x.

If f is continuous on $[-\pi, \pi]$, then it is bounded and therefore (4.21) holds for sufficiently large n and for all real x, and hence so does (4.19). This completes the proof of uniform convergence in this case. □

Remark Lebesgue proved that for a Lebesgue integrable function f, the Fourier series has almost everywhere Cesàro sum f [see Problem 5.8.P8].

Kolmogorov succeeded in finding an integrable function whose Fourier series diverges everywhere [34]. When f is square integrable, we shall see that the partial sums of the Fourier series converge to f in the L^2 norm [see Corollary 4.5.9]. For $1 \le p < \infty$ and $f \in L^p[-\pi, \pi]$, the Cesàro means converge to f in the L^p norm [see Problem 7.2.P4].

Recall from Sect. 1.5 that a finite sum

$$\alpha_0 + \sum_{k=1}^{n} (\alpha_k \cos kx + \beta_k \sin kx)$$

is called a *trigonometric polynomial.* Clearly, sums and constant multiples of trigonometric polynomials are again trigonometric polynomials. Therefore so are partial sums of Fourier series and Fejér sums. It should be borne in mind however, that although every trigonometric polynomial is the partial sum of a trigonometric series, a sequence of such polynomials can consist of partial sums of *different* series. Therefore being the limit of a sequence of trigonometric polynomials is not the same as being the sum of a trigonometric series. In particular, the next result does *not* claim that every continuous periodic function is the sum of a trigonometric *series*, which would be false. It is the same as Theorem 1.5.5, which was used in Proposition 3.4.4, but we now give a fresh proof based on Fourier series, without using any consequences of Proposition 3.4.4.

Theorem 4.3.5 (Weierstrass Approximation Theorem) *Every continuous real-valued function on* $\mathbb{R}$ *with period* 2π *can be approximated uniformly by trigonometric polynomials.*

Proof This is an immediate consequence of Fejér's Theorem 4.3.4 because the Fejér sums are trigonometric polynomials. □

Corollary 4.3.6 *If all Fourier coefficients of a continuous function of period* 2π *vanish, then the function vanishes everywhere.*

Proof If all Fourier coefficients of a continuous function of period 2π vanish, then its Fejér sums vanish as well and therefore $f(x) = \lim_{n\to\infty} \sigma_n(x) = 0$ for all real x. □

Problem Set 4.3

4.3.P1. If $f : [a, b] \to \mathbb{R}$ is continuous and $\varepsilon > 0$, then prove by using Fejér's Theorem 4.3.4 that there exists a polynomial $P(x)$ such that

$$\sup\{|f(x) - P(x)| : x \in [a, b]\} < \varepsilon.$$

[This is known as the Weierstrass Polynomial Approximation Theorem.]

4.3.P2. Use the Weierstrass Polynomial Approximation Theorem to prove that if f and g are continuous functions on $[a, b]$ such that

$$\int_a^b x^n f(x)dx = \int_a^b x^n g(x)dx \qquad \text{for all } n \geq 0,$$

then $f = g$.

4.3.P3. If a_n and b_n are the Fourier coefficients of f, show that the Fejér sums σ_n are

$$\sigma_n(x) = \frac{1}{2}a_0 + \sum_{k=1}^{n-1}\left(1 - \frac{k}{n}\right)(a_k \cos kx + b_k \sin kx).$$

Use this fact to show that

$$\frac{1}{\pi}\int_{-\pi}^{\pi} \sigma_n(x)^2 dx = \frac{1}{2}a_0^2 + \sum_{k=1}^{n-1}\left(1 - \frac{k}{n}\right)^2 (a_k^2 + b_k^2).$$

Deduce **Parseval's Theorem** that if f has period 2π and is continuous with Fourier coefficients a_n, b_n, then

$$\frac{1}{\pi}\int_{-\pi}^{\pi} f(x)^2 dx = \frac{1}{2}a_0^2 + \sum_{k=1}^{\infty}(a_k^2 + b_k^2).$$

4.3.P4. Two different periodic functions can have the same Fourier series (e.g. if they differ only on a finite set of points). Can this happen if the functions are continuous?

4.3.P5. If f and g are continuous functions of period 2π with Fourier coefficients a_n, b_n and α_n, β_n respectively, show that

$$\frac{1}{\pi}\int_{-\pi}^{\pi} f(x)g(x)dx = \frac{1}{2}a_0\alpha_0 + \sum_{n=1}^{\infty}(a_n\alpha_n + b_n\beta_n).$$

4.3.P6. Suppose $0 < \delta < \pi$ and $f(x) = 1$ if $|x| \leq \delta$, $f(x) = 0$ if $\delta < |x| < \pi$, and $f(x + 2\pi) = f(x)$ for all x.

(a) Compute the Fourier coefficients of f.

(b) Conclude that $\sum_{n=1}^{\infty} \frac{\sin n\delta}{n} = \frac{\pi - \delta}{2}$ when $0 < \delta < \pi$.

(c) Deduce from Parseval's Theorem [see Problem 4.3.P3] that $\sum_{n=1}^{\infty} \frac{\sin^2 n\delta}{n^2\delta} = \frac{\pi-\delta}{2}$ when $0 < \delta < \pi$.

(d) Let $\delta \to 0$ and prove that $\int_0^{\infty} \left(\frac{\sin x}{x}\right)^2 dx = \frac{\pi}{2}$.

4.3.P7. Show that $\sum_{n=1}^{\infty} \frac{\cos nx}{n^{3/2}}$ is the Fourier series of a continuous function and converges to it uniformly.

4.4 Even and Odd Functions

The standard form of a Fourier series is the one we have considered in the preceding sections. In this form, the function has period 2π and it is sufficient to consider it on $[-\pi, \pi]$, because its behaviour elsewhere is a repetition of its behaviour on this interval. In many applications, it is necessary to work with functions with some positive period $2l$ and it is therefore desirable to consider the function on $[-l, l]$ and adopt the corresponding form of the Fourier series. This is done by a change of variables.

Let $f : \mathbb{R} \to \mathbb{R}$ have period $2l$ and be Lebesgue integrable on some interval of length $2l$ (and hence on any bounded closed interval). Introduce a new variable t so that

$$\frac{t}{\pi} = \frac{x}{l}$$

and define $F(t) = f\left(\frac{l}{\pi}t\right) = f(x)$. Then F is defined on $\mathbb{R}$, is Lebesgue integrable and has period 2π, so that we may form its Fourier series. In the Fourier series of F, we replace t by $\frac{\pi}{l}x$. The resulting series is taken as the Fourier series of f,

$$f(x) \sim \frac{1}{2}a_0 + \sum_{n=1}^{\infty} \left(a_n \cos\frac{n\pi x}{l} + b_n \sin\frac{n\pi x}{l}\right),$$

where

$$a_n = \frac{1}{\pi}\int_{-\pi}^{\pi} F(t)\cos\, nt\, dt = \frac{1}{l}\int_{-l}^{l} f(x)\cos\frac{n\pi x}{l}dx, \qquad n = 0, 1, 2, \ldots$$

and

$$b_n = \frac{1}{\pi}\int_{-\pi}^{\pi} F(t)\sin\, nt\, dt = \frac{1}{l}\int_{-l}^{l} f(x)\sin\frac{n\pi x}{l}dx, \qquad n = 1, 2, 3, \ldots,$$

where the integrals have been transformed by using the change of variables made above and applying Remark 5.8.21 with $\frac{\pi}{l}$ as ‘γ’. We emphasise that the remark is independent on any results from this chapter.

A function f is said to be *odd* if

$$f(x) = -f(-x) \qquad \text{for every real } x$$

and is said to be *even* if

$$f(x) = f(-x) \qquad \text{for every real } x.$$

For example, sine is an odd function and cosine is an even function. Also, odd powers of x are odd functions and even powers are even functions. We note further that sums and constant multiples of even functions are even, and likewise for odd functions. The product of a pair of even or odd functions is even while the product of an even function with an odd function is odd.

The following facts may be familiar in the case of Riemann integrals from an elementary course on Calculus at least when the integrand f is a derivative. They involve breaking up the integral over $[-l, l]$ as a sum of integrals over $[-l, 0]$ and $[0, l]$, and then effecting the change of variables $t = -x$ in the integral over $[-l, 0]$. For this particular change of variables, the justification in the case of Lebesgue integration is given in Remark 5.8.21, which is independent of any results in this chapter.

(a) If a function f is Lebesgue integrable on $[-l, l]$ and satisfies $f(-x) = f(x)$ for $x \in (-l, l)$, then

$$\int_{-l}^{l} f(x)dx = 2\int_{0}^{l} f(x)dx.$$

(b) If a function f is Lebesgue integrable on $[-l, l]$ and satisfies $f(-x) = -f(x)$ for $x \in (-l, l)$, then

$$\int_{-l}^{l} f(x)dx = 0.$$

Proposition 4.4.1 *The Fourier series of an even function f of period $2l$ is a “Fourier cosine series”, which means every $b_n = 0$:*

$$f(x) \sim \frac{1}{2}a_0 + \sum_{n=1}^{\infty} a_n \cos\frac{n\pi x}{l}$$

with coefficients $a_n = \frac{2}{l}\int_0^l f(x)\cos\frac{n\pi x}{l}dx, \quad n = 0, 1, 2, \ldots.$

The Fourier series of an odd function of period 2l is a "Fourier sine series", which means every $a_n = 0$:

$$f(x) \sim \sum_{n=0}^{\infty} b_n \sin \frac{n\pi x}{l}$$

with coefficients $b_n = \frac{2}{l}\int_0^l f(x)\sin\frac{n\pi x}{l}dx, \quad n = 1, 2, 3, \ldots.$

Proof If f is an even function, then $f(x)\sin\frac{n\pi x}{l}$ is odd and $f(x)\cos\frac{n\pi x}{l}$ is even; so

$$b_n = \frac{1}{l}\int_{-l}^{l} f(x)\sin\frac{n\pi x}{l}dx = 0, \qquad n = 1, 2, \ldots$$

and

$$a_n = \frac{2}{l}\int_0^l f(x)\cos\frac{n\pi x}{l}dx, \qquad n = 0, 1, 2, \ldots.$$

If f is an odd function, then $f(x)\cos\frac{n\pi x}{l}$ is odd and $f(x)\sin\frac{n\pi x}{l}$ is even; so

$$a_n = \frac{1}{l}\int_{-l}^{l} f(x)\cos\frac{n\pi x}{l}dx = 0, \qquad n = 0, 1, 2, \ldots$$

and

$$b_n = \frac{2}{l}\int_0^l f(x)\sin\frac{n\pi x}{l}dx, \qquad n = 1, 2, \ldots.$$

The result now follows by substituting these values in the Fourier series of f as described in the second paragraph of this section. □

Examples 4.4.2

(a) For the odd periodic function f shown in Fig. 4.1, [which is $f(x) = x$ on $(-\pi, \pi]$, extended by periodicity], we have

$$\begin{aligned} b_n &= \frac{2}{\pi}\int_0^{\pi} x \sin nx\,dx = \frac{2}{\pi}[\frac{1}{n^2}\sin nx - \frac{x}{n}\cos nx]_0^{\pi} \\ &= -\frac{2}{\pi}\frac{\pi}{n}\cos n\pi = (-1)^{n+1}\frac{2}{n}. \end{aligned}$$

Fig. 4.1 Function in Example 4.4.2(a)

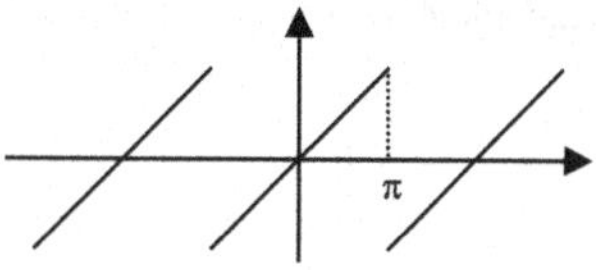

Thus the sine series in accordance with Proposition 4.4.1 is

$$\sum_{n=1}^{\infty}(-1)^{n+1}\frac{2}{n}\sin nx.$$

Since f is monotonic on $[0, \pi]$ and continuous at $\frac{\pi}{2}$, it follows from Corollary 4.2.11 that its Fourier series at $\frac{\pi}{2}$ converges to $f(\frac{\pi}{2})$. Therefore

$$\frac{\pi}{2}=\sum_{n=1}^{\infty}(-1)^{n+1}\frac{2}{n}\sin n\frac{\pi}{2}=2(1-\frac{1}{3}+\frac{1}{5}-+\cdots),$$

so that $\frac{\pi}{4}=1-\frac{1}{3}+\frac{1}{5}-+\cdots$.
Since the partial sums are continuous, if they were to converge uniformly on any neighbourhood of π, the limit would have to be continuous, which it obviously cannot be. Therefore they cannot converge uniformly on a neighbourhood of π.

(b) For the even periodic function f shown in Fig. 4.2, [which is $f(x) = |x|$ on $(-\pi, \pi]$, extended by periodicity], we have

$$a_n=\frac{2}{\pi}\int_0^{\pi}x\cos nx\,dx=\frac{2}{\pi}[\frac{1}{n^2}\cos nx+\frac{x}{n}\sin nx]_0^{\pi}=\frac{2}{\pi}\frac{\cos n\pi-1}{n^2}.$$

Thus $a_n=\begin{cases}-\frac{4}{\pi n^2} & \text{if } n \text{ is odd}\\ 0 & \text{if } n \text{ is even}, n\neq 0.\end{cases}$

For $n = 0$, we have

$$a_0=\frac{2}{\pi}\int_0^{\pi}xdx=\pi.$$

Therefore the cosine series in accordance with Proposition 4.4.1 is

$$\frac{\pi}{2}-\frac{4}{\pi}\sum_{n=0}^{\infty}\frac{\cos(2n+1)x}{(2n+1)^2},$$

which converges at each $x \in (0, \pi)$ in view of Corollary 4.2.11.
In particular, this series converges when $x=\frac{\pi}{4}$ or $\frac{\pi}{3}$

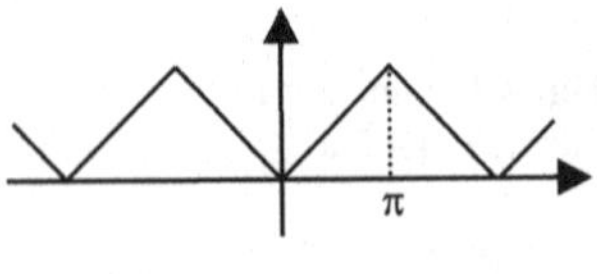

Fig. 4.2 Function in Example 4.4.2(b)

and therefore

$$\frac{\pi}{4} = \frac{\pi}{2} - \frac{4}{\pi}\frac{1}{\sqrt{2}}(1 - \frac{1}{3^2} - \frac{1}{5^2} + \frac{1}{7^2} + \frac{1}{9^2} - \frac{1}{11^2} - \frac{1}{13^2} + + - - \cdots),$$

which leads to $\frac{\pi^2}{16}\sqrt{2} = 1 - \frac{1}{3^2} - \frac{1}{5^2} + \frac{1}{7^2} + \frac{1}{9^2} - \frac{1}{11^2} - \frac{1}{13^2} + + - - \cdots,$
and

$$\frac{\pi}{3} = \frac{\pi}{2} - \frac{4}{\pi}\frac{1}{2}(1 - \frac{2}{3^2} + \frac{1}{5^2} + \frac{1}{7^2} - \frac{2}{9^2} + \frac{1}{11^2} + \frac{1}{13^2} - + + \cdots),$$

which leads to $\frac{\pi^2}{12} = 1 - \frac{2}{3^2} + \frac{1}{5^2} + \frac{1}{7^2} - \frac{2}{9^2} + \frac{1}{11^2} + \frac{1}{13^2} - + + \cdots.$

(c) Let $f : \mathbb{R} \to \mathbb{R}$ be the function with period 2π such that

$$f(x) = \frac{1}{2}(\pi - x) \qquad \text{for } 0 \le x < 2\pi.$$

See Fig. 4.3. With some effort one can check that this function is odd, with jumps at the points $2\pi k$ and is continuous as well as decreasing on the intervals $[2\pi k, 2\pi(k+1))$, $k = 0, \pm 1, \pm 2, \ldots$. Since it is odd, its Fourier coefficients a_n are all 0 while the coefficients b_n are given by

$$b_n = \frac{1}{\pi}\int_{-\pi}^{\pi} f(x) \sin\, nx\, dx = \frac{2}{\pi}\int_0^{\pi} f(x) \sin\, nx\, dx = \frac{1}{\pi}\int_0^{\pi} (\pi - x) \sin\, nx\, dx = \frac{1}{n}.$$

Therefore lits Fourier series is

$$\sum_{n=1}^{\infty} \frac{1}{n} \sin\, nx.$$

By Corollary 4.2.11, this series converges to $\frac{1}{2}(\pi - x)$ at each point of $(2\pi k, 2\pi(k+1))$, $k = 0, \pm 1, \pm 2, \ldots$, and at any point $2\pi k$, the sum of the series is 0. The latter can be verified directly.

In applications, we often want to employ a Fourier series for a (Lebesgue integrable) function f that is defined on an interval of the form $[0, l]$. One way of doing this is to extend the function to $[-l, l]$ to be even by setting

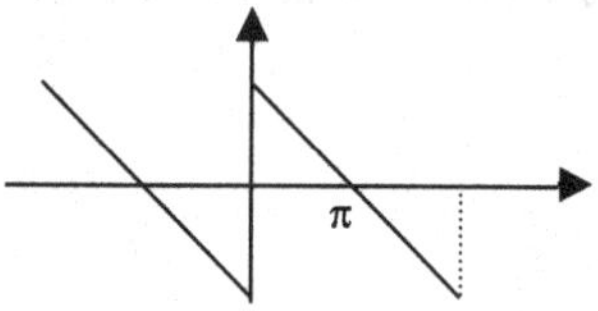

Fig. 4.3 Function in Example 4.4.2(c)

$$f(x) = f(-x) \qquad \text{for } -l \leq x < 0$$

and then extend it further to all of $\mathbb{R}$ so as to have period $2l$. The latter is possible since the extension to $[-l, l]$ satisfies $f(-l) = f(-l + 2l) = f(l)$. Moreover, the extension to $\mathbb{R}$ is even and Lebesgue integrable on $[-l, l]$, and therefore has a cosine series. We speak of this as the "cosine series" of the original function given on $[0, l]$.

Alternatively, we can first extend f to an odd function on $[-l, l]$ by setting

$$f(x) = -f(-x) \qquad \text{for } -l < x < 0$$

and

$$f(-l) = -f(l).$$

A periodic extension to $\mathbb{R}$ is now possible. Moreover, the extension to $\mathbb{R}$ is odd and Lebesgue integrable on $[-l, l]$, and therefore has a sine series. We speak of this as the "sine series" of the original function given on $[0, l]$.

Example 4.4.3 For $f(x) = x$, $0 \leq x \leq \pi$, we shall determine the sine series and cosine series below.

The odd and even periodic extensions are precisely the functions discussed in Examples 4.4.2(a) & (b) respectively. Therefore the sine and cosine series are none other than the Fourier series derived there, namely,

$$\sum_{n=1}^{\infty} (-1)^{n+1} \frac{2}{n} \sin nx$$

and

$$\frac{\pi}{2} - \frac{4}{\pi} \sum_{n=0}^{\infty} \frac{\cos(2n+1)x}{(2n+1)^2}.$$

Problem Set 4.4

4.4.P1. Find the Fourier cosine series for the function given on $[0, \pi]$ by $\sin x$. Deduce from the series that $\frac{1}{2} = \sum_{n=1}^{\infty} \frac{1}{4n^2-1}$.

4.4.P2. Find the Fourier series of the function

$$f(x) = \begin{cases} 0 & \text{if } -2 \leq x < -1 \\ k & \text{if } -1 \leq x \leq 1 \\ 0 & \text{if } 1 < x \leq 2, \end{cases}$$

extended periodically to the whole of $\mathbb{R}$ (period 4). Here k is some nonzero constant.

4.5 Orthonormal Expansions

In Sect. 3.3, the notion of the inner product of two functions in $L^2(X)$ was defined. The reader would have observed that it has properties analogous to those of the inner (or dot) product of ordinary vectors in $\mathbb{R}^n$. The notion of othogonality as is available between ordinary vectors in $\mathbb{R}^n$ was extended to elements of $L^2(X)$. It will yield rich dividends as the present section unfolds.

Definition 4.5.1 A sequence of elements $\phi_1, \phi_2, \ldots$ of $L^2[a, b]$ is said to be an **orthonormal sequence** if and only if every element is "normalised" and any two distinct elements are orthogonal to each other. In other words,

$$\langle \phi_i, \phi_j \rangle = \int_{[a,b]} (\phi_i \phi_j) dm = \begin{cases} 1 & \text{if } i = j \\ 0 & \text{if } i \neq j. \end{cases}$$

Here are some classical examples of orthonormal sequences.

Examples 4.5.2
(a) The trigonometric functions

$$\frac{1}{\sqrt{2\pi}}, \frac{\cos x}{\sqrt{\pi}}, \frac{\sin x}{\sqrt{\pi}}, \frac{\cos 2x}{\sqrt{\pi}}, \frac{\sin 2x}{\sqrt{\pi}}, \ldots$$

form an orthonormal sequence in $L^2[-\pi, \pi]$. This follows from Lemma 4.1.2 and the fact that their continuity ensures that all Lebesgue integrals concerned are Riemann integrals.

(b) Consider the sequence $\{P_n\}_{n \geq 0}$ of what are called "Legendre functions":

$$P_n(x) = \sqrt{\frac{2n+1}{2}} \frac{1}{2^n n!} \frac{d^n}{dx^n} (x^2 - 1)^n, \qquad -1 \leq x \leq 1, n = 0, 1, 2, \ldots.$$

It can be shown that they form an othonormal sequence in $L^2[-1, 1]$. Here,

$$P_0(x) = \frac{1}{\sqrt{2}}, \qquad P_1(x) = \sqrt{\frac{3}{2}} x, \qquad P_2(x) = \frac{\sqrt{10}}{4} (3x^2 - 1).$$

In the rest of this section, the symbol "m" will be used for an index of a term in a sequence (e.g. as in "k = 1, 2, …, m") except when it appears as "$m(E)$" or in an integral as "dm".

In the elementary geometry of $\mathbb{R}^3$, the reader will have encountered the problem of determining a point on a line (or plane) through the origin nearest to a given point in $\mathbb{R}^3$. It may be recalled that the nearest point turns out to be the orthogonal projection of the given point on the line (or plane) and can be computed by using the inner product. We consider an analogue of the problem in $L^2[a, b]$, where the concept of an inner product with similar properties has been introduced.

Let $\{\phi_k\}_{k\geq 1}$ be an orthonormal sequence in $L^2[a, b]$ and let f be an arbitrary function in the space. Consider the problem of approximating f in the mean, i.e. in the metric of $L^2[a, b]$, by a linear combination

$$c_1\phi_1 + c_2\phi_2 + \cdots + c_m\phi_m$$

of the first m elements of the orthonormal sequence. Here $c_1, \ldots, c_m$ are real numbers. In other words, we have to find $\sum_{k=1}^{m} c_k\phi_k$ such that the distance $\left\| f - \sum_{k=1}^{m} c_k\phi_k \right\|$ is minimum by choosing the coefficients c_k appropriately. Now,

$$\begin{aligned} \left\| f - \sum_{k=1}^{m} c_k\phi_k \right\|^2 &= \left\langle f - \sum_{k=1}^{m} c_k\phi_k, f - \sum_{k=1}^{m} c_k\phi_k \right\rangle = \langle f,f\rangle - 2\sum_{k=1}^{m} c_k\langle\phi_k,f\rangle + \sum_{k=1}^{m} c_k^2 \\ &= \|f\|^2 - \sum_{k=1}^{m} \langle\phi_k,f\rangle^2 + \sum_{k=1}^{m} (\langle\phi_k,f\rangle - c_k)^2. \end{aligned}$$

This equality implies that

$$\left\| f - \sum_{k=1}^{m} c_k\phi_k \right\| \text{ is minimum if and only if } c_k = \langle\phi_k,f\rangle \text{ for } k = 1, 2, \ldots, m.$$

This solves the approximation problem.

Upon substituting $c_k = \langle\phi_k,f\rangle$ in the same equality, we get the inequality

$$\sum_{k=1}^{m} \langle\phi_k,f\rangle^2 \leq \|f\|^2.$$

Since this inequality holds for all positive integers m, we may take the limit and obtain

$$\sum_{k=1}^{\infty} \langle\phi_k,f\rangle^2 \leq \|f\|^2.$$

This is known as **Bessel's Inequality** for square integrable functions with respect to an orthonormal sequence. [See also Problem 4.2.P2.]

We have thus proved the following:

Theorem 4.5.3 *If* $\{\phi_k\}_{k\geq 1}$ *is an orthonormal sequence in* $L^2[a, b]$, *then*

$$\sum_{k=1}^{\infty} \langle \phi_k, f\rangle^2 \leq \|f\|^2.$$

In particular, $\lim_{k\to\infty} \langle \phi_k, f\rangle = 0$.

Definition 4.5.4 Let $\{\phi_n\}_{n\geq 1}$ be an orthonormal sequence in $L^2[a, b]$. If $f \in L^2[a, b]$ and $c_k = \langle \phi_k, f\rangle$ for $k = 1, 2, \ldots$, then we call c_k, $k = 1, 2, \ldots$, the **generalised Fourier coefficients of** f. The series $\sum_{k=1}^{m} c_k\phi_k$ is called the **generalised Fourier series of** or the **orthonormal expansion of** f and we write

$$f \sim \sum_{k=1}^{\infty} c_k\phi_k.$$

For example, if $[a, b] = [-\pi, \pi]$ and we consider the orthonormal sequence

$$\left\{\frac{1}{\sqrt{2\pi}}, \frac{\cos x}{\sqrt{\pi}}, \frac{\sin x}{\sqrt{\pi}}, \frac{\cos 2x}{\sqrt{\pi}}, \frac{\sin 2x}{\sqrt{\pi}}, \ldots\right\},$$

then the generalised Fourier coefficients of $f \in L^2[-\pi, \pi]$ are

$$a_0\sqrt{\frac{\pi}{2}}, a_1\sqrt{\pi}, b_1\sqrt{\pi}, a_2\sqrt{\pi}, b_2\sqrt{\pi}, \ldots,$$

where

$$a_n = \frac{1}{\pi}\int_{[-\pi,\pi]} f(x)\cos nx\, dm(x), \qquad n = 0, 1, 2, \ldots$$

and

$$b_n = \frac{1}{\pi}\int_{[-\pi,\pi]} f(x)\sin nx dm(x), \qquad n = 1, 2, 3, \ldots.$$

In view of Theorem 4.5.3, it follows that the sum of the squares of these coefficients is less than or equal to the square of the norm of f, i.e.

$$\frac{1}{2}a_0^2 + \sum_{k=1}^{\infty} (a_k^2 + b_k^2) \leq \frac{1}{\pi}\int_{[-\pi,\pi]} f(x)^2 dm(x).$$

One consequence of Theorem 4.5.3 is that, if $\{c_k\}_{k\geq 1}$ is the sequence of generalized Fourier coefficients of $f \in L^2[a, b]$, then $\sum_{k=1}^{\infty} c_k^2$ is finite. We now prove the converse, namely, that if $\{\phi_k\}_{k\geq 1}$ is an orthonormal sequence in $L^2[a, b]$, then for any given sequence $\{c_k\}_{k\geq 1}$ satisfying $\sum_{k=1}^{\infty} c_k^2 < \infty$, there exists an $f \in L^2[a, b]$ for which the sequence of generalised Fourier coefficients $\langle \phi_k, f \rangle$ is none other than $\{c_k\}_{k\geq 1}$. This result is due to F. Riesz. That L^2 is a complete metric space was proved by E. Fischer. Since the arguments in the proof of Riesz' theorem are based on Fischer's theorem, the converse mentioned above is named after both of them.

Theorem 4.5.5 (Riesz–Fischer) *Let $\{\phi_k\}_{k\geq 1}$ be an orthonormal sequence in $L^2[a, b]$ and $\{c_k\}_{k\geq 1}$ be a sequence of real numbers satisfying $\sum_{k=1}^{\infty} c_k^2 < \infty$. Then there exists an $f \in L^2[a, b]$ such that $f \sim \sum_{k=1}^{\infty} c_k \phi_k$ and the sequence $\{s_n\}_{n\geq 1}$, where $s_n = \sum_{k=1}^{n} c_k \phi_k$, converges to f.*

Proof For $n > m$,

$$\|s_n - s_m\|^2 = \left\| \sum_{k=m+1}^{n} c_k \phi_k \right\|^2 = \sum_{k=m+1}^{n} c_k^2,$$

so that $\{s_n\}_{n\geq 1}$ is a Cauchy sequence in $L^2[a, b]$. By Theorem 3.3.10, there is a function $f \in L^2[a, b]$ such that

$$\lim_{n\to\infty} \|f - s_n\| = 0.$$

Again, for $n > k$,

$$\int_{[a,b]} (f\phi_k)dm - c_k = \int_{[a,b]} (f\phi_k)dm - \int_{[a,b]} (s_n\phi_k)dm$$

in view of Definition 4.5.1. Hence

$$\left| \int_{[a,b]} (f\phi_k)dm - c_k \right| = \left| \int_{[a,b]} ((f - s_n)\phi_k)dm \right| \leq \|f - s_n\| \|\phi_k\| = \|f - s_n\|,$$

where we have used the Cauchy–Schwarz Inequality 3.3.16. Letting $n \to \infty$, we obtain

$$c_k = \int_{[a,b]} (f\phi_k)dm = \langle \phi_k, f \rangle, \qquad \text{for } k = 1, 2, 3, \ldots. \qquad \square$$

Definition 4.5.6 An orthogonal sequence $\{\phi_k\}_{k \geq 1}$ of nonzero elements in $L^2[a, b]$ is said to be **complete** if the only element of $L^2[a, b]$ that is orthogonal to each ϕ_k is $[\![0]\!]$; in other words, if $\{\phi_k\}_{k \geq 1}$ satisfies the following condition:

whenever $\langle \phi_k, f \rangle = 0 \, \forall k \in \mathbb{N}$, we have $f = 0$ almost everywhere (i.e. $[\![f]\!] = [\![0]\!]$).

We shall show below in Theorem 4.5.8 that the trigonometric sequence

$$1, \cos\ x, \sin\ x, \cos 2x, \sin 2x, \ldots.$$

is complete in $L^2[-\pi, \pi]$.

The sequence obtained by deleting any term from a complete orthogonal sequence is never complete, because the deleted term is a nonzero element orthogonal to each term in it.

In Bessel's Inequality, the sign of equality holds whenever the orthonormal sequence in question is complete. This is, in fact, the content of the last part of our next result.

Theorem 4.5.7 *Given a complete orthonormal sequence $\{\phi_k\}_{k \geq 1}$ in $L^2[a, b]$, the orthonormal expansion of every element $f \in L^2[a, b]$ converges to it in the mean:*

$$f = \sum_{k=1}^{\infty} c_k \phi_k, \qquad \textit{where } c_k = \langle \phi_k, f \rangle \, \forall k \in \mathbb{N},$$

i.e.

$$\left\| f - \sum_{k=1}^{m} c_k \phi_k \right\| \to 0 \textit{ as } m \to \infty.$$

Furthermore, we have **Parseval's Theorem**:

$$\|f\|^2 = \sum_{k=1}^{\infty} c_k^2.$$

[This equality was proved in 4.3.P3 for continuous 2π-periodic functions with reference to the orthonormal sequence

$$\frac{1}{\sqrt{2\pi}}, \frac{\cos\ x}{\sqrt{\pi}}, \frac{\sin\ x}{\sqrt{\pi}}, \frac{\cos 2x}{\sqrt{\pi}}, \frac{\sin 2x}{\sqrt{\pi}}, \ldots.]$$

Proof For the complete orthonormal sequence $\{\phi_k\}_{k \geq 1}$ in $L^2[a, b]$, and $c_k = \langle \phi_k, f \rangle \, \forall k \in \mathbb{N}$, we have

$$\left\| \sum_{k=m+1}^{n} c_k \phi_k \right\|^2 = \left\langle \sum_{k=m+1}^{n} c_k \phi_k, \sum_{k=m+1}^{n} c_k \phi_k \right\rangle = \sum_{k=m+1}^{n} c_k^2. \tag{4.22}$$

Since by Bessel's Inequality, the series $\sum_{k=1}^{m} c_k^2$ is convergent, it follows that the right-hand side of (4.22) tends to 0 as m, $n \to \infty$. On using Theorem 3.3.10, we obtain a function $g \in L^2[a, b]$ such that

$$\left\| g - \sum_{k=1}^{m} c_k \phi_k \right\| \to 0 \text{ as } m \to \infty. \tag{4.23}$$

We next show that $\langle \phi_j, g \rangle = c_j$ for every j. In fact, for any given j and any $m \geq j$, we have

$$|\langle \phi_j, g \rangle - c_j| = \left| \left\langle \phi_j, g - \sum_{k=1}^{m} c_k \phi_k \right\rangle \right| \leq \left\| g - \sum_{k=1}^{m} c_k \phi_k \right\| \|\phi_j\| = \left\| g - \sum_{k=1}^{m} c_k \phi_k \right\|.$$

When $m \to \infty$, the right-hand side tends to 0 by virtue of (4.23) and so we get $\langle \phi_j, g \rangle = c_j$.

It is a consequence of what has just been proved that $\langle \phi_j, g - f \rangle = 0$ for every j. The completeness of the orthonormal sequence $\{\phi_k\}_{k \geq 1}$ implies $g = f$ a.e. We may now write (4.23) as

$$\left\| f - \sum_{k=1}^{m} c_k \phi_k \right\| \to 0 \text{ as } m \to \infty.$$

Also,

$$\begin{aligned} \left\| f - \sum_{k=1}^{m} c_k \phi_k \right\|^2 &= \left\langle f - \sum_{k=1}^{m} c_k \phi_k, f - \sum_{k=1}^{m} c_k \phi_k \right\rangle = \langle f, f \rangle - 2 \sum_{k=1}^{m} c_k \langle \phi_k, f \rangle + \sum_{k=1}^{m} c_k^2 \\ &= \|f\|^2 - 2 \sum_{k=1}^{m} c_k^2 + \sum_{k=1}^{m} c_k^2 = \|f\|^2 - \sum_{k=1}^{m} c_k^2. \end{aligned}$$

So,

$$\|f\|^2 - \sum_{k=1}^{m} c_k^2 = \left\| f - \sum_{k=1}^{m} c_k \phi_k \right\|^2 \to 0 \text{ as } m \to \infty.$$

Consequently,

$$\|f\|^2 = \sum_{k=1}^{\infty} c_k^2. \qquad \square$$

If for some orthonormal sequence, equality always holds in Bessel's Inequality, i.e. if Parseval's Theorem holds, then that orthonormal sequence is complete. This is because if an orthonormal sequence fails to be complete, then by definition of completeness, there must exist a nonzero f that is orthogonal to every term in the sequence, which implies $\sum_{k=1}^{\infty} c_k^2 = 0$ but $\|f\| > 0$.

Finally, we show that the orthogonal trigonometric sequence is complete in $L^2[-\pi, \pi]$.

Theorem 4.5.8 *If $f \in L^2[-\pi, \pi]$ satisfies the conditions*

$$\int_{[-\pi,\pi]} f(x)\, dm(x) = \int_{[-\pi,\pi]} f(x) \cos\, nx\, dm(x) = \int_{[-\pi,\pi]} f(x) \sin\, nx\, dm(x) = 0$$
$$\forall n \in \mathbb{N},$$

then $f = 0$ a.e.

Proof From the hypotheses, it immediately follows that

$$\int_{[-\pi,\pi]} f(x)T(x)dm(x) = 0 \text{ for an arbitrary trigonometric polynomial } T. \tag{4.24}$$

By Proposition 3.4.4, there exists a sequence $\{T_n\}_{n\geq 1}$ of trigonometric polynomials such that

$$\lim_{n\to\infty} \|f - T_n\| = 0. \tag{4.25}$$

But by Proposition 3.3.18,

$$\begin{aligned} 0 \leq \|f\|^2 &= \langle f, f\rangle \leq \langle f, f\rangle + \langle T_n, T_n\rangle \\ &= \langle f, f\rangle + \langle T_n, T_n\rangle - 2\langle f, T_n\rangle \qquad \text{by (4.24)} \\ &= \langle f - T_n, f - T_n\rangle. \end{aligned}$$

From this and (4.25), it follows upon letting $n \to \infty$ that $\|f\| = 0$, i.e. $f = 0$ a.e. $\square$

The following is an immediate corollary of Theorems 4.5.7 & 4.5.8.

Corollary 4.5.9 *With every $f \in L^2[-\pi, \pi]$ we associate its "Fourier series"*

$$f(x) = \frac{1}{2}a_0 + \sum_{n=1}^{\infty}(a_n \cos nx + b_n \sin nx),$$

where

$$a_n = \frac{1}{\pi}\int_{-\pi}^{\pi} f(x)\cos nx\,dx, \quad n = 0, 1, 2, \ldots$$

and

$$b_n = \frac{1}{\pi}\int_{-\pi}^{\pi} f(x)\sin nx\,dx, \quad n = 1, 2, 3, \ldots.$$

Then

$$\lim_{n\to\infty}\|f - s_n\| = 0,$$

where $s_n(x) = \frac{1}{2}a_0 + \sum_{k=1}^{n}(a_k \cos kx + b_k \sin kx)$, $n = 1, 2, \ldots$ *and Parseval's Theorem holds*:

$$\frac{1}{2}a_0^2 + \sum_{k=1}^{\infty}(a_k^2 + b_k^2) = \frac{1}{\pi}\int_{[-\pi,\pi]} f^2 dm.$$

Problem Set 4.5

4.5.P1. If $A \subseteq [-\pi, \pi]$ is measurable, prove that

$$\lim_{n\to\infty}\int_A \cos nx\,dm(x) = \lim_{n\to\infty}\int_A \sin nx\,dm(x) = 0.$$

4.5.P2. Let $n_1, n_2, \ldots$ be a sequence of positive integers and let

$$E = x \in [-\pi, \pi] : \lim_{k\to\infty}\sin n_k x \text{ exists}.$$

Prove that E has measure $m(E) = 0$.

4.5.P3. Show that each of the two sequences

$$1, \cos x, \cos 2x, \cos 3x, \ldots$$
$$\sin x, \sin 2x, \sin 3x, \ldots$$

is a complete orthogonal sequence in $L^2[0, \pi]$.

4.5.P4. Suppose E is a subset of $[-\pi, \pi]$ with positive measure and let $\delta > 0$ be given. Deduce from Bessel's Inequality that the number of positive integers k such that $\sin kx > \delta$ for all $x \in E$ is finite.

Appendix

The purpose of this appendix is to prove two results. One is a weaker version of Fejér's Theorem 4.3.4, in which continuity is assumed everywhere and which can be deduced from a theorem due to P. P. Korovkin. The latter is of independent interest and has been the subject of much research and generalisation. The second is the existence of a continuous function with a Fourier series that does not converge at 0. The construction of the function that we present is due to Fejér but du Bois-Reymond was the first to construct such a function in 1876.

Definition Let $C[a, b]$ be the space of all continuous real-valued functions on the interval $[a, b]$. A map $P:C[a, b] \to C[a, b]$ (also called an **operator in** $C[a, b]$) is said to be **positive** if $P(f) \geq 0$ wherever $f \geq 0$. It is said to be **linear** if $P(\alpha f + \beta g) = \alpha P(f) + \beta P(g)$ for all $\alpha, \beta \in \mathbb{R}$ and all $f, g \in C[a, b]$.

It may be observed that a linear positive operator P is **monotone**, in the sense that $f \leq g \Rightarrow P(f) \leq P(g)$. Also, one can have linear positive operators in the space $C_{2\pi}[-\pi, \pi]$ of all continuous functions on $[-\pi, \pi]$ with $f(-\pi) = f(\pi)$. The functions $f_0 = 1$, $f_1 = \cos$ and $f_2 = \sin$ belong to $C_{2\pi}[-\pi, \pi]$. Any $f \in C_{2\pi}[-\pi, \pi]$ can be uniquely extended to all of $\mathbb{R}$ by periodicity and the space of the extended functions is essentially the same as $C_{2\pi}[-\pi, \pi]$. In effect, we may assume every function in $C_{2\pi}[-\pi, \pi]$ to be so extended and speak of its Fourier series.

One often writes $P(f)$ as simply Pf, especially when the operator is denoted by an upper case letter and the function to which it is applied is denoted by a lower case letter.

Theorem (Korovkin's Theorem) *Let* $\{P_n\}_{n\geq 1}$ *denote a sequence of positive linear operators on* $C_{2\pi}[-\pi, \pi]$. *In order that* $P_n f \to f$ *uniformly for every* $f \in C_{2\pi}[-\pi, \pi]$, *it is necessary and sufficient that such convergence occur for* $f_0 = 1, f_1 = \cos$ *and* $f_2 = \sin$.

Proof We first prove sufficiency. Suppose the uniform convergence occurs for the three particular functions $f_0 = 1$, $f_1 = \cos$ and $f_2 = \sin$. Put

$$\varphi_y(x) = \sin^2\frac{y-x}{2} = \frac{1}{2}(1 - \cos(y-x)) = \frac{1}{2}(1 - \cos y \cos x - \sin y \sin x).$$

Then we have

$$\varphi_y = \frac{1}{2}(f_0 - f_1 \cos\, y - f_2 \sin\, y)$$

and

$$P_n(\varphi_y) = \frac{1}{2}(P_n(f_0) - P_n(f_1)\cos\, y - P_n(f_2)\sin\, y).$$

Thus

$$\begin{aligned}(P_n\varphi_y)(y) &= \frac{1}{2}\{[(P_nf_0)(y) - 1] - \cos\, y[(P_nf_1)(y) - \cos\, y] - \sin\, y[(P_nf_2)(y) - \sin\, y]\}\\ &\le \frac{1}{2}\{\|P_nf_0 - f_0\| + |\cos\, y|\|P_nf_1 - f_1\| + |\sin\, y|\|P_nf_2 - f_2\|\},\end{aligned}$$

where $\|f\|$ means sup $\{|f(z)| : z \in [-\pi, \pi]\}$ for any $f \in C_{2\pi}[-\pi, \pi]$. Since $|\cos\, y|$ and $|\sin\, y|$ are bounded on $[-\pi, \pi]$, it follows that $P_n\varphi_y(y)$ converges uniformly to 0 in y.

Now let f be an arbitrary element of $C_{2\pi}[-\pi, \pi]$ and let $\varepsilon > 0$ be given. Since $f \in C_{2\pi}[-\pi, \pi]$ is uniformly continuous, there is a positive $\delta < \pi$ such that

$$|x - y| < \delta \Rightarrow |f(x) - f(y)| < \varepsilon. \tag{4.26}$$

Also, the function $f \in C_{2\pi}[-\pi, \pi]$ is bounded and therefore there exists an $M > 0$ such that

$$-M \le f(x) \le M \qquad \forall x \in [-\pi, \pi]. \tag{4.27}$$

Let y be an arbitrary but fixed point of $[-\pi, \pi]$. If $y - \delta < x < y + \delta$, i.e., $x \in (y - \delta, y + \delta)$, then

$$|f(x) - f(y)| < \varepsilon$$

by (4.26) above and if $\delta \le x - y \le 2\pi - \delta$, i.e., $x \in [y + \delta, 2\pi + y - \delta]$ then

$$|f(x) - f(y)| \le 2M \le 2M\frac{\varphi_y(x)}{\sin^2\frac{\delta}{2}}.$$

by (4.27) above combined with the following argument: if $\delta \le x - y \le 2\pi - \delta$, then $\frac{\delta}{2} \le \frac{1}{2}(x - y) \le \pi - \frac{\delta}{2}$ and $\varphi_y(x) = \sin^2\frac{x-y}{2} \ge \sin^2\frac{\delta}{2}$. Thus, for $x \in (y - \delta, 2\pi + y - \delta] = (y - \delta, y + \delta) \cup [y + \delta, 2\pi + y - \delta]$, we have

$$-\varepsilon - \frac{2M}{\sin^2 \frac{\delta}{2}} \varphi_y(x) \le f(x) - f(y) \le \varepsilon + \frac{2M}{\sin^2 \frac{\delta}{2}} \varphi_y(x);$$

that is,

$$-\varepsilon f_0 - \frac{2M}{\sin^2 \frac{\delta}{2}} \varphi_y \le f - f(y) f_0 \le \varepsilon f_0 + \frac{2M}{\sin^2 \frac{\delta}{2}} \varphi_y \qquad \text{on } (y - \delta, 2\pi + y - \delta].$$

On using the periodicity of the functions involved, the above inequality is seen to hold on all of $\mathbb{R}$ and in particular on $[-\pi, \pi]$. Since the operators P_n are positive and linear, and hence monotone, we have

$$-\varepsilon (P_n f_0)(y) - \frac{2M}{\sin^2 \frac{\delta}{2}} (P_n \varphi_y)(y) \le (P_n f)(y) - f(y)(P_n f_0)(y)$$
$$\le \varepsilon (P_n f_0)(y) + \frac{2M}{\sin^2 \frac{\delta}{2}} (P_n \varphi_y)(y),$$

which implies

$$|(P_n f)(y) - f(y)(P_n f_0)(y)| \le \varepsilon \|P_n f_0\| + \frac{2M}{\sin^2 \frac{\delta}{2}} (P_n \varphi_y)(y).$$

Since $P_n f_0 \to f_0$ and $(P_n \varphi_y)(y) \to 0$ uniformly in y, it follows that $(P_n f)\ (y) \to f(y)$ uniformly in y. This completes the proof of sufficiency.

The necessity is trivial. □

Theorem (Fejér) *The Cesàro means of the Fourier series of a continuous function of period* 2π *converge uniformly to the function.*

Proof It is clear from Remark 4.3.2 that the Fejér sums (operators) σ_n are positive and linear. We may complete the proof by verifying that $\sigma_n f \to f$ uniformly when $f = 1$ or cos or sin. This requires only some simple computation:

$$\sigma_n 1 = \tfrac{1}{n}(1 + 1 + \cdots + 1) = 1 \to 1;$$
$$(\sigma_n \cos)(x) = \tfrac{1}{n}(0 + \cos x + \cos x + \cdots + \cos x) = \tfrac{n-1}{n} \cos x \to \cos x;$$
$$(\sigma_n \sin)(x) = \tfrac{1}{n}(0 + \sin x + \sin x + \cdots + \sin x) = \tfrac{n-1}{n} \sin x \to \sin x.$$ □

Theorem *There exists a continuous function with period* 2π *for which the Fourier series diverges to* ∞ *at* 0.

Proof For integers $m > n > 0$, set

$$T(x, m, n) = 2 \sin mx \sum_{k=1}^{n} \frac{\sin kx}{k}. \tag{4.28}$$

By Proposition 4.2.9, the sum on the right-hand side is bounded in absolute value (by $1 + \pi$) on $\mathbb{R}$ and therefore so is $T(x, m, n)$. Using elementary trigonometric identities, we can rewrite $T(x, m, n)$ as the sum of the sums

$$\frac{\cos(m-n)x}{n} + \frac{\cos(m-n+1)x}{n-1} + \cdots + \frac{\cos(m-1)x}{1} \tag{4.29}$$

and

$$-\left[\frac{\cos(m+1)x}{1} + \frac{\cos(m+2)x}{2} + \cdots + \frac{\cos(m+n)x}{n}\right]. \tag{4.30}$$

It will be convenient to refer to the sum of (4.29) and (4.30) as the *expansion* of $T(x, m, n)$. Observe that the multiples of x in (4.29) and in (4.30), *taken together*, are in increasing order, and consequently, the expansion of $T(x, m, n)$ consists of terms of a trigonometric series *in their proper order*.

Now let $\{n_k\}$ and $\{m_k\}$ be sequences of positive integers such that $n_k < m_k$ and let $\sum \alpha_k$ be a convergent series of positive terms. As $T(x, m, n)$ is bounded in absolute value, the series

$$\sum_{k=1}^{\infty} \alpha_k T(x, m_k, n_k) \tag{4.31}$$

converges uniformly and absolutely on $\mathbb{R}$. Therefore its sum is a continuous function $f(x)$ on $\mathbb{R}$ with period 2π. Each T is an even function and therefore the same is true of f.

We can choose the sequences $\{n_k\}$ and $\{m_k\}$ such that

$$m_k + n_k < m_{k+1} - n_{k+1} \qquad \forall k.$$

To do so, all we need to arrange for is that $n_{k+1} > 3n_k$ and $m_k = 2n_k$, so that $m_k + n_k = 3n_k < n_{k+1} = 2n_{k+1} - n_{k+1} = m_{k+1} - n_{k+1}$. This ensures that, when $k > j$, all the multiples of x in the expansion of $T(x, m_k, n_k)$ are higher multiples than all those in the expansion of $T(x, m_j, n_j)$. Therefore in view of the observation above, upon writing out the expansion of each $T(x, m_k, n_k)$ in the sum (4.31), we get a trigonometric series with terms in the proper order. Denote the series by $\sum a_j \cos jx$. The partial sums of (4.31) form a subsequence of the partial sums of the trigonometric series. Since the former converge uniformly to $f(x)$, it follows [see Problem 4.1.P7] that the Fourier series corresponding to f is in fact $\Sigma a_j \cos jx$.

Now consider the partial sum $s_{m_p}(x)$ of the Fourier series. [This will not be a partial sum of (4.31) because it stops midway in $T(x, m_p, n_p)$.]

$$
\begin{aligned}
s_{m_p}(x) &= \sum_{j=1}^{m_p} a_j \cos\, jx \\
&= \sum_{k=1}^{p-1} \alpha_k T(x, m_k, n_k) + \alpha_p \left[\frac{\cos(m_p - n_p)x}{n_p} \right. \\
&\quad \left. + \frac{\cos(m_p - n_p + 1)x}{n_p - 1} + \cdots + \frac{\cos(m_p - 1)x}{1} \right].
\end{aligned}
$$

Since $T(0, m, n) = 0$ by (4.28), we have

$$
s_{m_p}(0) = \alpha_p \sum_{k=1}^{n_p} \frac{1}{k}.
$$

By Problem 4.2.P5, $\sum_{k=1}^{n_p} \frac{1}{k} >$ ln n_p and therefore $s_{m_p}(0) < \alpha_p$ ln n_p. If we now show that the sequences $\{n_k\}$, $\{m_k\}$ and $\{\alpha_k\}$ can be so chosen that $\alpha_p \ln n_p \to \infty$ as $p \to \infty$, then it will follow that the partial sums of the Fourier series of f diverge to ∞ when $x = 0$.

To choose $\{n_k\}$, $\{m_k\}$ and $\{\alpha_k\}$ in the required manner, let $\alpha_k = 1/k^2$ and arrange not only that $m_k = 2n_k$ and $n_{k+1} > 3n_k$ as mentioned earlier, but also that $n_{k+1} > \exp(k + 1)^3$. □

Chapter 5
Differentiation

5.1 Background

In 1806, A.M. Ampère, a great scholar of his times, in a paper entitled '*Sur la théorie des fonctions derivées*', tried without success to establish the differentiability of an arbitrary function except at isolated points. The lack of success can safely be attributed to the fact that the concept of function had not fully evolved then.

During the nineteenth century, repeated attempts were made to establish the differentiability of arbitrary continuous functions. It was K. Weierstrass who, in the year 1861, put an end to these attempts by constructing a continuous function without a derivative at any point whatsoever. B. Bolzano had already obtained a function with this property in 1830. He was however debarred by an imperial decree from teaching and publishing. Later on, simpler examples of functions continuous everywhere but differentiable nowhere were constructed. Mathematicians of the era sought additional conditions under which a continuous function is at least differentiable at "very many points" of its domain. H. Lebesgue, in the year 1904, established as a consequence of his theory of integration (an equivalent version of which we have discussed in Chap. 3) that a continuous monotone function is differentiable except on a set of measure zero. Subsequently, in the year 1911, W. H. Young gave a proof of this striking and important theorem of Lebesgue without the assumption of continuity (Theorem 5.4.6). A simpler proof of the theorem, which is due to L. A. Rubel, as well as three other proofs, will be given in Sects. 5.3 and 5.4.

In this chapter, we shall consider a class of real-valued functions that have a derivative almost everywhere and the derivative is a measurable function. The derivative is then "eligible" to have an integral and we shall discuss how the integral is related to the original function.

S. Shirali and H. L. Vasudeva, *Measure and Integration*,
Springer Undergraduate Mathematics Series,
https://doi.org/10.1007/978-3-030-18747-7_5

5.2 Monotone Functions and Continuity

In this section, we take up some matters about discontinuity of monotone functions (see Sect. 1.6) and one-sided limits (see Sect. 1.3).

Remarks 5.2.1

(a) Since a decreasing function becomes an increasing function when multiplied by -1, we shall restrict our consideration of monotone functions to the class of increasing functions only.
(b) If $f{:}D \to \mathbb{R}$ is a nonconstant monotone function, and D is compact, then f is bounded; for, if $[a, b]$ is the smallest closed interval containing D, then the range of f is contained in an interval whose endpoints are $f(a)$ and $f(b)$. The result is of course trivial if the function is constant.

Examples

(a) The function f defined by $f(x) = \frac{x}{|x|}$ for $x \neq 0$ and $f(0) = k$, where $-1 \leq k \leq 1$, is increasing on $\mathbb{R}$, and is discontinuous at 0, because

$$f(0+) = \lim_{h\to 0+} f(h) = \lim_{h\to 0+} \frac{h}{|h|} = 1$$

and

$$f(0-) = \lim_{h\to 0-} f(h) = \lim_{h\to 0-} \frac{h}{|h|} = -1.$$

(b) The integer part function on $\mathbb{R}$ is increasing and has a discontinuity at each of the integral points.
(c) The function f defined on $\mathbb{R}$ by $f(x) = x$ is continuous and increasing.

We shall prove that functions which are monotone on compact intervals always have right-hand and left-hand limits at each interior point and a right [resp. left] limit at the left [resp. right] endpoint.

Proposition 5.2.2 *If f is increasing on* $[a, b]$, *then*

$$f(x+) = \lim_{h\to 0+} f(x+h) \textit{ exists and } f(x+) \geq f(x) \textit{ for every } x \in [a,\ b),$$
$$f(x-) = \lim_{h\to 0-} f(x+h) \textit{ exists and } f(x-) \leq f(x) \textit{ for every } x \in (a,\ b].$$

Furthermore, if $a \leq x < y \leq b$, then $f(x+) \leq f(y-)$.

Proof For the first two assertions, it is sufficient to prove for every $x \in (a, b]$ that $f(x-)$ exists and does not exceed $f(x)$, because the analogous statement for every $x \in [a, b)$ will follow in the same manner.

Consider any $x \in (a, b]$. For an arbitrary $\xi \in [a, x)$, the set of numbers

$$\{f(t) : \xi < t < x\}$$

is nonempty while being bounded above by the number $f(x)$, and therefore has a least upper bound A, say. Evidently, $A \leq f(x)$. We need only show that $A = f(x-)$ in order to conclude that the one-sided limit $f(x-)$ exists and $f(x-) \leq f(x)$. It will also follow that

$$f(x-) = \sup\{f(t) : \xi < t < x\} \quad \forall \xi \in [a, x). \tag{5.1}$$

Let $\varepsilon > 0$ be given. By definition of the least upper bound, we have

$$A - \varepsilon < f(y) \leq A \tag{5.2}$$

for some $y \in (\xi, x)$. In view of the monotonicity of f, the inequality

$$f(y) \leq f(t) \leq A$$

holds for all t such that $y < t < x$. Together with (5.2), this implies that

$$|f(t) - A| < \varepsilon$$

for all t satisfying the inequality $y < t < x$. Hence $A = f(x-)$, which is all that was needed to be shown for the existence of the one-sided limit $f(x-)$ and for the validity of the equality (5.1).

By a similar argument, we can establish for any $x \in [a, b)$ that $f(x+)$ exists, $f(x+) \geq f(x)$ and

$$f(x+) = \inf\{f(t) : x < t < \xi\} \quad \forall \xi \in (x, b]. \tag{5.3}$$

Now suppose that $a \leq x < y \leq b$. Then $f(x+) = \inf\{f(t) : x < t < y\}$ by (5.3) and $f(y-) = \sup\{f(t) : x < t < y\}$ by (5.1). It therefore follows that $f(x+) \leq f(y-)$. □

Remark 5.2.3 At a point in the interior of the interval of definition of a monotone function f, either $f(x-) = f(x+)$, in which case the function is continuous at x, or $f(x-) < f(x+)$ (see Proposition 5.2.2). In the latter case, there are two gaps in the range of the function if $f(x-) < f(x) < f(x+)$ and one gap if $f(x) = f(x-)$ or $f(x+)$. Consequently, when the range of a monotone function is an interval, there cannot be any discontinuity at an interior point of the interval of definition.

Moreover, there cannot be any discontinuity even at an endpoint. This is because the function can be extended beyond an endpoint to be constant, thereby preserving its monotonicity as well as its range but rendering that endpoint an interior point.

Proposition 5.2.2 has an obvious analogue for decreasing functions.

Definition 5.2.4 If $f:[a, b] \to \mathbb{R}$ is monotone and x is an interior point of $[a, b]$, the number $|f(x+) - f(x-)|$ is called the **jump** or **saltus** of f at x. The jump of f at a [*resp.b*] is defined to be $|f(a+) - f(a)|$ [*resp.* $|f(b) - f(b-)|$].

Theorem 5.2.5 *Let $f:[a, b] \to \mathbb{R}$ be increasing. Then the set of points at which f is discontinuous is at most countable.*

Proof Let E be the set of points in (a, b) at which f is discontinuous. With every point x of E we associate an interval $(f(x-), f(x+))$. Let $r(x)$ be a rational number such that

$$f(x-) < r(x) < f(x+).$$

If $x_1 < x_2$ are points of discontinuity, then for any ξ such that $x_1 < \xi < x_2$, we obviously have $f(x_1+) \le f(\xi) \le f(x_2-)$. Thus the intervals corresponding to different points of discontinuity do not overlap, although they may have an endpoint in common. We have thus shown that the set E is in one-to-one correspondence with a subset of the set of rational numbers, which is known to be countable. Adjoining the endpoints a and b, if they are points of discontinuity, does not alter its finiteness/countability. This completes the proof. □

Remark 5.2.6 The function f defined on $\mathbb{R}$ by $f(x) = [x]$ has a discontinuity at each $x \in \mathbb{Z}$, which is a countable discrete set. However the points of discontinuity of a monotone function need not be isolated. Indeed, it is possible to construct a monotone function with discontinuities at precisely the points of a given countable subset of $\mathbb{R}$, which may even be dense, as illustrated in the example below:

Example 5.2.7 Let E be the range of a sequence $\{x_n\}_{n \ge 1}$ of distinct terms in (a, b) and $\{\alpha_n\}_{n \ge 1}$ be a sequence of positive real numbers such that $\Sigma\alpha_n$ converges. Define

$$f_n(x) = 0 \text{ if } x < x_n \text{ and } f_n(x) = \alpha_n \text{ if } x \ge x_n. \tag{5.4}$$

Since $|f_n(x)| \le \alpha_n$ and $\Sigma\alpha_n$ converges, it follows by the Weierstrass M-test (see Proposition 1.5.6) that the series Σf_n converges uniformly to f, say. If x_0 does not coincide with any of the x_n, then it is a point of continuity for all the f_n and hence a point of continuity of f by Proposition 1.5.3.

On the other hand, if $x_0 = x_m$ for some m, then precisely one of the functions, namely f_m, is discontinuous at x_0. Then $\Sigma_{n \ne m} f_n = f - f_m$ is continuous at x_m. Hence the function $f = (f - f_m) + f_m$, which is the sum of two functions, one continuous at x_m and the other discontinuous, is itself discontinuous at x_m. Nevertheless, the function f is right continuous by the obvious analogue of Proposition 1.5.3.

We next show that f is increasing on (a, b). Indeed, if $\alpha < \beta$, then in view of (5.4),

$$f(\alpha) = \sum_{x_m \le \alpha} f_m(\alpha) = \sum_{x_m \le \alpha} \alpha_m \quad \text{and} \quad f(\beta) = \sum_{x_m \le \beta} f_m(\beta) = \sum_{x_m \le \beta} \alpha_m,$$

where the summations extend over all indices m for which $x_m \le \alpha$ and $x_m \le \beta$ respectively. As the terms of the convergent series are nonnegative, the order of summation is immaterial. Since all the x_m that are less than or equal to α are also less than or equal to β, it follows that $f(\alpha) \le f(\beta)$. Thus the function f has all the desired properties mentioned in Remark 5.2.6.

$$f(x_n+) = \lim_{x \to x_n+} f(x) = \lim_{x \to x_n+} (f - f_n)(x) + \lim_{x \to x_n+} f_n(x) = \lim_{x \to x_n+} (f - f_n)(x) + \alpha_n$$

and

$$f(x_n-) = \lim_{x \to x_n-} f(x) = \lim_{x \to x_n-} (f - f_n)(x) + \lim_{x \to x_n-} f_n(x) = \lim_{x \to x_n-} (f - f_n)(x) + 0.$$

Therefore, by continuity of $f - f_n$ at x_n, we have

$$f(x_n+) - f(x_n-) = \alpha_n.$$

Our next result will lead to the conclusion that the sum of the jumps of a monotone function defined on a closed bounded interval is bounded.

Proposition 5.2.8 *Let f:$[a, b] \to \mathbb{R}$ be an increasing function and suppose $a < x_1, x_2, \ldots, x_n < b$. The sum of the jumps of f at the points $x_1, x_2, \ldots, x_n$ does not exceed the difference $[f(b-) - f(a+)]$.*

Proof Without loss of generality, we may assume that $x_1, x_2, \ldots, x_n$ are in increasing order. Then by Proposition 5.2.2, $f(a+) \le f(x_1-)$, $f(x_n+) \le f(b-)$ and $f(x_i+) \le f(x_{i+1}-)$ for $i = 1, \ldots, n-1$. Hence we have

$$\sum_{i=1}^{n} [f(x_i+) - f(x_i-)] = \sum_{i=1}^{n-1} [f(x_i+) - f(x_{i+1}-)] + f(x_n+) - f(x_1-) \le 0 + f(b-) - f(a+),$$

as asserted. □

Theorem 5.2.5 can also be derived as a consequence of Proposition 5.2.8:

Corollary 5.2.9 *The set of points of discontinuity of an increasing function on $[a, b]$ is at most countable.*

Proof Let H be the set of all points in (a, b) at which f is discontinuous and H_n be the set of points x at which $[f(x+) - f(x-)] > \frac{1}{n}$, $n = 1, 2, \ldots$. Then $H = \cup_{n \ge 1} H_n$. We shall show that H_n is finite for each n and this implies that H is at most countable. If

H_n consists of infinitely many points, then for any finitely many $x_1, x_2,\ldots,x_k$ amongst them, we shall have

$$k\cdot\frac{1}{n}\leq\sum_{i=1}^{k}[f(x_i+)-f(x_i-)]\leq f(b-)-f(a+),$$

using Proposition 5.2.8. This will lead to a contradiction on letting k tend to ∞. This shows that H must be at most countable; adjoining the endpoints a and b, if they are points of discontinuity, does not alter the finiteness/countability. □

In view of Corollary 5.2.9, we may speak of the **sum of (all) jumps on [*a*, *b*]** meaning a sum

$$f(a+)-f(a)+\sum_{j=1}^{\infty}[f(x_j+)-f(x_j-)]+f(b)-f(b-)$$

where the numbers in the sequence $\{x_j\}_{j\geq 1}$ include all points of discontinuity in (a, b) but may also include other points.

Proposition 5.2.10 *Let* $f{:}[a, b]\to\mathbb{R}$ *be an increasing function. Then the sum of its jumps is bounded above by* $f(b) - f(a)$.

Proof If the number of discontinuities is finite, the result follows from Proposition 5.2.8. Otherwise, let the set of points of discontinuity of f in (a, b) be arranged in a sequence $\{x_j\}_{j\geq 1}$. Then

$$\sum_{j=1}^{\infty}[f(x_j+)-f(x_j-)]\leq f(b-)-f(a+). \tag{5.5}$$

Indeed, any partial sum of the series on the left-hand side here satisfies the same inequality by Proposition 5.2.8 and hence (5.5) follows by taking the limit.

Also,

$$f(a+)-f(a)+\sum_{j=1}^{\infty}[f(x_j+)-f(x_j-)]+f(b)-f(b-)\leq f(b)-f(a).$$

Hence the sum of jumps at the points of discontinuity of f in $[a, b]$ does not exceed $[f(b) - f(a)]$. □

Proposition 5.2.11 [Needed in Example 5.5.4] *Suppose* $\{f_n\}_{n\geq 1}$ *is a uniformly convergent sequence of increasing functions defined on an interval and* c *is a point in the domain such that the sequence formed by the jumps* $f_n(c+) - f_n(c-)$ *does not tend to* 0. *Then the limit function* f *is increasing and has a positive jump at* c.

Proof By passing to a subsequence if necessary, we may assume that there exists an $\alpha > 0$ such that $f_n(c+) - f_n(c-) > \alpha$ for every n. The given uniform convergence makes it possible to apply Proposition 1.5.7, which leads to $f(c+)-f(c-) = \lim_{n\to\infty}(f_n(c+)-f_n(c-)) \geq \alpha > 0$. Since f must be increasing by virtue of Proposition 1.6.7, the foregoing inequality means that f has a positive jump at c. □

Remark The hypothesis that the sequence formed by the jumps $f(c+)-f(c-)$ does not tend to 0 is essential, because each member of a sequence of increasing functions can have a nonzero jump at the same point and yet converge uniformly to a continuous function. An instance of this is $\phi_n(x) = 1-\frac{1}{n}$ for $x \leq x_0$ and $1-\frac{1}{2n}$ for $x > x_0$. The sequence $\{\phi_n\}$ converges uniformly to the constant function 1, although each ϕ_n has a nonzero jump at x_0.

Problem Set 5.2

5.2.P1. Give an example of a monotone function which is discontinuous at each rational number in [0, 1] (see Example 5.2.7).

5.2.P2. Let f be defined on an open interval (a, b) and assume that for each point x of the interval, there exists an open interval $\mathfrak{N}_x$ of x in which f is increasing. Prove that f is increasing throughout (a, b).

5.2.P3. Let f be continuous on a compact interval $[a, b]$ and assume that f does not have a local minimum or local maximum at any interior point. Prove that f must be monotonic on $[a, b]$.

5.2.P4. A function f increases on a closed interval $[a, b]$ and it is true that, if $f(a) < \lambda < f(b)$, there exists an $x \in [a, b]$ such $f(x) = \lambda$. Prove that f is continuous on $[a, b]$. Does the same conclusion hold if the hypothesis that f increases is dropped but it is still required that $f(a) < f(b)$?

A function f increases on an open interval I and, for each $a, b \in I$, it is true that, if $f(a) < \lambda < f(b)$, then there exists an $x \in I$ such $f(x) = \lambda$. Prove that f is continuous on I. Does the same conclusion hold if the hypothesis that f increases is dropped?

5.3 Monotone Functions and Differentiability (A)

Let f be an increasing function defined on $[a, b]$. Since the set of points of discontinuity is at most countable, there exist countably many points $x_k \in [a, b]$ that include all points of discontinuity (if there be any). Denote the jump of f at x_k by $j(x_k)$.

With the increasing function f on $[a, b]$, we associate a function S defined throughout the interval $[a, b]$ by means of the relations

$$S(a) = 0 \text{ and } S(x) = \sum_{x_k<x} j(x_k) + f(x) - f(x-) \text{ for } x \in (a, b], \tag{5.6}$$

where the summation extends over all indices k for which $x_k < x$. If the x_k are infinite in number, then the terms of the convergent series $\sum j(x_k)$ are nonnegative, and therefore the order of summation on the right-hand side of (5.6) is immaterial. Since $j(x_k) = 0$ when x_k is not a point of discontinuity, the sum is independent of how the set $\{x_k\} \subseteq [a, b]$ is chosen so long as it includes all points of discontinuity. So the function S is well defined on $[a, b]$ and is obviously increasing. By choosing $\{x_k\}$ so as to include a, we find from (5.6) that $S(x)$ is the sum of all jumps on $[a, x]$ when $x \in (a, b]$. This leads to the first part of the following double inequality, while Proposition 5.2.10 justifies the second part:

$$f(a+) - f(a) \leq S(x) \leq f(x) - f(a) \text{ for } x \in (a, b]. \tag{5.7}$$

The function S is called the **saltus function** or **jump function** associated with f.

Theorem 5.3.1 *Let f:$[a, b] \to \mathbb{R}$ be increasing and S denote the saltus function associated with f. The difference function $g = f - S$ is a continuous increasing function on $[a, b]$.*

Proof For $\alpha, \beta \in (a, b]$ satisfying $\alpha < \beta$, let $\{x_k\} \subseteq [a, b]$ be any countable set that contains all points of discontinuity and has infinitely many points in (a, α) as well as in (α, β). We have $\beta > a$ as well as $\alpha > a$, and hence the definition of S yields

$$S(\beta) = \sum_{a \leq x_k < \alpha} j(x_k) + j(\alpha) + \sum_{\alpha < x_k < \beta} j(x_k) + f(\beta) - f(\beta-),$$

$$S(\alpha) = \sum_{a \leq x_k < \alpha} j(x_k) + (f(\alpha) - f(\alpha-)).$$

Therefore,

$$\begin{aligned} S(\beta) - S(\alpha) &= j(\alpha) + \sum_{\alpha < x_k < \beta} j(x_k) + f(\beta) - f(\beta-) - (f(\alpha) - f(\alpha-)), \\ &= (f(\alpha+) - f(\alpha-)) + \sum_{\alpha < x_k < \beta} j(x_k) + f(\beta) - f(\beta-) - (f(\alpha) - f(\alpha-)) \\ &= (f(\alpha+) - f(\alpha-)) + (f(\alpha) - f(\alpha-)) \\ &\quad + \sum_{\alpha < x_k < \beta} j(x_k) + f(\beta) - f(\beta-) - (f(\alpha) - f(\alpha-)) \\ &= \sum_{\alpha < x_k < \beta} [f(x_k+) - f(x_k-)] + f(\alpha+) - f(\alpha) + f(\beta) - f(\beta-). \end{aligned} \tag{5.8}$$

The right-hand side is the sum of all the jumps of the function f restricted to the closed interval $[\alpha, \beta]$. It follows by Proposition 5.2.10 that

$$S(\beta) - S(\alpha) \leq f(\beta) - f(\alpha). \tag{5.9}$$

Though (5.9) has been proved for $\alpha > a$, it holds even for $\alpha = a$ in view of the second part of (5.7). So, $g(\beta) = f(\beta) - S(\beta) \geq f(\alpha) - S(\alpha) = g(\alpha)$ for $a \leq \alpha < \beta \leq b$, i.e. g is increasing on $[a, b]$.

Let $x_0 \in [a, b)$ be arbitrary. We shall show that g is right continuous at x_0. Let $x > x_0$. From (5.9), it follows that

$$S(x) - S(x_0) \leq f(x) - f(x_0).$$

On letting $x \to x_0+$, we obtain

$$S(x_0+) - S(x_0) \leq f(x_0+) - f(x_0). \tag{5.10}$$

On the other hand,

$$S(x) - S(x_0) \geq f(x_0+) - f(x_0),$$

using the first part of (5.7) if $x_0 = a$ and using (5.8) if $x_0 > a$. On letting $x \to x_0+$, we obtain

$$S(x_0+) - S(x_0) \geq f(x_0+) - f(x_0).$$

Comparing this with (5.10), we get

$$S(x_0+) - S(x_0) = f(x_0+) - f(x_0),$$

whence

$$g(x_0+) = f(x_0+) - S(x_0+) = f(x_0) - S(x_0) = g(x_0).$$

Thus g is right continuous at x_0, where $x_0 \in [a, b)$ is arbitrary. The left continuity of g on $(a, b]$ is proved analogously, except that the first part of (5.7) is not needed.

This completes the proof. □

In preparation for Lebesgue's Theorem on differentiation of monotone functions, namely, that an increasing function on a closed bounded interval has a finite derivative almost everywhere, we discuss below the definitions of one-sided limits superior and inferior of a function.

Let x be a point in the closure of an interval I other than its right endpoint (if any). In particular, for any $\delta > 0$, there exists a t in I such that $0 < t - x < \delta$. Suppose g is an extended real-valued function defined either on $I \backslash \{x\}$ or on I. We define

$$\begin{aligned}\limsup_{t \to x-} g(t) &= \inf_{\delta > 0}\{\sup g(t) : 0 < t - x < \delta\} \\ &= \inf\{\sup\{g(t) : 0 < t - x < \delta\} : \delta > 0\}.\end{aligned}$$

Since the sets $\{g(t): 0 < t - x < \delta\}$ decrease with δ, their sups also decrease with δ and it follows that the inf in the above definition can be replaced by a limit:

$$\limsup_{t \to x+} g(t) = \lim_{\delta \to 0} \sup\{g(t) : 0 < t - x < \delta\}.$$

Similarly, if x is a point in the closure of I other than its left endpoint (if any)

$$\begin{aligned}\limsup_{t \to x-} g(t) &= \inf_{\delta > 0} \{\sup g(t) : 0 < x - t < \delta\} \\ &= \inf\{\sup\{g(t) : 0 < x - t < \delta\} : \delta > 0\} \\ &= \lim_{\delta \to 0} \sup\{g(t) : 0 < x - t < \delta\}.\end{aligned}$$

We define $\liminf_{t \to x+} g(t)$ and $\liminf_{t \to x-} g(t)$ analogously:
If x is a point in the closure of I other than its right endpoint (if any),

$$\begin{aligned}\liminf_{t \to x+} g(t) &= \sup_{\delta > 0} \{\inf g(t) : 0 < t - x < \delta\} \\ &= \sup\{\inf\{g(t) : 0 < t - x < \delta\} : \delta > 0\} \\ &= \lim_{\delta \to 0} \inf\{g(t) : 0 < t - x < \delta\}.\end{aligned}$$

Similarly, if x is a point in the closure of I other than its left endpoint (if any),

$$\begin{aligned}\liminf_{t \to x-} g(t) &= \sup_{\delta > 0} \{\inf g(t) : 0 < x - t < \delta\} \\ &= \sup\{\inf\{g(t) : 0 < x - t < \delta\} : \delta > 0\} \\ &= \lim_{\delta \to 0} \inf\{g(t) : 0 < x - t < \delta\}.\end{aligned}$$

Since $\sup\{g(t): 0 < t - x < \delta\} \geq \inf\{g(t) : 0 < t - x < \delta\}$, it follows upon taking limits that

$$\limsup_{t \to x+} g(t) \geq \liminf_{t \to x+} g(t) \text{ provided that } x \text{ is not the right endpoint}$$

and similarly that

$$\limsup_{t \to x-} g(t) \geq \liminf_{t \to x-} g(t) \text{ provided that } x \text{ is not the left endpoint.}$$

For any function g, the above limits exist at the points indicated but they may be infinite.

We consider the following examples to illustrate the above definitions.

(1) We have $\limsup_{t \to 0^+} [t] = 0 = \liminf_{t \to 0^+} [t]$, $\limsup_{t \to 0-} [t] = -1 = \liminf_{t \to 0-} [t]$.

(2) Consider the function $f(x) = \sin(1/x)$, for $x \neq 0$, defined on $\mathbb{R}\backslash\{0\}$. In this case, $\limsup_{t\to 0+} f(t) = 1 = \limsup_{t\to 0-} f(t)$ and $\liminf_{t\to 0+} f(t) = -1 = \liminf_{t\to 0-} f(t)$.

(3) Let f be defined on $\mathbb{R}$ by means of the relations $f(t) = 1$ if t is rational and 0 if t is irrational. If $\alpha \in \mathbb{R}$ is arbitrary but fixed, $\limsup_{t\to\alpha+} f(t) = 1 = \limsup_{t\to\alpha} f(t)$, whereas $\liminf_{t\to\alpha+} f(t) = 0 = \liminf_{t\to\alpha} f(t)$. The reader will note that the one-sided limits $f(\alpha+)$ and $f(\alpha-)$ do not exist.

Remark The reader is invited to show that

$$\limsup_{t\to x+} g(t) = \inf\{\sup\{g(t) : 0<t-x<\delta\} : \delta_1 > \delta > 0\}$$

for an arbitrary $\delta_1 > 0$. Thus the values of g in an arbitrarily small interval to the right of x are the only relevant ones for purposes of the limsup as $t \to x+$.

It is easy to show that the right limit $g(x+)$ exists (possibly $\pm\infty$) if and only if $\limsup_{t\to x+} g(t) = \liminf_{t\to x+} g(t)$, in which case $\lim_{t\to x+} g(t) = \limsup_{t\to x+} g(t) = \liminf_{t\to x+} g(t)$. Similarly for $g(x-)$.

In order to prove Lebesgue's Theorem on differentiation of monotone functions, we shall need the concept of Dini derivatives, which involve replacing the limit in the definition of derivative by lim inf and lim sup from the left and from the right.

We consider real-valued functions whose domains are intervals in $\mathbb{R}$, containing more than one point. The four Dini derivatives that we are about to introduce have the advantage that they exist for functions that are not necessarily differentiable in the usual sense. They are defined as follows: For an arbitrary real-valued function f and any x which is not the right endpoint of the domain,

$$D^+f(x) = \limsup_{h\to 0+}\frac{f(x+h)-f(x)}{h},\quad D_+f(x) = \liminf_{h\to 0+}\frac{f(x+h)-f(x)}{h}$$

and for any x which is not the left endpoint of the domain,

$$D^-f(x) = \limsup_{h\to 0-}\frac{f(x+h)-f(x)}{h},\quad D_-f(x) = \liminf_{h\to 0-}\frac{f(x+h)-f(x)}{h}.$$

The values of these derivatives are in the extended real number system $\mathbb{R}^*$ and they are known as the **right upper** and **lower** and the **left upper** and **lower Dini derivatives** respectively.

Remarks 5.3.2

(a) The inequalities

$$D_+f(x) \le D^+f(x) \text{ and } D_-f(x) \le D^-f(x)$$

obviously hold wherever they make sense (i.e., except at endpoints).

(b) It is also easy to see that $D^+f(x)$ [resp. $D_+f(x)$] is the largest [resp. smallest] limit of a sequence $\left\{\frac{f(x+h_n)-f(x)}{h_n}\right\}_{n\ge 1}$ where $\{h_n\}_{n\ge 1}$ is a sequence of nonnegative numbers with limit 0 and such that each $x + h_n$ is in the domain of f. Similar statements hold for $D^-f(x)$ and $D_-f(x)$.

(c) If $D_+f(x) = D^+f(x)$ [resp. $D_-f(x) = D^-f(x)$], then f is said to have a **right derivative** [resp. **left derivative**] at x and we write $f'_+(x)$ [resp. $f'_-(x)$] for the common value of $D_+f(x)$ and $D^+f(x)$ [resp. $D_-f(x)$ and $D^-f(x)$]. If x is an interior point of the domain of f and $f'_+(x) = f'_-(x)$, then f is said to have a **derivative** at x and we write $f'(x)$ for the common value of $f'_+(x)$ and $f'_-(x)$, an obvious modification being needed if x is an endpoint. Note that our definition includes $\pm\infty$ as values of $f'(x)$. If $f'(x) \neq \pm\infty$, we say that f is **differentiable** at x. In this case, $f'(x)$ is the derivative in the usual sense, as reiterated in Sect. 1.6.

Examples 5.3.3

(a) Suppose f is defined by $f(x) = x\sin\frac{1}{x}$ for $x \neq 0$ and $f(0) = 0$. Then $D^+f(0) = D^-f(0) = 1$ and $D_+f(0) = D_-f(0) = -1$.
Indeed, the ratio $\frac{f(h)}{h} = \sin\frac{1}{h}$ assumes all values in $[-1, 1]$ as h varies in the interval $(\frac{1}{(2n+2)\pi}, \frac{1}{2n\pi})$, $n \in \mathbb{N}$. Therefore $D^+f(0) = 1$ and $D_+f(0) = -1$. Again by a similar argument, we have $D^-f(0) = 1$ and $D_-f(0) = -1$.

(b) Let f be defined by $f(x) = 1$ if x is rational and 0 otherwise. For rational x and $h > 0$, the ratio $\frac{f(x+h)-f(x)}{h}$ is either 0 or a negative number with large absolute value. So, $D^+f(x) = 0$ and $D_+f(x) = -\infty$. On the other hand, for $h < 0$, the same ratio is either 0 or a positive number with large absolute value. Hence $D^-f(x) = \infty$ and $D_-f(x) = 0$. If x is irrational, then similar considerations show that $D^+f(x) = \infty$ and $D_+f(x) = 0$, while $D^-f(x) = 0$ and $D_-f(x) = -\infty$. Thus the function does not have even one-sided derivatives.

(c) Let $f(x) = |x|$. Then $D^+f(0) = D_+f(0) = 1$ and $D^-f(0) = D_-f(0) = -1$.

(d) Let $f:\mathbb{R} \to \mathbb{R}$ be defined by $f(0) = 0$, $f(x) = x(1 + \sin(\ln x))$ if $x > 0$ and $x + \sqrt{-x}\,\sin^2(\ln(-x))$ if $x < 0$. Then $D^+f(0) = 2$, $D_+f(0) = 0$, $D^-f(0) = 1$, $D_-f(0) = -\infty$. Indeed, $\sin(\ln h)$ assumes all values between -1 and 1 in any interval from $\frac{1}{\exp(2n\pi+2\pi)}$ to $\frac{1}{\exp(2n\pi)}$, where $n \in \mathbb{N}$, and hence for positive values of h arbitrarily close to 0. Therefore $D^+f(0) = 2$ and $D_+f(0) = 0$. Again, by a similar argument, $D^-f(0) = 1$, $D_-f(0) = -\infty$.

F. Riesz' proof in the case of a continuous function, given in [21], is both elegant and simple. We shall borrow from Riesz' proof and combine it with an innovative idea of Rubel [18] to give a reasonably elementary proof of the theorem of Lebesgue. We begin by proving Riesz' Lemma.

Riesz' Lemma 5.3.4 *Let g be a continuous real-valued function on the interval* $[a, b]$ *and let*

$$E = \{x \in (a, b) : g(x) < g(y) \text{ for some } y > x\} \tag{5.11}$$

and

$$E' = \{x \in (a, b) : g(x) < g(y) \text{ for some } y < x\}. \tag{5.12}$$

Then the set E [resp. E′] is either empty or an open set; if E [resp. E′] is nonempty and (a_k, b_k) *is any component interval of it, then* $g(a_k) \le g(b_k)$ [resp. $g(a_k) \ge g(b_k)$].

Proof (F. Riesz) We first show that E is open. Let x_0 be in E. Then there exists $y > x_0$ with $g(y) > g(x_0)$. Let $\varepsilon = g(y) - g(x_0)$. By continuity of g, there exists a $\delta > 0$ such that $|x - x_0| < \delta$ implies $|g(x)-g(x_0)| < \varepsilon$. Let δ_1 be $\min\{\delta, y - x_0, x_0 - a\}$. Now $|x - x_0| < \delta_1$ implies $x < y$ as well as $g(x) < g(x_0) + \varepsilon = g(y)$, which in turn implies $x \in E$. Therefore E contains the open interval $(x_0 - \delta_1, x_0 + \delta_1)$ and is thus an open set.

The set E, if not empty, decomposes into an at most countable collection of disjoint nonempty open intervals (a_k, b_k), $k = 1, 2,\ldots$ (see Theorem 1.3.17); the points a_k, b_k do not belong to the set E. Let $a_k < x_0 < b_k$. We shall prove that $g(x_0) \le g(b_k)$; the desired result will follow upon letting x_0 tend to a_k. Let x_1 be a point in the interval $[x_0, b]$ at which g assumes a maximum value. Then

$$g(x_0) \le g(x_1). \tag{5.13}$$

Also, since $b_k \in [x_0, b]$, we have

$$g(b_k) \le g(x_1). \tag{5.14}$$

Furthermore, every point in $[x_1, b]$ satisfies $g(y) \le g(x_1)$, and it therefore follows that $x_1 \notin E$. Since the part of $[x_0, b]$ to the left of b_k lies entirely within E, we have $b_k \le x_1 \le b$. Again, $b_k \notin E$ and therefore the inequality $g(b_k) < g(x_1)$ cannot hold. By (5.14), it follows that $g(b_k) = g(x_1)$, and hence by (5.13), $g(x_0) \le g(b_k)$. This completes the proof for the set E. The proof for E' is similar. □

Lemma 5.3.5 *Let f be a continuous increasing function defined throughout* $[a, b]$. *For any positive number M, let* $E_M = \{x \in [a, b) : D^+f(x) > M\}$. *Then the measure* $m(E_M)$ *of* E_M *satisfies the inequality*

$$m(E_M) \le \frac{1}{M}(f(b) - f(a)).$$

Proof By definition of $D^+f(x)$, whenever $D^+f(x) > M$, there exists $y > x$ such that

$$\frac{f(y) - f(x)}{y - x} > M.$$

So,

$$f(y) - f(x) > My - Mx, \quad \text{i.e.,} \quad f(y) - My > f(x) - Mx. \tag{5.15}$$

If we set $g(t) = f(t) - Mt$ for $t \in [a, b]$, then g is continuous and (5.15) implies that $x \in E$, where E is the set as in Riesz' Lemma 5.3.4. Thus $E_M \subseteq E$. If E_M is empty, there is nothing to prove; so we may assume that E_M is nonempty, so that E is also nonempty. By Riesz' Lemma, E is open and is, therefore, representable as a countable disjoint union of component intervals:

$$E = \bigcup_{k=1}^{\infty} (a_k,\ b_k),$$

where $g(a_k) \leq g(b_k)$ for each k, i.e.

$$M(b_k - a_k) \leq f(b_k) - f(a_k),$$

whence

$$m(E_M) \leq m(E) = \sum\nolimits_k (b_k - a_k) \leq \frac{1}{M}\sum\nolimits_k (f(b_k) - f(a_k)) \leq \frac{1}{M}(f(b) - f(a)).$$

The last inequality follows in the following manner from the fact that f is increasing: Any n among the disjoint intervals (a_k, b_k) can be arranged so that $b_k \leq a_{k+1}$ for $k = 1, 2, \ldots, n - 1$ and

$$\begin{aligned}\sum_{k=1}^{n} (f(b_k) - f(a_k)) &= -f(a_1) + (f(b_1) - f(a_2)) + \cdots + (f(b_{n-1}) - f(a_n)) + f(b_n) \\ &\leq -f(a_1) + f(b_n) \leq f(b) - f(a).\end{aligned}$$

Taking the limit as $n \to \infty$ if there are infinitely many k yields the desired inequality. □

Corollary 5.3.6 *Let f be a continuous increasing function defined throughout* $[a, b]$. *Then* $m(E_\infty) = 0$, where $E_\infty = \{x \in (a, b) : D^+f(x) = \infty\}$.

Proof E_∞ is the intersection of the sequence of sets $\{E_n\}$, where

$$E_n = \{x \in [a,\ b] : D^+f(x) > n\}.$$

Since $E_1 \supseteq E_2 \supseteq \cdots$ and $m(E_1) < \infty$, we have $m(E_\infty) = \lim_{n\to\infty} m(E_n)$ by continuity of measure (see Proposition 2.3.21). But this limit is 0, in view of the inequality $m(E_n) \leq \frac{1}{n}(f(b) - f(a))$ assured by the foregoing lemma. □

Lemma 5.3.7 *Let f be a continuous increasing function defined throughout* $[a, b]$. *Then*

$$D^+f(x) \leq D_-f(x)$$

a.e. on (a, b).

Proof Step I. Let N and M be positive numbers such that $N < M$ and

$$E_{NM} = \{x \in (a,\ b) : D_-f(x) < N < M < D^+f(x)\}.$$

We shall show that $m(E_{NM}) = 0$.

We may assume that E_{NM} is nonempty, because otherwise there is nothing to prove. Let $x \in E_{NM}$. Since $D_-f(x) < N$, there exists a $y < x$ such that

$$\frac{f(y) - f(x)}{y - x} < N,$$

whence we get (using the fact that $y - x < 0$)

$$f(y) - f(x) > Ny - Nx, \quad \text{i.e.,} \quad f(y) - Ny > f(x) - Nx. \tag{5.16}$$

If we set $g(t) = f(t) - Nt$, $t \in [a, b]$, then g is continuous and (5.16) implies that $x \in E'$, where

$$E' = \{t \in (a,\ b) : g(t) < g(y) \text{ for some } y \in (a, t)\}$$

as in (5.12) of Riesz' Lemma. Thus E_{NM} is covered by an at most countable number of disjoint intervals (a_k, b_k) such that $g(b_k) \leq g(a_k)$ and therefore

$$f(b_k) - f(a_k) \leq N(b_k - a_k). \tag{5.17}$$

Next we consider inside each of the intervals (a_k, b_k) points x where $D^+f(x) > M$. That includes all the points of the nonempty set E_{NM}. Using arguments similar to the ones in the paragraph above, and applying Riesz' Lemma to the function $g_1(t) = f(t) - Mt$, $t \in [a_k, b_k]$, we see that these points x form an open nonempty set. There exists an at most countable number of disjoint intervals $\{(a_{k\ell}, b_{k\ell})\}_{\ell \geq 1}$ each contained in (a_k, b_k) and such that

$$M(b_{k\ell} - a_{k\ell}) \leq f(b_{k\ell}) - f(a_{k\ell})$$

and hence

$$M\sum_{\ell}(b_{k\ell} - a_{k\ell}) \leq \sum_{\ell}(f(b_{k\ell}) - f(a_{k\ell})) \leq f(b_k) - f(a_k).$$

On using (5.17), it follows that

$$M\sum_{k,\ell}(b_{k\ell} - a_{k\ell}) \leq \sum_{k}(f(b_k) - f(a_k)) \leq N\sum_{k}(b_k - a_k).$$

If $|\mathcal{S}_0|$, $|\mathcal{S}_1|$ and $|\mathcal{S}_2|$ denote the total lengths of the systems $\mathcal{S}_0 = \{(a, b)\}$, $\mathcal{S}_1 = \{(a_k, b_k)\}$ and $\mathcal{S}_2 = \{(a_{k\ell}, b_{k\ell})\}$ respectively, it follows that

$$|\mathcal{S}_2| \leq \frac{N}{M}|\mathcal{S}_1| \leq \frac{N}{M}|\mathcal{S}_0|.$$

Repeating the two arguments above alternately, we obtain a sequence $\mathcal{S}_0, \mathcal{S}_1, \mathcal{S}_2, \ldots$ of systems of disjoint open intervals, each having union contained in that of the preceding one, and satisfying the inequalities

$$|\mathcal{S}_{2n}| \leq \frac{N}{M}|\mathcal{S}_{2n-2}|, \quad n = 1, 2, \ldots,$$

which implies

$$|\mathcal{S}_{2n}| \leq \left(\frac{N}{M}\right)^n |\mathcal{S}_0| \to 0 \text{ as } n \to \infty.$$

Thus the set E_{NM} can be covered by a system of intervals of total length as small as we please; so

$$m(E_{NM}) = 0.$$

Step II. We next show that

$$m(\{x \in (a, b) : D^+f(x) > D_-f(x)\}) = 0.$$

Observe that

$$\{x \in (a, b) : D^+f(x) > D_-f(x)\} = \bigcup_{\substack{r, s \in \mathbb{Q} \\ r > s}} E_{sr},$$

where

$$E_{sr} = \{x \in (a,\ b) : D^+f(x) > r > s > D_-f(x)\}$$

and r and s are rational numbers. Since by Step I, $m(E_{sr}) = 0$ for every pair r and s of rational numbers ($r > s$), and since there are at most countably many such pairs, it follows on using Proposition 2.2.9 that

$$m(\{x \in (a,\ b) : D^+f(x) > D_-f(x)\}) \le \sum_{\substack{r,s \in \mathbb{Q} \\ r > s}} m(E_{sr}) = 0.$$

This completes the proof of Step II.

It now follows immediately that

$$D^+f(x) \le D_-(x)$$

a.e. on $[a,\ b]$, which is what the lemma claimed. □

We now present Lebesgue's Theorem for continuous monotone functions, which is the case he treated. The generalisation to all monotone function s, originally due to Young, will follow in the next section.

Theorem 5.3.8 (Lebesgue's Theorem (continuous case)). *Every continuous monotone function defined on* $[a,\ b]$ *possesses a finite derivative almost everywhere.*

Proof We need consider only increasing functions. The above lemma, when applied to the function $-f(-x)$, yields

$$D^-f(x) \le D_+(x)$$

a.e. on $(a,\ b)$. Since f is increasing on $[a,\ b]$, each of the four Dini derivatives is nonnegative. Combining this observation with the above inequality, Lemma 5.3.7 and Corollary 5.3.6 gives

$$0 \le D^+f(x) \le D_-f(x) \le D^-f(x) \le D_+f(x) \le D^+f(x) < \infty$$

a.e. on $(a,\ b)$, i.e. $f'(x)$ exists and is finite a.e. on $[a,\ b]$. □

Problem Set 5.3

5.3.P1. If the function f assumes its local maximum at an interior point c of its domain, then show that $D^+f(c) \le 0$ and $D_-f(c) \ge 0$.

5.3.P2. Suppose f is continuous on $[a,\ b]$ and $D^+f(x) > 0$ for all x in $[a,\ b)$. Show that $f(b) \ge f(a)$. Also, give a counterexample to show that the continuity hypothesis cannot be dropped.

5.3.P3. Show that f may be discontinuous at x_0 when all four Dini derivatives are equal to ∞.

5.3.P4. Let $f{:}[0,\ \infty) \to \ge \mathbb{R}$ be differentiable and suppose that $f(0) = 0$ and that f' is increasing. Prove that

$$g(x) = \begin{cases} \frac{f(x)}{x} & x > 0 \\ f'(0) & x = 0 \end{cases}$$

defines an increasing function of x.

5.3.P5.

(a) Show that if $f'(x)$ exists, then $D^+(f + g)(x) = f'(x) + D^+g(x)$ and similarly for other Dini derivatives.
(b) Give an example when $D^+(f + g)(x) \neq D^+f(x) + D^+g(x)$.

5.3.P6.

(a) Let f be continuous on $[a, b]$. If $\frac{f(b)-f(a)}{b-a} < C$, then $D^+f(x) \leq C$ for uncountably many $x \in [a, b]$.
(b) Let f be continuous on $[a, b]$. If $\frac{f(b)-f(a)}{b-a} > C$, then $D^-f(x) \geq C$ for uncountably many $x \in [a, b]$.

5.3.P7. Let f be a continuous function defined on $[a, b]$. Suppose that there exist real constants α, β such that $\alpha \leq D^+f(x) \leq \beta$ for all $x \in (a, b)$. Prove that $h\alpha \leq f(x + h) - f(x) \leq h\beta$ provided $a \leq x < x + h \leq b$.

5.3.P8. Let f be a continuous real-valued function defined on $[a, b]$ and let $x \in (a, b)$ be such that D^+f is finite in a neighbourhood of x and continuous at x. Prove that $f'(x)$ exists.

5.3.P9. If one of the Dini derivatives of a continuous function is zero everywhere in an interval, the function is constant there.

5.3.P10. Construct monotonic jump functions f on $[0, 1)$ whose discontinuities have 0 as a limit point and such that $f_+'(0)$ is
(a) zero (b) ∞ (c) positive and finite.
Also, compute the quantum of jump at each of the discontinuities.

5.3.P11. If all Dini derivatives of a function f satisfy $|Df(x)| \leq K$ everywhere on an interval, then the function satisfies the condition $|f(x) - f(y)| \leq K|x - y|$.

5.4 Monotone Functions and Differentiability (B)

In this section, we shall present three proofs of Lebesgue's Theorem on differentiation of arbitrary monotone functions.

We begin by proving an extension of Riesz' Lemma 5.3.4 that applies to the discontinuous case. It is our view that the extension is not so obvious and we spell out its proof in detail.

The reader may note that the functions in this section too are real-valued.

Suppose g is a function on a closed interval $[\xi, \eta]$ having finite one-sided limits at each point. Let G be the function on the domain $[\xi, \eta]$ defined by $G(z) = \max\{g(z+), g(z-), g(z)\}$ for $z \in (\xi, \eta)$ and $G(\xi) = \max\{g(\xi+), g(\xi)\}$, $G(\eta) = \max\{g(\eta-), g(\eta)\}$.

It is a trivial consequence that $g \leq G$ everywhere.

Suppose $z \in [\xi,\eta]$ and $\alpha > \beta > G(z)$. By definition of G, it follows that not only $g(z) < \beta$ but also $g(z+) < \beta$ (for $z \neq \eta$) and $g(z-) < \beta$ (for $z \neq \xi$), which further implies that $g \leq \beta$ on a neighbourhood of z, and this in turn implies $G < \alpha$ on a neighbourhood of z. Thus, $G(z) < \alpha$ implies $G < \alpha$ on a neighbourhood of z. This property of G is called **upper semicontinuity**. It is obvious for any upper semicontinuous function ϕ that a set of the form $\{z \in (\xi,\eta) \colon \phi(z) < \alpha\}$ is always open. Moreover, the same argument as for continuous functions in elementary analysis shows that an upper semicontinuous function on a closed bounded interval is bounded above and attains its supremum.

This property of G makes it possible to adapt the reasoning of Riesz' Lemma for the continuous case, where the fact that g had a maximum value on any closed subinterval of its domain played a crucial role.

Proposition 5.4.1 (Extended Riesz Lemma) *Let the notations be as in the paragraph above. Then the set*

$$E = \{x \in (\xi, \eta) : g(y) > G(x)\, for\, some\, y > x\}$$

is open; if $E \neq \varnothing$ *and* (a, b) *is a component interval of* E*, then* $g(x) \leq G(b)$ *for all* $x \in (a, b)$*; in particular,* $g(a+) \leq G(b)$*. The set*

$$F = \{x \in (\xi, \eta) : g(y) > G(x)\, for\, some\, y < x\}$$

is open, and if it is nonempty and (a', b') *is a component interval of* F*, then we have* $G(a') \geq g(x)$ *for all* $x \in (a', b')$*; in particular,* $g(b'-) \leq G(a')$*.*

Proof Let $x \in E$. By definition of E, there exists $y > x$ such that $g(y) > G(x)$. By upper semicontinuity of G, the inequality $G(z) < g(y)$ holds for all z in some neighbourhood of x. The intersection of this neighbourhood with (ξ, y) then lies in E. This shows that E is open.

By the defining condition of E, if $x \notin E$, then g is bounded above by $G(x)$ on the subinterval $[x,\eta]$, which implies that all one-sided limits on $(x, \eta]$ are also bounded by $G(x)$, which further implies that G is bounded by $G(x)$ on $[x, \eta]$. This observation will be used below.

The set E, if not empty, decomposes into an at most countable collection of disjoint open intervals (a_k, b_k), $k = 1, 2, \ldots$ (see Theorem 1.3.17); the points a_k, b_k do not belong to the set E. Let $a_k < x < b_k$. We shall prove that $g(x) \leq G(b_k)$; the desired result will follow upon letting x tend to a_k. Since G is upper semicontinuous, there exists some $x_1 \in [x, \eta]$ such that $G(x_1) = \sup\{G(z) \colon z \in [x, \eta]\}$. Then

$$g(x) \leq G(x) \leq G(x_1). \tag{5.18}$$

Since $x < b_k$, we have $b_k \in [x, \eta]$, which implies

$$G(b_k) \leq G(x_1). \tag{5.19}$$

Furthermore, since $[x_1, \eta] \subseteq [x, \eta]$, every point $y \in [x_1, \eta]$ satisfies $g(y) \leq G(y) \leq G(x_1)$, and it therefore follows that $x_1 \notin E$. Since the part of $[x,\eta]$ to the left of b_k lies entirely within E, we have $b_k \leq x_1 \leq \eta$. Again, $b_k \notin E$ and the observation above permits the conclusion that $G(x_1) \leq G(b_k)$. By (5.19), it follows that $G(b_k) = G(x_1)$, and hence by (5.18), $g(x) \leq G(b_k)$. This completes the proof for the set E.

The proof for the set F is similar. □

Remark

(a) The function G discussed above need not take the value $\gamma = \sup\{g(z): z \in I\}$ on I unless $I = [\xi, \eta]$. This is illustrated by the function g on $[\xi, \eta] = [0, 2]$ defined to satisfy $g(z) = z$ on $[0, 1)$, $g(z) = z + 2$ on $(1, 2]$ and $g(1) = 2$. Here $g(1+) = 3$ and therefore $G(z) = z$ on $[0, 1)$ and $G(z) = z + 2$ on $[1, 2]$, which means that when $I = [0, 1]$, the function G never takes the value $2 = \sup\{g(z): z \in I\}$. However, $G(1) = 3 > \sup\{g(z): z \in I\}$ and $1 \in I$.

(b) A function on a closed bounded interval having one-sided limits at each point can be discontinuous at every rational point (28, Problems 5.2.P5 and 6.2.P4). Nevertheless, such a function can be uniformly approximated by step functions (28, Corollary 6.2.7).

(c) It is a matter of peripheral interest that one can tweak the proof of Proposition 5.4.1 slightly to obtain $g(x) < G(b_k)$. All it takes is to note that one of the inequalities in (5.18) is actually strict, because otherwise g takes its maximum value on $[x, \eta]$ at x, making $g(x)$ an upper bound for g on $[x, \eta]$ and thus contradicting the fact that $x \in E$.

Proposition 5.4.2 *Let f be an increasing function on $[a, b]$ and let*

$$E_R = \{x \in (a, b) : f \text{ is continuous at } x \text{ and } D^+f(x) > R\},$$

where R is a fixed positive number. If E_R is nonempty, it can be covered by an at most countable set of intervals $\{(a_k, b_k)\}_{k \geq 1}$ whose total length $\Sigma(b_k - a_k)$ satisfies the inequality

$$\sum (b_k - a_k) \leq R^{-1}[f(b-) - f(a+)].$$

Proof If we alter the function at the single point b by resetting its value there to be $f(b-)$, the altered function will be increasing on $[a, b]$, have the same values of $f(a+)$, $f(b-)$ and D^+f, have the same set E_R, but will satisfy the extra condition that $f(b-) = f(b)$. It is therefore sufficient to prove the required inequality with $f(b-)$ replaced by $f(b)$.

Set $g(z) = f(z) - Rz$ for $z \in [a, b]$. Since f is increasing, not only does g have a one-sided limit at each point, but also, at any interior point, $f(z+) \geq f(z) \geq f(z-)$ and hence $f(z+) - Rz \geq f(z) - Rz \geq f(z-) - Rz$, so that

$$g(z+) \geq g(z) \geq g(z-).$$

Consequently, the function G defined on $[a, b]$ as

$$\begin{aligned} G(z) &= \max\{g(z+), g(z-), g(z)\} \quad \text{for every } z \in (a, b), \\ G(a) &= \max\{g(a+), g(a)\},\ G(b) = \max\{g(b-), g(b)\} \end{aligned}$$

actually satisfies

$$G(z) = g(z+) \text{ for every } z \in [a, b); \quad G(b) = g(b). \tag{5.20}$$

Now, consider any arbitrary $x \in E_R$. There exists $y > x$ satisfying $f(y) - f(x) > R(y - x)$, i.e.,

$$g(y) > g(x).$$

From the continuity of f at x, which implies that of g, we obtain $g(y) > g(x+)$, and hence by (5.20), also $g(y) > G(x)$. Since we have shown such a y to exist for an arbitrary $x \in E_R$, on applying Proposition 5.4.1 to g, we find that E_R is covered by an at most countable number of disjoint open intervals (a_k, b_k) such that

$$G(b_k) \geq g(a_k+).$$

By using (5.20), when $b_k < b$, this inequality can be rewritten as $g(b_k+) \geq g(a_k+)$ and when $b_k = b$ as $g(b) \geq g(a_k+)$. This can in turn be written as $f(b_k+) - Rb_k \geq f(a_k+) - Ra_k$, or equivalently, $R(b_k - a_k) \leq f(b_k+) - f(a_k+)$ when $b_k < b$ and as $R(b - a_k) \leq f(b) - f(a_k+)$ when $b_k = b$. It follows that

$$\sum (b_k - a_k) \leq R^{-1} \sum [f(b_k+) - f(a_k+)],$$

subject to the proviso that b_k+ is to be replaced by b if $b_k = b$. The fact that f is increasing and the intervals (a_k, b_k) are disjoint now leads to the inequality we seek. □

Corollary 5.4.3 *Let f be an increasing function on $[a, b]$. Then the set $\{x\colon D^+f(x) = \infty\}$ has measure zero.*

Proof Let H be the set of all points in $[a, b]$ at which f is discontinuous. We know that H is at most countable and hence has measure zero. Now, for any $R > 0$, we have

$$\{x : D^+f(x) = \infty\} \subseteq E_R \cup H$$

and so, the measure of the set $\{x : D^+f(x) = \infty\}$ does not exceed $m(E_R) + m(H) = m(E_R)$. If $E_R = \emptyset$, the proof is finished. If $E_R \neq \emptyset$, we know by the preceding Proposition 5.4.2 that

$$m(E_R) = \sum (b_k - a_k) \leq R^{-1}[f(b-) - f(a+)].$$

Since R can be chosen as large as we please, it follows that $m(\{x: D^+f(x) = \infty\}) = 0$. □

Proposition 5.4.4 *Let f be an increasing function on* $[a, b]$ *and let*

$$E_r = \{x \in (a,\ b) : f \text{ is continuous at } x \text{ and } D\_f(x) < r\},$$

where r is a fixed positive number. If E_r is nonempty, then E_r can be covered by an at most countable set of intervals $\{(a_k, b_k)\}_k$ whose total length $\Sigma(b_k - a_k)$ satisfies the inequality

$$\sum [f(b_k-) - f(a_k+)] \leq r \sum (b_k - a_k) \leq r(b - a).$$

Proof If $x \in E_r$, there exists a $y < x$ such that

$$\frac{f(y) - f(x)}{y - x} < r,$$

that is,

$$f(y) - f(x) > r(y - x),$$

since $y < x$. This implies

$$f(y) - ry > f(x) - rx. \tag{5.21}$$

Set $g(x) = f(x) - rx$. Since g is continuous at x, the function G used in Proposition 5.4.1 satisfies $G(x) = g(x)$. From (5.21), it follows that $g(y) > G(x)$. This means E_r is a subset of what has been called F in that proposition. Therefore we conclude that E_r can be covered by an at most countable set of disjoint intervals $\{(a_k, b_k)\}_k$ such that $g(b_k-) \leq G(a_k)$. Hence $f(b_k-) - rb_k \leq G(a_k)$. [To compute $G(a_k)$, note that, as in the proof of Proposition 5.4.2, we have

$$G(z) = g(z+) \text{ for every } z \in [a,\ b); \quad G(b) = g(b),$$

so that $G(a_k) = f(a_k+) - ra_k$.] So, $f(b_k-) - rb_k \leq f(a_k+) - ra_k$ and this implies that $f(b_k-) - f(a_k+) \leq r(b_k - a_k)$. Now the first part of the desired inequality follows upon summation. The second part follows from the fact that the intervals are disjoint and contained in $[a, b]$. □

Lemma 5.4.5 *Suppose that* $f:[a, b]\to\mathbb{R}$ *is an increasing function and* $0 < \alpha < \beta$. *If* $E_{\alpha,\beta} = \{x \in (a, b) : D_-f(x) < \alpha < \beta < D^+f(x)\}$, *then* $m(E_{\alpha,\beta}) = 0$.

Proof Let H be the at most countable set of points of discontinuity in $[a, b]$. It will be sufficient to show that $m(E_{\alpha,\beta}\backslash H) = 0$. If E_α denotes the set $\{x \in (a, b) : f$ is continuous at x and $D_-f(x) < \alpha\}$ and E_β the set $\{x \in (a, b) : f$ is continuous at x and $D^+f(x) > \beta\}$, then $E_{\alpha,\beta}\backslash H = E_\alpha \cap E_\beta$. It is therefore sufficient to show that $m(E_\alpha \cap E_\beta) = 0$.

Assume that $E_\alpha \cap E_\beta$ is nonempty, because otherwise there is nothing to prove. By Proposition 5.4.4, E_α can be covered by an at most countable family $\{(a_k, b_k)\}_k$ of open intervals such that

$$\sum [f(b_k-) - f(a_k+)] \le \alpha(b-a). \tag{5.22}$$

Next, we consider inside each of the intervals (a_k, b_k) the points of E_β, i.e. where $\beta < D^+f(x)$. The set F_k formed by such points satisfies $F_k = (a_k, b_k) \cap E_\beta \supseteq (a_k, b_k) \cap E_\alpha \cap E_\beta$ and their union taken over all k contains $E_\alpha \cap E_\beta$, considering that the family $\{(a_k, b_k)\}_k$ covers E_α. In view of the assumption that $E_\alpha \cap E_\beta$ is nonempty, F_k must be nonempty for some k. By Proposition 5.4.2, the set F_k, if nonempty, can be covered by disjoint open intervals $\{(a_{k\ell}, b_{k\ell})\}_\ell$ such that

$$\beta \sum\nolimits_\ell (b_{k\ell} - a_{k\ell}) \le [f(b_k-) - f(a_k+)]. \tag{5.23}$$

From (5.22) and (5.23), it follows that

$$\beta \sum\nolimits_{k,\ell} (b_{k\ell} - a_{k\ell}) \le \alpha(b-a),$$

where the summation is taken over only those k for which F_k is nonempty. Thus, if $|\mathcal{S}_1|$ denotes the total length of the system $\mathcal{S}_1 = \left\{(a_{k\ell}, b_{k\ell})_{k,\ell} : F_k \neq \varnothing\right\}$, it follows that

$$|\mathcal{S}_1| \le \frac{\alpha}{\beta}(b-a),$$

and that $\mathcal{S}_1$ covers the union of the sets F_k, which (as already noted) contains $E_\alpha \cap E_\beta$, and hence covers $E_\alpha \cap E_\beta$. Applying the same procedure to each of the intervals in $\mathcal{S}_1$, we get an at most countable family $\mathcal{S}_2$ of open intervals that covers $E_\alpha \cap E_\beta$ and satisfies $|\mathcal{S}_2| \le \frac{\alpha}{\beta}|\mathcal{S}_1|$. Continuing indefinitely in this manner, we obtain a sequence $\mathcal{S}_1, \mathcal{S}_2, \ldots$ of systems of intervals, each “imbedded” in the preceding one and such that

$$|S_{n+1}| \leq \frac{\alpha}{\beta}|S_n|, \ n = 1, 2, \ldots$$

and each S_n covers $E_\alpha \cap E_\beta$. It follows that

$$|S_n| \leq \left(\frac{\alpha}{\beta}\right)^n (b - a).$$

Thus the set $E_\alpha \cap E_\beta$ can be covered by an at most countable family of intervals whose total length is as small as we please; so, $m(E_\alpha \cap E_\beta) = 0$, which is all that was needed to be shown. □

We now restate Lebesgue's Theorem 5.3.8, this time without the assumption of continuity, and prove it.

Theorem 5.4.6 (Lebesgue's Theorem) *Let f:$[a, b] \to \mathbb{R}$ be an increasing function. Then f is differentiable almost everywhere.*

Proof Since f is increasing, each of the four Dini derivatives is nonnegative at each point of $[a, b]$. We assert that

$$m(\{x \in (a, b) : D_{-}f(x) < D^{+}f(x)\}) = 0.$$

Observe that

$$\{x \in (a, b) : D^{+}f(x) > D_{-}f(x)\} = \bigcup_{\substack{r, s \in \mathbb{Q} \\ r > s}} E_{r,s},$$

where

$$E_{r,s} = \{x \in (a, b) : D^{+}f(x) > r > s > D_{-}f(x)\}$$

and r and s are rational numbers. Consequently, by Lemma 5.4.5,

$$m(\{x \in (a, b) : D^{+}f(x) > D_{-}f(x)\}) \leq \sum_{\substack{r, s \in \mathbb{Q} \\ r > s}} m(E_{r,s}) = 0.$$

Hence

$$D^{+}f(x) \leq D_{-}f(x) \tag{5.24}$$

for almost all $x \in [a, b]$. The inequality (5.24), when applied to $-f(-x)$ yields

$$D^{-}f(x) \leq D_{+}f(x) \tag{5.25}$$

for almost all $x \in [a, b]$. Using Remark 5.3.2(a) with (5.24) and (5.25), we obtain

$$0 \leq D^{+}f(x) \leq D_{-}f(x) \leq D^{-}f(x) \leq D_{+}f(x) \leq D^{+}f(x)$$

for almost all $x \in [a, b]$. Hence the equality sign must hold a.e. on $[a, b]$. □

The next result is about the saltus function S of an increasing function f on an interval $[a, b]$. Recall that the definition of S (see beginning of Sect. 5.3) is that

$$S(a) = 0 \text{ and } S(x) = \sum_{x_k < x} j(x_k) + f(x) - f(x-) \text{ for } x \in (a, b],$$

where $x_1, x_2, \ldots$ form any countable set containing all the points of discontinuity of f, whether finite or infinite in number, and the summation extends over all indices k for which $x_k < x$. Here, $j(x)$, called the "jump of f at x", is understood in the sense that $j(x) = |f(x+) - f(x-)|$ if $x \in (a, b)$, whereas $j(a) = |f(a+) - f(a)|$ and $j(b) = |f(b) - f(b-)|$ (see Definition 5.2.4). The saltus function, as noted in Sect. 5.3, is increasing and bounded above by $f(b) - f(a)$.

The next proof follows the ideas of Boas [5, pp. 81–82]. In preparation for it, we note the following about saltus functions. Suppose $\alpha > a$ is a point of continuity. Then $S(\alpha) = \sum_{x_k < \alpha} j(x_k)$. Hence if $\beta > \alpha$ is also a point of continuity, $S(\beta) - S(\alpha) = \sum_{\alpha < x_k < \beta} j(x_k)$, which is the sum of the jumps at those points of discontinuity that lie in the open interval (α, β). The equality can be expressed as $S(\beta) - S(\alpha) = J((\alpha, \beta))$, where $J(G)$ means $\sum_{x_k \in G} j(x_k)$ for any $G \subseteq [a, b]$. This notation will continue to be used in the next proof. The obvious fact that J is a measure on $[a, b]$ with every subset taken as measurable will be used.

Theorem 5.4.7 *Let f be an increasing function defined on $[a, b]$ and S be its saltus function. Then $S'(x) = 0$ almost everywhere on $[a, b]$.*

Proof If there are only finitely many points of discontinuity, then S is constant between the points and there is nothing to prove. So, assume there are infinitely many, in which case, they must be countable. Let us enumerate them as $\{x_k\}_{k \geq 1}$.

The function S, being increasing, is measurable because $\{x \in [a, b]: S(x) < \gamma\}$ is always an interval. Hence $D^+S(x)$, being the limit superior from the right of a sequence of quotients of measurable functions [see Remark 5.3.2(b)], is also measurable. It follows that any set

$$E_n = \{x : D^{+}S(x) > \frac{2}{n}\}, \quad n = 1, 2, \ldots,$$

being the inverse image of the open interval $(\frac{2}{n}, \infty]$ under the measurable function D^+S, is measurable. We shall show that $m(E_n) = 0$ for $n = 1, 2, \ldots$. Since the set $\{x: D^+S(x) > 0\} = \cup_{n \geq 1} E_n$, it will follow that $m(\{x: D^+S(x) > 0\})$ is zero. It is sufficient to show that $m(E_n) = 0$ with the assumption that $a, b \notin E_n$.

Suppose for some n, E_n is not of measure zero. Let $d = m(E_n) > 0$ and consider a positive number η such that

$$0<\eta<\frac{d}{2}.$$

Since $\sum_{k=1}^{\infty} j(x_k) \leq S(b) < \infty$, there exists a k_0 such that

$$\sum\nolimits_{k > k_0} j(x_k) < \frac{d}{16n}. \tag{5.26}$$

For each $k \leq k_0$, choose a closed interval I_k of length $2^{-k}\eta$ and containing x_k, and denote the complement $[a, b]\backslash\cup_{k=1}^{k_0} I_k$ by A. This choice of intervals I_k has two consequences.

First, $x_k \in A$ implies $k > k_0$ and hence by (5.26),

$$G \subseteq A \text{ implies } J(G) = \sum_{x_k \in G} j(x_k) < \frac{d}{16n}. \tag{5.27}$$

Second,

$$m(\bigcup_{k=1}^{k_0} I_k) \leq \sum_{k=1}^{k_0} \ell(I_k) = \sum_{k=1}^{k_0} 2^{-k}\eta < \eta$$

and hence

$$m(E_n \backslash \bigcup_{k=1}^{k_0} I_k) \geq m(E_n) - m(\bigcup_{k=1}^{k_0} I_k) > d - \eta > d - \frac{d}{2} = \frac{d}{2}.$$

By Proposition 2.8.18, we can find a closed set $F \subseteq E_n \backslash \cup_{k=1}^{k_0} I_k$ such that

$$m((E_n \backslash \bigcup_{k=1}^{k_0} I_k) \backslash F) < \frac{d}{4};$$

so,

$$\begin{aligned} m(F) &= m(E_n \backslash \bigcup_{k=1}^{k_0} I_k) - m((E_n \backslash \bigcup_{k=1}^{k_0} I_k) \backslash F) \\ &> \frac{d}{2} - \frac{d}{4} = \frac{d}{4}. \end{aligned} \tag{5.28}$$

Since the bounded set F is closed and $\cup_{k=1}^{k_0} I_k$, being a finite union of closed intervals, is also closed, it follows that F is at a positive distance from $\cup_{k=1}^{k_0} I_k$. Keeping in mind that $a, b \notin E_n \supseteq F$, to each x belonging to F, we assign a positive δ_x such that

- $(x - \delta_x, x + \delta_x) \subseteq (a, b)$,
- neither $x - \delta_x$ nor $x + \delta_x$ is a point of discontinuity,
- $S(x + \delta_x) - S(x) > \frac{\delta_x}{n}$ and
- $(x - \delta_x, x + \delta_x) \subseteq A = [a, b]\backslash \cup_{k=1}^{k_0} I_k$.

As neither $x - \delta_x$ nor $x + \delta_x$ is a point of discontinuity, as noted before the statement of the theorem, $J((x - \delta_x, x + \delta_x)) = S(x + \delta_x) - S(x - \delta_x)$. Moreover, since f is increasing, we have

$$\frac{J((x-\delta_x, x+\delta_x))}{\delta_x} = \frac{S(x+\delta_x) - S(x-\delta_x)}{\delta_x} > \frac{S(x+\delta_x) - S(x)}{\delta_x} > \frac{1}{n}. \quad (5.29)$$

By the Heine–Borel Theorem, a finite collection $\{G_p\}$ of the intervals $(x - \delta_x, x + \delta_x)$ covers F. By Problem 5.4.P3, we can prune the collection $\{G_p\}$ of open intervals in such a manner that no three have a nonempty intersection but the union $\cup_p G_p$ remains the same. Then by Problem 5.4.P4(b),

$$\sum\nolimits_p J(G_p) \leq 2J\left(\bigcup\nolimits_p G_p\right), \quad (5.30)$$

considering that J is a measure on the σ-algebra of all subsets of $[a, b]$. Since $F \subseteq \cup_p G_p$, we have

$$m(F) \leq \sum \ell(G_p).$$

By (5.29), $J(G_p) \geq \frac{\ell(G_p)}{2n}$ and so, from the above inequality, we get

$$m(F) \leq 2n \sum\nolimits_p J(G_p). \quad (5.31)$$

But $\cup_p G_p \subseteq A$ and therefore by (5.27),

$$J\left(\bigcup\nolimits_p G_p\right) \leq \frac{d}{16n}.$$

Combining this with (5.30), we find that $\Sigma_p J(G_p) \leq 2(\frac{d}{16n})$ and hence by (5.31), $m(F) \leq 4n(\frac{d}{16n}) = \frac{d}{4}$. But $m(F) > \frac{d}{4}$ by (5.28) and we have obtained a contradiction, thereby establishing that $m(E_n) = 0$.

We have thus proved that $D^+S(x) = 0$ almost everywhere. Since $D^+S(x) \geq D_+S(x) \geq 0$, this shows that S has zero right-hand derivative almost everywhere. By considering the function $-f(-x)$, we see that S has zero left-hand derivative everywhere. This completes the proof. □

Remark Every increasing function f can be decomposed into the sum of its continuous part g and saltus part S (Theorem 5.3.1). By Theorem 5.3.8, the function g has finite derivative a.e. and, by Theorem 5.4.7, the function S has zero derivative almost everywhere. It therefore follows that the increasing function f has a finite derivative almost everywhere. We thus have a second proof of Lebesgue's Theorem 5.4.6. We now present a third proof due to Rubel [18]. But first, a lemma.

Lemma 5.4.8 *Let f be a strictly increasing function defined throughout* $[a, b]$. *Then f has a continuous increasing "left inverse" with domain* $[f(a), f(b)]$, *that is, there exists a continuous increasing function $F:[f(a), f(b)]\to\mathbb{R}$ such that $F(f(x)) = x$ for each $x \in [a, b]$.*

Proof Define $F:[f(a), f(b)]\to\mathbb{R}$ by setting

$$F(y) = \sup\{t : f(t) \leq y\}. \tag{5.32}$$

It is immediate that the range of F is contained in $[a, b]$.

First we show that $F(f(x)) = x$ for each $x \in [a, b]$. To this end, consider any $x \in [a, b]$. Since f is strictly increasing, we have $f(t) \leq f(x) \Leftrightarrow t \leq x$, so that $\{t: f(t) \leq f(x)\} = \{t: t \leq x\}$; hence $\sup\{t: f(t) \leq f(x)\} = x$. By (5.32), this is the same as saying that $F(f(x)) = x$.

Next, we show that F is increasing. If $y_1 \leq y_2$, then $f(t) \leq y_1 \Rightarrow f(t) \leq y_2$, and therefore $\{t: f(t) \leq y_1\} \subseteq \{t: f(t) \leq y_2\}$, so that $\sup\{t: f(t) \leq y_1\} \leq \sup\{t: f(t) \leq y_2\}$, i.e. $F(y_1) \leq F(y_2)$. Thus F is an increasing function.

It remains to prove continuity. Since we have shown that the function is increasing, its continuity will follow by Remark 5.2.3 if we can show that its range is the entire interval $[a, b]$. But this is a trivial consequence of the fact that $F(f(x)) = x$ for each $x \in [a, b]$. □

Example 5.4.9 Suppose f has domain $[0, 2]$ and $f(x) = x$ for $x \in [0, 1)$, $f(x) = x + 2$ for $x \in (1, 2]$ and $f(1) = \alpha$, where $1 \leq \alpha \leq 3$, then f is strictly increasing and $f(0) = 0$, $f(2) = 4$. A continuous left inverse F on $[0, 4]$ is given by $F(y) = y$ for $y \in [0,1)$, $F(y) = 1$ for $y \in [1, 3]$ and $F(y) = y - 2$ for $y \in (3, 4]$, regardless of the choice of $\alpha \in [1, 3]$.

If we alter the values of F at some points in $[1, 3]$ other than α, the resulting function is not monotone but is still a left inverse for f. We can alter the values in such a manner as to make it discontinuous as well, for example, by altering values at only a finite number of points.

Remark 5.4.10 The left inverse F of Lemma 5.4.8 is constant in a neighbourhood of every point that lies in the interior of the complement of the range of f. To see why, let y be such a point. Then y belongs to an open interval lying within the complement of the range and therefore the set $\{t: f(t) \leq z\}$ is the same for all z in the interval. It follows from the definition of the left inverse F in the statement of the lemma that the left inverse is constant on the interval.

Now we are ready to present Rubel's proof of the Lebesgue's Theorem 5.4.6 on differentiation of arbitrary monotone functions.

Without loss of generality, we may assume that f is strictly increasing, and also that it satisfies the condition

$$f(y) - f(x) \geq y - x \quad \text{whenever } y \geq x. \tag{5.33}$$

For, otherwise we could consider the function $g(x) = f(x) + x$. This function satisfies the assumed conditions, namely, g is strictly increasing and

$$g(y) - g(x) = f(y) - f(x) + y - x \geq y - x \quad \text{whenever } y \geq x.$$

Also, $g'(x) < \infty$ a.e. will imply $f'(x) < \infty$ a.e.

If F is the continuous left inverse of f as in Lemma 5.4.8, then $F'(y) < \infty$ a.e. by Theorem 5.3.8. Thus the set E_F where F' fails to exist has measure zero. We write

$$\frac{f(y) - f(x)}{y - x} = \frac{f(y) - f(x)}{F(f(y)) - F(f(x))} = \left(\frac{F(f(y)) - F(f(x))}{f(y) - f(x)}\right)^{-1}.$$

Thus, for every point x such that f is continuous at x and $f(x)$ does not belong to the exceptional set E_F, we see as in the elementary proof concerning the derivative of a two-sided inverse that

$$\lim_{y \to x} \frac{f(y) - f(x)}{y - x} = [F'(f(x))]^{-1},$$

i.e. $f'(x) \leq \infty$ exists. But the set of points of discontinuity of f is at most countable. Furthermore, $f^{-1}(E_F)$ has measure zero, because for any interval I, the set $f^{-1}(I)$ is an interval of length not exceeding that of I, using (5.33). Finally, in view of Corollary 5.4.3, the set of points x where $f'(x) = \infty$ has measure zero. Hence $f'(x) < \infty$ exists almost everywhere.

Lebesgue's Theorem may be proved in yet another way, using what is known as the Vitali Covering Theorem (see Problems 5.4.P1 and 5.4.P2). It does not use the continuous case, which two of the three previous proofs do.

Problem Set 5.4

Definition 5.4.P1 Let E be a subset of $\mathbb{R}$ of finite outer measure. A family $\mathcal{I}$ of closed intervals, each of positive length is called a **Vitali cover of E** if for each $x \in E$ and every $\varepsilon > 0$, there exists an $I \in \mathcal{I}$ such that $x \in I$ and $\ell(I) < \varepsilon$, i.e., each point of E belongs to an arbitrarily short interval of $\mathcal{I}$.

Example Let $\{r_n\}_{n \geq 1}$ be an enumeration of the rationals in $[a, b]$. Then the collection $\{I_{n,i}\}$, where $I_{n,i} = [r_n - \frac{1}{i}, r_n + \frac{1}{i}]$, $n, i \in \mathbb{N}$, forms a Vitali cover of $[a, b]$.

(Vitali Covering Theorem) Let E be a subset of $\mathbb{R}$ of finite outer measure and $\mathcal{I}$ be a Vitali cover of E. Then given any $\varepsilon > 0$, there is a finite disjoint collection $\{I_1, I_2, \ldots, I_N\}$ of intervals in $\mathcal{I}$ such that

$$m^*(E \backslash \bigcup_{i=1}^{N} I_i) < \varepsilon.$$

5.4.P2. **(Lebesgue's Theorem)** Let $[a, b]$ be a closed interval in $\mathbb{R}$ and let f be a real-valued monotone function on $[a, b]$. Using the Vitali Covering Theorem, prove that f is differentiable almost everywhere.

5.4.P3. If three open intervals have a nonempty intersection, then at least one among them is contained in the union of the other two, or equivalently, the union of some two among them is the same as the union of all three.

5.4.P4. **(a)** Let μ be a finite measure on a set X and $A_1,\ldots,A_n$ be finitely many distinct measurable subsets of X, where $n \geq 3$. If no three sets among the A_i have a nonempty intersection and each $\mu(A_i)$ is finite, show that

$$\mu(\bigcup_{i=1}^{n} A_i) = \sum_{i=1}^{n} \mu(A_i) - \sum_{i<j\leq n} \mu(A_i \cap A_j).$$

(b) Let μ be a measure on a set X and $A_1,\ldots,A_n$ be finitely many distinct measurable subsets of X, where $n \geq 3$. If no three sets among the A_i have a nonempty intersection, show that

$$\sum_{i=1}^{n} \mu(A_i) \leq 2\mu(\bigcup_{i=1}^{n} A_i).$$

5.5 Integral of the Derivative

Let f be Riemann integrable on $[a, b]$ and suppose that there exists a function F on the same interval such that $F'(x) = f(x)$ everywhere. Then

$$\int_a^b f(x)dx = F(b) - F(a). \tag{5.34}$$

This is the Second Fundamental Theorem of Calculus [see Theorem 1.7.2(b)]. It is the purpose of this section to investigate the relationship (5.34) in the case when f is monotone on $[a, b]$. It is proposed to replicate (5.34) for Lebesgue measurable functions under appropriate conditions. A weaker form of (5.34) is the content of the next theorem.

Theorem 5.5.1 *Let f be an increasing function defined on* $[a, b]$ *and E be the set of points in* $[a, b]$ *at which* f' *exists. Then the derivative* $f':E\to\mathbb{R}$ *is integrable in the sense of Lebesgue and*

$$\int_E f' dm = \int_{[a,\, b]} f' dm \le f(b) - f(a). \tag{5.35}$$

Proof Since the function f is increasing, it is continuous a.e. and is, therefore, measurable because $\{x \in [a, b]: S(x) < \gamma\}$ is always an interval. Let us extend it to the interval $[a, b + 1]$ by putting

$$f(x) = f(b) \text{ for } b < x \le b + 1.$$

We then define φ_n: $[a, b]\to\mathbb{R}$ by

$$\varphi_n(x) = n[f(x + \frac{1}{n}) - f(x)], \quad a \le x \le b, \quad n = 1, 2, \ldots.$$

Since f is increasing, the functions φ_n are nonnegative everywhere. It is clear that $\varphi_n(x) \to f'(x)$ if $x \in E$. Also, the functions φ_n are measurable and therefore f' is measurable by Proposition 2.6.1. Since $[a, b]\backslash E$ is of measure zero by Lebesgue's Theorem 5.4.6, we have

$$\begin{aligned}
\int_E \varphi_n(x)dx &= \int_a^b \varphi_n(x)dx = n[\int_a^b f(x + \frac{1}{n})dx - \int_a^b f(x)dx] \\
&= n[\int_{a+\frac{1}{n}}^{b+\frac{1}{n}} f(x)dx - \int_a^b f(x)dx] \\
&= n[\int_b^{b+\frac{1}{n}} f(x)dx - \int_a^{a+\frac{1}{n}} f(x)dx] \\
&= f(b) - n\int_a^{a+\frac{1}{n}} f(x)dx.
\end{aligned} \tag{5.36}$$

By monotonicity of f, we have $f(x) \ge f(a)$ and hence

$$n\int_a^{a+\frac{1}{n}} f(x)dx \ge f(a). \tag{5.37}$$

(5.36) and (5.37) together imply

$$\int_E \varphi_n(x)dx \le f(b) - f(a). \tag{5.38}$$

By Fatou's Lemma (Theorem 3.2.8) and (5.38), it follows that

$$\int_E f' dm \leq f(b) - f(a). \tag{5.39}$$

Once again using the fact that $m([a, b]\backslash E) = 0$, we obtain from (5.39)

$$\int_E f' dm = \int_{[a, b]} f' dm \leq f(b) - f(a).$$

Remark The inequality (5.35) obtained in the above theorem seems somewhat unsatisfactory when compared with (5.34). However, equality need not always hold in (5.34), e.g. the function given by

$$f(x) = \begin{cases} 0 & 0 \leq x < 1 \\ 1 & x = 1 \end{cases}$$

satisfies $f'(x) = 0$ a.e. but $f(1) - f(0) = 1 - 0 \neq \int_{[0, 1]} f'$.

We next construct a *continuous* monotone function called the *Cantor function* or *Lebesgue's singular function* which has the same feature.

Example 5.5.2 Recall that we defined a function g from the Cantor set C to $[0, 1]$ in Proposition 1.8.12. Recall also the terminology of *removed* and *component* intervals of C_n and that they are denoted by $I_{n,k}$ and $J_{n,k}$ respectively. It was proved that g is onto. We shall now show that the function assumes the same values at the endpoints of any interval that has been removed.

Indeed, for the pair $\frac{1}{3} = 0.0222\ \ldots$ and $\frac{1}{3} = 0.2000\ \ldots$, the endpoints of the interval $I_{1,1}$, we get $g\left(\frac{1}{3}\right) = 0.0111\ \ldots$ and $g\left(\frac{2}{3}\right) = 0.1000\ \ldots$, which represent the same number in base 2; for $(a, b) = I_{n,k}$ for some n and some k, $1 \leq k \leq 2^{n-1}$, where $a = \sum_{j=1}^{n-1} \frac{a_j}{3^j} + \sum_{j=n+1}^{\infty} \frac{2}{3^j}$ and $b = \sum_{j=1}^{n-1} \frac{a_j}{3^j} + \frac{2}{3^n}$ (see Remark 1.8.8, second paragraph), we find that $g(a) = \sum_{j=1}^{n-1} \frac{b_j}{2^j} + \sum_{j=n+1}^{\infty} \frac{1}{2^j}$ and $g(b) = \sum_{j=1}^{n-1} \frac{b_j}{2^j} + \frac{1}{2^n}$ and they represent the same number in base 2. Here each b_j is even and $b_j = \frac{1}{2} a_j$.

We extend the definition of g to the whole of $[0, 1]$ as follows. On any interval that has been removed, we define g by giving it the same value that g has at the endpoints of that interval. The function g has now been extended to all of $[0, 1]$ and the range of g is also $[0, 1]$.

We next show that g is monotone and continuous.

Let x and y be two points of C, where $x < y$, and suppose their (unique) ternary expansions with even digits (see Theorem 1.8.9) are

$$x = .\alpha_1 \alpha_2 \cdots \alpha_n \alpha_{n+1} \cdots,$$
$$y = .\beta_1 \beta_2 \cdots \beta_n \beta_{n+1} \cdots.$$

There exists a smallest integer i such that $\alpha_i \neq \beta_i$; call it n. Then $\beta_n > \alpha_n$ (see the last paragraph of Remark 1.8.1). In the binary expansions of $g(x)$ and $g(y)$, the nth digit of $g(y)$ will be 1 larger than the nth digit of $g(x)$, while the preceding digits (if any) will be the same. This implies $g(y) \geq g(x)$ (again by Remark 1.8.1). Thus the restriction of g to C is an increasing function. Also, on each of the removed intervals and their endpoints, the function is constant. Hence g is an increasing function on [0, 1]. We have already noted (two paragraphs ago) that the range of g is an interval, namely, [0, 1]. Since the function is increasing, it follows immediately from Remark 5.2.3 that it is continuous.

It was noted in Remark 1.8.13 that the unextended function g has the property that there exist arbitrarily small numbers at which it takes positive values. Since its extension is increasing, we know that it too has the property that $x > 0$ implies $g(x) > 0$.

Moreover, the equality $g'(x) = 0$ can be shown to hold for almost all $x \in [0, 1]$. In fact, $g'(x) = 0$ holds on $[0, 1]\backslash C$ and C is a set of measure zero [see Remark 2.2.11(c)]. Therefore

$$\int_{[0,\,1]} g' = 0 < 1 = g(1) - g(0).$$

The continuous monotone function g is called the **Cantor function** or **Lebesgue's singular function**.

It will be seen later in Theorem 5.8.2 that equality holds in Theorem 5.5.1 under suitable conditions.

We give below an alternative description of the Cantor function (notations as in Definition 1.8.7).

For any $x \in [0, 1]$, put

$$g_n = \left(\frac{2}{3}\right)^{-n} \chi_{C_n} \quad \text{and} \quad f_n(x) = \int_0^x g_n(t)dt.$$

The reader may check that $f_1(x) = \left(\frac{2}{3}\right)^{-1} x$ if $0 \leq x \leq \frac{1}{3}, \frac{1}{2}$ if $\frac{1}{5} < x < \frac{2}{5}$ and $\frac{1}{2} + \left(\frac{2}{3}\right)^{-1}\left(x - \frac{2}{3}\right)$ if $\frac{2}{3} \leq x \leq 1$ It will be found instructive to compute f_2 and f_3.

The functions f_n, $n = 1, 2, \ldots$ are such that $f_n(0) = 0$ and since each $J_{n,k}$ is of length $\frac{1}{3^n}$, we have

$$f_n(1) = \sum_{k=1}^{2^n} \int_0^1 \left(\frac{2}{3}\right)^{-n} \chi_{J_{n,k}}(t)dt = \left(\frac{2}{3}\right)^{-n} 2^n \left(\frac{1}{3^n}\right) = 1.$$

Since $g_n \geq 0$, it follows that f_n is increasing. Moreover, f_n is continuous [g_n is a step function and therefore its Lebesgue integral is the same as its Riemann integral]. If J is one of the component intervals of C_n, then

$$\int_J g_n(t)dt = \int_J \left(\frac{2}{3}\right)^{-n} dt = 2^{-n} \tag{5.40}$$

and, by the same computation with $n + 1$ in place of n,

$$\int_J g_{n+1}(t)dt = \int_{J_1} g_{n+1}(t)dt + \int_{J_2} g_{n+1}(t)dt = 2^{-n-1} + 2^{-n-1} = 2^{-n}, \tag{5.41}$$

where J_1 and J_2 are the parts of the closed interval J after removing its middle third.

Suppose $x \notin C_n$. Then x is in none of the component intervals of C_n and therefore in none of the component intervals of any C_m with $m > n$. So, the defining integrals for f_n and f_{n+1} are to be taken over only the intervals lying to the left of x. Hence, it follows from (5.40) and (5.41) that

$$f_{n+1}(x) = f_n(x) \quad \text{for } x \notin C_n.$$

Since $C_{n+1} \subseteq C_n$, it further follows that

$$f_m(x) = f_n(x) \quad \text{for} \quad m > n \text{ and } x \notin C_n. \tag{5.42}$$

Now suppose $x \in C_n$ and K be the component interval of C_n that x belongs to. Then $f_n(x) - f_{n+1}(x)$ is an integral over all the component intervals lying to the left of K and over the part of K that lies to the left of x. It follows from (5.40) and (5.41) that the integral over the component intervals lying to the left of K is 0. Consequently,

$$\begin{aligned} |f_n(x) - f_{n+1}(x)| &\leq \int_K |g_n(t) - g_{n+1}(t)|dt \\ &\leq \int_K |g_n(t)|dt + \int_K |g_{n+1}(t)|dt. \end{aligned}$$

Hence it follows from (5.40) and (5.41) that

$$|f_n(x) - f_{n+1}(x)| \leq 2^{-n} + 2^{-n} = 2^{-n+1} \text{ for } x \in C_n. \tag{5.43}$$

From (5.42) and (5.43), we see that $\{f_n\}$ converges uniformly on [0,1]. The uniform limit must then be a continuous increasing function f with $f(0) = 0, f(1) = 1$ and $f'(x) = 0$ for all $x \in C^c$. Indeed, the functions f_n being constant on each of the intervals removed up to and including the nth stage, the limit function f must be constant on each of the removed intervals and hence have derivative 0 at each point of it. In other words, $f'(x) = 0$ on C^c. Therefore

$$\int_0^1 f'(x)dx = 0 < 1 = f(1) - f(0).$$

It remains to show that the functions f and g are the same.

Consider any removed interval $I_{n,k}$ of some C_n and let $x \in I_{n,k}$. Then $x \notin C_n$, and as before, the defining integral for f_n is to be taken only over the component intervals lying to the left of x, which means component intervals lying to the left of the interval $I_{n,k}$. Hence, it follows from (5.40) that $f_n(x)$ equals 2^{-n} times the number of component intervals to the left of $I_{n,k}$. In fact, on the entire interval $I_{n,k}$, the function f_n takes the constant value equal to 2^{-n} times the number of component intervals to the left of $I_{n,k}$. On the basis of (5.42), we infer that the limit function f also takes that same value on the entire interval. Hence it follows from the continuity of the function (noted above), that f takes this value at the right endpoint of the removed interval under consideration. It is now a consequence of Remark 1.8.14 (f) that f and g agree there. But f and g are both constant on any removed interval and therefore they agree everywhere on every removed interval of C_n. Thus they agree on the complement in [0, 1] of the Cantor set C. But the complement is dense because C cannot contain an interval of positive length, as it has measure 0 [see Example 2.2.11(c)]. Since both f and g are continuous, it follows by Proposition 1.3.25 that they agree everywhere on [0, 1].

Lebesgue's Theorem 5.4.6 asserts that every monotone function is differentiable almost everywhere. Perhaps of no less importance is the following theorem of Fubini on the termwise differentiability of series with monotone terms, which Theorem 5.5.1 enables us to prove.

Fubini's Theorem. 5.5.3 *Let $f_n:[a, b]\to\mathbb{R}$ be increasing for $n = 1, 2, \ldots$ and assume that the series*

$$f(x) = \sum_{k=1}^{\infty} f_k(x) \tag{5.44}$$

converges pointwise on $[a, b]$. Then

$$f'(x) = \sum_{k=1}^{\infty} f'_k(x)$$

for almost all $x \in [a, b]$.

Proof Denote the nth partial sum of the series (5.44) by $s_n(x)$. We may then write

$$f(x) = s_n(x) + r_n(x),$$

where

$$r_n(x) = \sum_{k=n+1}^{\infty} f_k(x).$$

Let E denote the set of points in $[a, b]$ where each of the functions f_k, $k = 1, 2, \ldots$ and the function f are differentiable. By Lebesgue's Theorem 5.4.6, these functions are differentiable almost everywhere. So the set $[a, b]\backslash E$ has measure zero. If $x \in E$, then s_n is differentiable at x and therefore so is $r_n = f - s_n$. Also, $r_n' = f' - s_n'$. Since r_n is increasing, its derivative r_n' is nonnegative, and it follows that

$$f'(x) = s_n'(x) + r_n'(x) \geq s_n'(x) = \sum_{k=1}^{n} f_k'(x)$$

for all $x \in E$. On letting $n \to \infty$, we obtain

$$f'(x) \geq \sum_{k=1}^{\infty} f_k'(x). \tag{5.45}$$

An application of Theorem 5.5.1 to the increasing function r_n yields

$$0 \leq \int_a^b r_n'(x)dx \leq r_n(b) - r_n(a) = \sum_{k=n+1}^{\infty} [f_k(b) - f_k(a)].$$

By hypothesis, the series $\sum_{r=1}^{\infty} f_k(b)$ and $\sum_{k=1}^{\infty} f_k(a)$ converge and hence so does the series $\sum_{k=1}^{\infty} [f_k(b) - f_k(a)]$ Consequently,

$$\lim_{n\to\infty} \sum_{k=n+1}^{\infty} [f_k(b) - f_k(a)] = 0.$$

This implies

$$\lim_{n\to\infty} \int_a^b r_n'(x)dx = 0. \tag{5.46}$$

Now,

$$\begin{aligned} \int_a^b f'(x)dx &= \int_a^b \sum_{k=1}^{n} f_k'(x)dx + \int_a^b r_n'(x)dx \\ &\leq \int_a^b \sum_{k=1}^{\infty} f_k'(x)dx + \int_a^b r_n'(x)dx. \end{aligned}$$

Letting $n \to \infty$ and using (5.46), we get

$$\int_a^b f'(x)dx = \int_a^b \sum_{k=1}^{\infty} f_k'(x)dx. \tag{5.47}$$

From (5.45), we obtain

$$\int_a^b f'(x)dx \geq \int_a^b \sum_{k=1}^{\infty} f_k'(x)dx,$$

which, together with (5.47), implies

$$\int_a^b f'(x)dx = \int_a^b \sum_{k=1}^{\infty} f_k'(x)dx,$$

i.e.,

$$\int_a^b [f'(x) - \sum_{k=1}^{\infty} f_k'(x)]dx = 0.$$

As the integrand here is nonnegative by (5.45), it follows that $f'(x) = \sum_{k=1}^{\infty} f_k'(x)$ a.e. on $[a, b]$. □

As an application of the theorem we give an alternative proof of Theorem 5.4.7, according to which the saltus function S of an increasing function f on $[a, b]$ has derivative zero a.e. on $[a, b]$.

Let us enumerate all the points of discontinuity of f in a sequence $\{x_k\}_{k \geq 1}$. (As before, we may assume there are infinitely many.) To each $k \geq 1$ such that $a < x_k < b$, we assign a function f_k defined on $[a, b]$ as follows:

$$f_k(x) = \begin{cases} 0 & \text{for } x < x_k \\ f(x_k) - f(x_k-) & \text{for } x = x_k \\ f(x_k+) - f(x_k-) & \text{for } x > x_k. \end{cases} \tag{5.48–5.50}$$

If $x_p = a$, then we set

$$f_p(x) = 0 \text{ for } x = a, f_p(x) = f(a+) - f(a) \text{ for } x > a \tag{5.51}$$

and if $x_q = b$, then

$$f_q(x) = 0 \text{ for } x < b, f_q(x) = f(b) - f(b-) \text{ for } x = b. \tag{5.52}$$

It is possible that there is no such p and/or no such q, in which case, references to p and/or q in what follows are superfluous and are to be taken as expunged.

It is clear that each of the functions f_k ($k = 1, 2, \ldots$) is increasing and that $f_k'(x) = 0$ for all $x \neq x_k$. This ensures that

$$\sum_{k=1}^{\infty} f_k'(x) = 0 \text{ a.e.} \tag{5.53}$$

Furthermore,

$$S(x) = \sum_{k=1}^{\infty} f_k(x) \quad \text{for all } x \in [a, b], \tag{5.54}$$

as we shall soon show, starting from the next paragraph. It is worth keeping in mind that S was defined by using the notation $j(x)$, which stands for $f(x+) - f(x-)$ when $a < x < b$, $f(a+) - f(a)$ when $x = a$ and $f(b) - f(b-)$ when $x = b$. Thus $f_k(x) = j(x_k)$ when $p \neq k \neq q$ and $x > x_k$.

It is plain from the definition that $f_k(a) = 0$ for all k. Therefore $\sum_{k=1}^{\infty} f_k(a) = 0 = S(a)$ by definition of the latter.

Now let $x \in (a, b)$ and suppose also that $x \neq x_k$ for all k. Then $f(x) - f(x-) = 0$. When $p \neq k \neq q$, by (5.48) and (5.50), we have

$$f_k(x) = 0 \text{ for } x < x_k \text{ and } f_k(x) = j(x_k) \text{ for } x_k < x.$$

When $k = p$, by (5.51), we have

$$x_k = x_p = a < x \text{ and } f_k(x) = f_p(x) = f(a+) - f(a) = j(a) = j(x_p).$$

When $k = q$, by (5.52), we have

$$x_k = x_q = b > x \text{ and } f_k(x) = f_q(x) = 0.$$

It follows that

$$\begin{aligned}
\sum_{k=1}^{\infty} f_k(x) &= f_p(x) + f_q(x) + \sum_{p\neq k\neq q} f_k(x) = j(x_p) + 0 + \sum_{p\neq k\neq q} f_k(x) \\
&= j(x_p) + \sum\nolimits_{x_k < x, k\neq p} j(x_k) = \sum\nolimits_{x_k < x} j(x_k) \\
&= \sum\nolimits_{x_k < x} j(x_k) + f(x) - f(x-) \\
&= S(x) \quad \text{by definition of } S.
\end{aligned}$$

Continuing with $x \in (a, b)$, we drop the supposition that $x \neq x_k$ for all k. Let $x = x_r$. We know that $p \neq r \neq q$ because $a \neq x \neq b$. When $p \neq k \neq q$, by (5.48) and (5.50), we have

$$f_k(x) = 0 \text{ for } x < x_k \text{ and } f_k(x) = j(x_k) \text{ for } x_k < x, \text{ provided } k \neq r$$

and, by (5.49),

$$f_r(x) = f_r(x_r) = f(x_r) - f(x_r-) = f(x) - f(x-).$$

When $k = p$, we have (exactly as before)

$$x_k = x_p = a < x \text{ and } f_k(x) = f_p(x) = f(a+) - f(a) = j(a) = j(x_p).$$

When $k = q$, we have (exactly as before)

$$x_k = x_q = b > x \text{ and } f_k(x) = f_q(x) = 0.$$

It follows that

$$\begin{aligned}\sum_{k=1}^{\infty} f_k(x) &= f_p(x) + f_q(x) + f_r(x) + \sum_{p \neq k \neq q, k \neq r} f_k(x) \\ &= j(x_p) + 0 + f(x) - f(x-) + \sum_{x_k < x, k \neq p} j(x_k) \\ &= \sum\nolimits_{x_k < x} j(x_k) + f(x) - f(x-) \\ &= S(x) \text{ by definition of } S.\end{aligned}$$

So far, we have proved that $S(x) = \sum_{k=1}^{\infty} f_k(x)$ if $x \in [a, b)$.

Consider $x = b$. There are two cases to consider, $b > x_k$ for all k and $b = x_q$. In the first case, $f(b) - f(b-) = 0$ and, by (and (5.51), $f_k(5.50b) = j(x_k)$ for all k (including p). It follows that

$$\begin{aligned}\sum_{k=1}^{\infty} f_k(b) = \sum_{x_k < b} f_k(b) &= \sum_{x_k < b} j(x_k) \\ &= \sum_{x_k < b} j(x_k) + f(b) - f(b-) \\ &= S(b) \text{ by definition of } S.\end{aligned}$$

In the second case, $b > x_k$ for all $k \neq q$, and hence by (5.50), (5.51) and (5.52), $f_q(b) = f(b) - f(b\,-)$ and $f_k(b) = j(x_k)$ for all other $k \neq q$ (including p). It follows that

$$\begin{aligned}\sum_{k=1}^{\infty} f_k(b) &= f_q(b) + \sum_{k\neq q} f_k(b) = f(b) - f(b-) + \sum_{k\neq q} f_k(b)\\ &= f(b) - f(b-) + \sum_{k\neq q} j(x_k)\\ &= f(b) - f(b-) + \sum_{x_k<b} j(x_k)\\ &= S(b) \text{ by definition of } S.\end{aligned}$$

This completes the proof of (5.54).

By Fubini's Theorem 5.5.3, the equality (5.54) leads to

$$S'(x) = \sum_{k=1}^{\infty} f_k'(x) \text{ almost everywhere.}$$

In view of (5.53), this completes the alternative proof of Theorem 5.4.7 that the saltus function S of an increasing function f on $[a, b]$ has derivative zero a.e. on $[a, b]$.

A slight variation in the above proof would be to avoid proving $S(b) = \sum_{k=1}^{\infty} f_k(b)$ and to use Problem 5.5.P3 instead, after applying Fubini's Theorem 5.5.3 on intervals $[a, \beta]$ with $a < \beta < b$.

The following examples illustrate the use of Fubini's Theorem for producing nonconstant monotone functions, one with a dense set of discontinuities and the other without any discontinuities, and yet each having a zero derivative a.e.

Examples 5.5.4

(a) Let H be the function given by

$$H(x) = \begin{cases} 0 & x \leq 0 \\ 1 & x > 0. \end{cases}$$

That is, H is the characteristic function of the set $(0, \infty)$. Clearly, H is increasing and $H'(x) = 0$ for all $x \neq 0$. Let $\{r_k\}_{k\geq 1}$ be an enumeration of the rational numbers and set

$$f(x) = \sum_{k=1}^{\infty} \frac{1}{2^k} H(x - r_k). \tag{5.55}$$

Since $\frac{1}{2^k}H(x - r_k) \leq \frac{1}{2^k}$ and $\sum \frac{1}{2^k}$ is convergent, the series on the right of (5.55) is uniformly convergent. Observe that the function f is the limit of the sequence of partial sums of (5.55) and the nth partial sum has a nonzero jump at r_k with jump $\frac{1}{2^k}$, $k = 1, 2, \ldots, n$. The function f, being the sum of increasing functions, is itself an increasing function. Since

(a) the convergence of the partial sums is uniform and
(b) each partial sum from the kth one onward have the same jump at r_k, namely $\frac{1}{2^k}$, it follows from Proposition 5.2.11 that f has a positive jump at each r_k, $k = 1, 2, \ldots$. Applying Fubini's Theorem 5.5.3 to an arbitrary interval $[a, b]$ and then letting $a \to -\infty$ and $b \to \infty$, we obtain

$$f'(x) = \sum_{k=1}^{\infty} \frac{1}{2^k} H'(x - r_k) = 0$$

for almost all $x \in \mathbb{R}$. This function is strictly increasing, thus not constant on any interval.

We next give an example of a function defined on $\mathbb{R}$ which is continuous and strictly increasing, and yet has a zero derivative a.e.

(b) Let g be the Cantor function defined on $[0, 1]$ (see Example 5.5.2). The function g is continuous and increasing on $[0, 1]$ with $g(0) = 0$ and $g(1) = 1$ and is constant on each of the removed intervals which form the complement of the Cantor set. Also, $g'(x) = 0$ for all $x \in C$. Let

$$G(x) = \begin{cases} 0 & x<0 \\ g(x) & 0 \leq x \leq 1 \\ 1 & 1<x. \end{cases}$$

Then G has domain $\mathbb{R}$, is continuous and increasing and has $G'(x) = 0$ almost everywhere.

Let $\{r_k\}_{k \geq 1}$ be an enumeration of the rational numbers in $[0, 1]$ and set

$$F(x) = \sum_{k=1}^{\infty} \frac{1}{2^k} G(x - r_k).$$

Observe that F is continuous and increasing (being a limit of increasing functions), and using Fubini's Theorem 5.5.3, it follows that $F'(x) = 0$ a.e.

We shall argue that the function F is strictly increasing on $[0, 1]$. Suppose $x, y \in [0, 1]$ and $x < y$. Then $x - r_k < y - r_k$, so that

$$G(x - r_k) \leq G(y - r_k) \quad \text{for every } k.$$

As the sequence $\{r_k\}_{k \geq 1}$ goes through all the rational numbers in $[0, 1]$, we must have $x < r_j < y$ for some j. Then $x - r_j < 0 < y - r_j < 1$. It follows from the definition of G that $G(x - r_j) = 0$ and $G(y - r_j) = g(y - r_j) \geq 0$. But $g(y - r_j) > 0$ on account of the property of g noted in Example 5.5.2 that $x > 0$ implies $g(x) > 0$, and hence $G(x - r_j) < G(y - r_j)$. On the other hand, $G(x - r_k) \leq G(y - r_k)$ for every k. This implies $F(x) < F(y)$.

Problem Set 5.5

5.5.P1. Using the functions constructed in Examples 5.5.4 (a) and (b), show that strict inequality can hold in Theorem 5.5.1.

5.5.P2. Let f be Lebesgue's singular function. Compute all the four Dini derivatives of f at each point $x \in [0, 1]$. [Cf. Example 5.5.2.]

5.5.P3. Suppose $f:[a, b]\to\mathbb{R}$ has derivative 0 a.e. on $[a, \beta]$ whenever $\beta < b$. Show that f has derivative 0 a.e. on $[a, b]$.

5.6 Total Variation

The sum of two (or any finite number) of increasing functions is obviously an increasing function. However, their difference is, in general, not so. Indeed, the functions $f_1(x) = x$ and $f_2(x) = x^2$ defined on [0, 1] are increasing but the function $f(x) = f_1(x) - f_2(x) = x(1 - x), 0 \leq x \leq 1$, is such that $f(0) = 0 = f(1)$ and assumes its maximum value $\frac{1}{4}$ at $x = \frac{1}{2}$. The difference f of two increasing functions f_1 and f_2 defined on $[a, b]$ satisfies the following property: For every partition $a = x_0 < x_1 < \cdots < x_n = b$ of $[a, b]$, the sum

$$\begin{aligned}
\sum_{k=1}^{n} |f(x_k) - f(x_{k-1})| &\leq \sum_{k=1}^{n} |f_1(x_k) - f_1(x_{k-1})| + \sum_{k=1}^{n} |f_2(x_k) - f_2(x_{k-1})| \\
&= \sum_{k=1}^{n} [f_1(x_k) - f_1(x_{k-1})] + \sum_{k=1}^{n} [f_2(x_k) - f_2(x_{k-1})]
\end{aligned}$$

since $f_i(x_k) - f_i(x_{k-1}) \geq 0$ for $i = 1, 2$ and $k = 1, 2, \ldots, n$

$$= f_1(b) - f_1(a) + f_2(b) - f_2(a)$$

$$= \sum_{i=1}^{2} (f_i(b) - f_i(a)).$$

A function for which the sum

$$\sum_{k=1}^{n} |f(x_k) - f(x_{k-1})|$$

does not exceed some finite bound, independent of the partition, is called a function of bounded variation, or finite variation. The class of such functions plays a fundamental role in several branches of mathematical analysis, including the theory of Fourier series and the theory of integration. We shall study below the regularity (continuity and differentiability properties) of these functions.

Definition 5.6.1 Let P: $a = x_0 < x_1 < \cdots < x_n = b$ be a partition of $[a, b]$ and f be a function defined on $[a, b]$. We introduce the notation

$$T(P,f) = \sum_{k=1}^{n} |f(x_k) - f(x_{k-1})|.$$

The **total variation** of f on $[a, b]$ is defined to be

$$V([a,\ b],f) = \sup_P T(P,f),$$

the supremum being taken over all partitions P of $[a, b]$. *If* $V([a, b],f) < \infty$, we say that f is of **bounded (or finite) variation** on $[a, b]$.

Remarks 5.6.2

(a) A monotone function on a closed and bounded interval is always of bounded variation. For details, see the introductory paragraph of this section. In fact, $T(P, f) = |f(b) - f(a)|$ for every monotone function f and every partition P of $[a, b]$.

(b) Every function of bounded variation on $[a, b]$ is bounded on the interval. In fact, for $a \leq x \leq b$,

$$|f(x) - f(a)| + |f(b) - f(x)| \leq V([a,\ b],f).$$

Therefore

$$|f(x)| \leq |f(a)| + |f(x) - f(a)| \leq V([a,\ b],f) + |f(a)|.$$

(c) The sum, difference and product of functions of bounded variation are again of the same kind. To see why, let f and g be such functions on $[a, b]$ and $s = f + g$. Then for any partition P: $a = x_0 < x_1 < \cdots < x_n = b$, we have

$$\sum_{k=1}^{n} |s(x_k) - s(x_{k-1})| \leq \sum_{k=1}^{n} |f(x_k) - f(x_{k-1})| + \sum_{k=1}^{n} |g(x_k) - g(x_{k-1})|.$$

It follows that

$$V([a,\ b], s) \leq V([a,\ b],f) + V([a,\ b], g).$$

The proof that $f - g$ is of bounded variation is similar and is, therefore, not included.

Now suppose $p = fg$. Let

$$M = \sup\{|f(x)| : a \leq x \leq b\} \text{ and } N = \sup\{|g(x)| : a \leq x \leq b\}.$$

Then for any partition P: $a = x_0 < x_1 < \cdots < x_n = b$, we have

$$\begin{aligned}|p(x_k) - p(x_{k-1})| &\leq |f(x_k)g(x_k) - f(x_{k-1})g(x_k)| \\ &\quad + |f(x_{k-1})g(x_k) - f(x_{k-1})g(x_{k-1})| \\ &\leq N|f(x_k) - f(x_{k-1})| + M|g(x_k) - g(x_{k-1})|.\end{aligned}$$

This implies

$$V([a,\ b], p) \leq N \cdot V([a,\ b], f) + M \cdot V([a,\ b], g).$$

Thus $p = fg$ is of bounded variation whenever f and g are.

(d) If f and g are of bounded variation and there exists an $\alpha > 0$ such that $g(x) > \alpha$ for $a \leq x \leq b$, then $\frac{f}{g}$ is also of bounded variation.

It is enough to show that $\frac{1}{g}$ is of bounded variation. The result will then follow from (c) above. For any partition P: $a = x_0 < x_1 < \cdots < x_n = b$,

$$\sum_{k=1}^{n} \left| \frac{1}{g(x_k)} - \frac{1}{g(x_{k-1})} \right| \leq \frac{1}{\alpha^2} \sum_{k=1}^{n} |g(x_k) - g(x_{k-1})|$$

and this yields

$$V([a,\ b], \frac{1}{g}) \leq \frac{1}{\alpha^2} V([a,\ b], g).$$

(e) Recall that if f:$[a, b]\to\mathbb{R}$ is such that $|f(x) - f(y)| \leq M|x - y|$ for every x and y in $[a, b]$, where M is a fixed positive number, then f is said to satisfy a Lipschitz condition. Such a function is always of bounded variation, because for any partition P: $a = x_0 < x_1 < \cdots < x_n = b$,

$$\sum_{k=1}^{n} |f(x_k) - f(x_{k-1})| \leq M \sum_{k=1}^{n} (x_k - x_{k-1}) = M(b - a)$$

and so, $V([a, b], f) \leq M(b - a)$.

(f) If f:$[a, b]\to\mathbb{R}$ is continuous on $[a, b]$ and has a derivative $f'(x)$ at every point of (a, b) and if f' is bounded, then from the Mean Value Theorem, we have

$$|f(x) - f(y)| \le |f'(z)||x - y|,$$

where z lies between x and y. The function therefore satisfies a Lipschitz condition.

(g) If f_1 and f_2 are two continuous increasing functions, then $f = f_1 - f_2$ is a continuous function of bounded variation. This is an immediate consequence of parts (a) and (c).

We list below some examples of functions that are of bounded variation and some that are not.

Examples 5.6.3

(a) The function $f(x) = \sin x$ is of bounded variation on any compact interval $[a, b]$. This follows from Remark 5.6.2(f) above. In the paragraph just above Example 5.6.12, the notion of bounded variation will be extended to $\mathbb{R}$. We shall see that f is not of bounded variation on $\mathbb{R}$ [see Example 5.6.12 (b)].
(b) The function $f(x) = x - x^2$, $0 \le x \le 1$, being the difference of two increasing functions, is of bounded variation on [0, 1] [see Remark 5.6.2(c).]
(c) The function f:$[0, 1]\to\mathbb{R}$ defined by $f(0) = 0$ and $f(x) = x\sin\frac{\pi}{2x}$, $x \ne 0$, is not of bounded variation. However, it is continuous. Indeed, for the partition

$$P: 0 < \frac{1}{2n+1} < \frac{1}{2n} < \cdots < \frac{1}{3} < \frac{1}{2} < 1,$$

$$T(P,f) = \sum_{k=1}^{2n+1} |f(x_k) - f(x_{k-1})| = \frac{1}{2n+1} + \frac{1}{2n-1} + \cdots + \frac{1}{3} + 1.$$

These sums become infinite as $n \to \infty$. That the function is continuous on its domain is well known.

(d) The function $f(x) = x^p\sin\frac{1}{x}$, $0 < x \le 1$, and $f(0) = 0$ can be shown to be of bounded variation on [0, 1] for $p \ge 2$, as follows: The derivative $f'(x) = px^{p-1}\sin\frac{1}{x} - x^{p-2}\cos\frac{1}{x}$ is such that $|f'(x)| \le p + 1$, $0 < x < 1$, and f is continuous on [0, 1]. The assertion therefore follows by Remark 5.6.2(e) and (f).

Theorem 5.6.4 *Let f be any real-valued function on* $[a, b]$ (not necessarily of bounded variation) *and let $a < c < b$. Then*

$$V([a, b],f) = V([a,c],f) + V([c, b],f).$$

Proof If either $V([a, c],f)$ or $V([c, b],f)$ is infinite, then the equality is trivial. So, we may assume that both are finite.

If c is a point of the partition P: $a = x_0 < x_1 < \cdots < x_n = b$ of $[a, b]$, then P determines a partition P_1 of $[a, c]$ and a partition P_2 of $[c, b]$. It is clear that

$$T(P,f) = T(P_1,f) + T(P_2,f)$$

and therefore

$$T(P,f) \leq V([a,c],f) + V([c,\ b],f). \tag{5.56}$$

If c is not a point of the partition P, then the partition P' obtained by adding the point c is a refinement of P. Moreover, (5.56) holds with P' in place of P. Now, $x_{i-1} < c < x_i$ for some i. So,

$$|f(x_i) - f(x_{i-1})| \leq |f(x_i) - f(c)| + |f(c) - f(x_{i-1})|,$$

which implies

$$T(P,f) \leq T(P',f).$$

Since (5.56) holds with P' in place of P, the preceding inequality shows that (5.56) holds for P in this case as well (i.e. when c is not a point of P). Taking the supremum over all P, we obtain

$$V([a,\ b],f) \leq V([a,c],f) + V([c,\ b],f).$$

It remains to prove the reverse inequality.

Let $\varepsilon > 0$ be arbitrary. There exist partitions P_1 and P_2 of $[a,\ c]$ and $[c,\ b]$ respectively such that

$$T(P_1,f) > V([a,c],f) - \frac{\varepsilon}{2}$$

and

$$T(P_2,f) > V([c,\ b],f) - \frac{\varepsilon}{2}.$$

Let $P = P_1 \cup P_2$. Then P is a partition of $[a,\ b]$ and

$$T(P,f) = T(P_1,f) + T(P_2,f) > V([a,c],f) + V([c,\ b],f) - \varepsilon,$$

which implies

$$V([a,\ b],f) \geq V([a,\ c],f) + V([c,\ b],f) - \varepsilon.$$

Since $\varepsilon > 0$ is arbitrary, the required reverse inequality follows. □

Example Suppose that $f{:}[a,\ b]\to\mathbb{R}$ is an arbitrary step function. Then there exists a partition P: $a = x_0 < x_1 < \cdots < x_n = b$ of $[a,\ b]$ and numbers $\alpha_1, \alpha_2,\ldots,\alpha_n$ such that $f(x) = \alpha_i$, $x_{k-1} < x < x_k$, $k = 1, 2,\ldots, n$. We shall show that f is of bounded variation.

By Theorem 5.6.4, $V([a, b], f)$ is the sum of the variations over the n intervals $[x_{k-1}, x_k]$. If a partition P_k of $[x_{k-1}, x_k]$ consists of only the two endpoints, then $T(P_k, f) = |f(x_{k-1}) - f(x_k)|$, but if it consists of more than just the two endpoints, then $T(P_k, f) = |\alpha_k - f(x_{k-1})| + |\alpha_k - f(x_k)|$, because the function has the constant value α_k on the interval (x_{k-1}, x_k). In view of the inequality $|\alpha_k - f(x_{k-1})| + |\alpha_k - f(x_k)| \geq |f(x_{k-1}) - f(x_k)|$, it follows that $V([x_{k-1}, x_k], f) = |\alpha_k - f(x_{k-1})| + |\alpha_k - f(x_k)|$. Therefore the total variation of f on $[a, b]$ is

$$V([a,\ b],f) = \sum_{k=1}^{n} (|\alpha_k - f(x_{k-1})| + |\alpha_k - f(x_k)|) < \infty.$$

Corollary 5.6.5 *If it is possible to subdivide* $[a, b]$ *into a finite number of parts, on each of which* f *is monotone, then* f *has finite variation on the interval.*

Proof The assertion is immediate from Theorem 5.6.4. □

The following Theorem, due to C. Jordan, describes completely the class of functions of bounded variation.

Theorem 5.6.6 *A real-valued function on a compact interval is of bounded variation if and only if it can be expressed as the difference of two increasing functions.*

Proof The sufficiency of the condition follows from Remarks 5.6.2(a) and (c). To prove the necessity, we consider a function $f:[a, b] \to \mathbb{R}$ of bounded variation and set

$$v(x) = V([a,x],f), \quad a \leq x \leq b,$$

where $V([a, x], f)$ denotes the total variation of the function calculated on the interval $[a, x]$. For $\xi > x$, we have $v(\xi) - v(x) = V([x, \xi], f)$ by Theorem 5.6.4 and

$$V([x,\xi],f) \geq |f(\xi) - f(x)|, \tag{5.57}$$

since $|f(\xi) - f(x)|$ is the sum $T(P, f)$ for the particular partition P consisting only of x and ξ. So,

$$v(\xi) \geq v(x) \tag{5.58}$$

and

$$v(\xi) - v(x) \geq f(\xi) - f(x),$$

which implies

$$v(\xi) - f(\xi) \geq v(x) - f(x). \tag{5.59}$$

From (5.58) and (5.59), we conclude that v and $v - f$ are increasing functions and

$$f(x) = v(x) - (v(x) - f(x))$$

is a decomposition of f as the difference of increasing functions defined on $[a, b]$.

From (5.57), we also have

$$v(\xi) - v(x) \geq -(f(\xi) - f(x))$$

and this implies

$$v(\xi) + f(\xi) \geq v(x) + f(x). \tag{5.60}$$

It follows from (5.60) that $v + f$ is also an increasing function on $[a, b]$. We thus obtain a symmetric decomposition of f as the difference of increasing functions:

$$f(x) = \frac{1}{2}(v(x) + f(x)) - \frac{1}{2}(v(x) - f(x)).$$ □

Remark 5.6.7 As a consequence of Jordan's Theorem 5.6.6, a function $f:[a, b] \to \mathbb{R}$ of bounded variation inherits many properties of monotone functions:

(a) The limit from the right at x, denoted by $f(x+)$, exists if $a \leq x < b$; the limit from the left at x, denoted by $f(x-)$, exists if $a < x \leq b$.
(b) The set of points of discontinuity of f is at most countable.
(c) $f'(x)$ exists a.e. on $[a, b]$ and $f' \in L^1[a, b]$ by Theorem 5.5.1.
(d) The representation of a function of bounded variation as a difference of two increasing functions is not unique, because one can add the same increasing function to both of the latter.

We finally prove that $v(x) = V([a, x], f)$ is increasing and continuous at each point of continuity of f.

Theorem 5.6.8 *Let f be a function of bounded variation on $[a, b]$. The function v defined by*

$$v(x) = V([a, x], f) \text{ for } x \in [a, b]$$

is increasing and is continuous at each point of continuity of f.

Proof That the function v is increasing has been proved in Theorem 5.6.6 [statement (5.58) of its proof].

Suppose f is continuous at x, where $a < x \leq b$. We shall show that $v(x-) = v(x)$. Let $\varepsilon > 0$ be arbitrary. On using the definition of $v(x)$, we can find a partition P of $[a, x]$ such that

$$T(P, f) > v(x) - \frac{\varepsilon}{2}. \tag{5.61}$$

In view of the left continuity of f at x, we may choose a point y such that $a < y < x$ and

$$|f(y) - f(x)| < \frac{\varepsilon}{2} \tag{5.62}$$

and there is no point of subdivision between y and x. Let P' be the refinement of P obtained by adjoining y to the points of P. Then

$$T(P',f) \geq T(P,f) \tag{5.63}$$

(see the proof of Theorem 5.6.4). Let P'' be the partition of $[a, y]$ formed by using all the points of P'. Then

$$v(y) \geq T(P'',f) = T(P',f) - |f(y) - f(x)| > v(x) - \frac{\varepsilon}{2} - \frac{\varepsilon}{2} = v(x) - \varepsilon,$$

using (5.61), (5.63) and (5.62).

It may be noted that we have only used the left continuity of f at x. The use of right continuity will imply right continuity of v when $a \leq x < b$. □

Remark We note in passing that when f is of bounded variation, we can express it as $f = f_1 - f_2$, where f_1 and f_2 are increasing *nonnegative* functions. This is because in the decomposition of f as a difference $f = g_1 - g_2$ of increasing functions in accordance with Theorem 5.6.6, we have

$$g_1 = \frac{1}{2}(v+f), \quad g_2 = \frac{1}{2}(v-f).$$

By Remark 5.6.2 (b), both are bounded. Taking a common lower bound γ for them, we find that $f_1 = g_1 - \gamma$ and $f_2 = g_2 - \gamma$ are increasing *nonnegative* functions such that $f = f_1 - f_2$. Moreover, by Theorem 5.6.8, both f_1 and f_2 are left and/or right continuous wherever f is.

Theorem 5.6.6 of Jordan characterises a function of bounded variation as being decomposable into a difference of two increasing functions. In what follows, we shall obtain such a decomposition using positive and negative variations introduced below.

Let f be a real-valued function of bounded variation defined on $[a, b]$. If P: $a = x_0 < x_1 < \cdots < x_n = b$ is a partition of $[a, b]$, let

$$\mathcal{P}(P,f) = \sum_{k=1}^{n} [f(x_k) - f(x_{k-1})]^+$$

and

$$\mathcal{N}(P,f) = -\sum_{k=1}^{n} [f(x_k) - f(x_{k-1})]^-,$$

where r^+ means $\max\{r, 0\}$ and r^- means $\min\{r, 0\}$; i.e. $\mathcal{P}(P, f)$ is the sum of all those differences $f(x_k) - f(x_{k-1})$ that are nonnegative, while $\mathcal{N}(P, f)$ is the negative of the sum of those that are nonpositive. Note however that the definition makes $\mathcal{N}(P, f) \geq 0$.

We also define

$$\mathcal{P}([a, b], f) = \sup_P \mathcal{P}(P, f)$$
$$\mathcal{N}([a, b], f) = \sup_P \mathcal{N}(P, f),$$

where the supremum has been taken over all partitions P of $[a, b]$. $\mathcal{P}([a, b], f)$ and $\mathcal{N}([a, b], f)$ are called respectively the **positive and negative variations** of f over $[a, b]$.

Proposition 5.6.9 *If f is of bounded variation on $[a, b]$, then its positive and negative variations are finite. Moreover,*

$$2\mathcal{P}([a, b], f) = V([a, b], f) + f(b) - f(a), \tag{5.64}$$

$$2\mathcal{N}([a, b], f) = V([a, b], f) - f(b) + f(a), \tag{5.65}$$

$$\mathcal{P}([a, b], f) + \mathcal{N}([a, b], f) = V([a, b], f) \tag{5.66}$$

and

$$\mathcal{P}([a, b], f) - \mathcal{N}([a, b], f) = f(b) - f(a). \tag{5.67}$$

Proof The finiteness of the positive and negative variations follows from the fact that

$$0 \leq \mathcal{P}(P, f) \leq T(P, f) \quad \text{and} \quad 0 \leq \mathcal{N}(P, f) \leq T(P, f).$$

The following relations are evident:

$$\mathcal{P}(P, f) - \mathcal{N}(P, f) = f(b) - f(a)$$
$$T(P, f) = \mathcal{P}(P, f) + \mathcal{N}(P, f).$$

From these, we get

$$2\mathcal{P}(P, f) = T(P, f) + f(b) - f(a)$$
$$2\mathcal{N}(P, f) = T(P, f) - f(b) + f(a),$$

from which we further get (5.64) and (5.65). These then lead to (5.66) and (5.67). □

Remark If f is of bounded variation on $[a, b]$, it follows that f is of bounded variation on $[a, x]$, $a < x \leq b$. The symbols $\mathcal{P}([a, x], f)$, $\mathcal{N}([a, x], f)$ and $V([a, x], f)$ are all meaningful. Moreover, relations (5.64)–(5.67) of the above proposition hold if we replace b by x in them. We define

$$\mathcal{P}([a, a], f) = \mathcal{N}([a, a], f) = V([a, a], f) = 0.$$

We know by Theorem 5.6.8 that $v(x) = V([a, x], f)$ is an increasing function of x and is continuous wherever f is continuous.

Proposition 5.6.10 *The functions of* x *given by* $\mathcal{P}([a, x], f)$ *and* $\mathcal{N}([a, x], f)$ *are increasing.*

Proof Set $g(x) = V([a, x], f) - f(x)$ and $h(x) = V([a, x], f) + f(x)$. It has been shown in Theorem 5.6.6 that both are increasing.

From Proposition 5.6.9 (5.64), (5.65), with b replaced by x, we have

$$\mathcal{P}([a, x], f) = \frac{1}{2}(h(x) - f(a))$$

and

$$\mathcal{N}([a, x], f) = \frac{1}{2}(g(x) + f(a)).$$

So, $\mathcal{P}([a, x], f)$ and $\mathcal{N}([a, x], f)$ are increasing functions of x on $[a, b]$. □

Theorem 5.6.11 *If* f *is a function of bounded variation on* $[a, b]$*, there exist nonnegative increasing functions* f_1 *and* f_2 *on* $[a, b]$ *such that* $f = f_1 - f_2$*. The functions* $\mathcal{P}([a, x], f)$*,* $\mathcal{N}([a, x], f)$ *and* $V([a, x], f)$ *are continuous at any* $x \in [a, b]$ *where* f *is continuous. The functions* f_1 *and* f_2 *can be so chosen that if* f *is continuous at* x*, then so are* f_1 *and* f_2*.*

Proof Although the existence of the functions f_1 and f_2 is a consequence of Theorem 5.6.8, we give a fresh proof here, again using Theorem 5.6.8. Set

$$f_1(x) = \mathcal{P}([a, x], f) + \frac{1}{2}(f(a) + |f(a)|)$$

and

$$f_2(x) = \mathcal{N}([a, x], f) + \frac{1}{2}(-f(a) + |f(a)|).$$

In view of Proposition 5.6.10, f_1 and f_2 are increasing functions defined on $[a, b]$. Moreover, they are nonnegative and

$$f_1(x) - f_2(x) = \mathcal{P}([a,\ x],\ f) - \mathcal{N}([a,\ x],\ f) + f(a) = f(x),$$

where we have used (5.67) of Proposition 5.6.9 with x in place of b.

Assume that f is continuous at x_0. It follows from Theorem 5.6.8 that $V([a, x], f)$ is continuous at x_0. Consequently, the functions g and h of Proposition 5.6.10 are continuous at x_0 and so are $\mathcal{P}([a,\ x],\ f)$ and $\mathcal{N}([a,\ x],\ f)$. It follows that f_1 and f_2 are continuous at x_0. □

Remark Suppose g:$[a,\ b]\to\mathbb{R}^*$ is summable in the Lebesgue sense. Define f:$[a,\ b]\to\mathbb{R}$ by the formula

$$f(x) = \int_{[a,\ x]} g,$$

where it is understood that $f(a) = 0$. In view of Problem 3.2.P16, f is continuous on $[a,\ b]$. Moreover, it is of bounded variation on $[a,\ b]$ with $V([a,\ b], f) \leq \int_{[a,\ b]} |g|$. (That equality holds here is left as Problem 5.6.P9).

(i) Let $x_0 \in [a,\ b)$. Then for $x > x_0$, we have

$$f(x) - f(x_0) = \int_{[x_0, x]} g = \int_{[a,\ x]} \chi_{[x_0, x]} g.$$

Observe that $\left|\chi_{[x_0, x]} g\right| \leq |g|$ and the latter is Lebesgue integrable. Moreover, $\chi_{[x_0, x]} g \to 0$ as $x \to x_0$. By the Dominated Convergence Theorem 3.2.16, we obtain

$$|f(x) - f(x_0)| \leq \int_{[x_0, x]} |g| \to 0 \text{ as } x \to x_0.$$

Similarly, if $x_0 \in (a,\ b]$ and $x < x_0$, then $\int_{[x, x_0]} |g| \to 0$ as $x \to x_0$. This shows that f is continuous at x_0. Since x_0 is arbitrary, it follows that f is continuous on $[a,\ b]$.

(ii) Consider any partition P: $a = x_0 < x_1 < \cdots < x_n = b$ of $[a,\ b]$. Then

$$\sum_{k=1}^{n} |f(x_k) - f(x_{k-1})| = \sum_{k=1}^{n} \left| \int_{[x_{k-1}, x_k]} g \right| \leq \sum_{k=1}^{n} \int_{[x_{k-1}, x_k]} |g| \leq \int_{[a,\ b]} |g|.$$

Since by hypothesis, g is integrable on $[a,\ b]$, it follows that f is of bounded variation on $[a,\ b]$ with $V([a,\ b], f) \leq \int_{[a,\ b]} |g|$.

Obviously a function of bounded variation on $\mathbb{R}$ is of bounded variation on every compact subinterval of $\mathbb{R}$, where variation on $\mathbb{R}$ means $\sup\{V([a,\ b], f)\colon a < b\}$.

Examples 5.6.12

(a) The function $f(x) = x$ is of bounded variation on any compact interval. However it is not of bounded variation on $\mathbb{R}$. In fact, $V([a, b], f) = b - a$, and hence sup $\{V([a, b], f): a < b\} = \infty$.

(b) We have observed that the function $f(x) = \sin x$, $a \leq x \leq b$, is of bounded variation there [Example 5.6.3(a)]. However, it is not of bounded variation on $\mathbb{R}$. Indeed, for the partition

$$P : \frac{\pi}{2} < \frac{3\pi}{2} < \cdots < \frac{(2n-1)\pi}{2} < \frac{(2n+1)\pi}{2},$$

the sum

$$\sum_{k=1}^{n} \left| f\left(\frac{(2k+1)\pi}{2}\right) - f\left(\frac{(2k-1)\pi}{2}\right) \right| = 2n$$

exceeds every positive positive number as $n \to \infty$.

(c) Let $g \in L^1(\mathbb{R})$. Then the function $f(x) = \int_{[-\infty,x]} g$ is of bounded variation on $\mathbb{R}$. Indeed,

$$V([a, x], f) \leq \int_{[a, x]} |g| \leq \int_{[-\infty,\infty]} |g|;$$

consequently, $V(\mathbb{R}, f) \leq \int_{[-\infty,\infty]} |g|$.

Problem Set 5.6

5.6.P1. Show that

$$V([a, b], f+g) \leq V([a, b], f) + V([a, b], g) \text{ and } V([a, b], cf) = |c| V([a, b], f).$$

5.6.P2. Let $\phi:(a, b) \to \mathbb{R}$ have a (finite) left limit everywhere and $y \in [a, b)$ [resp. $y \in (a, b]$] be a point where the right limit $\phi(y+)$ [resp. left limit $\phi(y-)$] exists (and is finite). Then first,

$$\lim_{x \to y+} \phi(x-) = \phi(y+)$$

and second,

$$\lim_{x \to y-} \phi(x-) = \phi(y-).$$

In particular, $\lim_{x\to a+} \phi(x-) = \phi(a+)$ and $\lim_{x\to b-} \phi(x-) = \phi(b-)$.

5.6.P3. The function f defined by $f(x) = \sin\frac{\pi}{x}$, $0 < x \le 1$ and $f(0) = 0$, is not of bounded variation on [0, 1].

5.6.P4. Define f:[0, 1]→$\mathbb{R}$ by $f(0) = 0$ and $f(x) = x^2 \sin\frac{1}{x}$, $x \neq 0$. Show that f is of bounded variation on [0, 1].

5.6.P5. Show that the function defined on [0, 1] by

$$f(x) = \begin{cases} x^2 \sin\frac{1}{x^2} & \text{for } x \neq 0 \\ 0 & \text{for } x = 0 \end{cases}$$

is not of bounded variation.

5.6.P6. Suppose that f:[a, b] → [c, d] is monotone and that g is of bounded variation on [c, d]. Prove that $V([a, b], g \circ f) \le V([c, d], g)$.

5.6.P7. Let $\{f_n\}_{n\ge1}$ be a sequence of real-valued functions defined on [a, b] that converge pointwise to the function f. Prove that

$$V([a, b], f) \le \liminf_{n\to\infty} V([a, b], f_n).$$

5.6.P8. Let $\{f_n\}_{n\ge1}$ be a sequence of functions of bounded variation on [a, b] such that $\sum_n f_n(a)$ converges absolutely and $\sum_n V([a, b], f_n) < \infty$. Prove that

(i) $\sum_n f_n(x)$ converges absolutely for each $x \in [a, b]$,
(ii) $V([a, b], \Sigma_n f_n) \le \Sigma_n V([a, b], f_n)$.

5.6.P9. Suppose $g \in L^1[a, b]$ and set $f(x) = \int_{[a,x]} g$, $a \le x \le b$. Show that $V([a, b], f) = \int_{[a, b]} |g|$.

5.6.P10. Deduce from Problem 5.6.P9 that $\sin x$ is of bounded variation on $[0, 2\pi]$ and that its total variation is 4.

5.6.P11. Define f on [0, 1] by

$$f(0) = 0 \quad \text{and} \quad f(x) = x \sin\frac{1}{x}, x \neq 0.$$

Show that f is not an indefinite integral of a Lebesgue integrable function on [0, 1].

5.6.P12. Define

$$f(x) = \begin{cases} 1 & x \in \mathbb{Q} \\ 0 & x \notin \mathbb{Q}. \end{cases}$$

Show that f is not of bounded variation on any compact interval.

5.6.P13. Construct functions f_n, $n = 0, 1, 2, \ldots$ on an interval $[a, b]$ such that

(i) $f_n \to f$ uniformly,
(ii) each f_n is of bounded variation,
(iii) f is not of bounded variation.

5.6.P14. Let f be defined on [0, 1] by the following formulae (see Fig. 5.1):

$$f(x) = \begin{cases} x & x = \frac{1}{2n}, n \in \mathbb{N} \\ 0 & x = \frac{1}{2n+1}, n \in \mathbb{N} \\ \text{linear} & \frac{1}{n+1} \leq x \leq \frac{1}{n}, n \in \mathbb{N} \\ 0 & x = 0. \end{cases}$$

Show that f is not of bounded variation.

5.6.P15. If $f(x) = \int_{[a,\,x]} g$, where $g \in L^1[a, b]$, prove that the positive and negative variations of f, namely, $\mathcal{P}([a, b], f)$ and $\mathcal{N}([a, b], f)$, are given by $\int_{[a,\,b]} g^+$ and $\int_{[a,\,b]} g^-$ respectively.

5.6.P16. Let $x_1, x_2, \ldots$ be the points of discontinuity in (a, b) of the function f of bounded variation with domain $[a, b]$. We define j:$[a, b] \to \mathbb{R}$ by the following formulae:

$$j(a) = f(a) - f(a+) \text{ and } j(x) = f(x) - f(x-) + \sum_{x_i < x} [f(x_i+) - f(x_i)] \text{ if } a < x \leq b. \tag{5.68}$$

By the "jump" of f at x, we mean $|f(x) - f(x-)| + |f(x+) - f(x)|$ if $a < x < b$ and $|f(a+) - f(a)|$ if $x = a$ and $|f(b) - f(b-)|$ if $x = b$. [Note that this agrees with Definition 5.2.4 for a monotone function.]

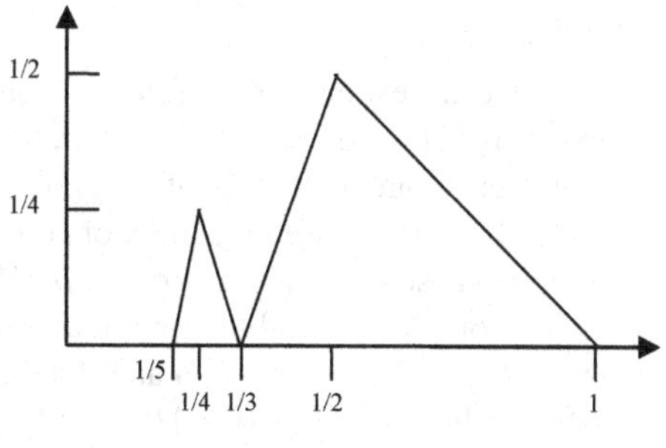

Fig. 5.1 Function in Problem 5.6.P14

Show without using Theorem 5.3.1 that

(a) The function j defined in (5.68) is of bounded variation on $[a, b]$.
(b) $j(y-) = \sum_{x_i<y} [f(x_i+) - f(x_i-)]$ for $y \in (a, b]$, $j(y+) = \sum_{x_i \le y} [f(x_i+) - f(x_i-)]$ for $y \in (a, b)$ and $j(a+) = 0$.
(c) If $F = f - j$, then F is continuous and of bounded variation.
(d) $j' = 0$ a.e.
(e) The representation of f as $F + j$, where F is continuous and $j' = 0$ a.e. is not unique (not even up to a constant) (Cf. Problem 5.7.P4).

5.7 Absolute Continuity

In this section, we shall study a special class of functions of bounded variation, known as *absolutely continuous* functions. These functions are representable as the integral of a summable function with a variable limit of integration (see Sect. 5.8). These functions are important for a number of applications and are, in addition, interesting in their own right.

We begin with a definition.

Definition 5.7.1 A real-valued function f defined on an interval $[a, b]$ or $\mathbb{R}$ or $(-\infty, b]$ *or* $[a, \infty)$ is said to be **absolutely continuous on its domain** if, given any $\varepsilon > 0$, there exists a $\delta > 0$ such that, for any finite system of disjoint intervals $\{(\alpha_k, \beta_k)\}_{1\le k\le n}$ contained in the domain, the sum of whose lengths is less than δ, *i.e.* $\sum_{k=1}^{n} (\beta_k - \alpha_k) < \delta$, there holds the inequality

$$\sum_{k=1}^{n} |f(\beta_k) - f(\alpha_k)| < \varepsilon.$$

Remarks 5.7.2

(a) We give an example of a function that is absolutely continuous on compact intervals but not on $\mathbb{R}$, $(-\infty, b]$ or $[a, \infty)$. The example is given by $f(x) = x^2$. For any open subinterval (α,β) of $[a, b]$, we have $|f(\beta) - f(\alpha)| = |\beta^2 - \alpha^2| \le 2\cdot\max\{|a|,|b|\}(\beta-\alpha)$ and we may therefore choose δ satisfying $\delta\cdot(2\max\{|a|,|b|\}) = \varepsilon$ to establish absolute continuity on $[a, b]$. However, if we take $\mathbb{R}$, $(-\infty, b]$ or $[a,\infty)$ as the domain, we find that, no matter what $\delta > 0$ is chosen, the single interval $(\alpha, \alpha + \gamma)$, where $0 < \gamma < \delta$ and $2\gamma|\alpha| > \varepsilon + \gamma^2$, has length less than δ and yet satisfies $|f(\alpha + \gamma) - f(\alpha)| = |2\alpha\gamma + \gamma^2| \ge |2\alpha\gamma| - \gamma^2 > \varepsilon$; note that such an α

can always be found in an unbounded interval, so that the criterion for absolute continuity is violated.

(b) It is readily seen that an absolutely continuous function is uniformly continuous (in the definition, take a system consisting of just one interval), and hence continuous. As the reader knows from elementary analysis, the function given by $f(x) = x^2$ is not uniformly continuous on $\mathbb{R}$, $(-\infty, b]$ or $[a, \infty)$, which is another way to see what we have noted in part (a), namely, that this function is not absolutely continuous there.
However, a uniformly continuous function need not be absolutely continuous, as we shall illustrate in Example 5.7.3(d). See also Problems 5.7.P2 and 5.7.P6.

(c) The inequality in Definition 5.7.1 can be replaced by the weaker condition

$$\left|\sum_{k=1}^{n} [f(\beta_k) - f(\alpha_k)]\right| < \varepsilon.$$

Indeed, given any $\varepsilon > 0$, let $\delta > 0$ be such that this inequality is fulfilled for any finite system of disjoint intervals $\{(\alpha_k, \beta_k)\}_{1\le k\le n}$ contained in $[a, b]$ and having sum of lengths less than δ. We divide the intervals into two disjoint classes A and B, where A consists of those for which $[f(\beta_k) - f(\alpha_k)] \geq 0$ and B consists of the rest. By hypothesis, the inequality in the weaker condition must be fulfilled by each of the classes A and B of intervals separately. Consequently,

$$\sum_{k=1}^{n} |f(\beta_k) - f(\alpha_k)| = \sum_{k\in A} |f(\beta_k) - f(\alpha_k)| + \sum_{k\in B} |f(\beta_k) - f(\alpha_k)|$$
$$= \left|\sum_{k\in A} [f(\beta_k) - f(\alpha_k)]\right| + \left|\sum_{k\in B} [f(\beta_k) - f(\alpha_k)]\right| < 2\varepsilon.$$

Since $\varepsilon > 0$ is arbitrary, it follows that f is absolutely continuous on $[a, b]$.

(d) Let f be a real-valued function defined on $[a, b]$. Then f is absolutely continuous if and only if for every $\varepsilon > 0$, there exists a $\delta > 0$ such that for any finite system of disjoint intervals $\{(\alpha_k, \beta_k)\}_{1\le k\le n}$ contained in $[a, b]$ and having sum of lengths less than δ, the inequality

$$\sum_{k=1}^{n} \omega_k < \varepsilon$$

holds, where

$$\omega_k = \sup\{|f(p) - f(q)| : p, q \in [\alpha_k, \beta_k]\}.$$

Let f be absolutely continuous on $[a, b]$ and let $\varepsilon > 0$ be given. Then there exists a $\delta > 0$ such that for any finite system $\{(\alpha_k, \beta_k)\}_{1\le k\le n}$ of disjoint open intervals contained in $[a, b]$ and having sum of lengths less than δ, the inequality

$$\sum_{k=1}^{n} |f(\beta_k) - f(\alpha_k)| < \varepsilon$$

holds. Let a_k, b_k in $[\alpha_k, \beta_k]$ be such that $f(a_k) = m_k$ and $f(b_k) = M_k$, where m_k and M_k are respectively the least and greatest values of the function f in $[\alpha_k, \beta_k]$. Then

$$\sum_{k=1}^{n} |(b_k - a_k)| \leq \sum_{k=1}^{n} (\beta_k - \alpha_k) < \delta$$

and so,

$$\sum_{k=1}^{n} \omega_k = \sum_{k=1}^{n} |f(b_k) - f(a_k)| < \varepsilon.$$

On the other hand, suppose the given condition holds. Consider $\varepsilon > 0$ and let $\delta > 0$ be as in the condition. Then for any finite system $\{(\alpha_k, \beta_k)\}_{1 \leq k \leq n}$ of disjoint open intervals contained in $[a, b]$ and having sum of lengths less than δ, we have $|f(\beta_k) - f(\alpha_k)| \leq \omega_k$, $k = 1, 2, \ldots, n$, and therefore

$$\sum_{k=1}^{n} |f(\beta_k) - f(\alpha_k)| \leq \sum_{k=1}^{n} \omega_k < \varepsilon.$$

Thus the function is absolutely continuous.

(e) A function f defined on $[a, b]$, $\mathbb{R}$, $(-\infty, b]$ or $[a, \infty)$ and satisfying a Lipschitz condition (see Definition 4.2.12)

$$|f(x) - f(y)| \leq M|x - y| \text{ for every } x \text{ and } y \text{ in the domain,}$$

where M is a fixed positive number, is absolutely continuous. This is so because, for any finite system $\{(\alpha_k, \beta_k)\}_{1 \leq k \leq n}$ of disjoint open intervals contained in the domain, we have

$$\sum_{k=1}^{n} |f(\beta_k) - f(\alpha_k)| \leq M \sum_{k=1}^{n} (\beta_k - \alpha_k),$$

from which it may be easily inferred that f is absolutely continuous.
As a particular case, if f' exists and is bounded in (a, b) or $\mathbb{R}$ and if $f:[a, b] \to \mathbb{R}$ is continuous, then it is absolutely continuous.

(f) If the functions f and g are absolutely continuous on $[a, b]$, then their sum, difference and product are also absolutely continuous. If g vanishes nowhere, then the quotient $\frac{f}{g}$ is absolutely continuous.
From the inequality

$$|(f \pm g)(x) - (f \pm g)(y)| \le |f(x) - f(y)| + |g(x) - g(y)|,$$

it may be easily inferred that $f \pm g$ are absolutely continuous provided f and g are absolutely continuous.
We shall next show that fg is absolutely continuous. Let $\varepsilon > 0$ be given. Then there exists a $\delta > 0$ such that, for any finite system $\{(\alpha_k, \beta_k)\}_{1 \le k \le n}$ of disjoint intervals contained in $[a, b]$ and having sum of lengths less than δ, we have

$$\sum_{k=1}^{n} |f(\beta_k) - f(\alpha_k)| < \frac{\varepsilon}{2(A+B)} \quad \text{and} \quad \sum_{k=1}^{n} |g(\beta_k) - g(\alpha_k)| < \frac{\varepsilon}{2(A+B)},$$

where $A \ge |f|$ and $B \ge |g|$ on $[a, b]$. Then

$$\begin{aligned}\sum_{k=1}^{n} |(fg)(\beta_k) - (fg)(\alpha_k)| &= \sum_{k=1}^{n} |f(\beta_k)g(\beta_k) - f(\alpha_k)g(\alpha_k)| \\ &\le \sum_{k=1}^{n} |[f(\beta_k) - f(\alpha_k)]g(\beta_k)| + \sum_{k=1}^{n} |f(\alpha_k)[g(\beta_k) - g(\alpha_k)]| \\ &\le B\sum_{k=1}^{n} |f(\beta_k) - f(\alpha_k)| + A\sum_{k=1}^{n} |g(\beta_k) - g(\alpha_k)| \\ &< \varepsilon.\end{aligned}$$

Finally, if g vanishes nowhere, then there exists an $m > 0$ such that $|g(x)| > m > 0$ for all $x \in [a, b]$. So,

$$\left|\frac{1}{g(x)} - \frac{1}{g(y)}\right| \le \frac{|g(x) - g(y)|}{m^2}.$$

Let $\varepsilon > 0$ be given. Then there exists a $\delta > 0$ such that, for any finite system $\{(\alpha_k, \beta_k)\}_{1 \le k \le n}$ of disjoint intervals contained in $[a, b]$ and having sum of lengths less than δ, we have

$$\sum_{k=1}^{n} |g(\beta_k) - g(\alpha_k)| < m^2\varepsilon.$$

Together with the inequality further above, this yields

$$\sum_{k=1}^{n}\left|\frac{1}{g(\beta_k)}-\frac{1}{g(\alpha_k)}\right|<\varepsilon.$$

Thus $\frac{1}{g}$ is absolutely continuous. Combining it with the result proved in the preceding paragraph for products, we get the desired conclusion.
If the domain of definition of the functions happens to be unbounded, then the sum and difference of absolutely continuous functions are absolutely continuous; however, the product and quotient can be shown to be absolutely continuous provided the functions involved are assumed to be bounded.

Examples 5.7.3

(a) If f is nonnegative and absolutely continuous on $[a, b]$, then so is $f^p, p \geq 1$. To prove this, consider any $y > x \geq 0$. Then first of all,

$$y^p - x^p = p(y-x)(x+\theta(y-x))^{p-1} \text{ for some } \theta \text{ such that } 0<\theta<1.$$

So,

$$\begin{aligned}|f(y)^p - f(x)^p| &= p|f(y)-f(x)||f(x)+\theta(f(y)-f(x))^{p-1}| \\ &\leq p(M+(2M)^{p-1})|f(y)-f(x)|,\end{aligned} \tag{5.69}$$

where $M \geq |f|$ on $[a, b]$. Now let $\varepsilon > 0$ be given. Since f is absolutely continuous, there exists a $\delta > 0$ such that, for any finite system $\{(\alpha_k, \beta_k)\}_{1\leq k\leq n}$ of disjoint intervals contained in $[a, b]$ and having sum of lengths less than δ, we have

$$\sum_{k=1}^{n}|f(\beta_k)-f(\alpha_k)|<\frac{\varepsilon}{p(M+(2M)^{p-1})},$$

which, in view of (5.69), implies

$$\sum_{k=1}^{n}|f(\beta_k)^p - f(\alpha_k)^p| \leq p(M+(2M)^{p-1})\sum_{k=1}^{n}|f(\beta_k)-f(\alpha_k)|<\varepsilon.$$

Boundedness of the domain is essential here, because x is absolutely continuous on every interval, bounded or not, but x^2 is not [see Remark 5.7.2(a)].

(b) Since the function $f(x) = x$, $0 \leq x \leq 1$, is trivially absolutely continuous, it follows in view of (a) above that $g(x) = x^p$, $p \geq 1$, $0 \leq x \leq 1$, is also absolutely continuous. Now, $g'(x)$ is bounded for $0 \leq x \leq A$ (any $A > 0$) and therefore Remark 5.7.2(e) also shows that g is absolutely continuous. We shall take up the case when $0 \leq p < 1$ in Problem 5.7.P6.

(c) The function $f(x) = \sin x$ is absolutely continuous on $\mathbb{R}$, and so is the function $g(x) = \sin^p x$, $p \geq 1$. The function $\sin x$ satisfies a Lipschitz condition with $M = 1$ and is, therefore, absolutely continuous by Remark 5.7.2(e). The function g can also be seen to be absolutely continuous by using Remark 5.7.2(e). Indeed, it satisfies a Lipschitz condition with $M = p$.

(d) Absolute continuity of f on $[a, b]$ does not imply that of $f^{\frac{1}{2}}$. For example, consider the function $f:[0, 1]\to\mathbb{R}$ defined by $f(0) = 0$ and $f(x)= x^2\sin^2 \frac{\pi}{2x}$, $x \neq 0$. Observe that $f'(x)=\sin^2 \frac{\pi}{2x} - \pi\sin \frac{\pi}{2x} \cos \frac{\pi}{2x}$ for $0 < x < 1$, so that $|f'(x)| \leq 2 + \pi$. Thus, f is absolutely continuous on $[0, 1]$ by Remark 5.7.2(e). However, $f(x)^{\frac{1}{2}} = x\left|\sin\frac{\pi}{2x}\right|$ when $x > 0$ and $f(0)^{\frac{1}{2}} = 0$. For the partition

$$0<\frac{1}{2n+1}<\frac{1}{2n}<\cdots<\frac{1}{3}<\frac{1}{2}<1,$$

$$\begin{aligned}
T(P,f^{\frac{1}{2}}) &= \sum_{k=1}^{2n+1} \left|f^{\frac{1}{2}}(x_k) - f^{\frac{1}{2}}(x_{k-1})\right| \\
&= \frac{1}{2n+1}\left|\sin(2n+1)\frac{\pi}{2}\right| + \left|\frac{1}{2n}|\sin n\pi| - \frac{1}{2n-1}\left|\sin(2n-1)\frac{\pi}{2}\right|\right| + \cdots \\
&\quad + \left|\frac{1}{3}\left|\sin\frac{3\pi}{2}\right| - \frac{1}{2}|\sin\pi|\right| + \left|\frac{1}{2}|\sin\pi| - \left|\sin\frac{\pi}{2}\right|\right| \\
&= \frac{1}{2n+1} + \frac{1}{2n-1} + \cdots + \frac{1}{3} + 1.
\end{aligned}$$

These sums become infinite as $n \to \infty$. Consequently, $f^{\frac{1}{2}}$ is not of bounded variation. As we shall see later (see Proposition 5.7.6), this implies that it cannot be absolutely continuous (despite being continuous on the closed and bounded interval $[0, 1]$ and hence uniformly continuous).

(e) A function ϕ defined on (a, b) is said to be *convex* if for every $x, y \in (a, b)$ and every $t \in [0, 1]$, the following inequality holds:

$$\phi((1-t)x+ty) \leq (1-t)\phi(x) + t\phi(y).$$

Graphically, the condition is that if $x < z < y$, then the point $(z, \phi(z))$ in the plane should lie below or on the line segment joining the points $(x, \phi(x))$ and $(y, \phi(y))$.

Consider any $x, y, z \in (a, b)$ such that $x < z < y$. Then

$$z = \frac{y-z}{y-x}x + \frac{z-x}{y-x}y = (1-t)x + ty, \quad \text{where } t = \frac{z-x}{y-x}.$$

Therefore, convexity of ϕ implies

$$\phi(z) \leq \frac{y-z}{y-x}\phi(x) + \frac{z-x}{y-x}\phi(y),$$

which leads to $(y - x)\phi(z) \leq (y - z)\phi(x) + (z - x)\phi(y)$ and hence to

$$(y-z+z-x)\phi(z) \leq (y-z)\phi(x) + (z-x)\phi(y).$$

It follows from here that

$$(z-x)(\phi(z) - \phi(y)) \leq (y-z)(\phi(x) - \phi(z)),$$

from which we obtain

$$\frac{\phi(z) - \phi(x)}{z-x} \leq \frac{\phi(y) - \phi(z)}{y-z}.$$

The fact that a convex function ϕ on (a, b) satisfies the above inequality for arbitrary $x, y, z \in (a, b)$ such that $x < z < y$ is about to be used.
A convex function ϕ defined on (a, b) is absolutely continuous on every closed subinterval $[c, d] \subset (a, b)$, as we show below:
Let $a < c' < c$ and $d < d' < b$. For $x, y \in [c, d]$, $x < y$, we have

$$\frac{\phi(c') - \phi(c)}{c'-c} \leq \frac{\phi(x) - \phi(c)}{x-c} \leq \frac{\phi(y) - \phi(x)}{y-x} \leq \frac{\phi(d) - \phi(y)}{d-y} \leq \frac{\phi(d') - \phi(d)}{d'-d},$$

using the fact that ϕ is convex on (a, b). The above inequalities imply

$$\left|\frac{\phi(y) - \phi(x)}{y-x}\right| \leq \max\left\{\left|\frac{\phi(d') - \phi(d)}{d'-d}\right|, \left|\frac{\phi(c') - \phi(c)}{c'-c}\right|\right\} = H, \quad \text{say}.$$

So, ϕ satisfies a Lipschitz condition on $[c, d]$ with Lipschitz constant H and is therefore absolutely continuous by Remark 5.7.2(e).

(f) The composition of absolutely continuous functions need not be absolutely continuous. (However, see Propositions 5.7.4 and 5.7.5.)
It is a consequence of Problem 5.7.P6 that the square root function is absolutely continuous on [0, 1]. Accordingly, it follows from the same example as in (d) above that the composition of absolutely continuous functions need not be absolutely continuous.

Proposition 5.7.4 *Let $f: I \to [c, d]$ be an absolutely continuous function, where I is any interval. If F is a function on $[c, d]$ which satisfies a Lipschitz condition, then the composite function $F \circ f$ is absolutely continuous.*

Proof If $|F(y) - F(x)| \leq M|y - x|$, then for an arbitrary finite system of disjoint open intervals $\{(\alpha_k, \beta_k)\}_{1 \leq k \leq n}$ contained in I, we have

$$\sum_{k=1}^{n} |F(f(\beta_k)) - F(f(\alpha_k))| \leq M \sum_{k=1}^{n} |f(\beta_k) - f(\alpha_k)|.$$

Since f is absolutely continuous on I, it follows that the right-hand side of the above inequality is small whenever $\sum_{k=1}^{n} (\beta_k - \alpha_k)$ is small. □

Proposition 5.7.5 *Let f be an absolutely continuous function defined on any interval I and suppose that it is increasing. If F is absolutely continuous on $f(I)$, then the composition $F \circ f$ is absolutely continuous on I.*

Proof Let $\varepsilon > 0$ be given. Since F is absolutely continuous on $f(I)$, there exists a $\delta > 0$ such that, for any finite system of disjoint open intervals $\{(A_k, B_k)\}_{1 \leq k \leq n}$ contained in $f(I)$ and having total length less than δ, we have

$$\sum_{k=1}^{n} |F(B_k) - F(A_k)| < \varepsilon.$$

For this $\delta > 0$, there exists an $\eta > 0$ such that $\sum_{k=1}^{m} (\beta_k - \alpha_k) < \eta$ implies $\sum_{k=1}^{m} |f(\beta_k) - f(\alpha_k)| < \delta$ whenever the intervals $\{(\alpha_k, \beta_k)\}_{1 \leq k \leq m}$ are disjoint. Choose any finite system $\{(c_k, d_k)\}_{1 \leq k \leq \ell}$ of disjoint open intervals with total length $\sum_{k=1}^{\ell} (d_k - c_k) < \eta$. Then the intervals $\{(f(c_k), f(d_k))\}_{1 \leq k \leq \ell}$ are disjoint, since f is increasing. Moreover the sum of their lengths, $\sum_{k=1}^{\ell} |f(d_k) - f(c_k)| < \delta$. Hence

$$\sum_{k=1}^{\ell} |F(f(d_k)) - F(f(c_k))| < \varepsilon.$$ □

Our next result shows that the class of absolutely continuous functions is contained in the class of functions of bounded variation.

Proposition 5.7.6 *Let f be an absolutely continuous function on $[a, b]$. Then f is of bounded variation.*

Proof Since f is absolutely continuous, choose $\delta > 0$ corresponding to $\varepsilon = 1$ such that, for every finite system $\{(\alpha_k, \beta_k)\}_{1 \leq k \leq n}$ of disjoint open intervals contained in $[a, b]$ and having sum of lengths less than δ, we have

$$\sum_{k=1}^{n} |f(\beta_k) - f(\alpha_k)| < 1.$$

Let $N > \frac{b-a}{\delta}$ be any integer, and let P: $a = x_0 < x_1 < \cdots < x_N = b$ be the partition of $[a, b]$ such that $x_k - x_{k-1} = \frac{b-a}{N}$, i.e., $x_k = a + k\frac{b-a}{N}$ for $1 \leq k \leq N$. Then $0 < x_k - x_{k-1} < \delta$ for $1 \leq k \leq N$. Therefore, for every subdivision of the subinterval $[x_{k-1}, x_k]$, the sum of the absolute increments of f on these intervals corresponding to the subdivision of $[x_{k-1}, x_k]$ is less than 1. Therefore $V([x_{k-1}, x_k], f) \leq 1$. In view of Theorem 5.6.4, it follows that $V([a, b], f) \leq N$. □

Remarks

(a) The converse of Proposition 5.7.6 is false. The Cantor ternary function is continuous and increasing, and hence is of bounded variation. However, it fails to be absolutely continuous by Problem 5.7.P2. Moreover, in contrast to the example in Remark 5.7.2(a), it is a continuous function on a *bounded* interval that is not absolutely continuous.

(b) The implication in Proposition 5.7.6 fails to be true if a or b is allowed to be infinite. In fact, the functions x and $\sin x$ are absolutely continuous but are not of bounded variation on $(-\infty, b]$ or on $[a, \infty)$.
By the Mean Value Theorem, $|\sin x - \sin y| \leq |x - y|$ for all $x, y \in \mathbb{R}$. Consequently, for any finite system $\{(\alpha_k, \beta_k)\}_{1 \leq k \leq n}$ of disjoint open intervals, we have

$$\sum_{k=1}^{n} |\sin \beta_k - \sin \alpha_k| \leq \sum_{k=1}^{n} (\beta_k - \alpha_k),$$

from which follows the absolute continuity of $\sin x$.
If we take the points

$$\frac{\pi}{2} < \frac{3\pi}{2} < \frac{5\pi}{2} < \cdots < \frac{(2n-1)\pi}{2} < \frac{(2n+1)\pi}{2}$$

for points of division, then the sum

$$\sum_{k=1}^{n} \left| \sin\frac{(2k-1)\pi}{2} - \sin\frac{(2k+1)\pi}{2} \right| = 2n$$

can be made large for large values of n. Hence the function $\sin x$, $x \in \mathbb{R}$, is not of bounded variation even on $[\frac{\pi}{2}, \infty)$ and *a fortiori* not on $\mathbb{R}$.

Corollary 5.7.7 *An absolutely continuous function on* $[a, b]$ *has a finite derivative a.e. on the domain and the derivative is Lebesgue integrable.*

Proof By Proposition 5.7.6, the function is of bounded variation on $[a, b]$. The Corollary now follows by using Theorem 5.5.1. □

Corollary 5.7.8 *There exist continuous functions on bounded intervals that are not absolutely continuous.* [The case of unbounded intervals has been taken care of in Remark 5.7.2(a).]

Proof The function $f(x) = x \sin \frac{\pi}{2x}, x \neq 0$, and $f(0) = 0$ is continuous on $[0, 1]$ but is not of bounded variation [see Example 5.6.3(c)], and hence by Proposition 5.7.6, cannot be absolutely continuous. □

Proposition 5.7.9 *Let* $f:[a, b]\to\mathbb{R}$ *be absolutely continuous and let*

$$v(a) = 0, v(x) = V([a,x],f) \text{ for } x \in [a, b]$$

(notations as in Theorem 5.6.8). *Then* v *is absolutely continuous on* $[a, b]$; f *can therefore be expressed as the difference of increasing absolutely continuous functions.*

Proof By Theorem 5.6.4, we see that $v(\beta) - v(\alpha) = V([\alpha, \beta], f)$ if $a \leq \alpha \leq \beta \leq b$. Let $\varepsilon > 0$ be given and $\delta > 0$ correspond to ε as in the definition of absolute continuity. Suppose $\{(\alpha_k,\beta_k)\}_{1\leq k\leq n}$ is a finite system of disjoint open intervals contained in $[a, b]$ and having sum of lengths less than δ, and

$$P_k : \alpha_k = \gamma_{k,1} < \gamma_{k,2} < \cdots < \gamma_{k,m_k} = \beta_k$$

is any partition of $[\alpha_k, \beta_k]$. Then $\{(\gamma_{k,j}, \gamma_{k,j+1}): 1 \leq j \leq m_k, 1 \leq k \leq n\}$ is a finite system of disjoint open intervals contained in $[a, b]$ with sum of lengths less than δ. By absolute continuity of f, we have

$$\sum_{k=1}^{n}\sum_{j=1}^{m_k} |f(\gamma_{k,j+1}) - f(\gamma_{k,j})| < \varepsilon.$$

Since the partitions P_k are arbitrary, the above implies

$$\sum_{k=1}^{n} |v(\beta_k) - v(\alpha_k)| = \sum_{k=1}^{n} V([\alpha_k, \beta_k],f) \leq \varepsilon.$$

Thus v is absolutely continuous.

Since the sum and difference of absolutely continuous functions are absolutely continuous, it follows that $\frac{1}{2}(v(x) \pm f(x))$, which are increasing functions (see proof of Theorem 5.6.6) and the difference of which is f, are also absolutely continuous.

Alternatively, one can argue that since the sum and difference of absolutely continuous functions are absolutely continuous, it follows that $\mathcal{P}([a, x], f)$, $\mathcal{N}([a, x], f)$ are also absolutely continuous and so are the functions f_1 and f_2 (notations as in Theorem 5.6.11). □

Problem Set 5.7

5.7.P1. If $f:[a, b]\to\mathbb{R}$ is absolutely continuous on $[a, c]$ as well as on $[c, b]$, then it is absolutely continuous on $[a, b]$.

5.7.P2. Let $f:[a, b]\to\mathbb{R}$ be an absolutely continuous function and let $E \subset [a, b]$ be a set of measure zero. Then show that $m(f(E)) = 0$. Use this to show that the Cantor function is not absolutely continuous. Show that on a closed bounded interval, the class of absolutely continuous functions is a proper subclass of continuous functions of bounded variation. [This is trivial if the domain is unbounded, because $\sin x$ will do the trick.]

5.7.P3. Let $f:[a, b]\to\mathbb{R}$ be an absolutely continuous function. Then f maps measurable sets into measurable sets.

5.7.P4. Let f be defined and measurable on the closed interval $[a, b]$ and let E be any measurable subset on which f' exists. Then prove that

$$m^*(f(E)) \leq \int_E |f'|.$$

So far, we have proved (see Remark 5.7.2(b), Proposition 5.7.6 and Problem 5.7.P2) that an absolutely continuous function is continuous, of bounded variation and maps a set of measure zero into a set of measure zero. The following result shows that these three properties characterise an absolutely continuous function [Banach–Zarecki Theorem].

5.7.P5. If f is continuous and of bounded variation on $[a, b]$, and if f maps a set of measure zero into a set of measure zero, then f is absolutely continuous on $[a, b]$.

5.7.P6. Give an example to show that continuity on $[0, 1]$, even when combined with absolute continuity on $[\varepsilon, 1]$ for each positive $\varepsilon < 1$, does not imply absolute continuity on $[0, 1]$. What if the function is also of bounded variation on $[0, 1]$? Show that x^p is absolutely continuous on $[0, 1]$ when $0 \leq p < 1$.

5.7.P7. Show that the function f given on $[0, \infty)$ by $f(x) = x^{\frac{1}{2}}$ for $0 \leq x < 1$ and satisfying $f(x + k) = f(x) + k$ for $0 \leq x < 1$ and $k \in \mathbb{N}$ is absolutely continuous on any bounded subinterval of its domain but not on the entire domain. (Note that the example of such a function given in Remark 5.7.2(a) satisfies a Lipschitz condition on every bounded subinterval of $[0, \infty)$ while the present one does not.)

5.7.P8. Let f be an increasing function defined on $[a, b]$. Show that f can be decomposed into a sum of increasing functions $f = g + h$, where g is absolutely continuous and h is increasing with $h' = 0$ a.e.

5.8 Differentiation and the Integral

In this section, we clarify the relation between differentiation and integration in the Lebesgue sense. In part (ii) of the remark preceding Example 5.6.12, we have seen that $f(x) = \int_{[a,x]} g$, where g is a Lebesgue integrable function on $[a, b]$, is a function of bounded variation. It turns out that f indeed is absolutely continuous. It will be shown in Theorem 5.8.2 and Corollary 5.8.3 that $f(x) - f(a) = \int_{[ax]} f'$ if and only if f is absolutely continuous. It generalises the Second Fundamental Theorem of Calculus [see Theorem 1.7.2 (b)]. Let g be an integrable function defined on $[a, b]$. The function

$$f(x) = c + \int_{[a,x]} g,$$

where c is an arbitrary constant is called an indefinite integral of g. We shall investigate whether the equality $f' = g$ holds if f is an indefinite integral of g. It will be shown in Theorem 5.8.5 that the equality $f'(x) = g(x)$ holds for almost all x for an arbitrary integrable function g. This generalises the First Fundamental Theorem of Calculus [see Theorem 1.7.2 (a)].

We begin by proving that an indefinite integral is absolutely continuous.

Proposition 5.8.1 *Let g be a Lebesgue integrable function defined on $[a, b]$. Then the indefinite integral*

$$f(x) = c + \int_{[a,x]} g$$

is absolutely continuous.

Proof In view of Problem 3.2.P15, for $\varepsilon > 0$, there is a $\delta > 0$ such that for every measurable set E with $m(E) < \delta$, we have

$$\int_E |g| < \varepsilon.$$

Now, suppose that $\{(\alpha_k, \beta_k)\}_{1 \le k \le n}$ is a finite system of disjoint open intervals contained in $[a, b]$ and having sum of lengths less than δ. Then

$$\sum_{k=1}^{n} |f(\beta_k) - f(\alpha_k)| = \sum_{k=1}^{n} \left| \int_{[\alpha_k,\beta_k]} g \right| \le \sum_{k=1}^{n} \int_{[\alpha_k,\beta_k]} |g| = \int_{\cup_k [\alpha_k,\beta_k]} |g| < \varepsilon$$

because $m(\cup_{1 \le k \le n} (\alpha_k, \beta_k)) = \sum_{t=1}^{n} (\beta_k - \alpha_k) < \delta$. Therefore f is absolutely continuous. □

Since the function given by $f(x) = x^p$ $(0 < p < 1)$ on a bounded interval $[0, A]$ is the indefinite integral of the function $g(t) = pt^{p-1}$, the above proposition shows that it is absolutely continuous. An independent derivation of this fact has already been seen in Problem 5.7.P6.

If f is absolutely continuous on $[a, b]$, then by Theorem 5.5.1 and Proposition 5.7.9, f is differentiable a.e. and the derivative f' is integrable in the Lebesgue sense. We shall show that its integral agrees with $f(b) - f(a)$.

Theorem 5.8.2 *Suppose f:$[a, b]\to\mathbb{R}$ is absolutely continuous. Let E be the set of points $x \in (a, b)$ such that f is differentiable at x. Then*

$$\int_{[a,\, b]} f' = \int_E f' = f(b) - f(a).$$

Proof The function f' is summable over $[a, b]$ by Theorem 5.5.1 and Proposition 5.7.9.

In view of the observation above, $m((a, b)\backslash E) = 0$. Consequently,

$$\int_{[a,\, b]} f' = \int_E f'.$$

It remains to prove that

$$\int_E f' = f(b) - f(a). \tag{5.70}$$

By Proposition 5.7.9, $f = f_1 - f_2$, where f_1 and f_2 are increasing and absolutely continuous. In view of this and the linearity of the integral, it suffices to prove (5.70) under the additional assumption that f is increasing on $[a, b]$. In the rest of the proof we make this assumption.

Since f is continuous on $[a, b]$, its Lebesgue integral over any interval $[a, a + \frac{1}{n}]$, $n > \frac{1}{b-a}$, is the same as its Riemann integral.

Extend the definition of f by setting $f(x) = f(b)$ for $x > b$. Then it is continuous on $[a, \infty)$ and therefore its Lebesgue integral over any bounded interval $[a, K]$, where $K > a$, is the same as its Riemann integral. Define ϕ_n:$[a, b]\to\mathbb{R}$ by the formula

$$\phi_n(x) = n[f(x + \frac{1}{n}) - f(x)].$$

Observe that, since f is increasing on $[a, b]$, we have

$$\phi_n(x) \geq 0 \text{ everywhere and } f'(x) \geq 0 \text{ wherever this derivative exists.}$$

Clearly, ϕ_n is continuous and at each $x \in E$, $\phi_n(x) \to f'(x)$. Now,

$$\begin{aligned}\int_E \phi_n &= \int_{[a,\,b]} \phi_n = n\int_{[a,\,b]} f\left(t+\frac{1}{n}\right)dt - n\int_{[a,\,b]} f(t)dt\\ &= n\int_{[a+\frac{1}{n},b+\frac{1}{n}]} f(t)dt - n\int_{[a,\,b]} f(t)dt\\ &= f(b) - n\int_{[a,a+\frac{1}{n}]} f(t)dt,\end{aligned}$$

where the integral on the right agrees with the corresponding Riemann integral for sufficiently large n. Since f is continuous, the First Fundamental Theorem of Calculus [see Theorem 1.7.2 (a)] yields

$$\lim_{n\to\infty} \int_E \phi_n = f(b) - f(a).$$

Therefore (5.70) will follow if we prove that

$$\lim_{n\to\infty} \int_E \phi_n = \int_E f'.$$

Let $\varepsilon > 0$ be given. Since the function f is constant on $[b, b + 1]$ and is absolutely continuous on $[a, b]$, it is absolutely continuous on $[a, b + 1]$. Hence there exists a $\delta_1 > 0$ such that, for any finite system $\{(\alpha_j, \beta_j)\}_{1 \le j \le p}$ of disjoint open intervals contained in $[a, b + 1]$ and having sum of lengths less than δ, we have

$$\sum_{j=1}^{p} |f(\beta_j) - f(\alpha_j)| < \frac{\varepsilon}{3}. \tag{5.71}$$

The function f' is summable over $[a, b]$, and so by Problem 3.2.P15, there exists a $\delta_2 > 0$ (which we choose less than δ_1) such that, for every measurable set $F \subseteq E$ with $m(F) < \delta_2$, we have

$$\int_F |f'| < \frac{\varepsilon}{3}. \tag{5.72}$$

Now by Egorov's Theorem 2.6.2, there is a measurable subset $F \subseteq E$ with $m(F) < \delta_2$ such that

$$\phi_n \to f'$$

uniformly on $E \backslash F$. Without loss of generality, we may assume that $F \subseteq (a, b)$. With F fixed, there exists an N such that $n \ge N$ implies

$$\int_{E\backslash F} |\phi_n - f'| < \frac{\varepsilon}{3}. \tag{5.73}$$

It then follows that

$$\left|\int_E \phi_n - \int_E f'\right| = \left|\int_{E\backslash F} (\phi_n - f') + \int_F (\phi_n - f')\right|$$
$$\leq \int_{E\backslash F} |\phi_n - f'| + \int_F \phi_n + \int_F f',$$

where we do not need to take the absolute value of the last two terms in view of the observation recorded above after defining ϕ_n. Upon using (5.72) and (5.73), it further follows that $n \geq N$ implies

$$\left|\int_E \phi_n - \int_E f'\right| \leq \int_F \phi_n + \frac{2}{3}\varepsilon \quad \text{for} \quad n \geq N. \tag{5.74}$$

Since $m(F) < \delta_2 < \delta_1$, there exists an open set V such that $F \subseteq V$ and $m(V) < \delta_1$. [This is because, by Proposition 2.3.24, there exists an open set $V \supseteq F$ such that $m(V\backslash F) < \delta_1 - \delta_2$, which implies $m(V) = m(V\backslash F) + m(F) < \delta_1 - \delta_2 + \delta_2 = \delta_1$]. As $F \subseteq (a, b)$, we may assume that $V \subseteq (a, b)$ by replacing it with $V \cap (a, b)$ if necessary. We can express V as a countable union of disjoint open intervals $(a_k, b_k) \subseteq (a, b) \subseteq [a, b]$, $k = 1, 2, \ldots$. In what follows, the obvious simplification for the case of finitely many disjoint intervals is left to the reader.

For any x such that $0 \leq x \leq 1$ and for any positive integer m, the intervals $(a_k + x, b_k + x) \subseteq [a, b + 1]$, $k = 1, \ldots, m$, are disjoint and have sum of lengths equal to

$$m\left(\bigcup_{k=1}^{m} (a_k + x, b_k + x)\right) \leq m(V) < \delta_1$$

and hence by (5.71),

$$\sum_{k=1}^{m} [f(b_k + x) - f(a_k + x)] < \frac{\varepsilon}{3}. \tag{5.75}$$

Now,

$$\begin{aligned}\int_{[a_k,b_k]}\phi_n &= n\Big[\int_{[a_k,b_k]} f(t+\frac{1}{n})dt - \int_{[a_k,b_k]} f(t)dt\Big]\\ &= n\Big[\int_{[a_k+\frac{1}{n},b_k+\frac{1}{n}]} f(t)dt - \int_{[a_k,b_k]} f(t)dt\Big]\\ &= n\Big[\int_{[b_k,b_k+\frac{1}{n}]} f(t)dt - \int_{[a_k,a_k+\frac{1}{n}]} f(t)dt\Big]\\ &= n\Big[\int_{[0,\frac{1}{n}]} f(t+b_k)dt - \int_{[0,\frac{1}{n}]} f(t+a_k)dt\Big].\end{aligned}$$

Therefore

$$\sum_{k=1}^{m}\int_{[a_k,b_k]}\phi_n = n\int_{[0,\frac{1}{n}]}\left(\sum_{k=1}^{m}[f(t+b_k)-f(t+a_k)]\right)dt \le \frac{\varepsilon}{3},$$

using (5.75) above. Consequently,

$$\int_F \phi_n \le \int_V \phi_n = \sum_{k=1}^{\infty}\int_{[a_k,b_k]}\phi_n \le \frac{\varepsilon}{3}.$$

Therefore by (5.74),

$$\left|\int_E \phi_n - \int_E f'\right| < \varepsilon \text{ for } n \ge N.$$

□

Corollary 5.8.3 *A function f:$[a, b]\to\mathbb{R}$ is absolutely continuous if and only if it can be expressed in the form*

$$f(x) = c + \int_{[a,\,x]} g,$$

where c is a constant and g:$[a, b]\to\mathbb{R}$ is integrable.

Proof Define $g(x) = f'(x)$ at those points x in (a, b) where f is differentiable and assign arbitrary values at the other points of $[a, b]$. The above theorem completes the argument in one direction. By Proposition 5.8.1, $\int_{[a,\,x]} g$ is absolutely continuous. Since a constant function is always absolutely continuous, it follows that f as defined above is absolutely continuous. □

Corollary 5.8.4 *If f:$[a, b]\to\mathbb{R}$ is absolutely continuous and $f'(x) = 0$ almost everywhere, then f is a constant function.*

Proof In fact, by Theorem 5.8.2, we have $0 = \int_{[a,x]} f' = f(x) - f(a)$ for any $x \in [a, b]$, so that $f(x) = f(a)$ everywhere on $[a, b]$. □

Theorem 5.8.5 *If* $g{:}[a, b]\to\mathbb{R}^*$ *is integrable and if* $f{:}[a, b]\to\mathbb{R}$ *is defined by*

$$f(x) = \int_{[a,x]} g,$$

then $f'(x) = g(x)$ *a.e. in* $[a, b]$.

Proof The function f is absolutely continuous by Proposition 5.8.1. Set $\phi(x) = f'(x)$ if $x \in (a, b)$ and f is differentiable at x; let $\phi(x)$ be arbitrary for other values of x. Then

$$\int_{[a,x]} \phi = f(x) - f(a) = f(x)$$

by Theorem 5.8.2. Hence by the given definition of f, we get

$$0 = \int_{[a,x]} g - \int_{[a,x]} \phi = \int_{[a,x]} (g - \phi),$$

which implies $g = \phi$ a.e. [see Problem 3-2.P14(b)]. So, $g(x) = f'(x)$ for almost all $x \in [a, b]$. □

Remark 5.8.6 If g is integrable, the above theorem may be formulated by stating that, for almost every $x \in (a, b)$, the expressions

$$\tfrac{1}{h}\textstyle\int_{[0,h]} [g(x+t) - g(x)]dt$$
and
$$\tfrac{1}{h}\textstyle\int_{[-h,0]} [g(x+t) - g(x)]dt$$

both tend to 0 as $h \to 0$. In what follows, we shall prove a stronger result (Theorem 5.8.9), namely, for almost every $x \in (a, b)$, the expressions

$$\tfrac{1}{h}\textstyle\int_{[0,h]} |g(x+t) - g(x)|dt = \tfrac{1}{h}\int_{[x,x+h]} |g(t) - g(x)|dt$$
and
$$\tfrac{1}{h}\textstyle\int_{[-h,0]} |g(x+t) - g(x)|dt = \tfrac{1}{h}\int_{[x-h,x]} |g(t) - g(x)|dt$$

both tend to 0 as $h \to 0$. We begin with a definition.

Definition 5.8.7 Let $g \in L^1[a, b]$. If

$$\lim_{h\to 0}\tfrac{1}{h}\textstyle\int_{[x,x+h]} |g(t) - g(x)|dt = 0$$
and
$$\lim_{h\to 0}\tfrac{1}{h}\textstyle\int_{[x-h,x]} |g(t) - g(x)|dt = 0,$$

the point $x \in (a, b)$ is said to be a **Lebesgue point** of the function g.

Remark 5.8.8

(a) Every interior point of continuity of an integrable function g is a Lebesgue point of the function.
Indeed, let $x \in (a, b)$ be a point of continuity of g. Then for $\varepsilon > 0$, there exists a $\delta > 0$ such that

$$|g(t) - g(x)| < \varepsilon \quad \text{whenever} \quad |t - x| < \delta.$$

For $0 < h < \delta$, we have

$$\frac{1}{h}\int_{[x,x+h]} |g(t) - g(x)|dt < \varepsilon$$
$$\text{and}$$
$$\frac{1}{h}\int_{[x-h,x]} |g(t) - g(x)|dt < \varepsilon.$$

This completes the proof.

(b) The function $g \in L^1[a, b]$ need not be continuous anywhere and yet almost every point in the domain of g may be a Lebesgue point.
To see how, consider the function g defined on [0, 1] which is 1 at every rational point and 0 at every irrational point. It is discontinuous everywhere and Lebesgue integrable on [0, 1] with $\int_{[0,1]} g = 0$ and, for every irrational number x,

$$\frac{1}{h}\int_{[x,x+h]} |g(t) - g(x)|dt = \frac{1}{h}\int_{[x,x+h]} |g(t)|dt = 0,$$

since the set $\{t : g(t) \neq 0\}$ has measure 0.

(c) We have proved that almost every point of the domain of the function g considered above is a Lebesgue point. In fact, the following theorem holds:

Theorem 5.8.9 *If $g \in L^1[a, b]$, then almost every point of $[a, b]$ is a Lebesgue point of g.*

Proof The function $|g(t) - r|$, where r is a rational number, is integrable on $[a, b]$ and hence, for almost all $x \in [a, b]$,

$$\lim_{h \to 0} \frac{1}{h}\int_{[x,x+h]} |g(t) - r|dt = |g(x) - r| \tag{5.76}$$

(see Theorem 5.8.5). Let $E(r)$ be the set of points of $[a, b]$ at which (5.76) does not hold. Let

$$E = \bigcup_{r \in \mathbb{Q}} (E(r) \cup \{t : |g(t)| = \infty\}).$$

Then

$$m(E) \leq \sum_{r \in \mathbb{Q}} m(E(r)) + m(\{t : |g(t)| = \infty\}) = 0.$$

We shall show that each point $x_0 \in (a, b) \backslash E$ is a Lebesgue point of g. Let $\varepsilon > 0$ be arbitrary and r be a rational number such that

$$|g(x_0) - r| < \frac{\varepsilon}{3}. \tag{5.77}$$

Then we have

$$||g(t) - r| - |g(t) - g(x_0)|| < \frac{\varepsilon}{3}$$

for all $t \in [a, b]$. It follows for $h > 0$ that

$$\left| \frac{1}{h} \int_{[x_0, x_0+h]} |g(t) - r| dt - \frac{1}{h} \int_{[x_0, x_0+h]} |g(t) - g(x_0)| dt \right| \leq \frac{1}{h} \int_{[x_0, x_0+h]} \frac{\varepsilon}{3} dt = \frac{\varepsilon}{3}.$$

Since $x_0 \notin E$, the equality (5.76) yields a $\delta > 0$ such that $0 < h < \delta$ implies

$$\left| \frac{1}{h} \int_{[x_0, x_0+h]} |g(t) - r| dt - |g(x_0) - r| \right| < \frac{\varepsilon}{3}$$

and hence by (5.77), further implies

$$\frac{1}{h} \int_{[x_0, x_0+h]} |g(t) - r| dt < \frac{2\varepsilon}{3}. \tag{5.78}$$

It is a consequence of (5.77) and (5.78) that for any $x_0 \notin E$, there exists a $\delta > 0$ such that $0 < h < \delta$ implies

$$\frac{1}{h} \int_{[x_0, x_0+h]} |g(t) - g(x_0)| dt \leq \frac{1}{h} \int_{[x_0, x_0+h]} |g(t) - r| dt + \frac{1}{h} \int_{[x_0, x_0+h]} |g(x_0) - r| dt < \varepsilon.$$

In other words, $\lim_{h \to 0} \frac{1}{h} \int_{[x_0, x_0+h]} |g(t) - g(x_0)| dt = 0$. Similarly, it can be shown that

$$\lim_{h \to 0} \frac{1}{h} \int_{[x_0 - h, x_0]} |g(t) - g(x_0)| dt = 0.$$

Thus x_0 is a Lebesgue point of g. □

Corollary 5.8.10 *Let $g \in L^1[a, b]$. Then we have*

$$\lim_{h\to 0}\frac{1}{h}\int_{[0,h]}|g(x+t)+g(x-t)-2g(x)|dt = 0$$

at every Lebesgue point x of g and hence for almost all $x \in (a, b)$.

Proof For fixed $x \in (a, b)$, write

$$\begin{aligned}&\frac{1}{h}\int_{[0,h]}|g(x+t)+g(x-t)-2g(x)|dt\\ &\le \frac{1}{h}\int_{[0,h]}|g(x+t)-g(x)|dt+\frac{1}{h}\int_{[0,h]}|g(x-t)-g(x)|dt\\ &=\frac{1}{h}\int_{[x,x+h]}|g(t)-g(x)|dt+\frac{1}{h}\int_{[x-h,x]}|g(t)-g(x)|dt.\end{aligned}$$

Since

$$\lim_{h\to 0}\frac{1}{h}\int_{[x,x+h]}|g(t)-g(x)|dt = 0 = \lim_{h\to 0}\frac{1}{h}\int_{[x-h,x]}|g(t)-g(x)|dt$$

at each Lebesgue point of g, the "almost all x" part of the result now follows by Theorem 5.8.9 above. □

Proposition 5.8.11 *Let x be a Lebesgue point of an integrable function g defined on $[a, b]$. The indefinite integral $f(x) = \int_{[a,x]} g$ is differentiable at the point x and $f'(x) = g(x)$.*

Proof It is easy to check that

$$\frac{f(x+h)-f(x)}{h} - g(x) = \frac{1}{h}\int_{[x,x+h]}[g(t)-g(x)]dt;$$

so,

$$\left|\frac{f(x+h)-f(x)}{h} - g(x)\right| \le \frac{1}{h}\int_{[x,x+h]}|g(t)-g(x)|dt \to 0 \text{ as } h\to 0+,$$

since x is a Lebesgue point of g. Thus $f_+'(x) = g(x)$. Similarly it can be shown that $f_-'(x) = g(x)$. □

Theorem 5.8.12 (Integration by Parts). *Let f and g be functions in $L^1[a, b]$ and*

$$F(x) = \int_{[a,\, x]} f + \alpha,$$

$$G(x) = \int_{[a,x]} g + \beta,$$

where α *and* β *are constants. Then*

$$\int_{[a,\, b]} Fg + \int_{[a,\, b]} Gf = G(b)F(b) - F(a)G(a).$$

Proof By Proposition 5.8.1, F and G are absolutely continuous functions on $[a, b]$ and hence so is FG by Remark 5.7.2(f), and

$$(FG)' = FG' + F'G \text{ a.e. on } [a,\ b]$$

as elementary calculations show. By Theorem 5.8.5, we have $F' = f$ and $G' = g$ a.e. Therefore the required equality follows upon applying Theorem 5.8.2. □

We turn our attention to change of variables in a Lebesgue integral. The discussionf below is based on Serrin and Varberg [24].

Theorem 5.8.13 *Let the function* $u{:}[a, b]\to\mathbb{R}$ *be differentiable at each point of a subset* $E \subseteq [a, b]$ *and* $u(E)$ *have Lebesgue measure* 0. *Then* $\{x \in E\colon u'(x) \neq 0\}$ *also has measure* 0.

Proof Denote the set $\{x \in E : u'(x) \neq 0\}$ by E_0. We have to show that E_0 has measure 0. For each $n \in \mathbb{N}$, let

$$E_n\{x \in E_0 : |u(x) - u(y)| \geq \frac{1}{n}|x - y| \text{ for all } y \in [a,\ b] \text{ such that } 0 < |x - y| < \frac{1}{n}\}. \tag{5.79}$$

We shall prove that

$$E_0 \subseteq \bigcup_{n\geq 1} E_n. \tag{5.80}$$

So, consider any $x \in E_0$. By definition of E_0, we have $|u'(x)| > 0$. By definition of derivative, there exists a $\delta > 0$ for which

$$y \in [a,\ b],\quad 0 < |x - y| < \delta \quad \Rightarrow \quad |u(x) - u(y)| > \frac{1}{2}|u'(x)||x - y|.$$

Now choose $n \in \mathbb{N}$ such that $\frac{1}{n} < \min\left\{\delta, \frac{1}{2}|u'(x)|\right\}$. For any such $n \in \mathbb{N}$, we obviously have $x \in E_n$. This completes the proof that (5.80) must hold.

The desired conclusion that $m(E_0) = 0$ will follow from (5.80) if for each $n \in \mathbb{N}$ we demonstrate that $m(E_n) = 0$.

For this purpose, fix some $n \in \mathbb{N}$. In order to demonstrate that $m(E_n) = 0$, we need only argue that, for any interval I of length less than $\frac{1}{n}$ the intersection

$$G_n = I \cap E_n \tag{5.81}$$

has measure 0. So, let $\varepsilon > 0$ be given. Since $G_n \subseteq E_n \subseteq E_0 \subseteq E$, we have $u(G_n) \subseteq u(E)$, which has measure 0 by hypothesis. Therefore $u(G_n)$ has measure 0. By definition of measure being 0, there must exist a sequence $\{I_k\}_{k \geq 1}$ of intervals such that $u(G_n) \subseteq \cup_{k \geq 1} I_k$ and $\sum_{k=1}^{\infty} \ell(I_k) < \varepsilon$. By breaking up each I_k further if necessary, we may assume that $\ell(I_k) < \frac{1}{n}$ for each k. Define

$$A_{n,k} = G_n \cap u^{-1}(I_k). \tag{5.82}$$

Considering that $u(G_n) \subseteq \cup_{k \geq 1} I_k$, we must have

$$G_n \subseteq \bigcup_{k \geq 1} A_{n,\,k}. \tag{5.83}$$

With a view to estimating the outer measure $m^*(A_{n,k})$, take any two distinct $x, y \in A_{n,k}$. By (5.81) and (5.82), we have first, $x, y \in I$ and second, $x, y \in E_n$. The first implies $|x - y| < \frac{1}{n}$ because I has length less than $\frac{1}{n}$. As a consequence of this inequality, the second implies $|u(x) - u(y)| \geq \frac{1}{n} |x - y|$ by virtue of (5.79). Since this has been proved for any two distinct $x, y \in A_{n,k}$, we arrive at

$$\begin{aligned} \sup\{|x - y| : x, y \in A_{n,k}\} &\leq n \cdot \sup\{|u(x) - u(y)| : x, y \in A_{n,k}\} \\ &\leq n \cdot \sup\{|\alpha - \beta| : \alpha, \beta \in I_k\} \quad \text{by (5.82)} \\ &\leq n \cdot \ell(I_k). \end{aligned}$$

By Problem 2.2.P13, the left-hand side here is greater than or equal to the outer measure $m^*(A_{n,k})$. Therefore we arrive at the estimate $m^*(A_{n,k}) \leq n \cdot \ell(I_k)$. It now follows upon using (5.83) that $m^*(G_n) \leq \sum_{k=1}^{\infty} m^*(A_{n,k}) \leq n \cdot \varepsilon$. Hence $m^*(G_n) = 0 = m(G_n)$. This completes the argument that for any interval I of length less than $\frac{1}{n}$, the intersection $I \cap E_n$ has measure 0. As already noted, this is all that we needed to establish. □

Theorem 5.8.14 (Chain Rule a.e.) *Let $F:[c, d] \to \mathbb{R}$ and $f:[c, d] \to \mathbb{R}$ be functions such that F maps sets of measure 0 into sets of measure 0 and $F' = f$ a.e. Suppose also that $u:[a, b] \to [c, d]$ is differentiable a.e. Then the equality*

$$(F \circ u)' = (f \circ u)u' \tag{5.84}$$

holds a.e. on $[a, b]$.

Proof According to the well-known Chain Rule (Proposition 1.6.1), if $t \in [a, b]$ has the property that u is differentiable at t and also F is differentiable at $u(t)$, then the composition $F \circ u$ is differentiable at t, and $(F \circ u)'(t) = (F' \circ u)(t)u'(t)$.

Let

$$U = \{t \in [a, b] : u'(t) \neq 0\}, \quad U_0 = \{t \in [a, b] : u'(t) = 0\}.$$

Since u is differentiable a.e., the complement $(U \cup U_0)^c$ has measure 0.

Since $F' = f$ a.e., there exists a set $Z \subseteq [c, d]$ of measure 0 such that $x \notin Z$ implies $F'(x) = f(x)$. In view of the hypothesis that F maps sets of measure 0 into sets of measure 0, the set $F(Z)$ has measure 0. Now, $U = A \cup B$, where

$$A = U \cap \{t \in [a, b] : u(t) \notin Z\},$$
$$B = U \cap \{t \in [a, b] : u(t) \in Z\}.$$

Also, $U_0 = A_0 \cup B_0 \cup C_0$, where

$$A_0 = U_0 \cap \{t \in [a, b] : u(t) \notin Z\},$$
$$B_0 = U_0 \cap \{t \in [a, b] : u(t) \notin Z\} \cap \{t \in [a, b] : (F \circ u)'(t) \neq 0\},$$
$$C_0 = U_0 \cap \{t \in [a, b] : u(t) \notin Z\} \cap \{t \in [a, b] : (F \circ u)'(t) = 0\}.$$

On the sets A as well as A_0, the equality (5.84) is seen to hold everywhere by virtue of the Chain Rule (Proposition 1.6.1). It holds on C_0 as well, because both sides of the equality are 0. Therefore the subset of $U \cup U_0$ on which it fails to hold is contained in $B \cup B_0$.

We shall demonstrate that both B and B_0 have measure 0. The definition of B implies that $B \subseteq U$ and $u(B) \subseteq Z$. Together with the fact that Z has measure 0, this immediately implies by Theorem 5.8.13 that B has measure 0. As regards B_0, its definition implies that $(F \circ u)(B_0) \subseteq F(Z)$, which has measure 0; since $(F \circ u)' \neq 0$ on B_0, Theorem 5.8.13 implies that B_0 has measure 0.

Thus the equality (5.84) can fail to hold only on a subset of the union of the sets

$$(U \cup U_0)^c, \; B \quad \text{and} \quad B_0,$$

each of which has measure 0. □

For the purpose of the Change of Variables Formula for Lebesgue integrals in one dimension, it will be convenient to introduce Lebesgue integrals from right to left, exactly as for Riemann integrals:

$$\int_b^a f\,dm = -\int_{[a,\,b]} f\,dm = -\int_a^b f\,dm \text{ when } a < b.$$

This enables us to restate Theorem 5.8.2 as

Theorem 5.8.15 *Suppose $f:[a, b] \to \mathbb{R}$ is absolutely continuous. Let E be the set of points $x \in (a, b)$ such that f is differentiable at x. Then*

$$\int_a^b f' dm = \int_E f' dm = f(b) - f(a)$$

and

$$\int_b^a f' dm = -\int_E f' dm = f(a) - f(b).$$

Theorem 5.8.16 (Change of Variables Formula 1). *Let f be an integrable function on $[c, d]$ and u be an absolutely continuous function on $[a, b]$ such that $u[a, b] \subseteq [c, d]$. Suppose $F \circ u$ is absolutely continuous on $[a, b]$, where*

$$F(x) = \int_c^x f\, dm.$$

Then $(f \circ u)u'$ is integrable on $[a, b]$ and, for any $\alpha, \beta \in [a, b]$,

$$\int_{u(\alpha)}^{u(\beta)} f\, dm = \int_\alpha^\beta (f \circ u)u' dm.$$

Proof By Proposition 5.8.1, the function F is absolutely continuous, and hence by Problem 5.7.P2, maps sets of measure 0 into sets of measure 0. Moreover, $F' = f$ a.e. on $[c, d]$ by Theorem 5.8.5. On the basis of the Chain Rule a.e. (Theorem 5.8.14), we therefore obtain

$$(F \circ u)' = (f \circ u)u' \text{ a.e. on } [a, b].$$

Since $F \circ u$ is given to be absolutely continuous, we now deduce from Corollary 5.7.7 that $(f \circ u)u'$ is integrable.

The absolute continuity of $F \circ u$ leads via Theorem 5.8.15 to

$$F(u(\beta)) - F(u(\alpha)) = \int_\alpha^\beta (F \circ u)' dm.$$

By definition of F, we have

$$F(u(\beta)) - F(u(\alpha)) = \int_{u(\alpha)}^{u(\beta)} f\, dm.$$

The above three equalities together yield the required conclusion. □

Theorem 5.8.17 (Change of Variables Formula 2) *Let f be an integrable function on* $[c, d]$ *and u be an absolutely continuous function on* $[a, b]$ *such that* $u[a, b] \subseteq [c, d]$. *Suppose* $(f \circ u)u'$ *is integrable on* $[a, b]$. *Then for any* $\alpha, \beta \in [a, b]$,

$$\int_{u(\alpha)}^{u(\beta)} f\,dm = \int_{\alpha}^{\beta} (f \circ u)u' dm,$$

and $F \circ u$ *is absolutely continuous on* $[a, b]$, *where*

$$F(x) = \int_{c}^{x} f\,dm.$$

Proof If f is bounded, then F satisfies a Lipschitz condition and $F \circ u$ is absolutely continuous in view of Proposition 5.7.4. It follows from Theorem 5.8.16 that the desired equality of integrals holds in the case of bounded f.

Now suppose f is not bounded. Consider the sequence $\{f_n\}_{n \geq 1}$ of functions on $[c, d]$ given by $f_n(x) = f(x)$ or $-n$ or n according as $|f(x)| < n$ or $f(x) \leq -n$ or $f(x) \geq n$. Then each f_n is bounded and measurable. By what has been observed in the paragraph above, we have

$$\int_{u(\alpha)}^{u(\beta)} f_n\,dm = \int_{\alpha}^{\beta} (f_n \circ u)u' dm \quad \text{for each } n.$$

Also, $f_n \to f$ everywhere and $(f_n \circ u)u' \to (f \circ u)u'$ a.e. Besides,

$$|f_n| \leq |f|, |(f_n \circ u)u'| \leq |(f \circ u)u'| \quad \text{and both } f \text{ and } (f \circ u)u' \text{ are integrable.}$$

Therefore, upon taking limits in the above equality of integrals and appealing to Problem 3-2.P20(a), we arrive at the required equality.

From the definition of F and the equality that has just been proved, it follows for any $t \in [a, b]$ that

$$(F \circ u)(t) - (F \circ u)(\alpha) = \int_{u(\alpha)}^{u(t)} f\,dm = \int_{\alpha}^{t} (f \circ u)u' dm.$$

Since $(f \circ u)u'$ has been assumed integrable, Proposition 5.8.1 shows that the function of t defined by the left-hand side must be absolutely continuous, and hence so is $F \circ u$. □

Corollary 5.8.18 *Let f be an integrable function on* $[c, d]$ *and u be an absolutely continuous function on* $[a, b]$ *such that* $u[a, b] \subseteq [c, d]$. *Then the function* $F \circ u$, *where*

$$F(x) = \int_{c}^{x} f\,dm,$$

is absolutely continuous if and only if $(f \circ u)u'$ *is integrable. When this is so,*

$$\int_{u(\alpha)}^{u(\beta)} f\,dm = \int_{\alpha}^{\beta} (f \circ u)u'dm$$

for any $\alpha, \beta \in [a, b]$.

Proof This is trivial from the previous two theorems. □

Remark 5.8.19 The following example, based on E. J. McShane [25, p. 214], shows that the hypothesis in Theorem 5.8.16 that $F \circ u$ be absolutely continuous is not redundant. Let $[a, b] = [c, d] = [0, 1]$ and

$$f(x) = x^{-\frac{2}{3}}, f(0) = 0, \quad \text{and} \quad u(t) = t^3 \cos^6\left(\frac{\pi}{t}\right), u(0) = 0.$$

Then $F(x) = 3x^{1/3}$ and $(F \circ u)(t) = 3t\cos^2(\frac{\pi}{t}) = \frac{3}{2}t(1 + \cos(\frac{2\pi}{t}))$. The function u is absolutely continuous because it has a bounded derivative; the function $F \circ u$ can be seen to be *not* absolutely continuous by modifying the argument of Example 5.6.3(c).

Moreover, it follows from Theorem 5.8.17 (or from Corollary 5.8.18) that $(f \circ u)u'$ is not integrable on $[u(0), u(1)] = [0, 1]$, although f is integrable on $[0, 1]$.

Corollary 5.8.20 *Let* $\gamma \neq 0$ *and let* f *be an integrable function on* $[a, b]$. *Then the function given by* $\Phi(x) = f(x/\gamma)$ *on* $[\gamma a, \gamma b]$ *or* $[\gamma b, \gamma a]$ *(according as* $\gamma > 0$ *or* < 0*) is integrable and*

$$\gamma \int_a^b f(x)dx = \int_{\gamma a}^{\gamma b} \Phi(t)dt = \int_{\gamma a}^{\gamma b} f(t/\gamma)dt.$$

Proof Consider the absolutely continuous function $u:[\gamma a, \gamma b] \to [a, b]$ or $u:[\gamma b, \gamma a] \to [a, b]$, as the case may be, for which $u(x) = x/\gamma$. The function F such that $F(x) = \int_a^x f\,dm$ is absolutely continuous by Corollary 5.8.3 and hence one can deduce directly from Definition 5.7.1 that $F \circ u$ is also absolutely continuous. It follows from Theorem 5.8.16 that $(f \circ u)u'$ is integrable on $[\gamma a, \gamma b]$ or $[\gamma b, \gamma a]$, as the case may be, and

$$\int_a^b f\,dm = \int_{\gamma a}^{\gamma b} (f \circ u)u'dm.$$

Since $f \circ u = \Phi$ and $u' = 1/\gamma$ everywhere, this is precisely what was to be proved. □

Remark 5.8.21 When $\gamma > 0$ and $[a, b] = [-l, l]$, where $l > 0$, the equality asserted by the above corollary takes the form

$$\gamma \int_{-l}^{l} f(x)dx = \int_{-\gamma l}^{\gamma l} \Phi(t)dt = \int_{-\gamma l}^{\gamma l} f(t/\gamma)dt.$$

When $\gamma = -1$ and $[a, b] = [-l, 0]$, where $l > 0$, it takes the form

$$\int_{-l}^{0} f(x)dx = \int_{0}^{l} \Phi(x)dx = \int_{0}^{l} f(-x)dx.$$

Problem Set 5.8

5.8.P1. Suppose f is absolutely continuous on $[a, b]$ and $f'(x) = 0$ a.e. on $[a, b]$. Then f is a constant function. Give a proof of this that is different from the one in Corollary 5.8.4.

5.8.P2. If f and g are absolutely continuous on $[a, b]$, $f'(x) = g'(x)$ a.e. on $[a, b]$ and $f(x_0) = g(x_0)$ for some $x_0 \in [a, b]$, then show that $f(x) = g(x)$ for every $x \in [a, b]$.

5.8.P3. Let $f:[a, b] \to \mathbb{R}$ be absolutely continuous and let

$$v(a) = 0,\ v(x) = V([a, x], f) \text{ for } x \in [a, b].$$

Then v is absolutely continuous on $[a, b]$ by Proposition 5.7.9. Show that $v'(x) = |f'(x)|$ for almost all $x \in [a, b]$.

5.8.P4. Show that, if g is an integrable function defined on $[a, b]$ with indefinite integral f, then $f'(x) = g(x)$ whenever x is a point of continuity of g.

5.8.P5. If $f(x) = \int_{[0,\, x]} g$, where g is an integrable function defined on $[0, 1]$, then $f' = g$ need not hold even when f' exists.

5.8.P6. (Cf. Theorem 5.5.1) Show that the inequality $\int_{[a,\, x]} f' \le f(b) - f(a)$ need not hold when f is not monotone increasing.

5.8.P7. (Lebesgue Decomposition Theorem for Functions) Suppose f is of bounded variation on $[a, b]$. Then there exists an absolutely continuous function g and a function h such that $f(x) = g(x) + h(x)$, $x \in [a, b]$, where $h'(x) = 0$ a.e. Up to constants, the decomposition is unique. [For the relation between this and Theorem 5.10.13, see Problems 5.11.P5 and 5.11.P8.]

5.8.P8. [Application to Fourier Series; due to Lebesgue] Let f be a function in $L^1([-\pi, \pi])$. Then show that

$$\lim_n \sigma_n(x) = f(x)$$

at every Lebesgue point x of f (and hence a.e. on $[-\pi, \pi]$ by Theorem 5.8.9). Here the notation is as in Fejér's Theorem 4.3.4.

Definition Let $f(x) = g(x) + ih(x)$ be a complex function on the closed bounded interval $[a, b]$, where $g(x)$ and $h(x)$ are real. For each partition P: $a = x_0 < x_1 < \cdots < x_n = b$ of $[a, b]$, we set

$$T(P,f) = \sum_{k=1}^{n} \|f(x_k) - f(x_{k-1})\|,$$

where $\|f(x) - f(y)\|^2 = (g(x) - g(y))^2 + (h(x) - h(y))^2$. The number

$$V([a, b], f) = \sup_P T(P, f)$$

is called the **total variation** of f over $[a, b]$.

Additivity over adjacent intervals as in Theorem 5.6.4 can be established by making obvious modifications in the proof of that theorem. It will be used in Problems 5.8.P9 and 5.8.P10.

5.8.P9. Show that the complex function $f = g + ih$ satisfies

$$V([a, b], f) \leq V([a, b], g) + V([a, b], h),$$

is of finite variation if and only if g and h are so, and is absolutely continuous if and only if g and h are so.

Definition Let g and h be continuous real-valued functions on the closed bounded interval $[a, b]$. The set of points $(g(x), h(x))$, $a \leq x \leq b$, in $\mathbb{R}^2$ is called a **continuous curve** and the total variation $V([a, b], f)$ of $f(x) = g(x) + ih(x)$ is called the **length of the curve**. Similarly, if

$$v(x) = V([a, x], f), \quad a \leq x \leq b,$$

then $v(x)$ is called the **length of the arc** of the curve **between the points** $(g(a), h(a))$ **and** $(g(x), h(x))$. If the curve has finite length, it is called a **rectifiable** curve.

5.8.P10. Show that the curve given by g and h is rectifiable if and only if the functions are of bounded variation on $[a, b]$. Show that in this case,

$$v(x) \geq \int_{[a,\, x]} \left[(g')^2 + (h')^2\right]^{\frac{1}{2}} \quad \text{for all } x$$

and equality holds for all x if and only if g and h are absolutely continuous.

In particular, if g is a function of bounded variation and $v(x)$ is its total variation over $[a, x]$, then $v(x) \geq \int_{[a,x]} |g'|$ for all x. Equality holds for all x if and only if g is absolutely continuous.

5.8.P11. Let g be a real-valued function on [0, 1] and $\gamma(t) = t + ig(t)$. The length of the graph of g is, by definition, the total variation of γ on [0, 1]. Show that the length is finite if and only if g is of bounded variation.

5.8.P12. Assume that g is continuous and increasing on [0, 1], $g(0) = 0$, $g(1) = \alpha$. By L_g we denote the length of the graph of g. Show that $L_g = 1 + \alpha$ if and only if $g'(x) = 0$ a.e. In particular, the length of the graph of the Cantor function constructed in Example 5.5.2 is 2.

5.8.P13. Let $f \in L^1[-\pi, \pi]$ and $\int_{-\pi}^{\pi} f(x)\,dx = 0$, $\int_{-\pi}^{\pi} f(x)\cos nx\,dx = 0$, $\int_{-\pi}^{\pi} f(x)\sin nx\,dx = 0$, $n = 1, 2, \ldots$. Then $f(x) = 0$ a.e. on $[-\pi, \pi]$.

5.9 Signed Measures

In this section, we consider a simple but rather useful generalisation of the concept of measure. It arises naturally if the measure is allowed to assume both positive and negative values.

Note that if μ_1 and μ_2 are measures defined on the same measurable space $(X, \mathcal{F})$, then $\alpha_1\mu_1 + \alpha_2\mu_2$, where α_1 and α_2 are nonnegative constants, is again a measure on $(X, \mathcal{F})$. However, the situation is different if we try to define a set function μ by

$$\mu(E) = \mu_1(E) - \mu_2(E) \text{ for } E \in \mathcal{F}.$$

It is obvious that μ is not always nonnegative. This is an interesting phenomenon which leads to the consideration of signed measures studied below. The difficulty, namely, that μ may not be defined ($\mu_1(E) = \mu_2(E) = \infty$ for some $E \in \mathcal{F}$) is overcome by assuming that either μ_1 or μ_2 is a finite measure.

In what follows, we shall study signed measures and their properties. The Hahn and Jordan Decomposition Theorems are discussed as well. The second of these says that a signed measure can only arise as a difference in the manner described above.

We begin with the following.

Definition 5.9.1 Let $(X, \mathcal{F})$ be a measurable space. By a **signed measure** we mean an extended real-valued set function μ defined on the class $\mathcal{F}$ of all measurable sets ($\mu: \mathcal{F} \to \mathbb{R}^*$) of the measurable space $(X, \mathcal{F})$ which satisfies the following:

(a) μ assumes at most one of the values ∞ and $-\infty$,
(b) $\mu(\emptyset) = 0$,
(c) μ is countably additive (as in Definition 2.2.7), that is, for every disjoint sequence of sets $\{A_j\}_{j \geq 1}$ such that each $A_j \in \mathcal{F}$, we have

$$\mu\left(\bigcup_{j=1}^{\infty} A_j\right) = \sum_{j=1}^{\infty} \mu(A_j).$$

A signed measure μ on $(X, \mathcal{F})$ is said to be **finite** if $|\mu(X)| < \infty$ and it is said to be **σ-finite** if there exists a sequence of sets $\{A_j\}_{j\geq 1}$ such that each $A_j \in \mathcal{F}$, $X = \cup_{j\geq 1} A_j$ and $|\mu(A_j)| < \infty$ for every j.

A signed measure which only takes nonnegative values is a measure in the sense of Chap. 3 and will sometimes be referred to in the context of signed measures as a *nonnegative measure.*

In the rest of this section, μ will be understood to be a signed measure on $(X, \mathcal{F})$ unless the context demands otherwise.

Remarks 5.9.2

(a) If the series in part (c) of Definition 5.9.1 were to be conditionally convergent, then it would be possible to rearrange it to converge to an arbitrary desired sum; however, the assertion of (c) implies that rearranging it does not alter the sum. Thus, it follows from (c) that the series cannot be conditionally convergent, which is to say, it must be either absolutely convergent or properly divergent to $\pm\infty$.

(b) A nonnegative measure is a special case of a signed measure but a signed measure is in general not a nonnegative measure.

(c) It follows from the definition that a signed measure, like any nonnegative measure, is finitely additive.
Let $E = \cup_{1\leq j\leq n} E_j$, $E_j \in \mathcal{F}$ for every j and $E_j \cap E_i = \varnothing$ whenever $i \neq j$. Then $E = \cup_{j\geq 1} E_j$ with $E_j = \varnothing$ for $i > n$ and $\mu(E) = \sum_{j=1}^{\infty} \mu(E_j) = \sum_{j=1}^{n} \mu(E_j)$.

(d) If $|\mu(E)| < \infty$ and $F \subseteq E$, then $|\mu(F)| < \infty$, $|\mu(E\backslash F)| < \infty$ and $\mu(E \backslash F) = \mu(E) - \mu(F)$. Indeed, the set-theoretic identities $E = (E\backslash F) \cup F$ and $(E \backslash F) \cap F = \varnothing$, in conjunction with finite additivity, lead to $\mu(E) = \mu(F) + \mu(E \backslash F)$. Therefore, when $|\mu(E)| < \infty$, we must have $|\mu(F)| < \infty$ and $|\mu(E\backslash F)| < \infty$. Moreover, it follows that $\mu(E\backslash F) = \mu(E) - \mu(F)$.

(e) μ is finite if and only if $|\mu(E)| < \infty$ for every $E \in \mathcal{F}$. Since $X \in \mathcal{F}$, it follows from this condition that $|\mu(X)| < \infty$. Conversely, if $|\mu(X)| < \infty$, it follows from (c) that $|\mu(E)| < \infty$ for every $E \in \mathcal{F}$.

(f) If $\{E_j\}_{j\geq 1}$ is a sequence of sets in $\mathcal{F}$ satisfying $E_1 \subseteq E_2 \subseteq \ldots$, then $\mu(\cup_{j\geq 1} E_j) = \lim_{n\to\infty} \mu(E_n)$. This follows by the same argument as Proposition 2.3.21 (a). The reader is reminded that $\lim E_n = \cup_{j\geq 1} E_j$ here.

(g) If $\{E_j\}_{j\geq 1}$ is a sequence of sets in $\mathcal{F}$ satisfying $E_1 \supseteq E_2 \supseteq \ldots$, and $|\mu(E_1)| < \infty$, then $\mu\left(\bigcap_{j\geq 1} E_j\right) = \lim_{n\to\infty} \mu(E_n)$. This follows by the same argument as Proposition 2.3.21 (b). The reader is reminded that $\lim E_n = \cap_{j\geq 1} E_j$ here.

(h) If $\{E_j\}_{j\geq 1}$ is a sequence of disjoint sets in $\mathcal{F}$ such that $|\mu(\cup_{j\geq 1}E_j)| < \infty$, then the series $\sum_{i=1}^{\infty}\mu(E_j)$ is absolutely convergent. This follows from (a) above.

(i) Suppose μ is finite and $\{E_i\}_{i\in I}$ is a family of disjoint sets in $\mathcal{F}$. Then $\{i \in I$: $\mu(E_i) \neq 0\}$ is at most countable, as we show below.
For each positive integer m, let

$$I_m = \{i \in I : \mu(E_i) > \frac{1}{m}\} \quad \text{and} \quad I'_m = \{i \in I : \mu(E_i) < -\frac{1}{m}\}.$$

We assert that I_m is finite. If not, then choose an infinite sequence $i_1, i_2,\ldots$ of distinct elements of I_m. Since E_{i_j} are disjoint, $\sum_{j=1}^{\infty}\mu(E_{i_j})$ converges. But this is a contradiction because $\mu(E_{i_j}) > \frac{1}{m}$ for all j. By a similar argument, $I_m{}'$ is also finite. It follows that $\cup_{m\geq 1}I_m$ and $\cup_{m\geq 1}I_m{}'$ are at most countable and hence so is their union. However, their union is the set we wanted to prove to be at most countable. In particular, when $\mu(E_i) \neq 0$ for every $i \in I$, we can conclude that I is at most countable.

(j) If $\mu(E) = \infty$ [resp. $-\infty$] for some set $E \in \mathcal{F}$, then $\mu(X) = \infty$ [resp. $-\infty$]. This follows from finite additivity: $\mu(X) = \mu(E) + \mu(X\backslash E)$.

(k) When μ_1 and μ_2 are signed measures, the sum $\mu_1(E) + \mu_2(E)$ fails to be defined if and only if $\mu_1(E) = -\mu_2(E) = \pm\infty$. By (j) above, this happens if and only if $\mu_1(X) = -\mu_2(X) = \pm\infty$. When this situation does not occur, we have a well-defined set function $\mu_1 + \mu_2$ on $\mathcal{F}$. We claim that $\mu_1 + \mu_2$ will be a signed measure. That $(\mu_1 + \mu_2)(\varnothing) = 0$ is trivial. If μ_1 is finite, then $\mu_1 + \mu_2$ fails to assume whichever of the values $\pm\infty$ that μ_2 fails to assume. Suppose μ_1 is not finite. If it fails to assume the value ∞, then it must take the value $-\infty$ and therefore $\mu_1(X) = -\infty$ by (j) above. It follows that $\mu_2(X) \neq \infty$ and hence, by (j) again, μ_2 also fails to assume the value ∞. It follows that $\mu_1 + \mu_2$ fails to assume the value ∞. An analogous argument shows that if μ_1 fails to assume the value $-\infty$, then $\mu_1 + \mu_2$ does the same. Thus, whether μ_1 is finite or not, $\mu_1 + \mu_2$ fails to assume one among the values $\pm\infty$. Lastly, countable additivity of $\mu_1 + \mu_2$ follows from the fact that two series can be added term by term except when one diverges to ∞ and the other one to $-\infty$.

The condition that $\mu_1 + \mu_2$ is well defined will be described by saying that $\mu_1 + \mu_2$ is a signed measure.

Examples 5.9.3

(a) Let μ_1 and μ_2 be signed measures on $\mathcal{F}$ such that at least one of them is finite. Then the set function μ defined by

$$\mu_0(E) = \mu_1(E) - \mu_2(E)$$

is a signed measure on $\mathcal{F}$.

Clearly, $\mu_0(\emptyset) = \mu_1(\emptyset) - \mu_2(\emptyset) = 0$. Let $E = \cup_{j \geq 1} E_j$, where the E_j are disjoint elements of $\mathcal{F}$. Then $\mu_1(E) = \sum_{j=1}^{\infty} \mu_1(E_j)$ and $\mu_2(E) = \sum_{j=1}^{\infty} \mu_2(E_j)$ Suppose μ_1 is finite, so that $\mu_1(A) < \infty$ for every $A \in \mathcal{F}$. If $\mu_2(E) < \infty$ as well, then the series $\sum_{j=1}^{\infty} (\mu_1(E_j) - \mu_2(E_j))$ is absolutely convergent to $\mu_1(E) - \mu_2(E)$. Hence

$$\mu_0(E) = \mu_1(E) - \mu_2(E) = \sum_{j=1}^{\infty} (\mu_1(E_j) - \mu_2(E_j)) = \sum_{j=1}^{\infty} \mu_0(E_j).$$

If $\mu_2(E) = \pm\infty$, then $\sum_{j=1}^{\infty} (\mu_1(E_j) - \mu_2(E_j))$ is divergent to $-\mu_2(E) = \mu_1(E) - \mu_2(E)$. Thus μ_0 is a signed measure on $\mathcal{F}$ when μ_1 is finite. The case when μ_2 is finite is argued analogously.

If μ_1 an μ_2 are both finite signed measures, then $|\mu_1(X)| < \infty$ and$|\mu_2(X)| < \infty$ and hence μ_0 is a finite signed measure. Similarly, if one among μ_1 and μ_2 is finite and the other is σ-finite, then μ_0 is σ-finite.

(b) Let ν be a nonnegative measure on $\mathcal{F}$ and f be an extended real-valued measurable function on X such that one among $\int_X f^+ d\nu$ and $\int_X f^- d\nu$ is finite, so that $\int_X f d\nu$ is defined as an extended real number in accordance with Definition 3.2.9. Set

$$\mu(E) = \int_E f\, d\nu \quad \text{for } E \in \mathcal{F}.$$

Then μ is a signed measure on $\mathcal{F}$, as we now show:

If $\int_X f^+ d\nu$ is finite, then $\int_E f^+ d\nu$ is finite for every $E \in \mathcal{F}$, so that $\int_E f\, d\nu$ cannot be ∞ for any $E \in \mathcal{F}$. Similarly, if $\int_E f^- d\nu$ is finite, then $\int_E f^- d\nu$ is finite for every $E \in \mathcal{F}$, so that $\int_E f\, d\nu$ cannot be $-\infty$ for any $E \in \mathcal{F}$. Thus, (a) of Definition 5.9.1 holds. It is trivial that (b) also holds. Let $\{E_j\}_{j \geq 1}$ be a sequence of disjoint sets in $\mathcal{F}$. For $E \in \mathcal{F}$, write

$$\mu^+(E) = \int_E f^+ d\nu \quad \text{and} \quad \mu^-(E) = \int_E f^- d\nu.$$

It follows by using Corollary 3.2.6 that μ^+ and μ^- are measures on $\mathcal{F}$. Moreover, $\mu = \mu^+ - \mu^-$. Therefore

$$\mu\Big(\bigcup_{j=1}^{\infty} E_j\Big) = \mu^+\Big(\bigcup_{j=1}^{\infty} E_j\Big) - \mu^-\Big(\bigcup_{j=1}^{\infty} E_j\Big) = \sum_{j=1}^{\infty} \mu^+(E_j) - \sum_{j=1}^{\infty} \mu^-(E_j) = \sum_{j=1}^{\infty} \mu(E_j),$$

as at no stage will "$\infty - \infty$" occur.

Definition 5.9.4 A set $A \subseteq X$ is a **positive set** (for μ) if A is measurable and every measurable subset $E \subseteq A$ satisfies $\mu(E) \geq 0$. Similarly, a set $B \subseteq X$ is a **negative set** (for μ) if B is measurable and every measurable subset $E \subseteq B$ satisfies $\mu(E) \leq 0$.

Note that every measurable subset of a positive set is a positive set. Let μ_A be defined on $\mathcal{F}$ by $\mu_A(E) = \mu(A \cap E)$ for all $E \in \mathcal{F}$. Then μ_A is a nonnegative measure. Indeed, $\mu_A(\emptyset) = \mu(\emptyset \cap A) = \mu(\emptyset) = 0$, and if $\{E_j\}_{j \geq 1}$ is a sequence of disjoint sets in $\mathcal{F}$, then

$$\mu_A\Big(\bigcup_{j=1}^{\infty} E_j\Big) = \mu\Big(\Big(\bigcup_{j=1}^{\infty} E_j\Big) \cap A\Big) = \mu\Big(\bigcup_{j=1}^{\infty} (E_j \cap A)\Big) = \sum_{j=1}^{\infty} \mu(E_j \cap A) = \sum_{j=1}^{\infty} \mu_A(E_j).$$

A set B is a negative set for μ if it is a positive set for $-\mu$. Every measurable subset of a negative set is again a negative set.

Finally, A is a **null set** for μ (or a **μ-null set**) if it is a positive as well as a negative set for μ. Equivalently, A is a null set if $A \in \mathcal{F}$ and $\mu(E) = 0$ for every $E \in \mathcal{F}$ such that $E \subseteq A$.

Remarks 5.9.5

(a) The empty set is a null set for every signed measure.
(b) A null set has measure zero. However, a set of measure zero may well be a union of two sets whose measures are not zero but are negatives of each other.
(c) The condition $\mu(A) \geq 0$ [resp. $\mu(A) \leq 0$] is necessary but not in general sufficient for A to be a positive [resp. negative] set.

The main result is the Hahn Decomposition Theorem 5.9.10, which states that X may be split into two disjoint measurable sets such that the respective restrictions of μ and $-\mu$ are nonnegative measures. The following results will be used to prove the main result.

Lemma 5.9.6 *The union of a countable or finite collection of positive sets is a positive set.*

Proof Let A be the union of a sequence $\{A_j\}_{j \geq 1}$ of positive sets. If $E \in \mathcal{F}$ and $E \subseteq A$, set

$$E_1 = E \cap A_1 \quad \text{and} \quad E_j = E \cap A_j \cap A_{j-1}^c \cap \cdots \cap A_1^c \text{ for } j > 1.$$

Then E_j is a measurable subset of A_j and so $\mu(E_j) \geq 0$. Since the sets E_j are disjoint and $E = \cup_{j \geq 1} E_j$, we have

$$\mu(E) = \sum_{j=1}^{\infty} \mu(E_j) \geq 0.$$

Thus A is a positive set.

A corresponding argument applies to a finite collection of sets. □

Corollary 5.9.7 *A countable union of negative [resp. null] sets is a negative [resp. null] set.*

Proposition 5.9.8 *Suppose that $E \in \mathcal{F}$ and $0 < \mu(E) < \infty$. Then there exists an $A \subseteq E$ such that A is a positive set and $\mu(A) > 0$.*

Proof If E contains no set of negative μ-measure, then E is a positive set and $A = E$ gives the result. Suppose no set A of the required sort exists. In particular, E is not a positive set. Let n_1 be the smallest positive integer for which there exists a measurable set $E_1 \subseteq E$ with $\mu(E_1) < -\frac{1}{n_1}$. Then we have

$$\mu(E \cap E_1^c) = \mu(E) - \mu(E_1) > \mu(E) > 0$$

and so by our assumption, $E \cap E_1^c$ is not a positive set for μ. As before, let n_2 be the smallest positive integer for which there exists a measurable set $E_2 \subseteq E \cap E_1^c$ with $\mu(E_2) < -\frac{1}{n_2}$. Then

$$\mu(E \cap (E_1 \cup E_2)^c) = \mu(E) - \mu(E_1) - \mu(E_2) > 0.$$

Continuing this process, we obtain a sequence $\{n_k\}_{k \geq 1}$ of minimal positive integers and a corresponding sequence $\{E_k\}_{k \geq 1}$ of sets in $\mathcal{F}$ such that $E_1 \subseteq E$ and $E_k \subseteq E \cap (\bigcup_{j=1}^{k-1} E_j)^c$ for $k > 1$and $\mu(E_k) < -\frac{1}{n_k}$ for each $k \in \mathbb{N}$. Note that the property $E_k \subseteq E \cap (\bigcup_{j=1}^{k-1} E_j)^c$ for $k > 1$ makes the E_k disjoint.

If we set $A = E \backslash \cup_{j \geq 1} E_j$, then $E = A \cup (\cup_{j \geq 1} E_j)$. Observe that

$$\infty > \mu(A) = \mu(E) - \mu(\bigcup_{j=1}^{\infty} E_j) = \mu(E) - \sum_{j=1}^{\infty} \mu(E_j) > \mu(E) + \sum_{j=1}^{\infty} \frac{1}{n_j} > 0.$$

In particular, $\sum_{i=1}^{\infty} \frac{1}{n_j} < \infty$;and hence $n_j \to \infty$ as $j \to \infty$; furthermore, A is not a positive set. Choose $F \subseteq A$ such that $\mu(F) < 0$ and choose $k > 1$ so large that $\mu(F) < -\frac{1}{n_k}$ and also $n_k > 2$. Then,

$$F \cup E_k \subseteq E \cap (\bigcup_{j=1}^{k-1} E_j)^c$$

and hence

$$\mu(F \cup E_k) = \mu(F) + \mu(E_k) < -\frac{1}{n_k} - \frac{1}{n_k}$$
$$< -\frac{1}{n_k - 1}, \text{ as } n_k > 2.$$

But this contradicts the minimality of n_k. Consequently, a set A of the required sort exists. □

Definition 5.9.9 A **Hahn decomposition** for a signed measure μ on $(X, \mathcal{F})$ is an ordered pair $\langle A, B\rangle \in \mathcal{F} \times \mathcal{F}$ such that $A \cup B = X$, $A \cap B = \varnothing$, A is a positive set for μ and B is a negative set for μ.

It is not required in the foregoing definition that A or B should be nonempty. In fact, if μ is a nonnegative measure, then $\langle X, \varnothing\rangle$ is a Hahn decomposition.

Theorem 5.9.10 (Hahn Decomposition Theorem) *Let μ be a signed measure on $(X, \mathcal{F})$. Then μ has a Hahn decomposition $\langle A, B\rangle$. It is unique in the sense that if $\langle A_1, B_1\rangle$ and $\langle A_2, B_2\rangle$ are both Hahn decompositions for μ, then $\mu(A_1 \cap E) = \mu(A_2 \cap E)$ and $\mu(B_1 \cap E) = \mu(B_2 \cap E)$ for any $E \in \mathcal{F}$.*

Proof Without loss of generality, we may assume that μ never takes the value ∞, for otherwise, we may switch to $-\mu$ and interchange the rôles of positive and negative sets.

Let $\lambda = \sup\{\mu(C) : C$ is a positive set for $\mu\}$. Since the empty set is positive, $\lambda \geq 0$. Choose a sequence $\{A_n\}_{n\geq 1}$ of positive sets such that $\lim_{n\to\infty} \mu(A_n) = \lambda$ and define $A = \cup_{n\geq 1}A_n$. By Lemma 5.9.6, A is itself a positive set and so $\lambda \geq \mu(A)$. But $A\backslash A_n \subseteq A$ and therefore $\mu(A\backslash A_n) \geq 0$ for each n. Thus, $\mu(A) = \mu(A_n) + \mu(A\backslash A_n) \geq \mu(A_n)$ for each n, so that $\mu(A) \geq \lim_{n\to\infty} A_n = \lambda$. It follows that $\lambda < \infty$.

Let $B = A^c$. Consider any positive subset E of B. Then $E \cap A = \varnothing$ and $E \cup A$ is a positive set. Hence

$$\lambda \geq \mu(E \cup A) = \mu(E) + \mu(A) = \mu(E) + \lambda,$$

which implies $\mu(E) \leq 0$, because $0 \leq \lambda < \infty$. Thus B contains no positive set of positive measure and hence contains no subset of positive measure by Proposition 5.9.8. Consequently, B is a negative set, so that $\langle A, B\rangle$ is a Hahn decomposition for μ.

Now suppose $\langle A_1, B_1\rangle$ and $\langle A_2, B_2\rangle$ are both Hahn decompositions for μ. Let $E \in \mathcal{F}$ be arbitrary. Since $E \cap A_1 \cap B_2$ is a subset of both A_1 and B_2, which are positive and negative sets for μ, we have $\mu(E \cap A_1 \cap B_2) = 0$. Similarly, $\mu(E \cap A_2 \cap B_1) = 0$. Observe that

$$E \cap A_1 = E \cap A_1 \cap X = E \cap A_1 \cap (A_2 \cup B_2) = (E \cap A_1 \cap A_2) \cup (E \cap A_1 \cap B_2)$$

and

$$(E \cap A_1 \cap A_2) \cap (E \cap A_1 \cap B_2) \subseteq A_2 \cap B_2 = \varnothing.$$

Since $\mu(E \cap A_1 \cap B_2) = 0$, this implies

$$\mu(E \cap A_1) = \mu(E \cap A_1 \cap A_2).$$

Similarly,

$$\mu(E \cap A_2) = \mu(E \cap A_1 \cap A_2).$$

Consequently, $\mu(E \cap A_1) = \mu(E \cap A_2)$ and a similar argument shows that $\mu(E \cap B_1) = \mu(E \cap B_2)$. This completes the proof. □

Remark 5.9.11 A Hahn decomposition need not be unique in the strict sense and there may actually be uncountably many Hahn decompositions. Indeed, any null set can be taken either as a part of A or of B. The following example illustrates this.

Let X be the interval $[0, 3]$ with Lebesgue measure m and $\mu(E) = \int_E f\,dm$, where the function $f:[0, 3] \to \mathbb{R}$ is $-\chi_{[0,1]} + \chi_{[2,3]}$. Then f vanishes on $(1, 2)$, so that $\mu(E) = 0$ for every measurable $E \subseteq (1, 2)$, which is to say, $(1, 2)$ is a null set. In particular, for an arbitrary $a \in (1, 2)$ we have $f \geq 0$ on $[a, 3]$ and $f \leq 0$ on $[0, a]$, whence it follows that $\langle [a, 3], [0, a] \rangle$ is a Hahn decomposition for μ whenever $a \in (1, 2)$. This exhibits uncountably many Hahn decompositions.

Example 5.9.12 Let $\mu(E) = \int_E xe^{-x^2}dx$, $E \in \mathfrak{M}$. We shall determine the positive, negative and null sets for μ.

$$\mu(E) = \int_{E \cap (0,\infty)} xe^{-x^2}dx + \int_{E \cap (-\infty,0)} xe^{-x^2}dx.$$

Observe that E is a positive [resp. negative] set for μ if and only if $m(E \cap (-\infty, 0)) = 0$ [resp. $m(E \cap (0, \infty)) = 0$] and E is a null set if and only if $m(E) = 0$.

The sets $[0, \infty)$ and $(-\infty, 0)$ form one of the uncountably many possible Hahn decompositions for μ.

A Hahn decomposition for a signed measure μ helps to obtain a decomposition of μ as a difference of measures. The idea of decomposing a signed measure into a difference of nonnegative measures is attributed to H. Lebesgue. However, earlier C. Jordan obtained a decomposition of a real function of bounded variation as a difference of increasing functions, which is a precursor to the decomposition of signed measures. Hence the name *Jordan Decomposition Theorem* (see Theorem 5.9.16).

Definition 5.9.13 Two nonnegative measures μ_1 and μ_2 on $(X, \mathcal{F})$ are said to be **mutually singular** [in symbols: $\boldsymbol{\mu_1 \perp \mu_2}$] if there are disjoint measurable sets A, B with $X = A \cup B$ such that $\mu_2(A) = 0 = \mu_1(B)$.

Remark 5.9.14

(a) In the above definition, one need not take A and B to be disjoint. If they were not disjoint, then $A \backslash B$ and B would constitute a pair of disjoint sets with the same properties.

(b) If μ_1, μ_2 and ν are nonnegative measures on $(X, \mathcal{F})$ and $\mu_i \perp \nu$ for $i = 1, 2$, then $\mu_1 + \mu_2 \perp \nu$. Indeed, there exists sets A_i, $B_i \in \mathcal{F}$ satisfying $A_i \cup B_i = X$ such that $\nu(A_i) = 0 = \mu_i(B_i)$ for $i = 1, 2$. Then $\nu(A_1 \cup A_2) \leq \nu(A_1) + \nu(A_2) = 0$ and

$$(\mu_1+\mu_2)(B_1\cap B_2)=\mu_1(B_1\cap B_2)+\mu_2(B_1\cap B_2)\leq\mu_1(B_1)+\mu_2(B_2)=0.$$

Moreover,

$$(A_1\cup A_2)\cup(B_1\cap B_2)=((A_1\cup A_2)\cup B_1)\cap((A_1\cup A_2)\cup B_2)=X.$$

Example 5.9.15 Let ν be a nonnegative measure on $(X, \mathcal{F})$ and A, B be measurable sets such that $\nu(A \cap B) = 0$. Define $\nu_1(E) = \nu(A \cap E)$ and $\nu_2(E) = \nu(B \cap E)$ for every $E \in \mathcal{F}$. Then $\nu_1 \perp \nu_2$. Indeed, $\nu_1(B) = \nu(A \cap B) = 0$ and $\nu_2(B^c) = \nu(B \cap B^c) = \nu(\emptyset) = 0$.

Theorem 5.9.16 (Jordan Decomposition Theorem) *Let μ be a signed measure on $(X, \mathcal{F})$. Then there exist nonnegative measures μ^+ and μ^- on $(X, \mathcal{F})$ such that $\mu = \mu^+ - \mu^-$ and $\mu^+ \perp \mu^-$. Moreover, the decomposition is unique.*

Proof Let $\langle A, B\rangle$ be a Hahn decomposition for μ. Define μ^+ and μ^- by

$$\mu^+(E)=\mu(E\cap A)\quad\text{and}\quad\mu^-(E)=-\mu(E\cap B)\quad\text{for}\, E\in\mathcal{F}.$$

It follows from the paragraph after Definition 5.9.4 that μ^+ and μ^- are nonnegative measures on $(X, \mathcal{F})$. Moreover $\mu^+(B) = \mu(B \cap A) = 0$ and $\mu^-(A) = -\mu(B \cap A) = 0$. So, $\mu^+ \perp \mu^-$. Also, for $E \in \mathcal{F}$,

$$\mu(E)=\mu(E\cap X)=\mu(E\cap A)+\mu(E\cap B)=\mu^+(E)-\mu^-(E).$$

Hence $\mu = \mu^+ - \mu^-$.

It remains to show that the decomposition is unique.

Let $\mu = \mu_1 - \mu_2$ with $\mu_1 \perp \mu_2$ be another decomposition of μ. Let A', $B' \in \mathcal{F}$ be such that $\mu_1(B') = 0 = \mu_2(A')$ with $A' \cap B' = \emptyset$ and $A' \cup B' = X$. If $D \in \mathcal{F}$ and $D \subseteq A'$, then

$$\mu(D)=\mu_1(D)-\mu_2(D)=\mu_1(D). \tag{5.85}$$

Also, if $D \in \mathcal{F}$ and $D \subseteq B'$, then

$$\mu_1(D)=0. \tag{5.86}$$

Since $\mu_1(D) \geq 0$, it follows from (5.85) that A' is a positive set for μ. Similarly, the analogues of (5.85) and (5.86) hold for μ_2. Since $\mu_2(D) \geq 0$, it follows from the analogue of (5.85) that B' is a negative set for μ. Thus $\langle A', B' \rangle$ is a Hahn decomposition for μ. By Theorem 5.9.10, it follows that for $E \in \mathcal{F}$, we have

$$\begin{aligned}\mu^+(E) &= \mu(E \cap A) = \mu(E \cap A') \\ &= \mu_1(E \cap A') \quad \text{by (5.85)} \\ &= \mu_1(E \cap A') + \mu_1(E \cap B') \quad \text{by (5.86)} \\ &= \mu_1(E).\end{aligned}$$

A similar argument using the analogues of (5.85) and (5.86) leads to $\mu^-(E) = \mu_2(E)$. □

Remark Since μ^+ and μ^- are nonnegative measures, their sum $|\mu| = \mu^+ + \mu^-$ is also a nonnegative measure. It satisfies $|\mu|(E) = \mu^+(E) + \mu^-(E) \geq |\mu^+(E) - \mu^-(E)| = |\mu(E)|$ for every $E \in \mathcal{F}$.

Definition 5.9.17 The decomposition of μ given by $\mu = \mu^+ - \mu^-$ as in Theorem 5.9.16 is called the **Jordan decomposition** of μ. The measures μ^+ and μ^- are called the **positive variation** and **negative variation** respectively of μ. The measure $|\mu| = \mu^+ + \mu^-$ is called the **total variation** of μ.

Example 5.9.18 Let $(X, \mathcal{F}, \nu)$ be a measure space, with ν a nonnegative measure and let f be an extended real-valued measurable function on X such that one among $\int_X f^+ d\nu$ and $\int_X f^- d\nu$ is finite. Define a set function μ on $\mathcal{F}$ by

$$\mu(E) = \int_E f\, d\nu \text{ for every } E \in \mathcal{F}.$$

As seen in Example 5.9.3 (b), μ is a signed measure. Let $A = \{x \in X: f(x) \geq 0\}$ and $B = \{x \in X: f(x) < 0\}$. Then $\langle A, B \rangle$ can be shown to be a Hahn decomposition of μ. Indeed, $A \cup B = X$, $A \cap B = \emptyset$ and for $E \in \mathcal{F}$ such that $E \subseteq A$, we have $\mu(E) \geq 0$, i.e., A is a positive set for μ. Similarly, B is a negative set for μ.

We next show that μ^+ and μ^- given by

$$\mu^+(E) = \int_E f^+\, d\nu \quad \text{and} \quad \mu^-(E) = \int_E f^- d\nu \quad \text{for } E \in \mathcal{F}$$

form the Jordan decomposition of μ. This is because clearly, for any $E \in \mathcal{F}$,

$$\mu(E) = \int_E f\, d\nu = \int_E f^+\, d\nu - \int_E f^- d\nu = \mu^+(E) - \mu^-(E)$$

and

$$\mu^+(B) = \int_B f^+ d\nu = 0, \quad \mu^-(A) = \int_A f^- d\nu = 0.$$

Furthermore, $|\mu|$ is given by

$$|\mu|(E) = \mu^+(E) + \mu^-(E) = \int_E f^+ d\nu + \int_E f^- d\nu = \int_E |f| d\nu \quad \text{for } E \in \mathcal{F}.$$

We note from the above definition of μ^+ and μ^- that

$$\mu^+(E) = \int_E f^+ \, d\nu = \int_{E \cap A} f \, d\nu = \mu(E \cap A),$$
$$\mu^-(E) = \int_E f^- d\nu = \int_{E \cap B} f \, d\nu = \mu(E \cap B),$$

which is an alternative way to see that μ^+ and μ^- form the Jordan decomposition of μ, keeping in view how the decomposition was obtained in the proof of the Jordan Decomposition Theorem 5.9.16.

Remark Define integration with respect to a signed measure μ by

$$\int_X f \, d\mu = \int_X f \, d\mu^+ - \int_X f \, d\mu^-,$$

where μ^+ and μ^- form the Jordan decomposition of μ. Then for $E \in \mathcal{F}$, we have

$$\left|\int_E f \, d\mu\right| \leq \left|\int_E f \, d\mu^+\right| + \left|\int_E f \, d\mu^-\right|.$$

If $|f| \leq M$ on X, then

$$\begin{aligned}\left|\int_E f \, d\mu\right| &\leq \int_E |f| d\mu^+ + \int_E |f| d\mu^- \\ &\leq M\left(\int_E d\mu^+ + \int_E d\mu^-\right) \\ &\leq M(\mu^+(E) + \mu^-(E)) \\ &= M|\mu|(E).\end{aligned}$$

Problem Set 5.9

5.9.P1. Let μ be a signed measure on $(X, \mathcal{F})$. Show that

$$|\mu|(E) = \sup \sum_{j=1}^{n} |\mu(E_j)|,$$

where the sets $E_j \in \mathcal{F}$ are disjoint and $E = \cup_{1 \leq j \leq n} E_j$.

5.9.P2. If $\mu = \mu_1 - \mu_2$, where μ_1 and μ_2 are nonnegative measures and either μ_1 or μ_2 is finite, then $\mu_1 \geq \mu^+$ and $\mu_2 \geq \mu^-$, where $\mu = \mu^+ - \mu^-$ is the Jordan decomposition of μ.

5.9.P3. Let $(X, \mathcal{F}, \mu)$ be a finite signed measure space. Show that there is an $M > 0$ such that $|\mu(E)| < M$ for every $E \in \mathcal{F}$.

5.9.P4. Show that a signed measure μ on $(X,\mathcal{F})$ is finite [resp. σ-finite] if and only if $|\mu|$ is finite [resp. σ-finite] if and only if μ^+ and μ^- are both finite [resp. σ-finite].

5.9.P5. Let μ be a signed measure μ on $(X,\mathcal{F})$. Then, for every $E \in \mathcal{F}$, the following hold:

(a) $\mu^+(E) = \sup\{\mu(F): F \subseteq E, F \in \mathcal{F}\}$;
(b) $\mu^-(E) = \sup\{-\mu(F): F \subseteq E, F \in \mathcal{F}\}$.

5.9.P6. Let μ_1 and μ_2 be nonnegative measures on $(X, \mathcal{F})$ such that for all $\alpha > 0$ and $\beta > 0$, there exist sets $A_{\alpha,\beta}, B_{\alpha,\beta} \in \mathcal{F}$ satisfying $A_{\alpha,\beta} \cup B_{\alpha,\beta} = X$, $\mu_1(A_{\alpha,\beta}) < \alpha$, $\mu_2(B_{\alpha,\beta}) < \beta$. Show that $\mu_1 \perp \mu_2$.

5.10 The Radon–Nikodým Theorem

In this section, a measure will be understood to be nonnegative unless specified as being a signed measure.

Let $(X, \mathcal{F}, \mu)$ be a measure space and f be an extended real-valued measurable function on X. For $E \in \mathcal{F}$, let

$$\nu(E) = \int_E f\,d\mu \tag{5.87}$$

provided one among $\int_X f^+ d\nu$ and $\int_X f^- d\nu$ is finite, so that the integral on the right-hand side is meaningful. Then ν is a signed measure on X [see Example 5.9.3 (b)]. This seems to be a natural generalisation of the indefinite integral, as it leads to such theorems as the Radon–Nikodým Theorem 5.10.8 and the Lebesgue Decomposition Theorem 5.10.13.

It is desired to determine when a signed measure is expressible as an integral, as is the case with ν in (5.87). Here the essential concept is *absolute continuity* and the main result is the Radon–Nikodým Theorem. We begin with the definition of absolute continuity.

Definition 5.10.1 Let $(X, \mathcal{F})$ be a measurable space and μ,ν be signed measures on X. We say that ν is **absolutely continuous with respect to μ**, in symbols, $\nu \ll \mu$ *if* $\nu(E) = 0$ for every $E \in \mathcal{F}$ such that $|\mu|(E) = 0$.

Remark 5.10.2 In view of the obvious fact that the total variation of $|\mu|$ is nothing but $|\mu|$, absolute continuity with respect to $|\mu|$ is the same thing as absolute continuity with respect to μ.

Examples 5.10.3

(a) Let $(X, \mathcal{F}, \mu)$ be a measure space and f be an extended nonnegative real-valued measurable function on X. Define

$$\nu(E) = \int_E f \, d\mu, \qquad E \in \mathcal{F}.$$

It is an immediate consequence of Corollary 3.2.6 that ν is a countably additive measure on $(X, \mathcal{F})$. Moreover, for all $E \in \mathcal{F}$, $\mu(E) = 0$ implies $\nu(E) = 0$, i.e., a μ-null set is also a ν-null set. Consequently, $\nu \ll \mu$.

(b) Let $(\mathbb{R}, \mathfrak{M}, m)$ be the Lebesgue measure space and μ be the counting measure on $\mathfrak{M}$. If $\mu(E) = 0$, then $E = \emptyset$ and hence $m(E) = 0$. This shows that $m \ll \mu$.

(c) Let $(\mathbb{R}, \mathfrak{M}, m)$ be the Lebesgue measure space and $\nu:\mathfrak{M} \to [0, \infty]$ be defined by $\nu(\emptyset) = 0$ and $\nu(E) = \infty$ for $E \in \mathfrak{M}$ and $E \neq \emptyset$. Clearly, $m \ll \nu$.

(d) Let $X = \mathbb{N}$ and $\mathcal{F} = \mathcal{P}(\mathbb{N})$. For $E \in \mathcal{F}$, define $\mu(E) = \sum_{n \in E} 2^n$ and $\nu(E) = \sum_{n \in E} 2^{-n}$. Obviously, μ and ν are measures on $\mathcal{P}(\mathbb{N})$. Observe that $\nu(E) = 0$ if and only if $\mu(E) = 0$ if and only if $E = \emptyset$. This shows that $\nu \ll \mu$ and also $\mu \ll \nu$.

Proposition 5.10.4 *Let μ and ν be signed measures on a measurable space $(X, \mathcal{F})$. Then the following conditions are equivalent*:

$$(\alpha) \quad \nu \ll \mu \quad (\beta) \quad \nu^+ \ll \mu, \nu^- \ll \mu \quad (\gamma) \quad |\nu| \ll |\mu| \quad (\delta) |\nu| \ll \mu.$$

Proof From Remark 5.10.2, we know that $\nu \ll \mu$ if and only if $\nu \ll |\mu|$. So, without loss of generality, we may assume that $\mu \geq 0$. Let (α) hold and $X = A \cup B$ be a Hahn decomposition for ν. Then, whenever $\mu(E) = 0$, we have

$$0 \leq \mu(E \cap A) \leq \mu(E) = 0$$

and

$$0 \leq \mu(E \cap B) \leq \mu(E) = 0,$$

and therefore

$$\nu^+(E) = \nu(E \cap A) = 0, \quad \nu^-(E) = \nu(E \cap B) = 0.$$

This proves that $(\alpha) \Rightarrow (\beta)$.

The implications $(\beta) \Rightarrow (\gamma)$ and $(\gamma) \Rightarrow (\alpha)$ follow from the relations

$$|\nu|(E) = \nu^+(E) + \nu^-(E) \quad \text{and} \quad 0 \le |\nu(E)| \le |\nu|(E)$$

(for the latter, see the Remark preceding Definition 5.9.17). Finally (γ) ⇔ (δ) by Remark 5.10.2. □

The assertions in the proposition below follow from the definitions of absolute continuity and mutual singularity. (Two signed measures λ_1 and λ_2 on $\mathcal{F}$ are said to be **mutually singular**, and we write $\lambda_1 \perp \lambda_2$, if there exists disjoint measurable sets A and B such that $A \cup B = X$ and $|\lambda_1|(A) = 0 = |\lambda_2|(B)$. Since $|(|\mu|)| = |\mu|$ always, the relation $\lambda_1 \perp \lambda_2$ is equivalent to $|\lambda_1| \perp \lambda_2$, $\lambda_1 \perp |\lambda_2|$ and to $|\lambda_1| \perp |\lambda_2|$. It is not required that A or B must be nonempty; consequently, the zero measure and any signed measure are always mutually singular.)

Proposition 5.10.5 *Let* μ_1, μ_2 *and* μ_3 *be signed measures on* $(X, \mathcal{F})$. *Then*

(a) *If* $\mu_1 \ll \mu_2$ *and* $\mu_2 \ll \mu_3$, *then* $\mu_1 \ll \mu_3$.
(b) *If* $\mu_1 \ll \mu_3$ *and* $\mu_2 \ll \mu_3$ *and if* $\mu_1 + \mu_2$ *is a signed measure, then* $\mu_1 + \mu_2 \ll \mu_3$.
(c) *If* $\mu_1 \ll \mu_3$ *and* $\mu_2 \perp \mu_3$, *then* $\mu_1 \perp \mu_2$.
(d) *If* $\mu_1 \ll \mu_2$ *and* $\mu_1 \perp \mu_2$, *then* $\mu_1 = 0$.

Proof

(a) Since $\mu_2 \ll \mu_3$, we know from Proposition 5.10.4 [the implication (α) ⇒ (δ) therein] that $|\mu_2| \ll \mu_3$. Let $E \in \mathcal{F}$ be such that $|\mu_3|(E) = 0$. Since $|\mu_2| \ll \mu_3$, we have $|\mu_2|(E) = 0$. The hypothesis that $\mu_1 \ll \mu_2$ now yields $\mu_1(E) = 0$.
(b) Obvious.
(c) Since $\mu_2 \perp \mu_3$, there are disjoint sets A and B in $\mathcal{F}$ such that $A \cup B = X$ and $|\mu_3|(A) = 0 = |\mu_2|(B)$. Since $\mu_1 \ll \mu_3$, we have $|\mu_1| \ll \mu_3$ by Proposition 5.10.4 [the implication (α) ⇒ (δ) therein] and hence $|\mu_1|(E) = 0$ for every $E \in \mathcal{F}$ for which $|\mu_3|(E) = 0$. In particular, $|\mu_1|(A) = 0$. We have thus proved that there are disjoint sets A and B in $\mathcal{F}$ satisfying $A \cup B = X$ and $|\mu_1|(A) = 0 = |\mu_2|(B)$, i.e., $\mu_1 \perp \mu_2$.
(d) By (c), the hypothesis of (d) implies $\mu_1 \perp \mu_1$ and this clearly implies $\mu_1 = 0$. □

The next proposition essentially says, if ν is a finite signed measure, then "ν μ" is equivalent to "ν is small when μ is small".

Proposition 5.10.6 *Let* ν *be a finite signed measure on* $(X, \mathcal{F})$. *Then the following conditions are equivalent*:

(α) $\nu \ll \mu$;
(β) *to every* $\varepsilon > 0$, *there corresponds a* $\delta > 0$ *such that* $|\nu(E)| < \varepsilon$ *for every measurable set* E *for which* $|\mu|(E) < \delta$.

Proof Assume (β). Consider any $F \in \mathcal{F}$ such that $|\mu|(F) = 0$. To show that $\nu(F) = 0$, take an arbitrary $\varepsilon > 0$. Then there exists a $\delta > 0$ such that $|\mu|(E) < \delta$

implies $|\nu(E)| < \varepsilon$ for any $E \in \mathcal{F}$. Since $|\mu|(F) = 0$, we have $|\mu|(F) < \delta$. It follows that $|\nu(F)| < \varepsilon$. Since ε is arbitrary, we conclude that $\nu(F) = 0$. Thus, (β) implies (α).

Now suppose (β) is false. Then there exists an $\varepsilon > 0$ and sets $E_n \in \mathcal{F}$, $n = 1, 2, \ldots$ such that $|\mu|(E_n) < 2^{-n}$ but $|\nu(E_n)| \geq \varepsilon$. Hence $|\nu|(E_n) \geq \varepsilon$. Put

$$A_n = \bigcup_{i=n}^{\infty} E_i, \quad A = \bigcap_{n=1}^{\infty} A_n.$$

Then

$$|\mu|(A_n) \leq \sum_{i=n}^{\infty} |\mu|(E_i) < 2^{-n+1}, \quad A_n \supseteq A_{n+1}$$

and so, Proposition 3.1.8 shows that $|\mu|(A) = 0$ and that, in view of the finiteness of ν,

$$|\nu|(A) = \lim_{n\to\infty} |\nu|(A_n) \geq \varepsilon > 0$$

since $|\nu|(A_n) \geq |\nu|(E_n)$.

It follows that we do not have, $|\nu| \ll \mu$ and hence by Proposition 5.10.4 that we do not have $\nu \ll \mu$, which means (α) is false. □

As noted in Example 5.9.3 (b), if $(X, \mathcal{F}, \mu)$ is a measure space and f an extended real-valued measurable function on X such that one among $\int_X f^+ d\mu$ and $\int_X f^- d\mu$ is finite, then the set function ν defined on $\mathcal{F}$ by

$$\nu(E) = \int_E f\, d\mu, \quad E \in \mathcal{F} \tag{5.88}$$

is a signed measure on $\mathcal{F}$. It is clear that $\nu \ll \mu$. The Radon–Nikodým Theorem asserts that if $\nu \ll \mu$ and both are σ-finite, then ν is of the form (5.88), where f is real-valued.

The following lemma will be needed in the proof of the Radon–Nikodým Theorem.

Lemma 5.10.7 *Let μ and ν be finite measures on $(X, \mathcal{F})$ such that $\nu \ll \mu$ and ν is not identically zero. Then there exists a set $E \in \mathcal{F}$ such that $\mu(E) > 0$ and an $\varepsilon > 0$ such that $\nu(F) - \varepsilon\mu(F) \geq 0$ for every $F \subseteq E$, $F \in \mathcal{F}$.*

Proof For every $n \in \mathbb{N}$, consider the signed measure $\nu - \frac{1}{n}\mu$. Let $X = A_n \cup B_n$ be a Hahn decomposition for the measure $\nu - \frac{1}{n}\mu$, $n = 1, 2, \ldots$. Then for every $n \in \mathbb{N}$, $(\nu - \frac{1}{n}\mu)(B_n) \leq 0$ and $(\nu - \frac{1}{n})(A_n) \geq 0$. Let

$$A_0 = \bigcup_{n=1}^{\infty} A_n \quad \text{and} \quad B_0 = \bigcap_{n=1}^{\infty} B_n.$$

Then $x \notin A_0$ implies $x \notin A_n$ for every n, which further implies $x \in B_n$ for every n, i.e., $x \in B_0$. Thus, $X = A_0 \cup B_0$. Moreover, since $B_0 \subseteq B_n$, we have

$$0 \leq \nu(B_0) \leq \nu(B_n) \leq \frac{1}{n}\mu(B_n) \leq \frac{1}{n}\mu(X).$$

Hence $\nu(B_0) = 0$. Since $X = A_0 \cup B_0$, we get $\nu(A_0) = \nu(X) > 0$. Thus, $\nu\,(A_{n_0}) > 0$ for some $n_0 \in \mathbb{N}$. Since $\nu \ll \mu$, we also have $\mu\,(A_{n_0}) > 0$.

Also, A_{n_0} is a positive set for $\nu - \frac{1}{n_0}\mu$. On choosing $\varepsilon = \frac{1}{n_0}$ and $E = A_{n_0}$, the required conclusion is seen to hold. □

Theorem 5.10.8 (Radon–Nikodým Theorem). *Let $(X,\mathcal{F},\mu)$ be a σ-finite measure space and ν be a σ-finite signed measure on $\mathcal{F}$ such that $\nu \ll \mu$. Then there exists a real-valued measurable function f on X such that*

$$\nu(E) = \int_E f\,d\mu, \quad E \in \mathcal{F}.$$

The function f is unique in the sense that if g is any extended real-valued measurable function on X with $\nu(E) = \int_E g\,d\mu$ for all $E \in \mathcal{F}$, then $f = g$ a.e.[μ].

Proof The uniqueness is easily seen from Problem 3.2.P14(a).

We proceed to prove the existence in three steps.

Step 1. Assume that both μ and ν are finite measures. Let $\mathcal{K}$ be the class of nonnegative measurable functions f integrable with respect to μ such that $\int_E f\,d\mu \leq \nu(E)$ for every $E \in \mathcal{F}$. Observe that $\mathcal{K}$ is nonempty, as $0 \in \mathcal{K}$. Let

$$\alpha = \sup\{\int_X f\,d\mu : f \in \mathcal{K}\}. \tag{5.89}$$

Since $\nu(X) < \infty$, it follows that $0 \leq \alpha < \infty$. Let $\{f_n\}_{n\geq 1}$ be a sequence in $\mathcal{K}$ such that

$$\lim_{n\to\infty} \int_X f_n d\mu = \alpha.$$

Let $g_n = \max\{f_1, f_2, \ldots ,f_n\}$. Clearly, $g_n \geq 0$ and g_n is integrable with respect to μ. We shall show that $g_n \in \mathcal{K}$.

Let $E \in \mathcal{F}$ be fixed and let

$$\begin{aligned} E_1 &= \{x : g_n(x) = f_1(x)\} \cap E, \\ E_2 &= \{x : g_n(x) = f_2(x)\} \cap (E \backslash E_1), \\ &\ldots \\ E_n &= \{x : g_n(x) = f_n(x)\} \cap (E \backslash \bigcup_{k=1}^{n-1} E_k). \end{aligned}$$

Then $E = \cup_{1 \le k \le n} E_k$, the sets E_k are disjoint and belong to $\mathcal{F}$. Besides, $g_n(x) = f_j(x)$ on E_j, $j = 1, 2, \ldots, n$. Consequently, we have

$$\int_E g_n \, d\mu = \sum_{j=1}^{n} \int_{E_j} f_j \, d\mu \le \sum_{j=1}^{n} \nu(E_j) = \nu(E). \tag{5.90}$$

Therefore $g_n \in \mathcal{K}$. Now, let $g = \sup_n f_n$. Clearly, $g_n \le g_{n+1}$ and $g_n \to g$. Since $\int_E g_n \, d\mu \le \nu(X)$, we see that g is integrable. By the Monotone Convergence Theorem 3.2.4, $\int_E g_n \, d\mu \to \int_E g \, d\mu$ and, by (5.90), it follows that $g \in \mathcal{K}$, i.e.,

$$\int_E g \, d\mu \le \nu(E), \quad E \in \mathcal{F}.$$

Since $f_n \le g_n$, we see that $\alpha = \int_X g \, d\mu$. We define

$$f(x) = \begin{cases} g(x) & \text{if } g(x) < \infty \\ 0 & \text{if } g(x) = \infty. \end{cases}$$

Then $f(x) = g(x)$ a.e.[μ], so that f is integrable and $\int_E f \, d\mu = \int_E g \, d\mu$ for all $E \in \mathcal{F}$. Note that it follows that $f \in \mathcal{K}$. We shall prove that if $\nu_0(E) = \nu(E) - \int_E f \, d\mu$, then the measure ν_0 is identically zero. This will complete Step 1.

Suppose, if possible, that ν_0 is not identically zero. Observe that ν_0 is a finite measure and $\nu_0 \ll \mu$. Using Lemma 5.10.7, we obtain an $\varepsilon > 0$ and an $F \in \mathcal{F}$ such that $\mu(F) > 0$, and for every set $E \subseteq F$, $E \in \mathcal{F}$, we have $\nu_0(E) - \varepsilon\mu(E) \ge 0$, i.e.,

$$\nu(E) - \int_E f \, d\mu - \varepsilon\mu(E) \ge 0.$$

In particular,

$$\varepsilon\mu(E \cap F) \le \nu(E \cap F) - \int_{E \cap F} f \, d\mu, \quad E \in \mathcal{F}.$$

If $h = f + \varepsilon\chi_F$, then

$$\int_E h\,d\mu = \int_E f\,d\mu + \varepsilon\mu(E\cap F)$$
$$\leq \int_{E\backslash F} f\,d\mu + \nu(E\cap F) \leq \nu(E\backslash F) + \nu(E\cap F),$$

using the fact that $f \in \mathcal{K}$.

Consequently,

$$\int_E h\,d\mu \leq \nu(E\backslash F) + \nu(E\cap F) = \nu(E)$$

for every measurable set E, so that $h \in \mathcal{K}$. However, since

$$\int_X h\,d\mu = \int_X f\,d\mu + \varepsilon\mu(F) > \alpha,$$

equality (5.89) is violated, showing that ν_0 is identically zero, and Step 1 is therefore complete.

Step 2. Assume that μ and ν are σ-finite measures on $(X, \mathcal{F})$. With the assumption of σ-finiteness, we can write $X = \cup_{j\geq 1}A_j$, $\mu(A_j) < \infty$ and $X = \cup_{k\geq 1}B_k$, $\nu(B_k) < \infty$ and $\{A_j\}$, $\{B_k\}$ may be supposed to be sequences of disjoint sets. So, setting $X = \cup_{k\geq 1, j\geq 1}(A_j \cap B_k)$, we obtain X as the union of disjoint sets on which μ and ν are finite, say $X = \cup_{n\geq 1}X_n$. Let $\mathcal{F}_n = \{E \cap X_n : E \in \mathcal{F}\}$, a σ-algebra in X_n. Considering μ and ν restricted to $\mathcal{F}_n$, we obtain f_n such that every $E \in \mathcal{F}_n$ satisfies $\nu(E) = \int_E f_n\,d\mu$. So, if $E \in \mathcal{F}$, we have $E = \cup_{n\geq 1}E_n$, where $E_n = E \cap X_n \in \mathcal{F}_n$, and therefore upon defining $f = f_n$ on X_n, we obtain a measurable function on X and $\nu(E) = \sum_{n=1}^{\infty} \int_{E_n} f_n\,d\mu = \int_E f\,d\mu$. This completes Step 2.

Step 3. Now suppose that μ is a σ-finite measure on $(X, \mathcal{F})$ and ν a σ-finite signed measure. Let $\nu = \nu^+ - \nu^-$ be the Jordan decomposition of ν. Since $|\nu| \ll \mu$ by Proposition 5.10.4, the nonnegative measures ν^+ and ν^-, which are obviously both σ-finite, satisfy $\nu^+ \ll \mu$ and $\nu^- \ll \mu$. By the result of Step 2, there exist nonnegative real-valued measurable f_1 and f_2 such that

$$\nu^+(E) = \int_E f_1\,d\mu, \qquad \nu^-(E) = \int_E f_2\,d\mu$$

for every $E \in \mathcal{F}$.

Since f_1 and f_2 are real-valued, their difference $f = f_1 - f_2$ is also real-valued and we claim that the inequalities $f^+ \leq f_1$ and $f^- \leq f_2$ hold. Consider any $x \in X$. If $f(x) \geq 0$ then $f_1(x) = f(x) + f_2(x) \geq f(x) = f^+(x)$, while if $f(x) \leq 0$ then $f_1(x) \geq 0 = f^+(x)$. Hence $f^+ \leq f_1$ for all $x \in X$. We deduce from here that $f_2(x) = f_1(x) - f(x) \geq f^+(x) - f(x) = f^-(x)$ for all $x \in X$. This justifies our claim.

Now, either $\nu^+(X) < \infty$ or $\nu^-(X) < \infty$, so that either $\int_X f_1 d\mu < \infty$ or $\int_X f_2 d\mu < \infty$. In conjunction with the inequalities proved in the preceding paragraph, this implies that the function $f = f_1 - f_2$ has the property that either $\int_X f^+ d\mu < \infty$ or $\int_X f^- d\mu < \infty$, and hence that $\int_E f\, d\mu$ is well defined (though not necessarily finite). Consequently, the equality $\int_E f\, d\mu = \int_E f_1 d\mu - \int_E f_2\, d\mu$ holds for every $E \in \mathcal{F}$, which validates the computation

$$\nu(E) = \nu^+(E) - \nu^-(E) = \int_E f_1 d\mu - \int_E f_2 d\mu$$
$$= \int_E f\, d\mu.$$

This completes Step 3 and thereby, also the proof of the theorem. □

Remarks 5.10.9

(a) The function f given by the Radon–Nikodým Theorem is called the **Radon–Nikodým derivative** of ν with respect to μ and is sometimes denoted by $[\frac{d\nu}{d\mu}]$.
(b) The condition that μ is σ-finite is essential in the Radon–Nikodým Theorem.

Let $(\mathbb{R}, \mathfrak{M}, m)$ be the Lebesgue measure space and μ be the counting measure on $\mathfrak{M}$. So, μ is not σ-finite. Also, $m \ll \mu$, as seen in Example 5.10.3 (b). Suppose f exists such that $m(E) = \int_E f\, d\mu$ whenever $E \in \mathcal{F}$. Then for any $x \in \mathbb{R}$, we have

$$0 = m\{x\} = \int_{\{x\}} f\, d\mu = f(x)\mu\{x\} = f(x).$$

Hence f is identically 0. This means for every $E \in \mathfrak{M}$, we have

$$m(E) = \int_E 0\, d\mu = 0.$$

This contradicts the fact that m is Lebesgue measure.

(c) The condition that ν is σ-finite is also essential in the Radon–Nikodým Theorem. Consider the set X consisting of a single point x_0. Let μ be the counting measure on X and ν be the measure that $\nu(X) = \infty$. Clearly, μ is σ-finite and $\nu \ll \mu$. If f is any real-valued function on X, then $\int_X f\, d\mu$ is the real number $f(x_0)$ but $\nu(X) = \infty$, so that $\nu(X) \neq \int_X f\, d\mu$.

Remark 5.10.10 Suppose $E \in \mathcal{F}$. Then $\{F \cap E: F \in \mathcal{F}\} = \{F \in \mathcal{F}: F \subseteq E\}$ and is a σ-algebra of subsets of E. For a signed measure μ on $(X, \mathcal{F})$, the restriction to this σ-algebra is called the **restriction of the measure** to E and we shall denote it by $\mu\restriction_E$. The following facts about restrictions of measures are easy to justify in the order in which they are stated.

(a) If $\langle A, B\rangle$ is a Hahn decomposition of X for μ, then $\langle A \cap E, B \cap E\rangle$ is a Hahn decomposition of E for $\mu 1_E$.
(b) $(\mu 1_E)^+ = \mu^+ 1_E$, $(\mu 1_E)^- = \mu^- 1_E$, $|(\mu 1_E)| = (|\mu|)1_E$.
(c) $\nu \ll \mu \Rightarrow \nu 1_E \ll \mu 1_E$.
(d) $\nu \perp \mu \Rightarrow \nu 1_E \perp \mu 1_E$.

Let $(X, \mathcal{F}, \mu)$ be a σ-finite measure space. In what follows, we shall show that all σ-finite signed measures on $(X, \mathcal{F})$ can be analysed by considering only those that are absolutely continuous or singular with respect to μ. More specifically, we have the following:

Definition 5.10.11 Given a measure μ on $(X, \mathcal{F})$ and a signed measure ν on $(X, \mathcal{F})$, a **Lebesgue decomposition** of ν **with respect to** μ is a pair of measures ν_a and ν_s on $(X, \mathcal{F})$ with the property that

$$\nu = \nu_a + \nu_s, \quad \text{where} \quad \nu_a \ll \mu \text{ and } \nu_s \perp \mu.$$

Remark 5.10.12 Suppose $E \in \mathcal{F}$, μ is a measure on $(X, \mathcal{F})$ and ν a signed measure on $(X, \mathcal{F})$. If $\nu = \nu_a + \nu_s$ is a Lebesgue decomposition of ν, then by parts (c) and (d) of Remark 5.10.10, $\nu 1_E = \nu_a 1_E + \nu_s 1_E$ is a Lebesgue decomposition of $\nu 1_E$ with respect to $\mu 1_E$.

Theorem 5.10.13 (Lebesgue Decomposition Theorem for Measures) *Let $(X, \mathcal{F}, \mu)$ be a σ-finite measure space and ν be a σ-finite signed measure on $(X, \mathcal{F})$. Then ν has a unique Lebesgue decomposition with respect to μ consisting of a pair of σ-finite measures; i.e., there exists a unique pair of σ-finite measures ν_a and ν_s on $(X, \mathcal{F})$ with the property that*

$$\nu = \nu_a + \nu_s, \quad \textit{where} \quad \nu_a \ll \mu \quad \textit{and } \nu_s \perp \mu.$$

Proof We shall prove the existence in two steps.

Step 1. Let ν be nonnegative and set $\lambda = \mu + \nu$. Since both μ and ν are absolutely continuous with respect to λ, by the Radon–Nikodým Theorem 5.10.8, there exist nonnegative measurable functions f and g such that, for $E \in \mathcal{F}$,

$$\mu(E) = \int_E f \, d\lambda \quad \text{and} \quad \nu(E) = \int_E g \, d\lambda.$$

Let $A = \{x: f(x) > 0\}$ and $B = \{x: f(x) = 0\}$. Then $X = A \cup B$, where $A \cap B = \emptyset$ and $\mu(B) = 0$. If we define ν_s on $\mathcal{F}$ by

$$\nu_s(E) = \nu(E \cap B),$$

we have $\nu_s(A) = \nu(A \cap B) = 0$ and so, $\nu_s \perp \mu$. Define ν_a on $\mathcal{F}$ by

$$\nu_a(E) = \nu(E \cap A) = \int_{E \cap A} g\, d\lambda.$$

Then, for any $E \in \mathcal{F}$, we have

$$\nu(E) = \nu(E \cap A) + \nu(E \cap B) = \nu_a(E) + \nu_s(E),$$

i.e., $\nu = \nu_a + \nu_s$. It remains to show that $\nu_a \ll \mu$.

Let $E \in \mathcal{F}$ be such that $\mu(E) = 0$. Then

$$0 = \mu(E) = \int_E f\, d\lambda,$$

which implies $f = 0$ a.e.[λ] on E. Since $f > 0$ on $A \cap E$, we must have $\lambda(A \cap E) = 0$. Hence $\nu(A \cap E) = 0$ (recalling that $\nu \ll \lambda$) and so, $\nu_a(E) = \nu(E \cap A) = 0$.

Observe that if ν does not take the value ∞, the same is true of ν_a and ν_s.

Step 2. Suppose ν is a σ-finite signed measure. By the Jordan Decomposition Theorem 5.9.16, $\nu = \nu^+ - \nu^-$. We treat ν^+ and ν^- as in Step 1.

We consider the case when ν does not take the value ∞; then ν^+ does not take the value ∞. Also, the measures ν^+ and ν^- are σ-finite by Problem 5.9.P4. So, let $\nu^+ = \nu_a^+ + \nu_s^+$ and $\nu^- = \nu_a^- + \nu_s^-$ be any Lebesgue decompositions of ν^+ and ν^-. From the observation at the end of Step 1, it follows that both ν_a^+ and ν_s^+ do not take the value ∞. The set functions $\nu_a^+ - \nu_a^-$ and $\nu_s^+ - \nu_s^-$ are thus well defined and therefore must be signed measures by Remark 5.9.2(k). This permits us to write

$$\nu = \nu^+ - \nu^- = (\nu_a^+ - \nu_a^-) + (\nu_s^+ - \nu_s^-).$$

In view of Proposition 5.10.5(b), we have $(\nu_a^+ - \nu_a^-) \ll \mu$. The σ-finiteness of $\nu_a^+ - \nu_a^-$ and $\nu_s^+ - \nu_s^-$ follows from Problem 5.9.P4. It remains only to demonstrate that $(\nu_s^+ - \nu_s^-) \perp \mu$. Since ν_s^+ does not take the value ∞, we have

$$(\nu_s^+ - \nu_s^-)^+ \leq \nu_s^+ \quad \text{and} \quad (\nu_s^+ - \nu_s^-)^- \leq \nu_s^- \tag{5.91}$$

by Problem 5.9.P2. It is trivial to see that for (nonnegative) measures μ_1, μ_2, μ_3,

$$\mu_1 \leq \mu_2 \text{ and } \mu_2 \perp \mu_3 \quad \Rightarrow \quad \mu_1 \perp \mu_3.$$

Therefore, (5.91) and fact that $\nu_s^+ \perp \mu$ and $\nu_s^- \perp \mu$ together yield $(\nu_s^+ - \nu_s^-)^+ \perp \mu$ and $(\nu_s^+ - \nu_s^-)^- \perp \mu$. By Remark 5.9.14 (b), it follows that $((\nu_s^+ - \nu_s^-)^+ + (\nu_s^+ - \nu_s^-)^-) \perp \mu$. However $(\nu_s^+ - \nu_s^-)^+ + (\nu_s^+ - \nu_s^-)^- = |\nu_s^+ - \nu_s^-|$. Thus we have demonstrated that $|\nu_s^+ - \nu_s^-| \perp \mu$, which is equivalent to $(\nu_s^+ - \nu_s^-) \perp \mu$.

The case when ν does not take the value $-\infty$ is analogous.

This completes the proof of existence.

We next prove the uniqueness of the decomposition.

First suppose that ν is finite. Then in any Lebesgue decomposition of it, both measures must be finite, as is evident from the requirement in Definition 5.10.11 that their sum is ν. If $\nu = \nu_a + \nu_s$ and $\nu = \bar{\nu}_a + \bar{\nu}_s$ are two Lebesgue decompositions of ν, then the finiteness of all measures concerned permits the assertion that $\nu_s - \bar{\nu}_s = \bar{\nu}_a - \nu_a$. Since the left-hand side, namely, $\nu_s - \bar{\nu}_s$ can be shown to be singular in the same manner as in Step 2, and since the right-hand side, namely, $\bar{\nu}_a - \nu_a$ is absolutely continuous by Prop. 5.10.5(b), it follows on the basis of Proposition 5.10.5 (d) that $\nu_s - \bar{\nu}_s = \bar{\nu}_a - \nu_a = 0$. This proves uniqueness in the finite case.

For a general σ-finite ν, we can express X as a countable disjoint union of measurable sets X_n with each $\nu(X_n) < \infty$. By Remark 5.10.12, for any n, the restrictions to X_n of any two Lebesgue decompositions provide Lebesgue decompositions of the restriction of ν to X_n. It follows from the finite case that the restrictions of the Lebesgue compositions agree and it is then easy to deduce by "piecing together" the restrictions that the Lebesgue decompositions (on X) agree. □

For the relation between the above theorem and Problem 5.8.P7, see Problems 5.11.P5 and 5.11.P8.

The following example shows that the condition of σ-finiteness cannot be dropped in the Lebesgue Decomposition Theorem.

Example 5.10.14 Let μ be the Lebesgue measure m on $\mathfrak{M}$, the σ-algebra of Lebesgue measurable subsets of [0, 1] and suppose ν is the counting measure on $\mathfrak{M}$. We argue that ν cannot be written as $\nu = \nu_a + \nu_s$, where $\nu_a \ll \mu$ and $\nu_s \perp \mu$.

Suppose the contrary. Then for each $x \in [0, 1]$, we have

$$1 = \nu\{x\} = \nu_a\{x\} + \nu_s\{x\} = 0 + \nu_s\{x\}.$$

Since $\nu_s \perp \mu$, there exists disjoint measurable sets A, B with $A \cup B = [0, 1]$ such that $\mu(A) = 0 = \nu_s(B)$. Now if $B \neq \varnothing$, take $x \in B$. Then

$$0 = \nu_s(B) \geq \nu_s\{x\} = 1,$$

which is a contradiction. Consequently, $B = \varnothing$. Since $A \cap B = \varnothing$ and $A \cup B = [0, 1]$, it follows that $A = [0, 1]$, which implies $\mu(A) = m(A) = m[0, 1] = 1$. This contradicts the stipulation that $\mu(A) = 0$.

Problem Set 5.10

5.10.P1. Let $\mu = m + \delta$, where m is Lebesgue measure on [0, 1] and δ is the "Dirac" measure, which is defined by setting $\delta(E) = 1$ if $0 \in E$ and 0 if $0 \notin E$. Determine $[\frac{dm}{d\mu}]$.

5.10.P2. Let

$$f(x) = \begin{cases} \sqrt{1-x} & \text{if } x \le 1 \\ 0 & \text{if } x > 1 \end{cases}, \quad g(x) = \begin{cases} x^2 & \text{if } x \ge 0 \\ 0 & \text{if } x < 0 \end{cases}$$
$$\nu(E) = \int_E f\, dx \quad \text{and} \quad \mu(E) = \int_E g\, dx, \quad E \in \mathfrak{M}.$$

Find the Lebesgue decomposition of ν with respect to μ.

5.10.P3. Let $\mu = m + \delta$, where m is the usual Lebesgue measure on $\mathbb{R}$ and δ is the "Dirac" measure defined by

$$\delta(E) = \begin{cases} 1 & \text{if } 0 \in E \\ 0 & \text{if } 0 \notin E, \end{cases} \quad E \in \mathfrak{M}.$$

Determine the Lebesgue decomposition of μ with respect to m.

5.10.P4. Let μ_1,μ_2 and ν be σ-finite measures on $(X, \mathcal{F})$. Prove the following:

(a) If $\mu_i \ll \nu$, $i = 1, 2$, then $\mu_1 + \mu_2 \ll \nu$ and

$$\left[\frac{d(\mu_1+\mu_2)}{d\nu}\right] = \left[\frac{d\mu_1}{d\nu}\right] + \left[\frac{d\mu_2}{d\nu}\right] \quad \text{a.e.}[\nu].$$

(b) If $\mu_1 \ll \mu_2$ and $\mu_2 \ll \mu_1$, then $\left[\frac{d\mu_1}{d\mu_2}\right]\left[\frac{d\mu_2}{d\mu_1}\right] = 1$ a.e.$[\mu_1]$ and a.e.$[\mu_2]$.

5.10.P5. $(X, \mathcal{F}, \nu)$ be a σ-finite measure space and μ_1,μ_2 be σ-finite signed measures such that $\mu_1 + \mu_2$ is also a signed measure on $(X, \mathcal{F})$. Prove the following:

(a) If $\mu_i \ll \nu$, $i = 1, 2$, then $\mu_1 + \mu_2 \ll \nu$ and

$$\left[\frac{d(\mu_1+\mu_2)}{d\nu}\right] = \left[\frac{d\mu_1}{d\nu}\right] + \left[\frac{d\mu_2}{d\nu}\right] \quad \text{a.e.}[\nu].$$

(b) If μ is a σ-finite signed measure on $(X, \mathcal{F})$ such that $\mu \ll \nu$, then

$$\left[\frac{d|\mu|}{d\nu}\right] = \left|\left[\frac{d\mu}{d\nu}\right]\right| \quad \text{a.e.}[\nu].$$

5.10.P6. Let $(X, \mathcal{F}, \mu)$ be a σ-finite measure space. Let ν be a measure on $(X, \mathcal{F})$ for which the conclusion of the Radon–Nikodým Theorem holds. Prove that ν is σ-finite.

5.11 The Lebesgue–Stieljes Measure

In what follows, we shall study the connection between bounded increasing functions and finite measures—the functions and measures will both be on $\mathbb{R}$, the domain of the measure being the σ-algebra of Borel subsets of $\mathbb{R}$ (see Definition 2.3.13).

Let $\mathfrak{B}$ denote the σ-algebra of Borel subsets of $\mathbb{R}$ and μ be a finite (nonnegative) measure on $\mathfrak{B}$. Define a function f_μ on $\mathbb{R}$ by the rule

$$f_\mu(x) = \mu((-\infty, x)) = \mu(\{y \in \mathbb{R} : y<x\}).$$

Theorem 5.11.1

(a) *The function f_μ is a bounded increasing function on $\mathbb{R}$ which is left continuous. Moreover* $\lim_{x\to-\infty} f_\mu(x) = 0$.

(b) *The function f_μ is continuous at x if and only if* $\mu(\{x\}) = 0$.

Proof

(a) Observe that for $a < b$,

$$\mu([a,\ b)) = \mu((-\infty, b)\backslash(-\infty, a)) = f_\mu(b) - f_\mu(a).$$

Since μ is nonnegative, $f_\mu(b) \geq f_\mu(a)$, i.e., f_μ is an increasing function on $\mathbb{R}$. It follows from the fact that μ is finite that f_μ is also bounded.

Also, $[a, b)$ being the intersection of the sets $[a - \frac{1}{n}, b)$, outer continuity of measure (see Proposition 3.1.8) implies

$$\mu([a,\ b)) = \lim_{n\to\infty} \mu([a - \frac{1}{n}, b))$$

and so,

$$\begin{aligned} f_\mu(b) - f_\mu(a) &= f_\mu(b) - \lim_{n\to\infty} f_\mu(a - \frac{1}{n}) \\ &= f_\mu(b) - f_\mu(a-) \text{ by monotonicity of } f_\mu. \end{aligned}$$

Consequently, $f_\mu(a) = f_\mu(a-)$, i.e., f_μ is left continuous.

Since $\cap_{n\geq1}(-\infty, -n) = \varnothing$, by outer continuity of measure again, we have $f_\mu(-n) = 0$ and hence $\lim_{x\to-\infty} f_\mu(x) = 0$ by monotonicity of f_μ.

(b) Since $\mu(\{x\}) = \mu(\cap_{n\geq1}[x, x + \frac{1}{n})) = \lim_{n\to\infty} f_\mu(x + \frac{1}{n}) - f_\mu(x) = f_\mu(x+) - f_\mu(x)$, the function f_μ is continuous at x if and only if the set $\{x\}$ consisting of x alone has measure 0. □

The finite measure μ on B and the monotone increasing function f_μ are closely related. Indeed, we have the following theorem.

Theorem 5.11.2 If μ is a finite measure on the σ-algebra $\mathfrak{B}$ of Borel subsets of $\mathbb{R}$, then the monotone function f_μ is absolutely continuous if and only if $\mu \ll m$.

Proof Suppose $\mu \ll m$. By Proposition 5.10.6, to every $\varepsilon > 0$, there corresponds a $\delta > 0$ such that $\mu(E) < \varepsilon$ for every Borel set E with $m(E) < \delta$. Let $\{(a_i, b_i): i = 1,\ldots, n\}$ be finite disjoint collection of bounded open intervals for which

$$m(\bigcup_{i=1}^{n}[a_i, b_i)) = \sum_{i=1}^{n}(b_i - a_i) < \delta.$$

Then

$$\sum_{i=1}^{n}|f_\mu(b_i) - f_\mu(a_i)| = \sum_{i=1}^{n}\mu([a_i, b_i)) = \mu(\bigcup_{i=1}^{n}[a_i, b_i)) < \varepsilon;$$

this proves that f_μ is absolutely continuous. Suppose conversely that f_μ is absolutely continuous. Let $\varepsilon > 0$ be arbitrary and $\delta > 0$ be such that

$$\sum_{i=1}^{n}(b_i - a_i) < \delta \quad \text{implies} \quad \sum_{i=1}^{n}|f_\mu(b_i) - f_\mu(a_i)| < \varepsilon,$$

where $\{(a_i, b_i) : i = 1,\ldots,n\}$ is any finite disjoint collection of open intervals. Let E be a Borel set of Lebesgue measure 0. By using Proposition 2.3.24, we can find a pairwise disjoint sequence $\{[a_i, b_i)\}_{i\geq 1}$ of bounded intervals such that

$$E \subseteq \bigcup_{i=1}^{\infty}[a_i, b_i) \quad \text{and} \quad \sum_{i=1}^{\infty}(b_i - a_i) < \delta.$$

Since it follows that

$$\sum_{i=1}^{n}|f_\mu(b_i) - f_\mu(a_i)| < \varepsilon$$

for every positive integer n, we have

$$\mu(E) \leq \sum_{i=1}^{\infty}\mu([a_i, b_i)) = \sum_{i=1}^{\infty}|f_\mu(b_i) - f_\mu(a_i)| \leq \varepsilon.$$

Since $\varepsilon > 0$ is arbitrary, we have $\mu(E) = 0$. □

The intervals of the form $[a, b)$, where $a \leq b$, will be referred to in the present section as *left closed* intervals.

Conversely, let f be a bounded increasing function on $\mathbb{R}$ which is left continuous. We shall show that there is a unique measure μ_f on $\mathfrak{B}$ such that

$$\mu_f([a, b)) = f(b) - f(a)$$

for all left closed intervals $[a, b)$. See Theorem 5.11.22. Indeed, it will be shown in Remark 5.11.24 (b) that, if $\mu = \mu_f$, then f_μ will be equal to f + constant. We will call

μ_f as the *Lebesgue–Stieltjes measure induced by f*. To begin with, let μ be defined on the family of all left closed intervals $[a, b)$ by

$$\mu([a, b)) = f(b) - f(a).$$

Observe that μ assumes only nonnegative values and $\mu(\varnothing) = 0$. Indeed, $\varnothing = [a, b)$ if and only if $a = b$.

We begin with the following lemmas.

Lemma 5.11.3 *Let* $\{E_i\}_{1 \le i \le n}$ *be disjoint left closed intervals* ($E_i = [a_i, b_i)$, $1 \le i \le n$) *such that* $\cup_{1 \le i \le n} E_i \subseteq I$, *where* I *is a left closed interval. Then* $\sum_{i=1}^{n} \mu(E_i) \le \mu(I)$.

Proof Let $I = [a, b)$. Without loss of generality, we may assume that $a_1 < a_2 < \cdots < a_n$. Since $\cup_{1 \le i \le n} E_i \subseteq I$, we know that $a \le a_1$ and $b_n \le b$. The disjointness of the intervals then tells us that we also have $b_i \le a_{i+1}$ for $1 \le i \le n-1$. Since f is increasing, these inequalities imply that

$$\begin{aligned}\sum_{i=1}^{n} (f(b_i) - f(a_i)) &\le \sum_{i=1}^{n} (f(b_i) - f(a_i)) + \sum_{i=1}^{n-1} (f(a_{i+1}) - f(b_i)) = f(b_n) - f(a_1)\\ &\le f(b) - f(a).\end{aligned}$$

Up to this stage, the argument is valid for any kind of bounded intervals I and E_i with $\cup_{1 \le i \le n} E_i \subseteq I$ without even a continuity hypothesis about the increasing function f. But when I and E_i are left closed, the inequality above asserts precisely what the lemma claims. This completes the proof. □

Lemma 5.11.4 *If* $I = [a_0, b_0]$ *is contained in the union of a finite number of bounded open intervals* $U_i = (a_i, b_i)$, $1 \le i \le n$, *then*

$$f(b_0) - f(a_0) \le \sum_{i=1}^{n} (f(b_i) - f(a_i)).$$

Proof Let k_1 be such that $a_0 \in U_{k_1}$. Then $a_0 < b_{k_1}$, and if also $b_{k_1} \le b_0$, then $b_{k_1} \in I$; so, let k_2 be such that $b_{k_1} \in U_{k_2}$; similarly, if $b_{k_2} \le b_0$, then let k_3 be such that $b_{k_2} \in U_{k_3}$ and so on, inductively. The process stops with k_m if $b_{k_m} > b_0$. Without loss of generality, we assume that $m = n$ and that $U_{k_i} = U_i = (a_i, b_i)$ for $1 \le i \le n$. This is easily achieved by omitting superfluous U_i's and changing the notation if necessary. In other words, we may assume that

$$a_1 < a_0 < b_1, \quad a_n < b_0 < b_n$$

and, if $n > 1$, then $a_{i+1} < b_i < b_{i+1}$ for $1 \le i \le n-1$. It follows that

$$
\begin{aligned}
f(b_0) - f(a_0) &\le f(b_n) - f(a_1) = f(b_1) - f(a_1) + \sum_{i=1}^{n-1} (f(b_{i+1}) - f(b_i)) \\
&\le f(b_1) - f(a_1) + \sum_{i=1}^{n-1} (f(b_{i+1}) - f(a_{i+1})) \\
&= \sum_{i=1}^{n} (f(b_i) - f(a_i)).
\end{aligned}
$$

This completes the proof. □

Lemma 5.11.5

(a) *If the left closed interval* $[a_0, b_0)$ *is contained in the union of a sequence of left closed intervals* $[a_i, b_i)$, $i = 1, 2, \ldots$, *then*

$$\mu([a_0, b_0)) \le \sum_{i=1}^{\infty} \mu([a_i, b_i)).$$

(b) *If the intervals in the sequence* $[a_i, b_i)$, $i = 1, 2, \ldots$, *are disjoint and* $[a_0, b_0) = \cup_{i \ge 1}[a_i, b_i)$, *then*

$$\mu([a_0, b_0)) = \sum_{i=1}^{\infty} \mu([a_i, b_i)).$$

Proof

(a) If $a_0 = b_0$, then the result is trivial. Let ε be a positive number such that $\varepsilon < b_0 - a_0$. Since f is left continuous at a_i, to every number δ and positive integer i, there corresponds a positive number ε_i such that

$$f(a_i) - f(a_i - \varepsilon_i) < \frac{\delta}{2^i}, \quad i = 1, 2, \ldots. \tag{5.92}$$

If $F_0 = [a_0, b_0 - \varepsilon]$ and $U_i = (a_i - \varepsilon_i, b_i)$, $i = 1, 2, \ldots$, then $F_0 \subseteq \cup_{i \ge 1} U_i$ and therefore by the Heine–Borel Theorem, there is an integer n such that F_0 is contained in the union of n among the intervals U_i. By renumbering the U_i if necessary, we may assume that

$$F_0 \subseteq \bigcup_{1 \le i \le n} U_i.$$

It follows from Lemma 5.11.4 that

$$\begin{aligned} f(b_0 - \varepsilon) - f(a_0) &\le \sum_{i=1}^{n} (f(b_i) - f(a_i - \varepsilon)) \\ &= \sum_{i=1}^{n} (f(b_i) - f(a_i)) + \sum_{i=1}^{n} (f(a_i) - f(a_i - \varepsilon)) \\ &\le \sum_{i=1}^{n} (f(b_i) - f(a_i)) + \delta \quad \text{by (5.92)}. \end{aligned}$$

Since ε and δ are arbitrary, the proof of (a) is complete.

(b) It follows from Lemma 5.11.3 that $\sum_{i=1}^{n} \mu([a_i, b_i)) \le \mu([a_0, b_0))$ for $n = 1, 2, \ldots$. So, $\sum_{i=1}^{\infty} \mu([a_i, b_i)) \le \mu([a_0, b_0))$. The proof is completed by using (a) above. □

Remark 5.11.6

(a) Let $[a, b)$ and $[\alpha, \beta)$ be left closed intervals. Then x belongs to their intersection if and only if $\max\{a, \alpha\} \le x < \min\{b, \beta\}$. Thus the intersection is empty if $\max\{a, \alpha\} \ge \min\{b, \beta\}$ and is the left closed interval

$$[\max\{a, \alpha\}, \min\{b, \beta\})$$

otherwise. Therefore the intersection is always a left closed interval.

(b) The set-theoretic difference $[a, b)\backslash[\alpha, \beta)$ is a union of at most two disjoint left closed intervals. This is trivial if $a = b$. Suppose $a < b$. Clearly, the complement of $[\alpha, \beta)$ is the union $(-\infty, \alpha) \cup [\beta, \infty)$ and therefore $[a, b)\backslash[\alpha, \beta)$ is the union of the intersections

$$[a,\ b) \cap (-\infty, \alpha) \text{ and } [a,\ b) \cap [\beta, \infty).$$

The first of these is empty if $a \ge \alpha$ and is the left closed interval $[a, \min\{b, \alpha\})$ otherwise; the second is empty if $b \le \beta$ and is the left closed interval $[\max\{a, \beta\}, b)$ otherwise.

(c) The union of two left closed intervals $[a, b)$ and $[\alpha, \beta)$ that have a nonempty intersection can be shown to be a left closed interval by reasoning as follows. First of all, the union of any family of intervals with a nonempty intersection is an interval. This is easy to prove by means of the characterisation of an interval as being a set that contains any number that lies between two of its numbers. It follows that the union of $[a, b)$ and $[\alpha, \beta)$ is an interval. It obviously contains $\min\{a, \alpha\}$ and nothing smaller; it also contains numbers arbitrarily close to $\max\{b, \beta\}$ and nothing bigger or equal. It follows that the union is precisely the left closed interval

$$[\min\{a, \alpha\}, \max\{b, \beta\}).$$

(d) A union of finitely many left closed intervals is also a union of finitely many disjoint left closed intervals, as we now prove by induction. If there is only one interval in the union, there is nothing to prove. Suppose the claim is true when there are k intervals in the union and let $E = \cup_{1 \le i \le k+1} E_i$, where E_i, $1 \le i \le k+1$ are left closed intervals. If all the E_i are disjoint, we are done. Suppose instead that there are distinct p and q such that E_p and E_q have a nonempty intersection. By (c) their union is a left closed interval F. Then F and the remaining E_i $(p \ne i \ne q)$ constitute k left closed intervals whose union is E. Therefore by the induction hypothesis, E is a union of finitely many disjoint left closed intervals.

Definition 5.11.7 A collection $\mathfrak{R}$ of subsets of a nonempty set X is called a **ring** if whenever E, $F \in \mathfrak{R}$, we have E $\cup$ F, $E \backslash F \in \mathfrak{R}$. A function $\mu:\mathfrak{R} \to \mathbb{R}$ is called a **measure (on the ring** $\mathfrak{R}$) if it is nonnegative-valued, $\mu(\emptyset) = 0$ and is countably additive [in the sense of Definition 2.2.7].

$$\text{Since } E \cap F = E \backslash (E \backslash F), \text{ any ring } \mathfrak{R} \text{ satisfies } E, F \in \mathfrak{R} \Rightarrow E \cap F \in \mathfrak{R}.$$

Definition 5.11.8 The family of all finite unions of left closed intervals will be denoted by $\tilde{I}$.

Remark 5.11.9 It is obvious that E, $F \in \tilde{I}$ implies $E \cup F \in \tilde{I}$. It therefore follows from Remark 5.11.6 (b) that $\tilde{I}$ is a ring.

Furthermore, by Remark 5.11.6 (d), each set belonging to this ring is a union of finitely many left closed intervals that are disjoint.

Proposition 5.11.10 *There exists a smallest ring containing a given collection of subsets of any nonempty set X.*

Proof The proof of Proposition 2.3.17 applies with the minor modification that the word "σ-algebra" is to be replaced by "ring". □

The smallest ring containing a given collection of subsets is called the ring **generated by** that collection. The σ-algebra generated by a family $\mathfrak{J}$ of subsets of $\mathbb{R}$ (as defined in Sect. 2.3) will be denoted by $S(\mathfrak{J})$. It is easy to check that the ring generated by the collection of left closed intervals is precisely $\tilde{I}$.

Proposition 5.11.11 $S(\tilde{I}) = \mathfrak{B}$.

Proof Since any nonempty bounded open interval (a, b) is a countable union $\cup_{n \ge N} [a + \frac{1}{n}, b)$, where $N > 1/(b - a)$, it must belong to $S(\tilde{I})$. But every open subset of $\mathbb{R}$ is a countable union of bounded open intervals, and therefore must also belong to $S(\tilde{I})$. Since $\mathfrak{B}$ is defined to be the σ-algebra generated by the family of open subsets, it follows that $\mathfrak{B} \subseteq S(\tilde{I})$. On the other hand, any left closed interval $[a, b)$ is a countable intersection $\cap_{n \ge 1} (a - \frac{1}{n}, b)$, and must therefore belong to $\mathfrak{B}$, which implies that $\mathfrak{B} \supseteq S(\tilde{I})$. Thus, $S(\tilde{I}) = \mathfrak{B}$. □

Theorem 5.11.12 *There exists a unique measure $\tilde{\mu}$ on $\tilde{I}$ such that if $I = [a, b)$, where $a \leq b$, then $\tilde{\mu}(I) = \mu(I)$.*

Proof Let $E \in \tilde{I}$. Then E can be written as $E = \cup_{1 \leq i \leq n} E_i$, where E_i, $1 \leq i \leq n$, are disjoint left closed intervals [Remark 5.11.9 justifies taking the E_i to be disjoint]. Define

$$\tilde{\mu}(E) = \sum_{i=1}^{n} \mu(E_i).$$

This $\tilde{\mu}$ is uniquely defined on $\tilde{I}$, because if $E = \cup_{1 \leq j \leq m} F_j$ is another decomposition of E into disjoint left closed intervals, then

$$E = \bigcup_{i,j} (E_i \cap F_j),$$

the intervals $E_i \cap F_j$ are disjoint and also left closed, and moreover,

$$\sum_{i=1}^{n} \mu(E_i) = \sum_{i=1}^{n} \sum_{j=1}^{m} \mu(E_i \cap F_j) = \sum_{j=1}^{m} \sum_{i=1}^{n} \mu(E_i \cap F_j) = \sum_{j=1}^{m} \mu(F_j),$$

using the additivity of μ [Lemma 5.11.5(b)]. So, $\tilde{\mu}$ is uniquely defined on $\tilde{I}$.

It is clear that $\tilde{\mu}$ is finitely additive.

Let $\{E_i\}_{n \geq 1}$ be a sequence of disjoint sets of $\tilde{I}$ such that $E = \cup_{i \geq 1} E_i \in \tilde{I}$. Then for each i, we have $E_i = \cup_{1 \leq j \leq m_i} E_{i,j}$, where the sets $E_{i,j}$ are disjoint left closed intervals. It follows that

$$\tilde{\mu}(E_i) = \sum_{j=1}^{m_i} \mu(E_{i,j}).$$

If E is any left closed interval, then by Lemma 5.11.5 (b), we have

$$\tilde{\mu}(E) = \mu(E) = \sum_{i=1}^{\infty} \sum_{j=1}^{m_i} \mu(E_{i,j}) = \sum_{i=1}^{\infty} \tilde{\mu}(E_i) \tag{5.93}$$

as the intervals $E_{i,j}$ are disjoint. In general, we can write any $E \in \tilde{I}$ as $E = \cup_{1 \leq k \leq m} F_k$, where the F_k are disjoint left closed intervals. Then we have

$$\tilde{\mu}(E) = \sum_{k=1}^{m} \tilde{\mu}(F_k) \quad \text{by finite additivity of } \tilde{\mu}$$
$$= \sum_{k=1}^{m}\sum_{i=1}^{\infty} \tilde{\mu}(F_k \cap E_i) \quad \text{by (5.93), because } F_k \cap E_i \in \tilde{I}$$
$$= \sum_{i=1}^{\infty}\sum_{k=1}^{m} \tilde{\mu}(F_k \cap E_i)$$
$$= \sum_{i=1}^{\infty} \tilde{\mu}(E_i) \quad \text{by finite aditivity of } \tilde{\mu} \text{ again.}$$

Thus $\tilde{\mu}$ is countably additive. Since $\tilde{\mu}(\emptyset) = \mu(\emptyset) = 0$, $\tilde{\mu}$ is a measure.

Clearly, any measure on $\tilde{I}$ that extends μ, by definition of $\tilde{\mu}$, equals $\tilde{\mu}$ on each set of $\tilde{I}$. Thus, the extension is unique. □

Remark 5.11.13 We may now drop the notation $\tilde{\mu}$ and write $\mu(E)$ for any $E \in \tilde{I}$.

Outer measures arise naturally in the attempt to extend measures from rings to larger classes of sets.

Definition 5.11.14 For an arbitrary $E \subseteq \mathbb{R}$,

$$\mu^*(E) = \inf\{\sum_{i=1}^{\infty} \mu(E_i) : E_i \in \tilde{I} \, for \, i = 1, 2, \ldots \, and \, E \subseteq \bigcup_{i=1}^{\infty} E_i\}.$$

The set function μ^* is called the **outer measure induced by** μ.

Proposition 5.11.15 *For an arbitrary* $E \subseteq \mathbb{R}$,

$$\mu^*(E) = \inf\{\sum_{i=1}^{\infty} \mu(E_i) : E_i, i$$
$$= 1, 2, \ldots, \textit{are disjoint left closed intervals such that } E \subseteq \bigcup_{i=1}^{\infty} E_i\}.$$

Proof Let

$$\mu^{**}(E) = \inf\{\sum_{i=1}^{\infty} \mu(E_i) : E_i, i = 1, 2, \ldots,$$
$$\text{are disjoint left closed intervals such that } E \subseteq \bigcup_{i=1}^{\infty} E_i\}.$$

The assertion of the proposition is that $\mu^{**}(E) = \mu^*(E)$.

In the definition of μ^{**}, the infimum is taken over a subset of the set occurring in the definition of μ^* and therefore the infimum $\mu^{**}(E)$ satisfies $\mu^{**}(E) \geq \mu^*(E)$. We claim that actually $\mu^{**}(E) = \mu^*(E)$.

In order to justify the claim, it is enough to show that whenever $E_i \in \tilde{I}$ for $i = 1, 2, \ldots$ and $E \subseteq \cup_{i \geq 1} E_i$, there exist disjoint left closed intervals F_n for $n = 1, 2, \ldots$ such that

$$E \subseteq \bigcup_{n=1}^{\infty} F_n \quad \text{and} \quad \sum_{n=1}^{\infty} \mu(F_n) \leq \sum_{i=1}^{\infty} \mu(E_i).$$

The first step is to show that there exist disjoint $G_i \in \tilde{I}$ such that $E \subseteq \cup_{i \geq 1} G_i$ and $\sum_{i=1}^{\infty} \mu(G_i) \leq \sum_{i=1}^{\infty} \mu(E_i)$.We do this by setting $G_1 = E_1$ and $G_{i+1} = E_{i+1} \backslash \cup_{1 \leq j \leq i} E_j$ for $i = 1, 2, \ldots$. Since $\tilde{I}$ is a ring, we have $G_i \in \tilde{I}$. It is clear that $G_i \subseteq E_i$ and $\cup_{i \geq 1} G_i = \cup_{i \geq 1} E_i$ and therefore

$$E \subseteq \bigcup_{i=1}^{\infty} G_i. \tag{5.94}$$

Since μ is a measure on $\tilde{I}$ by Theorem 5.11.22, it satisfies $A \subseteq B \Rightarrow \mu(A) \leq \mu(B)$ for $A, B \in \tilde{I}$ and therefore

$$\sum_{i=1}^{\infty} \mu(G_i) \leq \sum_{i=1}^{\infty} \mu(E_i). \tag{5.95}$$

As the next step, we note that by definition of $\tilde{I}$, each G_i can be written as a finite union

$$G_i = \bigcup_{j=1}^{m_i} E_{i,j}, \tag{5.96}$$

where $E_{i,j}$, $1 \leq j \leq m_i$, are disjoint [Remark 5.11.9 justifies taking the $E_{i,j}$ to be disjoint] left closed intervals. By finite additivity of μ, we have $\sum_{i=1}^{\infty} \mu(G_i) = \sum_{i=1}^{\infty} \sum_{j=1}^{m_i} \mu(E_{i,j})$ Therefore

$$\sum_{i=1}^{\infty} \mu(G_i) = \sum_{n=1}^{\infty} \mu(F_n), \tag{5.97}$$

where the sequence $\{F_n\}_{n \geq 1}$ is an enumeration of the countably many left open intervals $E_{i,j}$. It follows from (5.97) and (5.95) that

$$\sum_{n=1}^{\infty} \mu(F_n) \leq \sum_{i=1}^{\infty} \mu(E_i).$$

Now, not only are the $E_{i,j}$ disjoint for each i, the but also the sets $G_i = \cup_{1 \le j \le mi} E_{i,j}$ are disjoint; therefore the left open intervals $E_{i,j}$, and hence the F_n, are disjoint. Moreover, by (5.94) and (5.96),

$$E \subseteq \bigcup_{i=1}^{\infty} G_i = \bigcup_{i=1}^{\infty} \bigcup_{j=1}^{m_i} E_{i,j} = \bigcup_{n=1}^{\infty} F_n.$$

This justifies the claim. □

Remarks 5.11.16

(a) If E is a left closed interval, then $\mu^*(E) = \mu(E)$:

We have $E = E \cup \emptyset \cup \emptyset \cup \ldots$ and therefore $\mu^*(E) \le \mu(E)$; on the other hand, if E_i are left closed intervals for $i = 1, 2, \ldots$ and $E \subseteq \cup_{i \ge 1} E_i$, then $\mu(E) \le \sum_i \mu(E_i)$ by Lemma 5.11.5 (a), so that, on the basis of Proposition 5.11.15, we have $\mu(E) \le \mu^*(E)$.

Actually, the equality $\mu^*(E) = \mu(E)$ holds for all $E \in \tilde{I}$, but this will be proved later in Theorem 5.11.22.

(b) If $E \subseteq F \subseteq \mathbb{R}$, then $\mu^*(E) \le \mu^*(F)$; i.e., μ^* is monotone:

If $E_i \in \tilde{I}$ for $i = 1, 2, \ldots$ and $F \subseteq \cup_{i \ge 1} E_i$, then $E \subseteq \cup_{i \ge 1} E_i$ and therefore $\mu^*(E) \le \mu^*(F)$.

(c) If $E \subseteq \mathbb{R}$, $E_i \subseteq \mathbb{R}$ for $i = 1, 2, \ldots$ and $E \subseteq \cup_{i \ge 1} E_i$, then $\mu^*(E) \le \sum_{i=1}^{\infty} \mu^*(E_i)$; i.e. μ^* is countably subadditive:

Let $\varepsilon > 0$. For each i, use the definition of μ^* to choose a sequence $\{E_{i,j}\}_{j \ge 1}$ of sets belonging to $\tilde{I}$ such that

$$E_i \subseteq \bigcup_{j=1}^{\infty} E_{i,j} \text{ and } \sum_{j=1}^{\infty} \mu(E_{i,j}) \le \mu^*(E_i) + \frac{\varepsilon}{2^i}.$$

Note that $E \subseteq \cup_{i \ge 1} E_i \subseteq \cup_{i \ge 1, j \ge 1} E_{i,j}$. So,

$$\mu^*(E) \le \sum_{i=1}^{\infty} \left(\sum_{j=1}^{\infty} \mu(E_{i,j}) \right) \le \sum_{i=1}^{\infty} \left(\mu^*(E_i) + \frac{\varepsilon}{2^i} \right) = \sum_{i=1}^{\infty} \mu^*(E_i) + \varepsilon.$$

Since $\varepsilon > 0$ is arbitrary, it follows that $\mu^*(E) \le \sum_{i=1}^{\infty} \mu^*(E_i)$.

An outer measure is not necessarily countably or even finitely additive. We single out a family of subsets of $\mathbb{R}$ on which μ^* is countably, and hence also finitely, additive. The precise formulation uses Carathéodory's condition of measurability along the lines of Definition 2.3.1.

Definition 5.11.17 Let μ^* be the outer measure induced by μ (see Definition 5.11.14). A set $E \subseteq \mathbb{R}$ is called **μ^*-measurable** if

$$\mu^*(A) = \mu^*(A \cap E) + \mu^*(A \cap E^c) \textit{ for all } A \subseteq \mathbb{R}.$$

Remark 5.11.18 Since the outer measure is subadditive [see Remark 5.11.16 (c)], we have $\mu^*(A) \leq \mu^*(A \cap E) + \mu^*(A \cap E^c)$, and it follows that E is μ^*-measurable if

$$\mu^*(A) \geq \mu^*(A \cap E) + \mu^*(A \cap E^c) \text{ for all } A \subseteq \mathbb{R}.$$

This observation is frequently used in the discussion that follows.

Remark 5.11.19 By Remark 5.11.16 (a), we know that $\mu^*(\varnothing) = 0$. Therefore it is immediate from Definition 5.11.17 that $\varnothing$ and $\mathbb{R}$ are μ^*-measurable.

Next, we have the following theorem.

Theorem 5.11.20 *Let μ^* be the outer measure induced by μ and let $\mathcal{S}^*$ denote the class of μ^*-measurable sets. Then $\mathcal{S}^*$ is a σ-algebra and μ^* restricted to $\mathcal{S}^*$ is a complete measure.*

Proof The argument of Theorems 2.3.12 and 2.3.13 carries over upon replacing m^* there by what we have called μ^* here. The argument of Remark 2.3.2 (d) carries over to show that μ^* restricted to $\mathcal{S}^*$ is complete. □

Theorem 5.11.21 *Every set in $\mathfrak{B}$ is μ^*-measurable. In particular, the restriction of μ^* to $\mathfrak{B}$ is a measure. Moreover, μ^* coincides with μ on $\tilde{I}$ and it is therefore an extension of μ.*

Proof First consider $E \in \tilde{I}$ and let $A \subseteq \mathbb{R}$ be arbitrary. By definition of μ^*, for any $\varepsilon > 0$, there exists a sequence $\{E_i\}_{i \geq 1}$ of sets in $\tilde{I}$ such that $A \subseteq \cup_{i \geq 1} E_i$ and

$$\mu^*(A) + \varepsilon \geq \sum_{i=1}^{\infty} \mu(E_i).$$

Since μ is a measure on $\tilde{I}$ (see Remark 5.11.13 and Theorem 5.11.12),

$$\sum_{i=1}^{\infty} \mu(E_i) = \sum_{i=1}^{\infty} (\mu(E_i \cap E) + \mu(E_i \cap E^c)$$
$$\geq \mu^*(A \cap E) + \mu * (A \cap E^c) \quad \text{by definition of } \mu^*.$$

Combining this with the previous inequality, we get

$$\mu^*(A) + \varepsilon \geq \mu^*(A \cap E) + \mu^*(A \cap E^c).$$

Since this holds for any $\varepsilon > 0$, the sufficient condition for μ^*-measurability noted in Remark 5.11.18 is found to hold. We thus conclude that E is μ^*-measurable. Since E is an arbitrary element of $\tilde{I}$, this proves $\tilde{I} \subseteq \mathcal{S}^*$. Now, $\mathcal{S}^*$ is a σ-algebra by Theorem 5.11.20. Therefore $S(\tilde{I}) \subseteq \mathcal{S}^*$. However, $S(\tilde{I}) = \mathfrak{B}$ by Proposition 5.11.11 and hence $\mathfrak{B} \subseteq \mathcal{S}^*$. This means every set in $\mathfrak{B}$ is μ^*-measurable.

By Theorem 5.11.20, μ^* is a measure on $\mathcal{S}^*$ and is now seen to be a measure on $\mathfrak{B}$ because $\mathfrak{B} \subseteq \mathcal{S}^*$.

To see why μ^* coincides with μ on $\tilde{I}$, consider an arbitrary $E \in \tilde{I}$. Then E can be written as $E = \cup_{1 \le i \le n} E_i$, where E_i, $1 \le i \le n$, are disjoint left closed intervals. Recall from Remark 5.11.13 and the proof of Theorem 5.11.12 that $\mu(E) = \sum_{i=1}^{n} \mu(E_i)$. By Remark 5.11.16 (a), this leads to $\mu(E) = \sum_{i=1}^{n} \mu^*(E_i)$. Now that μ^* has been shown to be a measure on $\mathfrak{B}$, which contains $\tilde{I}$ by Prop. 5.11.11, we also have $\sum_{i=1}^{n} \mu^*(E_i) = \mu^*(E)$. Thus $\mu(E) = \mu^*(E)$. □

Theorem 5.11.22 *Let f be a bounded increasing function on $\mathbb{R}$ which is left continuous. Then there exists a unique measure μ_f on $\mathfrak{B}$ such that*

$$\mu_f([a, b)) = f(b) - f(a)$$

for all left closed intervals $[a, b)$.

Proof Denote the restriction of μ^* to the Borel σ-algebra $\mathfrak{B}$ by μ_f. By Theorem 5.11.21, μ_f coincides with μ on $\tilde{I}$ and it therefore satisfies

$$\mu_f([a, b)) = \mu([a, b)) = f(b) - f(a),$$

for all left closed intervals $[a, b)$. This proves the existence of μ_f.

Suppose ν is a measure on $\mathfrak{B}$ such that $\nu = \mu$ on all left closed intervals. Since every $E \in \tilde{I}$ can be written as a finite union of disjoint left closed intervals [Remark 5.11.6 (d)], it follows from the additivity of ν and the definition of μ on $\tilde{I}$ that $\nu = \mu$ on $\tilde{I}$. It follows that whenever $A = \cup_{i \ge 1} E_i$ and E_i are disjoint sets in $\tilde{I}$, the definition of μ_f and Theorem 5.11.21 lead to

$$\mu_f(A) = \sum_{i=1}^{\infty} \mu_f(E_i) = \sum_{i=1}^{\infty} \mu^*(E_i) = \sum_{i=1}^{\infty} \mu(E_i) = \sum_{i=1}^{\infty} \nu(E_i) = \nu(A). \qquad (5.98)$$

We wish to show that $\nu = \mu_f$ on $\mathfrak{B}$.

If $E \in \mathfrak{B}$ and $\varepsilon > 0$, Proposition 5.11.15 shows that there exists a sequence $\{E_i\}_{i \ge 1}$ of disjoint left closed intervals such that $E \subseteq \cup_{i \ge 1} E_i$ and

$$\mu_f(E) + \varepsilon \ge \sum_{i=1}^{\infty} \mu(E_i) = \sum_{i=1}^{\infty} \nu(E_i).$$

Since the E_i are disjoint, we further get

$$\mu_f(E) + \varepsilon \ge \nu\left(\bigcup_{i=1}^{\infty} E_i\right) \ge \nu(E),$$

which implies

$$\mu_f(E) \geq \nu(E),$$

because $\varepsilon > 0$ is arbitrary.

Conversely, suppose that $E \in \mathfrak{B}$. Then $\mu_f(E) < \infty$. Let $\varepsilon > 0$ be given. Then there exists an $A \supseteq E$ such that $\mu_f(A) < \mu_f(E) + \varepsilon$, where $A = \cup_{i \geq 1} E_i$ and E_i are disjoint sets in $\tilde{I}$. Observe that $\mu_f(A) = \nu(A)$ on the basis of (5.98). So,

$$\mu_f(E) \leq \mu_f(A) = \nu(A) = \nu(E) + \nu(A \backslash E).$$

But by what has been proved in the preceding paragraph, $\nu(A \backslash E) \leq \mu_f(A \backslash E)$. Also, since $\mu_f(E) < \infty$, we have $\mu_f(A \backslash E) = \mu_f(A) - \mu_f(E) < \varepsilon$. This implies

$$\mu_f(E) \leq \nu(E) + \varepsilon.$$

Since $\varepsilon > 0$ is arbitrary, we have $\mu_f(E) \leq \nu(E)$. □

Definition 5.11.23 If f is a bounded increasing function on $\mathbb{R}$ which is left continuous, the unique measure μ_f on $\mathfrak{B}$ such that

$$\mu_f([a,\ b)) = f(b) - f(a)$$

for all left closed intervals [a, b) is called the **Lebesgue–Stieltjes measure induced by** f.

Remark 5.11.24

(a) Since two bounded increasing left continuous functions on $\mathbb{R}$ that differ by a constant obviously lead to the same $\tilde{\mu}$ and μ^*, they induce the same Lebesgue–Stieltjes measure.
(b) Let f be a bounded increasing function on $\mathbb{R}$ which is left continuous and μ_f be the Lebesgue–Stieltjes measure on $\mathfrak{B}$ induced by f, i.e., $\mu_f([a, b)) = f(b) - f(a)$ for all left closed intervals $[a, b)$. Then μ_f is bounded. Conversely, starting with a bounded measure μ on $\mathfrak{B}$, we can generate a bounded increasing function f_μ on $\mathbb{R}$ which is left continuous. Although there is no one-to-one correspondence here, the following is true:

Starting with the bounded measure μ_f on $\mathfrak{B}$ induced by a bounded increasing function f on $\mathbb{R}$, if we generate the bounded increasing function f_{μ_f} on $\mathbb{R}$, then f_{μ_f} differs from f by a constant. It is sufficient to show that when F is also a bounded increasing function and $\mu_F = \mu_f$, the functions F and f differ by a constant. Since f and F are both bounded and increasing, the limits $\lim_{x \to -\infty} f(x)$ and $\lim_{x \to -\infty} F(x)$ must both exist. It follows for any real b that

$$F(b) - \lim_{x\to-\infty} F(x) = \lim_{x\to-\infty} (F(b) - F(x)) = \lim_{x\to-\infty} \mu_F([x,b))$$
$$= \lim_{x\to-\infty} \mu_f([x,b)) = \lim_{x\to-\infty} (f(b) - f(x)) = f(b) - \lim_{x\to-\infty} f(x).$$

Hence,

$$F(b) - f(b) = \lim_{x\to-\infty} F(x) - \lim_{x\to-\infty} f(x).$$

Since this holds for any real b, the functions F and f differ by the constant $\lim_{x\to-\infty} F(x) - \lim_{x\to-\infty} f(x)$.

(c) Observe the consequence of part (b) that Theorem 5.11.2 has the companion result that if f is a bounded increasing function on $\mathbb{R}$ then $\mu_f \ll m$ if and only if f is absolutely continuous.
(d) For any $x \in \mathbb{R}$, we have $\mu_f(\{x\}) = f(x+) - f(x)$. This is because $\{x\} = \cap_{n\geq 1}[x, x + \frac{1}{n})$, which leads via outer continuity of measure (see Proposition 3.1.8) to $\mu_f(\{x\}) = \lim_{n\to\infty} \mu f\left(\left[x, x+\frac{1}{n}\right)\right) = \lim_{n\to\infty} f\left(x+\frac{1}{n}\right) - f(x) = f(x+) - f(x)$.

Remark 5.11.25 The relation between the Riemann–Stieltjes integral and the integral with respect to Lebesgue–Stieltjes measure is as described below.

If φ is a Borel measurable function and f is a monotone increasing function which is left continuous, we define the Lebesgue–Stieltjes integral of φ as

$$\int \varphi\, df = \int \varphi\, d\mu,$$

where $\mu = \mu_f$ is the Lebesgue–Stieltjes measure induced by f as in this section. It is clear what is meant by Lebesgue–Stieltjes integral $\int_{[a,\, b]} \varphi\, df$, where $a \leq b$.

We shall show for a certain class of functions φ on $[a, b]$, which includes continuous functions, that

$$\int_{[a,\, b)} \varphi df = \int_a^b \varphi(x) df(x),$$

where the integral on the right is the Riemann–Stieltjes integral.

If $\varphi = \chi_{[\alpha,\beta)}$, where $[\alpha, \beta) \subseteq [a, b)$, then φ is right continuous and hence by Theorem 11.2.10 of [28],

$$\int_a^b \varphi(x)\, df(x) = f(\beta) - f(\alpha).$$

Also,

$$\int_{[a,\,b)} \varphi\, df = \mu_f[\alpha, \beta) = f(\beta) - f(\alpha).$$

The equality in question holds therefore for linear combinations of characteristic functions of left closed intervals.

Let φ be any continuous function on $[a, b]$. It is uniformly continuous and therefore there exists a sequence of functions φ_n, each of which is a linear combination of characteristic functions of left closed intervals, such that φ_n converges uniformly to φ. By Theorem 11.3.1 of [28], the uniform convergence yields

$$\int_a^b \varphi_n(x)\, df(x) \to \int_a^b \varphi(x)\, df(x).$$

Since μ_f is a finite measure, the uniform convergence also yields

$$\int_{[a,\,b)} \varphi_n\, df \to \int_{[a,\,b)} \varphi\, df.$$

Since the equality in question has been shown to hold for every φ_n, it follows from the above limits that the equality holds for φ.

Let μ be a finite nonnegative measure on $(\mathbb{R},\mathfrak{B})$ and f_μ be the related function as defined at the beginning of this section:

$$f_\mu(x) = \mu((-\infty, x)) = \mu(\{y \in \mathbb{R} : y < x\}).$$

Recall that by Theorem 5.11.1, the function f_μ is bounded, increasing and left continuous. Moreover, $\lim_{x \to -\infty} f_\mu(x) = 0$ Being an increasing function, it has a derivative a.e. (see Lebesgue's Theorem 5.4.6). In the result that follows, we shall compute the derivative $f_\mu{}'$ wherever it exists and motivate a definition of the derivative of a measure on $\mathbb{R}$, independently of the Radon–Nikodým derivative. Its relation to the latter will be described later in Proposition 5.11.28.

Theorem 5.11.26 *Let* μ *and* $f = f_\mu$ *be as in the paragraph above and let* $x_0 \in \mathbb{R}$. *Then the following statements are equivalent*:

(a) *f is differentiable at x_0 with $f'(x_0) = k$;*
(b) *for every $\varepsilon > 0$, there exists a $\delta > 0$ such that*

$$\left|\frac{\mu(I)}{m(I)} - k\right| < \varepsilon$$

for every open interval I with $x_0 \in I$ and $\ell(I) = m(I) < \delta$.

Proof (a) $\Rightarrow$ (b). Suppose that (a) holds.

Let $\varepsilon > 0$ be given. There exists a $\delta > 0$ such that

$$\left|\frac{f(t)-f(x_0)}{t-x_0}-k\right|\leq\varepsilon$$

provided $|t-x_0|<\delta$. Suppose $x_0\in I=(s,t)$ and $t-s<\delta$. Choose $\{s_n\}_{n\geq 1}$ such that $x_0>s_1>s_2>\cdots>s_n\rightarrow s$. Then

$$\left|\frac{\mu([s_n,t))}{t-s_n}-k\right|=\left|\frac{f(t)-f(s_n)}{t-s_n}-k\right|=\left|\frac{f(t)-f(x_0)}{t-s_n}+\frac{f(x_0)-f(s_n)}{t-s_n}-k\right|$$
$$=\left|\frac{t-x_0}{t-s_n}\frac{f(t)-f(x_0)}{t-x_0}+\frac{x_0-s_n}{t-s_n}\frac{f(x_0)-f(s_n)}{x_0-s_n}-\frac{t-x_0}{t-s_n}k-\frac{x_0-s_n}{t-s_n}k\right|$$
$$\leq\frac{t-x_0}{t-s_n}\left|\frac{f(t)-f(x_0)}{t-x_0}-k\right|+\frac{x_0-s_n}{t-s_n}\left|\frac{f(x_0)-f(s_n)}{x_0-s_n}-k\right|$$
$$\leq\frac{t-x_0}{t-s_n}\varepsilon+\frac{x_0-s_n}{t-s_n}\varepsilon=\varepsilon.$$

Since $I=\cup_{n\geq 1}[s_n,t)$, it follows by using inner continuity of measure (see Proposition 3.1.8) that $\mu(I)=\lim_{n\to\infty}\mu([s_n,t))$. But $m(I)=t-s=\lim_{n\to\infty}(t-s_n)$. Therefore

$$\left|\frac{\mu(I)}{m(I)}-k\right|\leq\varepsilon.$$

An alternative presentation of the above argument that makes the use of convexity more explicit is as follows.

Let I be an open interval (α,β) with $x_0\in I$. i.e. $\alpha<x_0<\beta$. The simple equality

$$\frac{f(\beta)-f(\alpha)}{\beta-\alpha}=\frac{x_0-\alpha}{\beta-\alpha}\frac{f(x_0)-f(\alpha)}{x_0-\alpha}+\frac{\beta-x_0}{\beta-\alpha}\frac{f(\beta)-f(x_0)}{\beta-x_0}$$

shows that $\frac{f(\beta)-f(\alpha)}{\beta-\alpha}$ is a convex combination of $\frac{f(x_0)-f(\alpha)}{x_0-\alpha}$ and $\frac{f(\beta)-f(x_0)}{\beta-x_0}$. Therefore, if the latter two lie within some positive distance η of $f'(x_0)$, so does the first.

For any $\varepsilon>0$, there exists a $\delta>0$ such that $0<x_0-\alpha<\delta$ and $0<\beta-x_0<\delta$ imply that the latter two numbers mentioned in the preceding paragraph both lie within $\frac{1}{2}\varepsilon$ of $f'(x_0)$.

Now, consider $x_0\in I$, i.e. $\alpha<x_0<\beta$ and suppose $m(I)<\delta$, i.e., $\beta-\alpha<\delta$. Let α_1 such that $\alpha<\alpha_1<x_0$ be arbitrarily chosen. Then $\beta-\alpha_1<\delta$ and the inequalities $0<x_0-\alpha<\delta$, $0<x_0-\alpha_1<\delta$ and $0<\beta-x_0<\delta$ all hold, and hence by the convexity noted above,

$$\frac{f(\beta)-f(\alpha)}{\beta-\alpha}\quad\text{and}\quad\lim_{\alpha_1\to\alpha}\frac{f(\beta)-f(\alpha_1)}{\beta-\alpha_1}$$

both lie within a distance ε of $f'(x_0)$. Now,

$$\frac{\mu(I)}{m(I)} = \frac{\mu((\alpha, \beta))}{\beta - \alpha} \leq \frac{\mu([\alpha, \beta))}{\beta - \alpha} = \frac{f(\beta) - f(\alpha)}{\beta - \alpha}$$

and

$$\frac{\mu(I)}{m(I)} = \frac{\mu((\alpha, \beta))}{\beta - \alpha} \geq \frac{\lim_{\alpha_1 \to \alpha} \mu([\alpha_1, \beta))}{\beta - \alpha} = \lim_{\alpha_1 \to \alpha} \frac{\mu([\alpha_1, \beta))}{\beta - \alpha_1} = \lim_{\alpha_1 \to \alpha} \frac{f(\beta) - f(\alpha_1)}{\beta - \alpha_1}.$$

It follows that $\frac{\mu(I)}{m(I)}$ lies within a distance ε of $f'(x_0)$.

(b) $\Rightarrow$ (a). On the other hand, suppose (b) holds. Since $\frac{\mu(I)}{m(I)} \geq 0$ always, we must have $k \geq 0$.

First we prove that f is continuous at x_0. By taking $\varepsilon = 1$ in (b), we obtain an increasing sequence $\{s_n\}$ converging to x_0 and a decreasing sequence $\{t_n\}$ converging to x_0 such that $\mu((s_n, t_n)) \leq (1 + k)(t_n - s_n)$. Since $\{x_0\} = \cap_{n \geq 1}(s_n, t_n)$, we use outer continuity of measure (see Proposition 3.1.8) to arrive at the equality $\mu(\{x_0\}) = 0$. We deduce from this and Theorem 5.11.1 that f is continuous at x_0. The same theorem also tells us that it is an increasing function.

Next, for $\varepsilon > 0$, choose δ as in (b). If $s < x_0 < t$ and $t - s < \delta$, then

$$\left| \frac{\mu((s - \frac{1}{n}, t))}{t - s + \frac{1}{n}} - k \right| < \varepsilon$$

for all sufficiently large n. Since $[s, t) = \cap_{n \geq 1}(s - \frac{1}{n}, t)$ and $f(t) - f(s) = \mu([s, t))$, it follows that

$$\left| \frac{f(t) - f(s)}{t - s} - k \right| \leq \varepsilon \quad \text{for } s < x_0 < t < s + \delta, \tag{5.99}$$

using outer continuity of measure once again. Now we would like to take limits as $s \to x_0$ and as $t \to x_0$ separately in (5.99); however, the range of values of s or that of t, as described in (5.99), is not explicitly an interval. Therefore we resort to an artifice. The inequalities

$$x_0 - \frac{1}{2}\delta < s < x_0 \quad \text{and} \quad x_0 < t < x_0 + \frac{1}{2}\delta$$

together imply

$$s < x_0 < t < s + \delta.$$

In conjunction with (5.99), this leads to

$$\left|\frac{f(t) - f(s)}{t - s} - k\right| \leq \varepsilon \quad \text{for } x_0 - \frac{1}{2}\delta < s < x_0 \quad \text{and } x_0 < t < x_0 + \frac{1}{2}\delta.$$

This permits us to take limits as $s \to x_0$ and as $t \to x_0$ separately. Taking the limits and keeping in mind the continuity of f at x_0, we obtain respectively

$$\left|\frac{f(t) - f(x_0)}{t - x_0} - k\right| \leq \varepsilon \quad \text{for } x_0 < t < x_0 + \frac{1}{2}\delta \tag{5.100}$$

and

$$\left|\frac{f(x_0) - f(s)}{x_0 - s} - k\right| \leq \varepsilon \quad \text{for } x_0 - \frac{1}{2}\delta < s < x_0. \tag{5.101}$$

Since we have shown that, for every $\varepsilon > 0$, there exists a $\delta > 0$ such that (5.100) and (5.101) hold, it follows that the right and left derivatives of f at x_0 both exist and are both equal to k. □

Definition 5.11.27 Let μ be a finite nonnegative measure on $(\mathbb{R}, \mathfrak{B})$. We say that μ is differentiable at $x \in \mathbb{R}$ if the limit

$$D\mu(x) = \lim_{m(I)\to 0} \frac{\mu(I)}{m(I)}, \quad x \in \mathbb{R}$$

exists in the sense of part (b) of Theorem 5.11.26. The limit $D\mu(x)$ is called the **derivative of μ at x**.

Proposition 5.11.28 *Let μ_f be the Lebesgue–Stieltjes measure induced by an absolutely continuous bounded increasing function f on $\mathbb{R}$. Then*

$$\left[\frac{d\mu_f}{dm}\right] = f' = D\mu_f \quad a.e.$$

Proof Since f is absolutely continuous, Remark 5.11.24(c) allows us to infer that $\mu_f \ll m$. Therefore the Radon–Nikodým derivative in question exists and is characterised by the equality

$$\mu_f(E) = \int_E \left[\frac{d\mu_f}{dm}\right] dm \quad \text{for all Borel sets } E.$$

Since μ_f is finite, the Radon–Nikodým derivative is integrable. It follows by Theorem 5.5.1 that f' is integrable and by Theorem 5.8.2 that whenever $a < b$, we have

$$\int_{[a,\,b]} f'dm = f(b) - f(a) = \mu_f([a,\,b)) = \int_{[a,\,b]} \left[\frac{d\mu_f}{dm}\right] dm.$$

The first of the desired equalities is now a consequence of Problem 3.2.P14(b) and the integrability of both the Radon–Nikodým derivative and f'. The second follows upon applying Theorem 5.11.26 and Definition 5.11.27. □

Remark 5.11.29 Suppose the bounded increasing left continuous function f on $\mathbb{R}$ is constant on some interval (α,∞). Then it is trivial for every left closed interval $E \subseteq (\alpha, \infty)$ that $\mu_f(E) = \mu(E) = 0$. It follows in two obvious steps that $\mu^*(E) = 0$ for an arbitrary set $E \subseteq (\alpha, \infty)$ and hence $\mu(E) = 0$ for every Borel set $E \subseteq (\alpha, \infty)$. Similarly, if f is constant on some interval $(-\infty, \alpha)$, then $\mu(E) = 0$ for every Borel set $E \subseteq (-\infty, \alpha)$.

Problem Set 5.11

5.11.P1. Suppose μ is a measure on $\tilde{I}$ and μ^* is the outer measure induced by μ. Let I_σ denote the family of countable unions of sets of $\tilde{I}$. Given any set E and any $\varepsilon > 0$, show that there is a set $A \in I_\sigma$ such that $E \subseteq A$ and

$$\mu^*(A) \le \mu^*(E) + \varepsilon.$$

5.11.P2. Suppose μ is a measure on $\tilde{I}$ and μ^* is the outer measure induced by μ. Let I_σ denote the family of countable unions of sets of $\tilde{I}$ and $I_{\sigma\delta}$ denote the family of countable intersections of sets of I_σ. Given any set E, show that there is a set $A \in I_{\sigma\delta}$ such that $E \subseteq A$ and $\mu^*(E) = \mu^*(A)$.

5.11.P3. Let F be a real-valued bounded increasing right continuous function on $\mathbb{R}$ and for any right closed interval $(a, b]$, define $\mu((a, b]) = F(b) - F(a)$. Show that:

(i) Let $\{E_i\}_{1\le i\le n}$ be disjoint right closed intervals ($E_i = (a_i, b_i]$, $1 \le i \le n$) such that $\cup_{1\le i\le n}E_i \subseteq I$, where I is a right closed interval. Then $\sum_{i=1}^{n} \mu(E_i) \le \mu(I)$.

(ii) If the right closed interval $(a_0, b_0]$ is contained in the union of a sequence of right closed intervals $(a_i, b_i]$, $i = 1, 2, \ldots$, then

$$\mu((a_0, b_0]) \le \sum_{i=1}^{\infty} \mu((a_i, b_i]);$$

moreover, if the intervals in the sequence $(a_i, b_i]$, $i = 1, 2, \ldots$, are disjoint and $(a_0, b_0] = \cup_{i\ge 1}(a_i, b_i]$, then

$$\mu((a_0, b_0]) = \sum_{i=1}^{\infty} \mu((a_i, b_i]).$$

5.11.P4. Let h be a real-valued bounded increasing left continuous function on $\mathbb{R}$ such that $h' = 0$ a.e. and $\mu = \mu_h$ be the Lebesgue–Stieltjes measure induced by h. Suppose also that there is a closed bounded interval I such that h is constant on each of the unbounded open intervals that constitute the complement I^c. Show that μ and Lebesgue measure m (on Borel sets) are mutually singular (see Definition 5.9.13).

5.11.P5. Suppose f is an increasing function on $[a, b]$ that is left continuous and let $f = g + h$ be a Lebesgue decomposition in accordance with Problem 5.8.P7. Extend each of the three functions to $\mathbb{R}$ so as to take the same value on $(-\infty, a)$ as at a and to take any constant value on (b, ∞) not less than the value at b, but so chosen that the equality $f = g + h$ holds on (b, ∞). Prove the following:

(a) There is some constant c such that $g(x) = c + \int_{[a,\, x]} f'$ for all $x \in \mathbb{R}$.
(b) The functions f, g and h on $\mathbb{R}$ are also bounded, increasing and left continuous; moreover, $f = g + h$ on $\mathbb{R}$.
(c) If μ_f, μ_g and μ_h are the Lebesgue–Stieltjes measures induced by f, g and h respectively, then $\mu_f = \mu_g + \mu_h$ and this equality constitutes the unique Lebesgue decomposition of μ_f with respect to Lebesgue measure m.

5.11.P6. Let μ be a finite nonnegative measure on $(\mathbb{R}, \mathfrak{B})$ and $f = f_\mu$. Suppose $[a, b]$ is an arbitrary interval and ϕ is the function on $\mathbb{R}$ obtained by extending the restriction of f to $[a, b]$ so as to be equal to $f(a)$ on $(-\infty, a)$ and equal to $f(b+)$ on (b, ∞). Then obviously, ϕ is bounded, increasing and left continuous. Show that the Lebesgue–Stieltjes measure μ_ϕ induced by ϕ is given by $\mu_\phi(E) = \mu(E \cap [a, b])$ for every Borel set E.

5.11.P7. Let μ be a finite nonnegative measure on $(\mathbb{R}, \mathfrak{B})$ and $f = f_\mu$. If μ and Lebesgue measure m (on $\mathfrak{B}$) are mutually singular, show that $f' = 0$ a.e.

5.11.P8. Let μ be a finite nonnegative measure on $(\mathbb{R}, \mathfrak{B})$ and $\mu = \nu + \lambda$ be its Lebesgue decomposition with respect to Lebesgue measure m. Then f_ν is absolutely continuous on any interval $[a, b]$ and $f_\lambda' = 0$ a.e. Also, $f_\mu = f_\nu + f_\lambda$. [Remark: This means the equality $f_\mu = f_\nu + f_\lambda$ is a Lebesgue decomposition of f_μ on any interval $[a, b]$ in accordance with Problem 5.8.P7.]

Chapter 6
Lebesgue Spaces and Modes of Convergence

6.1 The Spaces L^p as Normed Linear Spaces

The theory of integration developed in Chap. 3 enables us to define certain spaces of functions, namely, L^p spaces, $1 \leq p \leq \infty$, whose importance in analysis is hard to overemphasise. We considered the L^p norm to define the distance between two points of the space. With this notion of distance, a satisfying and useful property of the space, namely, completeness, was proved (Theorems 3.3.10 and 3.3.12). Certain dense subsets of the L^p spaces, $1 \leq p < \infty$, were also listed. In what follows, we shall be concerned with a characterisation of what are known as bounded linear functionals on L^p spaces, $1 \leq p < \infty$.

A **real linear** (or **vector**) **space** is a set V, whose elements are called *vectors* and, in which two operations called *addition* and *scalar multiplication* are defined with the following familiar algebraic properties:

(a) To each pair of vectors x and y, there corresponds a vector $x + y$ satisfying the following:

$$x + y = y + x;$$
$$x + (y + z) = (x + y) + z;$$

V contains a unique vector 0 (the *zero vector* or the *origin of* V) such that

$$x + 0 = x \quad \text{for}\, x \in V$$

and to each $x \in V$, there corresponds a unique vector $-x$ such that

$$x + (-x) = 0.$$

S. Shirali and H. L. Vasudeva, *Measure and Integration*,
Springer Undergraduate Mathematics Series,
https://doi.org/10.1007/978-3-030-18747-7_6

(b) To each pair (α, x), where $x \in V$ and $\alpha \in \mathbb{R}$ (called a *scalar* in this context), there corresponds a vector $\alpha x \in V$ in such a way that

$$1x = x, \ \alpha(\beta x) = (\alpha\beta)x, \ \alpha(x+y) = \alpha x + \alpha y, \ (\alpha+\beta)x = \alpha x + \beta x$$

for $x, y \in V$ and scalars α, β.

The concept defined above makes sense when the scalars are allowed to be rational or complex numbers, but we shall be concerned only with real numbers as scalars. So, we shall drop the word "real" and speak of simply a "linear space".

Examples

(i) Let $V = \mathbb{R}^n$. For $x = (x_1, x_2, \ldots, x_n), y = (y_1, y_2, \ldots, y_n)$ and $\alpha \in \mathbb{R}$, define $x + y = (x_1 + y_1, x_2 + y_2, \ldots, x_n + y_n)$ and $\alpha x = (\alpha x_1, \alpha x_2, \ldots, \alpha x_n)$. Then V is a linear space. In particular, $\mathbb{R}$ is seen to be a linear space when $n = 1$.

(ii) Let V be the collection of all functions $f : X \to \mathbb{R}$ on a nonempty set X. Define $(f + g)(x) = f(x) + g(x)$ and $(\alpha f)(x) = \alpha f(x)$ for all $x \in X$ and $\alpha \in \mathbb{R}$. The preceding example is just a special case with X chosen to be the finite set $\{1, 2, \ldots, n\}$.

(iii) Let V be the collection of all *bounded* functions $f : X \to \mathbb{R}$ on a nonempty set X. Define $(f + g)(x) = f(x) + g(x)$ and $(\alpha f)(x) = \alpha f(x)$ for all $x \in X$ and $\alpha \in \mathbb{R}$. Then V is a linear space, as can be easily verified. The first step is to observe that sums and constant multiples of bounded functions are again bounded.

(iv) Let $X = [a, b]$, a closed bounded interval, and $C[a, b]$ be the set of all continuous real-valued functions on $[a, b]$. A sum of continuous functions and a constant multiple of a continuous function are always continuous and hence the operations $f + g$ and αf can be set up in $C[a, b]$ in the same manner as in (ii) above. It is then a trivial verification that $C[a, b]$ becomes a linear space.

(v) By virtue of Theorem 3.3.8, operations $f + g$ and αf can be set up in the space L^p of functions, where $1 \le p \le \infty$. It is as easy as in the case of $C[a, b]$ to verify that the space L^p of functions thereby becomes a linear space.

The L^p spaces studied so far are examples of a *complete normed linear space*, also known as a *Banach space*. This concept, and some related ones that we shall work with, are defined below:

Definition 6.1.1 Let V be a linear space. A function $\|\cdot\| : V \to \mathbb{R}$, whose value at $x \in V$ is written as $\|x\|$, is said to be a **norm** on V if it satisfies the following conditions for all $x, y \in V$ and all $\alpha \in \mathbb{R}$: $\|x\| \ge 0$ with equality if and only if $x = 0$; $\|\alpha x\| = |\alpha|\|x\|$; and $\|x + y\| \le \|x\| + \|y\|$.

If $\|\cdot\|$ is a norm on V, then V is called a **normed linear space**.

When V is a normed linear space, $d(x, y) = \|x - y\|$ defines a metric on V, called the **metric induced by** the norm. If V is a complete metric space with this metric, then V is called a **complete normed linear space** or a **Banach space**.

Examples

(a) $\mathbb{R}^n$ is a normed linear space. The following are examples of norms on $\mathbb{R}^n$:

$$\|x\|_p = \left(\sum_{i=1}^{n} |x_i|^p\right)^{\frac{1}{p}}, \text{ where } x = (x_1, x_2, \ldots, x_n) \in \mathbb{R}^n \text{ and } 1 \le p < \infty;$$
$$\|x\|_\infty = \sup\{|x_i| : 1 \le i \le n\}, \text{ where } x = (x_1, x_2, \ldots, x_n) \in \mathbb{R}^n.$$

In the particular case when $n = 1$, the norm $\|x\|_p$ agrees with the absolute value $|x|$ for $1 \le p \le \infty$.

(b) The space $C(I)$ of continuous real-valued functions defined on a closed bounded interval $I \subseteq \mathbb{R}$ may be normed by $\|f\| = \|f\|_\infty = \sup_{x \in I} |f(x)|$. It is understood that I contains more than one point. It can be verified that $\|\cdot\|_\infty$ is a norm on $C(I)$ and $C(I)$ is a complete normed linear space, the latter being essentially a reformulation of the well-known fact that a uniformly Cauchy sequence of continuous functions on I is uniformly convergent with a continuous limit. However, $C(I)$ with $\|f\|_1 = \int_I |f|\, d\mu$ is a normed space that is not complete. Lack of completeness is a consequence of Proposition 3.3.14.

(c) The list of properties of $\|f\|_p$ in Remark 3.3.9 says that $L^p(X)$, $1 \le p \le \infty$, is a normed linear space. By Theorems 3.3.10 and 3.3.12, it is complete and therefore a Banach space.

Definition A **linear transformation** of a linear space V into a linear space V_1 is a mapping F of V into V_1 such that

$$F(\alpha x + \beta y) = \alpha F(x) + \beta F(y)$$

for $x, y \in V$ and scalars α, β.

It is trivial to verify that $F(0) = 0$. In the particular case when $V = \{0\}$, there is only one linear transformation possible.

In the special case in which V_1 is $\mathbb{R}$, the field of scalars, F is called a **linear functional**.

Examples

(i) Let X be an arbitrary nonempty set and V be the linear space of all real-valued functions on X. Given any fixed $x_0 \in X$, the map F_0 of V into $\mathbb{R}$ defined by $F_0(f) = f(x_0)$ is a linear functional on V. More generally, given finitely many $x_1, x_2, \ldots, x_n \in X$ and corresponding real numbers $\lambda_1, \lambda_2, \ldots, \lambda_n$, the mapping F of V into $\mathbb{R}$ defined by $F(f) = \sum_{j=1}^{n} \lambda_j f(x_j)$ is a linear functional on V.

(ii) Let $V = \mathbb{R}^n$. Recall that this is just a special case of (i) above, wherein X has been chosen to be the finite set $\{1, 2, \ldots, n\}$. Thus, given real numbers

$\lambda_1, \lambda_2, \ldots, \lambda_n$, the mapping F such that $F(x_1, x_2, \ldots, x_n) = \sum_{j=1}^{n} \lambda_j x_j$ is a linear functional. In fact, every linear functional F is of this kind: Given F, put $\lambda_j = F(0, 0, \ldots, 0, 1, 0, \ldots, 0)$, where the only "1" is in the jth position, and one can then check by computation that the equality $F(x_1, x_2, \ldots, x_n) = \sum_{j=1}^{n} \lambda_j x_j$ holds. Thus all linear functionals look alike and we know exactly how they look.

(iii) The map $F : C[a,b] \to \mathbb{R}$ defined by $F(f) = \int_{[a,b]} f\, dm$ is a linear functional on the linear space $C[a, b]$. As before, given finitely many $x_1, x_2, \ldots, x_n \in [a, b]$ and corresponding real numbers $\lambda_1, \lambda_2, \ldots, \lambda_n$, the mapping F of $C[a, b]$ into $\mathbb{R}$ defined by $F(f) = \sum_{j=1}^{n} \lambda_j f(x_j)$ is a linear functional on $C[a, b]$.

The **norm** of any linear transformation T from a normed linear space $V \neq \{0\}$ into a normed linear space V_1, denoted by $\|\boldsymbol{T}\|$, is defined to be

$$\|T\| = \sup\left\{\frac{\|Tx\|}{\|x\|} : x \neq 0\right\}.$$

In the event that $V = \{0\}$, there is only one linear transformation T possible and its norm is 0. A linear transformation is said to be **bounded** if the norm is finite; when $V \neq \{0\}$, this is equivalent to boundedness of the set mentioned above. When we speak of the norm of a linear transformation or linear functional, it is often presumed to be finite unless the context demands otherwise.

Examples In the course of the preceding two examples, we described infinitely many norms on $\mathbb{R}^n$ and all linear functionals on it. The linear functionals can be shown to be bounded no matter which norm one takes on $\mathbb{R}^n$. The norm of a linear functional naturally depends on what norm one takes on $\mathbb{R}^n$ and the focus here is on the different norms of the same linear functional when one takes the different norms $\|\cdot\|_1, \|\cdot\|_2$ and $\|\cdot\|_\infty$ on $\mathbb{R}^n$. A generalisation will be established in Propositions 6.1.2(v) and 6.1.3. Recall that a linear functional F on $\mathbb{R}^n$ is given by $\lambda_1, \lambda_2, \ldots, \lambda_n$ via the equality $F(x_1, x_2, \ldots, x_n) = \sum_{i=1}^{n} \lambda_j x_j$.

(i) First consider $\mathbb{R}^n$ with the norm $\|\cdot\|_1$. We have

$$|F(x_1, x_2, \ldots, x_n)| = \left|\sum_{j=1}^{n} \lambda_j x_j\right| \leq \sum_{j=1}^{n} |x_j| \cdot \max\{|\lambda_i| : 1 \leq i \leq n\},$$

which means the value

$$M = \max\{|\lambda_i| : 1 \leq i \leq n\}$$

fulfils the inequality

$$\frac{|F(x)|}{\|x\|_1} \leq M \quad \text{whenever } 0 \neq x \in \mathbb{R}^n.$$

Furthermore, M has to be $|\lambda_k|$ for some k. Consider the vector $z = (z_1, z_2, \ldots, z_n) \in \mathbb{R}^n$ such that $z_k = 1$ but all other z_j are 0. This z has the property that $\|z\|_1 = 1$ and

$$|F(z)| = \left|\sum_{j=1}^{n} \lambda_j z_j\right| = |\lambda_k| = |\lambda_k| \cdot 1 = M\|z\|_1.$$

Since $\|z\|_1 > 0$, this shows that the value of M that we have given is the largest possible value of $|F(x)|/\|x\|_1$, $\|x\|_1 \neq 0$. It is therefore the supremum involved in the definition of $\|F\|$. Thus max $\{|\lambda_i| : 1 \leq i \leq n\}$ is the norm of the linear functional given by λ_1, λ_2, ..., λ_n when $\mathbb{R}^n$ is taken with the norm $\|\cdot\|_1$. In other words, if we think of the linear functional F as given by $(\lambda_1, \lambda_2, \ldots, \lambda_n) \in \mathbb{R}^n$, then its norm as a linear functional on $\mathbb{R}^n$ taken with the norm $\|\cdot\|_1$ is $\|(\lambda_1, \lambda_2, \ldots, \lambda_n)\|_\infty$. This is generalised in Proposition 6.1.2(v).

(ii) Now consider $\mathbb{R}^n$ with the norm $\|\cdot\|_2$. We have

$$|F(x_1, x_2, \ldots, x_n)| = \left|\sum_{j=1}^{n} \lambda_j x_j\right| \leq (\sum_{j=1}^{n} \lambda_j^2)^{\frac{1}{2}} (\sum_{j=1}^{n} x_j^2)^{\frac{1}{2}}$$

which means the value

$$M = (\sum_{j=1}^{n} \lambda_j^2)^{\frac{1}{2}}$$

fulfils the inequality

$$\frac{|F(x)|}{\|x\|_2} \leq M \text{ whenever } 0 \neq x \in \mathbb{R}^n.$$

In the case when $\lambda_j = 0$ for every j, it is trivial to see that $F(x) = 0$ for every x and hence that $\|F\| = 0 = (\sum_{j=1}^{n} \lambda_j^2)^{1/2}$. Suppose $\lambda_j \neq 0$ for some j. Consider the vector $z = (\lambda_1, \lambda_2, \ldots, \lambda_n) \in \mathbb{R}^n$. This z has the property that $\|z\|_2 = (\sum_{j=1}^{n} \lambda_j^2)^{1/2}$ and

$$|F(z)| = \left|\sum_{j=1}^{n} \lambda_j z_j\right| = \sum_{j=1}^{n} \lambda_j^2 = (\sum_{j=1}^{n} \lambda_j^2)^{\frac{1}{2}} (\sum_{j=1}^{n} \lambda_j^2)^{\frac{1}{2}} = M\|z\|_2.$$

Since $\|z\|_2 > 0$, this shows that the value of M that we have given is the largest possible value of $|F(x)|/\|x\|_2$, $\|x\|_2 \neq 0$. It is therefore the supremum involved in the definition of $\|F\|$. Thus $(\sum_{j=1}^{n} \lambda_j^2)^{1/2}$ is the norm of the linear functional given by $\lambda_1, \lambda_2, \ldots, \lambda_n$ when $\mathbb{R}^n$ is taken with the norm $\|\cdot\|_2$.
In other words, if we think of the linear functional F as given by $(\lambda_1, \lambda_2, \ldots, \lambda_n) \in \mathbb{R}^n$, then its norm as a linear functional on $\mathbb{R}^n$ taken with the norm $\|\cdot\|_2$ is $\|(\lambda_1, \lambda_2, \ldots, \lambda_n)\|_2$. This is generalised in Proposition 6.1.3.

(iii) It is left to the reader to check that the norm of the linear functional given by $\lambda_1, \lambda_2, \ldots, \lambda_n$ is $\sum_{j=1}^{n} |\lambda_j|$ when $\mathbb{R}^n$ is taken with the norm $\|\cdot\|_\infty$. That is to say, if we think of the linear functional F as given by $(\lambda_1, \lambda_2, \ldots, \lambda_n) \in \mathbb{R}^n$, then its norm as a linear functional on $\mathbb{R}^n$ taken with the norm $\|\cdot\|_\infty$ is $\|(\lambda_1, \lambda_2, \ldots, \lambda_n)\|_1$. This too is generalised in Proposition 6.1.3 below.

Remark The above ideas extend to the case when the linear space admits complex scalars. However, we shall not be concerned with the extension.

Hölder's Inequality suggests a way of defining bounded linear functionals on L^p spaces. We begin by doing this in the case $p = 1$.

Proposition 6.1.2 *Let $g \in L^\infty(X)$ and let F_g be the mapping on $L^1(X)$ defined by*

$$F_g(f) = \int_X fg \, d\mu, \qquad f \in L^1(X). \tag{6.1}$$

Then F_g has the following properties:

(i) *F_g is well defined*;
(ii) *F_g is linear, which is to say, for $\alpha, \beta \in \mathbb{R}$ and f, $h \in L^1(X)$, we have $F_g(\alpha f + \beta h) = \alpha F_g(f) + \beta F_g(h)$*;
(iii) $|F_g(f)| \leq \|g\|_\infty \|f\|_1, f \in L^1(X)$;
(iv) *F_g is bounded and hence continuous everywhere on $L^1(X)$: If $\|f_n - f\|_1 \to 0$ as $n \to \infty$, then $F_g(f_n) \to F_g(f)$ as $n \to \infty$*;

(v) $\sup\left\{\frac{|F_g(f)|}{\|f\|_1} : \|f\|_1 \neq 0\right\} = \|g\|_\infty$ *provided that every set of positive measure contains a subset of finite positive measure; in particular,* $\|F_g\| = \|g\|_\infty$ *under the proviso.*

Proof (i) The mapping F_g is well defined because $f = f_1$ a.e. and $g = g_1$ a.e. implies $\int_X fg\, d\mu = \int_X f_1 g_1 d\mu$, so that the value of F_g given by (6.1) depends only on the equivalence classes that f and g represent. Since

$$\left|\int_X fg\, d\mu\right| \leq \int_X |fg| d\mu \leq \|g\|_\infty \|f\|_1,$$

$F_g(f)$ is a real number.

(ii) and (iii) are trivial.

(iv) follows from (ii) and (iii).

(v) It follows from (iii) that the supremum here does not exceed the right-hand side. To obtain the reverse inequality, we proceed as follows: Since the matter is trivial when $\|g\|_\infty = 0$, we may assume $\|g\|_\infty > 0$. Given any positive $\varepsilon < \|g\|_\infty$, let $E \subseteq X$ be a measurable subset of positive measure on which $|g| > \|g\|_\infty - \varepsilon > 0$. Since every set of positive measure contains a subset of finite positive measure, we may take E to have finite positive measure, so that $\chi_E \in L^1(X)$. Consequently, the function $f = (\text{sgn } g)\chi_E \in L^1(X)$ and is nonzero. Besides,

$$F_g(f) = \int_E ((\text{sgn } g)g) d\mu = \int_E |g| d\mu \geq (\|g\|_\infty - \varepsilon)\|f\|_1.$$

This shows that

$$\frac{F_g(f)}{\|f\|_1} \geq \|g\|_\infty - \varepsilon,$$

and since the positive number $\varepsilon < \|g\|_\infty$ is arbitrary, the desired inequality is obtained. □

In (v) of the above proposition, the proviso cannot be omitted. Let X consist of two points called a and b. Consider the measure which is 1 on $\{a\}$ and ∞ on $\{b\}$. Then L^1 consists of functions that are 0 at b. The function g defined to be 0 at a and 1 at b belongs to L^∞ with $\|g\|_\infty = 1$. However, $F_g = 0$.

F_g is the prototype of a bounded linear functional defined on $L^p(X)$.

Remarks

(a) If $L^p(X)$ contains only the zero element, then any real-valued function on it is trivially continuous.

Suppose $L^p(X)$ contains a nonzero element. Then a linear functional F that is defined on $L^p(X)$ is continuous if and only if it is bounded. Suppose that F is continuous but unbounded. Then for each n, there is an $f_n \in L^p(X)$ such that $|F(f_n)| \geq n\|f_n\|_p \neq 0$, or

$$\frac{|F(f_n)|}{n\|f_n\|_p} = \left|F\left(\frac{f_n}{n\|f_n\|_p}\right)\right| \geq 1.$$

Now, the sequence $\left\{\frac{f_n}{n\|f_n\|_p}\right\}$ is such that $\left\|\frac{f_n}{n\|f_n\|}\right\| = \frac{1}{n} \to 0$ as $n \to \infty$, contradicting the continuity of F at 0 in view of the above inequality. On the other hand, if $f_n, f \in L^p(X), n = 1, 2, \ldots$, then by linearity and boundedness of F, we have

$$|F(f_n) - F(f)| = |F(f_n - f)| \leq M\|f_n - f\|_p.$$

If the right-hand side of the foregoing inequality tends to zero as n tends to infinity, so does the left-hand side. Thus the continuity of F is proved. Note that this argument also works in a general normed linear space that contains a nonzero element.

(b) Let ℓ^p, $0 < p < \infty$, be the set of all sequences $\{x_n\}_{n\geq 1}$ such that $\sum_{n=1}^{\infty} |x_n|^p < \infty$. When $p \geq 1$, it is easy to see by using Problem 3.3.P21 that ℓ^p is a linear space and $\|\{x_n\}_{n\geq 1}\|_p = (\sum_{n=1}^{\infty} |x_n|^p)^{1/p}$ defines a norm on it. In Problem 3.3.P22, it has been proved that it is a complete metric space. When $p < 1$, it can be proved that ℓ^p is a linear space on the basis of the inequality $(a + b)^p \leq a^p + b^p$ for $a, b \geq 0$.

To see why the inequality $(a + b)^p \leq a^p + b^p$ holds for $a, b \geq 0$ when $0 < p \leq 1$, take $f(t) = (b + t)^p - b^p - t^p$, $t \geq 0$, and note that $f'(t) = p(b+t)^{p-1} - pt^{p-1} \leq 0$ for $t \geq 0$. Consequently, $f(t) \leq 0$ for $t \geq 0$ and hence for $t = a$.

The reader may note that, when $p < 1$, $(\sum_{n=1}^{\infty} |x_n|^p)^{1/p}$ does not provide a norm on ℓ^p. Indeed, take $x = (1, 0, 0, \ldots)$ and $y = (0, 1, 0, 0, \ldots)$; $p = \frac{1}{2}$. Then $\|x\|_p = \|y\|_p = 1$. So, $\|x\|_p + \|y\|_p = 2$. But $\|x+y\|_p = 4$.

Let ℓ^∞ be the set of all bounded sequences $\{\sigma_n\}_{n\geq 1}$ of real numbers, i.e. all sequences $\{\sigma_n\}_{n\geq 1}$ such that $\sup \{|\sigma_n| : n \geq 1\} < \infty$. The proof that ℓ^∞ is a linear space with norm given by $\|\{\sigma_n\}_{n\geq 1}\|_\infty = \sup\{|\sigma_n| : n \geq 1\}$ is as below. Suppose $\rho = \{\rho_n\}_{n\geq 1}$ and $\sigma = \{\sigma_n\}_{n\geq 1}$ are sequences in ℓ^∞. Then

$$\begin{aligned}\|\rho + \sigma\|_\infty &= \sup\{|\rho_n + \sigma_n| : n \geq 1\} \\ &\leq \sup\{|\rho_n| : n \geq 1\} + \sup\{|\sigma_n| : n \geq 1\} \\ &= \|\rho\|_\infty + \|\sigma\|_\infty.\end{aligned}$$

Also, $\|\rho\|_\infty = \sup_n|\rho_n| = 0$ if and only if $\rho_n = 0$, $n = 1, 2, \ldots$, and $\|\alpha\rho\|_\infty = \sup_n|\alpha\rho_n| = |\alpha|\|\rho\|_\infty$, where α is any real number.

The above results about $\ell^p (1 \leq p < \infty)$, including those of Problems 3.3.P21 and 3.3.P22, can also be obtained from the corresponding results about L^p by using Problem 3.2.P31, according to which the sum $\sum_{n=1}^{\infty} |x_n|^p$ is the integral with respect to the counting measure of the pth power of the absolute value of the function on $\mathbb{N}$ that is given by $\{x_n\}_{n\geq 1}$. The result about ℓ^∞ can also be obtained from the corresponding result about L^∞ because the supremum is the same as the essential supremum for the counting measure.

For $\sigma = \{\sigma_n\}_{n\geq 1} \in \ell^\infty$, define a linear mapping F on ℓ^p, $0 < p \leq 1$, by the following rule:

$$F(c) = \sum_{n=1}^{\infty} \sigma_n c_n, \qquad c = \{c_n\}_{n\geq 1} \in \ell^p.$$

It is obvious that, if F is real-valued (i.e. is a functional), then it is linear. Now,

$$|F(c)| \leq \|\sigma\|_\infty \sum_{n=1}^{\infty} |c_n|.$$

Since

$$\left(\sum_{n=1}^{k} |c_n|\right)^p \leq \sum_{n=1}^{k} |c_n|^p \leq \sum_{n=1}^{\infty} |c_n|^p \text{ for } 0 < p \leq 1,$$

using $(a + b)^p \leq a^p + b^p$, for $a, b \geq 0$, it follows that $|F(c)| \leq \|\sigma\|_\infty \|c\|_p$. Thus F is real-valued and also a *bounded* linear mapping on ℓ^p. Observe that F is not a bounded linear functional in the sense defined earlier, because ℓ^p need not be a normed linear space when $p < 1$.

It is easy to check that the linear space $L^p(X)$, where $1 \leq p \leq \infty$, contains a nonzero element if and only if there exists a measurable set $E \subseteq X$ such that $0 < \mu(E) < \infty$. In what follows in this section, whatever we have to say about an L^p space is trivial if the space consists of only the zero element; so, we shall consider only L^p spaces containing a nonzero element.

(c) There are plenty of bounded linear functionals on $L^p(X)$, $1 \leq p < \infty$, and it is possible to characterise them as we do below.
Let p and q be a pair of conjugate exponents. If $1 \leq q < \infty$ (in which case it is also true that $1 < p \leq \infty$) and $g \in L^q(X)$, we define a linear functional F_g on $L^p(X)$ by

$$F_g(f) = \int_X fg\, d\mu, \qquad f \in L^p(X).$$

Then F_g is bounded. In fact, we have the following:

Proposition 6.1.3 *Let p and q be a pair of conjugate exponents,* $1 \leq q < \infty$, $1 < p \leq \infty$, *and* $g \in L^q(X)$. *Then the linear functional* F_g *defined on* $L^p(X)$ *by*

$$F_g(f) = \int_X fg\, d\mu, \qquad f \in L^p(X). \tag{6.2}$$

is bounded and $\|F_g\| = \|g\|_q$.

Proof The mapping F_g is well defined because $f = f_1$ a.e. and $g = g_1$ a.e. implies $\int_X fg\, d\mu = \int_X f_1 g_1 d\mu$, so that the value of F_g given by (6.2) above depends only on the equivalence classes that f and g represent.

If $\|g\|_q = 0$, then $g = 0$ a.e. and there is nothing to prove. So, assume $\|g\|_q \neq 0$. Observe that by Hölder's Inequality, we have

$$|F_g(f)| \leq \|g\|_q \|f\|_p$$

and indeed F_g is a bounded, hence continuous, linear functional on $L^p(X)$, satisfying $\|F_g\| \leq \|g\|_q$.

To show that $\|F_g\| \leq \|g\|_q$, we first suppose that $p < \infty$, so that $q > 1$. Select an f defined by

$$f = (\operatorname{sgn} g)|g|^{\frac{q}{p}}$$

and note that

$$\|f\|_p^p = \int_X |g|^q d\mu = \|g\|_q^q < \infty,$$

i.e.

$$f \in L^p(X) \quad \text{and} \quad fg = (\operatorname{sgn} g)|g|^{\frac{q}{p}} g = |g|^{\frac{q}{p}+1} = |g|^q.$$

Thus for this particular $f \in L^p(X)$, we have

$$F_g(f) = \int_X fg\, d\mu = \int_X |g|^q d\mu = \|g\|_q^q.$$

Consequently,

$$\|F_g\| \geq \frac{|F_g(f)|}{\|f\|_p} = \frac{\|g\|_q^q}{\|g\|_q^{\frac{q}{p}}} = \|g\|_q.$$

Now consider the case when $p = \infty$ and $q = 1$. Select $f = \operatorname{sgn} g$. Since $\|g\|_q \neq 0$, the function cannot be 0 a.e. and hence $\operatorname{sgn} g \neq 0$ on a set of positive measure, which implies $\|f\|_p = \|f\|_\infty = 1$. Also, $fg = |g|$. Proceeding as before, we obtain $\|F_g\| \geq \|g\|_1 = \|g\|_q$. □

The next theorem shows that $L^q(X)$ is, in some sense, the set of all bounded linear functionals on $L^p(X)$ when $X \subseteq \mathbb{R}$ is of finite Lebesgue measure and $p < \infty$. We begin with the following lemma, which will prove useful in the sequel.

Lemma 6.1.4 *Let $(X, \mathcal{F}, \mu)$ be a measure space with $\mu(X) < \infty$ and g an integrable function on X. Suppose M is a nonnegative real number such that*

$$\left| \int_X fg \, d\mu \right| \leq M \|f\|_p$$

for all bounded measurable functions f. Then $g \in L^q(X)$ and $\|g\|_q \leq M$.

Proof Define a sequence of bounded measurable functions as follows:

$$g_n(x) = \begin{cases} g(x) & \text{if } |g(x)| \leq n \\ 0 & \text{if } |g(x)| > n. \end{cases}$$

Since g_n is bounded and $\mu(X) < \infty$, we have $\|g_n\|_q < \infty$.

First we assume $p > 1$, so that $q < \infty$. Set

$$f_n = (\operatorname{sgn} g_n)|g_n|^{\frac{q}{p}}, \qquad n = 1, 2, \ldots$$

Now,

$$\int_X |f_n|^p d\mu = \int_X |g_n|^q d\mu = \|g_n\|_q^q$$

and

$$|g_n|^q = |g_n|^{\frac{q}{p}} |g_n| = f_n g_n = f_n g.$$

Hence

$$\|g_n\|_q^q = \int_X f_n g \, d\mu \leq M \|f_n\|_p = M(\|g_n\|_q)^{\frac{q}{p}},$$

using the hypothesis. Since $q - q/p = 1$ and $\|g_n\|_q < \infty$, we have $\|g_n\|_q \le M$ and $\int_X |g_n|^q d\mu \le M^q$. Since $|g_n|^q \to |g|^q$ as $n \to \infty$, it follows by Fatou's Lemma 3.2.8 that

$$\int_X |g|^q d\mu \le \liminf \int_X |g_n|^q d\mu \le M^q.$$

Thus $g \in L^q(X)$ and $\|g\|_q \le M$, proving the result for $p > 1$.

Now, let $p = 1$, so that $q = \infty$. Let the real number $M_0 > M$ be arbitrary and $A = X(|g| > M_0)$. If we can show that $\mu(A) = 0$, it will follow that $\|g\|_\infty \le M_0$, with the consequence that not only $g \in L^\infty(X)$ but also $\|g\|_\infty \le M$, because $M_0 > M$ is arbitrary. Suppose otherwise: $\mu(A) > 0$. Write

$$E_n = X(|g| > (1 + \frac{1}{n})M_0).$$

Then $A = \cup_{n \ge 1} E_n$ and therefore $\mu(E_n) > 0$ for some n. Since $|g(x)| > (1 + \frac{1}{n})M_0$ for all $x \in E_n$, we have

$$\int_{E_n} |g| d\mu \ge (1 + \frac{1}{n})M_0 \mu(E_n). \tag{6.3}$$

Set $f = (\operatorname{sgn} g)\chi_{E_n}$. Then f is a bounded measurable function such that $\|f\|_1 = \mu(E_n) \le \mu(X) < \infty$, so that $0 < \|f\|_1 < \infty$. From (6.3), we obtain

$$\begin{aligned} \int_X fg\, d\mu = \int_{E_n} |g| d\mu &\ge (1 + \frac{1}{n})M_0 \mu(E_n) = (1 + \frac{1}{n})M_0 \|f\|_1 \\ &> M_0 \|f\|_1, \text{ considering that } 0 < M_0 < \infty \text{ and } 0 < \|f\|_1 < \infty \\ &\ge M \|f\|_1, \end{aligned}$$

in violation of the hypothesis about M. This contradiction shows that $\mu(A) = 0$, which is what we needed to prove. □

It is convenient in certain parts of the next proof to assume that $X = [0, 1]$, and we make this assumption. The necessary changes when $X \subseteq \mathbb{R}$ is an arbitrary measurable subset are cumbersome but minor. The theorem provides an instance when all bounded linear functionals look alike and we know how they look.

Theorem 6.1.5 (Riesz Representation) *Let F be a bounded linear functional on $L^p(X)$, $1 \le p < \infty$. Then there exists a unique element $g \in L^q(X)$, where $\frac{1}{p} + \frac{1}{q} = 1$, such that*

$$F(f) = \int_X fg\, dm, \qquad f \in L^p(X). \tag{6.4}$$

Moreover,

$$\|F\| = \|g\|_q. \tag{6.5}$$

Proof The uniqueness of g is clear, for if g and g_1 both satisfy (6.4), then

$$\int_X f(g - g_1)dm = 0, \qquad f \in L^p(X).$$

In particular, for $f = \chi_E$, where $E \subseteq X$ is measurable, we have $\int_X \chi_E(g - g_1)dm = 0$. So, $g = g_1$ a.e. by Problem 3.2.P14(a). Next, if (6.4) holds, then (6.5) follows in the case $p > 1$ by Proposition 6.1.3 and in the case $p = 1$ by Proposition 6.1.2(v). So, it remains to prove that g exists such that equality (6.4) holds. If $\|F\| = 0$, then (6.4) and (6.5) hold with $g = 0$. Therefore we may assume that $\|F\| > 0$.

Let $f = \chi_{[0,s]}$, where $\chi_{[0,s]}$ denotes the characteristic function of the interval $[0, s]$. Let $\phi(s) = F(\chi_{[0,s]})$. Clearly, ϕ defines a real-valued function on $[0, 1]$ and $\phi(0) = 0$. We first show that ϕ is absolutely continuous. Suppose $\varepsilon > 0$ is given. Let $\{(x_i, x_i'): 1 \le i \le n\}$ be a finite collection of disjoint subintervals of $[0, 1]$ such that $\sum_{i=1}^{n} (x_i' - x_i) < \delta$, where δ will be appropriately chosen later. Then, for

$$f = \sum_{i=1}^{n} (\chi_{[0,x_i']} - \chi_{[0,x_i]})\mathrm{sgn}(\phi(x_i') - \phi(x_i)),$$

we have

$$\sum_{i=1}^{n} |\phi(x_i') - \phi(x_i)| = F(f).$$

Since $\|f\|_p^p < \delta$, we have

$$\sum_{i=1}^{n} |\phi(x_i') - \phi(x_i)| = F(f) \le \|F\|\|f\|_p < \delta^{\frac{1}{p}}\|F\|.$$

If we choose $\delta = \varepsilon^p/\|F\|^p$, then it follows that $\sum_{i=1}^{n} |\phi(x_i') - \phi(x_i)| < \varepsilon$ for any finite collection of disjoint intervals of total length less than $\delta = \varepsilon^p/\|F\|p$, that is, ϕ is absolutely continuous on $[0, 1]$. By Corollary 5.8.3, there exists an integrable function g defined on X such that

$$\phi(s) = \int_{[0,s]} g \quad \text{for all } s \in [0, 1],$$

that is,

$$F(\chi_{[0,s]}) = \int_X g\chi_{[0,s]} dm.$$

Since every step function on [0, 1] is equal (except at finitely many points) to a linear combination $\phi = \sum_{j=1}^{n} c_j\chi_{[0,s_j]}$, we have

$$F(\phi) = \int_X g\phi \, dm$$

for every step function Φ by linearity of F and that of the integral.

Let f be any nonnegative bounded measurable function on [0, 1] with bound M. By Proposition 3.4.3, there exists a sequence $\{\Phi_n\}_{n\geq 1}$ of nonnegative step functions such that $\lim_n \|f - \phi_n\|_p = 0$. By passing to a subsequence if necessary, we may assume that $\Phi_n \to f$ a.e. (see Corollary 3.3.11). Now, the functions $\Psi_n = \min\{\Phi_n, M\}$ are also nonnegative step functions and satisfy $\Psi_n \to f$ a.e. as well as $\lim_n \|f - \Psi_n\|_p = 0$ in view of the simple fact about real numbers that

$$\alpha \leq M \Rightarrow |\min\{\sigma, M\} - \alpha| \leq |\sigma - \alpha|.$$

Moreover, they have a common bound M. Then $F(\Psi_n) \to F(f)$ because F is continuous and $\int_X \Psi_n g \to \int_X fg$ by the Dominated Convergence Theorem, the applicability of which follows from the existence of a common bound for the Ψ_n. So, the equality (6.4) holds for an arbitrary nonnegative bounded measurable function f and hence for an arbitrary bounded measurable function f. Together with the fact that $|F(f)| \leq \|F\|\|f\|_p$, this implies that $\left|\int_X fg\right| \leq \|F\|\|f\|_p$ for an arbitrary bounded measurable function f. By appealing to Lemma 6.1.4, we can conclude that $g \in L^q(X)$.

We proceed to prove that g satisfies (6.4). Let $\varepsilon > 0$ be given and $f \in L^p(X)$ be arbitrary. By Proposition 3.4.3, there is a step function Φ such that $\|f - \phi\|_p < \varepsilon$. Since Φ is a step function, we have

$$F(\phi) = \int_X \phi g \, dm.$$

Hence

$$\begin{aligned}
\left|F(f) - \int_X fg\, dm\right| &= \left|F(f) - F(\phi) + \int_X \phi g\, dm - \int_X fg\, dm\right| \\
&\leq |F(f) - F(\phi)| + \left|\int_X (\phi - f)g\, dm\right| \\
&\leq \|F\|\|f - \phi\|_p + \|g\|_q\|f - \phi\|_p \\
&\leq (\|F\| + \|g\|_q) \cdot \varepsilon.
\end{aligned}$$

Since $\varepsilon > 0$ is arbitrary, it follows that (6.4) holds.

The equality (6.5) now follows upon applying Proposition 6.1.3 if $p > 1$ and Proposition 6.1.2(v) if $p = 1$. □

The next result shows that the translates of a function f in $L^p(\mathbb{R})$ are uniformly continuous in the norm. More precisely, we have the following:

Proposition 6.1.6 (Cf. Problem 3.3.P4) *If $f \in L^p(\mathbb{R}), 1 \leq p < \infty$ and $h \in \mathbb{R}$, then the function $f_h : \mathbb{R} \to \mathbb{R}$ such that $f_h(x) = f(x + h)$ for all $x \in \mathbb{R}$ belongs to $L^p(\mathbb{R})$. Moreover, the function from $\mathbb{R}$ to $L^p(\mathbb{R})$ that maps $h \in \mathbb{R}$ into $f_h \in L^p(\mathbb{R})$ is uniformly continuous.*

Proof That f_h belongs to $L^p(\mathbb{R})$ is a consequence of Problem 6.1.P4. We shall prove the uniform continuity of the map $h \to f_h$.

Fix $\varepsilon > 0$. Since $f \in L^p(\mathbb{R})$, there exists some positive number A and a continuous function $g : \mathbb{R} \to \mathbb{R}$ whose support lies in $[-A, A]$ such that $\|f - g\|_p < \varepsilon$ (see Proposition 3.4.6). It follows from the uniform continuity of g that there exists a positive $\delta < A$ such that

$$|s - t| < \delta \text{ implies } |g(s) - g(t)| < (3A)^{-\frac{1}{p}} \cdot \varepsilon.$$

Note that the provision that $\delta < A$ ensures that $(3A)^{-1} \cdot (2A + \delta) < 1$.

Consider any s and t such that $|s - t| < \delta$. Then

$$\int_{\mathbb{R}} |g(x+s) - g(x+t)|^p dm < (3A)^{-1} \cdot \varepsilon^p \cdot (2A + \delta),$$

from which we obtain

$$\|g_s - g_t\|_p < \varepsilon,$$

bearing in mind that $(3A)^{-1} \cdot (2A + \delta) < 1$.

Observe that $\|f\|_p = \|f_s\|_p$ for all $s \in \mathbb{R}$ (see Problem 6.1.P4). Consequently,

$$\begin{aligned} \|f_s - f_t\|_p &\leq \|f_s - g_s\|_p + \|g_s - g_t\|_p + \|g_t - f_t\|_p \\ &\leq \|(f - g)_s\|_p + \|(g - f)_t\|_p + \varepsilon \\ &< 3\varepsilon. \end{aligned}$$

□

Proposition 6.1.7 *Let p and q be conjugate exponents, where $1 \leq p < \infty$, and let $f \in L^p(\mathbb{R})$, $g \in L^q(\mathbb{R})$. Then $F(t) = \int_{\mathbb{R}} f(x+t)g(x)dm$ is continuous in t.*

Proof Indeed, given t, $h \in \mathbb{R}$, note that

$$\begin{aligned}|F(t+h)-F(t)| &= \left|\int_{\mathbb{R}} [f(x+t+h)g(x)-f(x+t)g(x)]dm\right| \\ &\le \int_{\mathbb{R}} |f(x+t+h)-f(x+t)||g(x)|dm \\ &\le \|(f_t)_h - f_t\|_p \|g\|_q,\end{aligned}$$

which tends to zero as $h \to 0$ in view of the proposition above. □

As an application of the Radon–Nikodým Theorem 5.10.8, we discuss the Riesz Representation Theorem 6.1.5 on a general finite measure space, not just a measurable subset of $\mathbb{R}$. We present only that portion of the proof which differs significantly from the proof of the case above. As the rest of the arguments are no different from those in Theorem 6.1.5 above, they are not included.

Theorem 6.1.8 *Let* $(X, \mathcal{F}, \mu)$ *be a measure space with* $\mu(X) < \infty$ *and* F *be a bounded linear functional on* $L^p(X) = L^p(X, \mathcal{F}, \mu), 1 \le p < \infty$. *Then there exists a unique element* $g \in L^q(X) = L^q(X, \mathcal{F}, \mu)$, *where* $\frac{1}{p} + \frac{1}{q} = 1$, *such that*

$$F(f) = \int_X fg\, d\mu, \qquad f \in L^p(X). \tag{6.6}$$

Moreover,

$$\|F\| = \|g\|_q. \tag{6.7}$$

Proof If $\|F\| = 0$, then (6.6) and (6.7) hold with $g = 0$. Therefore we may assume that $\|F\| > 0$.

For $E \in \mathcal{F}$, define

$$\nu(E) = F(\chi_E).$$

Observe that ν is a signed measure:

Clearly, $\nu(\varnothing) = F(\chi_\varnothing) = F(0) = 0$. Since $\chi_{A \cup B} = \chi_A + \chi_B$ for disjoint A and B, we have $\nu(A \cup B) = F(\chi_{A \cup B}) = F(\chi_A + \chi_B) = F(\chi_A) + F(\chi_B) = \nu(A) + \nu(B)$. It follows that ν is finitely additive.

Let $E = \cup_{i \ge 1} E_i$, where the measurable sets E_i are disjoint, and $F_n = \cup_{1 \le k \le n} E_k$. We have $\|\chi_{F_n} - \chi_E\|_p = (\mu(E \backslash F_n))^{1/p} \to 0$ as $n \to \infty$. Since F is continuous, we have $\nu(F_n) \to \nu(E)$. Observe that $\nu(F_n) = \sum_{k=1}^{n} \nu(E_k)$ because the sets E_i are disjoint and we have proved that ν is finitely additive. So, the convergence just proved says that ν is countably additive.

Also, if $\mu(E) = 0$ then $\|\chi_E\|_p = 0$ and so, $\nu(E) = 0$, which shows that $\nu \ll \mu$. By the Radon–Nikodým Theorem 5.10.8, there exists a $g \in L^1(\mu)$ such that, for every $E \in \mathcal{F}$,

$$F(\chi_E) = \nu(E) = \int_E g\, d\mu = \int_X \chi_E g\, d\mu.$$

It follows from the linearity of F and of the integral that, for every simple function s,

$$F(s) = \int_X sg\, d\mu.$$

Let f be a nonnegative bounded measurable function. By Theorem 2.5.9, there is a sequence $\{s_n\}_{n \geq 1}$ of simple functions, which converges uniformly to f, and satisfies $0 \leq s_n \leq f$. Since $\mu(X) < \infty$, we have $\|f - s_n\|_p \to 0$ as $n \to \infty$ and hence

$$F(f) = \lim_{n \to \infty} F(s_n).$$

Moreover, $\{s_n g\}_{n \geq 1}$ converges pointwise to fg and satisfies $|s_n g| \leq |fg|$. Since f is bounded and $g \in L^1(\mu)$, the function $|fg|$ is integrable. Hence by the Dominated Convergence Theorem,

$$\lim_{n \to \infty} \int_X s_n g\, d\mu = \int_X fg\, d\mu.$$

However, each n satisfies $F(s_n) = \int_X s_n g\, d\mu$ because each s_n is a simple function. Combining this with the previous two equalities established, we obtain

$$F(f) = \int_X fg\, d\mu$$

on the assumption that f be a nonnegative bounded measurable function. By taking differences, we can deduce the same for any bounded measurable function f. It now follows exactly as in the proof of Theorem 6.1.5 that $g \in L^q(X)$. Moreover, $\int_X |fg|\, d\mu < \infty$ for every $f \in L^p(X)$, a fact that will be used below.

We shall now prove that (6.6) holds. For every n, let

$$\phi_n(x) = \begin{cases} n & \text{if } f(x) > n \\ f(x) & \text{if } |f(x)| \leq n \\ -n & \text{if } f(x) < -n. \end{cases}$$

Then each Φ_n is a bounded measurable function and hence

$$F(\phi_n) = \int_X \phi_n g\, d\mu.$$

Also, $|\phi_n - f|^p \leq |f|^p$ and tends to 0 pointwise. Since $|f|^p$ is integrable, it follows by the Dominated Convergence Theorem that $\{\Phi_n\}_{n\geq 1}$ converges to f in the L^p norm, so that

$$F(f) = \lim_{n\to\infty} F(\phi_n).$$

Moreover, $\phi_n g \to fg$, $\{|\phi_n g|\}_{n\geq 1}$ is dominated by $|fg|$ and $\int_X |fg| d\mu < \infty$, so that by the Dominated Convergence Theorem,

$$\int_X fg\, d\mu = \lim_{n\to\infty} \int_X \phi_n g\, d\mu.$$

From the three preceding equalities, we conclude that $F(f) = \int_X fg\, d\mu$. □

Problem Set 6.1

6.1.P1. If $f \in L^p(X)$, $\infty > p > 1$, then $\|f\|_p = \sup\{\int_X |fg|\, d\mu : g \in L^q(X), \|g\|_q = 1\}$, where q is given by $\frac{1}{p} + \frac{1}{q} = 1$.
If $f \in L^1(X)$, then $\|f\|_1 = \sup\{\int_X |fg|\, d\mu : g \in L^\infty(X), \|g\|_\infty = 1\}$.
[Note that this differs from Proposition 6.1.3 in that the absolute value is inside the integral rather than outside it.]
6.1.P2. If $f \in L^\infty(X)$, then $\|f\|_\infty = \sup\{\int_X |fg|\, d\mu : g \in L^1(X), \|g\|_1 = 1\}$, provided that every set of positive measure contains a subset of finite positive measure. Show also that this is false if the proviso about the measure is omitted. [Note that this differs from Proposition 6.1.2(v) in that the absolute value is inside the integral rather than outside it.]
6.1.P3. Let $\{f_n\}_{n\geq 1}$ be a sequence of functions in $L^\infty(X)$. Prove that $\{f_n\}_{n\geq 1}$ converges to $f \in L^\infty(X)$ if and only if there is a set E of measure zero such that $f_n \to f$ uniformly on E^c.
6.1.P4. Suppose f is integrable on $\mathbb{R}$ and for fixed $h \in \mathbb{R}$, let $f_h(x) = f(x + h)$ be a translate of f. Show that f_h is also integrable and that

$$\int_{\mathbb{R}} f_h dm = \int_{\mathbb{R}} f\, dm.$$

6.1.P5. Show that if $|f_n| \leq M$ a.e. and $f_n \to f$ in $L^p(X)$, where $\mu(X) < \infty$ and $p \geq 1$, then $f_n \to f$ in $L^{p'}(X)$ for $1 \leq p' < \infty$.

6.1.P6. Let $I = [a, b] \subset \mathbb{R}$ and $1 < p < \infty$. If $f \in L^p(I)$ and $F(x) = C + \int_{[a,x]} f\,dm$, then show that

$$\sup_{\mathcal{P}} \sum_{k=0}^{n-1} \frac{|F(x_{k+1}) - F(x_k)|^p}{(x_{k+1} - x_k)^{p-1}} \leq \int_I |f|^p dm, \tag{6.8}$$

where $\mathcal{P} : a = x_0 < x_1 < \cdots < x_n = b$ is a partition of $[a, b]$.

Conversely, if

$$K = \sup_{\mathcal{P}} \sum_{k=0}^{n-1} \frac{|F(x_{k+1}) - F(x_k)|^p}{(x_{k+1} - x_k)^{p-1}} < \infty,$$

where $\mathcal{P}$ is a partition of I, then there exists a $\phi \in L^p(I)$ such that $F(x) = C + \int_{[a,x]} \phi\,dm$; show also that $\int_I |\phi|^p dm \leq K$.

Use this to show that equality holds in (6.8).

6.1.P7. (Chebychev's Inequality) Let $f \in L^p(X)$, where $1 \leq p < \infty$. Show that, for every $\lambda > 0$,

$$\lambda^p \mu(\{x : |f(x)| > \lambda\}) \leq \int_X |f|^p d\mu.$$

Moreover,

$$\lim_{\lambda \to \infty} \lambda^p \mu(\{x : |f(x)| > \lambda\}) = 0.$$

6.1.P8. Let $\{f_n\}_{n \geq 1}$ be a sequence in $L^p(X)$, $1 \leq p \leq \infty$, which converges to a function f in $L^p(X)$. Then show that, for each g in $L^q(X)$, where $\frac{1}{p} + \frac{1}{q} = 1$, we have

$$\int_X fg\,dm = \lim_{n \to \infty} \int_X f_n g\,dm.$$

6.2 Modes of Convergence

Up to this point we have encountered various types of convergence of a sequence of measurable functions, such as pointwise, pointwise a.e., uniform and L^p convergence. There are other modes of convergence that are of importance in dealing with measurable functions, namely, almost uniform convergence and convergence in measure.

The definitions of various modes of convergence that we have already encountered are restated for the sake of convenience and ready reference. Throughout this section, we shall consider only real-valued functions defined on a set X with a σ-algebra $\mathcal{F}$ of subsets and a measure μ on it. We shall soon restrict the functions to be measurable as well.

Suppose that we have a sequence of functions f_k on X. We may fix our attention either on the sequence $\{f_k\}_{k\geq 1}$ whose terms are the functions themselves or on the sequence $\{f_k(x)\}_{k\geq 1}$ whose terms are the values of the functions at the individual points x of the domain X. Let f be a real-valued function defined on X.

(i) The sequence $\{f_k\}_{k\geq 1}$ *converges uniformly* to f on X [in symbols, $\lim_k f_k = f$ (unif) or $f_k \xrightarrow{unif} f$] if for every $\varepsilon > 0$, there exists a natural number N such that

$$|f_k(x) - f(x)| < \varepsilon \quad \text{whenever} \quad k \geq N \text{ and } x \in X.$$

(ii) The sequence $\{f_k\}_{k\geq 1}$ *converges pointwise* to f on X [in symbols, $\lim_k f_k(x) = f(x)$, $x \in X$ or $f_k \to f$] if the sequence $\{f_k(x)\}_{k\geq 1}$ converges to $f(x)$ for each $x \in X$, that is, if for every $\varepsilon > 0$ and every $x \in X$, there exists a natural number N such that

$$|f_k(x) - f(x)| < \varepsilon \quad \text{whenever} \quad k \geq N.$$

Uniform and pointwise convergence make sense even when there is no σ-algebra. However, the other modes of convergence to be discussed only make sense if $(X, \mathcal{F}, \mu)$ is a measure space.

(iii) The sequence $\{f_k\}_{k\geq 1}$ *converges almost everywhere* to f on X [in symbols, $\lim_k f_k(x) = f(x)$ for almost all x or $f_k \xrightarrow{ae} f$] if there exists a set $S \in \mathcal{F}$ with $\mu(S) = 0$ such that the sequence $\{f_k(x)\}_{k\geq 1}$ converges to $f(x)$ for each $x \in X\backslash S$, that is, for every $\varepsilon > 0$ and every $x \in S^c$, there exists a natural number N such that

$$|f_k(x) - f(x)| < \varepsilon \quad \text{whenever} \quad k \geq N.$$

(iv) The sequence $\{f_k\}_{k\geq 1}$ in $L^p(X)$, $1 \leq p < \infty$, converges in the L^p norm to $f \in L^p(X)$ [in symbols, $\lim_k f_k = f$ in $L^p(X)$ or $f_k \xrightarrow{L^p} f$] if for every $\varepsilon > 0$, there exists a natural number N such that

$$\|f_k - f\|_p = \left(\int_X |f_k - f|^p d\mu\right)^{\frac{1}{p}} < \varepsilon \quad \text{whenever} \quad k \geq N.$$

L^p norm convergence is also called convergence in the mean of order p or simply L^p convergence. This kind of convergence applies only to a sequence of L^p functions while the other three modes of convergence just described apply to any kinds of real-valued functions, whether measurable or not.

Besides the above four modes of convergence, we discuss two other modes of convergence—alluded to in the opening paragraph—making six in all. The first of the remaining two modes of convergence applies to functions that need not be measurable.

Definition 6.2.1 Let $\{f_k\}_{k\geq 1}$ be a sequence of functions on X and f be a function on X. Then we say that f_k **converges to** f **almost uniformly** [in symbols, $\lim_k f_k = f$ a.u. or $f_k \xrightarrow{au} f$] if for every $\eta > 0$, there exists a set $E \in \mathcal{F}$ with $\mu(E) < \eta$ such that $f_k \xrightarrow{unif} f$ on $X\backslash E$.

Definition 6.2.2 Let $\{f_k\}_{k\geq 1}$ be a sequence of measurable functions on X and f be a measurable function on X. Then we say that f_k **converges to** f **in measure** [in symbols, $\lim_k f_k = f$ in measure or $f_k \xrightarrow{meas} f$] if

$$\lim_{k\to\infty} \mu(\{x \in X : |f_k(x) - f(x)| > a\}) = 0$$

for every $a > 0$.

Using the notation explained at the beginning of Sect. 2.4, the condition can be abbreviated as

$$\lim_{k\to\infty} \mu(X(|f_k - f| > a)) = 0.$$

We shall continue to use this shorter notation in this section whenever possible.

Definition 6.2.3 The sequence $\{f_k\}_{k\geq 1}$ of measurable functions on X is said to be **Cauchy in measure** if $\lim_{p,k}\mu(X(|f_k - f_p| > a)) = 0$ for every $a > 0$.

Definition 6.2.4 The sequence $\{f_k\}_{k\geq 1}$ of real-valued functions on X is said to be **almost uniformly Cauchy** if, for every $\eta > 0$, there exists a set of measure less than η, on the complement of which, the sequence is uniformly Cauchy.

Remarks 6.2.5

(a) If a sequence of functions converges a.e. or almost uniformly, then the limit function is unique a.e.

For almost everywhere convergence, this is no different from Problem 3.2.P30 (b). Nonetheless, we discuss the matter here ab initio. Let $f_k \xrightarrow{ae} f$ and $f_k \xrightarrow{ae} g$. Then there exist sets $S \in \mathcal{F}$, $T \in \mathcal{F}$ with $\mu(S) = \mu(T) = 0$ such that the sequence $\{f_k(x)\}_{k\geq 1}$ converges to $f(x)$ for each $x \in X\backslash S$ and to $g(x)$ for each $x \in X\backslash T$. It follows that it converges to $f(x)$ as well as $g(x)$ for each $x \in X\backslash(S \cup T)$. Therefore $f(x) = g(x)$ for each $x \in X\backslash(S \cup T)$. Since $\mu(S \cup T) = 0$, we conclude that $f = g$ a.e. The following converse is also true: If $f_k \xrightarrow{ae} f$ and $f = g$ a.e., then

$f_k \overset{ae}{\to} g$. This is easy to see because there exist two sets of measure zero, outside one of which $f_k \to f$ pointwise and outside the other, $f = g$ everywhere; their union provides a set of measure zero, outside which $f_k \to f$ pointwise and $f = g$ everywhere, so that $f_k \to g$ pointwise. Thus, $f_k \overset{ae}{\to} g$.

Next, suppose $f_k \overset{au}{\to} f$ and $f_k \overset{au}{\to} g$. Then, for each $p \in \mathbb{N}$, there exist sets E_p, F_p with $\mu(E_p) < 1/2p$, $\mu(F_p) < 1/2p$ and $f_k \overset{unif}{\to} f$ on the complement E_p^c and $f_k \overset{unif}{\to} g$ on the complement F_p^c; in particular, $f = g$ on $E_p^c \cap F_p^c = (E_p \cup F_p)^c$. Thus $G_p = E_p \cup F_p$ is a measurable set with $\mu(G_p) < 1/p$ such that $f = g$ on its complement G_p^c. The set $S = \cap_{p \geq 1} G_p$ has measure 0 and $x \in S^c \Rightarrow x \in G_p^c$ for some $p \Rightarrow f(x) = g(x)$. Thus $f = g$ a.e. Here again, a converse is true: If $f_k \overset{au}{\to} f$ and $f = g$ a.e., then $f_k \overset{au}{\to} g$. The proof proceeds along the same lines as in the preceding paragraph, using the fact the union of a set of measure zero and a set of measure less than ε is a set of measure less than ε.

However, it may happen that f is measurable and g is not. Example: $X = \{a, b, c\}$, $\mathcal{F} = \{\varnothing, \{a\}, \{b, c\}, X\}$, $\mu(\varnothing) = 0$, $\mu(\{a\}) = 1$, $\mu(\{b, c\}) = 0$, $\mu(X) = 1$; $f_k = f = 1$ everywhere, $g(a) = 1$, $g(b) = 2$, $g(c) = 3$. Then $f_k \overset{ae}{\to} f$ and $f_k \overset{ae}{\to} g$ as well as $f_k \overset{au}{\to} f$ and $f_k \overset{au}{\to} g$ but only f is measurable. This cannot happen when the measure is what will later be called "complete" (see Definition 7.2.1), because in that event, a function that agrees with a measurable function a.e. can be shown to be measurable (see Problem 3.2.P30(e)).

(b) If a sequence of measurable functions converges in measure, then the limit function (measurable by the very definition of convergence in measure) is unique a.e.

Let $f_k \overset{meas}{\to} f$ and $f_k \overset{meas}{\to} g$. Since $|f - g| \leq |f - f_k| + |f_k - g|$, it follows that, for any $a > 0$,

$$X(|f - g| > a) \subseteq X(|f_k - f| > \frac{a}{2}) \cup X(|f_k - g| > \frac{a}{2}).$$

So,

$$\mu(X(|f - g| > a)) \leq \mu(X(|f_k - f| > \frac{a}{2})) + \mu(X(|f_k - g| > \frac{a}{2})).$$

Since the right-hand side of the above inequality tends to 0 as $k \to \infty$, it follows that $f = g$ a.e. It is left to the reader to show that, if $f_k \overset{meas}{\to} f$ and $f = g$ a.e., then $f_k \overset{meas}{\to} g$.

(c) If $f_k \overset{meas}{\to} f$ and $g_k \overset{meas}{\to} g$, then

(i) $f_k + g_k \overset{meas}{\to} f + g$;

(ii) $\alpha f_k \overset{meas}{\to} \alpha f$ for any real α;

(iii) $|f_k| \overset{meas}{\to} |f|$;

(iv) $\max\{f_k, g_k\} \overset{meas}{\to} \max\{f, g\}, \quad \min\{f_k, g_k\} \overset{meas}{\to} \min\{f, g\};$

(v) $f_k^+ \overset{meas}{\to} f^+, \quad f_k^- \overset{meas}{\to} f^-;$

(vi) If $\mu(X) < \infty$, then $f_k^2 \overset{meas}{\to} f^2$ and $f_k g_k \overset{meas}{\to} fg$.

Part (i) follows from the relation

$$X(|(f_k + g_k) - (f + g)| > a) \subseteq X(|f_k - f| > \frac{a}{2}) \cup X(|g_k - g| > \frac{a}{2})$$

and the arguments in (b) above.

Part (ii) follows from the relation

$$X(|\alpha f_k - \alpha f| > a) = X(|f_k - f| > \frac{a}{|\alpha|}).$$

Part (iii) is a consequence of the relation $||f_k| - |f|| \leq |f_k - f|$.

One now deduces part (iv) from the fact that

$$\max\{\beta, \gamma\} = \frac{1}{2}[(\beta + \gamma) + |\beta - \gamma|] \quad \text{and}$$
$$\min\{\beta, \gamma\} = \frac{1}{2}[(\beta + \gamma) - |\beta - \gamma|]$$

for any real numbers β, γ.

For part (v), one just notes that $f^+ = \max\{f, 0\}$ and $f^- = \max\{-f, 0\}$.

We shall establish (vi) in several steps.

Step 1. If $f = 0$, then $f_k^2 \overset{meas}{\to} 0$. This is because $X(|f_k^2 - 0| > a) = X(|f_k - 0| > a^{1/2})$.

Step 2. If ϕ is any measurable real-valued function, then for any $\delta > 0$, there exists a measurable set $E \subseteq X$ and a constant $M > 0$ such that $\mu(E) < \delta$ and $|\phi| \leq M$ on the complement $X \backslash E = E^c$. To see why, let $E_n = X(|\phi| > n)$, $n = 1, 2, \ldots$. The sets E_n form a descending sequence with empty intersection because ϕ is real-valued. By Proposition 3.1.8, and the hypothesis that $\mu(X) < \infty$, we have $\lim_{n\to\infty} \mu(E_n) = 0$. Choose n large enough to make $\mu(E_n) < \delta$. Then $E = E_n$ and $M = n$ have the required properties.

Step 3. If ϕ is any measurable real-valued function, then $f_k\phi \overset{meas}{\to} f\phi$. To prove this, consider any $a > 0$. We have to prove that $\lim_{k\to\infty} \mu(X(|f_k\phi - f\phi| > a)) = 0$. So, let $\varepsilon > 0$ be arbitrary and we need to produce N such that $k \geq N \Rightarrow \mu(X(|f_k\phi - f\phi| > a)) < \varepsilon$. As warranted by Step 2, there exists some measurable E and a constant $M > 0$ such that $\mu(E) < \frac{\varepsilon}{2}$ and $|\phi| \leq M$ on the complement E^c. Observe that this complement must then have the property that

$$X(|f_k\phi - f\phi| > a) \cap E^c \subseteq X(|f_k - f| > \frac{a}{M}).$$

Now choose the integer N to be so large that

$$k \geq N \Rightarrow \mu(X(|f_k - f| > \frac{a}{M})) < \frac{\varepsilon}{2}.$$

We claim that this N has the requisite property. Indeed,

$$X(|f_k\phi - f\phi| > a) = X(|f_k\phi - f\phi| > a) \cap (E^c \cup E) \subseteq (X(|f_k\phi - f\phi| > a) \cap E^c) \cup E$$
$$\subseteq X(|f_k - f| > \frac{a}{M}) \cup E \text{ by the observation above.}$$

In conjunction with the fact that $\mu(E) < \frac{\varepsilon}{2}$ and the manner in which N was chosen, this inclusion shows that $k \geq N \Rightarrow \mu(X(|f_k\phi - f\phi| > a)) < \varepsilon$, as claimed.

Step 4. The convergence $f_k^2 \xrightarrow{meas} f^2$ holds. This follows from the identity

$$f_k^2 = (f_k - f)^2 + 2f_kf - f^2$$

upon applying Step 1, Step 3, (i) and (ii).
Step 5. The convergence $f_k g_k \xrightarrow{meas} fg$ holds. The identity

$$f_k g_k = \frac{1}{4}[(f_k + g_k)^2 - (f_k - g_k)^2]$$

yields the convergence when taken in conjunction with Step 4, (i) and (ii).

The analogues of (i)–(v) hold for almost uniform convergence, but obviously not (vi), which is false even for uniform convergence.

(d) In part (vi) of (c) above, the condition that $\mu(X) < \infty$ cannot be dispensed with, as the following example shows:
Let $X = (0, \infty)$ and $f_k(x) = x$ for each $k \in \mathbb{N}$ and each $x \in X$. Then $f(x) = x$. Let $g_k(x) = \alpha_k$, where $\{\alpha_k\}_{k \geq 1}$ is a sequence of positive real numbers converging to 0. Here $\mu(X) = \infty$. Also,

$$\mu(X(|f_k g_k - fg| > a)) = \mu(X(\alpha_k x > a)) = \infty$$

for each k and each $a > 0$. Consequently, $f_k g_k$ does not converge to fg in measure.
The space of measurable functions is "complete" with respect to convergence in measure:

Theorem 6.2.6 *A sequence of measurable functions is convergent in measure if and only if it is Cauchy in measure.*

Proof Suppose $\{f_k\}_{k\geq 1}$ converges in measure to some (measurable) function f. It follows from the relation

$$X(|f_k - f_j| > a) \subseteq X(|f_k - f| > \frac{a}{2}) \cup X(|f_j - f| > \frac{a}{2})$$

that

$$\mu(X(|f_k - f_j| > a)) \leq \mu(X(|f_k - f| > \frac{a}{2})) + \mu(X(|f_j - f| > \frac{a}{2})).$$

This implies that $\{f_k\}_{k\geq 1}$ is Cauchy in measure.

For the converse, suppose $\{f_k\}_{k\geq 1}$ is Cauchy in measure. We can, for each natural number j, choose a natural number k_j such that

$$\mu(X(|f_r - f_s| > \frac{1}{2^j})) < \frac{1}{2^j} \text{ whenever } r, s \geq k_j.$$

Furthermore, the integers k_j can be selected to be strictly increasing. Now, let

$$E_j = X(|f_{k_j} - f_{k_{j+1}}| > \frac{1}{2^j}).$$

Observe that $\mu(E_j) < \frac{1}{2^j}$. Let $F_J = \cup_{j\geq J} E_j$, so that $F_J \in \mathcal{F}$ and $\mu(F_J) < 2^{-(J-1)}$. If $i > j > J$ and $x \notin F_J$, then

$$\begin{aligned} |f_{k_i}(x) - f_{k_j}(x)| &\leq |f_{k_i}(x) - f_{k_{i-1}}(x)| + \cdots + |f_{k_{j+1}}(x) - f_{k_j}(x)| \\ &\leq \frac{1}{2^{i-1}} + \cdots + \frac{1}{2^j} < \frac{1}{2^{j-1}}. \end{aligned} \tag{6.9}$$

Let $F = \cap_{J\geq 1} F_J$. Then $F \in \mathcal{F}$ and $\mu(F) \leq \mu(F_J) < \frac{1}{2^{J-1}}$ for all J, so that $\mu(F) = 0$. By (6.9), $\{f_{k_j}\}_{j\geq 1}$ converges on $X \backslash F$. If we define

$$f(x) = \begin{cases} \lim_{j\to\infty} f_{k_j}(x) & \text{if } x \notin F \\ 0 & \text{if } x \in F, \end{cases}$$

then $\{f_{k_j}\}_{j\geq 1}$ converges a.e. to the measurable function f. Letting $i \to \infty$ in (6.9), we infer that, if $j \geq J$ and $x \notin F_J$, then

$$|f(x) - f_{k_j}(x)| \leq \frac{1}{2^{j-1}} \leq \frac{1}{2^{J-1}}. \tag{6.10}$$

We shall show by using (6.10) that $\{f_{k_j}\}_{j\geq 1}$ converges uniformly to f on the complement of each set F_J and hence, for every positive a,

$$\mu(X(|f_{k_j}-f|>\frac{a}{2}))\to 0 \text{ as } j\to\infty. \tag{6.11}$$

Indeed, given $a>0$, choose J such that $\frac{1}{2^J}<\frac{a}{2}$. Then $X\left(|f_{k_j}-f|>\frac{a}{2}\right)\subseteq X\left(|f_{k_j}-f|>\frac{1}{2^j}\right)$ for all j, which implies by (6.10) that $X(|f_{k_j}-f|>\frac{a}{2})\subseteq F_{J+1}$ for $j\geq J+1$, thus ensuring that $\mu(X(|f_{k_j}-f|>\frac{a}{2}))\leq\mu(F_{J+1})<2^{-J}$ for $j\geq J+1$.

Now,

$$X(|f_k-f|>a)\subseteq X(|f_k-f_{k_j}|>\frac{a}{2})\cup X(|f_{k_j}-f|>\frac{a}{2}).$$

If k and k_j are sufficiently large, the measure of the first set on the right-hand side is arbitrarily small because $\{f_k\}_{k\geq 1}$ is Cauchy in measure, and the measure of the second set is arbitrarily small because of (6.11). □

Corollary 6.2.7 *If a sequence of measurable functions converges in measure to some limit, then it has a subsequence which converges a.e. to the same limit.*

Proof Follows from the argument of Theorem 6.2.6. □

If a sequence converges in measure to some limit, the sequence itself need not converge to it a.e. or almost uniformly. This is illustrated in Example 6.2.20 below.

Corollary 6.2.8 *If a sequence of measurable functions converges in measure to some limit, then it has a subsequence which converges almost uniformly to the same limit.*

Proof Follows from the argument of Theorem 6.2.6. □

We note in passing that Corollary 6.2.7 is also a consequence of Corollary 6.2.8 in conjunction with Theorem 6.2.11.

In what follows, we study the interrelationships between the six types of convergence described in this section up to Definition 6.2.2. We begin with some that the reader is perhaps familiar with.

In order to ensure transparency in our discussions, we shall consider the interrelationship under:

(A) no restriction on the total measure $\mu(X)$,
(B) $\mu(X)<\infty$,
(C) the sequence $\{f_k\}_{k\geq 1}$ dominated by a suitable function.

We begin by considering (A).

It is obvious that uniform convergence implies pointwise convergence, which in turn, implies almost everywhere convergence.

Uniform convergence	$\Rightarrow$ Trivial	Almost uniform convergence
$\Downarrow$ Trivial		Theorem 6-2.11 $\Downarrow$
Pointwise convergence	$\Rightarrow$ Trivial	Almost everywhere convergence

Also, uniform convergence obviously implies almost uniform convergence. The reverse implications all fail, as Examples 6.2.9(a)–(c) and 6.2.12 show.

Examples 6.2.9

(a) On [0, 1], the sequence given by $f_k(x) = x^k$ converges pointwise to the limit function f that vanishes on [0, 1) but has value 1 when $x = 1$. However, the convergence is not uniform, because otherwise the limit would have to be continuous.

(b) The sequence $\{f_k\}_{k\geq 1}$ defined on [0, 1] by

$$f_k(x) = \begin{cases} 0 & \text{if } x \notin \mathbb{Q} \\ (-1)^k & \text{if } x \in \mathbb{Q} \end{cases}$$

converges to the identically zero function a.e. but has no pointwise limit at all. If the only set of measure zero is the empty set (as is the case with the counting measure), almost everywhere convergence is the same as pointwise convergence.

(c) Let $X = [0, 2]$, $\mathcal{F}$ denote the σ-algebra of Lebesgue measurable subsets of X and m be Lebesgue measure on $\mathcal{F}$. Define f_k to be $\chi_{[\frac{1}{k},\frac{2}{k}]}$ for $k = 1, 2, \ldots$ and set $f(x) = 0$ for $x \in X$. Then $f_k \to f$ uniformly on the complement of $[0, \delta]$, where δ is any positive number less than 2. In fact, if $k_0 > \frac{2}{\delta}$, then f_k vanishes on $(\delta, 2]$ for $k \geq k_0$. Thus $f_k \overset{au}{\rightarrow} f$. However, f_k does not converge uniformly to f because f_k takes the value 1 on $\left[\frac{1}{k}, \frac{2}{k}\right]$.

(d) The sequence in part (a) above converges almost uniformly to the function which is 0 everywhere on [0, 1], but does not converge to it pointwise. Thus almost uniform convergence does not imply pointwise convergence. However, almost uniform convergence does imply almost everywhere convergence (Theorem 6.2.11). It will be seen later (Theorem 6.2.14) that it implies convergence in measure if all functions in the sequence are measurable. That almost everywhere convergence does not imply almost uniform convergence is demonstrated in Example 6.2.12.

The notion of almost uniformly Cauchy is equivalent to that of almost uniform convergence.

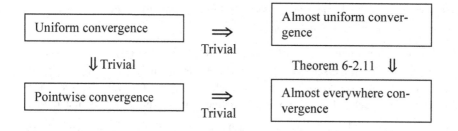

Theorem 6.2.10 *Let $\{f_k\}_{k\geq 1}$ be an almost uniformly Cauchy sequence. Then there exists a real-valued function f such that $f_k \xrightarrow{au} f$.*

Proof For each n, there exists an E_n such that $\mu(E_n) \leq \frac{1}{n}$ and $\{f_k\}_{k\geq 1}$ is uniformly Cauchy on E_n^c. Then the set $E = \cap_{n\geq 1} E_n$ is measurable and $\mu(E) \leq \mu(E_n) \leq \frac{1}{n}$ for all n, so that $\mu(E) = 0$. Moreover, for each $x \in E^c = \cup_{n\geq 1} E_n^c$, the sequence $\{f_k(x)\}_{k\geq 1}$ is Cauchy. For each $x \in E^c$, define $f(x) = \lim_{k\to\infty} f_k(x) \in \mathbb{R}$ and define $f(x) = 0$ for $x \in E$. Then for every n, we have $f_k(x) \to f(x)$ on E_n^c and $\{f_k\}_{k\geq 1}$ is uniformly Cauchy on E_n^c. We shall show that $\{f_k\}_{k\geq 1}$ converges almost uniformly to f.

With this in view, consider an arbitrary $\eta > 0$. On account of $\{f_k\}_{k\geq 1}$ being almost uniformly Cauchy, there exists a set F of measure less than η such that $\{f_k\}_{k\geq 1}$ is uniformly Cauchy on F^c. Considering that $\mu(E) = 0$, the set $G = F \cup E$ has the property that not only is $\mu(G) < \eta$ but also $f_k(x) \to f(x)$ at each $x \in G^c$. But $\{f_k\}_{k\geq 1}$ is uniformly Cauchy on F^c and *a fortiori* on its subset G^c. It follows that $\{f_k\}_{k\geq 1}$ converges uniformly on G^c to some limit function g. However, $G^c \subseteq E^c$ and we already know that $\{f_k\}_{k\geq 1}$ converges to f on E^c. Hence $g = f$ on G^c. Thus, $\{f_k\}_{k\geq 1}$ converges uniformly on G^c to f, while at the same time, $\mu(G) < \eta$. □

We go on to consider the relation between convergence a.e. and almost uniform convergence. The result may be summarised as: *Almost uniform convergence implies convergence a.e.*

Theorem 6.2.11 *Let $\{f_k\}_{k\geq 1}$ be a sequence of functions converging to f almost uniformly. Then it converges to f almost everywhere.*

Proof For each k, let $E_k \in \mathcal{F}$ be such that $\mu(E_k) < \frac{1}{k}$ and $\{f_k\}_{k\geq 1}$ converges uniformly to f on E_k^c. Then, for $x \in \cup_{k\geq 1} E_k^c$, we have $x \in E_j^c$ for some j, and so, $f_k(x) \to f(x)$. But $(\cup_{k\geq 1} E_k^c)^c = \cap_{k\geq 1} E_k$ is in $\mathcal{F}$ and $\mu(\cap_{k\geq 1} E_k) = 0$. □

Example 6.2.12 The converse of the theorem is false (see the example after Egorov's Theorem 2.6.2): The sequence $\{f_k\}_{k\geq 1}$, where $f_k(x) = \chi_{[k,k+1]}$ on $[0,\infty)$ converges pointwise to the function $f = 0$. To see this, fix $x \in [0, \infty)$ and choose $k_0 > x$. Then $|f(x) - f_k(x)| = 0$ for $k \geq k_0$. However, the sequence does not converge almost uniformly. If it did, it would have to converge to f (and also to functions equal a.e. to it). But for $\varepsilon = \frac{1}{2}$, any measurable set E_ε with $m(E_\varepsilon) < \varepsilon$ satisfies $m([k,k+1] \cap E_\varepsilon^c) > \frac{1}{2}$, which has the consequence that $[k,k+1] \cap E_\varepsilon^c$ is nonempty, and any point x in it satisfies $|f(x) - f_k(x)| = 1$. This is true for all k and thus f_k cannot converge uniformly to f on E_ε^c.

It should be noted that f_k does not converge to f either in measure or in the L^p norm. This is because $X(|f(x) - f_k(x)| > \frac{1}{2}) = [k, k+1]$ has measure 1 and

$$\left(\int_{[0,\infty)} |f_k - f|^p\right)^{\frac{1}{p}} = 1,$$

neither of which tends to 0. This shows that pointwise convergence does not imply convergence in measure or in L^p norm, and hence neither does convergence a.e. However, it will be seen in Theorem 6.2.22 that when $\mu(X) < \infty$, convergence a.e. implies convergence in measure. On the other hand, pointwise convergence does not imply convergence in the L^p norm even when $\mu(X) < \infty$. This will be shown in Example 6.2.16.

Recall from the last paragraph in Remark 6.2.5(a) that a sequence $\{f_k\}_{k\geq 1}$ of measurable functions can converge almost uniformly to a nonmeasurable limit f. However, as with convergence a.e., there always exists a measurable limit function to which the sequence converges.

Proposition 6.2.13 *Suppose a sequence $\{f_k\}_{k\geq 1}$ of measurable functions converges almost uniformly to f. Then there exists a measurable function g such that $g = f$ a.e. and any such g has the property that $f_k \overset{au}{\to} g$.*

Proof By Theorem 6.2.11, $f_k \overset{ae}{\to} f$. Using the results of part (c) and then part (b) of Problem 3.2.P30, there exists some measurable g such that $f_k \overset{ae}{\to} g$ and $g = f$ a.e. It now follows from Remark 6.2.5(a) that $f_k \overset{au}{\to} g$. □

It is trivial to check that uniform convergence of a sequence of measurable functions implies convergence in measure. We shall prove the corresponding result for almost uniform convergence, which may be summarised as: *Almost uniform convergence implies convergence in measure.*

Theorem 6.2.14 *If a sequence $\{f_k\}_{k\geq 1}$ of measurable functions converges almost uniformly to f, then some measurable function g is equal to f a.e. and the sequence converges in measure to any such function g.*

Proof By Proposition 6.2.13, some measurable function g is equal to f a.e. and $f_k \overset{au}{\to} g$ for any such g. In order to prove that $f_k \overset{meas}{\to} g$, consider any $a > 0$. Since $f_k \overset{au}{\to} g$, for any $\varepsilon > 0$, there exists a set $X_\varepsilon \in \mathcal{F}$ such that $\mu(X \backslash X_\varepsilon) < \varepsilon$ and $f_k \overset{unif}{\to} g$ on X_ε. So, for sufficiently large k, the set

$$X(|f_k - g| > a)$$

must be contained in $X \backslash X_\varepsilon$. Since f_k and g are both measurable functions, the set is measurable and satisfies

$$\mu(X(|f_k - g| > a)) \leq \mu(X \backslash X_\varepsilon) < \varepsilon.$$

Thus $f_k \overset{meas}{\to} f$. □

Corollary 6.2.8 is a partial converse of the preceding theorem.

L^p norm convergence implies convergence in measure.

Theorem 6.2.15 *Let* $\{f_k\}_{k\geq 1}$ *be a sequence of functions in* $L^p(X)$ *such that* $f_k \xrightarrow{L^p} f \in L^p(X)$, *i.e.,* $\|f_k - f\|_p \to 0$ *as* $k \to \infty$. *Then* $f_k \xrightarrow{meas} f$.

Proof First suppose $p < \infty$. For $a > 0$, set

$$E_k = X(|f_k - f| > a);$$

then

$$\left(\int_X |f_k - f|^p d\mu\right)^{\frac{1}{p}} \geq \left(\int_{E_k} |f_k - f|^p d\mu\right)^{\frac{1}{p}} > a\mu(E_k)^{\frac{1}{p}}.$$

Since $\|f_k - f\|_p \to 0$ as $k \to \infty$, it follows that $\mu(E_k) \to 0$ as $k \to \infty$.

Now suppose $p = \infty$. Consider any $a > 0$ and any $\varepsilon > 0$. Since $\lim_{k\to\infty} \|f_k - f\|_\infty = 0$, there exists an N such that $k \geq N$ implies $\|f_k - f\|_\infty < a$. On the basis of the definition of the norm $\|\cdot\|_\infty$, this means $k \geq N$ implies $\mu(X(|f_k - f| > a)) = 0 < \varepsilon$. So, $\lim_{k\to\infty} \mu(X(|f_k - f| > a)) = 0$. □

Example 6.2.16 The converse of the above theorem fails. In fact, none among almost uniform convergence, convergence in measure, pointwise convergence and almost everywhere convergence implies convergence in the L^p norm. To see why, let $X = [0, 2]$, $\mathcal{F}$ be the σ-algebra of Lebesgue measurable subsets of X and μ be the Lebesgue measure m on $\mathcal{F}$. The sequence $\{f_k\}_{k\geq 1}$, where

$$f_k = k\chi_{[\frac{1}{k},\frac{2}{k}]}, \qquad k = 1, 2, \ldots,$$

converges to the zero function f pointwise as well as almost uniformly and in measure. However,

$$\begin{aligned}\|f_k - f\|_p &= \left(\int_{[0,2]} \left|k\chi_{[\frac{1}{k},\frac{2}{k}]}\right|^p dm\right)^{\frac{1}{p}}\\ &= k^{1-\frac{1}{p}} \geq 1 \text{ for all } k \text{ and all } p \geq 1,\end{aligned}$$

that is, f_k does not converge to f in the L^p norm, $p < \infty$. Since $\|f_k\|_\infty = k$, neither does the sequence converge to f in the L^∞ norm.

We shall show in Theorem 6.2.23 that uniform convergence implies convergence in the L^p norm, provided that $\mu(X) < \infty$. Here we record an example to show that this is not so when $\mu(X) = \infty$.

Example 6.2.17 Let $X = \mathbb{R}$, $\mathcal{F} = \mathfrak{M}$ and μ be the Lebesgue measure m. The sequence $\{f_k\}_{k\geq 1}$, where $f_k = k^{-1/p}\chi_{[0,k]}$, $k = 1, 2, \ldots$ converges to the zero function f uniformly. Indeed, $|f_k - f|$ equals $k^{-1/p}$ on $[0, k]$ and 0 elsewhere. So, $|f_k(x) - f(x)|$ is arbitrarily small for all large k and all $x \in \mathbb{R}$. But

$$\|f_k - f\|_p = \left(\int_{\mathbb{R}} |f_k - f|^p dm\right)^{\frac{1}{p}} = 1 \text{ for every } k,$$

that is, f_k does not converge to f in the L^p norm.

In the reverse direction, we have the following two theorems:

Theorem 6.2.18 *Let $\{f_k\}_{k \geq 1}$ be a sequence of functions in $L^p(X)$ such that $f_k \xrightarrow{L^p} f \in L^p(X)$. Then it has a subsequence $\{f_{k_j}\}_{j \geq 1}$ such that $f_{k_j} \xrightarrow{au} f$.*

Proof An immediate consequence of Theorem 6.2.15 and Corollary 6.2.8. □

Theorem 6.2.19 *Let $\{f_k\}_{k \geq 1}$ be a sequence of functions in $L^p(X)$ such that $f_k \xrightarrow{L^p} f \in L^p(X)$. Then it has a subsequence $\{f_{k_j}\}_{j \geq 1}$ such that $f_{k_j} \xrightarrow{ae} f$.*

Proof See Corollary 3.3.11. Also an immediate consequence of Theorem 6.2.15 and Corollary 6.2.7 as well as of Theorems 6.2.11 and 6.2.18. □

We shall now present an example to show that neither L^p norm convergence nor convergence in measure implies convergence a.e. It would be pertinent to note first that convergence in measure implies the existence of a subsequence that converges to the same limit function almost uniformly (Corollary 6.2.8) and hence by Theorem 6.2.11 also a.e. Moreover, by Theorem 6.2.19, L^p norm convergence also implies the same.

Example 6.2.20 Let $X = [0, 1]$ with Lebesgue measure on the family of all Lebesgue measurable subsets of it. Set $E_{k,j} = \left[\frac{j-1}{k}, \frac{j}{k}\right], j = 1, 2, \ldots, k$ and enumerate all these intervals as follows: $E_{1,1} = [0, 1]$, $E_{2,1} = \left[0, \frac{1}{2}\right]$, $E_{2,2} = \left[\frac{1}{2}, 1\right]$, $E_{3,1} = \left[0, \frac{1}{3}\right]$, $E_{3,2} = \left[\frac{1}{3}, \frac{2}{3}\right]$, $E_{3,3} = \left[\frac{2}{3}, 1\right]$, …. Let $\{F_k\}_{k \geq 1}$ denote the above sequence of intervals. Define $f_k = \chi_{F_k}$, $k = 1, 2, \ldots$. Since $\mu(F_k) \to 0$ as $k \to \infty$, we have $f_k \xrightarrow{Lp} 0$ and also $f_k \xrightarrow{meas} 0$ [the latter convergence can be seen directly but also as a consequence of the former in view of Theorem 6.2.15]. On the other hand, we observe that $X(|f_k| > 0) = F_k$. So, for each $x \in X = [0, 1]$, $f_k(x) = 1$ for infinitely many values of k. Hence $f_k(x)$ does not converge to 0 for any $x \in X$. Thus f_k does not converge a.e. to 0 and hence also not almost uniformly (Theorem 6.2.11). However, the subsequence $\{\chi_{E_{n,1}}\}_{k \geq 1}$ does converge to 0 almost uniformly and (hence) a.e., which confirms Corollaries 6.2.7 and 6.2.8.

Example 6.2.20 leads to the (converse) question: Is convergence in (a) L^p norm or (b) measure (c) pointwise (d) almost uniformly implied by convergence a.e.? The answer to (a) is in the negative even if $\mu(X) < \infty$ [see Example 6.2.16; this example also shows that pointwise $\nRightarrow L^p$ and a.u. $\nRightarrow L^p$]. However, the answer to (b), though negative in general (Example 6.2.12), is in the affirmative when $\mu(X) < \infty$, as we demonstrate in the theorem below. The answers to (c) and (d) are both in the negative, as demonstrated in Examples 6.2.9(b) and 6.2.12 respectively.

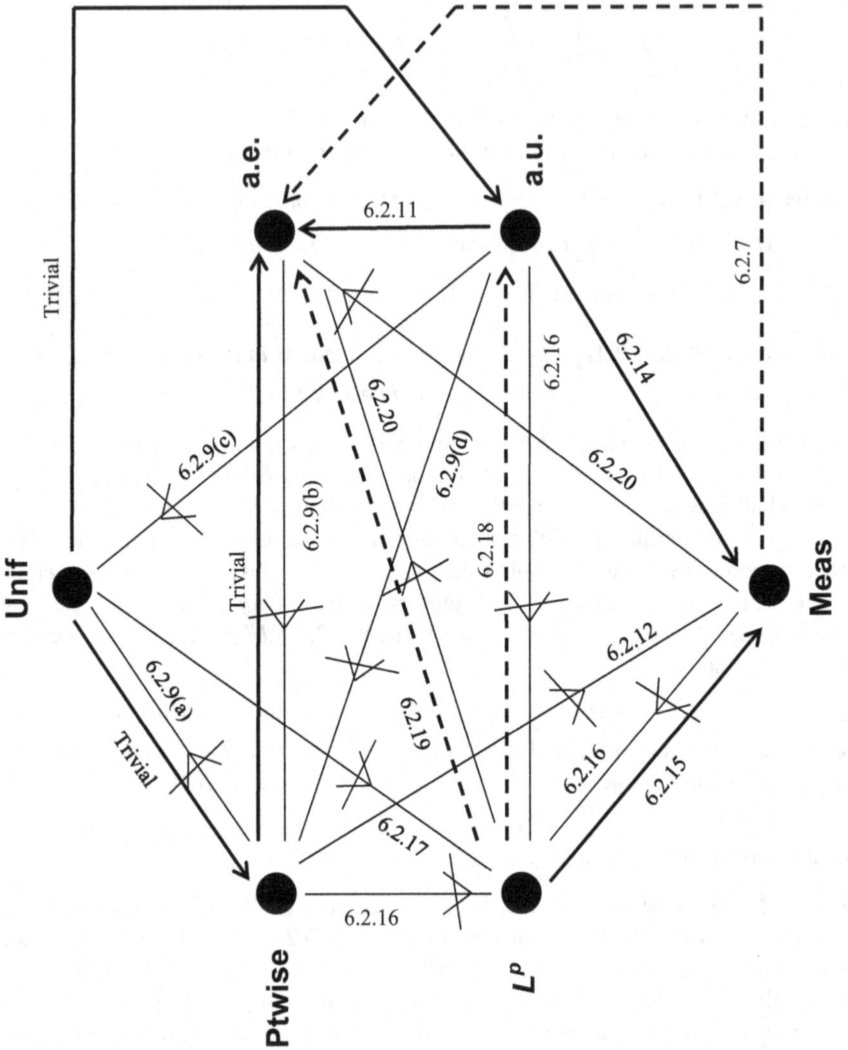

Fig. 6.1 Relations between various modes of convergence with no restriction on $\mu(X)$

The relations between various modes of convergence with no restriction on $\mu(X)$ have been indicated in Fig. 6.1 on the next page. The dotted arrows mean that convergence of a suitable subsequence is implied. In order to avoid cluttering, some non-implications that follow from those that have been shown have been left out. For instance it follows from what has been shown that convergence a.e. does not imply convergence in measure (not shown) and hence that it also does not imply almost uniform convergence (also not shown).

We go on to consider the situation in (B), namely, $\mu(X) < \infty$.

The finiteness hypothesis plays a significant role in determining the relation between modes of convergence. For instance, we saw in an example above that uniform convergence does not imply L^p norm convergence; in that example $\mu(X)$ was not finite. We shall see below (Theorem 6.2.23) that when $\mu(X) < \infty$, uniform convergence indeed implies L^p norm convergence. There are other instances of a similar nature. Almost uniform convergence does not follow from pointwise convergence (see Example 6.2.12), but if $\mu(X) < \infty$, the situation can be retrieved. Recall that almost everywhere convergence (hence pointwise convergence) implies almost uniform convergence when the total measure is finite (see Egorov's Theorem 2.6.2). Although the theorem was proved only for Lebesgue measure m, the argument carries over to a general measure μ. It may be noted that the argument need not use any analogue of Proposition 2.6.1, because Proposition 6.2.13 allows us to replace the limit by a measurable function and then Remark 6.2.5(a) allows us to switch back to the given limit. Moreover, Proposition 2.3.21 and Problem 2.3.P11, which were used in the proof of Egorov's Theorem 2.6.2, are valid for a general measure μ (see Proposition 3.1.8). We restate the theorem below for the sake of completeness.

Theorem 6.2.21 (Egorov) *Suppose that a sequence $\{f_k\}_{k\geq 1}$ of measurable functions converges a.e. to f and $\mu(X) < \infty$. Then for any $\varepsilon > 0$, there is a measurable subset X_ε such that $\mu(X \backslash X_\varepsilon) < \varepsilon$ and the sequence $\{f_k\}_{k\geq 1}$ converges to f uniformly on X_ε. In other words, when the measure of the space is finite, convergence a.e. implies almost uniform convergence.*

Remark The requirement that $\mu(X) < \infty$ cannot be dropped: see the example provided soon after Theorem 2.6.2 and repeated in Example 6.2.12.

Theorem 6.2.22 *Let $\{f_k\}_{k\geq 1}$ be a sequence of measurable functions such that $f_k \xrightarrow{ae} f$. Assume that $\mu(X) < \infty$. Then there exists a measurable function g such that $g = f$ a.e. and any such g has the property that $f_k \xrightarrow{meas} g$.*

Proof Immediate from Theorems 6.2.14 and 6.2.21. □

The converse of Theorem 6.2.22 is false in view of the Example 6.2.20, in which the measure was finite.

We have already seen in Example 6.2.17 that, in general, uniform convergence does not imply L^p norm convergence. We shall now show that the situation is different when $\mu(X) < \infty$.

Theorem 6.2.23 *Suppose that $\mu(X) < \infty$ and that the sequence $\{f_k\}_{k\geq 1}$ in L^p converges uniformly to f on X. Then $f \in L^p$ and $f_k \xrightarrow{L^p} f$.*

Proof The case $p = \infty$ is trivial; so we consider only $1 \leq p < \infty$. Let $\varepsilon > 0$ be given. Since $f_k \xrightarrow{unif} f$, there exists a natural number $k_0(\varepsilon)$ such that

$$|f_k(x) - f(x)| < \varepsilon \quad \text{for } k \geq k_0(\varepsilon) \text{ and } x \in X.$$

For $k \geq k_0(\varepsilon)$,

$$\|f_k - f\|_p = \left(\int_X |f_k - f|^p d\mu\right)^{\frac{1}{p}} \leq \left(\int_X \varepsilon^p d\mu\right)^{\frac{1}{p}} = \varepsilon\mu(X)^{\frac{1}{p}},$$

so that

$$\lim_{k\to\infty} \|f_k - f\|_p = 0. \qquad \square$$

Remarks 6.2.24

(a) The converse of the above theorem does not hold even if the measure is finite. The sequence with $f_k = \chi_{[\frac{1}{k},\frac{2}{k}]}$ on [0, 1] illustrates this, but Example 6.2.20 shows this and more.
(b) In Example 6.2.12, we showed that pointwise convergence does not imply convergence in measure when the measure is *infinite*. When the measure is finite however, convergence a.e. implies almost uniform convergence as already noted in Theorem 6.2.21, which then implies convergence in measure by Theorem 6.2.14.

The relations between various modes of convergence with $\mu(X) < \infty$ have been indicated in Fig. 6.2 on the next page. As before, the dotted arrows mean that convergence of a suitable subsequence is implied, and some implications that follow from those that have been shown are left out. For instance it follows from what has been shown that pointwise convergence implies convergence in measure (not shown) but not vice versa (also not shown).

Lastly, we take up case (C), when there is a dominating function.

The next result, also known as the Dominated Convergence Theorem, is restated below without proof, as it follows easily from Theorem 3.2.16 and Problem 3.3. P16(b).

Theorem 6.2.25 *Let $\{f_k\}_{k\geq 1}$ be a sequence of functions in L^p such that $f_k \xrightarrow{ae} f$. Assume that $|f_k| \leq g$ a.e. for every $k \in \mathbb{N}$, where $g \in L^p$. Then $f \in L^p$ and $f_k \xrightarrow{L_p} f$.*

That the condition $|f_k| \leq g$ cannot be dropped follows from the remark following the proof of Theorem 3.2.16.

When $\mu(X)$ is not necessarily finite, almost uniform convergence, and hence also convergence in measure, does not imply L^p norm convergence (Example 6.2.16). However, this implication does hold when the convergence is dominated.

Theorem 6.2.26 *Let $\{f_k\}_{k\geq 1}$ be a sequence of functions in L^p such that $f_k \xrightarrow{meas} f$. Assume that $|f_k| \leq g$ a.e. for every $k \in \mathbb{N}$, where $g \in L^p$. Then $f \in L^p$ and $f_k \xrightarrow{L^p} f$.*

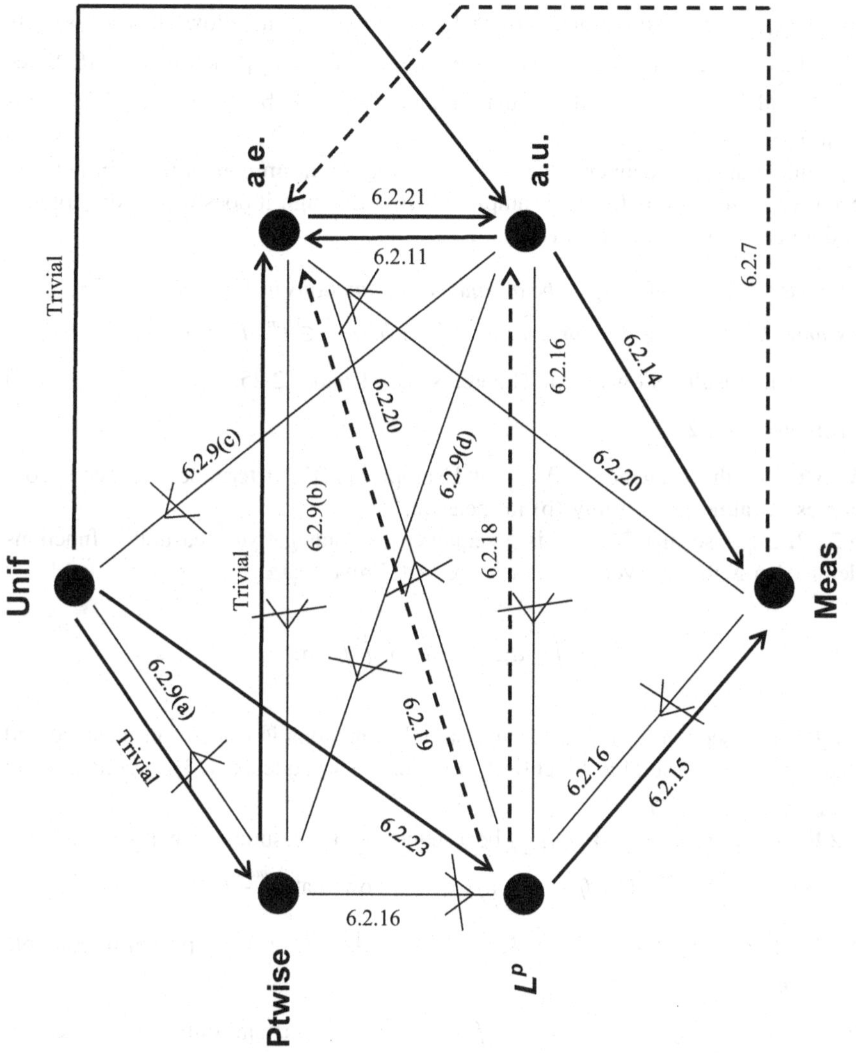

Fig. 6.2 Relations between various modes of convergence with $\mu(X) < \infty$

Proof Since the hypothesis implies that $|f| \leq g$ a.e. and it is given that $g \in L^p$, it follows that $f \in L^p$.

Suppose $\lim_{k\to\infty} \|f_k - f\|_p \neq 0$. Then there exists $\varepsilon > 0$ and a subsequence $\{g_j\}_{j\geq 1}$ of $\{f_k\}_{k\geq 1}$ such that

$$\|g_j - f\|_p \geq \varepsilon \quad \text{for every } j \in \mathbb{N}. \tag{6.12}$$

Since $\{g_j\}_{j\geq 1}$ is a subsequence of $\{f_k\}_{k\geq 1}$ and $f_k \overset{meas}{\longrightarrow} f$, it follows that $g_j \overset{meas}{\longrightarrow} f$. By Corollary 6.2.7, there is a subsequence $\{h_i\}_{i\geq 1}$ of $\{g_j\}_{j\geq 1}$ such that $h_i \overset{ae}{\longrightarrow} f$. Since $h_i \overset{ae}{\longrightarrow} f$ and $|h_i| \leq g$, it follows that $\lim_i \|h_i - f\|_p = 0$ by Theorem 6.2.25. This contradicts (6.12). □

Almost uniform convergence does not imply L^p norm convergence in general, even if the measure is finite (Example 6.2.16), although it does if the convergence is dominated by a function in L^p.

Theorem 6.2.27 *Let $\{f_k\}_{k\geq 1}$ be a sequence of functions in L^p such that $f_k \overset{au}{\longrightarrow} f \in L^p$. Assume that $|f_k| \leq g$ a.e. for every $k \in \mathbb{N}$, where $g \in L^p$. Then $f_k \overset{L^p}{\longrightarrow} f$.*

Proof The result follows from Theorems 6.2.11 and 6.2.25. □

Problem Set 6.2

6.2.P1. For the sequence $\{f_k\}_{k\geq 1}$ of Example 6.2.17, determine whether it converges (a) almost uniformly (b) in measure.

6.2.P2. Suppose that $\{f_k\}_{k\geq 1}$ is a sequence of nonnegative measurable functions defined on X that converges in measure to f. Prove that

$$\int_X f\, d\mu \leq \liminf \int_X f_k\, d\mu.$$

6.2.P3. If a sequence $\{f_k\}_{k\geq 1}$ of measurable functions is Cauchy in measure and there exists a measurable function f to which a subsequence $\{f_j\}_{j\geq 1}$ converges in measure, then $f_k \overset{meas}{\longrightarrow} f$.

6.2.P4. Let $\mu(X) < \infty$ and $\{f_k\}_{k\geq 1}$ be a sequence of measurable functions such that, for every $\varepsilon > 0$, $\sum_{n=1}^{\infty} \mu(X(|f_k - f| \geq \varepsilon)) < \infty$. Show that $f_k \overset{meas}{\longrightarrow} f$.

6.2.P5. Let $\mu(X) < \infty$. Define $\rho(f, g) = \int_X \frac{|f-g|}{1+|f-g|}\, d\mu$ for every pair of measurable functions f and g. Show that

(a) $\infty > \rho(f, g) \geq 0$, $\rho(f, g) = \rho(g, f)$ and $\rho(f, g) = 0$ if and only if $f = g$ a.e.;
(b) $\rho(f, h) \leq \rho(f, g) + \rho(g, h)$ [triangle inequality];

Remark This means ρ is a pseudometric in the sense of Definition 1.3.3.

(c) $f_k \overset{meas}{\longrightarrow} f$ if and only if $\rho(f_k, f) \to 0$ as $k \to \infty$;
(d) A sequence $\{f_k\}_{k\geq 1}$ of measurable functions is Cauchy in measure if and only if it is Cauchy in the sense of the pseudometric ρ;
(e) setting $\|f\| = \rho(f, 0)$ does not provide a norm on the space of equivalence classes of measurable functions that agree a.e., except in the trivial case when $\mu(X) = 0$.

6.2.P6. For any two measurable functions f and g on X, define

$$\rho(f,g) = \inf\{c + \mu(X(|f-g| > c)) : c > 0\}.$$

Show that, if either $\mu(X) < \infty$ or $f, g, f_k \in L^\infty$, then

(a) $\infty > \rho(f, g) \geq 0$, and $\rho(f, g) = 0$ if and only if $f = g$ a.e.; also, $\rho(f, g) = \rho(g, f)$;
(b) $\rho(f, h) \leq \rho(f, g) + \rho(g, h)$ [triangle inequality].

Remark This means ρ is a pseudometric in the sense of Definition 1.3.3.

(c) $f_k \overset{meas}{\to} f$ if and only if $\rho(f_k, f) \to 0$ as $k \to \infty$;
(d) a sequence $\{f_k\}_{k\geq 1}$ of measurable functions is Cauchy in measure if and only if it is Cauchy in the sense of the pseudometric ρ;
(e) if $A \subseteq X$ is measurable and $a > 0$, then $\rho(a\chi_A, 0) = \min\{a, \mu(A)\}$;
(f) setting $\|f\| = \rho(f, 0)$ does not provide a norm on the space of equivalence classes of measurable functions that agree a.e., except in the trivial case that no subset $A \subseteq X$ satisfies $0 < \mu(A) < \infty$.

6.2.P7. If $f_k \overset{meas}{\to} f$ and $g \in L^\infty(X)$, then show that $f_k g \overset{meas}{\to} fg$.

6.2.P8. If f and $\{f_k\}_{k\geq 1}$ are measurable functions defined on X with $\mu(X) < \infty$, show that the following are equivalent:

(i) $f_k \overset{meas}{\to} f$,
(ii) every subsequence of $\{f_k\}_{k\geq 1}$ has a subsequence converging to f a.e.

6.2.P9. For Lebesgue measure on [0, 1], show that there cannot exist a ρ satisfying (a) and (b) of Problems 6.2.P5 and 6.2.P6, but satisfying

$$f_k \overset{ae}{\to} f \text{ if and only if } \rho(f_k, f) \to 0 \text{ as } k \to \infty$$

instead of (c).

6.2.P10.

(a) If $f_k \overset{meas}{\to} f$, and $f_k \overset{ae}{\to} g$, then $g = f$ a.e.
(b) If $f_k \overset{meas}{\to} f$, and $f_k \leq f_{k+1}$ a.e. for each k, then $f_k \overset{ae}{\to} f$.

6.2.P11. Show that if $X = \mathbb{Z}$ with the counting measure, then convergence in measure is equivalent to uniform convergence. Does this equivalence hold for the counting measure on an arbitrary set?

6.2.P12. Let $(X, \mathcal{F}, \mu)$ be a measure space with $\mu(X) < \infty$ and suppose f and $\{f_k\}_{k \geq 1}$ are measurable functions on X. For $\varepsilon > 0$ and integer $k \geq 1$, put

$$E_k^\varepsilon = X(|f_k - f| \geq \varepsilon).$$

Prove that $f_k \to f$ (pointwise) if and only if, for all $\varepsilon > 0$, $\left\{ \cup_{k \geq n} E_k^\varepsilon \right\}_{n \geq 1}$ decreases to $\varnothing$.

6.2.P13. Let $(X, \mathcal{F}, \mu)$ be a measure space with $\mu(X) < \infty$ and suppose f and $\{f_k\}_{k \geq 1}$ are measurable functions on X. For $\varepsilon > 0$ and integer $k \geq 1$, put

$$E_k^\varepsilon = X(|f_k - f| \geq \varepsilon).$$

Prove that $f_k \xrightarrow{ae} f$ if and only if $\lim_{n \to \infty} \mu\left(\cup_{k \geq n} E_k^\varepsilon \right) = 0$ for all $\varepsilon > 0$.

6.2.P14. Let $(X, \mathcal{F}, \mu)$ be a measure space with $\mu(X) < \infty$ and suppose f and $\{f_k\}_{k \geq 1}$ are measurable functions on X. For $\varepsilon > 0$ and integer $m \geq 1$, let

$$E_m^\varepsilon = X(|f_m - f| \geq \varepsilon).$$

Prove that $f_k \xrightarrow{au} f$ if and only if $\lim_{n \to \infty} \mu\left(\cup_{m \geq n} E_m^\varepsilon \right) = 0$ for all $\varepsilon > 0$.

Chapter 7
Product Measure and Completion

7.1 Product Measure

In Chap. 2, the intuitive notion of volume defined for cuboid s in $\mathbb{R}^n$ was extended to the class of Lebesgue measurable sets. The class of Lebesgue measurable sets includes, amongst others, open and closed subsets of $\mathbb{R}^n$, their countable unions as well as countable intersections. In fact, the σ-algebra of Lebesgue measurable subsets of $\mathbb{R}^n$ contains the σ-algebra of Borel subsets of $\mathbb{R}^n$. Since $\mathbb{R}^{n+p} = \mathbb{R}^n \times \mathbb{R}^p$, it is natural to ask (i) whether the σ-algebra $\mathfrak{M}_{n+p}$ of Lebesgue measurable subsets of $\mathbb{R}^{n+p}$ is the "Cartesian product" $\mathfrak{M}_n \times \mathfrak{M}_p$ of the σ-algebras $\mathfrak{M}_n$ and $\mathfrak{M}_p$ of Lebesgue measurable subsets of $\mathbb{R}^n$ and $\mathbb{R}^p$ respectively; (ii) the Lebesgue measure m_{n+p} in $\mathbb{R}^{n+p}$ is the "Cartesian product" of the Lebesgue measure m_n in $\mathbb{R}^n$ with Lebesgue measure m_p in $\mathbb{R}^p$. In general, we shall be concerned with the following problem. Given two measure spaces $(X, \mathcal{F}, \mu)$ and $(Y, \mathcal{G}, \nu)$, how to define an appropriate measure space $(X \times Y, \mathcal{H}, \lambda)$, where $\mathcal{H}$ and λ are related to $\mathcal{F}$, $\mathcal{G}$ and μ, ν in some reasonable way, so as to include the following:

$$A \in \mathcal{F}, B \in \mathcal{G} \quad \Rightarrow \quad A \times B \in \mathcal{H} \text{ and } \lambda(A \times B) = \mu(A)\nu(B). \tag{7.1}$$

We shall address these problems in reverse order, starting with the problem raised in the general context.

In order to consider multiple integrals, we need to deal with measure and integration on the Cartesian product of measure spaces. To help motivate such a definition of measure, we consider the following concrete examples:

(i) Let I and J be intervals in $\mathbb{R}$. Then the Cartesian product $I \times J$ is a rectangle in $\mathbb{R}^2$, whose area can be expressed as

$$\text{area}(I \times J) = \ell(I)\ell(J) = m(I)m(J),$$

S. Shirali and H. L. Vasudeva, *Measure and Integration*,
Springer Undergraduate Mathematics Series,
https://doi.org/10.1007/978-3-030-18747-7_7

where m denotes the Lebesgue or the Borel measure on $\mathbb{R}$. This is a restatement of (7.1) in the present context and the product measure λ will be a generalisation of area to $\mathfrak{M} \times \mathfrak{M}$-measurable sets.

(ii) Let γ be the counting measure on $\mathcal{P}(\mathbb{N})$. If $A, B \in \mathcal{P}(\mathbb{N})$, then the number of elements in $A \times B$ can be expressed as

$$|A \times B| = |A||B| = \gamma(A)\gamma(B),$$

or

$$\gamma'(A \times B) = \gamma(A)\gamma(B),$$

where γ' is the counting measure on $\mathcal{P}(\mathbb{N} \times \mathbb{N})$.

In what follows, $(X, \mathcal{F}, \mu)$ and $(Y, \mathcal{G}, \nu)$ are understood to be given measure spaces.

Definition 7.1.1 A **rectangle** is a subset $A \times B \subseteq X \times Y$, where $A \in \mathcal{F}$, $B \in \mathcal{G}$. An **elementary set** is a finite union of disjoint rectangles. The family of elementary sets will be denoted by $\mathcal{E}$.

Some authors prefer the name "measurable rectangle", but we avoid it, because it suggests that we already have a σ-algebra of subsets of $X \times Y$.

A part of what (7.1) says is that the σ-algebra $\mathcal{H}$, which we are yet to construct, must contain all rectangles, and therefore, by virtue of being a σ-algebra, must contain all elementary sets as well. This can be arranged for by using Proposition 2.3.17 [see the terminology "generated by" mentioned just after that Proposition].

The representation of a rectangle in the form $A \times B$ need not be unique. Indeed,

$$\varnothing = X \times \varnothing = \varnothing \times Y.$$

However, if $P \times Q = R \times S$ is nonempty, then $P = R$ and $Q = S$. If $y \in Q$, then for $x \in P$, $(x, y) \in P \times Q = R \times S$, so that $x \in R$ and $y \in S$, which implies $P \subseteq R$ and $Q \subseteq S$. Similarly, one may show that $R \subseteq P$ and $S \subseteq Q$.

Let A be a measurable subset of $\mathbb{R}$. Then $A \times \varnothing = \varnothing \times \varnothing = \varnothing$.

Definition 7.1.2 The **product σ-algebra** of $\mathcal{F}$ and $\mathcal{G}$ is the σ-algebra generated by elementary sets and is denoted by the same symbol $\mathcal{F} \times \mathcal{G}$ as the Cartesian product (there will be no confusion on this account).

Next, we wish to extend the set function defined on the family of all rectangles $A \times B$ as $\mu(A)\nu(B)$ to the product σ-algebra $\mathcal{F} \times \mathcal{G}$ so as to be a measure. This will be facilitated if we establish that the collection of elementary sets is an algebra. To this end, we begin with the following:

Lemma 7.1.3 *A finite disjoint union of elementary sets is an elementary set.*

Proof Let $E_1, E_2, \ldots, E_n$ be disjoint elementary sets. Then their union F is a union of rectangles. The rectangles occurring in the union forming E_i are disjoint not only from each other but also from the rectangles occurring in the union forming E_j, $j \neq i$. This means all the rectangles occurring in all the n unions forming the respective elementary sets E_i are disjoint from each other. Hence F is an elementary set. □

Lemma 7.1.4 *If $P_1, P_2, \ldots, P_m$ are disjoint rectangles and Q is a rectangle, then $(P_1 \backslash Q) \cup (P_2 \backslash Q) \cup \cdots \cup (P_m \backslash Q)$ is an elementary set.*

Proof It is sufficient to prove that if P and Q are rectangles, then $P \backslash Q$ is an elementary set. The rest will follow by Lemma 7.1.3.

Let $P = A \times B$ and $Q = C \times D$, where $A, C \in \mathcal{F}$ and $B, D \in \mathcal{G}$. Then

$$P \backslash Q = ((A \backslash C) \times B) \cup (A \times (B \backslash D)) = ((A \backslash C) \times B) \cup ((A \cap C) \times (B \backslash D)),$$

which is an elementary set. □

Proposition 7.1.5 *$\mathcal{E}$ is closed under taking finite intersections and differences. In symbols,*

$$P \in \mathcal{E}, Q \in \mathcal{E} \quad \Rightarrow \quad P \cap Q \in \mathcal{E}, P \backslash Q \in \mathcal{E}. \tag{7.2}$$

Proof Let

$$P = P_1 \cup P_2 \cup \cdots \cup P_p \quad \text{and} \quad Q = Q_1 \cup Q_2 \cup \cdots \cup Q_q,$$

where $P_1, P_2, \ldots, P_p$ are disjoint rectangles and so are $Q_1, Q_2, \ldots, Q_q$. Then

$$P \cap Q = \bigcup_{i=1}^{p} \bigcup_{j=1}^{q} (P_i \cap Q_j)$$

and $P_i \cap Q_j$ ($1 \leq i \leq p$, $1 \leq j \leq q$) are disjoint rectangles. So, $P \cap Q \in \mathcal{E}$. To complete the proof of (7.2), we must show that $P \backslash Q \in \mathcal{E}$. We accomplish this by induction on n. For $n = 1$, it is a consequence of Lemma 7.1.4 that $P \backslash Q \in \mathcal{E}$. Now suppose $P \backslash Q \in \mathcal{E}$ whenever Q is a finite disjoint union of k rectangles. For a finite disjoint union Q of $k + 1$ rectangles Q_j ($1 \leq j \leq k + 1$), we have

$$\begin{aligned} P \backslash Q &= P \backslash (Q_1 \cup Q_2 \cup \cdots \cup Q_{k+1}) = P \cap (Q_1^c \cap Q_2^c \cap \cdots \cap Q_k^c \cap Q_{k+1}^c) \\ &= (P \cap (Q_1^c \cap Q_2^c \cap \cdots \cap Q_k^c)) \cap Q_{k+1}^c = (P \backslash (Q_1 \cup Q_2 \cup \cdots \cup Q_k)) \cap Q_{k+1}^c. \end{aligned}$$

By the induction hypothesis, the set $P \backslash (Q_1 \cup Q_2 \cup \ldots \cup Q_k)$ on the right-hand side here is an elementary set:

$$P\backslash(Q_1 \cup Q_2 \cup \cdots \cup Q_k) = R_1 \cup R_2 \cup \cdots \cup R_r,$$

where $R_1, \ldots, R_r$ are disjoint rectangles. Therefore we can further recast $P\backslash Q$ as

$$\begin{aligned} P\backslash Q &= (R_1 \cup R_2 \cup \cdots \cup R_r) \cap Q_{k+1}^c \\ &= (R_1\backslash Q_{k+1}) \cup (R_2\backslash Q_{k+1}) \cup \cdots \cup (R_r\backslash Q_{k+1}). \end{aligned}$$

By Lemma 7.1.4, $P\backslash Q$ is an elementary set. Induction is complete and (7.2) is established.

Corollary 7.1.6 *$\mathcal{E}$ is closed under taking finite unions and under complementation. In symbols,*

$$P \in \mathcal{E}, Q \in \mathcal{E} \quad \Rightarrow \quad P \cup Q \in \mathcal{E}, P^c \in \mathcal{E}.$$

Thus $\mathcal{E}$ is an algebra.

Proof Since $P \cup Q = P \cup (Q\backslash P)$, the corollary follows immediately from Lemma 7.1.3 and Proposition 7.1.5.

Corollary 7.1.7 $\mathcal{F} \times \mathcal{G} = \mathfrak{M}_0(\mathcal{E})$, where $\mathfrak{M}_0(\mathcal{E})$ denotes the monotone class generated by $\mathcal{E}$.

Proof The result follows from Corollary 7.1.6 and Problem 2.3.P19(b).

Definition 7.1.8 The ***x*-section of a subset** $E \subseteq X \times Y$, where $x \in X$, is the subset $\boldsymbol{E_x} = \{y \in Y: (x, y) \in E\}$ of Y. Similarly for the ***y*-section** $\boldsymbol{E^y}$, where $y \in Y$.

Examples 7.1.9

(a) Let $A = S \times T$, where S and T are subsets of X and Y respectively. Then

$$A_x = \begin{Bmatrix} T & x \in S \\ \varnothing & x \notin S \end{Bmatrix} \text{ and } A^y = \begin{Bmatrix} S & y \in T \\ \varnothing & y \notin T \end{Bmatrix}.$$

(b) Suppose $A = \{(x, y) : x^2 + 4y^2 \leq 4\}$. Then

$$A_x = \begin{cases} [-\frac{1}{2}(4 - x^2)^{\frac{1}{2}}, \frac{1}{2}(4 - x^2)^{\frac{1}{2}}] & |x| \leq 2 \\ \varnothing & \text{otherwise} \end{cases}$$

and

$$A^y = \begin{cases} [-2(1 - y^2)^{\frac{1}{2}}, 2(1 - y^2)^{\frac{1}{2}}] & |y| \leq 1 \\ \varnothing & \text{otherwise.} \end{cases}$$

(c) Let $A = \{(x, y) : 0 \leq y \leq x^2 \text{ and } x \geq 0\}$. Then

$$A_x = \begin{cases} [0, x^2] & x \geq 0 \\ \varnothing & \text{otherwise} \end{cases}$$

and

$$A^y = \begin{cases} [\sqrt{y}, \infty) & y \geq 0 \\ \varnothing & \text{otherwise.} \end{cases}$$

Lemma 7.1.10 *$(E_x)^c = (E^c)_x$ and $(E^y)^c = (E^c)^y$. Moreover, for any sequence of sets $E_1, E_2, \ldots,$ we have $\cup_{j\geq 1}(E_j)_x = (\cup_{j\geq 1}E_j)_x$ and $\cap_{j\geq 1}(E_j)^y = (\cap_{j\geq 1}E_j)^y$. Similarly for $\cap_{j\geq 1}(E_j)_x$ and $\cup_{j\geq 1}(E_j)^y$.*

Proof $y \in (E_x)^c \Leftrightarrow y \notin E_x \Leftrightarrow (x, y) \notin E \Leftrightarrow (x, y) \in E^c \Leftrightarrow y \in (E^c)_x$. So, $(E_x)^c = (E^c)_x$. An analogous argument shows that $(E^y)^c = (E^c)^y$.

$$y \in \bigcup_{j=1}^{\infty}(E_j)_x \Leftrightarrow y \in (E_j)_x \quad \text{for some } j \Leftrightarrow (x, y) \in E_j \quad \text{for some } j$$
$$\Leftrightarrow (x, y) \in (\bigcup_{j=1}^{\infty} E_j) \Leftrightarrow y \in (\bigcup_{j=1}^{\infty} E_j)_x.$$

This means $\cup_{j\geq 1}(E_j)_x = (\cup_{j\geq 1}E_j)_x$.

An analogous argument shows that $\cap_{j\geq 1}(E_j)^y = (\cap_{j\geq 1}E_j)^y$. □

Lemma 7.1.11 *If E is a rectangle $A \times B$, then $E_x = B$ for $x \in A$ and $E_x = \varnothing$ for $x \notin A$; also, $E^y = A$ for $y \in B$ and $E^y = \varnothing$ for $y \notin B$. In particular, if $A \in \mathcal{F}$ and $B \in \mathcal{G}$, then $\nu(E_x) = \nu(B)\chi_A(x)$ and $\mu(E^y) = \mu(A)\chi_B(y)$.*

Proof By definition of an x-section of a set, we have $y \in E_x \Leftrightarrow (x, y) \in E$. Let $x \in A$. Since $E = A \times B$, we have $(x, y) \in E \Leftrightarrow y \in B$. Thus $y \in E_x \Leftrightarrow y \in B$. This means $E_x = B$. Now let $x \notin A$. Then $(x, y) \notin E = A \times B$ for every $y \in Y$, which means $E_x = \varnothing$. *The proof for* E^y *is similar. The last part now follows trivially.* □

Proposition 7.1.12 *Let $E \in \mathcal{F} \times \mathcal{G}$. Then $E_x \in \mathcal{G}$ for every $x \in X$, and $E^y \in \mathcal{F}$ for every $y \in Y$.*

Proof Let $\mathcal{H}$ be the family of all $E \in F \times \mathcal{G}$ which satisfy the condition that $E_x \in \mathcal{G}$ for every $x \in X$, and $E^y \in \mathcal{F}$ for every $y \in Y$. It is an immediate consequence of Lemma 7.1.11 that every rectangle belongs to $\mathcal{H}$. Also, it is an immediate consequence of Lemma 7.1.10 that $E \in \mathcal{H} \Rightarrow E^c \in \mathcal{H}$ and that for any sequence of sets $E_1, E_2, \ldots$ in $\mathcal{H}$ we have $\cup_{j\geq 1}E_j \in \mathcal{H}$. Thus $\mathcal{H}$ is a σ-algebra containing all rectangles. By definition of the σ-algebra $\mathcal{F} \times \mathcal{G}$, it follows that $\mathcal{F} \times \mathcal{G} \subseteq \mathcal{H}$. On the other hand, it follows by the definition of $\mathcal{H}$ that $\mathcal{H} \subseteq \mathcal{F} \times \mathcal{G}$. Hence $\mathcal{H} = \mathcal{F} \times \mathcal{G}$. □

Definition 7.1.13 Given $x \in X$, the ***x*-section of** an extended real-valued function f on $X \times Y$ is the extended real-valued function f_x on the domain Y given by $f_x(y) = f(x, y)$ for all $y \in Y$. Similarly for the ***y*-section** f^y, where $y \in Y$.

Note: The reader is cautioned that the *y*-section and the *y*th power are denoted by the same symbol and judgement has to be exercised to determine which is intended in any particular instance.

When f_x is integrable or has integral $\pm\infty$, we shall denote $\int_Y f_x$ by the more elaborate symbol $\int_Y f(x,y)d\nu(y)$, as it will maintain greater clarity. Similarly, the integral $\int_X f^y$, when meaningful, will be denoted by $\int_X f(x,y)d\mu(x)$. This notation obviates the need to introduce the symbols for the sections f_x and f^y.

Example 7.1.14 Let $f : (0,\infty) \times (0,\infty) \to \mathbb{R}$ be defined by $f(x,y) = x^y + \frac{x^2}{y}$. Then

$$f_{\frac{1}{2}} : (0,\infty) \to \mathbb{R} \text{ is given by } \frac{1}{2^y} + \frac{1}{4y} \text{ and}$$
$$f^2 : (0,\infty) \to \mathbb{R} \text{ is given by } x^2 + \frac{x^2}{2} = \frac{3x^2}{2}.$$

Proposition 7.1.15 *If $E \subseteq X \times Y$, then*

$$\chi_{E_x}(y) = (\chi_E)_x(y) = (\chi_E)^y(x) = \chi_{E^y}(x) = \chi(x,y) \quad \textit{for all } x \in X \textit{ and } y \in Y.$$

If moreover $E \in \mathcal{F} \times \mathcal{G}$, then

$$\int_Y \chi_E(x,y)d\nu(y) = \int_Y (\chi_E)_x = \int_Y \chi_{E_x} = \nu(E_x) \quad \textit{for all } x \in X$$

and

$$\int_X \chi_E(x,y)d\mu(x) = \int_X (\chi_E)^y = \int_X \chi_{E^y} = \mu(E^y) \quad \textit{for all } y \in Y.$$

Proof By definition of section, we have

$$(\chi_E)_x(y) = (\chi_E)^y(x) = \chi_E(x,y) = 1 \text{ or } 0 \text{ according as } (x,y) \in E \text{ or } (x,y) \notin E$$

and

$$\begin{aligned}\chi_{E_x}(y) &= 1 \text{ or } 0 \text{ according as } y \in E_x \text{ or } y \notin E_x \\ &= 1 \text{ or } 0 \text{ according as } (x,y) \in E \text{ or } (x,y) \notin E.\end{aligned}$$

Similarly,

$$\chi_{E^y}(x) = 1 \text{ or } 0 \text{ according as } (x,y) \in E \text{ or } (x,y) \notin E.$$

This proves the string of equalities regarding characteristic functions.

Since the integral of the characteristic function of a measurable set is always equal to measure of that set, we have $\int_Y \chi_{E_x} = \nu(E_x)$ and $\int_X \chi_{E^y} = \mu(E^y)$. The equalities concerning the remaining integrals now follow from the string of equalities regarding characteristic functions. □

Proposition 7.1.16 *If E is a rectangle A × B, then*

$$(\chi_E)_x = \chi_{E_x} = \chi_B \quad \textit{and} \quad (\chi_E)^y = \chi_{E^y} = \chi_A.$$

Proof Simple consequence of Proposition 7.1.15 and Lemma 7.1.11. □

Proposition 7.1.17 *Let the extended real-valued function f on X × Y be measurable. Then its sections f_x and f^y are also measurable.*

Proof Let $E = \{(x, y) \in X \times Y : f(x, y) > \alpha\}$, where α is a real number. Then for fixed $x \in X$, the set $E_x = \{y \in Y : f_x(y) > \alpha\}$ belongs to $\mathcal{G}$ by Proposition 7.1.12. That is, f_x is a measurable function. Similarly, f^y is also a measurable function. □

Recall the concept of σ-finiteness from Definition 5.9.1. We restate it formally here for ready reference.

Definition 7.1.18 A measure space $(X, \mathcal{F}, \mu)$ is called a **σ-finite measure space** if there is a sequence $\{X_n\}_{n\geq 1}$ of $\mathcal{F}$-measurable subsets such that $\cup_{j\geq 1} X_n = X$ and $\mu(X_n) < \infty$ for each n.

$(\mathbb{R}, \mathfrak{M}, m)$ is σ-finite. Indeed, the sets $X_n = [-n, n]$, $n \in \mathbb{N}$, satisfy $\cup_{j\geq 1} X_n = \mathbb{R}$ and $m(X_n) < \infty$ for each n. Let γ be the counting measure on $\mathcal{P}(\mathbb{N})$. Then the sets $X_n = \{n\}$, consisting of a single point each, satisfy the requirements: $\cup_{j\geq 1} X_n = \mathbb{N}$ and $\gamma(X_n) = 1$ for each n. Thus, $(\mathbb{N}, \mathcal{P}(\mathbb{N}), \gamma)$ is a σ-finite measure space.

Lemma 7.1.19 *Suppose that $(X, \mathcal{F}, \mu)$ and $(Y, \mathcal{G}, \nu)$ are σ-finite measure spaces. Then for each $E \in \mathcal{F} \times \mathcal{G}$,*

(a) *the function $x \to \nu(E_x)$ defined on X is measurable,*
(b) *the function $y \to \mu(E^y)$ defined on Y is measurable,*
(c) $\int_X \nu(E_x) d\mu(x) = \int_Y \mu(E^y) d\nu(y)$.

Proof To begin with, we suppose that μ and ν are finite measures. Let Ω be the family of all subsets $E \in \mathcal{F} \times \mathcal{G}$ such that the assertions of the lemma hold. In what follows, we shall show that $\Omega = \mathcal{F} \times \mathcal{G}$. This will be accomplished by showing that Ω contains every measurable rectangle and every elementary set, that is, Ω contains the algebra $\mathcal{E}$, and that it is a monotone class contained in $\mathcal{F} \times \mathcal{G}$. The desired equality $\Omega = \mathcal{F} \times \mathcal{G}$ will then follow by Corollary 7.1.7.

If $E = A \times B$, where $A \in \mathcal{F}$ and $B \in \mathcal{G}$, then by Lemma 7.1.11, $\nu(E_x) = \nu(B)\chi_A(x)$ and therefore $\nu(E_x)$ is a measurable function on X satisfying

$$\int_X \nu(E_x)d\mu(x) = \nu(B)\mu(A). \tag{7.3}$$

Similarly, $\mu(E^y)$ is a measurable function on Y and its integral over Y equals the right-hand side of (7.3). Therefore the assertions of the lemma are true for a measurable rectangle.

If E is a finite union of disjoint rectangles, that is, if $E = \cup_{1 \leq n \leq N}(A_n \times B_n)$ is an elementary set, then

$$\chi_{E_x}(y) = \sum_{n=1}^{N} \chi_{A_n}(x)\chi_{B_n}(y),$$

from which it readily follows that

$$\nu(E_x) = \sum_{n=1}^{N} \left(\chi_{A_n}(x) \int_Y \chi_{B_n}(y)d\nu(y) \right) = \sum_{n=1}^{N} \nu(B_n)\chi_{A_n}(x).$$

Thus $\nu(E_x)$, being a sum of measurable functions, is itself measurable. Moreover,

$$\int_X \nu(E_x)d\mu(x) = \sum_{n=1}^{N} \nu(B_n)\mu(A_n).$$

Similarly, $\mu(E^y)$ is a measurable function on Y and

$$\int_Y \mu(E^y)d\nu(y) = \sum_{n=1}^{N} \nu(B_n)\mu(A_n).$$

Thus the assertions of the lemma hold for all elementary sets.

It remains to show that Ω is a monotone class. Let $E_n \in \Omega$, $n \geq 1$, be such that $E_n \subseteq E_{n+1}$ for every n and $E = \cup_{n \geq 1} E_n$. Then $(E_n)_x \subseteq (E_{n+1})_x$ and $(E_n)^y \subseteq (E_{n+1})^y$ for every $x \in X$ and $y \in Y$. Thus $\{\nu((E_n)_x)\}_{n \geq 1}$ and $\{\mu((E_n)^y)\}_{n \geq 1}$ are increasing sequences of nonnegative measurable functions, and $\lim \nu((E_n)_x) = \nu(E_x)$, $\lim \mu((E_n)^y) = \mu(E^y)$. Consequently, $\nu(E_x)$ and $\mu(E^y)$, being limits of measurable functions, are themselves measurable. Moreover, the functions $\nu((E_n)_x)$ and $\mu((E_n)^y)$ are nonnegative. By the Monotone Convergence Theorem 3.2.4, we have

$$\int_X \nu(E_x)d\mu(x) = \lim_{n\to\infty} \int_X \nu((E_n)_x)d\mu(x)$$

and

$$\int_Y \mu(E^y)d\nu(y) = \lim_{n\to\infty} \int_Y \mu((E_n)^y)d\nu(y).$$

Since $E_n \in \Omega$ for each $n \geq 1$, we have

$$\int_X \nu((E_n)_x)d\mu(x) = \int_Y \mu((E_n)^y)d\nu(y).$$

Therefore, it follows from the preceding two equalities that

$$\int_X \nu(E_x)d\mu(x) = \int_Y \mu(E^y)d\nu(y),$$

that is, $E = \cup_{n\geq 1}E_n \in \Omega$. Similarly, if $E_n \in \Omega$ and $E_n \supseteq E_{n+1}$, $n \geq 1$, we can conclude that $E = \cap_{n\geq 1}E_n \in \Omega$ on using the Dominated Convergence Theorem 3.2.16 and the hypothesis that μ and ν are both finite measures.

To handle the general case, we may write $X = \cup_{n\geq 1}X_n$, $Y = \cup_{k\geq 1}Y_k$, where $\{X_n\}$ and $\{Y_n\}$ are disjoint sequences of sets of finite measure. In the paragraph above, we have proved that the assertions of the lemma hold for each rectangle $X_n \times Y_k$ when we are considering μ and ν restricted to measurable subsets of X_n and Y_k respectively. Let $E \in \mathcal{F} \times \mathcal{G}$ and write $E_{n,k} = E \cap (X_n \times Y_k)$. Then for each x, $E_x = \cup_{n,k\geq 1}(E_{n,k})_x$. By the finite case proved above, $\nu((E_{n,k})_x)$ is a measurable function of x defined on X_n for each k; so, $\sum_{k=1}^{\infty} \nu((E_{n,k})_x)$ is a measurable function defined on X_n. Hence

$$\nu(E_x) = \sum_{n,k=1}^{\infty} \nu((E_{n,k})_x)$$

is $\mathcal{F}$-measurable. Similarly, $\mu(E^y)$ is $\mathcal{G}$-measurable. An application of the Monotone Convergence Theorem shows that

$$\begin{aligned}\int_X \nu(E_x)d\mu(x) &= \sum_{n=1}^{\infty}\int_{X_n}\sum_{k=1}^{\infty}\nu((E_{n,k})_x)d\mu(x) = \sum_{n=1}^{\infty}\sum_{k=1}^{\infty}\int_{X_n}\nu((E_{n,k})_x)d\mu(x)\\ &= \sum_{n=1}^{\infty}\sum_{k=1}^{\infty}\int_{Y_k}\mu((E_{n,k})^y)d\nu(y) = \int_Y \mu(E^y)d\nu(y).\end{aligned}$$

This completes the proof. □

Remark Problem 7.1.P8 shows that the σ-finiteness condition in the above lemma is essential.

Definition 7.1.20 Let $(X, \mathcal{F}, \mu)$ and $(Y, \mathcal{G}, \nu)$ be σ-finite measure spaces. Then the set function $\mu \times \nu$ defined on $\mathcal{F} \times \mathcal{G}$ by

$$(\mu \times \nu)(E) = \int_X \nu(E_x)d\mu(x) = \int_Y \mu(E^y)d\nu(y)$$

is called the **product** *of* μ *and* ν. The equality of the integrals is assured by Lemma 7.1.19.

It is an immediate consequence of this definition and (7.3) of Lemma 7.1.19 that $(\mu \times \nu)(A \times B) = \mu(A)\nu(B)$, where $A \in \mathcal{F}$ and $B \in \mathcal{G}$.

The next result shows that the set function $\mu \times \nu$ defined on $\mathcal{F} \times \mathcal{G}$ is indeed a measure which is σ-finite.

Theorem 7.1.21 *Let* $(X, \mathcal{F}, \mu)$ *and* $(Y, \mathcal{G}, \nu)$ *be* σ*-finite measure spaces. Then the set function* $\mu \times \nu$ *defined on the* σ*-algebra* $\mathcal{F} \times \mathcal{G}$ *as*

$$(\mu \times \nu)(E) = \int_X \nu(E_x)d\mu(x) = \int_Y \mu(E^y)d\nu(y)$$

is a σ*-finite measure.*

Proof Since the function $\nu(E_x)$ defined on X is nonnegative and measurable, it has an integral, the value of which may be infinite. We shall only show that $\mu \times \nu$ is a measure on $\mathcal{F} \times \mathcal{G}$, because the σ-finiteness will then follow easily.

Clearly, $(\mu \times \nu)(E) \geq 0$ for all $E \in \mathcal{F} \times \mathcal{G}$ and $(\mu \times \nu)(\emptyset) = 0$. Assume that $\{E_n\}_{n\geq 1}$ is a sequence of disjoint members of $\mathcal{F} \times \mathcal{G}$. Then $\{(E_n)_x\}_{n\geq 1}$ is a sequence of disjoint members of $\mathcal{G}$. Consequently,

$$\begin{aligned}(\mu \times \nu)(\bigcup_{n=1}^{\infty} E_n) &= \int_X \nu((\bigcup_{n=1}^{\infty} E_n)_x)d\mu(x) = \int_X \nu((\bigcup_{n=1}^{\infty} (E_n)_x))d\mu(x) \\ &= \int_X \sum_{n=1}^{\infty} \nu((E_n)_x)d\mu(x) = \sum_{n=1}^{\infty} \int_X \nu((E_n)_x)d\mu(x) = \sum_{n=1}^{\infty} (\mu \times \nu)(E_n),\end{aligned}$$

using the Monotone Convergence Theorem. Hence $\mu \times \nu$ is a measure. □

The measure $\mu \times \nu$ of the above theorem is called the **product measure** of μ and ν.

Remark 7.1.22 In view of Proposition 7.1.15, $\int_Y \chi_E(x, y)d\nu(y) = \int_Y (\chi_E)_x = \nu(E_x)$ while $\int_X \chi_E(x, y)d\mu(x) = \int_X (\chi_E)^y = \mu(E^y)$. The conclusions (a) and (b) of Lemma 7.1.19 can be expressed by saying that and these two functions of x and y respectively are measurable; conclusion (c) can be written as

$$\int_X (\int_Y (\chi_E)_x) = \int_Y (\int_X (\chi_E)^y)$$

or in alternative notation as

$$\int_X d\mu(x) \int_Y \chi_E(x,y)d\nu(y) = \int_Y d\nu(y) \int_X \chi_E(x,y)d\mu(x).$$

Also, one can write $(\mu \times \nu)(E)$ as the integral $\int_{X\times Y} \chi_E d(\mu \times \nu)$ or alternatively, as $\int_{X\times Y} \chi_E(x,y)d(\mu \times \nu)(x,y)$. Therefore by Definition 7.1.20, we further have

$$\int_X \Big(\int_Y (\chi_E)_x\Big) = \int_{X\times Y} \chi_E d(\mu \times \nu) = \int_Y \Big(\int_X (\chi_E)^y\Big)$$

or in alternative notation,

$$\begin{aligned}\int_X d\mu(x) \int_Y \chi_E(x,y)d\nu(y) &= \int_{X\times Y} \chi_E(x,y)d(\mu \times \nu)(x,y)\\ &= \int_Y d\nu(y) \int_X \chi_E(x,y)d\mu(x).\end{aligned}$$

Another way to say this is that the "double" integral of the characteristic function of a measurable subset of the product space equals both "iterated" integrals of that function. It is understood here that integrability of the functions involved is part of the assertion. We go on to prove it for a larger class of measurable functions on the product space.

Theorem 7.1.23 (Tonelli) *Let $(X, \mathcal{F}, \mu)$ and $(Y, \mathcal{G}, \nu)$ be σ-finite measure spaces and let $(X \times Y,\ \mathcal{F} \times \mathcal{G},\ \mu \times \nu)$ be the product measure space. If f is a nonnegative extended real-valued $\mathcal{F} \times \mathcal{G}$-measurable function on $X \times Y$, then*

(a) *the function f^y is $\mathcal{F}$-measurable,*
(b) *the function f_x is $\mathcal{G}$-measurable,*
(c) *the function $x \to \int_Y f(x,y)d\nu(y)$ is $\mathcal{F}$-measurable,*
(d) *the function $y \to \int_X f(x,y)d\mu(x)$ is $\mathcal{G}$-measurable, and the equation*

(e)
$$\begin{aligned}\int_X d\mu(x) \int_Y f(x,y)d\nu(y) &= \int_{X\times Y} f(x,y)d(\mu \times \nu)(x,y)\\ &= \int_Y d\nu(y) \int_X f(x,y)d\mu(x)\end{aligned}$$

holds.

Proof It follows from Proposition 7.1.17 that (a) and (b) hold. We next deal with (c), (d) and (e). Because of the symmetry between x and y, it is enough to prove (c) and the first half of (e).

As recorded in Remark 7.1.22, the result is valid when f is the characteristic function of an $(\mathcal{F} \times \mathcal{G})$-measurable set. Therefore it is also valid for any simple function [the fact that $(f + g)_x = f_x + g_x$ is relevant here]. Now consider an arbitrary extended real-valued nonnegative $(\mathcal{F} \times \mathcal{G})$-measurable function f on $X \times Y$. There exists an increasing sequence of nonnegative simple functions s_k, $k \in \mathbb{N}$, converging to f [see Theorem 2.5.9]. Since the required conclusion is valid for simple functions, we know that each $\int_Y s_k(x,y)d\nu(y)$ is an $\mathcal{F}$-measurable function on X and

$$\int_X d\mu(x) \int_Y s_k(x,y)d\nu(y) = \int_{X\times Y} s_k d(\mu\times\nu)(x,y). \tag{7.4}$$

Now, the Monotone Convergence Theorem 3.2.4 leads to the equality

$$\lim_{k\to\infty} \int_Y s_k(x,y)d\nu(y) = \int_Y f(x,y)d\nu(y),$$

showing that $\int_Y f(x,y)d\nu(y)$ is an $\mathcal{F}$-measurable function on X, thereby establishing (c). Moreover, the $\mathcal{F}$-measurable functions $\int_Y s_k(x,y)d\nu(y)$ form an increasing sequence, and hence another application of the Monotone Convergence Theorem 3.2.4 leads to the equality

$$\lim_{k\to\infty} \int_X d\mu(x) \int_Y s_k(x,y)d\nu(y) = \int_X d\mu(x) \int_Y f(x,y)d\nu(y),$$

which is the same as

$$\lim_{k\to\infty} \int_{X\times Y} s_k d(\mu\times\nu)(x,y) = \int_X d\mu(x) \int_Y f(x,y)d\nu(y),$$

in view of (7.4). Finally, yet another application of the Monotone Convergence Theorem 3.2.4 leads to the equality

$$\lim_{k\to\infty} \int_{X\times Y} s_k(x,y)d(\mu\times\nu)(x,y) = \int_{X\times Y} f(x,y)d(\mu\times\nu)(x,y).$$

The required equality, that is the first half of (e), is now immediate from the preceding two. □

Remark 7.1.24

(a) Suppose F is a function defined almost everywhere and there exists an integrable function Φ defined everywhere that agrees with F a.e. Then Φ agrees a.e. with any integrable function that agrees with F a.e. and therefore both have the same integral. We speak of F as being integrable and the integral of Φ as being the integral of F. The statements (c) and (d) in the next theorem are to be understood in the light of this observation.

(b) Let f be a complex-valued $\mathcal{F}\times\mathcal{G}$-measurable function. Then it follows from Tonelli's Theorem that the three integrals $\int_{X\times Y} |f(x,y)| d(\mu\times\nu)(x,y)$, $\int_X d\mu(x) \int_Y |f(x,y)| d\nu(y)$, $\int_Y d\nu(y) \int_X |f(x,y)| d\mu(x)$ are equal.

Theorem 7.1.25 (Fubini) *Let $(X, \mathcal{F}, \mu)$ and $(Y, \mathcal{G}, \nu)$ be σ-finite measure spaces and let $(X \times Y,\ \mathcal{F} \times \mathcal{G},\ \mu \times \nu)$ be the product measure space. Suppose f is a complex-valued $\mathcal{F} \times \mathcal{G}$-measurable function on $X \times Y$ such that $\int_{X\times Y} |f(x,y)|\, d(\mu \times \nu)(x,y) < \infty$. Then*

(a) *the function $f^y \in L^1(X, \mathcal{F}, \mu)$ for almost all $y \in Y$,*
(b) *the function $f_x \in L^1(Y, \mathcal{G}, \nu)$ for almost all $x \in X$,*
(c) *the a.e. defined function $x \to \int_Y f(x,y)d\nu(y)$ belongs to $L^1(X, \mathcal{F}, \mu)$,*
(d) *the a.e. defined function $y \to \int_X f(x,y)d\mu(x)$ belongs to $L^1(Y, \mathcal{G}, \nu)$, and the equation*

(e)
$$\int_X d\mu(x) \int_Y f(x,y)d\nu(y) = \int_{X\times Y} f(x,y)d(\mu\times\nu)(x,y) = \int_Y d\nu(y) \int_X f(x,y)d\mu(x)$$

holds.

Proof It follows from Proposition 7.1.17 (the proposition being evidently true for complex-valued functions) that the functions in (a) and (b) are measurable.

From the finiteness in the hypothesis and Remark 7.1.24(b), we have

$$\int_Y d\nu(y) \int_X |f(x,y)|d\mu(x) < \infty$$

and hence

$$\int_X |f(x,y)|d\mu(x) < \infty \quad \text{for almost all } y \in Y.$$

The latter inequality means the same thing as (a). The assertion (b) is proved in like manner.

We shall first prove the result for real-valued f.

Since f^+ is measurable and is nonnegative, Tonelli's Theorem 7.1.23 shows that $\int_Y f^+(x,y)d\nu(y)$ is a measurable function on X satisfying

$$\int_X d\mu(x) \int_Y f^+(x,y)d\nu(y) = \int_{X\times Y} f^+ d(\mu\times\nu)(x,y).$$

Now, f is integrable and therefore

$$\int_X d\mu(x) \int_Y f^+(x,y)d\nu(y) < \infty. \tag{7.5}$$

It follows that the measurable function on X given by $\int_Y f^+(x,y)d\nu(y)$ is finite almost everywhere, which is the same as $(f^+)_x$ being integrable for almost all x. Similarly,

$$\int_X d\mu(x) \int_Y f^-(x,y)d\nu(y) < \infty, \tag{7.6}$$

and the measurable function on X given by $\int_Y f^-(x,y)d\nu(y)$ is finite almost everywhere, which is the same as $(f^-)_x$ being integrable for almost all x. By virtue of the obvious equality

$$f_x = (f^+ - f^-)_x = (f^+)_x - (f^-)_x, \tag{7.7}$$

we can assert that f_x is integrable for almost all x, thereby justifying (b). Part (a) can be justified on the basis of an analogous argument.

By (7.5) and (7.6), the functions defined on the set X by $\int_Y f^+(x,y)d\nu(y)$ and $\int_Y f^-(x,y)d\nu(y)$ are integrable. Therefore they may have ∞ as a value only on a set of measure 0, so that the function Φ given by

$$\Phi(x) = \int_Y f^+(x,y)d\nu(y) - \int_Y f^-(x,y)d\nu(y) \tag{7.8}$$

is defined almost everywhere. Upon extending it to be 0 on the set where it may fail to be defined, we obtain a measurable function Φ such that (7.8) holds for almost all x. The functions Φ^+ and Φ^- then agree a.e. with the respective functions occurring on the right-hand side of (7.8), both of which are already known to be integrable, and accordingly, Φ is integrable with

$$\int_X \Phi = \int_X d\mu(x) \int_Y f^+(x,y)d\nu(y) - \int_X d\mu(x) \int_Y f^-(x,y)d\nu(y). \tag{7.9}$$

As a consequence of (7.7),

$$\int_Y f_x = \int_Y (f^+)_x - \int_Y (f^-)_x \quad \text{for almost all } x.$$

In other words,

$$\int_Y f(x,y)d\nu(y) = \int_Y f^+(x,y)d\nu(y) - \int_Y f^-(x,y)d\nu(y) \quad \text{for almost all } x.$$

Since (7.8) holds for almost all x, this leads to

$$\Phi(x) = \int_Y f(x,y)d\nu(y) \quad \text{for almost all } x. \tag{7.10}$$

Thus there exists an integrable function on X that agrees with $\int_Y f(x,y)d\nu(y)$ almost everywhere. This proves (c). A similar argument proves (d).

We obtain from (7.9) and (7.10) that

$$\int_X d\mu(x)\int_Y f(x,y)d\nu(y) = \int_X d\mu(x)\int_Y f^+(x,y)d\nu(y) - \int_X d\mu(x)\int_Y f^-(x,y)d\nu(y). \tag{7.11}$$

Now, by definition of integral,

$$\int_{X\times Y} f(x,y)d(\mu\times\nu)(x,y) = \int_{X\times Y} f^+(x,y)d(\mu\times\nu)(x,y) - \int_{X\times Y} f^-(x,y)d(\mu\times\nu)(x,y).$$

By Tonelli's Theorem 7.1.23, the right-hand side here agrees with the one in (7.11). Therefore so does the left-hand side, which is what the first equality of part (e) says. The second equality is also valid for analogous reasons.

This completes the proof for real-valued f.

The case when f is complex-valued now follows by splitting into real and imaginary parts. □

Problem Set 7.1

7.1.P1. Let $X = Y = [0, 1]$, $\mathcal{F} = \mathcal{G}$ be the σ-algebra of measurable subsets of [0, 1] and $\mu = \nu$ be Lebesgue measure on [0, 1]. Let

$$f(x,y) = \frac{x^2-y^2}{(x^2+y^2)^2}, \quad (x,y)\in(0,1)\times(0,1).$$

Prove that each of the iterated integrals of f exists. The function f is, however, not in $L^1([0, 1] \times [0, 1])$.

7.1.P2. Let $X = Y = [-1, 1]$ and $\mathcal{F} = \mathcal{G}$ be the σ-algebra of measurable subsets of [–1,1] and $\mu = \nu$ be Lebesgue measure on [–1, 1]. Let

$$f(x,y) = \begin{cases} \frac{xy}{(x^2+y^2)^2} & \text{if } (x,y)\neq(0,0) \\ 0 & \text{otherwise.} \end{cases}$$

Show that

$$\int_{[-1,1]}\left(\int_{[-1,1]} f(x,y)d\nu(y)\right)d\mu(x) = 0 = \int_{[-1,1]}\left(\int_{[-1,1]} f(x,y)d\mu(x)\right)d\nu(y),$$

but the function is not Lebesgue integrable over $[-1, 1] \times [-1, 1]$.

7.1.P3. Let $f \in L^1(X, \mathcal{F}, \mu)$ and $g \in L^1(Y, \mathcal{G}, \nu)$, and suppose $\phi(x, y) = f(x)g(y)$, $x \in X$, $y \in Y$. Show that $\phi \in L^1(X \times Y, \mathcal{F} \times \mathcal{G}, \mu \times \nu)$ and

$$\int_{X\times Y} \phi(x,y)d(\mu\times\nu) = \left(\int_X f(x)d\mu(x)\right)\left(\int_Y g(y)d\nu(y)\right).$$

7.1.P4. Let $X = Y = [0, 1]$, $\mathcal{F} = \mathcal{G} =$ the σ-algebra of Lebesgue measurable subsets of $[0, 1]$ and $\mu = \nu =$ Lebesgue measure. Suppose that either $f \in L^1([0, 1] \times [0, 1])$ or f is a measurable nonnegative function on $[0, 1] \times [0, 1]$. Prove that

$$\int_{[0,1]}\left(\int_{[0,x]} f(x,y)dy\right)dx = \int_{[0,1]}\left(\int_{[y,1]} f(x,y)dx\right)dy.$$

7.1.P5. Let $f \in L^1((0, a))$ and $g(x) = \int_{[x,a]} \frac{f(t)}{t}dt$, $0 < x \le a$. Show that $g \in L^1((0, a))$ and that $\int_{[0,a]} g(x)dx = \int_{[0,a]} f(t)dt$.

7.1.P6. Let $X = Y = [0, 1]$, $\mathcal{F} = \mathcal{G} =$ the σ-algebra of Lebesgue measurable subsets of $[0, 1]$ and $\mu = \nu =$ Lebesgue measure. Define for $x, y \in [0, 1]$,

$$f(x,y) = \begin{cases} 1 & x \in \mathbb{Q} \\ 2y & x \notin \mathbb{Q}. \end{cases}$$

Compute

$$\int_{[0,1]}\left(\int_{[0,1]} f(x,y)d\nu(y)\right)d\mu(x) \quad \text{and} \quad \int_{[0,1]}\left(\int_{[0,1]} f(x,y)d\mu(x)\right)d\nu(y).$$

Does $f \in L^1([0, 1] \times [0, 1])$?

7.1.P7. Suppose that $\{a_{m,n}\}_{m,n\ge 1}$ is a double sequence of nonnegative real numbers. Then show that

$$\sum_{m=1}^{\infty}\sum_{n=1}^{\infty} a_{m,n} = \sum_{n=1}^{\infty}\sum_{m=1}^{\infty} a_{m,n}.$$

7.1.P8. Let $X = Y = [0, 1]$, $\mathcal{F} = \mathcal{G} =$ the σ-algebra of Borel subsets of $[0, 1]$; suppose μ is Lebesgue measure on Borel subsets of X and let γ be the counting measure on Y. Show that $V = \{(x, y) \in [0, 1] \times [0, 1] : x = y\}$ is $\mathcal{F} \times \mathcal{G}$-measurable and

$$\int_Y d\gamma \int_X \chi_V d\mu = 0 \quad \text{but} \quad \int_Y d\mu \int_X \chi_V d\gamma = 1.$$

7.1.P9. Let $X = Y = \mathbb{N}$ and $\mu = \nu =$ counting measure on $\mathbb{N}$. Let

$$f(x,y) = \begin{cases} 2 - 2^{-x} & \text{if } x = y \\ -2 + 2^{-x} & \text{if } x = y + 1 \\ 0 & \text{otherwise.} \end{cases}$$

Show that the iterated integrals of f are not equal and $f \notin L^1(\mu \times \nu)$. [This shows that neither the integrability condition in Fubini's Theorem nor the nonnegativity condition in Tonelli's Theorem can be omitted.]

7.1.P10. Let $(X, \mathcal{F}, \mu)$ and $(Y, \mathcal{G}, \nu)$ be measure spaces with $\mu(X) = 1$ and ν σ-finite. If E is a measurable subset of $X \times Y$ and if $\nu(E_x) \leq \frac{1}{2}$ for almost all $x \in X$, then

$$\nu(\{y \in X : \mu(E^y) = 1\}) \leq \frac{1}{2}.$$

7.1.P11. Let $X = Y = [0, 1]$, equipped with the σ-algebra of Lebesgue measurable subsets and Lebesgue measure. Consider a real sequence $\{\alpha_n\}_{n \geq 1}$, $0 < \alpha_1 < \alpha_2 < \ldots$, satisfying $\lim_{n \to \infty} \alpha_n = 1$. For each n, choose a continuous function g_n such that $\{t : g_n(t) \neq 0\} \subseteq (\alpha_n, \alpha_{n+1})$ and also $\int_{[0,1]} g_n(t)dt = 1$. Define

$$f(x,y) = \sum_{n=1}^{\infty} [g_n(x) - g_{n+1}(x)]g_n(y).$$

Note that for each (x, y), at most two terms in the sum can be nonzero. Thus no convergence problem arises in the definition of f. Show that f is not integrable and that its repeated integrals do not agree.

7.1.P12. This problem deals with a measurable function on $[0, 1] \times [0, 1] \subseteq \mathbb{R}^2$ for which the iterated integrals exist and are equal though the function fails to be integrable. Let $I = [0, 1] \times [0, 1]$; divide I into four equal squares. Let $I_1 = \left[0, \frac{1}{2}\right] \times \left[0, \frac{1}{2}\right]$. Next, divide the square $\left[\frac{1}{2}, 1\right] \times \left[\frac{1}{2}, 1\right]$ into four equal squares and set $I_2 = \left[\frac{1}{2}, \frac{3}{2^2}\right] \times \left[\frac{1}{2}, \frac{3}{2^2}\right]$, and so on. On each square I_k, $k = 1, 2, \ldots$, define a function φ_k as follows; divide I_k into four equal squares and let φ_k be -1 on the interiors of the left bottom and right upper squares, 1 on the interiors of the right bottom and left upper squares and zero elsewhere. Set

$$f(x,y) = \sum_{k=1}^{\infty} \frac{1}{|I_k|} \varphi_k(x,y),$$

where $|I_k|$ is the area of the square I_k. The function $f(x, y)$ is well defined since each point $(x, y) \in I$ belongs to the interior of at most one of the I_k. Show that the repeated integrals exist and are equal but the function is not integrable, though it is measurable.

7.1.P13. This is another example (besides 7.1.P2 and 7.1.P12) which shows that the integrability condition cannot be omitted from Fubini's Theorem. Let $X = Y = \mathbb{Z}$, the set of integers, $\mathcal{P}(\mathbb{Z})$ the σ-algebra of all subsets of $\mathbb{Z}$ and γ be the counting measure on $\mathcal{P}(\mathbb{Z})$. Let

$$f(x,y) = \begin{cases} x & \text{if} & y = x \\ -x & \text{if} & y = x+1 \\ 0 & \text{otherwise.} & \end{cases}$$

Show that the repeated integrals exist and are unequal. Also, show by a direct computation that the function is not integrable.

7.1.P14. Let $(X, \mathcal{F}, \mu)$ and $(Y, \mathcal{G}, \nu)$ be measure spaces, $A \subseteq X$ and $B \subseteq Y$. If $A \times B \in \mathcal{F} \times \mathcal{G}$ show that, if $A \neq \emptyset$, then B is measurable and if $B \neq \emptyset$, then A is measurable. Hence show that $A \times B \in \mathcal{F} \times \mathcal{G}$ if and only if $A \neq \emptyset \neq B$ and either A or B is nonmeasurable.

7.1.P15. [Cf. Problem 3.2.P13(d)] Let $(X, \mathcal{F}, \mu)$ be a measure space. Verify for every real-valued measurable function f that

$$\int_X |f|^p d\mu = \int_{(0,\infty)} pt^{p-1} \mu(\{|f| > t\}) dt, \quad 0<p<\infty.$$

7.2 The Completion of a Measure

Let $(X, \mathcal{F}, \mu)$ be a measure space. A measurable set of measure zero is called a *null set*. If N is a null set and $A \subseteq N$, then by monotonicity of μ, it follows that $\mu(A) = 0$ provided that $A \in \mathcal{F}$. But it may happen that some null set may have a subset that is not measurable. Consider for instance the following simple example: let $X = \{a, b, c\}$, $\mathcal{F} = \{\emptyset, \{b\}, \{a, c\}, X\}$, and $\mu(X) = \mu(\{b\}) = 1$, $\mu(\emptyset) = \mu(\{a, c\}) = 0$. Then μ is a measure on $\mathcal{F}$. The set $N = \{a, c\}$ is a null set, but its subsets $\{a\}$ and $\{c\}$ are not $\mathcal{F}$-measurable. The reader will recall that there is a subset of a Borel set of Lebesgue measure zero which is not a Borel set [Problem 2.3.P20].

If N is a null set and f is a nonnegative measurable function defined on X, then $\int_N f = 0$, i.e., null sets are negligible in integration. It is desirable that a subset of a negligible set is also negligible, i.e., a subset of a null set be measurable—in which case, its measure must necessarily be 0.

The product $m \times m$ of Lebesgue measure m on $\mathbb{R}$ with itself is a measure on $\mathbb{R}^2$, which turns out to be different from Lebesgue measure on $\mathbb{R}^2$ but closely related to it. To clarify the relation between $m \times m$ on $\mathbb{R}^2$ and Lebesgue measure on $\mathbb{R}^2$, we shall need the following concept, which has already occurred in Problem 3.2.P30(e):

Definition 7.2.1 A measure space $(X, \mathcal{F}, \mu)$ is said to be **complete** if whenever $N \in \mathcal{F}$ is a null set and $A \subseteq N$, A is measurable (in which case it has to be a null set). It means the same thing to say that the measure μ is complete.

It is trivial that a measure space $(X, \mathcal{P}(X), \mu)$ is complete no matter what μ is.

The Lebesgue measure space $(\mathbb{R}^n, \mathfrak{M}_n, m_n)$ is a complete measure space. Indeed, if $N \in \mathfrak{M}_n$ is a null set and $A \subseteq N$, then A is a Lebesgue measurable null set [see Remark 2.8.2(d)]. Since a subset of a Borel set of measure zero need not be a Borel set [see 2.3.P20], the Borel measure space $(\mathbb{R}, \mathfrak{B}, m)$ is not complete. In particular, $\mathfrak{B} \subset \mathfrak{M}$.

It is worth noting that when $(X, \mathcal{F}, \mu)$ is complete, a function on X that agrees with a measurable function almost everywhere is itself measurable. This is essentially Problem 3.2.P30(e).

By *extending* a given measure on a σ-algebra, we mean obtaining a measure on a bigger σ-algebra of subsets of the same space, which agrees with the given measure when restricted to the given σ-algebra. One also speaks of extending a measure space. We next consider whether an incomplete measure can be extended to a complete measure. An affirmative answer is provided in the theorem below. We begin with some simple examples of the relation between the σ-algebras on which the incomplete measure and its complete extension are defined.

The incomplete measure space described in the opening paragraph of this section can be extended to a complete measure space by enlarging $\mathcal{F}$ to $\mathcal{P}(X)$, the σ-algebra of all subsets of X, and setting $\bar{\mu} = \mu$ on those sets that are in $\mathcal{F}$ and $\bar{\mu}(\{a\}) = \bar{\mu}(\{c\}) = 0$, $\bar{\mu}(\{a, b\}) = 1 = \bar{\mu}(\{c, b\})$. Note that, like any measure on $\mathcal{P}(X)$, $\bar{\mu}$ is complete. Observe that every set in $\mathcal{P}(X)$ is the union of a set belonging to $\mathcal{F}$ and a subset of a null set belonging to $\mathcal{F}$.

The other incomplete space we have encountered is the Borel measure space $(\mathbb{R}, \mathfrak{B}, m)$. In contrast, $(\mathbb{R}, \mathfrak{M}, m)$ is a complete measure space. Whereas $\mathfrak{B} \subset \mathfrak{M}$, there is a relation between $\mathfrak{B}$ and $\mathfrak{M}$ that is analogous to the one between $\mathcal{F}$ and $\mathcal{P}(X)$ in the previous example, as we now show. A subset E of $\mathbb{R}$ is Lebesgue measurable, i.e., $E \in \mathfrak{M}$, if and only if there exists an $\mathcal{F}_\sigma$ set F and a $\mathcal{G}_\delta$-set G such that $F \subseteq E \subseteq G$ with $m(G \backslash F) = 0$ [Remark 2.3.25]. Now $E \backslash F \subseteq G \backslash F$, a Borel set of measure zero and $E = F \cup (E \backslash F)$. So, E is the union of the Borel set F and a subset $E \backslash F$ of a Borel set of measure zero. Thus every set of $\mathfrak{M}$ is the union of a set belonging to $\mathfrak{B}$ and a subset of a null set belonging to $\mathfrak{B}$.

Theorem 7.2.2 *Let $(X, \mathcal{F}, \mu)$ be a measure space and let $\mathfrak{N} = \{N \in \mathcal{F} : \mu(N) = 0\}$. The collection $\overline{\mathcal{F}}$ of all sets of the form $E \cup F$, where $E \in \mathcal{F}$ and $F \subseteq N \in \mathfrak{N}$, is a σ-algebra containing $\mathcal{F}$. For such sets, define $\bar{\mu}(E \cup F) = \mu(E)$. Then $\bar{\mu}$ is a measure on $\overline{\mathcal{F}}$, $(X, \overline{\mathcal{F}}, \bar{\mu})$ is a complete measure space and the restriction of $\bar{\mu}$ to $\mathcal{F}$ agrees with μ, i.e. $\bar{\mu}|_{\mathcal{F}} = \mu$. Moreover, $\bar{\mu}$ is the unique extension of μ to $\overline{\mathcal{F}}$.*

(Note: If $(X, \mathcal{F}, \mu)$ is complete, then obviously $\overline{\mathcal{F}} = \mathcal{F}$ and $\bar{\mu} = \mu$.)

Proof The collection

$$\overline{\mathcal{F}} = \{E \cup F : E \in \mathcal{F} \text{ and } F \subseteq N \in \mathfrak{N}\}$$

has X as an element. We shall show that it is a σ-algebra.

(i) $\overline{\mathcal{F}}$ is closed under complementation, i.e., $A \in \overline{\mathcal{F}} \Rightarrow A^c \in \overline{\mathcal{F}}$.

Let $A = E \cup F$, where $E \in \mathcal{F}$ and $F \subseteq N \in \mathfrak{N}$. Then

$$\begin{aligned} A^c &= (A^c \cap N^c) \cup (A^c \cap N) = ((E \cup F)^c \cap N^c) \cup (A^c \cap N) \\ &= ((E^c \cap F^c) \cap N^c) \cup (A^c \cap N) = (E^c \cap (F^c \cap N^c)) \cup (A^c \cap N) \\ &= (E^c \cap N^c) \cup (A^c \cap N) \text{because } F \subseteq N. \end{aligned}$$

Now, $(E^c \cap N^c) \in \mathcal{F}$ and $A^c \cap N \subseteq N$. Therefore $A^c \in \overline{\mathcal{F}}$.

(ii) $\overline{\mathcal{F}}$ is closed under countable unions, i.e., if $\{A_k\}_{k \geq 1}$ is a sequence of subsets belonging to $\overline{\mathcal{F}}$ and $A = \cup_{k \geq 1} A_k$, then $A \in \overline{\mathcal{F}}$.

By definition of $\overline{\mathcal{F}}$, for each k, we have $A_k = E_k \cup F_k$, where $E_k \in \mathcal{F}$ and $F_k \subseteq N_k \in \mathfrak{N}$. Therefore $A = \cup_{k \geq 1} A_k = (\cup_{k \geq 1} E_k) \cup (\cup_{k \geq 1} F_k) = E \cup F$, say. Clearly, $E = \cup_{k \geq 1} E_k \in \mathcal{F}$ and $F \subseteq \cup_{k \geq 1} N_k$ is a subset of a set of measure zero. Thus $A \in \overline{\mathcal{F}}$.

Since $\overline{\mathcal{F}}$ has X as an element (as already noted), it follows from (i) and (ii) that $\overline{\mathcal{F}}$ is a σ-algebra.

Next, we shall prove that $\bar{\mu}$ is a measure on $\overline{\mathcal{F}}$ and $(X, \overline{\mathcal{F}}, \bar{\mu})$ is a complete measure space.

If $(X, \mathcal{F}, \mu)$ is complete, then $\overline{\mathcal{F}} = \mathcal{F}$ and $\bar{\mu} = \mu$; so there is nothing to prove. Suppose $(X, \mathcal{F}, \mu)$ is not complete. We first show that $\bar{\mu}$ is well defined. Suppose for $i = 1$ and 2, we have $A = E_i \cup F_i$, where $E_i \in \mathcal{F}$ and $F_i \subseteq N_i \in \mathfrak{N}$. Then $\mu(E_1) = \mu(E_2)$, because the fact that $F_2 \subseteq N_2$ implies

$$E_1 \subseteq E_1 \cup F_1 = E_2 \cup F_2 \subseteq E_2 \cup N_2 \in \mathcal{F}$$

and the fact that $N_2 \in \mathfrak{N}$ further implies

$$\mu(E_1) \leq \mu(E_2 \cup N_2) \leq \mu(E_2) + \mu(N_2) = \mu(E_2);$$

reversing the roles of E_1 and E_2, we get $\mu(E_2) \leq \mu(E_1)$. Consequently, $\bar{\mu}$ is well defined.

Since it is obvious that $\bar{\mu}$ is nonnegative-valued and $\bar{\mu}(\varnothing) = 0$, it remains only to check its countable additivity. To this end, let $\{A_k\}_{k\geq 1}$ be a sequence of disjoint sets in $\overline{\mathcal{F}}$, so that $A_k = E_k \cup F_k$, where $E_k \in \mathcal{F}$ and $F_k \subseteq N_k \in \mathfrak{N}$, $k = 1, 2, \ldots$. Then $F = \cup_{k\geq 1} F_k \subseteq \cup_{k\geq 1} N_k \in \mathfrak{N}$. Since the sets E_k must also be disjoint, we have

$$\bar{\mu}\Big(\bigcup_{k=1}^{\infty} A_k\Big) = \bar{\mu}\Big(\Big(\bigcup_{k=1}^{\infty} E_k\Big) \cup F\Big) = \mu\Big(\bigcup_{k=1}^{\infty} E_k\Big) = \sum_{k=1}^{\infty} \mu(E_k) = \sum_{k=1}^{\infty} \bar{\mu}(A_k).$$

This proves the countable additivity that was claimed.

Next, we show that $\bar{\mu}|_{\mathcal{F}} = \mu$ and that is complete. Suppose $E \in \mathcal{F}$. Then we have $E = E \cup \varnothing$ and $\varnothing \in \mathfrak{N}$, and it is clear that $\bar{\mu}(E) = \mu(E)$. Thus $\bar{\mu}$ extends μ.

To show that $\bar{\mu}$ is complete, let $A \subseteq N \in \overline{\mathcal{F}}$ and $\bar{\mu}(N) = 0$. Then $N = E_0 \cup F_0$, where $F_0 \subseteq N_0$ and $E_0 \in \mathcal{F}$, $\mu(N_0) = 0$. Since $\bar{\mu}\ (N) = \mu(E_0)$ by definition of $\bar{\mu}$ and $\bar{\mu}(N) = 0$, it follows that $\mu(E_0) = 0$. It follows that $A \subseteq E_0 \cup N_0$ and $\mu(E_0 \cup N_0) = 0$. Therefore $A = \varnothing \cup A$, where $\varnothing \in \mathcal{F}$ and $A \subseteq E_0 \cup N_0 \in \mathfrak{N}$, so that $A \in \overline{\mathcal{F}}$.

Finally, we show that $\bar{\mu}$ is the only measure on $\overline{\mathcal{F}}$ that extends μ. Suppose $\bar{\mu}_1$ is any extension of μ to $\overline{\mathcal{F}}$, which may or may not be complete. For $F \subseteq N \in \mathfrak{N}$, we have $F = \varnothing \cup F \in \overline{\mathcal{F}}$, $N = \varnothing \cup N \in \overline{\mathcal{F}}$ and $\bar{\mu}_1(F) \leq \bar{\mu}_1(N) = \mu(N) = 0$. Thus if $A = E \cup F \in \overline{\mathcal{F}}$, it readily follows that

$$\mu(E) = \bar{\mu}_1(E) \leq \bar{\mu}_1(A) \leq \bar{\mu}_1(E) + \bar{\mu}_1(F) = \bar{\mu}_1(E) = \mu(E).$$

Therefore $\bar{\mu}_1(A) = \mu(E) = \bar{\mu}(A)$. This completes the proof of uniqueness. □

As the following example shows, not every complete extension is of the kind described in the theorem. The space $(X, \mathcal{F}, \mu)$, where $X = \{a, b\}$, $\mathcal{F} = \{\varnothing, X\}$, $\mu(\varnothing) = 0$, $\mu(X) = 1$, is complete. Nevertheless it has several extensions which are again complete: Let $\mathcal{G} = \mathcal{P}(X)$ and set $\nu(\varnothing) = 0$, $\nu(X) = 1$, $\nu(\{a\}) = \alpha$, $\nu(\{b\}) = 1 - \alpha$, where $0 \leq \alpha \leq 1$. Note that in this instance, $\mathcal{F} = \overline{\mathcal{F}} \neq \mathcal{G}$.

Definition 7.2.3 The measure $\bar{\mu}$ of the above theorem is called the **completion *of* μ** and $\overline{\mathcal{F}}$ is called the **completion of $\mathcal{F}$** with respect to μ. The measure space $(X, \overline{\mathcal{F}}, \bar{\mu})$ is called the **completion of $(X, \mathcal{F}, \mu)$**.

Let $(X, \mathcal{F}, \mu)$ be a measure space and let $(X, \overline{\mathcal{F}}, \bar{\mu})$ be its completion. The following theorem describes in another manner the relationship between the members of $\mathcal{F}$ and those of $\overline{\mathcal{F}}$.

Theorem 7.2.4 *Let $(X, \mathcal{F}, \mu)$ be a measure space and let $(X, \overline{\mathcal{F}}, \bar{\mu})$ be its completion. Then $A \in \overline{\mathcal{F}}$ if and only if there exist sets F and G in $\mathcal{F}$ such that*

$$F \subseteq A \subseteq G \, and \, \mu(G \backslash F) = 0.$$

Proof Suppose A satisfies the condition stated in the theorem. That is, there exist sets F and G in $\mathcal{F}$ such that $F \subseteq A \subseteq G$ and $\mu(G \backslash F) = 0$. We may write $A = F \cup (A \backslash F)$. Then $F \in \mathcal{F}$, $A \backslash F \subseteq G \backslash F$ and $\mu(G \backslash F) = 0$.

On the other hand let $A \in \overline{\mathcal{F}}$. Then $A = E \cup F$, where $E \in \mathcal{F}$ and $F \subseteq N$ with $\mu(N) = 0$, i.e., F is a subset of a null set. Observe that

$$E \subseteq A \subseteq E \cup N,$$

where $E \in \mathcal{F}$, $E \cup N \in \mathcal{F}$ (as $\mathcal{F}$ is a σ-algebra) and $\mu((E \cup N) \backslash E) \leq \mu(N) = 0$. □

The example discussed just before Theorem 7.2.2 showed that the Lebesgue measure space $(\mathbb{R}, \mathfrak{M}, m)$ is the completion of the Borel measure space $(\mathbb{R}, \mathfrak{B}, m|_B)$, although the term "completion" had not been introduced then. The same is true in n dimensions and we record it as a theorem, without prejudice to whether the n-dimensional Borel measure space $(\mathbb{R}^n, \mathfrak{B}_n, m_n|_{B_n})$ is complete or not. In fact, we shall use it to demonstrate in Theorem 7.2.17 later that $(\mathbb{R}^n, \mathfrak{B}_n, m_n|_{B_n})$ is indeed not complete. A somewhat different demonstration without using the next theorem directly will be called for in a problem.

Theorem 7.2.5 *The Lebesgue measure space* $(\mathbb{R}^n, \mathfrak{M}_n, m_n)$ *is the completion of the Borel measure space* $(\mathbb{R}^n, \mathfrak{B}_n, m_n|_{B_n})$.

Proof The argument is exactly as in the one-dimensional case above but using Remark 2.8.19 instead of Remark 2.3.25. □

Remark 7.2.6 Let $\mathcal{F}$ be a σ-algebra generated by a class of sets $\mathcal{S}$. Then the σ-algebra generated by any class of sets $\mathcal{T}$ such that $\mathcal{S} \subseteq \mathcal{T} \subseteq \mathcal{F}$ is obviously the same as $\mathcal{F}$. In what follows, this will be of significance when $\mathcal{T}$ is the class of countable (or finite) unions of countable (or finite) intesections of sets belonging to $\mathcal{S}$ and, in particular, finite unions of disjoint sets belonging to $\mathcal{S}$.

A rephrasing of Definition 2.7.1 is that a cuboid in $\mathbb{R}^q (q \in \mathbb{N})$ is either the empty set or a Cartesian product of q bounded intervals, each of positive length. (Recall the requirement "$a_i, b_i \in \mathbb{R}$ and $a_i < b_i$" stated there.) For the purposes of the present section, it will be convenient to extend the notion to include Cartesian products of intervals of any kind, including those having length ∞ as well as those having length 0, such as single point intervals and the empty set. From the elementary fact that such intervals are countable unions of countable intersections of intervals of finite positive length, it is straightforward to deduce that a Cartesian product of intervals of any nonnegative length is a countable union of countable intersections of cuboids. We shall call them "extended cuboids" or "e-cuboids" for short. Note that $\mathbb{R}^q$ is an e-cuboid, though not a cuboid in the sense of Definition 2.7.1.

Definition 7.2.7 An **e-cuboid** *in* $\mathbb{R}^q$ is a Cartesian product of q intervals, called its **edges**.

A nonempty e-cuboid has uniquely determined edges and is a cuboid if and only if all its edges are of finite positive length.

Proposition 7.2.8 *The σ-algebra generated by finite unions of disjoint e-cuboids is the same as the one generated by cuboids.*

Proof In the light of the observations preceding the above definition, the σ-algebra generated by cuboids contains all e-cuboids and hence also all finite unions of disjoint e-cuboids. The result now follows from Remark 7.2.6. □

Proposition 7.2.9 *The Borel σ-algebra $\mathfrak{B}_q$ is generated by the class of all cuboids in $\mathbb{R}^q$ and also by the class of all finite unions of disjoint e-cuboids in $\mathbb{R}^q$. In particular, every e-cuboid is a Borel set.*

Proof Denote by $\mathfrak{C}$ the σ-algebra generated by open cuboids in $\mathbb{R}^q$. Now, every open set is a countable union of open cuboids; indeed, for each x in an open set $O \subseteq \mathbb{R}^q$, there exists an open cuboid with rational endpoints and containing x and contained in O. It therefore follows by Remark 7.2.6 that the σ-algebra generated by open sets is also $\mathfrak{C}$. Thus $\mathfrak{B}_q = \mathfrak{C}$. As every cuboid is a countable intersection of open cuboids, it follows by Remark 7.2.6 that the σ-algebra generated by all cuboids is also $\mathfrak{C}$. Since we know that $\mathfrak{B}_q = \mathfrak{C}$, we conclude that the σ-algebra generated by all cuboids is $\mathfrak{B}_q$. By Proposition 7.2.8, $\mathfrak{B}_q$ is also the σ-algebra generated by finite unions of disjoint e-cuboids.

It is now trivial that every e-cuboid is a Borel set. □

Remark 7.2.10 In the context of Cartesian products in general, it is obvious that

- $(E_1 \cup E_2) \times F = (E_1 \times F) \cup (E_2 \times F)$, $(E_1 \cap E_2) \times F = (E_1 \times F) \cap (E_2 \times F)$,
- $(E_1 \backslash E_2) \times F = (E_1 \times F) \backslash (E_2 \times F)$,
- if E_1 and E_2 are disjoint, then $E_1 \times F$ and $E_2 \times F$ are disjoint

and that the assertions in the first line hold also for infinite unions and intersections.

Proposition 7.2.11 *For any $q \in \mathbb{N}$, finite unions of disjoint e-cuboids in $\mathbb{R}^q$ form an algebra of sets.*

Proof Denote the family of finite unions of disjoint e-cuboids in $\mathbb{R}^q$ by $\mathcal{E}_q$.

Step 1. A finite union of disjoint sets belonging to $\mathcal{E}_q$ belongs to $\mathcal{E}_q$.

This is proved exactly like Lemma 7.1.3 with sets of $\mathcal{E}_q$ playing the role of "elementary" sets of Sect. 7.1 and e-cuboids playing the role of "rectangles".

Step 2. (Analogue of Lemma 7.1.4) If P_1, P_2, …, P_k are disjoint e-cuboids and Q is an e-cuboid, then $(P_1 \backslash Q) \cup (P_2 \backslash Q) \cup \ldots \cup (P_k \backslash Q) \in \mathcal{E}_q$.

As in the proof of Lemma 7.1.4, it is sufficient to prove that if P and Q are e-cuboids, then $P \backslash Q \in \mathcal{E}_q$. The rest will follow by Step 1. We proceed by induction on q.

Suppose $q = 1$. Then a cuboid is nothing but an interval. It is an elementary case by case argument that a difference $P\backslash Q$ of intervals P and Q is a finite union of disjoint intervals.

Now assume as induction hypothesis that the assertion in question holds for e-cuboids in $\mathbb{R}^q$ and consider e-cuboids P and Q in $\mathbb{R}^{q+1}$. We can represent P and Q as $P = A \times B$ and $Q = C \times D$, where A, C are intervals and B,D are e-cuboids in $\mathbb{R}^q$. We have (as in Proposition 7.1.4).

$$P\backslash Q = ((A\backslash C) \times B) \cup (A \times (B\backslash D)) = ((A\backslash C) \times B) \cup ((A \cap C) \times (B\backslash D)). \tag{7.12}$$

Since $A\backslash C$ is a finite union of disjoint intervals and B is an e-cuboid in $\mathbb{R}^q$, it follows from Remark 7.2.10 that $(A\backslash C) \times B \in \mathcal{E}_{q+1}$. Since $A \cap C$ is an interval and $B\backslash D \in \mathcal{E}_q$ by the induction hypothesis, it follows from Remark 7.2.10 that $(A \cap C) \times (B\backslash D) \in \mathcal{E}_{q+1}$. Thus we find that both sets in the union on the right-hand side of (7.12) belong to $\mathcal{E}_{q+1}$. Since the two sets are disjoint, Step 1 leads to the required conclusion that $P\backslash Q \in \mathcal{E}_{q+1}$. This completes Step 2.

Step 3. (Analogue of Proposition 7.1.5) $\mathcal{E}_q$ is closed under taking finite intersections and differences. In symbols,

$$P \in \mathcal{E}_q, Q \in \mathcal{E}_q \Rightarrow \quad P \cap Q \in \mathcal{E}_q, P\backslash Q \in \mathcal{E}_q.$$

This is proved exactly like Proposition 7.1.5 with sets of $\mathcal{E}_q$ playing the role of "elementary" sets of Sect. 7.1 and e-cuboids playing the role of "rectangles".

Step 4. (Analogue of Corollary 7.1.6) $\mathcal{E}_q$ is closed under taking finite unions and under complementation. In symbols,

$$P \in \mathcal{E}_q, Q \in \mathcal{E}_q \quad \Rightarrow \quad P \cup Q \in \mathcal{E}_q, P^c \in \mathcal{E}_q.$$

This is proved exactly like Corollary 7.1.6 with sets of $\mathcal{E}_q$ playing the role of "elementary" sets of Sect. 7.1 and e-cuboids playing the role of "rectangles". □

By Proposition 7.2.9, an e-cuboid $I_1 \times \cdots \times I_q$ is a Borel set and therefore has q-dimensional measure $m_q(I_1 \times \cdots \times I_q)$. If it is a nonempty cuboid, i.e., each edge I_k is a bounded interval of positive length, we know from Proposition 2.7.8 that $m_q(I_1 \times \cdots \times I_q)$ equals the volume, which is defined as the product of the lengths of the edges, $\Pi_{1 \le k \le q} \ell(I_k)$.

Proposition 7.2.12 *For any e-cuboid with edges $I_1, \ldots, I_q$, the q-dimensional measure is the product of the lengths of the edges*:

$$m_q(I_1 \times \cdots \times I_q) = \prod_{k=1}^{q} \ell(I_k).$$

Proof For $q = 1$, this is trivial because the Lebesgue measure of an interval, whether bounded or not, is its length. We need argue the matter only for $q > 1$.

If the e-cuboid is a nonempty cuboid, then the required equality is essentially Proposition 2.7.8. We may assume that none of the edges is $\varnothing$ because otherwise the e-cuboid reduces to $\varnothing$ and there is nothing to argue. So, suppose $I_1 \times \cdots \times I_q$ is not a cuboid. Then one of the edges has length 0 or ∞. We consider two cases.

Case 1. One of the edges has length 0.

The product of the lengths of the edges is then 0 and we therefore have to show that $m_q(I_1 \times \cdots \times I_q) = 0$. We break up the matter into two subcases.

Subcase 1(a). All the other edges also have finite length. Now, any edge of length 0 consists of a single point. By replacing every such edge by an interval having arbitrarily small positive length and containing the single point, we get a cuboid containing the given e-cuboid $I_1 \times \cdots \times I_q$ and also having arbitrarily small positive volume, which is also its m_q-measure. By monotonicity of Lebesgue measure, $m_q(I_1 \times \cdots \times I_q) = 0$.

Subcase 1(b). One of the other edges has infinite length. Any edge of infinite length, which means an unbounded interval, is a countable union of disjoint intervals of finite length. Consequently, Remark 7.2.10 permits us to represent the entire e-cuboid $I_1 \times \cdots \times I_q$ as a countable union of disjoint e-cuboids of the kind discussed in subcase 1a. Each e-cuboid in the countable union must then have m_q-measure 0 and hence so must the entire e-cuboid.

This completes the discussion of Case 1.

Case 2. None of the edges has length 0.

Since $I_1 \times \cdots \times I_q$ is not a cuboid, one of the edges must have infinite length. The product of the lengths of the edges is then ∞ and we therefore have to show that $m_q(I_1 \times \cdots \times I_q) = \infty$.

If there are any edges of finite length, let α be the product of those lengths. If there are no edges of finite length, then let $\alpha = 1$. In either event, $\alpha > 0$. Now, any unbounded interval is a countable union of disjoint intervals of length 1. Consequently, Remark 7.2.10 permits us to represent the entire e-cuboid $I_1 \times \cdots \times I_q$ as a countable union of disjoint cuboids of volume α each. Since $\alpha > 0$, it follows that $m_q(I_1 \times \cdots \times I_q) = \infty$.

This completes the discussion of Case 2. □

Theorem 7.2.13 *Let $(X, \mathcal{F}, \mu)$ be a σ-finite measure space. Suppose that $\mathcal{A}$ is an algebra of subsets of X such that the σ-algebra generated by $\mathcal{A}$ is $\mathcal{F}$. If ν is a measure on $\mathcal{F}$ such that $\nu(A) = \mu(A)$ for all $A \in \mathcal{A}$, then $\nu(A) = \mu(A)$ for all $A \in \mathcal{F}$.*

Proof Case (i): $\mu(X) < \infty$. Let

$$\mathcal{D} = \{E \in \mathcal{F} : \nu(E) = \mu(E)\}$$

Observe that $\mathcal{A} \subseteq \mathcal{D} \subseteq \mathcal{F}$. We shall show that $\mathcal{D}$ is a monotone class. Let $\{E_j\}_{j \geq 1}, E_j \in \mathcal{D}$ for each j, be such that $E_j \subseteq E_{j+1}$. Since μ and ν are measures, using Proposition 3.1.8(a), we get

$$\nu\left(\bigcup_{j=1}^{\infty} E_j\right) = \lim_{j\to\infty} \nu(E_j) = \lim_{j\to\infty} \mu(E_j) = \mu\left(\bigcup_{j=1}^{\infty} E_j\right).$$

Thus $\cup_{j\geq 1} E_j \in \mathcal{D}$.

Similarly, if $\{E_j\}_{j\geq 1}$, $E_j \in \mathcal{D}$ for each j, are such that $E_j \supseteq E_{j+1}$, then we have $\cap_{j\geq 1} E_j \in \mathcal{D}$, using Proposition 3.1.8 (b) and the fact that $\nu(E_1) = \mu(E_1) < \infty$. Thus $\mathcal{D}$ is a monotone class. By 2.3.P19 (b), it follows that $\mathcal{F} \subseteq D$. This completes the proof in this case.

Case (ii): $\mu(X) = \infty$. Since the measure space is σ-finite, there must exist a sequence $\{X_k\}_{k\geq 1}$ of measurable subsets X_k such that each $\mu(X_k)$ is finite and $X = \cup_{k\geq 1} X_k$. By replacing each X_k by $X_k \setminus \cup_{k'<k} X_{k'}$ for $k > 1$, we may assume that the sets X_k are disjoint. Now, $A = A \cap X = \cup_{k\geq 1}(A \cap X_k)$. Because $\mu(X_k) < \infty$, it follows that Case (i) is applicable to X_k, keeping in view Problem 2-3.P24. Hence $\nu(A \cap X_k) = \mu(A \cap X_k)$ for each n, and consequently,

$$\nu(A) = \sum_{k=1}^{\infty} \nu(A \cap X_k) = \sum_{k=1}^{\infty} \mu(A \cap X_k) = \mu(A). \qquad \square$$

In what follows, it is understood that $r, s \in \mathbb{N}$ and we shall be concerned with the Euclidean spaces $\mathbb{R}^r$, $\mathbb{R}^s$ and $\mathbb{R}^{r+s} = \mathbb{R}^r \times \mathbb{R}^s$.

Theorem 7.2.14

$$\mathfrak{B}_r \times \mathfrak{B}_s = \mathfrak{B}_{r+s}.$$

Proof For any $q \in \mathbb{N}$, let $\mathcal{I}_q$ denote the class of all cuboids in $\mathbb{R}^q$. It follows from the associativity of the Cartesian product that a cuboid in $\mathbb{R}^{r+s}$ is exactly the same as a Cartesian product $I \times J : I \in \mathcal{I}_r, J \in \mathcal{I}_s$ of a cuboid in $\mathbb{R}^r$ and a cuboid in $\mathbb{R}^s$. Thus,

$$\begin{aligned} K \in \mathcal{I}_{r+s} &\Rightarrow K = I \times J \text{ for some } I \in \mathcal{I}_r, J \in \mathcal{I}_s \\ &\Rightarrow K = I \times J \text{ for some } I \in \mathfrak{B}_r, J \in \mathfrak{B}_s \\ &\Rightarrow K \in \mathfrak{B}_r \times \mathfrak{B}_s, \end{aligned}$$

$$\text{i.e., } \mathcal{I}_{r+s} \subseteq \mathfrak{B}_r \times \mathfrak{B}_s, \tag{7.13}$$

and also

$$K = I \times J \text{ for some } I \in \mathcal{I}_r, J \in \mathcal{I}_s \Rightarrow K \in \mathcal{I}_{r+s}. \tag{7.14}$$

By Proposition 7.2.9, $\mathfrak{B}_{r+s}$ is the σ-algebra generated by $\mathcal{I}_{r+s}$. Therefore it follows from (7.13) that

$$\mathfrak{B}_{r+s} \subseteq \mathfrak{B}_r \times \mathfrak{B}_s.$$

For the reverse inclusion, fix any $I \in \mathcal{I}_r$ and consider

$$\mathcal{D} = \{T \subseteq \mathbb{R}^s : I \times T \in \mathfrak{B}_{r+s}\}.$$

Since

$$I \times (\bigcup_{i=1}^{\infty} T_i) = \bigcup_{i=1}^{\infty} (I \times T_i)$$

and

$$I \times (\mathbb{R}^s \backslash T) = (I \times \mathbb{R}^s) \backslash (I \times T),$$

it follows that $\mathcal{D}$ is a σ-algebra. By Proposition 7.2.9, we have $I_{r+s} \subseteq \mathfrak{B}_{r+s}$ and hence it follows from (7.14) that the σ-algebra $\mathcal{D}$ contains $\mathcal{I}_s$; we therefore conclude, again on the strength of Proposition 7.2.9, that $\mathcal{D}$ contains $\mathfrak{B}_s$. So, if $I \in \mathcal{I}_r$ and $J \in \mathfrak{B}_s$, then $I \times J \in \mathfrak{B}_{r+s}$.

Similarly, fix $J \in \mathfrak{B}_s$ and consider the collection of all $S \subseteq \mathbb{R}^r$ such that $S \times J \in \mathfrak{B}_{r+s}$; this collection is a σ-algebra and it contains $\mathcal{I}_r$ because of what was proved in the preceding paragraph. Hence it contains $\mathfrak{B}_r$. So, if $I \in \mathfrak{B}_r$ and $J \in \mathfrak{B}_s$, then $I \times J \in \mathfrak{B}_{r+s}$. Thus $\mathfrak{B}_{r+s}$ contains every Cartesian product of Borel sets ("rectangles" in the terminology of Sect. 7.1) and hence also a union of finitely many disjoint Cartesian products of Borel sets ("elementary sets" in the terminology of Sect. 7.1). By Definition 7.1.2 of a product σ-algebra, it follows that

$$\mathfrak{B}_r \times \mathfrak{B}_s \subseteq \mathfrak{B}_{r+s},$$

which is the required reverse inclusion. □

Theorem 7.2.15 *The measures* m_{r+s} and $m_r \times m_s$ *agree on* $\mathfrak{B}_{r+s}$.

Proof It follows from the associativity of the Cartesian product that an e-cuboid in $\mathbb{R}^{r+s}$ exactly the same as a Cartesian product $I \times J$ of an e-cuboid I in $\mathbb{R}^r$ and an e-cuboid J in $\mathbb{R}^s$. Moreover, it follows from Proposition 7.2.12 that

$$m_{r+s}(I \times J) = m_r(I) m_s(J).$$

However, $m_r(I)m_s(J) = (m_r \times m_s)(I \times J)$ by definition of the product measure. Therefore m_{r+s} and $m_r \times m_s$ agree on all e-cuboids and hence also on all finite unions of disjoint e-cuboids. Now, finite unions of disjoint e-cuboids form an algebra according to Proposition 7.2.11. Moreover, the σ-algebra generated by the algebra is seen to be $\mathfrak{B}_{r+s}$ on the basis of Proposition 7.2.9. As m_{r+s} is σ-finite, Theorem 7.2.13 yields the desired conclusion. □

Theorem 7.2.16 *Let* $n > 1$ *and* $1 \leq r,s \leq n - 1$, *where* $r + s = n$. *Then*

$$\mathfrak{B}_n \subseteq \mathfrak{M}_r \times \mathfrak{M}_s \subseteq \mathfrak{M}_n.$$

Proof Since $\mathfrak{B}_r \subseteq \mathfrak{M}_r$ and $\mathfrak{B}_s \subseteq \mathfrak{M}_s$, we have $\mathfrak{B}_r \times \mathfrak{B}_s \subseteq \mathfrak{M}_r \times \mathfrak{M}_s$. But $\mathfrak{B}_r \times \mathfrak{B}_s = \mathfrak{B}_n$ by Theorem 7.2.14.

Next, suppose that $A \in \mathfrak{M}_r$ and $B \in \mathfrak{M}_s$. Then both $A \times \mathbb{R}^s$ and $\mathbb{R}^r \times B$ belong to $\mathfrak{M}_n$, as we now show: Since $A \in \mathfrak{M}_r$, there exists an $\mathcal{F}_\sigma$-set $E \subseteq \mathbb{R}^r$ and a $\mathcal{G}_\delta$-set $F \subseteq \mathbb{R}^r$ such that $E \subseteq A \subseteq F$ and $m_r(F \backslash E) = 0$ by Remark 2.8.19; so $E \times \mathbb{R}^s \subseteq A \times \mathbb{R}^s \subseteq F \times \mathbb{R}^s$, and

$$(m_r \times m_s)((F \times \mathbb{R}^s) \backslash (E \times \mathbb{R}^s)) = m_r(F \backslash E) m_s(\mathbb{R}^s) = 0.$$

However, $(F \times \mathbb{R}^s) \backslash (E \times \mathbb{R}^s) \in \mathfrak{B}_r \times \mathfrak{B}_s = \mathfrak{B}_{r+s}$ and hence by Theorem 7.2.15, we have

$$m_n((F \times \mathbb{R}^s) \backslash (E \times \mathbb{R}^s)) = 0.$$

Since $E \times \mathbb{R}^s$ is an $\mathcal{F}_\sigma$-set in $\mathbb{R}^{r+s} = \mathbb{R}^n$ and $F \times \mathbb{R}^s$ a $\mathcal{G}_\delta$-set in $\mathbb{R}^{r+s} = \mathbb{R}^n$, it follows on using Remark 2.8.19 that $A \times \mathbb{R}^s \in \mathfrak{M}_n$. By a similar argument, we can show that $\mathbb{R}^r \times B \in \mathfrak{M}_n$. The same is therefore true of their intersection $A \times B$. It follows that $\mathfrak{M}_r \times \mathfrak{M}_s \subseteq \mathfrak{M}_n$. □

Theorem 7.2.17 *The Borel measure space* $(\mathbb{R}^n, \mathfrak{B}_n, m_n|_{B_n})$ *is not complete.*

Proof If $(\mathbb{R}^n, \mathfrak{B}_n, m_n|_{B_n})$ were to be complete, it would be its own completion. Now, we know from Theorem 7.2.5 that its completion is $(\mathbb{R}^n, \mathfrak{M}_n, m_n)$. Therefore it suffices to show that $\mathfrak{B}_n \subset \mathfrak{M}_n$. We do so by induction.

For $n = 1$, we already know the result to be true. Assume it to be true for some $n \in \mathbb{N}$. Then there exists an $E \in \mathfrak{M}_n$ such that $E \notin \mathfrak{B}_n$. Consider the set $F = E \times \mathbb{R} \subseteq \mathbb{R}^{n+1}$. Trivially, $F \in \mathfrak{M}_n \times \mathfrak{M}$ and hence $F \in \mathfrak{M}_{n+1}$ by Theorem 7.2.16. Now, fix any $y \in \mathbb{R}$. Since $\{x \in \mathbb{R}^n : (x, y) \in F\} = E$ and $E \notin \mathfrak{B}_n$, we have $F \notin \mathfrak{B}_n \times \mathfrak{B}$ by Proposition 7.1.12. But $\mathfrak{B}_n \times \mathfrak{B} = \mathfrak{B}_{n+1}$ by Theorem 7.2.14. Therefore it follows that $F \notin \mathfrak{B}_{n+1}$, although $F \in \mathfrak{M}_{n+1}$. The existence of such a set F shows that $\mathfrak{B}_{n+1} \subset \mathfrak{M}_{n+1}$. □

Theorem 7.2.18 *Let* $n > 1$ *and* $1 \leq r,s \leq n-1$, *where* $r + s = n$. *Then* m_n *is the completion of the product measure* $m_r \times m_s$.

Proof Let $A \in \mathfrak{M}_r \times \mathfrak{M}_s$. By Theorem 7.2.16, $A \in \mathfrak{M}_n$. So, by Remark 2.8.19, there exists an $\mathcal{F}_\sigma$-set $E \subseteq \mathbb{R}^n$ and a $\mathcal{G}_\delta$-set $F \subseteq \mathbb{R}^n$ such that $E \subseteq A \subseteq F$ and $m_n(F \backslash E) = 0$. Since $E, F \in \mathfrak{B}_n$ and $\mathfrak{B}_r \times \mathfrak{B}_s = \mathfrak{B}_n$ by Theorem 7.2.14, it follows that $E, F \in \mathfrak{M}_r \times \mathfrak{M}_s$ and

$$(m_r \times m_s)(A \backslash E) \leq (m_r \times m_s)(F \backslash E) \tag{7.15}$$

because $E \subseteq A \subseteq F$. But the measures m_n and $m_r \times m_s$ agree on $\mathfrak{B}_n$ by Theorem 7.2.15. Therefore

$$(m_r \times m_s)(E) = m_n(E) \tag{7.16}$$

and also $(m_r \times m_s)(F\backslash E) = m_n(F\backslash E) = 0$. The latter equality and (7.15) together lead to

$$(m_r \times m_s)(A\backslash E) = 0.$$

Consequently,

$$\begin{aligned}(m_r \times m_s)(A) &= (m_r \times m_s)(E \cup (A\backslash E)) \\ &= (m_r \times m_s)(E) \\ &= m_n(E) \text{ by (7.16)} \\ &= m_n(A).\end{aligned}$$

So, $m_r \times m_s$ agrees with m_n on $\mathfrak{M}_r \times \mathfrak{M}_s$, which is to say $m_r \times m_s$ is the restriction of m_n to $\mathfrak{M}_r \times \mathfrak{M}_s$. On the other hand, $\mathfrak{B}_n \subseteq \mathfrak{M}_r \times \mathfrak{M}_s \subseteq \mathfrak{M}_n$ by Theorem 7.2.16 while $\mathfrak{M}_n$ is the completion of $\mathfrak{B}_n$ [see Theorem 7.2.5]. Therefore Problem 7.2.P3 leads to the required conclusion. □

Remark 7.2.19 From the argument above, we know independently of Theorem 7.2.5 that $m_r \times m_s$ agrees with m_n on $\mathfrak{M}_r \times \mathfrak{M}_s$.

A somewhat different proof of incompleteness of $(\mathbb{R}^n, \mathfrak{B}_n, m_n|_{B_n})$ that uses Remark 7.2.19 but not Theorem 7.2.5 is as follows.

By Remark 2.8.19, a set of Lebesgue measure 0 is always contained in a Borel set of measure 0. Therefore incompleteness of $(\mathbb{R}^n, \mathfrak{B}_n, m_n|_{B_n})$ will follow from the existence of a set of Lebesgue measure 0 that is not a Borel set. As before, we proceed by induction on n. For $n = 1$, we already know that there exists such a set. Assume the same for some $n \in \mathbb{N}$. That is to say, there exists an $E \in \mathfrak{M}_n$ such that $m_n(E) = 0$ and $E \notin \mathfrak{B}_n$. Consider the set $F = E \times \mathbb{R} \subseteq \mathbb{R}^{n+1}$. Trivially, $F \in \mathfrak{M}_n \times \mathfrak{M}$ and hence $F \in \mathfrak{M}_{n+1}$ by Theorem 7.2.16. Moreover, $(m_n \times m)(F) = m_n(E)m(\mathbb{R}) = 0$, which implies $m_{n+1}(F) = 0$ in view of Remark 7.2.19. Now, fix any $y \in \mathbb{R}$. Since $\{x \in \mathbb{R}^n : (x, y) \in F\} = E$ and $E \notin \mathfrak{B}_n$, we have $F \notin \mathfrak{B}_n \times \mathfrak{B}$ by Proposition 7.1.12. But $\mathfrak{B}_n \times \mathfrak{B} = \mathfrak{B}_{n+1}$ by Theorem 7.2.14. Therefore it follows that $F \notin \mathfrak{B}_{n+1}$. Thus $F \subseteq \mathbb{R}^{n+1}$ is a set of Lebesgue measure 0 that is not a Borel set.

Problem Set 7.2

7.2.P1. Show that if μ is not complete, then f measurable and $f = g$ a.e. do not imply that g is measurable.

7.2.P2. Suppose $\mathcal{F}$ and $\mathcal{G}$ are σ-algebras of subsets of X with measures μ and ν respectively. If $\mathcal{F} \subseteq \mathcal{G}$ and $\mu = \nu|_{\mathcal{F}}$, show that $\overline{\mathcal{F}} \subseteq \overline{\mathcal{G}}$.

7.2.P3. Suppose $\mathcal{F}$ and $\mathcal{G}$ are σ-algebras of subsets of X with measures μ and ν respectively. If $\mathcal{F} \subseteq \mathcal{G} \subseteq \overline{\mathcal{F}}$ and $\nu = \bar{\mu}|_{\mathcal{G}}$, show that $\overline{\mathcal{G}} = \overline{\mathcal{F}}$ and $\bar{\nu} = \bar{\mu}$.

7.2.P4. [Application to Fourier Series] Prove the following: Let f be a function in $L^p[-\pi, \pi]$, where $1 \le p < \infty$. Then the Cesàro means of the Fourier series for f converge to f in the L^p norm.

Chapter 8
Hints

Problem Set 2.1

2.1.P1. From properties (i) and (iii) of m, show that $m(\varnothing) = 0$, provided that at least one set C in the domain of m satisfies $m(C) < \infty$.
Hint: Let $E_1 = C$ and $E_j = \varnothing$ for each $j > 1$. Then $\{E_j\}_{j\geq 1}$ is a sequence of disjoint subsets in the domain of m. Also, $\cup_{j\geq 1}E_j = C$ and property (iii) implies $m(C) = m(C) + \sum_{i=2}^{\infty} m(\varnothing)$. Since $0 \leq m(C) < \infty$, this leads to $m(\varnothing.) = 0$.

2.1.P2. From properties (i) and (iii) of m, show that, if $A \cap B = \varnothing$, where A, B, $A \cup B$ are in the domain of m, then $m(A \cup B) = m(A) + m(B)$.
Hint: If $m(C) = \infty$ for every C in the domain of m, then there is nothing to prove. So, assume $m(C) < \infty$ for some C. Let $E_1 = A$, $E_2 = B$ and $E_j = \varnothing$ for $j > 2$. Then $\{E_j\}_{j\geq 1}$ is a sequence of disjoint subsets in the domain of m. Also, $\cup_{j\geq 1}E_j = A \cup B$ is in the domain of m. The result now follows from property (iii) and 2.1.P1.

2.1.P3. From properties (i) and (iii) of m, show that, if $A \backslash B$, $B \backslash A$, A, B, $A \cup B$ and $A \cap B$ are in the domain of m, then $m(A \cup B) + m(A \cap B) = m(A) + m(B)$.
Hint: A is the disjoint union of $A \backslash B$ and $A \cap B$. Therefore by 2.1.P2, $m(A) = m(A \backslash B) + m(A \cap B)$. Similarly, $m(B) = m(B \backslash A) + m(A \cap B)$. Now, $A \cup B$ is the disjoint union of $A \backslash B$, $B \backslash A$ and $A \cap B$. Therefore $m(A \cup B) = m(A \backslash B) + m(B \backslash A) + m(A \cap B)$. The three equalities proved, when taken together, imply the required equality.

Problem Set 2.2

2.2.P1. Use the results of this section to show that [0, 1] is uncountable.
Hint: If it were countable or finite, its outer measure would be 0 by Example 2.2.5, but the outer measure must be 1 by Proposition 2.2.6.

S. Shirali and H. L. Vasudeva, *Measure and Integration*,
Springer Undergraduate Mathematics Series,
https://doi.org/10.1007/978-3-030-18747-7_8

2.2.P2. If the domain of a countably subadditive set function ν includes ∅ and if ν(∅) = 0, show that ν is finitely subadditive.
Hint: Imitate the argument of Corollary 2.2.10.

2.2.P3. Show that for any two sets A and B with union [0, 1], the outer measure satisfies

$$m^*(A) \geq 1 - m^*(B).$$

Hint: $m^*(A \cup B) = m^*([0, 1]) = \ell([0, 1]) = 1$. Also, $m^*(A \cup B) \leq m^*(A) + m^*(B)$ by Corollary 2.2.10.

2.2.P4. Let $\{I_j\}_{1 \leq j \leq n}$ be a finite sequence of open intervals covering the rationals in [0, 1]. Show that $\sum_{j=1}^{n} \ell(I_j) \geq 1$. [Note: With a little extra effort, it can be shown that the sum of lengths is strictly greater than 1; however, we shall not need this fact.]
Hint: First assume that each I_j ($1 \leq j \leq n$) has rational endpoints. Since 0 belongs to $\cup_{1 \leq j \leq n} I_j$, it must be in one of the I_j. Let this interval be (a_1, b_1). Then we have $a_1 < 0 < b_1$. If $b_1 \leq 1$, then $b_1 \in [0, 1]$ and since $b_1 \notin (a_1, b_1)$, there must be an interval (a_2, b_2) in the collection $\{I_j: 1 \leq j \leq n\}$ such that $b_1 \in (a_2, b_2)$, i.e. $a_2 < b_1 < b_2$. Continuing in this manner, we obtain a sequence of intervals (a_i, b_i) belonging to the collection $\{I_j: 1 \leq j \leq n\}$ such that $a_i < b_{i-1} < b_i$. The second part of this double inequality ensures that the intervals in our sequence are distinct. Since $\{I_j: 1 \leq j \leq n\}$ is a finite collection, our sequence is also finite and therefore terminates with some (a_k, b_k). However the sequence terminates with (a_k, b_k) only if $b_k > 1$. Thus $a_1 < 0 < 1 < b_k$ and $a_i < b_{i-1}$ for $1 \leq i \leq k$. Now

$$\sum_{j=1}^{n} \ell(I_j) \geq \sum_{i=1}^{k} \ell(a_i, b_i) = (b_k - a_k) + (b_{k-1} - a_{k-1}) + \cdots + (b_1 - a_1).$$

Therefore, in view of the inequalities noted just earlier, we have

$$\begin{aligned}\sum_{j=1}^{n} \ell(I_j) &> b_k - (a_k - b_{k-1}) - (a_{k-1} - b_{k-2}) - \cdots - (a_2 - b_1) - a_1 \\ &> b_k - a_1 > 1 - 0 = 1.\end{aligned}$$

If some of the I_j do not have rational endpoints, replace each I_j by I_j', where I_j' has rational endpoints, $I_j' \supseteq I_j$ and $\ell(I_j') < \ell(I_j) + \frac{\varepsilon}{n}$. Then, from what has already been proved, we have

$$\sum_{j=1}^{n} \ell(I_j) > \sum_{j=1}^{n} \ell(I'_j) - \varepsilon > 1 - \varepsilon.$$

Since $\varepsilon > 0$ is arbitrary, it follows that $\sum_{j=1}^{n} \ell(I_j) \geq 1$.

2.2.P5. Show that if we were to define outer measure as approximation by finitely many open intervals, i.e. if $m^*(A)$ were to be

$$\inf\{\sum_{i=1}^{n} \ell(I_i) : A \subseteq \bigcup_{i=1}^{n} I_i, \text{each } I_i \text{ an open interval}\},$$

then it would not be countably subadditive.
Hint: The total length of finitely many open intervals covering the rationals in [0, 1] is at least 1 by 2.2.P4. Thus the outer measure of the rationals in [0, 1] would turn out to be at least 1. On the other hand, the outer measure of a set consisting of a single point would still be zero. Therefore we would not have countable subadditivity.

2.2.P6. Show that if we were to define outer measure as approximation from within, i.e. if $m^*(A)$ were to be

$$\sup\{\sum_{i=1}^{\infty} \ell(I_i) : A \supseteq \bigcup_{i=1}^{\infty} I_i, \text{ each } I_i \text{ an open interval and } i \neq j \Rightarrow I_i \cap I_j = \emptyset\},$$

then it would not be finitely subadditive.
Hint: This procedure would assign outer measure 0 to the set of rationals in [0, 1] as well as to the set of irrationals in [0, 1], because neither of them contains a nonempty open interval [see Remark 2.2.2(a)]. But the union of the two sets, which is [0, 1] would be assigned outer measure at least 1 [it contains the open interval (0, 1)]. So, finite subadditivity would not hold.

2.2.P7. Prove that, if the open set A is the union of a sequence $\{I_n\}_{n \geq 1}$ of disjoint open intervals, then

$$m^*(A) = \sum_{n=1}^{\infty} \ell(I_n).$$

Hint: Since $A = \cup_{n \geq 1} I_n$, we have $m^*(A) = m^*(\cup_{n \geq 1} I_n) \leq \sum_{n=1}^{\infty} m^*(I_n) = \sum_{n=1}^{\infty} \ell(I_n)$. To prove the reverse inequality, consider an arbitrary $\varepsilon > 0$. There exists a sequence $\{I'_n\}_{n>1}$ of open intervals such that $\cup_{n \geq 1} I'_n \supseteq A$ and

$\sum_{n=1}^{\infty} \ell(I'_n) \leq m^*(A) + \varepsilon$. As $\cup_{n \geq 1} I'_n \supseteq A = \cup_{n \geq 1} I_n$ and $i \neq j \Rightarrow I_i \cap I_j = \varnothing$, it follows that

$$\ell(I_n) = m^*(I_n) = m^*(\bigcup_{k=1}^{\infty} (I_n \cap I'_k)) \leq \sum_{k=1}^{\infty} m^*(I_n \cap I'_k) = \sum_{k=1}^{\infty} \ell(I_n \cap I'_k).$$

Therefore

$$\sum_{n=1}^{\infty} \ell(I_n) \leq \sum_{n=1}^{\infty}\sum_{k=1}^{\infty} \ell(I_n \cap I'_k) \leq \sum_{k=1}^{\infty}\sum_{n=1}^{\infty} \ell(I_n \cap I'_k) \leq \sum_{k=1}^{\infty} \ell(I'_k) \leq m^*(A) + \varepsilon.$$

The last but one inequality is a consequence of the fact that if a countable number of disjoint open intervals are contained in a single open interval, then the sum of their lengths is less than or equal to the length of the single interval. This follows from the corresponding fact for finitely many intervals, for which the reader can supply an argument suggested by Fig. 8.1 for three intervals. Since $\varepsilon > 0$ is arbitrary, the required reverse inequality now follows. Remark: The above result is an easy consequence of Proposition 2.2.6, 2.3.13, 2.3.16 of the next section.

2.2.P8. Show that using closed intervals instead of open intervals in the definition of outer measure does not change the evaluation of $m^*(A)$.
Hint: Denote the outer measure obtained by using closed intervals by m^*_C. From the definition, $m^*_C(A) \leq m^*(A)$. Let $\varepsilon > 0$ be given. There exists a sequence $\{I_n\}_{n \geq 1}$ of closed intervals such that $\cup_{n \geq 1} I_n \supseteq A$ and $\sum_{n=1}^{\infty} \ell(I_n) \leq m^*_C(A) + \varepsilon$. For each closed interval I_n, let I'_n be an open interval containing I_n with $\ell(I'_n) = (1+\varepsilon)\ell(I_n)$. Then $m^*_C(A) + \varepsilon \geq \sum_{n=1}^{\infty} \ell(I_n) = (1+\varepsilon)^{-1} \sum_{n=1}^{\infty} \ell(I'_n)$. So, $\sum_{n=1}^{\infty} \ell(I'_n) \leq (1+\varepsilon)\left[m^*_C(A) + \varepsilon\right]$. Clearly, $\cup_{n \geq 1} I'_n \supseteq A$, and therefore the preceding inequality implies $m^*(A) \leq m^*_C(A)$, considering that $\varepsilon > 0$ is arbitrary.

2.2.P9. Show that using intervals closed only on the left [or on the right, or a mixture of various types of intervals] rather than open intervals does not change the evaluation of $m^*(A)$.
Hint: Let $\{I_n\}_{n \geq 1}$ be a sequence of intervals which cover A. Then $\{I_n{}^{\circ}\}_{n \geq 1}$ is the sequence of corresponding open intervals and let $\{r_i\}_{i \geq 1}$ be an enumeration of the points of $\cup_{n \geq 1}(I_n \backslash I_n{}^{\circ})$. Enclose these points in open intervals of lengths $\varepsilon/2^{i+1}$.

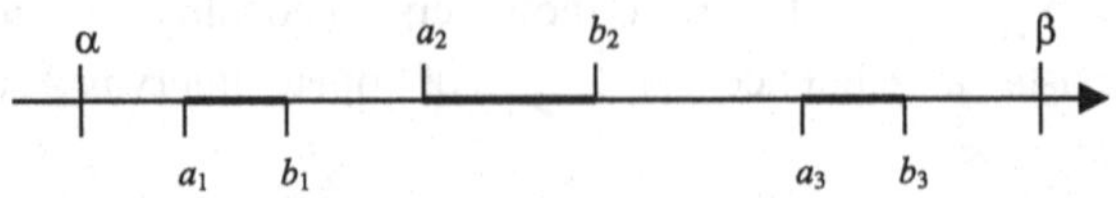

Fig. 8.1 For Problem 2.2.P7

These intervals together with I_n° form an open cover of A of total length less than $\sum_{n=1}^{\infty} \ell(I_n) + \varepsilon$.

2.2.P10. **(a)** For $k > 0$ and $A \subseteq \mathbb{R}$, let kA denote $\{x\colon k^{-1}x \in A\}$. Show that $m^*(kA) = k \cdot m^*(A)$.
(b) For $A \subseteq \mathbb{R}$, let $-A$ denote $\{x\colon -x \in A\}$. Show that $m^*(-A) = m^*(A)$.
Hint: **(a)** For $\varepsilon > 0$, there exists a sequence $\{I_n\}_{n\geq 1}$ of open intervals such that $kA \subseteq \cup_{n\geq 1} I_n$ and $m^*(kA) + \varepsilon \geq \sum_{n=1}^{\infty} \ell(I_n)$. But then $A \subseteq \cup_{n\geq 1} k^{-1} I_n$. So, $m^*(A) \leq \sum_{n=1}^{\infty} m^*(k^{-1}I_n) = \sum_{n=1}^{\infty} \ell(k^{-1}I_n) = k^{-1}\sum_{n=1}^{\infty} \ell(I_n) \leq k^{-1}(m^*(kA) + \varepsilon)$. Hence $m^*(kA) \geq k \cdot m^*(A)$. By a similar argument, $m^*(A) = m^*(k^{-1}(kA)) \geq k^{-1}(m^*(kA))$.
(b) For $\varepsilon > 0$, there exists a sequence $\{I_n\}_{n\geq 1}$ of open intervals such that $-A \subseteq \cup_{n\geq 1} I_n$ and $m^*(-A) + \varepsilon \geq \sum_{n=1}^{\infty} \ell(I_n)$. But then $A \subseteq \cup_{n\geq 1}(-I_n)$. So,

$$m^*(A) \leq \sum_{n=1}^{\infty} m^*(-I_n) = \sum_{n=1}^{\infty} \ell(-I_n) = \sum_{n=1}^{\infty} \ell(I_n) \leq m^*(-A) + \varepsilon.$$

Since $\varepsilon > 0$ is arbitrary, we have $m^*(A) \leq m^*(-A)$. Replace A by $-A$ to obtain the reverse inequality.
2.2.P11. Let $A = \{x \in [0, 1]\colon x$ has a decimal expansion not containing the digit $5\}$. Note that $0.5 \in A$, because $0.5 = 0.4999 \ldots$, but $0.51 \notin A$. Then show that $m^*(A) = 0$.
Hint: In fact, $A = \cap_{n\geq 1} A_n$, where $A_0 = [0, 1]$, $A_1 = [0, \frac{5}{10}] \cup [\frac{6}{10}, 1]$, which is obtained from A_0 on removing the open interval $(\frac{5}{10}, \frac{6}{10})$; A_2 is obtained from A_1 on removing the open intervals

$$(\tfrac{5}{10^2}, \tfrac{6}{10^2}), (\tfrac{15}{10^2}, \tfrac{16}{10^2}), (\tfrac{25}{10^2}, \tfrac{26}{10^2}), (\tfrac{35}{10^2}, \tfrac{36}{10^2}), (\tfrac{45}{10^2}, \tfrac{46}{10^2}), (\tfrac{65}{10^2}, \tfrac{66}{10^2}), (\tfrac{75}{10^2}, \tfrac{76}{10^2}),$$
$$(\tfrac{85}{10^2}, \tfrac{86}{10^2}), (\tfrac{95}{10^2}, \tfrac{96}{10^2}).$$

Thus A_2 is a union of 10 closed intervals of total length $(\frac{9}{10})^2$. In general, A_n is the union of 10^n closed intervals of total length $(\frac{9}{10})^n$. Since $A \subseteq A_n$, it follows that $m^*(A) \leq (\frac{9}{10})^n \to 0$ as $n \to \infty$.
2.2.P12. Suppose $m^*(A \cap I) \leq \frac{1}{2}m^*(I)$ for every interval I. Prove that $m^*(A) = 0$.
Hint: Suppose $0 < m^*(A) < \infty$. Then we may take $\varepsilon = \frac{1}{2}m^*(A)$. From the definition of outer measure, there exists a cover $\{I_n\}_{n\geq 1}$ consisting of open intervals such that

$$\sum_{n=1}^{\infty}\ell(I_n) < \frac{3}{2}m^*(A) = \frac{3}{2}m^*(A\cap\bigcup_{n=1}^{\infty}I_n) \le \frac{3}{2}\sum_{n=1}^{\infty}m^*(A\cap I_n).$$

For at least one n, we must have $0<\ell(I_n)\le\frac{3}{2}m^*(A\cap I_n)$, which is a contradiction.

Now drop the hypothesis that $m^*(A) < \infty$. Let J be an arbitrary bounded interval and put $A' = A \cap J$. Then for any interval I, the intersection $J \cap I$ is an interval and hence we have $m^*(A'\cap I) = m^*(A\cap(J\cap I)) \le \frac{1}{2}m^*(J\cap I) \le \frac{1}{2}m^*(I)$. Since $m^*(A') \le m^*(J) < \infty$, it follows from what has already been proved that $m^*(A') = 0$. This means $m^*(A \cap J) = 0$ for every bounded interval J. In particular, $m^*(A \cap [n, n + 1]) = 0$ for every integer n. This implies $m^*(A) = 0$.

2.2.P13. For nonempty $E \subseteq \mathbb{R}$, the diameter is diam $E = \sup\{|x-y|: x, y \in E\}$. Show that $m^*(E) \le$ diam$E$.

Hint: Let d denote diam$E$. If $d = 0$, E consists of a single point and there is nothing to prove. So, suppose $d > 0$ and consider an arbitrary positive $\varepsilon < 2d$. By definition, there exist $x, y \in E$ such that $|x - y| > d - \frac{\varepsilon}{2} > 0$. Without loss of generality, we may assume that $x < y$. Note that $E\subseteq\left(x-\frac{\varepsilon}{2}, y+\frac{\varepsilon}{2}\right)$. For, if $z \in E$ but $z\notin\left(x-\frac{\varepsilon}{2}, y+\frac{\varepsilon}{2}\right)$, then $y+\frac{\varepsilon}{2}-z > y-x+\varepsilon$ if z lies to the left of $x-\frac{\varepsilon}{2}$ and $z-x+\frac{\varepsilon}{2} > y-x+\varepsilon$ if z lies to the right of $y+\frac{\varepsilon}{2}$. In the first case, $y-z > y-x+\frac{\varepsilon}{2} > d$ and in the second case, $z - x > d$, a contradiction in either case. So, $m^*(E) \le y - x + \varepsilon \le d + \varepsilon$. Since this is true for each $\varepsilon > 0$, it follows that $m^*(E) \le d$.

Problem Set 2.3

2.3.P1. Let A, B be subsets of $\mathbb{R}$ and $A \subseteq B$. Show that

(a) If $m^*(B\backslash A) = 0$, then $m^*(B) = m^*(A)$.
(b) If $m^*(B) = m^*(A) < \infty$ and $A \in \mathfrak{M}$, then $m^*(B\backslash A) = 0$.

Hint: **(a)** By the subadditivity of outer measure, $m^*(B) = m^*(A \cup (B\backslash A)) \le m^*(A) + m^*(B\backslash A) = m^*(A)$ since $m^*(B\backslash A) = 0$. The reverse inequality follows from Proposition 2.2.3(c).

(b) Since A is measurable, $m^*(B) = m^*(B \cap A) + m^*(B \cap A^c) = m^*(A) + m^*(B\backslash A)$.

2.3.P2. Suppose that A is a subset of $\mathbb{R}$ with the property that, for $\varepsilon > 0$, there exist measurable sets B and C such that $B \subseteq A \subseteq C$ and $m(C \cap B^c) < \varepsilon$. Show that A is measurable.

Hint: For $n \in \mathbb{N}$, there exist measurable sets B_n and C_n such that $B_n \subseteq A \subseteq C_n$ and $m(C_n\cap B_n^c) < \frac{1}{n}$. Set $B = \cup_{n\ge1}B_n$ and $C = \cap_{n\ge1}C_n$. Then B and C are measurable (the collection $\mathfrak{M}$ is a σ-algebra) and $m(C \cap B^c) \le m(C_n \cap B_n^c) \le \frac{1}{n}$ for each $n \in \mathbb{N}$; so, $m(C \cap B^c) = 0$. Now, $A \cap B^c \subseteq C \cap B^c$ and hence $m^*(A \cap B^c) \le m^*(C \cap B^c) = 0$. This implies $A \cap B^c \in \mathfrak{M}$ [Remark 2.3.2(d)] and hence $A = (A \cap B^c) \cup B \in \mathfrak{M}$.

2.3.P3. Let $\{E_n\}_{n\geq 1}$ be a sequence of sets such that $E_1 \subseteq E_2 \subseteq \cdots$. Show that $m^*(\lim E_n) = \lim_{n\to\infty} m^*(E_n)$.

Hint: Since the sequence $\{m^*(E_n)\}_{n\geq 1}$ is monotone, $\lim_{h\to\infty} m^*(E_n)$ exists, although it may be infinite. Since $m^*(E_n) \leq m^*(\cup_{n\geq 1}E_n)$, we have $\lim_{n\to\infty} m^*(E_n) \leq m^*(\cup_{n\geq 1}E_n)$. Choose measurable $F_n \supseteq E_n$ such that $m(F_n) = m^*(E_n)$ (see Proposition 2.2.13). Writing $B_n = \cap_{j\geq n}F_j$, we have $B_1 \subseteq B_2 \subseteq \cdots$, with each B_n measurable and $E_n \subseteq B_n \subseteq F_n$. So, $m^*(E_n) \leq m(B_n) \leq m(F_n)$, which implies $m(B_n) = m^*(E_n)$. Then by Proposition 2.3.21(a), $m^*(\cup_{n\geq 1}E_n) \leq m(\cup_{n\geq 1}B_n) = \lim_{n\to\infty} m(B_n) = \lim_{n\to\infty} m^*(E_n)$.

2.3.P4. Given a subset A of $\mathbb{R}$, let $A_n = A \cap [-n, n]$ for $n \in \mathbb{N}$. Show that $m^*(A) = \lim_{n\to\infty} m^*(A_n)$.

Hint: Note that $A_1 \subseteq A_2 \subseteq \cdots$, and $A = \cup_{n\geq 1}A_n = \lim A_n$. Apply Problem 2.3.P3.

2.3.P5. The **symmetric difference** of sets A, B is defined to be $A\Delta B = (A\backslash B) \cup (B\backslash A)$, consisting of points belonging to one of the two sets but not to both. Show that if $A \in \mathfrak{M}$ and $m^*(A\Delta B) = 0$, then $B \in \mathfrak{M}$.

Hint: From Remark 2.3.2(d), we have $A\Delta B \in \mathfrak{M}$ and so also the subsets $A\backslash B$ and $B\backslash A$. Since $\mathfrak{M}$ is a σ-algebra, it follows that $A \cap B = A\backslash(A\backslash B) \in \mathfrak{M}$ and hence also that $B = (A \cap B) \cup (B\backslash A) \in \mathfrak{M}$.

2.3.P6. Show that every nonempty open subset O has positive measure.

Hint: Since O is nonempty, there is some $a \in O$. Since O is open, it is measurable, so that $m(O) = m^*(O)$, and furthermore, $(a - \delta, a + \delta) \subseteq O$ for some $\delta > 0$. Hence, $m(O) = m^*(O) \geq m^*(a - \delta, a + \delta) = \ell(a - \delta, a + \delta)$ by Proposition 2.2.6. But $\ell(a - \delta, a + \delta) = 2\delta > 0$.

2.3.P7. Let $O = \cup_{n\geq 1}(x_n - \frac{1}{n^2}, x_n + \frac{1}{n^2})$, where $x_1, x_2, \ldots$ is an enumeration of all the rationals. Prove that $m(O\Delta F) > 0$ for any closed set F.

Hint: If $m(O\backslash F) > 0$, then $m(O\Delta F) = m(O\backslash F) + m(F\backslash O) \geq m(O\backslash F) > 0$. Suppose $m(O\backslash F) = 0$. Since $O\backslash F$ is open, it follows from Problem 2.3.P6 that it is empty and hence $O \subseteq F$. Now, O contains $\mathbb{Q}$, which has closure $\mathbb{R}$. So, $F = \mathbb{R}$ and $m(F) = \infty$. But $m(O) \leq 2\sum_{n=1}^{\infty}\frac{1}{n^2} < \infty$. So, $m(F\backslash O) = \infty$ and hence $m(O\Delta F) = m(O\backslash F) + m(F\backslash O) = \infty$.

2.3.P8. The number of elements in a σ-algebra generated by n given sets cannot exceed 2^{2^n}.

Hint: Let $\mathcal{A}$ be a family of n distinct subsets $A_1, A_2, \ldots, A_n$ of a nonempty set X. Set up the family $\mathcal{B}$ of all possible intersections $A^*_1 \cap A^*_2 \cap \cdots \cap A^*_n$, where each A^*_k is either A_k or its complement A_k^c. Such intersections are at most 2^n in number, so that the cardinality N of $\mathcal{B}$ is at most 2^n. Now let $\mathcal{C}$ be the family of all possible finite unions of sets in $\mathcal{B}$, including $\emptyset$. Trivially, $\mathcal{C} \supseteq \mathcal{B}$. Also, any such union is a union of at most N sets. Since there are $\binom{N}{m}$ ways to choose m sets to include in a union of m sets ($m \geq 1$), the cardinality of $\mathcal{C}$ is at most $1 + \sum_{m=1}^{N}\binom{N}{m} = 2^N$.

We record the observation here that $\mathcal{C}$ is the smallest family of sets that (i) contains $\mathcal{B}$ (ii) is closed under unions and (iii) has $\varnothing$ as an element.

We shall argue that $\mathcal{C}$ is an algebra containing $\mathcal{A}$. Since any algebra containing $\mathcal{A}$ must necessarily contain $\mathcal{C}$, it will follow immediately that $\mathcal{C}$ is the algebra generated by $\mathcal{A}$. Being finite, it must then be also the σ-algebra generated by $\mathcal{A}$.

It is obvious that $\mathcal{C}$ is closed under (finite) unions and has $\varnothing$ as an element. It is sufficient therefore to show that $\mathcal{C}$ contains $\mathcal{A}$, and is closed under intersections and complements.

In order to show that $\mathcal{C} \supseteq \mathcal{A}$, consider an arbitrary set in $\mathcal{A}$. For ease of notation, we shall take it to be A_1. We have to produce sets in $\mathcal{B}$ whose union is precisely A_1. Consider the family $\mathcal{B}_1$ of intersections $A_1 \cap A^*_2 \cap \cdots \cap A^*_n$, where each A^*_k $(k > 1)$ is either A_k or its complement A_k^c. Any such intersection belongs to $\mathcal{B}$ by definition of the latter. We shall demonstrate that the union of all the sets belonging to $\mathcal{B}_1$ is precisely A_1. Since each of the sets in question is a subset of A_1, their union is a subset of A_1. In order to show that that A_1 is a subset of their union, consider any $x \in A_1$. For each $k > 1$, select A^*_k to be A_k if $x \in A_k$ and A_k^c if $x \notin A_k$. This selection ensures $x \in A^*_k$ for all $k > 1$, so that $x \in A_1 \cap A^*_2 \cap \cdots \cap A^*_n$. Thus each element of A_1 is seen to belong to at least one of the sets belonging to $\mathcal{B}_1$, which means that A_1 is a subset of their union. This completes the demonstration that the union of all the sets belonging to $\mathcal{B}_1$ is precisely A_1, thereby showing that $\mathcal{C} \supseteq \mathcal{A}$.

We proceed to argue that $\mathcal{C}$ is closed under intersections.

To start with, we observe that the intersection of any two distinct sets in $\mathcal{B}$ is empty. This is because in the respective representations of the distinct sets as intersections $A^*_1 \cap A^*_2 \cap \cdots \cap A^*_n$, the kth sets must differ for some k, which means the kth sets must be complements of each other, thereby rendering the intersections disjoint. In particular, the intersection of any two sets of $\mathcal{B}$ (whether distinct or not) is in $\mathcal{C}$. Since $\mathcal{C}$ is closed under unions, we further conclude that the intersection of a set of $\mathcal{B}$ with any set of $\mathcal{C}$ belongs to $\mathcal{C}$.

Let $\mathcal{D}$ be the family of all sets in $\mathcal{C}$ whose intersection with any set of $\mathcal{C}$ belongs to $\mathcal{C}$. It is immediate that $\varnothing \in \mathcal{D}$ and that $\mathcal{D}$ is closed under unions (because $\mathcal{C}$ is). Also, $\mathcal{C} \supseteq \mathcal{D}$. In view of the conclusion of the preceding paragraph, $\mathcal{D} \supseteq \mathcal{B}$. We have thus shown that $\mathcal{D}$ (i) contains $\mathcal{B}$ (ii) is closed under unions and (iii) has $\varnothing$ as an element. In conjunction with the observation recorded earlier, this implies that $\mathcal{C} \subseteq \mathcal{D}$. Thus $\mathcal{C} = \mathcal{D}$ and the definition of $\mathcal{D}$ now instantly yields the consequence that $\mathcal{C}$ is closed under intersections.

It remains to prove that $\mathcal{C}$ is closed under complements.

To begin with, observe that the argument given above for establishing that $\mathcal{A} \subseteq \mathcal{C}$ can be easily modified to establish that the complement of any set in $\mathcal{A}$ is also in $\mathcal{C}$: all one has to do is to replace $A_1 \cap A^*_2 \cap \cdots \cap A^*_n$ by $A_1^c \cap A^*_2 \cap \cdots \cap A^*_n$. Since $\mathcal{C}$ is closed under unions, the foregoing observation implies that any union $A^*_1 \cup A^*_2 \cup \cdots \cup A^*_n$ belongs to $\mathcal{C}$. But the complement of any set in $\mathcal{B}$ is a union of this form and therefore belongs to $\mathcal{C}$. Now, the complement of any set in $\mathcal{C}$ is a finite intersection of complements of sets in $\mathcal{B}$. Since $\mathcal{C}$ has been shown to be closed under intersections, it is now plain that $\mathcal{C}$ is closed under complements.

2.3.P9. Show that a σ-algebra consisting of infinitely many distinct sets contains an uncountable number of sets.

Hint: Let $\mathcal{F}$ be an infinite σ-algebra and $A_1, A_2, \cdots$ be an enumeration of a countable subfamily $\mathcal{A} \subseteq \mathcal{F}$. Set up the family $\mathcal{B}$ of all possible countable intersections $A^*_1 \cap A^*_2 \cap \cdots$, where each A^*_k is either A_k or its complement A_k^c. Clearly, $\mathcal{F} \supseteq \mathcal{B}$.

We shall demonstrate that an arbitrary set of $\mathcal{A}$ is an intersection of sets belonging to $\mathcal{B}$. Consider an arbitrary set in $\mathcal{A}$. For ease of notation, we shall take it to be A_1. We shall produce a subfamily of $\mathcal{B}$ whose union is precisely A_1. Consider the family $\mathcal{B}_1$ of intersections $A_1 \cap A^*_2 \cap A^*_3 \cap \cdots$, where each A^*_k $(k > 1)$ is either A_k or its complement A_k^c. Any such intersection belongs to $\mathcal{B}$ by definition of the latter, and so $\mathcal{B}_1 \subseteq \mathcal{B}$. We shall demonstrate that the union of all the sets belonging to $\mathcal{B}_1$ is precisely A_1. Since each of the sets in question is a subset of A_1, their union is a subset of A_1. In order to show that that A_1 is a subset of their union, consider any $x \in A_1$. For each $k > 1$, select A^*_k to be A_k if $x \in A_k$ and A_k^c if $x \notin A_k$. This selection ensures $x \in A^*_k$ for all $k > 1$, so that $x \in A_1 \cap A^*_2 \cap A^*_3 \cap \cdots$. Thus each element of A_1 is seen to belong to at least one of the sets belonging to $\mathcal{B}_1$, which means that A_1 is a subset of their union. This completes the demonstration that an arbitrary set of $\mathcal{A}$ is an intersection of sets belonging to $\mathcal{B}$. It is a consequence of this that if $\mathcal{B}$ were to be finite, then the same would be true of $\mathcal{A}$. Therefore $\mathcal{B}$ is seen to be infinite.

Next, we observe that the intersection of any two distinct sets in $\mathcal{B}$ is empty. This is because in the respective representations of the distinct sets as intersections $A^*_1 \cap A^*_2 \cap \cdots$, the kth sets must differ for some k, which means the kth sets must be complements of each other, thereby rendering the intersections disjoint.

Since $\mathcal{B}$ is infinite, there exists a sequence $\{B_n\}_{n \geq 1}$ of distinct nonempty sets belonging to $\mathcal{B}$. The set of all sequences in the two-element set $\{0, 1\}$ has cardinality c of $\mathbb{R}$ (see Sect. 1.2). Therefore, if we can exhibit an injective map from the set into $\mathcal{F}$, it will follow that $\mathcal{F}$ has cadinality at least c. We claim that the map Θ from this set into $\mathcal{F}$ given by

$$\Theta(\alpha) = \bigcup_{n=1}^{\infty} B^*_n, \text{ where } B^*_n \text{ means } B_n \text{ if } \alpha(n) = 1 \text{ and } B_n^c \text{ otherwise,}$$

is injective.

Let $\alpha \neq \beta$ be sequences in the two-element set $\{0, 1\}$. Then $\alpha(k) \neq \beta(k)$ for some k and hence one among $\Theta(\alpha)$ and $\Theta(\beta)$ contains the nonempty set B_k and the other is a union of B_k^c with sets of $\mathcal{B}$ that are distinct from B_k. However, distinct sets of $\mathcal{B}$ have been shown to be disjoint, and therefore sets of $\mathcal{B}$ that are distinct from B_k are disjoint from B_k and hence so is their union with B_k^c. Thus, one among $\Theta(\alpha)$ and $\Theta(\beta)$ contains the nonempty set B_k and the other is disjoint from it. Now, a set that is disjoint from a nonempty set must be distinct from any set containing the latter. Therefore $\Theta(\alpha) \neq \Theta(\beta)$. This justifies the claim that Θ is injective, thereby completing the proof that $\mathcal{F}$ has cadinality at least c.

2.3.P10. Show that, if E_1 and E_2 are measurable, then

$$m(E_1 \cup E_2) + m(E_1 \cap E_2) = m(E_1) + m(E_2).$$

Hint: Observe that $E_1 \cup E_2$ as well as $E_1 \cap E_2$ are measurable. Now, $m(E_1 \cup E_2) = m((E_1 \cup E_2) \cap E_1) + m((E_1 \cup E_2) \cap E_1^c) = m(E_1) + m((E_1 \cup E_2) \cap E_1^c)$, since E_1 is measurable and $(E_1 \cup E_2) \cap E_1 = E_1$. Adding $m(E_1 \cap E_2)$ to both sides, we get

$$\begin{aligned} m(E_1 \cup E_2) + m(E_1 \cap E_2) &= m(E_1) + m(E_2 \cap E_1^c) + m(E_1 \cap E_2) \\ &= m(E_1) + m(E_2), \end{aligned}$$

since E_1 is measurable.

Note that the measurability of E_2 is required only for obtaining the measurability of $E_1 \cup E_2$ and $E_1 \cap E_2$.

Alternatively (using finite additivity; Proposition 2.3.8): The sets $E_1 \cap E_2^c$, $E_1^c \cap E_2$, $E_1 \cap E_2$ are disjoint. Moreover,

$$E_1 = (E_1 \cap E_2^c) \cup (E_1 \cap E_2), \quad E_2 = (E_1^c \cap E_2) \cup (E_1 \cap E_2)$$

and

$$E_1 \cup E_2 = \left(E_1 \cap E_2^c\right) \cup (E_1 \cap E_2) \cup \left(E_1^c \cap E_2\right).$$

From finite additivity of m, we get the three equalities

$$m(E_1) = m\left(E_1 \cap E_2^c\right) + m(E_1 \cap E_2), \quad m(E_2) = m\left(E_1^c \cap E_2\right) + m(E_1 \cap E_2)$$

and

$$m(E_1 \cup E_2) = m\left(E_1 \cap E_2^c\right) + m(E_1 \cap E_2) + m\left(E_1^c \cap E_2\right).$$

Upon adding $m(E_1 \cap E_2)$ to both sides of the third equality and using the first two, we get the desired result.

2.3.P11. If $E_1 \in \mathfrak{B}$ and $E_2 \in \mathfrak{B}$ are such that $E_1 \supseteq E_2$ and $m(E_2) < \infty$, then $m(E_1 \backslash E_2) = m(E_1) - m(E_2)$. [Since $\mathfrak{B} \subseteq \mathfrak{M}$, it follows trivially that this is true also when $E_1 \in \mathfrak{B}$ and $E_2 \in \mathfrak{B}$.]

Hint: In fact, $E_1 = (E_1 \backslash E_2) \cup E_2$. The result follows on using the hypothesis $m(E_2) < \infty$ and the fact that m is finitely additive. To do this without finite additivity, use Problem 2.3.P10 in the following manner:

$$\begin{aligned} m(E_1\backslash E_2)+m(E_2) &= m((E_1\backslash E_2)\cup E_2)+m((E_1\backslash E_2)\cap E_2) \\ &= m(E_1\cup E_2) = m(\varnothing) \\ &\text{because}(E_1\backslash E_2)\cup E_2 = E_1\cup E_2 \text{ and } (E_1\backslash E_2)\cup E_2 = \varnothing \\ &= m(E_1) \text{ because } E_1 \supseteq E_2. \end{aligned}$$

2.3.P12. Let $0 < \alpha < 1$. Construct a measurable set $E \subseteq [0, 1]$ of measure $1 - \alpha$ and containing no interval of positive length.

Hint: Let $\{a_n\}_{n\geq 1}$ be a sequence of positive real numbers such that $\sum_{n=1}^{\infty} a_n = \alpha$ ($a_n = \alpha/2^n$ is one such sequence). Remove from the middle of the interval [0, 1] the open interval of length a_1, leaving behind two closed intervals of length $(1-a_1)/2$ each. Note that $(1-a_1)/2 > (\alpha-a_1)/2 > a_2/2$. From the middle of each of these intervals, remove an open interval of length $a_2/2$, leaving behind 2^2 closed intervals of length $(1-a_1-a_2)/2^2$ each. Note that $(1-a_1-a_2)/2^2 > (\alpha-a_1-a_2)/2^2 > a_3/2^2$. Repeating this process, at the $(n + 1)$st stage, from the middle of each of the 2^n closed intervals, remove an open interval of length $a_{n+1}/2^n$, leaving behind 2^{n+1} closed intervals of length $(1-a_1-\cdots-a_{n+1})/2^{n+1}$ each. Note that $(1-a_1-\cdots-a_{n+1})/2^{n+1} > (\alpha-a_1-\cdots-a_{n+1})/2^{n+1} > a_{n+2}/2^{n+1}$. The total length of the open intervals so removed is

$$a_1+2(a_2/2)+2^2(a_3/2^2)+\cdots+2^n(a_{n+1}/2^n)+\cdots=\sum_{n=1}^{\infty}a_n=\alpha.$$

Thus the set E which is left behind after successively removing open intervals as described, being the difference of the measurable sets [0,1] and the countable union of intervals removed, is measurable, and its measure is $m(E) = m([0, 1])-\alpha = 1-\alpha$. Observe that, at the nth stage, each remaining interval has length less than $\frac{1}{2^n}$, which tends to 0. Consequently, E contains no interval of positive length.

Remark The set constructed above is called the **generalised Cantor set.**

2.3.P13. Show that, if E is such that $0 < m^*(E) < \infty$ and $0 < \alpha < 1$, then there exists an open interval I such that $m^*(I \cap E) > \alpha\ell(I)$.

Hint: By Proposition 2.2.12 (second part), there exists an open set $O \supseteq E$ such that $m(O) < \alpha^{-1}m^*(E)$. But $O = \cup_{i\geq 1}I_i$, where I_i are disjoint open intervals. So, $\alpha\sum_{i=1}^{\infty}\ell(I_i) < m^*(E) \leq \sum_{i=1}^{\infty} m^*(E \cap I_i)$. But then there must exist some n such that $\alpha\ell(I_n) < m^*(E \cap I_n)$.

2.3.P14. **(a)** Suppose $m^*(E) < \infty$ and, for every $\varepsilon > 0$, there exists an open set U such that $m^*(E\Delta U) < \varepsilon$. Show that $E \in \mathfrak{M}$.

(b) If E is measurable and $m(E) < \infty$, then show that, for every $\varepsilon > 0$, there exists a finite union U of disjoint open intervals such that $m^*(E\Delta U) < \varepsilon$.

Hint: **(a)** First assume that $m^*(E) < \infty$.

Let $\varepsilon > 0$. By definition of outer measure, there exists an open set $V \supseteq E$ such that

$$m^*(V) < m^*(E) + \varepsilon. \tag{8.1}$$

Set $S = U \cap V$. Then $S \subseteq U$, $S \subseteq V$ and hence

$$\begin{aligned} S\Delta E &= (S\backslash E) \cup (E\backslash S) \\ &\subseteq (U\backslash E) \cup (E\backslash S) \quad \text{because } S \subseteq U \\ &\subseteq (U\backslash E) \cup (E \cap (U \cap V)^c) \quad \text{because } S = U \cap V \\ &= (U\backslash E) \cup (E \cap (U^c \cup V^c)) = (U\backslash E) \cup ((E \cap U^c) \cup (E \cap V^c)) \\ &= (U\backslash E) \cup (E \cap U^c) \quad \text{because } E \subseteq V \\ &= (U\backslash E) \cup (E\backslash U) = E\Delta U. \end{aligned}$$

Consequently,

$$m^*(S\Delta E) \leq m^*(E\Delta U) < \varepsilon. \tag{8.2}$$

We claim that $E \subseteq S \cup (S\Delta E)$.

$$\begin{aligned} S \cup (S\Delta E) &= S \cup (S\backslash E) \cup (E\backslash S) \\ &\supseteq (S \cap E) \cup (S\backslash E) \cup (E\backslash S) \quad \text{because } S \supseteq S \cap E \\ &= (S \cap E) \cup (S \cap E^c) \cup (E \cap S^c) \\ &\supseteq (S \cap (E \cup E^c)) \cup (E \cap S^c) \supseteq S \cup (E \cap S^c) \\ &= (S \cup E) \cap (S \cup S^c) = S \cup E \supseteq E. \end{aligned}$$

It follows that

$$m^*(E) \leq m^*(S) + \varepsilon. \tag{8.3}$$

Now,

$$\begin{aligned} V\backslash E &= ((V\backslash S) \cup S)\backslash E \quad \text{because } S \subseteq V \\ &= ((V \cap S^c) \cup S) \cap E^c = ((V \cap S^c) \cap E^c) \cup (S \cap E^c) \\ &\subseteq ((V \cap S^c) \cup (S \cap E^c)) \cup (E\backslash S) \\ &= (V\backslash S) \cup (S\Delta E). \end{aligned}$$

Therefore by (8.2),

$$m^*(V\backslash E) \leq m^*(V\backslash S) + \varepsilon$$

By (8.1) and (8.3), we further obtain

$$m^*(V) < m^*(S) + 2\varepsilon.$$

Since $S \subseteq V$ and $m^*(V) < \infty$, so that $m^*(S) < \infty$, it follows that

$$m^*(V \backslash S) = m^*(V) - m^*(S) < 2\varepsilon$$

and hence that

$$m^*(V \backslash E) < 3\varepsilon$$

By Proposition 2.3.24, we conclude that E is measurable.

(b) By Proposition 2.3.24, for every $\varepsilon > 0$, there exists an open set $O \supseteq E$ such that $m^*(O \backslash E) < \varepsilon$. Let $O = \cup_{i \geq 1} I_i$, where I_i are disjoint open intervals. Then $m^*(O) = \sum_{i=1}^{\infty} \ell(I_i) < \infty$. Since this series is convergent, there exists an $N \in \mathbb{N}$ such that $\sum_{i=N+1}^{\infty} \ell(I_i) < \varepsilon$. Let $U = \cup_{1 \leq i \leq N} I_i$. Then $O \backslash U \subseteq \cup_{i \geq N+1} I_i$, so that $m^*(O \backslash U) \leq \sum_{i=N+1}^{\infty} \ell(I_i) < \varepsilon$. Now, $E \Delta U = (E \backslash U) \cup (U \backslash E) \subseteq (O \backslash U) \cup (O \backslash E)$. Therefore $m^*(E \Delta U) \leq m^*(O \backslash U) + m^*(O \backslash E) < 2\varepsilon$.

Remark Part (b) is false without the condition $m(E) < \infty$. To see why, let $E = \cup_{n \geq 2} E_n$, where $E_n = [n, n + \frac{1}{n})$, $n \geq 2$. Then $m(E) = m^*(E) = \sum_{n=2}^{\infty} \ell(E_n) = \sum_{n=2}^{\infty} \frac{1}{n} = \infty$. There can be no finite union U of disjoint open intervals for which $m^*(E \Delta U) < \varepsilon$ for arbitrary ε. Indeed, if U contains an infinite interval, then $m^*(U \backslash E) = \infty$, and if each of the intervals comprising U is bounded, then $m^*(E \backslash U) = \infty$.

2.3.P15. Give an example of a nonmeasurable set.

Hint: Define an equivalence relation in $\mathbb{R}$ by setting

$$x \sim y \Leftrightarrow x - y \in \mathbb{Q}.$$

Clearly, the relation '$\sim$' partitions $\mathbb{R}$ into disjoint equivalence classes E_α. Any two elements of the same class differ by a rational number while those of two different classes differ by an irrational number. Moreover $\cup_\alpha E_\alpha = \mathbb{R}$. For any real number x, let $[x]$ be its integer part. Then

$$x \sim (x - [x]) \quad \text{and} \quad (x - [x]) \in (0, 1].$$

We construct a set A using the Axiom of Choice 1.1.1 by choosing one element from each set $E_\alpha \cap [0, 1)$. Observe that $\cup_\alpha E_\alpha \cap [0, 1) = [0, 1)$. Let $x_1, x_2, \ldots$ be an enumeration of the rationals in $(-1, 1)$ and set

$$A_n = A + x_n.$$

The sets $\{A_n\}_{n\geq 1}$ are disjoint. In fact, if there is a point $x \in A_m \cap A_n$, then $x = x' + x_m = x'' + x_n$, where x' and x'' are points in A. This implies $x' - x'' = x_n - x_m$ is a rational number, which forces $x' \sim x''$ and hence $x_m = x_n$; this implies $m = n$. Moreover, the following inclusions hold:

$$(0,\ 1) \subseteq \bigcup_{n=1}^{\infty} A_n \subseteq (-1,\ 2)$$

To justify the first inclusion, consider any $x \in (0, 1)$. Then $x \in [0, 1) = \cup_\alpha E_\alpha \cap [0, 1)$, as observed above. Therefore $x - x'$ is a rational number for a unique $x' \in A \subseteq [0, 1)$; so $x - x' = x_n$ for some n and hence $x \in A_n$. The second inclusion follows from the fact that each A_n is obtained by translating the elements of $A \subseteq [0, 1)$ by $x_n \in (-1, 1)$.

If A were to be measurable, then it would follow from the preceding paragraph that

$$1 = m((0,1)) \leq m\left(\bigcup_{n=1}^{\infty} A_n\right) = \sum_{n=1}^{\infty} m(A_n) \leq m((-1,2)) = 3$$

and also that

$$m\left(\bigcup_{n=1}^{\infty} A_n\right) = \sum_{n=1}^{\infty} m(A_n) = \sum_{n=1}^{\infty} m(A) = 0 \text{ or } \infty,$$

a contradiction.

2.3.P16. Show that the outer measure m^* is not finitely additive.

Hint: Let $E \subseteq \mathbb{R}$ be a nonmeasurable set. Then there exists an $A \subseteq \mathbb{R}$ for which $m^*(A) < m^*(A \cap E) + m^*(A \cap E^c)$. Here $A \cap E$ and $A \cap E^c$ are disjoint and have union A. Thus m^* is not finitely additive.

2.3.P17. Let f be defined on $[0, 1]$ by $f(0) = 0$ and $f(x) = x\sin\frac{1}{x}$ for $x > 0$. Show that the measure of the set $\{x: f(x) > 0\}$ is $1-(\ln 2)/\pi$.

Hint: The function f is positive on the set $S = \left(\frac{1}{\pi}, 1\right] \cup \left(\cup_{n\geq 1}\left(\frac{1}{(2n+1)\pi}, \frac{1}{2n\pi}\right)\right) \subseteq [0, 1]$. This is a union of countably many disjoint intervals and so,

$$m(S) = 1 - \frac{1}{\pi} + \sum_{n=1}^{\infty} \frac{1}{\pi}\left(\frac{1}{2n} - \frac{1}{2n+1}\right) = 1 - \frac{1}{\pi} + \frac{1}{\pi}\sum_{n=1}^{\infty} \frac{1}{(2n)(2n+1)}$$
$$= 1 - \frac{1}{\pi}\ln 2.$$

2.3.P18. Let E be Lebesgue measurable with $0 < m(E) < \infty$ and let $\varepsilon > 0$ be given. Then there exists a compact set $K \subseteq E$ such that $m(E\backslash K) = m(E) - m(K) < \varepsilon$.
Hint: Let $E_n = E \cap [-n, n]$ for $n \geq 1$. Then $\{E_n\}_{n \geq 1}$ is an increasing sequence of Lebesgue measurable sets. Therefore $m(E_n) \leq m(E) < \infty$. Also, $\cup_{n \geq 1} E_n$ is E and hence the continuity property (Proposition 2.3.21) yields $m(E) = \lim_n m(E_n)$. It follows that there exists an $N \in \mathbb{N}$ such that $m(E) - m(E_N) < \min\{\frac{\varepsilon}{2}, m(E)\}$. Then $m(E_N) > 0$. By Proposition 2.3.24, there exists a closed set $K \subseteq E_N \subseteq [-N, N]$ such that
$m(E_N\backslash K) = m(E_N) - m(K) < \min\{\frac{\varepsilon}{2}, m(E_N)\}$. Note that $m(K) > 0$, so that $K \neq \varnothing$ and also, K is compact, $K \subseteq E$, and $m(E) < \infty$, so that

$$m(E\backslash K) = m(E) - m(K) = m(E) - m(E_N) + m(E_N) - m(K) < \varepsilon.$$

Remark 1 Let $(\mathbb{R}, \mathfrak{B}, m)$ be the Borel measure space. It has been proved that m is an extended real-valued nonnegative function defined on $\mathfrak{B}$ which is translation invariant and is both inner and outer regular, i.e. (i) for $E \in \mathfrak{B}$ with $0 < m(E) < \infty$ and for $\varepsilon > 0$, there exists a compact set $K \subseteq E$ with $m(E\backslash K) < \varepsilon$ (2.3.P18) and (ii) for $E \in \mathfrak{B}$ and $\varepsilon > 0$, there exists an open set $O \supseteq E$ such that $m(O\backslash E) < \varepsilon$ (Proposition 2.3.24). Therefore m is often referred to as a **regular Borel measure**. Moreover, it is translation invariant (Proposition 2.3.23).

We next show that m is determined uniquely up to a constant multiple. Indeed, we have the following result.

Let μ be a σ-finite measure on the Borel σ-algebra $\mathfrak{B}$ of subsets of $\mathbb{R}$, satisfying the following properties:

(i) $\mu(U) > 0$ for an arbitrary nonempty open subset $U \subseteq \mathbb{R}$;
(ii) $\mu(K) < \infty$ for an arbitrary compact $K \subseteq \mathbb{R}$;
(iii) $\mu(E + x) = \mu(E)$ for an arbitrary $E \in \mathfrak{B}$ and arbitrary $x \in \mathbb{R}$.

Then there exists a positive $c \in \mathbb{R}$ such that $\mu(E) = cm(E)$ for all $E \in \mathfrak{B}$.

Step 1. *Let μ be a measure on the Borel sets of $\mathbb{R}$ which is translation invariant and let $\mu(0, 1) < \infty$. Then $\mu\{x\} = 0$ for any real number x, and $\mu(0, 1) = \mu[0, 1) = \mu(0, 1] = \mu[0, 1]$.*
Proof: Suppose if possible that there is some real number x_0 such that $\mu\{x_0\} \neq 0$. Then $\mu\{x_0\} > 0$. It follows by translation invariance that $\mu\{x\} = \mu\{x_0\} > 0$ for all x. Now the interval $(0, 1)$ contains infinitely many real numbers and it follows by countable additivity that any $\mu(0, 1) = \infty$, contrary to hypothesis.

It now follows that $\mu(0, 1) = \mu[0, 1) = \mu(0, 1] = \mu[0, 1]$.

Step 2. *Let* μ *be a measure on the Borel sets of* $\mathbb{R}$ *which is translation invariant and let* $\mu(0, 1) < \infty$. *Then* $\mu(I) = \mu(0, 1)m(I)$ *for any interval* I *and* $\mu(O) = \mu(0, 1)$ $m(O)$ *for any open set* O.
Proof: By Step 1, $\mu(0, 1) = \mu[0, 1) = \mu(0, 1] = \mu[0, 1]$. Call this nonnegative real number c. We have to prove that $\mu(I) = c \bullet m(I)$ for any interval I and $\mu(O) = c \bullet m$ (O) for any open set O. Since an open set is a countable union of disjoint open intervals, we need prove the equality only for intervals.

Since $\mu\{x\} = 0 = m\{x\}$ for any real number x, we need consider only open intervals.

By (Proposition 2.3.23), any two intervals of the same length have the same μ-measure as well as m-measure. It follows that the μ-measure of an interval of rational length is c times the length, i.e. c times the m-measure.

For an arbitrary bounded interval (a, b), we choose a rational strictly increasing sequence $\{a_n\}$ with limit a and a rational strictly decreasing sequence $\{b_n\}$ with limit b. Then $\mu[a, b] = \lim_n \mu(a_n, b_n) = c(b_n - a_n) = c(b - a)$. It follows from the fact that the measure of a singleton set is 0 that $\mu(a, b) = \mu[a, b) = \mu(a, b] = \mu[a, b] = c(b - a)$. Since an unbounded interval is a countable disjoint union of bounded intervals, the same equality must hold for unbounded intervals as well.
Step 3. According to Theorems 5.11.12 and 5.11.21, it can be concluded that for every Borel set E, the equality $\mu(E) = cm(E)$ holds.

In what follows, we provide a characterisation of the σ-algebra generated by an algebra $\mathcal{A}$ in terms of what are called "monotone classes".

A family $\mathfrak{M}$ of subsets of a nonempty set X is called a **monotone class** if it satisfies the following two conditions:

(i) if $A_1 \subseteq A_2 \subseteq \cdots$ and each $A_j \in \mathfrak{M}$, then $\cup_{j \geq 1} A_j \in \mathfrak{M}$;
(ii) if $B_1 \supseteq B_2 \supseteq \cdots$ and each $B_j \in \mathfrak{M}$, then $\cap_{j \geq 1} B_j \in \mathfrak{M}$.

Any σ-algebra is a monotone class. Let $A \subseteq X$, where X is any nonempty set and $\mathfrak{M} = \{A\}$. Then $\mathfrak{M}$ is a monotone class which is not a σ-algebra.

2.3.P19. (a) If $\mathcal{Y}$ is any class of subsets of a nonempty set X, then show that there exists a smallest monotone class containing $\mathcal{Y}$. We shall denote it by $\mathfrak{M}_0(\mathcal{Y})$.
(b) If $\mathcal{A}$ is an algebra, show that $\mathcal{S}(\mathcal{A}) = \mathfrak{M}_0(\mathcal{A})$; that is, the σ-algebra generated by an algebra is also the smallest monotone class containing the algebra.
Hint: **(a)** Let $\{\mathfrak{M}_\alpha\}$ be the collection of all those monotone classes of subsets of X that contain $\mathcal{Y}$. Since the collection of all subsets of X is a monotone class containing $\mathcal{Y}$, the collection $\{\mathfrak{M}_\alpha\}$ is nonempty. Also, any intersection of monotone classes is a monotone class. So, $\cap_\alpha \mathfrak{M}_\alpha$ is the required monotone class $\mathfrak{M}_0(\mathcal{Y})$.
(b) Let us write $\mathfrak{M}_0$ in place of $\mathfrak{M}_0(\mathcal{A})$. Since every σ-algebra is a monotone class, it is sufficient to show that $\mathfrak{M}_0$ is a σ-algebra.

We first show that $\mathfrak{M}_0$ is closed under complementation. To that end, let $\mathfrak{M}'_0 = \{A \in \mathfrak{M}_0 : A^c \in \mathfrak{M}_0\}$. Because $\mathcal{A}$ is an algebra and $\mathfrak{M}_0 \supseteq \mathcal{A}$, it follows that $\mathfrak{M}'_0 \supseteq \mathcal{A}$. Also, because $\mathfrak{M}_0$ is a monotone class, it is easy to see that $\mathfrak{M}'_0$ is a monotone class. Therefore $\mathfrak{M}'_0 \supseteq \mathfrak{M}_0$, which implies that $\mathfrak{M}_0$ is closed under complementation.

We next show that $\mathfrak{M}_0$ is closed under finite unions.

Suppose $A \in \mathcal{A}$ and let $K(A) = \{F \in \mathfrak{M}_0 \colon A \cup F \in \mathfrak{M}_0\}$. We shall show that $K(A)$ is a monotone class containing $\mathcal{A}$. Since $\mathcal{A}$ is an algebra and $\mathfrak{M}_0 \supseteq \mathcal{A}$, it follows that $K(A) \supseteq \mathcal{A}$. Now suppose that $E_1 \subseteq E_2 \subseteq \cdots$ and each $E_j \in K(A)$. Since $K(A) \subseteq \mathfrak{M}_0$ by definition and the latter is a monotone class, we have $\cup_{j \geq 1} E_j \in \mathfrak{M}_0$. Again by definition of $K(A)$, we have $A \cup E_j \in \mathfrak{M}_0$ for each j; also, $A \cup E_1 \subseteq A \cup E_2 \subseteq \cdots$. Since $\mathfrak{M}_0$ is a monotone class, we have $\cup_{j \geq 1}(A \cup E_j) \in \mathfrak{M}_0$, i.e. $A \cup (\cup_{j \geq 1} E_j) \in \mathfrak{M}_0$. Therefore $\cup_{j \geq 1} E_j \in K(A)$. Similarly, $K(A)$ is closed under intersections of nonincreasing sequences. Thus $K(A)$ is a monotone class containing $\mathcal{A}$ and, consequently, $K(A) \supseteq \mathfrak{M}_0$. But, $K(A) \subseteq \mathfrak{M}_0$ by definition. Thus $K(A) = \mathfrak{M}_0$. In other words, $A \cup F \in \mathfrak{M}_0$ for all $A \in \mathcal{A}$ and $F \in \mathfrak{M}_0$.

Now suppose $B \in \mathfrak{M}_0$, and let $\mathrm{K}(B) = \{F \in \mathfrak{M}_0 \colon B \cup F \in \mathfrak{M}_0\}$. We shall show that $\mathrm{K}(B)$ is a monotone class containing $\mathcal{A}$. From the previous paragraph, we know that $\mathrm{K}(B) \supseteq \mathcal{A}$ and using the same argument as in that paragraph, we find that $\mathrm{K}(B)$ is a monotone class. This implies that $\mathrm{K}(B) = \mathfrak{M}_0$. In other words, $F \cup G \in \mathfrak{M}_0$ for all $F, G \in \mathfrak{M}_0$. Thus $\mathfrak{M}_0$ is closed under finite unions.

It remains to prove that $\mathfrak{M}_0$ is a σ-algebra. Let $F_n \in \mathfrak{M}_0$ for every n. For each n, define $E_n = \cup_{1 \leq k \leq n} F_k$. Then $E_1 \subseteq E_2 \subseteq \cdots$ and $\cup_{n \geq 1} E_n = \cup_{n \geq 1} F_n$. Since $\mathfrak{M}_0$ is an algebra, each E_n belongs to it and, therefore, since $\mathfrak{M}_0$ is a monotone class, $\cup_{n \geq 1} E_n = \cup_{n \geq 1} F_n \in \mathfrak{M}_0$.

2.3.P20. Show that (a) the cardinality of $\mathfrak{M}$ is greater than c and (b) the Cantor set has a subset which is not a Borel set.

Hint: As usual, denote the Cantor set by C. Since the cardinality of C is c (see Proposition 1.8.12), the power set $\mathcal{P}(C)$ has cardinality 2^c, which is greater than c. Now, we know from Remark 2.2.11(c) that $m(C) = 0$ and hence by Remark 2.3.2 (d) that $\mathcal{P}(C) \subseteq \mathfrak{M}$. Therefore the cardinality of $\mathfrak{M}$ is also greater than c, which proves (a). On the other hand, the cardinality of the family of all Borel subsets of C is at most c. Since the cardinality of $\mathcal{P}(C)$ is greater than c, it follows that there exists a subset of C which is not a Borel set, which proves (b).

2.3.P21. Let $\mathcal{F}$ be an algebra of subsets of a set X. If A and B are subsets of X such that B, $B \cap A^c$, $A \cap B^c$ all belong to $\mathcal{F}$, show that A also belongs to $\mathcal{F}$.

Hint: The set

$$A \cap B = (B \cap B^c) \cup (A \cap B) = B \cap (B^c \cup A) = B \cap (B \cap A^c)^c$$

belongs to $\mathcal{F}$, because both B and $B \cap A^c$ do. Hence

$$A = A \cap (B \cup B^c) = (A \cap B) \cup (A \cap B^c)$$

belongs to $\mathcal{F}$, because both $A \cap B$ and $A \cap B^c$ do.

2.3.P22. Let f be a real-valued function on $[a, b]$. Suppose $E \subseteq [a, b]$, f' exists and satisfies $|f'(x)| \leq M$ for all $x \in [a, b]$. Prove that $m^*(f(E)) \leq Mm^*(E)$.
Hint: Assume that E does not contain the points a and b, so that $E \subseteq (a, b)$. There exists an open set $G \supseteq E$ such that $m(G) < m^*(E) + \frac{\varepsilon}{M}$. Without loss of generality, suppose that $G \subseteq (a, b)$. $G = \cup_{n\geq 1} I_n$, where I_n are disjoint open intervals. Let I_n' denote I_n with its endpoints adjoined to it. Observe that $I_n' \subseteq [a, b]$. Since f is continuous on I_n', there exist α and β such that $f(\alpha) = \min\{f(x): x \in I_n'\}$ and $f(\beta) = \max\{f(x): x \in I_n'\}$. By the Mean Value Theorem, $|f(\beta) - f(\alpha)| \leq Mm(I_n') = M\ell(I_n')$. This implies $\ell(f(I_n)) \leq \ell(f(I_n')) \leq M(\ell(I_n))$. So, $m^*(f(E)) \leq m^*(f(G)) = m^*(f(\cup_{n\geq 1} I_n)) = m^*(\cup_{n\geq 1} f(I_n)) \leq \Sigma_n m^*(f(I_n)) \leq M\Sigma_n \ell(I_n) = Mm(G) \leq M(m^*(E) + \frac{\varepsilon}{M}) = Mm^*(E) + \varepsilon$. Note that the removal of points a and b from the set E implies the removal at most two points $f(a)$ and $f(b)$ from $f(E)$.
2.3.P23. **(a)** Let F and Y be subsets of a set X. Then show that $f^c \cap Y$ is the complement in Y of $F \cap Y$.
(b) Let $Y \subseteq X$ and $\mathcal{G}$ be a σ-algebra of subsets of Y. Show that the family

$$\mathcal{F}_0 = \{F \subseteq X : F \cap Y \in \mathcal{G}\}$$

of subsets of X is a σ-algebra.
Hint: **(a)** $f^c \cap Y = (f^c \cap Y) \cup (Y^c \cap Y) = (f^c \cup Y^c) \cap Y = (F \cap Y)^c \cap Y$.
(b) It is trivial that $X \in \mathcal{F}_0$. If $\{F_n\}_{n\geq 1}$ is any sequence of sets belonging to $\mathcal{F}_0$, then the equality $\cup_{n\geq 1}(F_n \cap Y) = (\cup_{n\geq 1} F_n) \cap Y$ shows that $\cup_{n\geq 1} F_n \in \mathcal{F}_0$. As for complements, suppose $F \in \mathcal{F}_0$. Then $F \cap Y \in \mathcal{G}$ and $f^c \cap Y$ is the complement in Y of $F \cap Y$, as assured by part (a). It follows that $f^c \cap Y \in \mathcal{G}$ and thus $f^c \in \mathcal{F}_0$.
2.3.P24. Let $\mathcal{A}$ be a family of subsets of a set X and $\mathcal{F}$ the σ-algebra generated by it. Suppose $Y \subseteq X$ and

$$\mathcal{A}_Y = \{A \cap Y : A \in \mathcal{A}\}, \quad \mathcal{F}_Y = \{F \cap Y : F \in \mathcal{F}\}.$$

Show that $\mathcal{F}_Y$ is the σ-algebra of subsets of Y generated by $\mathcal{A}_Y$.
Hint: It is plain that $\mathcal{F}_Y$ contains $\mathcal{A}_Y$. That it is a σ-algebra follows (as in the preceding proof) by using the equality $\cup_{n\geq 1}(F_n \cap Y) = (\cup_{n\geq 1} F_n) \cap Y$ and Problem 2.3.P23(a).

Suppose $\mathcal{G}$ is a σ-algebra of subsets of Y that contains $\mathcal{A}_Y$. All we need to show is that $\mathcal{F}_Y \subseteq \mathcal{G}$.

Let $\mathcal{F}_0 = \{F \subseteq X: F \cap Y \in \mathcal{G}\}$. By Problem 2.3.23(b), $\mathcal{F}_0$ is a σ-algebra. Any set $A \in \mathcal{A}$ satisfies $A \cap Y \in \mathcal{A}_Y \subseteq \mathcal{G}$, so that it also satisfies $A \in \mathcal{F}_0$. Hence the σ-algebra $\mathcal{F}_0$ contains $\mathcal{A}$ and consequently, $\mathcal{F} \subseteq \mathcal{F}_0$. So, any $F \in \mathcal{F}$ satisfies $F \cap Y \in \mathcal{G}$. This means $\mathcal{F}_Y \subseteq \mathcal{G}$.

Problem Set 2.4

2.4.P1. Let the function f: $[0, 1] \to \mathbb{R}$ be defined by $f(x) = \frac{1}{x}$ if $0 < x \leq 1$, $f(0) = 0$. Show that f is measurable.
Hint: For any real number α,

$$\{x \in [0, 1] : f(x) > \alpha\} = \begin{cases} [0, 1] & \alpha < 0 \\ (0, 1] & 0 \leq \alpha < 1 \\ (0, \frac{1}{\alpha}) & \alpha \geq 1. \end{cases}$$

2.4.P2. Show that, if f is a real-valued function on a measurable subset $X \subseteq \mathbb{R}$ such that $X(f \geq r)$ is measurable for every rational number r, then f is measurable.
Hint: Let α be any real number. There is a decreasing sequence $\{r_n\}_{n \geq 1}$ of rational numbers such that $\lim r_n = \alpha$ and $r_n > \alpha$ for each n. Then

$$X(f > \alpha) = \bigcup_{n=1}^{\infty} X(f \geq r_n).$$

Since each of the sets on the right is measurable, their countable union is measurable too. So, f is measurable.
2.4.P3. Let $X \subseteq \mathbb{R}$ be measurable. Without using Theorem 2.4.6, show that if f and g are measurable functions defined on X, then the set $\{x \in X: f(x) > g(x)\}$ is measurable.
Hint: For each $x \in X$ such that $f(x) > g(x)$, we can find a rational number r such that $f(x) > r > g(x)$. Since

$$\{x \in X : f(x) > r > g(x)\} = \{x \in X : f(x) > r\} \cap \{x \in X : g(x) < r\},$$

it follows that the set on the left is measurable, being the intersection of two measurable sets. Now,

$$\{x \in X : f(x) > g(x)\} = \bigcup_{r \in \mathbb{Q}} \{x \in X : f(x) > r > g(x)\}.$$

Since $\mathbb{Q}$ is countable and a countable union of measurable sets is measurable, it follows that the set $\{x \in X: f(x) > g(x)\}$ is measurable.
2.4.P4. Let f be a real-valued function defined on a measurable subset $X \subseteq \mathbb{R}$. Then f is measurable if and only if for every open set $V \subseteq \mathbb{R}$,

$$f^{-1}(V) = \{x \in X : f(x) \in V\}$$

is measurable.

Hint: Suppose that f is measurable. Being an open subset of $\mathbb{R}$, V can be written as $V = \cup_{n \geq 1} I_n$, where $I_n = (a_n, b_n)$, $n = 1, 2, \ldots$, are disjoint open intervals (see Theorem 1.3.17). Then

$$\begin{aligned} f^{-1}(V) &= \bigcup_{n=1}^{\infty} f^{-1}(I_n) = \bigcup_{n=1}^{\infty} \{x \in X : a_n < f(x) < b_n\} \\ &= \bigcup_{n=1}^{\infty} [\{x \in X : f(x) > a_n\} \cap \{x \in X : f(x) < b_n\}]. \end{aligned}$$

It follows from the measurability of f that $f^{-1}(V)$ is measurable. The reverse conclusion follows on taking $V = (\alpha, \infty)$, where α is an arbitrary real number.

2.4.P5. Let f be a measurable function defined on a measurable subset $X \subseteq \mathbb{R}$ and ϕ be defined and continuous on the range of f. Then $\phi \circ f$ is a measurable function on X.

Hint: Let α be an arbitrary real number.

$$X(\phi \circ f > \alpha) = \{x \in X : f(x) \in U\},$$

where $U = \{u: \phi(u) > \alpha\}$. But U is open, since ϕ is continuous. Hence in view of 2.4.P4, the set $X(\phi \circ f > \alpha)$ is measurable. This proves that $\phi \circ f$ is measurable.

2.4.P6. Show that any function f defined on a set X of Lebesgue measure zero is Lebesgue measurable.

Hint: For an arbitrary real number α, $\{x \in X: f(x) > \alpha\} \subseteq X$ and so $m^*(\{x \in X: f(x) > \alpha\}) = 0$. Consequently, $\{x \in X: f(x) > \alpha\}$ is Lebesgue measurable [see Remark 2.3.2(d)].

2.4.P7. Let $X \subseteq \mathbb{R}$ and $f: X \to \mathbb{R}$ be any function. Show that the family of subsets of X given by $\mathcal{F} = \{f^{-1}(V): V \in \mathfrak{B}\}$ is a σ-algebra. (The same is true if $\mathfrak{B}$ is replaced by $\mathfrak{M}$.)

Hint: It is easy to verify the set-theoretic identities

$$f^{-1}(\mathbb{R}) = X, f^{-1}(V^c) = f^{-1}(V)^c \text{ and } f^{-1}\left(\bigcup_{n=1}^{\infty} V_n\right) = \bigcup_{n=1}^{\infty} f^{-1}(V_n).$$

This immediately implies that $\mathcal{F} = \{f^{-1}(V): V \in \mathfrak{B}\}$ is a σ-algebra.

2.4.P8. Let $X \subseteq \mathbb{R}$ and $f: X \to \mathbb{R}$ be any function. If $\mathcal{F}$ is a σ-algebra of subsets of X, show that the family of subsets of $\mathbb{R}$ given by $\mathcal{G} = \{V \subseteq \mathbb{R}: f^{-1}(V) \in \mathcal{F}\}$ is a σ-algebra. Hence show that, if f is Borel measurable, then $V \in \mathfrak{B} \Rightarrow f^{-1}(V) \in$

$\{X \cap U: U \in \mathfrak{B}\}$. Is it true that, if f is Lebesgue measurable, then $V \in \mathfrak{B} \Rightarrow f^{-1}(V) \in \{X \cap U: U \in \mathfrak{M}\}$?

Hint: The first part follows from the same set-theoretic identities as 2.4.P7. Now, $\mathfrak{B}_X = \{X \cap U: U \in \mathfrak{B}\}$ is a σ-algebra of subsets of X (easy verification), and hence by the first part, $\mathcal{G} = \{V \subseteq \mathbb{R}: f^{-1}(V) \in \mathfrak{B}_X\}$ is also a σ-algebra. By Corollary 2.4.3, all open sets of the form (α, ∞) as well as $(-\infty, \alpha)$, with $\alpha \in \mathbb{R}$ are in $\mathcal{G}$. Since the latter is a σ-algebra, all countable unions of finite intersections of such sets are in it. This means all open sets are in it. By definition of $\mathfrak{B}$ and the fact that $\mathcal{G}$ a σ-algebra, it follows that $\mathfrak{B} \subseteq \mathcal{G}$. Thus $V \in \mathfrak{B} \Rightarrow V \in \mathcal{G}$. By definition of $\mathcal{G}$, this means that $V \in \mathfrak{B} \Rightarrow f^{-1}(V) \in \mathfrak{B}_X = \{X \cap U: U \in \mathfrak{B}\}$. The last part is true, because the same argument works with $\mathfrak{M}_X$ in place of $\mathfrak{B}_X$ but *not* $\mathfrak{M}$ in place of $\mathfrak{B}$, the details being as below.

Now, $\mathfrak{M}_X = \{X \cap U: U \in \mathfrak{M}\}$ is a σ-algebra of subsets of X (easy verification), and hence by the first part, $\mathcal{G} = \{V \subseteq \mathbb{R}: f^{-1}(V) \in \mathfrak{M}_X\}$ is also a σ-algebra. By Corollary 2.4.3, all open sets of the form (α, ∞) as well as $(-\infty, \alpha)$, with $\alpha \in \mathbb{R}$ are in $\mathcal{G}$. Since the latter is a σ-algebra, all countable unions of finite intersections of such sets are in it. This means all open sets are in it. By definition of $\mathfrak{B}$ and the fact that $\mathcal{G}$ a σ-algebra, it follows that $\mathfrak{B} \subseteq \mathcal{G}$. Thus $V \in \mathfrak{B} \Rightarrow V \in \mathcal{G}$. By definition of $\mathcal{G}$, this means that $V \in \mathfrak{B} \Rightarrow f^{-1}(V) \in \mathfrak{M}_X = \{X \cap U: U \in \mathfrak{M}\}$.

2.4.P9. Let f be a real-valued measurable function defined on a measurable set $X \subseteq \mathbb{R}$. Prove that

(a) if $\alpha > 0$, then the function $|f|^\alpha$ is measurable;
(b) if $f(x) \neq 0$ on X and $\alpha < 0$, then $|f|^\alpha$ is measurable.

Hint: **(a)** For $\alpha > 0$, the function $\phi(t) = |t|^\alpha \ \forall\, t \in \mathbb{R}$ is continuous on $\mathbb{R} \supseteq$ range (f) and hence, $|f|^\alpha = \phi \circ f$ is measurable by Problem 2.4.P5.
(b) If $\alpha < 0$, the function $\phi(t) = |t|^\alpha \ \forall$ real $t \neq 0$ is continuous on $\mathbb{R}\backslash\{0\} \supseteq$ range (f) and hence, $|f|^\alpha = \phi \circ f$ is measurable by Problem 2.4.P5.

2.4.P10. A complex-valued function f with domain a measurable set $X \subseteq \mathbb{R}$ is said to be measurable if its real and imaginary parts, which are real-valued functions on X, are measurable. Prove that a complex-valued function is measurable if and only if $f^{-1}(V)$ is measurable for every open set $V \subseteq \mathbb{C}$ (the complex plane).

Hint: Suppose $f = g + ih$ is measurable. Any open $V \subseteq \mathbb{C}$ is a union of a countable family $\{I_n\}_{n \geq 1}$ of Cartesian products of open intervals (see Theorem 1.3.18). Let $I_n = (a_n, b_n) \times (c_n, d_n)$. Then

$$f^{-1}(V) = \bigcup_{n=1}^{\infty} f^{-1}(I_n) = \bigcup_{n=1}^{\infty} \left[g^{-1}((a_n, b_n)) \cap h^{-1}((c_n, d_n))\right]$$

is measurable. Conversely, suppose $f^{-1}(V)$ is measurable for every open set $V \subseteq \mathbb{C}$. Define α: $\mathbb{C} \to \mathbb{R}$ by $\alpha(z) = \mathfrak{R}(z)$, the real part of z. Then α is a real-valued continuous function defined on $\mathbb{C}$. Also, $g = \alpha \circ f$ and

$$g^{-1}((a,\infty)) = (\alpha \circ f)^{-1}((a,\infty)) = f^{-1}(\alpha^{-1}((a,\infty))) = f^{-1}((a,\infty) \times \mathbb{R})$$

is a measurable set. So g is measurable. Similarly, one may show that h is measurable.

Problem Set 2.5

2.5.P1. Let $g(x) = f'(x)$ exist for every $x \in [a, b]$. Prove that g is Lebesgue measurable.
Hint: Define $f(x) = f(b)$ for $x > b$. Let

$$g_n(x) = n\left(f\left(x + \frac{1}{n}\right) - f(x)\right) \quad \text{for } a \leq x \leq b \text{ and } n \in \mathbb{N}.$$

The functions g_n ($n \in \mathbb{N}$), being continuous on $[a, b]$, are measurable. Moreover, $\lim_n g_n(x) = f'(x) = g(x)$ for $x \in [a, b)$ and 0 when $x = b$. On using Corollary 2.5.5 and Proposition 2.5.13, it follows that g is measurable.
2.5.P2. Prove that the set of points at which a sequence of measurable functions converges or diverges to $\pm\infty$ is a measurable set.
Hint: The set $E = \{x \in \mathbb{R}: \limsup f_n - \liminf f_n = 0\}$ is clearly the set where the sequence converges. Since $\limsup f_n$ and $\liminf f_n$ are measurable by Theorem 2.5.4, it follows upon using the (easily proven) analogue of Proposition 2.4.5 for extended real-valued functions that the set E is measurable. The set on which the sequence diverges is the complement of E and therefore measurable.
2.5.P3. Show that $\sup\{f_\alpha: \alpha \in \Lambda\}$ is not necessarily measurable even if each f_α is.
Hint: Let E be a nonmeasurable set, $\Lambda = E$ and let $f_\alpha = \chi_{\{\alpha\}}$ for each $\alpha \in \Lambda = E$. Each f_α, being the characteristic function of a singleton set, is measurable. However, $\sup\{f_\alpha: \alpha \in \Lambda\} = \chi_E$ is nonmeasurable.
2.5.P4. If f is a real-valued Lebesgue measurable function defined on a Borel set $X \subseteq \mathbb{R}$, then there exists a Borel measurable function g on X such that $f(x) = g(x)$ a.e.
Hint: First suppose $X = \mathbb{R}$. Let $f = \chi_E$, where E is Lebesgue measurable. By Proposition 2.3.24, there exists an $\mathcal{F}_\sigma$-set $F \subseteq E$ such that $m(E \backslash F) = 0$ and hence also a G_δ-set $G \supseteq E \backslash F$ such that $m(G) = 0$. Since F is Borel measurable, χ_F is a Borel measurable function such that $\{x \in \mathbb{R}: \chi_E \neq \chi_F\} = \{x \in \mathbb{R}: \chi_{E \backslash F} \neq 0\} = E \backslash F \subseteq G$, which is a Borel set of measure zero. So, the characteristic function of a Lebesgue measurable set E has the property that it agrees with the characteristic function of a certain Borel set F on the complement of some Borel set of measure zero (which we called G above). It follows that a Lebesgue measurable simple function has the same property, because a finite union of Borel measurable sets of measure zero is again a set of the same kind. If f is Lebesgue measurable and nonnegative, there exists (see Theorem 2.5.9) a sequence $\{s_n\}_{n \geq 1}$ of Lebesgue measurable simple functions converging to f. Therefore there exists a sequence $\{t_n\}_{n \geq 1}$ of Borel measurable simple functions and a sequence $\{G_n\}_{n \geq 1}$ of Borel

sets of measure zero such that $t_n = s_n$ on the complement of G_n. Now $G = \cup_{n\geq 1}G_n$ is a Borel set of measure zero and $t_n(x) = s_n(x)\ \forall\, n \in \mathbb{N}\ \forall\, x \in G^c$. Reset each t_n to be 0 on G^c. Then $g(x) = \lim_n t_n(x)$ exists for every $x \in \mathbb{R}$, and g is Borel measurable (because each t_n is still Borel measurable). Moreover, $x \in G^c \Rightarrow g(x) = \lim_n s_n(x) = f(x)$. Since G has measure zero, this means $g = f$ a.e. To remove the assumption that $X = \mathbb{R}$, now consider any Borel set $X \subseteq \mathbb{R}$. Extend f to $\mathbb{R}$ by setting it equal to 0 outside X. Then the extended function is also Lebesgue measurable and, by what has already been proved, there exists a Borel measurable g on $\mathbb{R}$ which agrees with f a.e. The restriction of this g to X serves our purpose.

2.5.P5. Let f and g be extended real-valued measurable functions defined on a measurable subset $X \subseteq \mathbb{R}$. Then their product fg is also measurable.

Hint: Let $n \in \mathbb{N}$ and let the "truncated" function f_n be defined by

$$f_n(x) = \begin{cases} f(x) & \text{if } |f(x)| \leq n \\ n & \text{if } f(x) > n \\ -n & \text{if } f(x) < -n. \end{cases}$$

Let g_n be defined similarly. It follows from Remark 2.5.3(a) that f_n and g_n are measurable. Since they are real-valued (do not take $\pm\infty$ as values), it follows by Theorem 2.4.6 that each $f_n g_n$ is measurable. Since $f(x)g_m(x) = \lim_n f_n(x)g_m(x)$, it follows from Corollary 2.5.5 that fg_m is measurable. The result is now a consequence of the identity

$$(fg)(x) = f(x)g(x) = \lim_{m\to\infty} f(x)g_m(x) \quad \text{for } x \in X$$

and another application of Corollary 2.5.5.

2.5.P6. Let s be a simple function on a measurable set X such that $s = \Sigma_{1\leq i\leq p}\alpha_i\chi_{A_i}$, where

$$i \neq i' \Rightarrow A_i \cap A_{i'} = \varnothing \quad \text{and} \quad \bigcup_{i=1}^{p} A_i = X.$$

(a) If $\gamma_1, \gamma_2, \ldots, \gamma_n$ are the distinct elements of the range of s, show that

$$\{\gamma_j : 1 \leq j \leq n\} = \{\alpha_i : 1 \leq i \leq p, A_i \neq \varnothing\}.$$

(b) Show that, for any j, $1 \leq j \leq n$, there must exist some index i $(1 \leq i \leq p)$ for which the coefficient α_i equals γ_j and, at the same time, $A_i \neq \varnothing$. If for each j $(1 \leq j \leq n)$, N_j is the set of all such indices i corresponding to a given j, namely,

$$N_j = \{i : 1 \le i \le p, \alpha_i = \gamma_j \quad \text{and} \quad A_i \ne \emptyset\},$$

show that $\cup_{i \in N_j} A_i = X(s = \gamma_j)$.

(c) Finally, show that the canonical representation of s is $\Sigma_{1 \le j \le n} \gamma_j \chi_{B_j}$, where $B_j = \cup_{i \in N_j} A_i$.

Hint: **(a)** Observe that $x \in A_i \Rightarrow s(x) = \alpha_i$, because the A_i are disjoint. It follows from this observation that $A_i \ne \emptyset \Rightarrow \alpha_i \in s(X)$, i.e.

$$\{\alpha_i : 1 \le i \le p, A_i \ne \emptyset\} \subseteq s(X) = \{\gamma_1, \gamma_2, \ldots, \gamma_n\}.$$

Since $\cup A_i = X$ it follows from the same observation that the reverse inclusion also holds.

(b) Consider any j, $1 \le j \le n$. By (a), there must exist some index i ($1 \le i \le p$) for which the coefficient α_i equals γ_j and $A_i \ne \emptyset$. Therefore the set of all such indices i corresponding to a given j, namely,

$$N_j = \{i : 1 \le i \le p, \alpha_i = \gamma_j \quad \text{and} \quad A_i \ne \emptyset\}$$

is nonempty for $1 \le j \le n$. Moreover, $s(x) = \gamma_j$ for precisely those points x that belong to some A_i with $i \in N_j$, which is precisely the assertion in question.

(c) Immediate from (b) and Remark 2.5.8(c).

2.5.P7. Let $\Sigma_{1 \le i \le p} \alpha_i \chi_{A_i}$ be the canonical representation of a simple function s. If $0 < \alpha < \infty$, show that $\Sigma_{1 \le i \le p} (\alpha\alpha_i) \chi_{A_i}$ is the canonical representation of αs.

Hint: Clearly, $\Sigma_{1 \le i \le p} (\alpha\alpha_i) \chi_{A_i} = \alpha s$. We need show only that it is a canonical representation. Since $\Sigma_{1 \le i \le p} \alpha_i \chi_{A_i}$ is a canonical representation, the sets A_j are nonempty, disjoint with $\cup_{1 \le j \le n} A_j = X$ and are measurable; moreover the real numbers $\alpha_1, \alpha_2, \ldots, \alpha_n$ are distinct and nonnegative and hence so are the products $\alpha\alpha_i$ (keeping in view that $0 < \alpha < \infty$). This means $\Sigma_{1 \le i \le p} (\alpha\alpha_i) \chi_{A_i}$ is a canonical representation.

2.5.P8. Give an example when $(f + g)^+$ is not the same as $f^+ + g^+$.

Hint: Take f to be a nonnegative function with at least one positive value, and $g = -f$.

2.5.P9. If f and g are extended real-valued functions and $g \ge 0$, show that $(fg)^+ = f^+g$ and $(fg)^- = f^-g$.

Hint: Consider any x in the domain and let $f(x) \ge 0$. Then $f^+(x) = f(x)$, and therefore $f(x)g(x) = f^+(x)g(x) = (f^+g)(x)$. Also, $f(x)g(x) \ge 0$ and so $((fg)^+)(x) = f(x)g(x)$. Thus $(fg)^+(x) = (f^+g)(x)$ when $f(x) \ge 0$. The second equality holds at x because both sides are 0. A similar argument works when $f(x) \le 0$.

2.5.P10. Let f be a bounded measurable function defined on a bounded closed interval $[a, b]$ and let $\varepsilon > 0$ be arbitrary. Show that there exists a step function g on $[a, b]$ such that

$$m(\{x \in [a, b] : |f(x) - g(x)| \geq \varepsilon\}) < \varepsilon.$$

Hint: Step I. Let $f = \chi_E$, where $E \subseteq [a, b]$ is measurable. By Problem 2.3.P14(b), there exist disjoint open intervals $I_1, \ldots, I_k$ such that[1] $m(E \Delta \cup_{1 \leq r \leq k} I_r) < \eta$. Let g be the restriction to $[a, b]$ of the characteristic function of $\cup_{1 \leq r \leq k} I_r$. Then g is a step function on $[a, b]$. Furthermore, $\{x \in [a, b]: |f(x)-g(x)| \geq \varepsilon\} = \emptyset$ if $\varepsilon > 1$ and for $0 < \varepsilon \leq 1$, $\{x \in [a, b]: |f(x)-g(x)| \geq \varepsilon\} \subseteq E \Delta \cup_{1 \leq r \leq k} I_r$. Hence

$$m(\{x \in [a, b] : |f(x) - g(x)| \geq \varepsilon\}) < \eta.$$

Step II. Let f be any simple function on $[a, b]$ with representation $f = \sum_{i=1}^{n} a_i \chi_{E_i}$, where $E_1, \ldots, E_n$ are disjoint measurable sets with $\cup_{1 \leq i \leq n} E_i = [a, b]$. For each $i = 1, \ldots, n$, let g_i be the step function on $[a, b]$ (Step I) such that $m(\{x \in [a, b]: |a_i \chi_{E_i}(x) - g_i(x)| \geq \frac{1}{n}\varepsilon\}) < \frac{1}{n}\eta$. Let $g = \sum_{i=1}^{n} g_i$. Then

$$m(\{x \in [a, b] : |f(x) - g(x)| \geq \varepsilon\}) < \eta.$$

This is because $|f(x) - g(x)| = \left|\sum_{i=1}^{n} a_i \chi_{E_i} - \sum_{i=1}^{n} g_i\right| \leq \sum_{i=1}^{n} |a_i \chi_{E_i}(x) - g_i(x)|$, so that $|f(x) - g(x)| \geq \varepsilon$ implies $|a_i \chi_{E_i}(x) - g_i(x)| \geq \frac{1}{n}\varepsilon$ for some i, and hence

$$\{x \in [a, b] : |f(x) - g(x)| \geq \varepsilon\} \subseteq \bigcup_{i=1}^{n} \left\{x \in [a, b] : |a_i \chi_{E_i}(x) - g_i(x)| \geq \frac{1}{n}\varepsilon\right\},$$

which implies

$$m(\{x \in [a, b] : |f(x) - g(x)| \geq \varepsilon\}) \leq \sum_{i=1}^{n} m\left(\left\{x \in [a, b] : |a_i \chi_{E_i}(x) - g_i(x)| \geq \frac{1}{n}\varepsilon\right\}\right).$$

---

[1]The fact that the intervals can be chosen to be disjoint is of no consequence here.

Step III. Let f be a bounded measurable function defined on $[a, b]$. By Theorem 2.5.9, there exists a simple function s such that $|f - s| < \varepsilon/2$. By Step II, corresponding to the simple function s, there is a step function g such that

$$m\left(\left\{x \in [a, b] : |s(x) - g(x)| \geq \frac{\varepsilon}{2}\right\}\right) < \eta,$$

Since $|f(x) - s(x)| < \varepsilon/2$, the inequality $|s(x) - g(x)| < \varepsilon/2$ implies $|f(x) - g(x)| < \varepsilon$, which means $\{x \in [a, b]: |f(x)-g(x)| \geq \varepsilon\} \subseteq \{x \in [a, b]: |s(x)-g(x)| \geq \varepsilon/2\}$. It follows that

$$m(\{x \in [a, b] : |f(x) - g(x)| \geq \varepsilon\}) \leq m\left(\left\{x \in [a, b] : |s(x) - g(x)| \geq \frac{\varepsilon}{2}\right\}\right) < \eta.$$

This completes the argument.

Problem Set 2.7

2.7.P1. Prove that $|m_n^*(A) - m_n^*(B)| \leq \max\{m_n^*(A\backslash B), m_n^*(B\backslash A)\} \leq m_n^*(A\Delta B)$ provided $m_n^*(A)$ and $m_n^*(B)$ are finite.
Hint: Since $A = (A\backslash B) \cup (A \cap B)$, we have $m_n^*(A) \leq m_n^*(A\backslash B) + m_n^*(A \cap B) \leq m_n^*(A\backslash B) + m_n^*(B)$. So, $m_n^*(A) - m_n^*(B) \leq m_n^*(A\backslash B)$. Similarly, $m_n^*(B) - m_n^*(A) \leq m_n^*(B\backslash A)$.
2.7.P2. An *outer measure* on $\mathbb{R}^n$ is an extended real-valued, nonnegative, monotone and countably subadditive set function μ^* defined on all subsets of $\mathbb{R}^n$ and satisfying $\mu^*(\emptyset) = 0$. If $\{\mu_n^*\}_{k\geq 1}$ is a sequence of outer measures and $\{a_k\}_{k\geq 1}$ a sequence of positive real numbers, then show that the set function defined by $\mu^*(E) = \sum_{k=1}^{\infty} a_k \mu_k^*(E)$ is an outer measure.

Hint: $\mu^*(\emptyset) = \sum_{k=1}^{\infty} a_k \mu_k^*(\emptyset) = 0$. $E \subseteq F \Rightarrow \mu_k^*(E) \leq \mu_k^*(F)$. Monotonicity now follows.

$$\mu_k^*\left(\bigcup_{j=1}^{\infty} E_j\right) \leq \sum_{j=1}^{\infty} \mu_k^*(E_j)$$

$$\sum_{k=1}^{\infty} a_k \mu_k^*\left(\bigcup_{j=1}^{\infty} E_j\right) \leq \sum_{k=1}^{\infty} a_k \sum_{j=1}^{\infty} \mu_k^*(E_j) = \sum_{j=1}^{\infty} \sum_{k=1}^{\infty} a_k \mu_k^*(E_j) = \sum_{j=1}^{\infty} \mu^*(E_j).$$

Problem Set 2.8

2.8.P1 Let $\{E_i\}_{i\geq 1}$ be a sequence of measurable sets. Then

(a) if $E_i \subseteq E_{i+1}$, then $m_n(\lim E_n) = \lim\limits_k m_n(E_k)$;

(b) if $E_i \supseteq E_{i+1}$ and $m_n(E_1) < \infty$, then $m_n(\lim E_n) = \lim\limits_k m_n(E_k)$.

Hint: Same as Proposition 2.3.21.

2.8.P2. Prove that for subsets A, B, C of any nonempty set X,

$$A\Delta C \subseteq (A\Delta B) \cup (B\Delta C).$$

Hint: $A\Delta C = (A\backslash C) \cup (C\backslash A)$. Consider any $x \in A\Delta C$. Suppose $x \in A\backslash C$; then $x \in A$ and $x \notin C$. If $x \in B$, then $x \in B\backslash C$ and if $x \notin B$, then $x \in A\backslash B$. Thus $A\backslash C \subseteq (A\Delta B) \cup (B\Delta C)$. Upon interchanging A and C, we get $C\backslash A \subseteq (C\Delta B) \cup (B\Delta A)$.

2.8.P3. Let A, B be subsets of $\mathbb{R}^n$. Show that

$$m_n^*(A\Delta B) \leq m_n^*(A) + m_n^*(B).$$

Hint: Since $A\Delta B = (A\backslash B) \cup (B\backslash A)$, it follows that

$$\begin{aligned} m_n^*(A\Delta B) &\leq m_n^*(A\backslash B) + m_n^*(B\backslash A) \\ &\leq m_n^*(A) + m_n^*(B) \end{aligned}$$

because $A\backslash B \subseteq A$ and $B\backslash A \subseteq B$.

2.8.P4. If μ^* is an outer measure on $\mathbb{R}^n$ and if A and B are subsets of $\mathbb{R}^n$, of which at least one is μ^*-measurable, then show that

$$\mu^*(A) + \mu^*(B) = \mu^*(A \cup B) + \mu^*(A \cap B).$$

Remark Under the additional hypothesis that at least one among A and B has finite outer measure, the above equality can be written as

$$\mu^*(A \cup B) = \mu^*(A) + \mu^*(B) - \mu^*(A \cap B).$$

Hint: We may assume that A is μ^*-measurable. Then

$$\mu^*(B) = \mu^*(A \cap B) + \mu^*(A^c \cap B)$$

and

$$\begin{aligned}\mu^*(A \cup B) &= \mu^*(A \cap (A \cup B)) + \mu^*(A^c \cap (A \cup B)) \\ &= \mu^*(A) + \mu^*(A^c \cap B).\end{aligned}$$

Hence

$$\mu^*(A \cup B) + \mu^*(A \cap B) = \mu^*(A) + \mu^*(A^c \cap B) + \mu^*(A \cap B) = \mu^*(A) + \mu^*(B).$$

2.8.P5. Let $A \subseteq E$, where E is measurable and $m_n(E) < \infty$. Show that A is measurable provided

$$m_n^*(E) = m_n^*(A) + m_n^*(E \backslash A).$$

Hint: By Propositions 2.8.18 and 2.8.14, we have $A \subseteq A'$ and $E \backslash A \subseteq B'$, where A' and B' are measurable and $m_n{}^*(A) = m_n{}^*(A')$ and $m_n{}^*(E \backslash A) = m_n{}^*(B')$. Replacing A' by $A' \cap E$ and B' by $B' \cap E$, we may assume that $A' \subseteq E$ and $B' \subseteq E$. Since A' and B' are measurable and $A' \cup B' = E$, we have

$$m_n(A' \Delta B') + m_n(A' \cap B') = m_n(A' \cup B') = m_n(E). \tag{8.4}$$

Also,

$$m_n(A' \backslash B') + m_n(A' \cap B') = m_n(A'), \tag{8.5}$$

$$m_n(B' \backslash A') + m_n(A' \cap B') = m_n(B'). \tag{8.6}$$

On adding (8.5) and (8.6), we get

$$\begin{aligned}m_n(A' \backslash B') + m_n(B' \backslash A') + 2m_n(A' \cap B') &= m_n(A') + m_n(B') \\ &= m_n^*(A) + m_n^*(E \backslash A) = m_n(E).\end{aligned} \tag{8.7}$$

From (8.4) and (8.7), it follows that

$$m_n(A'\backslash B') = 0.$$

Since $A' \subseteq E$ and $A \subseteq E$, we have $A'\backslash A = A' \cap (E\backslash A) \subseteq A' \cap B'$. So,

$$m_n^*(A'\backslash A) \le m_n^*(A' \cap B') = m_n(A' \cap B') = 0,$$

which implies $A'\backslash A$ is measurable. Consequently, $A = A'\backslash(A'\backslash A)$ is measurable.

2.8.P6. If $F \in \mathfrak{M}_n$ and $m_n{}^*(F\Delta G) = 0$, then show that G is measurable.
Hint: By Remark 2.8.2(d), $F\Delta G$ is measurable and so are the subsets $F\backslash G$ and $G\backslash F$. Now, $F = (F\backslash G) \cup (F \cap G)$. Therefore $F \cap G = F\backslash(F\backslash G)$ is measurable. Consequently, $G = (G\backslash F) \cup (F \cap G)$ is measurable.

2.8.P7. Show that every nonempty open set has positive measure.
Hint: A nonempty open set must contain a nonempty cuboid, the volume of which is always positive.

2.8.P8. Let $q_1, q_2, \ldots$ be an enumeration of points in $\mathbb{R}^n$ with rational coordinates and let $G = \cup_{k\geq 1} I_k$, where I_k is an open cuboid centred at q_k with volume $1/k^2$. Prove that for any closed set F, $m_n(G\Delta F) > 0$.
Hint: If $m_n(G\backslash F) > 0$, there is nothing to prove. Suppose $m_n(G\backslash F) = 0$ and since $G\backslash F$ is open, we must have $G \subseteq F$ by Problem 2.8.P7. But G contains the set of all points with rational coordinates, whose closure is $\mathbb{R}^n$. So, $F = \mathbb{R}^n$ and $m_n(F) = \infty$. But $m_n(G) \le \Sigma(1/k^2) < \infty$. Therefore $m_n(F\backslash G) = \infty$ and hence

$$m_n(G\Delta F) = m_n(G\backslash F) + m_n(F\backslash G) = \infty > 0.$$

Remark In fact, we have proved more than what is demanded in the problem, namely, either $m_n(G\backslash F) > 0$ or $m_n(F\backslash G) = \infty$.

2.8.P9. Let E be Lebesgue measurable with $0 < m_n(E) < \infty$ and let $\varepsilon > 0$ be given. Then there exists a compact set $K \subseteq E$ such that $m_n(E\backslash K) = m_n(E) - m_n(K) < \varepsilon$.
Hint: Follow the ideas of Problem 2.3.P18.

Problem Set 3.1

3.1.P1. Let $X = [0, 4]$ and $s = 2\chi_{[0,2]} + 3\chi_{(2,4]}$ and $t = 6\chi_{(1,3]}$. Find $\int_X s\,dm$ and $\int_X t\,dm$. Also, find the canonical form of $s + t$ and use it to compute $\int_X (s+t)dm$ from the definition.
Hint: $\int_X s\,dm = 2(2 - 0) + 3(4 - 2) = 10$, $\int_X t\,dm = 6(3 - 1) = 12$. In canonical form,

$$s + t = 2\chi_{[0,\,1]} + 8\chi_{(1,\,2]} + 9\chi_{(2,\,3]} + 3\chi_{(3,\,4]}.$$

So, $\int_X (s+t)\,dm = 2(1-0) + 8(2-1) + 9(3-2) + 3(4-3) = 22.$

3.1.P2. Let $X = (0, 4]$ and $A_n = [\frac{1}{n}, 4]$, so that $A_1 \subseteq A_2 \subseteq \cdots$ and $\cup_{n \geq 1} A_n = X$. For $s = \chi_{(0,2]}$, find $\int_X (s\chi_{A_n})dm$, $\int_X s\,dm$ and $\lim_{n\to\infty} \int_X (s\chi_{A_n})dm$.
Hint: $s\chi_{A_n} = \chi_{[\frac{1}{n},2]}$. Therefore $\int_X (s\chi_{A_n})dm = 1 \cdot (2 - \frac{1}{n})$; hence $\lim_{n\to\infty} \int_X (s\chi_{A_n}) dm = 2$. Also, $\int_X s\,dm = 1 \cdot (2 - 0) = 2$.

3.1.P3. Let X be a measurable subset of $\mathbb{R}$ and $\{A_n\}_{n\geq 1}$ be a sequence of measurable subsets such that $A_1 \subseteq A_2 \subseteq \cdots$. If $X \backslash \cup_{n\geq 1} A_n$ has positive measure, show that there exists a simple function s such that $\lim_{n\to\infty} \int_X (s\chi_{A_n})d\mu \neq \int_X s\,d\mu$.
Hint: Take $s = \chi_E$, where $E = X \backslash \cup_{n\geq 1} A_n$. Then $\int_X s\,d\mu > 0$ but $\int_X (s\chi_{A_n})d\mu = 0 \forall n \in \mathbb{N}$.

3.1.P4. Let X be a measurable set, A_i $(1 \leq i \leq p)$ disjoint measurable subsets of it and α_i $(1 \leq i \leq p)$ nonnegative real numbers.
(a) If the sets A_i do *not* form a partition of X, show that there are infinitely many simple functions taking the respective values α_i on A_i.
(b) If the sets A_i form a partition of X, show that there is a unique simple function s taking the respective values α_i on A_i and that it is given by $s = \Sigma_{1\leq i\leq p}\alpha_i\chi_{A_i}$.
Hint: **(a)** Since the A_i are disjoint, their not forming a partition means that their union has a nonempty complement. The specification that a function takes the respective values α_i on A_i allows any nonnegative real number to be taken as a value on the nonempty complement.
(b) Here the union is the whole of X and the specification that a function takes the respective values α_i on A_i determines the function completely. Therefore such a function must be unique. Since $s = \Sigma_{1\leq i\leq p}\alpha_i\chi_{A_i}$ fulfils the specification (in view of the given disjointness), the unique function must be none other than s.
3.1.P5. Is Proposition 3.1.5 valid without the hypothesis that the A_i and the B_j form partitions of X?
Hint: Yes, by Proposition 3.1.7.

3.1.P6. [Needed in 3.2.P13] Let s be a simple function on X. Define ϕ_s: $[0, \infty) \to [0, \infty]$ as $\phi_s(u) = \mu(X(s > u))$. If $\mu(X) < \infty$, then ϕ_s takes values in $[0, \infty)$. Show that
(a) $\phi_s(u) = 0$ if $u \geq M = \max\{s(x): x \in X\}$.
(b) ϕ_s is a step function.
(c) The Riemann integral $\int_0^M \phi_s(u)du$, which is equal to the improper integral $\int_0^\infty \phi_s(u)du$ in view of (a), is also equal to the measure space integral $\int_X s d\mu$.
(d) What happens if $\mu(X) = \infty$?
Hint: **(a)** If $u \geq M$, then $X(s > u) = \varnothing$ and hence $\phi_s(u) = \mu(\varnothing) = 0$.
(b) Let the values of s be α_i, $1 \leq i \leq p$, in increasing order, i.e. $0 \leq \alpha_1 < \alpha_2 < \cdots < \alpha_p = M$. Then $s = \Sigma_{1\leq i\leq p}\alpha_i\chi_{Ai}$, where A_i is the set on which s takes the value α_i. It will be convenient to introduce the notation $S_i = \Sigma_{i\leq j\leq p} \mu(A_j)$. Suppose $0 < \alpha_1$ and consider any u such that $0 \leq u < \alpha_1$. Then $s(x) > u$ for all x. Therefore $X(s > u) = X$ and hence $\phi_s(u) = \mu(X) = \Sigma_{1\leq i\leq p} \mu(A_i)$ by the additivity of measure. This shows that, if $0 \leq u < \alpha_1$, then $\phi_s(u) = S_1$. Therefore, unless the interval $[0, \alpha_1)$ is empty, ϕ_s has the constant value S_1 on it. Now suppose $p > 1$ and consider any u such that $\alpha_{i-1} \leq u < \alpha_i$, where $i > 1$. Then $s(x) > u \Leftrightarrow$

$s(x) \geq \alpha_i \Leftrightarrow s(x) = \alpha_j$ for some $j \geq i$. This means $X(s > u) = \cup_{i \leq j \leq p} A_j$ and hence $\phi_s(u) = S_i$ by the additivity of measure. This shows that ϕ_s has the constant value S_i on $[\alpha_{i-1}, \alpha_i)$. Finally, consider any u such that $M = \alpha_p \leq u$. Since $s(x) \leq M$ everywhere, $X(s > u)$ is now $\emptyset$ and therefore $\phi_s(u) = 0$. This shows that ϕ_s has the constant value 0 on $[\alpha_p, \infty)$. Thus ϕ_s is the following step function:

$$\phi_s(u) = \begin{cases} S_1 & \text{if} 0 \leq u < \alpha_1 \\ S_i & \text{if} \alpha_{i-1} \leq u < \alpha_i \\ 0 & \text{if} \alpha_p \leq u. \end{cases}$$

The only difference if $\alpha_1 = 0$ is that the case $0 \leq u < \alpha_1$ is to be omitted. If $p = 1$, then the case $\alpha_{i-1} \leq u < \alpha_i$ has to be omitted. If $p = 1$ and also $\alpha_1 = 0$, then both the aforementioned cases are vacuous and s as well as ϕ_s are zero everywhere.
(c) If $p = 1$, there are two subcases, namely, $\alpha_1 = 0$ and $\alpha_1 > 0$. Both are trivial and we go on to consider the case $p \geq 2$. Since ϕ_s is a step function as described above, its Riemann integral over $[0, M] = [0, \alpha_p]$ is (regardless of whether $\alpha_1 = 0$ or not)

$$\begin{aligned} \int_0^M \phi_s(u)du &= \alpha_1 S_1 + \sum_{i=2}^{p} S_i(\alpha_i - \alpha_{i-1}) \\ &= \sum_{i=1}^{p-1} \alpha_i(S_i - S_{i+1}) + \alpha_p S_p = \sum_{i=1}^{p} \alpha_i \mu(A_i). \end{aligned}$$

(d) If $\mu(X) = \infty$, then ϕ_s may have ∞ as a value on some interval of positive length. It is nonetheless a nonnegative extended real-valued step function and has Riemann integral ∞. But in that event, the measure space integral of s is also ∞ and the equality proved in **(c)** is still valid.

3.1.P7. Let $\{s_n\}_{n \geq 1}$ be the sequence of simple functions given by $s_n(x) = 1/n$ for $|x| \leq n$ and 0 for $|x| > n$. Show that $s_n \to 0$ uniformly on $X = \mathbb{R}$, but $\int_X s_n dm = 2$ for every $n \in \mathbb{N}$. [Uniform convergence on a set X of finite measure to a bounded limit function does imply convergence of integrals; this will be seen to be true in the next section even for more general functions in the light of the Dominated Convergence Theorem 3.2.16]
Hint: Let $\varepsilon > 0$. Choose $n_0 \in \mathbb{N}$ such that $n_0 > 1/\varepsilon$. For $n \geq n_0$, we have $s_n(x) = 1/n < 1/n_0 < \varepsilon$ provided $|x| \leq n$. Since $|x| > n \Rightarrow s_n(x) = 0$, it follows that $|s_n(x)| < \varepsilon$ for $n \geq n_0$ and all $x \in X = \mathbb{R}$. Also, s_n is a simple function with canonical representation $s_n = \frac{1}{n} \cdot \chi_{[-n,n]} + 0 \cdot \chi_{(-\infty,-n) \cup (n,\infty)}$ so that $\int_X s_n = \frac{1}{n} 2n = 2$.

3.1.P8. *Optional.* [The principle that permits grouping of terms in a sum according to any scheme whatsoever can be expressed in this manner: Suppose we have a sum $\Sigma_{1 \leq i \leq p} b_i$ to evaluate and that the set $\{i: 1 \leq i \leq p\}$ of indices has been partitioned into n subsets N_j, $1 \leq j \leq n$, meaning thereby that these subsets are nonempty and disjoint with union equal to $\{i: 1 \le i \le p\}$. Then the required sum can be evaluated as the grand total of the n subtotals $\Sigma_{i\in N_j} b_i$. In other words,

$$\sum_{i=1}^{p} b_i = \sum_{j=1}^{n}\left(\sum_{i\in N_j} b_i\right)$$

provided $\bigcup_{j=1}^{n} N_j = \{1,2,\ldots,p\}$, every $N_j \neq \emptyset$ and $j \neq j' \Rightarrow N_j \cap N_{j'} = \emptyset$. This is valid even if one or more of the b_i are ∞ as long as each b_i is nonnegative. We shall refer to it as the *grouping principle*.]

Let s be a simple function on a measurable set X such that

$$s = \sum_{i=1}^{p} \alpha_i \chi_{A_i},$$

where

$$i \neq i' \Rightarrow A_i \cap A_{i'} = \emptyset \quad \text{and} \quad \text{each } A_i \text{ is measurable}$$

and

$$\bigcup_{i=1}^{p} A_i = X.$$

Using Problem 2.5.P6 and the *grouping principle*, but not Proposition 3.1.5, show that

$$\sum_{i=1}^{p} \alpha_i \mu(A_i) = \int_X s\, d\mu. \tag{8.8}$$

Hint: On the left-hand side of (8.8), the terms with $A_i = \emptyset$ can be omitted: (*) $\Sigma_{1\le i\le p}\, \alpha_i\mu(A_i) = \Sigma_{A_i\neq\emptyset}\, \alpha_i\mu(A_i)$. Now let $\gamma_1, \gamma_2, \ldots, \gamma_n$ be the distinct elements of $s(X)$ and $N_j = \{i: 1 \le i \le p,\ \alpha_i = \gamma_j \text{ and } A_i \neq \emptyset\}$. Then the N_j are disjoint because the γ_j are distinct. Moreover, by Problem 2.5.P6, we have $\{\gamma_j: 1 \le j \le n\} = \{\alpha_i: 1 \le i \le p, A_i \neq \emptyset\}$ and $\cup_{i\in N_j} A_i = X(s = \gamma_j)$. The first of these two equalities shows that $\cup_{1\le j\le n} N_j = \{i: 1 \le i \le p, A_i \neq \emptyset\}$ and therefore the sets N_j provide the kind of partition of $\{i: 1 \le i \le p,\ A_i \neq \emptyset\}$ needed for applying the *grouping principle*. The second of the equalities leads to: (**) $\Sigma_{i\in N_j}\, \mu(A_i) = \mu(X(s = \gamma_j))$. Therefore

$$\begin{aligned}\sum\nolimits_{A_i \neq \varnothing} \alpha_i \mu(A_i) &= \sum_{j=1}^{n} \left[\sum_{i \in N_j} \alpha_i \mu(A_i) \right] \text{by the } \textit{grouping principle} \\ &= \sum_{j=1}^{n} \left[\sum_{i \in N_j} \gamma_j \mu(A_i) \right] \text{by definition of } N_j \\ &= \sum_{j=1}^{n} \left(\gamma_j \sum_{i \in N_j} \mu(A_i) \right) \\ &= \sum_{j=1}^{n} (\gamma_j \mu(X(s = \gamma_j))) \text{ by } (**) \\ &= \int_X s\, d\mu \text{ by definitionof the integral.}\end{aligned}$$

In conjunction with (*), this leads immediately to the required equality (8.8).

3.1.P9. If in a measure space, $\{A_n\}_{n \geq 1}$ is a sequence of sets of measure 0 and $\{B_n\}_{n \geq 1}$ is a descending sequence of sets of finite measure such that their measure tends to 0, show that $\cap_{n \geq 1}(A_n \cup B_n)$ has measure 0.
Hint: $\cap_{n \geq 1}(A_n \cup B_n) \subseteq (\cup_{n \geq 1} A_n) \cup (\cap_{n \geq 1} B_n)$ and both $\cup_{n \geq 1} A_n$ and $\cap_{n \geq 1} B_n$ have measure 0 (the latter by Proposition 3.1.8).

3.1.P10. Let $\mathcal{F}$ be a σ-algebra of subsets of a set X and f: $X \to \mathbb{R}$ be any function whatsoever. Show that the family $\mathcal{G} = \{A \subseteq \mathbb{R}: f^{-1}(A) \in \mathcal{F}\}$ of subsets of $\mathbb{R}$ is a σ-algebra. Hence show that if f is $\mathcal{F}$-measurable, then $f^{-1}(A) \in \mathcal{F}$ whenever $A \subseteq \mathbb{R}$ is any Borel subset of $\mathbb{R}$.
Hint: Clearly, $\mathbb{R} \in \mathcal{G}$. Since $f^{-1}(A)^c = f^{-1}(A^c)$, we have $A \in \mathcal{G} \Rightarrow A^c \in \mathcal{G}$. If $\{A_j\}_{j \geq 1}$ is a sequence of sets in $\mathcal{G}$, then $f^{-1}(A_j) \in \mathcal{F}$ for each j and hence $f^{-1}(\cup_{j \geq 1} A_j) = \cup_{j \geq 1} f^{-1}(A_j) \in \mathcal{F}$. Thus, $\mathcal{G}$ is a σ-algebra.

Now suppose f is $\mathcal{F}$-measurable. The result of Problem 2.4.P4, which is easily seen to carry over to any set X with a σ-algebra of subsets, shows that every open subset of $\mathbb{R}$ is in $\mathcal{G}$. From what has been proved in the paragraph above, it therefore follows that the entire σ-algebra of Borel subsets of $\mathbb{R}$ is contained in $\mathcal{G}$, which is what we needed to prove.

3.1.P11. Suppose f:$\mathbb{R} \to \mathbb{R}$ and g:$\mathbb{R} \times \mathbb{R} \to \mathbb{R}$ are both Borel measurable. Show that the composition $f \circ g$:$\mathbb{R} \times \mathbb{R} \to \mathbb{R}$ is Borel measurable.
Hint: Let $V = \{x \in \mathbb{R}: x > \alpha\}$, where $\alpha \in \mathbb{R}$ is arbitrary. Then $f^{-1}(V)$ is a Borel subset of $\mathbb{R}$ because f is given to be Borel measurable. Since g is also given to be Borel measurable, it follows by Problem 3.1.P10 that $g^{-1}(f^{-1}(V))$ is a Borel subset of $\mathbb{R} \times \mathbb{R}$. Since $(f \circ g)^{-1}(V) = g^{-1}(f^{-1}(V))$, and $\alpha \in \mathbb{R}$ is arbitrary, we have shown that the composition $f \circ g$:$\mathbb{R} \times \mathbb{R} \to \mathbb{R}$ is Borel measurable.

3.1.P12. Let μ be a measure on the Borel σ-algebra $\mathfrak{B}$ of subsets of $\mathbb{R}$, satisfying the following properties:

(i) $\mu(0, 1) > 0$;
(ii) $\mu(0, 1) < \infty$;
(iii) for $E \in \mathfrak{B}$ and $\varepsilon > 0$, there exists an open set $O \supseteq E$ such that $\mu(O\backslash E) < \varepsilon$;
(iv) $\mu(E + x) = \mu(E)$ for an arbitrary $E \in \mathfrak{B}$ and arbitrary $x \in \mathbb{R}$ (translation invariance).

Show that there exists a positive $\alpha \in \mathbb{R}$ such that $\mu(E) = \alpha \cdot m(E)$ for all $E \in \mathfrak{B}$.
Hint: This can be accomplished in three steps.
Step 1. *Let* μ *be a measure on the Borel sets of* $\mathbb{R}$ *which is translation invariant and let* $\mu(0, 1) < \infty$. *Then* $\mu\{x\} = 0$ *for any real number* x, *and* $\mu(0, 1) = \mu[0, 1) = \mu(0, 1] = \mu[0, 1]$.
Proof: Suppose if possible that there is some real number x_0 such that $\mu\{x_0\} \neq 0$. Then $\mu\{x_0\} > 0$. It follows by translation invariance that $\mu\{x\} = \mu\{x_0\} > 0$ for all x. Now the interval (0, 1) contains infinitely many real numbers and it follows by countable additivity that any $\mu(0, 1) = \infty$, contrary to hypothesis.

It now follows that $\mu(0, 1) = \mu[0, 1) = \mu(0, 1] = \mu[0, 1]$, whereby the proof of Step 1 is complete.
Step 2. *Let* μ *be a measure on the Borel sets of* $\mathbb{R}$ *which is translation invariant and let* $\mu(0, 1) < \infty$. *Then* $\mu(I) = \mu(0, 1)m(I)$ *for any interval* I *and* $\mu(O) = \mu(0, 1)\cdot m(O)$ *for any open set* O.
Proof: By Step 1, $\mu(0, 1) = \mu[0, 1) = \mu(0, 1] = \mu[0, 1]$. Call this nonnegative real number α. Then, for the particular interval $I = [0, 1]$ of length 1, the equality $\mu(I) = \alpha\cdot\cdot m(I)$ holds because $m(I) = 1$. We have to prove that $\mu(I) = \alpha\bullet m(I)$ for any interval I and $\mu(O) = \alpha\cdot\cdot m(O)$ for any open set O. Since an open set is a countable union of disjoint open intervals, we need prove the equality only for intervals.

Since $\mu\{x\} = 0 = m\{x\}$ for any real number x, we need consider only open intervals.

By translation invariance of m (Proposition 2.3.23) and of μ, any two intervals of the same length have the same m-measure as well as μ-measure. Upon combining this with fact that for the particular interval $I = [0, 1]$ of length 1, the equality $\mu(I) = \alpha\cdot\cdot m(I)$ holds, it follows that the μ-measure of an interval of rational length is α times the length, i.e. α times its m-measure

For an arbitrary bounded interval (a, b), we choose a rational strictly increasing sequence $\{a_n\}$ with limit a and a rational strictly decreasing sequence $\{b_n\}$ with limit b. Upon appealing to the continuity property of an arbitrary measure [noted just before Proposition 3.1.8], we arrive at $\mu[a, b] = \lim_{n\to\infty} \mu(a_n, b_n) = \lim_{n\to\infty} \alpha(b_n - a_n) = \alpha(b - a)$. It follows from the fact that the measure of a singleton set is 0 that $\mu(a, b) = \mu[a, b) = \mu(a, b] = \mu[a, b] = \alpha(b - a)$. Since an unbounded interval is a countable disjoint union of bounded intervals, the same equality must hold for unbounded intervals as well.

Since α is defined to mean $\mu(0, 1)$, the proof of Step 2 is now complete.

Step 3. *Let μ be a measure on the Borel sets of $\mathbb{R}$ which is translation invariant and let $0 < \mu(0, 1) < \infty$. Suppose also that, for $E \in \mathfrak{B}$ and $\varepsilon > 0$, there exists an open set $O \supseteq E$ such that $\mu(O \backslash E) < \varepsilon$. Then there exists a positive $\alpha \in \mathbb{R}$ such that $\mu(E) = \alpha \bullet m(E)$ for all $E \in \mathfrak{B}$.*

Proof: By Step 2, the number $\alpha = \mu(0, 1)$ satisfies $\mu(O) = \alpha \bullet m(O)$ for any open set O. Since it is assumed in this Step that $0 < \mu(0, 1)$, we have $\alpha > 0$.

Consider an arbitrary $E \in \mathfrak{B}$ and an arbitrary $\varepsilon > 0$. By hypothesis, there exists an open set $O_\mu \supseteq E$ such that $\mu(O_\mu \backslash E) < \varepsilon$. By Proposition 2.3.24, there exists an open set $O_m \supseteq E$ such that $m(O_m \backslash E) < \varepsilon$. It follows that the open set $O = O_\mu \cap O_m$ satisfies

$$O \supseteq E, \quad m(O \backslash E) < \varepsilon \quad \text{and} \quad \mu(O \backslash E) < \varepsilon.$$

Hence

$$m(O) \geq m(E) = m(O) - m(O \backslash E) \geq m(O) - \varepsilon$$

and

$$\mu(O) \geq \mu(E) = \mu(O) - \mu(O \backslash E) \geq \mu(O) - \varepsilon.$$

The former leads to

$$\alpha \cdot m(O) \geq \alpha \cdot m(E) \geq \alpha \cdot m(O) - \alpha \cdot \varepsilon$$

and the latter leads to

$$\alpha \cdot m(O) \geq \mu(E) \geq \alpha \cdot m(O) - \varepsilon.$$

If $m(O) = \infty$, then we have $\mu(E) = \infty = m(E)$, so that $\mu(E) = \alpha \bullet m(E)$. So, suppose $m(O) < \infty$. Then it follows from the preceding two inequalities that

$$\varepsilon \geq \alpha\, m(E) - \mu(E) \geq -\alpha \cdot \varepsilon.$$

Since $\varepsilon > 0$ is arbitrary, it follows that $\mu(E) = \alpha \bullet m(E)$. Since this has been established for arbitrary $E \in \mathfrak{B}$, the proof of Step 3 is complete.

Remark The result holds with $\mathfrak{M}$ in place of $\mathfrak{B}$.

Problem Set 3.2

3.2.P1. (a) If $\int_X f\,d\mu > 0$, where f is a measurable extended real-valued function on X, show that $X(f > 0)$ has positive measure.
(b) [Needed in Theorem 3.2.20, Problems 3.2.P6 and 3.2.P14 and Theorem 3.3.5] When f is nonnegative, show that if $X(f > 0)$ has positive measure, then $\int_X f\,d\mu > 0$.
(c) When f and g are measurable, $f \geq g$ and $\int_X f\,d\mu = \int_X g\,d\mu$, does it follow that $f = g$ a.e.?
Hint: **(a)** Some simple function s such that $0 \leq s \leq f^+$ must satisfy $\int_X s\,d\mu > 0$. Let $s = \sum_{1 \leq i \leq p} \alpha_i \chi_{A_i}$ canonically, so that $\int s = \sum_{1 \leq i \leq p} \alpha_i \mu(A_i)$. Then some i must satisfy $\alpha_i > 0$ as well as $\mu(A_i) > 0$. For any $x \in A_i$, we have $f^+(x) \geq s(x) = \alpha_i > 0$. Thus $X(f > 0) = X(f^+ > 0) \supseteq A_i$ and hence $\mu(X(f > 0)) \geq \mu(A_i) > 0$.
(b) $f(x) > 0 \Leftrightarrow f(x) > \frac{1}{n}$ for some $n \in \mathbb{N}$. This means $X(f > 0) = \cup_{n\in\mathbb{N}} E_n$, where $E_n = X\left(f > \frac{1}{n}\right)$. Since $E_n \subseteq E_{n+1}$, continuity of measure (see Proposition 3.1.8) implies that $\mu(E_n) > 0$ for some n. When f is nonnegative, we have $f \geq f\chi_A$ for any $A \subseteq X$ and by definition of E_n, we have $f\chi_{E_n} \geq \frac{1}{n}\chi_{E_n}$. So, $f \geq \frac{1}{n}\chi_{E_n}$, which leads to $\int f \geq \frac{1}{n}\mu(E_n) > 0$.
(c) No. But if $\int_X f\,d\mu = \int_X g\,d\mu < \infty$, then it does follow, because we can infer that $\int_X (f - g)d\mu = 0$.

3.2.P2. Let $\{f_n\}_{n \geq 1}$ be the sequence of functions on $X = [0, 2]$ defined by $f_n = \chi_{[0,1]}$ if n is even and $\chi_{(1,2]}$ if n is odd. Find $\int_X (\liminf f_n)d\mu$ and $\liminf \int_X f_n d\mu$.
Hint: For each $x \in X$, the sequence $\{f_n(x)\}_{n \geq 1}$ alternates between 0 and 1, starting with 0 or 1 according as $x \in [0, 1]$ or $x \in (1, 2]$. Therefore $\liminf f_n$ is 0 everywhere and has integral 0. But $\int_X f_n d\mu = 1$ for every n, with the consequence that lim inf $\int_X f_n d\mu = 1$.
3.2.P3. Let L be the class of all Lebesgue integrable functions on X. Define $\rho(f, g)$ to be $\int_X |f - g|d\mu$. Show that $\rho(f, g)$ can be 0 when $f \neq g$ but that ρ has all the other properties of a metric (i.e. ρ is a "pseudometric"). In what way is this different if $X = [a, b]$ and we set $\rho(f, g) = \int_a^b |f(x) - g(x)|dx$ in the class of Riemann integrable functions?
Hint: If f and g differ at only finitely many points, then $f \neq g$. However, $\rho(f, g) = 0$. Also, $0 \leq \int_X |f - g|d\mu = \int_X |g - f|d\mu$, i.e. $0 \leq \rho(f, g) = \rho(g, f)$. For Lebesgue integrable functions f, g, h, we have $|f - h| \leq |f - g| + |g - h|$ almost everywhere on X and hence, by monotonicity and linearity, $\int_X |f - h|d\mu \leq \int_X |f - g|d\mu + \int_X |g - h|d\mu$, i.e. $\rho(f, h) \leq \rho(f, g) + \rho(g, h)$. Not different.
3.2.P4. Give an example of a nonnegative real-valued function on [0, 1] which is unbounded on every subinterval of positive length (so that it cannot have an improper Riemann integral) but has Lebesgue integral 0.

Hint: Let $r_1, r_2, \ldots$ be an enumeration of the rationals in [0, 1] and define $f(x)$ to be 0 if x is irrational and $f(r_k) = k$. Since every interval of positive length contains infinitely many rationals, it must contain r_k with arbitrarily large k; this means that f is unbounded on it. Now consider the sequence of simple functions s_n defined as $s_n(r_k) = k$ if $k \leq n$ and $s_n(x) = 0$ for all other x. This is an increasing sequence converging to f and each function having integral 0. It follows by the Monotone Convergence Theorem 3.2.4 that f has integral 0.

3.2.P5. **(a)** [Needed in 3.2.P8(a) & 3.2.P9] Let Y be a measurable subset of X. If f is a measurable function on X, then its restriction to Y, denoted by $f|_Y$, is a measurable function on Y. Show that the product $f\chi_Y$ (defined on X) has an integral if and only if the restriction $f|_Y$ has an integral, and that when this is so, $\int_X (f\chi_Y) = \int_Y (f|_Y)$. [Note: It is customary to denote this common value by $\int_Y f$ or $\int_Y f d\mu$.]
(b) Suppose f is a measurable extended nonnegative real-valued function on $\mathbb{R}$. Let $E \in \mathfrak{M}$ or $\mathfrak{B}$, and let $\int_E f = 0$. Show that f vanishes a.e. on E, i.e. $\{x \in E: f(x) > 0\}$ has measure zero.
Hint: **(a)** A nonnegative measurable function always has an integral, though it may be ∞. First consider a characteristic function $f = \chi_A$, where $A \subseteq X$ is measurable. Then $f\chi_Y = \chi_{A \cap Y}$ and $\int_X (f\chi_Y) = \mu(A \cap Y)$. On the other hand, $f|_Y$ is the function on Y that is 1 on $A \cap Y$ and 0 elsewhere, so that $\int_X (f|_Y) = \mu(A \cap Y)$. This proves the equality in question for characteristic functions. Since $(\alpha f + \beta g)\chi_Y = \alpha(f\chi_Y) + \beta(g\chi_Y)$ and $(\alpha f + \beta g)|_Y = \alpha(f|_Y) + \beta(g|_Y)$, it follows by the linearity of the Lebesgue integral that the equality holds for all simple functions. Next, let f be a measurable nonnegative extended real-valued function and consider any increasing sequence $\{s_n\}$ of simple functions converging to it pointwise. Then $\{s_n|_Y\}$ is an increasing sequence of simple functions on Y converging to $f|_Y$ pointwise, while $\{s_n\chi_Y\}$ is an increasing sequence of simple functions on X converging to $f\chi_Y$ pointwise. Since the equality in question holds for simple functions, it follows by the Monotone Convergence Theorem 3.2.4 that it holds for f. Finally, consider any measurable f. Observe that

$$f^+\chi_Y = (f\chi_Y)^+, f^-\chi_Y = (f\chi_Y)^- \text{ and } (f^+|_Y) = (f|_Y)^+, (f^-|_Y) = (f|_Y)^-.$$

The equality just established for the nonnegative case shows that $\int_X (f^+\chi_Y) = \int_X (f^+|_Y)$ and $\int_X (f^-\chi_Y) = \int_Y (f^-|_Y)$. In view of the above observation, we therefore have

$$\int_X (f\chi_Y)^+ = \int_Y (f|_Y)^+ \text{ and } \int_X (f\chi_Y)^- = \int_Y (f|_Y)^-.$$

The rest is now clear.

(b) By definition, $\int_E f = \int_{\mathbb{R}} (f\chi_E)$. Now, $\mathbb{R}(f\chi_E > 0) = \{x \in E: f(x) > 0\}$. Apply Problem 3.2.P1(b).

3.2.P6. [Theorem due to Lebesgue] Show that a bounded function $f:[a, b]\to\mathbb{R}$ is Riemann integrable if and only if it is continuous a.e.

Hint: Let ϕ, ψ be as in Proposition 3.2.19. If f is continuous a.e. then by (iv) we have $\phi = \psi$ a.e. and hence by (iii), f is Riemann integrable. Conversely, assume f is Riemann integrable. Then by (iii), we have $\int_{[a, b]}(\psi - \phi)dx = 0$, and by (i) and (ii), we also have $\psi - \phi \geq 0$. By Problem 3.2.P1(b), it follows that $\phi = \psi$ a.e. By (v) of (a), f is continuous at all points except perhaps those where $\phi \neq \psi$ or which belong to one of the partitions P_n. But this exceptional set has measure zero.

3.2.P7. Show that there does not exist a nonnegative Lebesgue integrable function g on [0, 1] such that $g(x) \geq n^2x^n(1 - x)$ everywhere.

Hint: $\lim_{n\to\infty} n^2x^n(1 - x) = 0$ everywhere on [0, 1]. By Theorem 3.2.20,

$$\int_{[0,\,1]} n^2x^n(1-x)dm(x) = \int_0^1 n^2x^n(1-x)dx = \frac{n^2}{(n+1)(n+2)} \to 1$$
$$\neq 0 = \int_{[0,1]} \lim_{n\to\infty} n^2x^n(1-x)dm(x).$$

By the Dominated Convergence Theorem 3.2.16, no such g can exist.

3.2.P8. (a) Let I denote the interval $\left[\frac{1}{(2k+1)\pi}, \frac{1}{2k\pi}\right]$, where $k \in \mathbb{N}$. Using Problem 3.2. P5 and Theorem 3.2.20, show that the Lebesgue integral

$$\int_{[0,1]} \left(\frac{1}{x}\sin\frac{1}{x}\right)\chi_I(x)dm(x)$$

is no less than $\frac{2}{(2k+1)\pi}$.

(b) Hence show that, for $f(x) = \frac{1}{x}\sin\frac{1}{x}$, we have $\int_{[0,1]} f^+dm = \infty$.

(c) Does $\int_{[0,\,1]} f\,dm$ exist?

Hint: **(a)** By Problem 3.2.P5(a), the given Lebesgue integral equals the Lebesgue integral $\int_I \frac{1}{x}\sin\frac{1}{x}dm(x)$, and by Theorem 3.2.20, this in turn equals the Riemann integral

$$\int_{1/(2k+1)\pi}^{1/2k\pi} \frac{1}{x}\sin\frac{1}{x}dx = \int_{1/(2k+1)\pi}^{1/2k\pi} x\left(\frac{1}{x^2}\sin\frac{1}{x}\right)dx.$$

Now, for $x \in \left[\frac{1}{(2k+1)\pi}, \frac{1}{2k\pi}\right]$, we have $2k\pi \leq \frac{1}{x} \leq (2k+1)\pi$ and hence $\sin\frac{1}{x} \geq 0$, so that $x\left(\frac{1}{x^2}\sin\frac{1}{x}\right) \geq \frac{1}{(2k+1)\pi}\left(\frac{1}{x^2}\sin\frac{1}{x}\right)$. Hence

$$\int_{1/(2k+1)\pi}^{1/2k\pi} x\left(\frac{1}{x^2}\sin\frac{1}{x}\right)dx \geq \frac{1}{(2k+1)\pi}\int_{1/(2k+1)\pi}^{1/2k\pi}\left(\frac{1}{x^2}\sin\frac{1}{x}\right)dx.$$

But

$$\begin{aligned}\frac{1}{(2k+1)\pi}\int_{1/(2k+1)\pi}^{1/2k\pi}\left(\frac{1}{x^2}\sin\frac{1}{x}\right)dx &= \frac{1}{(2k+1)\pi}(\cos(2k\pi)-\cos((2k+1)\pi))\\ &= \frac{2}{(2k+1)\pi}.\end{aligned}$$

(b) Let $I_k = \left[\frac{1}{(2k+1)\pi}, \frac{1}{2k\pi}\right]$ for each $k \in \mathbb{N}$. Then $f(x) \geq 0 \Leftrightarrow x \in \cup_{k\in\mathbb{N}} I_k = K$, say. Therefore $f^+ = f^+\chi_K$ and $f^+\chi_{I_k} = f\chi_{I_k}\ \forall k \in \mathbb{N}$. Set $J_n = \cup_{1\leq k\leq n} I_k$. Since the intervals I_k are disjoint, we have $\chi_{J_n} = \Sigma_{1\leq k\leq n}\chi_{I_k}$. Since $J_n \subseteq J_{n+1}$, we also have $f^+\chi_{J_n} \leq f^+\chi_{J_{n+1}}$. Also, $\cup_{n\in\mathbb{N}} J_n = K$; so, $\lim_{n\to\infty}\chi_{J_n} = \chi_K$ and hence $\lim_{n\to\infty} f^+\chi_{J_n} = f^+\chi_K = f^+$. By the Monotone Convergence Theorem 3.2.4, it follows that $\int_{(0,1]} f^+ dm = \lim_{x\to\infty}\int_{(0,1]}(f^+\chi_{J_n})dm$. Now,

$$\begin{aligned}\int_{(0,1]} f^+\left(\sum_{k=1}^{n}\chi_{I_k}\right)dm &= \sum_{k=1}^{n}\int_{(0,1]}(f^+\chi_{I_k})dm\\ &= \sum_{k=1}^{n}\int_{(0,1]}(f\chi_{I_k})dm\\ &\geq \sum_{k=1}^{n}\frac{2}{(2k+1)\pi}\text{ by part(a)}.\end{aligned}$$

But this is the partial sum of a divergent series.

(c) No, because similar reasoning shows that $\int_{(0,1]} f^-\, dm = \infty$.

3.2.P9. Suppose that the function f:(0, 1]→$\mathbb{R}$ is Riemann integrable on $[c,1]$ whenever $0 < c \leq 1$, and that $\lim_{c\to 0}\int_c^1 |f(x)|dx$ and $\lim_{r\to 0}\int_c^1 f(x)dx$ exist. Then by definition, these limits are respectively equal to $\int_0^1|f(x)|dx$ and $\int_0^1 f(x)dx$. Show that f is measurable and that its Lebesgue integral $\int_{(0,1]} f$ is the same as $\int_0^1 f(x)dx$. [*Remark* The corresponding assertion is true when f:$[A, \infty)\to\mathbb{R}$ is Riemann integrable over $[A, B]$ for any finite $B > A$ and the improper integral $\int_A^\infty f(x)dx$ converges absolutely.]

Hint: The restriction of $f\chi_{[1/n,1]}$ to $[1/n, 1]$ is Riemann integrable and hence measurable on $[1/n, 1]$; since $f\chi_{[1/n,1]}$ vanishes outside $[1/n, 1]$, it is measurable on (0, 1].

Now, $f = \lim_{n\to\infty} f\chi_{[1/n,1]}$ and is therefore measurable. Using Problem 3.2.P5(a) and Theorem 3.2.20, we get $\int_{(0,1]}(|f|\chi_{[1/n,1]}) = \int_{[1/n,1]}|f| = \int_{1/n}^{1}|f(x)|dx$. Since $|f| = \lim_{h\to\infty}\left(|f|\chi_{[1/n,1]}\right)$, the Monotone Convergence Theorem 3.2.4 now yields $\int_{(0,1]}|f| = \lim_{n\to\infty}\int_{1/n}^{1}|f(x)|dx = \int_0^1|f(x)|dx$. Thus $|f|$ is Lebesgue integrable. A similar argument using the Dominated Convergence Theorem 3.2.16 shows that $\int_{(0,1]}f = \lim_{n\to\infty}\int_{1/n}^{1}f(x)dx = \int_0^1 f(x)dx$.

3.2.P10. Suppose f: [0, 1]→$\mathbb{R}$ is Lebesgue integrable and is also Riemann integrable on [ε, 1] for every positive ε < 1. Show that $\lim_{\varepsilon\to n}\int_{\varepsilon}^{1}f(x)dx$ exists and equals $\int_{[0,1]}f\,dm$. In other words, the Riemann integral $\int_0^1 f(x)dx$ exists, possibly as an improper integral, and whether improper or not, equals the Lebesgue integral of f over [0, 1]. [*Remark* The corresponding assertion is true when f: [A, ∞)→$\mathbb{R}$ is Lebesgue integrable over [A, ∞).]

Hint: Let $\{a_n\}_{n\geq 1}$ be a sequence of positive real numbers less than 1 such that $\lim_{n\to\infty} a_n = 0$. Denote $f\chi_{[a_n,1]}$ by f_n. By Problem 3.2.P5 and Theorem 3.2.20, we get: (*) $\int_{[0,1]}f_n = \int_{[a_n,1]} = \int_{a_n}^{1}f(x)dx$. Also, $|f_n| \leq |f|$ and f is given to be Lebesgue integrable. By the Dominated Convergence Theorem 3.2.16, it follows that $\int_{[0,1]}f_n \to \int_{[0,1]}f$ as $n\to\infty$. In view of (*), this is precisely what was required to be proved.

3.2.P11. Let f_n: [0, 1]→$\mathbb{R}$ be defined by $f_n(x) = nx/(1+n^{10}x^{10})$. Find the pointwise limit f of $\{f_n\}_{n\geq 1}$ and show that the convergence is neither uniform nor monotone. Also, establish the convergence of Riemann integrals $\int_0^1 f_n(x)dx \to \int_0^1 f(x)dx$.

Hint: It is elementary that f is 0 everywhere. Since $f\left(\frac{1}{n}\right) = \frac{1}{2}\forall n$, convergence is not uniform. Also, $f_n(x) \geq f(x)$ on [0, 1] and therefore the convergence, if monotone, must be decreasing. But this is not so, because $f_{n+1}(x) > f_n(x)$ when $x = 1/n(n+1)$. It is easy to verify that $0 \leq f_n(x) < 1$ everywhere. By the Dominated Convergence Theorem 3.2.16, $\int_{[0,1]}f_n(x)dm(x) \to \int_{[0,1]}f(x)dm(x)$. But all functions involved are continuous and therefore, by Theorem 3.2.20, the Lebesgue integrals are the same as Riemann integrals.

3.2.P12. Let f: [0, 1] × [0, 1]→$\mathbb{R}$ satisfy $|f(x, y)| \leq 1$ everywhere. Suppose that for each fixed $x \in [0, 1]$, $f(x, y)$ is a measurable function of y, and also that for each fixed $y \in [0, 1]$, $f(x, y)$ is a continuous function of x. If g: [0, 1]→$\mathbb{R}$ is defined as $g(x) = \int_{[0,1]}f(x, y)dm(y)$, show that g is continuous.

Hint: Let $\{x_n\}_{n\geq 1}$ be any sequence in [0, 1] converging to $x_0 \in [0, 1]$. We have to show $g(x_n) \to g(x_0)$. Consider the sequence of functions f_n on [0, 1] defined by $f_n(y) = f(x_n, y)$. According to the hypothesis, the sequence $\{f_n\}$ converges pointwise to f_0, where $f_0(y) = f(x_0, y)$, and $|f_n(y)| \leq 1$, $|f_0(y)| \leq 1$ everywhere. By the Dominated Convergence Theorem 3.2.16,

$$\int_{[0,1]} f_n(y)dm(y) \to \int_{[0,1]} f_0(y)dm(y).$$

This says precisely that $g(x_n) \to g(x_0)$.
Note The improper Riemann integral $\int_0^\infty g(u)du$ is the sum of limits of Riemann integrals

$$\lim_{a\to 0} \int_a^c g(u)du + \lim_{b\to\infty} \int_c^b g(u)du,$$

where c can be taken as any positive real number without affecting the value of the sum. The improper integral is finite if and only if each of the limits is finite. This will be needed in the next problem.
3.2.P13. Let f be a measurable extended nonnegative real-valued function on X. Define ϕ_f: $[0, \infty) \to [0, \infty]$ as $\phi_f(u) = \mu(X(f > u))$. Show that

(a) ϕ_f is a decreasing function;
(b) $f \leq g \Rightarrow \phi_f \leq \phi_g$;
(c) For an increasing sequence $\{h_n\}$ of measurable functions converging to h, we have $\phi_{h_n} \to \phi_h$.
(d) $\int_0^\infty \phi_f(u)du$ is equal to the integral $\int_X f\, d\mu$.
(e) $\int_0^\infty \mu(X(f > u))pu^{p-1}du = \int_X f^p d\mu$, where $0 < p < \infty$.

Hint: **(a)** Let $0 \leq u < v$. Then $f(x) > v \Rightarrow f(x) > u$, which means $X(f > v) \subseteq X(f > u)$, which implies $\mu(X(f > u)) \geq \mu(X(f > v))$, i.e. $\phi_f(u) \geq \phi_f(v)$.
(b) $f(x) > u \Rightarrow g(x) > u$. So, $X(f > u) \subseteq X(g > u)$, which implies $\mu(X(f > u)) \leq \mu(X(g > u))$, i.e. $\phi_f(u) \leq \phi_g(u)$.
(c) By definition, $\phi_{h_n} v) = \mu(X(h_n > v)) \to \mu(X(h > v)) = \phi_h(v)$. Indeed, from the hypothesis that $\{h_n\}$ is an increasing sequence of measurable functions converging to h, it follows that the sets $X(h_n > v)$ form an increasing sequence with union $X(h > v)$.
(d) Let $\{s_n\}_{n\geq 1}$ be a sequence of simple functions increasing to f. By Problem 3.1. P6, $\int_X s_n d\mu = \int_0^\infty \phi_{s_n}(u)du$ and by the Monotone Convergence Theorem 3.2.4, $\int_X s_n d\mu \to \int_X f\, d\mu$. So, it is enough to show that $\int_0^\infty \phi_{s_n}(u)du \to \int_0^\infty \phi_f(u)du$. In view of (a), (b) and (c), this is a consequence of the following result about improper Riemann type integrals: *If* $\{g_n\}_{n\geq 1}$ *is an increasing sequence of nonnegative extended real-valued decreasing functions on* $[0, \infty)$, *so that* $g = \lim_{n\to\infty} g_n$ *exists everywhere on the domain and is decreasing there, then* $\int_0^\infty g_n(u)du \to \int_0^\infty g(u)du$. We give the proof of this assertion below.

Suppose that $g(a) = \infty$ for some $a > 0$. Since g is decreasing, we have $g = \infty$ on $[0, a]$. Therefore $\int_0^\infty g(u)du \geq \int_0^a g(u)du = \infty$. So we need to show that $\int_0^\infty g_n(u)du \to \infty$. To this end, consider any $K > 0$. Since $g_n(a)$ increases to

$g(a) = \infty$, there exists a natural number N such that $g_n(a) > \frac{K}{a}$ for $n \geq N$. For such n, we have $g_n(u) > \frac{K}{a}$ for $0 \leq u \leq a$ and hence $\int_0^\infty g_n(u)du \geq \int_0^a g_n(u)du > a\frac{K}{a} = K$. This proves that $\int_0^\infty g_n(u)du \to \infty = \int_0^\infty g(u)du$ if $g(a) = \infty$ for some $a > 0$.

In view of what has just been proved, we may assume that $g(a) < \infty$ for every $a > 0$. This permits us when $0 < a < b < \infty$ to form differences such as $g_n(a) - g_n(b)$ and $g(a) - g(b)$, and we know that the integrals $\int_a^b g_n(u)du$ and $\int_a^b g(u)du$ are finite.

To begin with, we shall show for $0 < a < b < \infty$ that

$$\int_a^b g_n(u)du \to \int_a^b g(u)du. \tag{8.9}$$

Let $\varepsilon > 0$ be arbitrary. Since $g_n(a) - g_n(b) \to g(a) - g(b)$, there exists an $M > 0$ such that $0 \leq g(a) - g(b) < M$ as well as $0 \leq g_n(a) - g_n(b) < M$ for all n. Let P: $a = u_0 < u_1 < \cdots < u_p = b$ be a partition of $[a, b]$ that satisfies: $0 < u_i - u_{i-1} < \frac{\varepsilon}{M}$ for each i. Then for any decreasing G: $[a, b] \to [0,\infty)$ with $0 \leq G(a) - G(b) < M$, we have

$$\begin{aligned}
0 \leq U(G, P) - L(G, P) &= \sum_{i=1}^{p} (G(u_{i-1}) - G(u_i))(u_i - u_{i-1}) \\
&\leq \frac{\varepsilon}{M}\sum_{i=1}^{p} (G(u_{i-1}) - G(u_i)) \\
&= \frac{\varepsilon}{M}(G(a) - G(b)) < \frac{\varepsilon}{M}M = \varepsilon.
\end{aligned}$$

In particular,

$$0 \leq U(G, P) - \int_a^b G(u)du < \varepsilon \tag{8.10}$$

and this holds for $G = g$ as well as for $G = g_n$ (any n).

Since $g_n(u_i) \to g(u_i)$ for each of the finitely many points of P, there exists an $N \in \mathbb{N}$ such that $0 \leq g(u_i) - g_n(u_i) < \frac{\varepsilon}{b-a}$ for $0 \leq i \leq p$ whenever $n \geq N$. For such n, we have

$$\begin{aligned}
0 \leq U(g, P) - U(g_n, P) &= \sum_{i=1}^{p} (g(u_{i-1}) - g_n(u_{i-1}))(u_i - u_{i-1}) \\
&< \frac{\varepsilon}{(b-a)}\sum_{i=1}^{P} (u_i - u_{i-1}) = \varepsilon.
\end{aligned}$$

Since (8.10) holds for $G = g$ as well as for $G = g_n$ (any n), it now follows that

$$\left|\int_a^b g_n(u)du - \int_a^b g(u)du\right| < 3\varepsilon \quad \text{whenever } n \geq N.$$

This proves (8.9).

To prove the same convergence for the corresponding improper integrals (as required), first assume that

$$\int_0^\infty g(u)du < \infty.$$

Then $\lim\limits_{a\to 0} \int_a^c g(u)du$ and $\lim\limits_{b\to\infty} \int_c^b g(u)du$ are both finite. Consider any $\eta > 0$. In view of the finiteness of the aforementioned limits, there exist a and b such that $0 < a < b < \infty$ and

$$0 \leq \int_0^a g(u)du < \frac{\eta}{4}, \quad 0 \leq \int_b^\infty g(u)du < \frac{\eta}{4}.$$

Since $0 \leq g_n \leq g$ for all n, the above two inequalities are satisfied by every g_n as well. Now an improper integral from 0 to ∞ is a sum of three integrals in the obvious way, of which two are governed by the inequalities just mentioned. It follows that

$$\left|\int_0^\infty g_n(u)du - \int_0^\infty g(u)du\right| \leq \eta + \left|\int_a^b g_n(u)du - \int_a^b g(u)du\right|.$$

By (8.9), the second term on the right-hand side approaches 0 as $n \to \infty$. Since $\int_0^\infty g_n(u)du$ is an increasing sequence and must therefore have a limit, it follows from the above inequality that

$$\left|\lim_{n\to\infty} \int_0^\infty g_n(u)du - \int_0^\infty g(u)du\right| \leq \eta.$$

This holds for all $\eta > 0$ and hence $\int_0^\infty g_n(u)du \to \int_0^\infty g(u)du$, on the assumption that $\int_0^\infty g(u)du < \infty$.

It remains to prove that the same convergence obtains even if $\int_0^\infty g(u)du = \infty$. In this situation, for any $K > 0$, there exist a and b such that $0 < a < b < \infty$ and $\int_a^b g(u)du > K$. Since $\int_a^b g_n(u)du \to \int_a^b g(u)du$ by (8.9), there exists an $N \in \mathbb{N}$ such that $\int_a^b g_n(u)du > K$ whenever $n \geq N$. Now, $\int_0^\infty g_n(u)du \geq \int_a^b g_n(u)du$ because each g_n is nonnegative. So, $\int_0^\infty g_n(u)du > K$ whenever $n \geq N$. Thus $\int_0^\infty g_n(u)du \to \infty = g(u)du$ as $n \to \infty$.

(e) If $\mu(X(f^p > u)) = \infty$ for some finite positive u, then both sides of the equality are ∞. Suppose this does not happen. Then from part (d), we have

$\int_X f^p d\mu = \int_0^\infty \mu(X(f^p > u))du$. Now, for $0 < a < b < \infty$, we have $\int_a^b \mu(X(f^p > u))du = \int_a^b \mu(X(f > u^{1/p}))du$. Since this is a Riemann integral, we may use the substitution $t = u^{1/p}$ (see Proposition 1.7.3) to obtain $\int_a^b \mu(X(f > u^{1/p}))du = \int_{a^{1/p}}^{b^{1/p}} \mu(X(f > t))pt^{p-1}dt$. We arrive at the desired result by taking limits as $a \to 0$ and $b \to \infty$.

3.2.P14. **(a)** Let f be an integrable function on a set X such that, for every measurable $E \subseteq X$, we have $\int_E f\,d\mu = 0$ (which means $\int_X f\chi_E\,d\mu = 0$, as explained in 3.2.P5). Show that $f = 0$ a.e.

(b) Let f be a Lebesgue integrable function on $[a, b]$. If $\int_{[a,x]} f\,dm = 0$ for all $x \in [a, b]$, show that $f = 0$ a.e. on $[a, b]$.

Hint: **(a)** Let $E = X(f \geq 0)$. Then $f^+ = f\chi_E$ and it follows from the hypothesis that $\int_X f^+ d\mu = 0$. From Problem 3.2.P1(b), it follows that $f^+ = 0$ a.e. A similar argument shows that $f^- = 0$ a.e.

(b) Suppose not. Then there exists a measurable set $E \subseteq [a, b]$ with $m(E) > 0$ such that $f(t) \neq 0$ for $t \in E$. Obviously, we may suppose $f > 0$ on E. By Proposition 2.3.24 [the part that says $(\alpha) \Rightarrow (\delta)$], there exists a closed set $F \subseteq E$ with $m(F) > 0$. Set $A = [a, b]\backslash F$. Then A is open and $A = \cup_i I_i$, a countable union of disjoint intervals I_i. Now,

$$0 = \int_{[a,b]} f\,dm = \int_{A \cup F} f\,dm = \int_A f\,dm + \int_F f\,dm,$$

which implies $\int_A f\,dm = -\int_F f\,dm \neq 0$, since $f > 0$ on $E \supseteq F$. Therefore $\int_A f\,dm \neq 0$. Consequently, there exists an i such that $\int_{I_i} f\,dm \neq 0$. This implies that either $\int_{[a,\alpha]} f\,dm \neq 0$ or $\int_{[a,\beta]} f\,dm \neq 0$, where α and β are the endpoints of I_i. This contradicts the hypothesis.

3.2.P15. Let f be an integrable function on a set E. Show that for every $\varepsilon > 0$, there exists a $\delta > 0$ such that whenever $A \subseteq E$ with $\mu(A) < \delta$, we have $|\int_A f d\mu| < \varepsilon$ (which means $|\int_E (f\chi_A) d\mu| < \varepsilon$, as explained in Problem 3.2.P5).

Hint: Since $|\int_A f| \leq \int_A |f|$, it is sufficient to consider $f \geq 0$. Suppose, if possible, that there exists some $\eta > 0$ such that we can find subsets $A_n \subseteq E$ for which $\mu(A_n) < 2^{-n}$ and $\int_{A_n} f \geq \eta$, i.e. $\int_E (f\,\chi_{A_n}) \geq \eta$. Let $g_n = f\,\chi_{A_n}$. Then $g_n \to 0$ except on the set $\cap_{n \geq 1}(\cup_{i \geq n} A_i)$. Since $\mu(\cup_{i \geq n} A_i) < 2^{-n+1}$, we have $g_n \to 0$ a.e. Let $f_n = f - g_n = f(1 - \chi_{A_n})$. Then $\{f_n\}_{n \geq 1}$ is a sequence of measurable nonnegative functions and $f_n \to f$ a.e. Hence by Fatou's Lemma 3.2.8, $\int_E f \leq \liminf \int_E f_n = \int_E f - \limsup \int_E g_n \leq \int_E f - \eta$. This inequality cannot hold unless $\int_E f = \infty$, contrary to the hypothesis.

3.2.P16. Let f be an integrable function on $[a, b]$. If $F(x) = \int_{[a,x]} f\,dm$, prove that F is continuous on $[a, b]$. If f is an integrable function on $[a, \infty)$, is it true that F is uniformly continuous on $[a, \infty)$?

Hint:

$$F(x+h) - F(x) = \int_{[a,b]} \left(f\chi_{[a,x+h]}\right) - \int_{[a,b]} \left(f\chi_{[a,x]}\right)$$
$$= \int_{[a,b]} f \cdot \left(\chi_{[a,x+h]} - \chi_{[a,x]}\right) = \int_{[a,b]} \left(f\chi_{(x,x+h]}\right) = \int_{[x,x+h]} f,$$

assuming $h > 0$ and $x < b$. Now, let $\varepsilon > 0$ be given. By 3.2.P15, there exists a $\delta > 0$ such that $|\int_{[x,x+h]} f| < \varepsilon$ provided $h = m((x, x+h)) < \delta$, i.e. such that $0 < h < \delta \Rightarrow |F(x+h) - F(x)|$. This proves right continuity at any $x \in [a, b)$. Left continuity at an arbitrary $x \in (a, b]$ is proved analogously. Since δ does not depend on x, the preceding argument proves uniform continuity and is valid even if $[a, b]$ is replaced by $[a, \infty)$.

3.2.P17. Let $\{f_n\}_{n\geq 1}$ be a sequence of measurable nonnegative functions on $\mathbb{R}$ such that $f_n \to f$ a.e. and suppose that $\int_{\mathbb{R}} f_n dm \to \int_{\mathbb{R}} f\, dm < \infty$. Show that for each measurable set $E \subseteq \mathbb{R}$, $\int_E f_n dm \to \int_E f\, dm$. [The result also holds on a general measure space.]

Hint: Suppose there is a measurable subset A for which the convergence fails. Then there exists a $\sigma > 0$ for which there are infinitely many n such that $\int_A f_n$ lies outside the interval $(\int_A f - 2\sigma, \int_A f + 2\sigma)$. If $\int_A f_n \leq \int_A f - 2\sigma$ for infinitely many n, then there is a subsequence for which the inequality holds. However, this contradicts Fatou's Lemma 3.2.8 and therefore the inequality can hold only for finitely many n. It must therefore be the case that $\int_A f_n \geq \int_A f + 2\sigma$ for infinitely many n. Again there must be a subsequence for which the inequality holds. But by hypothesis, $|\int_{\mathbb{R}} f_n - \int_{\mathbb{R}} f| < \sigma$, and hence $\int_{\mathbb{R}} f_n < \int_{\mathbb{R}} f + \sigma$, for all sufficiently large n, in particular for all terms of the last mentioned subsequence. In terms of $B = \mathbb{R} \backslash A$, this means that the subsequence satisfies $\int_B f_n < \int_B f - \sigma$. This leads to a contradiction to Fatou's Lemma 3.2.8 again.

3.2.P18. Let f be a measurable nonnegative function on $\mathbb{R}$ and E be a measurable subset of finite measure. Show that

(a) $\int_E f = \lim_{N\to\infty} \int_E f_N$, where $f_N(x) = \min\{f(x), N\}$;
(b) $\int_{\mathbb{R}} f = \lim_{N\to\infty} \int_{[-N,N]} f$.

Hint: **(a)** $\{f_N\}_{N\geq 1}$ is an increasing sequence of measurable nonnegative functions converging to f pointwise. The required conclusion follows by the Monotone Convergence Theorem 3.2.4.

(b) Set $f_N(x) = f(x)$ if $|x| \leq N$ and 0 otherwise. Then $\{f_N\}_{N\geq 1}$ is an increasing sequence of measurable nonnegative functions converging to f pointwise. Again, the required conclusion follows by the Monotone Convergence Theorem 3.2.4.

3.2.P19. Let $E \subseteq \mathbb{R}$ be measurable and g an integrable function on E. Suppose $\{f_n\}_{n\geq 1}$ is a sequence of measurable functions on E such that $|f_n| \leq g$ a.e. Show that

$$\int_E \liminf f_n \leq \liminf \int_E f_n \leq \limsup \int_E f_n \leq \int_E \limsup f_n.$$

Hint: $\liminf f_n = \lim_n(\inf_{k \geq n} f_k)$. Define the sequence $\{p_n\}_{n \geq 1}$ by the following rule:

$$p_n(x) = \inf\{f_n(x), f_{n+1}(x), \ldots\}.$$

Observe that $\{p_n\}_{n \geq 1}$ is an increasing sequence of measurable functions such that $|p_n| \leq g$ a.e. and that $\lim_{n \to \infty} p_n = \liminf f_n$ a.e. The sequence $g + p_n$ is therefore an increasing sequence of measurable nonnegative functions. By the Monotone Convergence Theorem 3.2.4, we have $\lim_{n \to \infty} \int_E (g + p_n) = \int_E \lim_{n \to \infty} (g + p_n)$. Since g is integrable, this leads to $\lim_{n \to \infty} \int_E p_n = \int_E (\lim_{n \to \infty} p_n) = \int_E (\liminf f_n)$. Since $f_n \geq p_n$ everywhere, we have $\liminf \int_E f_n \geq \lim_{n \to \infty} \int_E p_n = \int_E (\liminf f_n)$. This proves the first of the three inequalities in question. The third follows from the first by considering $-f_n$ and the second is true for any sequence.

3.2.P20(a) Let $\{f_n\}_{n \geq 1}$ and $\{g_n\}_{n \geq 1}$ be sequences of measurable functions on X such that $|f_n| \leq g_n$ for every n. Let f and g be measurable functions such that $\lim_n f_n = f$ a.e. and $\lim_n g_n = g$ a.e. If $\lim_n \int_X g_n = \int_X g < \infty$, show that

$$\lim_{n \to \infty} \int_X f_n = \int_X f.$$

(b) Let $\{f_n\}_{n \geq 1}$ be a sequence of integrable functions on X such that $f_n \to f$ a.e., where f is also integrable. Show that $\int_X |f_n - f| d\mu \to 0$ if and only if $\int_X |f_n| d\mu \to \int_X |f| d\mu$.

Hint: **(a)** Since $|f_n| \leq g_n$, it follows first, that $g_n \pm f_n$ are both nonnegative and second, that $|f| \leq g$, so that, by Fatou's Lemma and the hypothesis, $\int_X (g \pm f) \leq \liminf \int_X (g_n \pm f_n)$. On using the two facts that (i) $\lim_n \int_X g_n$ exists and (ii) $\int_X g < \infty$ in that order, we obtain

$$\int_X f \leq \liminf \int_X f_n \leq \limsup \int_X f_n \leq \int_X f.$$

(b) The "only if" part follows from the simple inequality

$$||f_n| - |f|| \leq |f_n - f|.$$

For the "if" part, observe that

$$|f_n - f| \leq |f_n| + |f|.$$

Since $|f_n - f| \to 0$, $|f_n| + |f| \to 2|f|$ as $n \to \infty$, and $\int_X |f| < \infty$, the conclusion follows from the result of part (a).

3.2.P21. Let the function $f: I_1 \times I_2 \to \mathbb{R}$, where I_1 and I_2 are intervals, satisfy the following conditions:

(i) The function $f(x, y)$ is measurable for each fixed $y \in I_2$ and $f(x, y_0)$ is Lebesgue integrable on I_1 for some $y_0 \in I_2$;

(ii) $\frac{\partial}{\partial y} f(x,y)$ exists for each interior point $(x, y) \in I_1 \times I_2$;

(iii) There exists a nonnegative integrable function g on I_1 such that $\left|\frac{\partial}{\partial y} f(x,y)\right| \leq g(x)$ for each interior point $(x, y) \in I_1 \times I_2$.

Show that the Lebesgue integral $\int_{I_1} f(x, y)dx$ exists for every $y \in I_2$ and that the function F on I_2 given by

$$F(y) = \int_{I_1} f(x,y)dx$$

is differentiable at each interior point of I_2, the derivative being

$$F'(y) = \int_{I_1} \frac{\partial}{\partial y} f(x,y)dx.$$

Hint: Fix $y \in I_2$. By the Mean Value Theorem, we have $f(x, y) - f(x, y_0) = (y - y_0) \frac{\partial}{\partial y} f(x,y')$, where y' lies between y_0 and y. By (iii), it follows that $|f(x, y)| \leq |f(x, y_0)| + |y - y_0| g(x)$. Hence by (i) and (iii), $f(x, y)$ is integrable on I_1, i.e. the Lebesgue integral $\int_{I_1} f(x, y)dx$ exists for every $y \in I_2$. For differentiability, consider any sequence $\{y_n\}_{n\geq 1}$ in I_2 such that $y_n \neq y$ $(\forall\, n)$ and $y_n \to y$. The functions $h_n(x) = (f(x, y_n) - f(x, y))/(y_n - y)$ are integrable on I_1 and $h_n(x) \to \frac{\partial}{\partial y} f(x,y)$ pointwise. Moreover, $|h_n(x)| = \left|\frac{\partial}{\partial y} f(x,y'')\right| \leq g(x)$ by the Mean Value Theorem and hypothesis (iii). Now the Dominated Convergence Theorem 3.2.16 implies that $\lim_n \int_{I_1} h_n = \int_{I_1} \lim_n h_n = \int_{I_1} \frac{\partial}{\partial y} f(x, y)dx$.

But

$$\int_{I_1} h_n = \int_{I_1} \frac{f(x,y_n) - f(x,y)}{y_n - y} dx = \frac{F(y_n) - F(y)}{y_n - y} \to F'(y).$$

3.2.P22. Use the result of Problem 3.2.P21 to prove that $\int_0^\infty x^n \exp(-x)dx = n!$

Hint: By the remark in Problem 3.2.P9, $\int_0^\infty x^n \exp(-x)dx = \int_{[0,\infty)} x^n \exp(-x)dx$ and $\exp(-x/2)$ is Lebesgue integrable on $[0, \infty)$. Consider the function $f(x, y) = \exp(-xy)$

on $I_1 \times I_2$, where $I_1 = [0, \infty)$ and $I_2 = (0, \alpha)$, $\alpha > 1$. The function f is continuous in x for each fixed $y \in I_2$ and is therefore a measurable function of x. Moreover, as noted above, $f\left(x, \frac{1}{2}\right) = \exp(-x/2)$ is Lebesgue integrable on I_1. For each n, $\frac{\partial^n}{\partial y^n} f(x, y) = (-1)^n x^n \exp(-xy)$ exists at each interior point $(x, y) \in I_1 \times I_2$. Observe that, firstly, $0 \leq x^n \exp(-xy) \leq x^n \exp(-xy_0) < \exp\left(-\frac{1}{2}xy_0\right)$, where $y \geq y_0 > 0$, for all sufficiently large x, and secondly, $\exp\left(-\frac{1}{2}xy_0\right)$ is integrable. [The last inequality follows from the fact that $x^n \exp(-xy_0) \exp\left(\frac{1}{2}xy_0\right) = x^n \exp\left(-\frac{1}{2}xy_0\right) < 1$, as the exponential tends to ∞ faster than any power of x.] So, we have

$$F^{(n)}(y) = (-1)^n \int_{[0,\infty]} x^n \exp(-xy) dx, \tag{8.11}$$

where $F(y) = \int_{[0,\infty]} \exp(-xy) dx$, using Problem 3.2.P21. Also, $F(y) = 1/y$ and therefore

$$F^{(n)}(y) = (-1)^n n!/y^{n+1}. \tag{8.12}$$

From (8.11) and (8.12), we obtain $\int_{[0,\infty]} x^n \exp(-xy) dm(x) = n!/y^{n+1}$. Substitute $y = 1$.

[*Remark* This can be more easily worked out using induction and the Riemann theory.]

3.2.P23. Prove that the improper integral $\int_0^\infty e^{-xy} \frac{\sin x}{x} dx$, where it is understood that $\frac{\sin x}{x}$ is to be replaced by 1 when $x = 0$, converges absolutely for $y > 0$, and evaluate it.

Hint: $\left|\frac{\sin x}{x}\right|$ is bounded above on $[0, \infty)$; so let K be an upper bound for it. Then $\left|e^{-xy} \frac{\sin x}{x}\right| \leq K e^{-xy}$; now, for each $y > 0$, e^{-xy} is Lebesgue integrable (as a function of x) on $[0, \infty)$, and therefore so is $e^{-xy} \frac{\sin x}{x}$. By the remark in Problem 3.2.P10, it follows that the given improper integral converges absolutely for $y > 0$ and agrees with the Lebesgue integral $\int_{[0,\infty)} e^{-xy} \frac{\sin x}{x} dm(x) = F(y)$, say. Let $I_1 = [0, \infty)$ and $I_2 = (\alpha, \infty)$, where α is any positive real number. Then $f(x, y) = e^{-xy} \frac{\sin x}{x}$ satisfies the hypotheses of Problem 3.2.P21 with $g(x) = e^{-x\alpha}$. Therefore $F'(y) = -\int_{[0,\infty]} e^{-xy} \sin x \, dm(x)$. By Theorem 3.2.20, the integrals involved are Riemann integrals, permitting integration by parts. Doing this twice, for any $y \in I_2$ and $b > 0$, we have

$$\int_{[0,b]} e^{-xy} \sin x \, dm(x) = \frac{e^{-by}(-y \sin b - \cos b)}{1 + y^2} + \frac{1}{1 + y^2}.$$

Letting $b \to \infty$, we find that $\int_0^\infty e^{-xy} \sin x dx = \frac{1}{1+y^2}$, keeping in mind that $y > 0$. By the remark in Problem 3.2.P9, this integral equals the corresponding Lebesgue integral. Hence $F'(y) = -\frac{1}{1+y^2}$ for $y \in I_2$, i.e. for $y > \alpha$. Since $\alpha > 0$ is arbitrary, the equality holds for all $y > 0$. Upon integrating both sides, we find that $F(y) - F(b) = \arctan b - \arctan y$. Therefore, for any sequence $b_n \geq 1$ tending to ∞, it follows that

$$F(y) - \lim_{n\to\infty} F(b_n) = \frac{\pi}{2} - \arctan y. \tag{8.13}$$

But $\lim_n F(b_n) = 0$, because on $(0, \infty)$, $e^{-xb_n} \frac{\sin x}{x} \to 0$ pointwise, $\left|e^{-xb_n} \frac{\sin x}{x}\right| \leq Ke^{-x}$, where K is an upper bound of $\left|\frac{\sin x}{x}\right|$, and Ke^{-x} is integrable, so that the Dominated Convergence Theorem 3.2.16 is applicable. Therefore (8.13) shows that $F(y) = (\pi/2) - \arctan y$. It may be noted that the argument here does not work when $y = 0$, which is the case in the next problem.

3.2.P24. Assuming that $\int_0^\infty e^{-xy} \frac{\sin x}{x} dx$ converges for $y = 0$, i.e. $\int_0^\infty \frac{\sin x}{x} dx$ converges, find its value by using the Dominated Convergence Theorem 3.2.16. It is understood that $\frac{\sin x}{x}$ is to be replaced by 1 when $x = 0$. [In Problem 4.2.P10, we shall obtain the value without employing Lebesgue integration.] Show that $\frac{\sin x}{x}$ is nevertheless not Lebesgue integrable over $[1, \infty)$.

Hint: Since the integral is assumed to exist, it is enough to compute $\lim_n \int_0^n \frac{\sin x}{x} dx$. Consider the functions f_n on $[0, \infty)$ given by $f_n(y) = \int_0^n e^{-xy} \frac{\sin x}{x} dx$. Then

$$\int_0^\infty \frac{\sin x}{x} dx = \lim_{n\to\infty} f_n(0). \tag{8.14}$$

By Theorem 3.2.20, the Riemann integral that defines $f_n(y)$ equals the corresponding Lebesgue integral $\int_{[0,n]} e^{-xy} \frac{\sin x}{x} dm(x)$. Since $\frac{\partial}{\partial y}\left(e^{-xy} \frac{\sin x}{x}\right) = -e^{-xy} \sin x$ and e^{-xy} is Lebesgue integrable over $[0, n]$, it follows by using Problem 3.2.P21 with $g(x) = 1$ on $[0, n]$ that $f_n'(y) = -\int_{[0,n]} e^{-xy} \sin x \, dm(x)$. By making use of the Dominated Convergence Theorem 3.2.16, we can show that f_n' is continuous: indeed, $-f_n'(y+h) + f_n'(y) = \int_{[0,n]} \left(e^{-xy} \sin x(e^{-xh} - 1)\right) dm(x) \to 0$ as $h \to 0$ because the integrand is bounded in absolute value by e^n for $|h| \leq 1$ and tends to 0 pointwise as $h \to 0\left[\text{ Reason: } x \in [0, n], |h| \leq 1 \Rightarrow |xh| \leq n \Rightarrow 0 < e^{-xh} \leq e^n\right]$. Being continuous, f_n' has a Riemann integral, and by Theorem 3.2.20,

$$\int_{[0,n]} f_n'(y) dm(y) = \int_0^n f_n'(y) dy = f_n(n) - f_n(0). \tag{8.15}$$

Since $|f_n(n)| \leq \int_0^n Ke^{-nx}dx \leq K/n$, where K is an upper bound of $\left|\frac{\sin x}{x}\right|$ on $[0, \infty)$, we have $\lim_n f_n(n) = 0$. Using this limit in (8.15) leads to

$$-\lim_{n\to\infty} \int_0^n f_n'(y)dy = \lim_{n\to\infty} f_n(0). \tag{8.16}$$

Using Theorem 3.2.20 again and then integrating by parts twice, we have

$$-f_n'(y) = \int_{[0,n]} e^{-xy} \sin x \, dm(x) = \int_0^n e^{-xy} \sin x dx = \frac{1 - e^{-ny}(y \sin n + \cos n)}{1+y^2},$$

so that $-\chi_{[0,n]}(y)f_n'(y) \to \frac{1}{1+y^2}$ as $n \to \infty$.
Now,

$$\left|\chi_{[0,n]}(y)f_n'(y)\right| \leq \frac{1+e^{-y}(1+y)}{1+y^2} \leq \frac{1}{1+y^2} + e^{-y}\frac{1+y}{1+y^2} \leq \frac{1}{1+y^2} + \frac{3}{2}e^{-y},$$

which is Lebesgue integrable over $[0, \infty)$ in view of Theorem 3.2.20 and the Remark in Problem 3.2.P9. On using the Dominated Convergence Theorem 3.2.16, we get

$$\lim_{n\to\infty} \int_{[0,n]} (-f_n'(y))dm(y) = \int_{[0,\infty)} \frac{1}{1+y^2} dm(y).$$

Together with Theorem 3.2.20 and the Remark in Problem 3.2.P10, the above equality yields

$$\lim_{n\to\infty} \int_0^n (-f_n'(y))dy = \int_0^\infty \frac{1}{1+y^2} dy = \frac{\pi}{2}. \tag{8.17}$$

By (8.14), (8.16) and (8.17), we have $\int_0^\infty \frac{\sin x}{x} dx = \pi/2$.
If $\frac{\sin x}{x}$ were Lebesgue integrable over $[1, \infty)$, then by the remark in Problem 3.2. P10, we would have

$$\int_{[1,\infty)} \left|\frac{\sin x}{x}\right| dm(x) = \int_0^\infty \left|\frac{\sin x}{x}\right| dx = \lim_{B\to\infty} \int_0^B \left|\frac{\sin x}{x}\right| dx.$$

Now, $\int_1^B \left|\frac{\sin x}{x}\right| dx = \int_1^{1/B} \left|x \sin\frac{1}{x}\right| \left(-\frac{1}{x^2}\right) dx = -\int_1^{1/B} \left|\frac{1}{x}\sin\frac{1}{x}\right| dx$. Hence $\lim_{c\to 0} \int_c^1 \left|\frac{1}{x}\sin\frac{1}{x}\right| dx$ would exist, in which case, by Problem 3.2.P9, $\int_{(0,1]} \left|\frac{1}{x}\sin\frac{1}{x}\right| dm(x)$ would also exist, contradicting Problem 3.2.P8(c).

3.2.P25. Let f be a nonnegative integrable function on $[a, b]$. For each $n \in \mathbb{N}$, let $E_n = X(n - 1 \leq f < n)$. Prove that $\sum_{k=1}^{\infty} (k-1) \cdot m(E_k) < \infty$.

Hint: Let χ_k, χ be the characteristic functions of E_k and $\cup_{k \geq 1} E_k$ respectively. Since $f \geq 0$, we have $f = f\chi$; since the E_k are disjoint, we further have $f = f\chi = \sum_{k=1}^{\infty} (f\chi_k)$, where the partial sums are increasing and pointwise convergent. By the Monotone Convergence Theorem 3.2.4, it follows that

$$\infty > \int_{[a,b]} f = \sum_{k=1}^{\infty} \int_{[a,b]} (f\chi_k) \geq \sum_{k=1}^{\infty} (k-1) \cdot m(E_k).$$

3.2.P26. Let $x^\gamma f(x)$ be integrable over $(0, \infty)$ for $\gamma = \alpha$ and $\gamma = \beta$, where $0 < \alpha < \beta$. Show that for each $\gamma \in (\alpha, \beta)$, the integral $\int_{(0,\infty)} x^\gamma f(x) dm(x)$ exists and is a continuous function of γ.

Hint: Consider any $\gamma \in (\alpha, \beta)$. Then $x^\gamma \leq x^\alpha$ for $0 < x \leq 1$, while $x^\gamma \leq x^\beta$ for $x > 1$. So, $|x^\gamma f(x)| \leq (x^\alpha + x^\beta)|f(x)|$ for all $x \in (0,\infty)$. Since the right-hand side is integrable, so is the left-hand side. For continuity, choose h so small that $\gamma + h \in (\alpha, \beta)$. Now, $|(x^{\gamma+h} - x^\gamma) f(x)| \leq 2(x^\alpha + x^\beta)|f(x)|$. The right-hand side being integrable, the Dominated Convergence Theorem 3.2.16 implies that

$$\int_{(0,\infty)} (x^{\gamma+h} - x^\gamma) f(x) dm(x) \to 0 \text{ as } h \to 0.$$

3.2.P27. Suppose f is an integrable real-valued function on $\mathbb{R}$. Define $\mu_f(E) = \int_E f d\mu$ for every measurable $E \subseteq \mathbb{R}$. Show that: $\mu_f(E) = 0\ \forall\ E \Leftrightarrow f = 0$ a.e. on $\mathbb{R}$.

Hint: $\Rightarrow$. Put $f = f^+ - f^-$. Let $E_+ = \{x \in \mathbb{R}: f(x) \geq 0\}$. Then $0 = \int_{E_+} f = \int_{E_+} f^+$. By 3.2.P1(b), f^+ vanishes a.e. on E_+. It follows that f vanishes a.e. on E_+. Similarly, f^- vanishes a.e. on $E_- = \{x \in \mathbb{R}: f(x) \leq 0\}$ and hence so does f. But $E_+ \cup E_- = \mathbb{R}$. $\Leftarrow$. Suppose f vanishes a.e. on $\mathbb{R}$. Then so does $|f|$. Moreover, $\int_E |f| = \int_{\mathbb{R}} (|f|\chi_E) = 0$ by 3.2.P1(b), considering that $|f|\chi_E$ vanishes a.e. and is nonnegative on $\mathbb{R}$. Hence $|\int_E f| \leq \int_E |f| = 0$.

3.2.P27. [Complement to Fatou's Lemma] Let $\{f_n\}_{n\geq 1}$ be a sequence of measurable functions on X such that $|f_n| \leq g$, where g is an integrable function on X. Prove that $\int_X \limsup f_n d\mu \geq \limsup \int_X f_n d\mu$ and give an example to show that the condition about the function g cannot be dropped.

Hint: Set $g_n = g - f_n$. Then $\{g_n\}_{n\geq 1}$ is a sequence of nonnegative measurable functions. By Fatou's Lemma 3.2.8, $\int_X \liminf g_n \leq \liminf \int_X g_n$, which implies by Theorem 3.2.15 and the assumed finiteness of $\int_X g$ that $\int_X \limsup f_n \geq \limsup \int_X f_n$. For the example, take $X = \mathbb{R}$ and $f_n = n^2\chi_{[-1/n,1/n]}$. Then $\int_X f_n = n^2(2/n) = 2n$ and so $\limsup \int_X f_n = \infty$. On the other hand, $\limsup f_n(x) = 0$ if $x \neq 0$ and ∞ if $x = 0$, so that $\int_X \limsup f_n = 0$. Alternatively, take $f_n = \chi_{[n,n+1]}$ instead. Then the f_n are also bounded.

3.2.P28. Show that the improper Riemann integral $\int_0^\infty \left(1+\frac{x}{n}\right)^{-n} \sin\frac{x}{n}\,dx$ has limit 0 as $n \to \infty$.

Hint: For $n > 1$ and $x > 0$, we have $\left(1+\frac{x}{n}\right)^n = 1 + x + \frac{n(n-1)}{n^2}\frac{x^2}{2} + \cdots > \frac{x^2}{4}$. We define $g(x)$ to be $\frac{4}{x^2}$ for $x \geq 1$ and $x^{-1/2}$ for $0 < x < 1$. Then by an elementary computation, the improper Riemann integral $\int_0^\infty e^{-x/2}dx$ is absolutely convergent and hence by the remark in Problem 3.2.P9, g is Lebesgue integrable over $[0, \infty)$. Also, $\left|\left(1+\frac{x}{n}\right)^{-n}\sin\frac{x}{n}\right| \leq g(x)$ for $n > 1$ for $n > 1$ and all $x > 0$. [Indeed, for $0 < x < 1$ and for all n, we have $x^{1/2} < 1 < 1 + x + \frac{n(n-1)}{n^2}\frac{x^2}{2} + \cdots = \left(1+\frac{x}{n}\right)^n$]. Therefore $\left(1+\frac{x}{n}\right)^{-n}\sin\frac{x}{n}$ is also Lebesgue integrable on $[0, \infty)$. Now, $0 \leq \left|\left(1+\frac{x}{n}\right)^{-n}\sin\frac{x}{n}\right| \leq \frac{4}{x^2}\left|\sin\frac{x}{n}\right|$, so that $\left(1+\frac{x}{n}\right)^{-n}\sin\frac{x}{n} \to 0$ pointwise as $n \to \infty$. By the Dominated Convergence Theorem 3.2.16, it follows that $\int_0^\infty \left(1+\frac{x}{n}\right)^{-n}\sin\frac{x}{n}\,dm(x) \to 0$ as $n \to \infty$. However, the Lebesgue integral here is the same as the improper Riemann integral in question in virtue of the remark in Problem 3.2.P10.

3.2.P29. Evaluate the limit as $n \to \infty$ of the Riemann integral $\int_0^n \left(1-\frac{x}{n}\right)^n e^{x/2}dx$.

Hint: Set $f_n(x) = \chi_{[0,n]}\left(1-\frac{x}{n}\right)^n e^{x/2}$. Observe that $f_n(x) \leq e^{-x/2}$. Indeed, $0 < x < n \Rightarrow 0 < \left(1-\frac{x}{n}\right)^n < 1$. By an elementary computation, the improper Riemann integral $\int_0^\infty e^{-x/2}dx$ is absolutely convergent with value 2 and hence by the remark in Problem 3.2.P9, $g(x) = e^{-x/2}$ has Lebesgue integral 2 over $[0, \infty)$. Moreover, $f_n(x) \to e^{-x/2}$. Therefore by the Dominated Convergence Theorem 3.2.16, it follows that $\int_{[0,\infty)} f_n \to \int_{[0,\infty)} g = 2$. But $\int_{[0,\infty)} f_n = \int_{[0,n)} f_n$ by Problem 3.2.P5 and this equals the Riemann integral in question by virtue of Theorem 3.2.20.

3.2.P30. Let $(X, \mathcal{F}, \mu)$ be a measure space and $\{f_k\}_{k\geq 1}$ be a sequence of functions on X which converges almost everywhere to a limit function f. In other words, there exists some $N \in \mathcal{F}$ such that $\mu(N) = 0$ and $f_k(x) \to f(x)$ for every $x \notin N$. In symbols, $f_k \xrightarrow{ae} f$.

(a) Give an example to show that f need not be measurable even if each f_k is.
(b) Show that $f_k \xrightarrow{ae} g$ if and only if $f = g$ a.e.
(c) If each f_k is measurable, show that there always exists a measurable g such that $f_k \xrightarrow{ae} g$.
(d) State the Monotone and Dominated Convergence Theorems for the case when the inequalities and the convergence in the hypotheses are assumed to hold only almost everywhere.

(e) If $f_k \overset{ae}{\to} f$ and each f_k is measurable, then under the additional hypothesis that $X, \mathcal{F}, \mu$ is *complete*, which is to say,

$$F \subseteq E, \mu(E) = 0 \quad \text{implies} \quad F \in \mathcal{F},$$

show that f has to be measurable.

Hint: **(a)** Let $X = \{a, b, c\}$, $\mathcal{F} = \{\emptyset, \{a\}, \{b, c\}, X\}$, $\mu(\emptyset) = 0$, $\mu(\{a\}) = 1$, $\mu(\{b, c\}) = 0$, $\mu(X) = 1$; $f_k = 1$ everywhere, $f(a) = 1, f(b) = 2, f(c) = 3$. Then $f_k \overset{ae}{\to} f$ but f is not measurable.
(b) Suppose $f_k \overset{ae}{\to} f$ and $f_k \overset{ae}{\to} g$. There exists sets $N_1 \in \mathcal{F}$ and $N_2 \in \mathcal{F}$ such that $\mu(N_1) = 0 = \mu(N_2)$, $f_k(x) \to f(x)$ for every $x \in N_1^c$ and $f_k(x) \to g(x)$ for every $x \in N_2^c$. It follows that $f_k(x) \to f(x)$ as well as $f_k(x) \to g(x)$ for every $x \in N_1^c \cap N_2^c$. Thus $f(x) = g(x)$ for every $x \in N_1^c \cap N_2^c$. Since $N_1^c \cap N_2^c = (N_1 \cup N_2)^c$ and $\mu(N_1 \cup N_2) = 0$, this means that $f = g$ a.e. Conversely, suppose $f_k \overset{ae}{\to} f$ and $f = g$ a.e. Then there exists sets $N_1 \in \mathcal{F}$ and $N_2 \in \mathcal{F}$ such that $\mu(N_1) = 0 = \mu(N_2)$, $f_k(x) \to f(x)$ for every $x \in N_1^c$ and $f(x) = g(x)$ for every $x \in N_2^c$. Therefore $f_k(x) \to g(x)$ for every $x \in N_1^c \cap N_2^c$. Once again, since $N_1^c \cap N_2^c = (N_1 \cup N_2)^c$ and $\mu(N_1 \cup N_2) = 0$, this means that $f_k \overset{ae}{\to} g$.
(c) Let $N \in \mathcal{F}$, $\mu(N) = 0$ and $f_k(x) \to f(x)$ for every $x \notin N$. Then

$$(f_k\chi_{N^c})(x) \to (f\chi_{N^c})(x) \text{for every } x \in X.$$

Since each function $f_x\chi_{N^c}$ is measurable, it follows that $g = f\chi_{N^c}$ is measurable. Now, $(f_k\chi_{N^c})(x) = f_k(x)$ and $g(x) = (f\chi_{N^c})(x) = f(x)$ for every $x \in N^c$. Therefore $f_k(x) \to g(x)$ for every $x \in N^c$. Since g is measurable and $\mu(N) = 0$, the function g has the required property

(d) Monotone Convergence Theorem: *Let $\{f_n\}_{n\geq 1}$ be a sequence of measurable nonnegative extended real-valued functions on X such that*

$$0 \leq f_1 \leq f_2 \leq \cdots \quad a.e.$$

and

$$f_n \to f \; a.e. \, as \, n \to \infty.$$

Then some measurable function equals f a.e. and for any such function F,

$$\lim_{n\to\infty}\int_X f_n d\mu = \int_X F\, d\mu.$$

Dominated Convergence Theorem: *Let* $\{f_n\}_{n\geq 1}$ *be a sequence of measurable extended real-valued functions on X, converging almost everywhere to a function f and let g be an integrable function on X such that*

$$|f_n(x)| \leq g(x) \quad \textit{almost everywhere.}$$

Then some measurable function equals f a.e. and for any such function F,

$$\lim_{n\to\infty}\int_X |f_n - F|d\mu = 0 \quad \text{and} \quad \lim_{n\to\infty}\int_X f_n d\mu = \int_X F\, d\mu.$$

(e) By part (c) above, $f_k \xrightarrow{ae} g$ for some measurable g. By part (b), $f = g$ a.e. So, we need only show that, if $f = g$ a.e. and g is measurable, then f is measurable. Consider a real number α and let $A = X(f > \alpha)$, $B = X(g > \alpha)$. Then B is measurable and we must argue that A has to be measurable as well. Since $f = g$ a.e., there exists some N with $\mu(N) = 0$ such that $\{x \in X: f(x) \neq g(x)\} \subseteq N$. Since $A \cap B^c$ and $B \cap A^c$ are both subsets of $\{x \in X: f(x) \neq g(x)\}$, they are measurable by virtue of the completeness hypothesis. It follows by the argument of Problem 2.3.P5 that A is measurable.

3.2.P31. Let γ be the counting measure [see Example 3.1.2(a)] on either $\mathbb{N}$ or $\{1, 2, \ldots, n\}$. Show for any nonnegative extended real-valued function f that the integral is given by the sum $\Sigma_k f(k)$.

Hint: The conclusion holds trivially if f is permitted to take the value ∞. So we consider only real-valued f.

Let γ be the counting measure on $\mathbb{N}$. Then any function is measurable. It is given that $f{:}\mathbb{N}\to\mathbb{R}$ is nonnegative. Define

$$f_n(k) = \begin{cases} f(k) & 1 \leq k \leq n \\ 0 & \text{otherwise.} \end{cases}$$

Clearly, for each $k \in \mathbb{N}$, we have $f_n(k) \to f(k)$ and $0 \leq f_1(k) \leq f_2(k) \leq \cdots$, the latter because f is nonnegative. By the Monotone Convergence Theorem 3.2.4,

$$\lim_{n\to\infty}\int_X f_n d\gamma = \int_X f\, d\gamma.$$

Now, $f_n = f(1)\chi_{\{1\}} + f(2)\chi_{\{2\}} + \cdots + f(n)\chi_{\{n\}} + 0\cdot\chi_{\{n,n+1,\cdots\}}$ and therefore by Proposition 3.1.5, $\int_{\mathbb{N}} f_n d\gamma = \sum_{j=1}^{n} f(j)\gamma(\{j\})$, which is the same as $\sum_{j=1}^{n} f(j)$ by definition of the counting measure. Hence

$$\int_X f\, d\gamma = \lim_{n\to\infty} \sum_{j=1}^{n} f(j) = \sum_{k=1}^{\infty} f(k).$$

If instead of $\mathbb{N}$, the domain is $\{1, 2, \ldots, n\}$, then all nonnegative real-valued functions are simple and a similar conclusion with the obvious modification that the summation is finite can be derived; the argument then does not need the Monotone Convergence Theorem and we merely need to represent f as $f(1)\chi_{\{1\}} + f(2)\chi_{\{2\}} + \cdots + f(n)\chi_{\{n\}}$ before applying Proposition 3.1.5.

Problem Set 3.3

3.3.P1. Give an example of a nonnegative extended real-valued function on [0, 1] which has Lebesgue integral 1 and is (a) the limit of an increasing sequence of continuous functions (b) unbounded on every subinterval of positive length (so that it cannot have an improper Riemann integral).
Hint: The limit of the sequence of functions in the proof of Proposition 3.3.14 has the required features.

3.3.P2. Give an example when $\int_X |f_n - f|\,d\mu \to 0$ but f_n does not converge pointwise to f anywhere.
Hint: Let $X = [0, 1]$ and consider the intervals $\left[\frac{j-1}{k}, \frac{j}{k}\right], 1 \le j \le k, k \in \mathbb{N}$, arranged in this order. The first five intervals are [0, 1], $\left[0, \frac{1}{2}\right]$, $\left[\frac{1}{2}, 1\right]$, $\left[0, \frac{1}{3}\right]$, $\left[\frac{1}{3}, \frac{2}{3}\right]$. Let f_n be the characteristic function of the nth interval and let f be zero everywhere. If $n \ge k(k+1)/2$, then f_n is the characteristic function of an interval of measure at most $1/k$. Therefore $\int_X |f_n - f| \le 1/k \to 0$. For any given $x \in [0, 1]$, any subdivision of [0, 1] into 3 or more subintervals must include a subinterval containing x and also one not containing x, which means $\{f_n(x)\}_{n\ge 1}$ has a subsequence with each term 1 and also a subsequence with each term 0. So the sequence $\{f_n\}_{n\ge 1}$ converges nowhere on [0, 1].

3.3.P3. Let $\{f_k\}_{k\ge 1}$ be a sequence of integrable functions on a measurable set X such that $\sum_{k=1}^{\infty} \int_X |f_k| < \infty$. Show that the series $\sum_{k=1}^{\infty} f_k$ converges a.e. to an integrable sum f and that $\int_X f = \sum_{k=1}^{\infty} \int_X f_k$.

Hint: Let $g = \sum_{k=1}^{\infty} |f_k|$. By the Monotone Convergence Theorem 3.2.4 and the given finiteness of the sum of integrals, it follows that $\int_X g < \infty$ and so g is finite a.e. (see Remark 3.2.2). Thus there exists a measurable set $H \subseteq X$ such that g is finite on H and $\mu(H^c) = 0$. Then the series $\sum_{k=1}^{\infty} f_k$ is convergent on H, and denoting its sum on H by f_0, we have $|f_0| \le g$ on H. Since H is measurable and restrictions of the partial sums of $\sum_{k=1}^{\infty} f_k$ to H are also measurable, it follows that f_0 is a measurable function on H. Therefore its extension f to X obtained by setting it equal to 0 on H^c is measurable (the reader is invited to supply the simple justification). Since the extension satisfies $|f| \leq g$ on X, it is integrable. Write $g_n(x) = \sum_{k=1}^{n} f_k$. Then $|g_n(x)| \leq g(x)$ and $\lim_{n\to\infty} g_n(x) = f(x) \forall x \in H$. By the Dominated Convergence Theorem 3.2.16 and the fact that $\mu(H^c) = 0$, we obtain $\lim_{n\to\infty} \sum_{k=1}^{n} \int_X f_k = \lim_{n\to\infty} \sum_{k=1}^{n} \int_H f_k = \lim_{n\to\infty} \int_H \sum_{k=1}^{n} f_k = \lim_{n\to\infty} \int_H g_n = \int_H f = \int_X f$

3.3.P4. For $f \in L^1(\mathbb{R})$, show that

$$\lim_{h\to 0} \int_{\mathbb{R}} |f(x+h) - f(x)| dx = 0, \tag{8.18}$$

i.e., the mapping $\varphi:\mathbb{R}\to L^1(\mathbb{R})$ defined by $\varphi(h) = f_h$, where $f_h(x) = f(x + h)$, $x \in \mathbb{R}$, is continuous.

Hint: Let $g \in C_c(\mathbb{R})$, the space of continuous functions on $\mathbb{R}$ with compact support, be such that $\|f - g\|_1 < \frac{\varepsilon}{3}$. Observe that

$$|f(x+h) - f(x)| \leq |f(x+h) - g(x+h)| + |g(x+h) - g(x)| + |g(x) - f(x)|.$$

On integrating both sides over $\mathbb{R}$, we find that the integral of the left-hand side of (8.18) does not exceed

$$\int_{\mathbb{R}} |f(x+h) - g(x+h)| dx + \int_{\mathbb{R}} |g(x+h) - g(x)| dx + \int_{\mathbb{R}} |g(x) - f(x)| dx$$

$= A + B + C$, say. Observe that $A = C$ for all h (translation invariance of measure) and by our choice of g, we have $A, C \leq \frac{\varepsilon}{3}$.

Note that, g, being a continuous function with compact support, is actually uniformly continuous. Let I be any interval containing the support of g. Now, for any $\eta > 0$, there exists a $\delta > 0$ such that

$$|g(x+h) - g(x)| < \eta \text{ for all } |h| < \delta \text{ and } x \in \mathbb{R}.$$

Then if $|h| < \delta$, it follows that $\int_{\mathbb{R}} |g(x+h) - g(x)| dx < 2\eta\ell(I) < 2\eta\ell(I) < \frac{\varepsilon}{3}$, by choosing η sufficiently small. This shows that the left-hand side of (8.18) does not exceed ε when $|h| < \delta$.

3.3.P5. Suppose that $f \in L^1(\mathbb{R})$. Show that

$$\lim_{|h|\to\infty} \int_{\mathbb{R}} |f(x+h)+f(x)|dx = 2\int_{\mathbb{R}} |f(x)|dx.$$

Hint: Assume that supp$f \subseteq I \subseteq \mathbb{R}$, where I is a bounded interval. In this case, the sum $f(x)+f(x+h)$ equals either $f(x)$, $f(x+h)$ or zero, provided $|h| > \ell(I)$. Therefore, for these values of h, we have

$$\begin{aligned}\int_{\mathbb{R}} |f(x+h)+f(x)|dx &= \int_{I-h} |f(x+h)|dx + \int_I |f(x)|dx \\ &= \int_{\mathbb{R}} |f(x+h)|dx + \int_{\mathbb{R}} |f(x)|dx = 2\int_{\mathbb{R}} |f(x)|dx.\end{aligned}$$

Consequently,

$$\lim_{|h|\to\infty} \int_{\mathbb{R}} |f(x+h)+f(x)|dx = 2\int_{\mathbb{R}} |f(x)|dx.$$

So far the result has been shown to hold for an integrable function whose support is bounded. Now, let f be an arbitrary integrable function. Define

$$f_n(x) = \begin{cases} f(x) & -n \le x \le n \\ 0 & |x| > n. \end{cases}$$

Clearly, $|f - f_n| \le |f|$ and $|f - f_n| \to 0$ pointwise. Therefore, $\int_{\mathbb{R}} |f(x) - f_n(x)|dx \to 0$ by the Dominated Convergence Theorem 3.2.16. Let $\varepsilon > 0$ be given. There exists an n_0 such that $n \ge n_0$ implies $\int_{\mathbb{R}} |f(x) - f_n(x)|dx < -\frac{\varepsilon}{4}$. It follows that $n \ge n_0$ also implies $\int_{\mathbb{R}} |f_n(x)|dx > \int_{\mathbb{R}} |f(x)|dx - \frac{\varepsilon}{4}$ The inequality

$$|f(x+h)+f(x)| \ge |f_{n_0}(x+h)+f_{n_0}(x)| - |f_{n_0}(x+h)+f_{n_0}(x) - f(x+h) - f(x)|,$$

on integration, yields

$$\begin{aligned}\int_{\mathbb{R}} |f(x+h)+f(x)|dx &\ge \int_{\mathbb{R}} |f_{n_0}(x+h)+f_{n_0}(x)|dx - 2\cdot\frac{\varepsilon}{4} \\ &= 2\int_{\mathbb{R}} |f_{n_0}(x)|dx - 2\cdot\frac{\varepsilon}{4} \text{for} |h| > 2n_0 \\ &\ge 2\int_{\mathbb{R}} |f(x)|dx - 4\cdot\frac{\varepsilon}{4} \text{for} |h| > 2n_0.\end{aligned}$$

Thus we have shown that for any $\varepsilon > 0$, there exists an $M > 0$, namely, $M = 2n_0$, such that

$$\int_{\mathbb{R}} |f(x+h)+f(x)|dx \geq 2\int_{\mathbb{R}} |f(x)|dx - \varepsilon \quad \text{for } |h| > M.$$

On the other hand,

$$\int_{\mathbb{R}} |f(x+h)+f(x)||dx \leq \int_{\mathbb{R}} |f(x+h)|dx + \int_{\mathbb{R}} |f(x)|dx = 2\int_{\mathbb{R}} |f(x)|dx.$$

This proves the required result.

3.3.P6. If $f \in L^2[0, 1]$, show that $f \in L^1[0, 1]$. However, if $f \in L^2[0, \infty)$ then f need not belong to $L^1[0, \infty)$.
Hint: Let $g = 1$ everywhere on $[0, 1]$. Then $\int_{[0,1]}g^2 = 1$ and by the Cauchy–Schwarz Inequality, we have $(\int_{[0,1]}|f|)^2 = (\int_{[0,1]}(|f|\cdot g))^2 \leq (\int_{[0,1]}|f|^2)(\int_{[0,1]}g^2) = (\int_{[0,1]}|f|^2) < \infty$ when $f \in L^2[0, 1]$. However, $f(x) = \frac{1}{1+x}$ defines a function in $L^2[0, \infty)$ that does not belong to $L^1[0, \infty)$. Indeed, $\int_{[0,\infty]}\left(\frac{1}{1+x}\right)^2 dm(x) = 1$, whereas $\int_{[0,\infty]}\frac{1}{1+x}dm(x) = \infty$.

3.3.P7. The functions in $L^2[-\pi, \pi]$ given by $f_n(x) = \sin nx$, $n \in \mathbb{N}$, form a bounded and closed subset which is not compact.
Hint: By an elementary computation, for $n \neq k$, we have $\|f_n - f_k\|^2 = \int_{[-\pi,\pi]}(\sin nx - \sin kx)^2 dm(x) = \int_{[-\pi,\pi]}(\sin^2 nx + \sin^2 kx - 2\sin nx \sin kx)dm(x) = 2\pi$. Thus $\|f_n - f_k\| = \sqrt{2\pi}$ when $n \neq k$. So, the sequence $\{f_n\}_{n\geq 1}$ cannot have a convergent subsequence and the set $\{f_n: n \in \mathbb{N}\}$ therefore cannot be compact. As the distance between any two distinct points of the set has been shown to be $\sqrt{2\pi}$, it has no limit points, which means that it is vacuously true that it contains all its limit points; consequently, the set is closed. It is also bounded, because all its points lie in a closed ball of radius $\sqrt{2\pi}$ around any one of its points.

3.3.P8. Let f and g be positive measurable functions on $[0, 1]$ such that $fg \geq 1$. Prove that

$$\left(\int_{[0,1]} f\, dm\right)\left(\int_{[0,1]} g\, dm\right) \geq 1.$$

Hint: If one of the integrals on the left is infinite, there is nothing to prove. Assume both to be finite. Then $f^{1/2} \in L^2[0, 1]$ and $g^{1/2} \in L^2[0, 1]$. Also, $f^{1/2}g^{1/2} \geq 1$, so that $\int_{[0,1]}(f^{1/2}g^{1/2}) \geq 1$. But by the Cauchy–Schwarz Inequality, we have $\left(\int_{[0,1]}f\right)^{1/2}\left(\int_{[0,1]}g\right)^{1/2} \geq \int_{[0,1]}(f^{1/2}g^{1/2})$.

3.3.P9. Show that the following inequalities are inconsistent for a function $f \in L^2[0, \pi]$:

$$\|f - \cos\| \leq \frac{2}{3}, \quad \|f - \sin\| \leq \frac{1}{3}.$$

Hint: The given inequalities imply that $\|\cos - \sin\| \leq \|f - \cos\| + \|f - \sin\| \leq \frac{2}{3} + \frac{1}{3} = 1$. However, this is false because $\|\cos - \sin\|^2 = \int_{[0,\pi]}(\cos x - \sin x)^2 dm(x) = \int_0^{\pi}(1 - \sin 2x)dx = \pi$ by an elementary computation.

3.3.P10. Let $f_n(x) = \frac{n}{1+n\sqrt{x}}$ for $x \in [0, 1]$ and $n \in \mathbb{N}$.

(a) Show that the sequence $\{f_n\}_{n\geq 1}$ has a pointwise limit a.e.;
(b) Does f_n belong to $L^2[0, 1]$ for each (or some) $n \in \mathbb{N}$?
(c) Does the pointwise limit (a.e.) of f_n belong to $L^2[0, 1]$?

Hint: Proceeding as in Example 3.3.13, we find that **(a)** the pointwise limit is $\frac{1}{\sqrt{x}}$ when $x \neq 0$ **(b)** f_n belongs to $L^2[0, 1]$ for every $n \in \mathbb{N}$. For **(c)**, we note that

$$\int_{[0,1]} \left(\frac{1}{\sqrt{x}}\right)^2 dm(x) = \int_0^1 \frac{1}{x} dx = \infty.$$

3.3.P11. [Cf. Problem 3.3.P14] Let $(X, \mathcal{F}, \mu)$ be a measure space with $\mu(X) < \infty$. Show that $L^\infty(X) \subseteq L^p(X)$ for all p, $0 < p < \infty$. Also, show that, if f is measurable, then

$$\lim_{p\to\infty} \|f\|_p = \|f\|_\infty.$$

Hint: Let $f \in L^\infty(X)$. Then $|f(x)|^p \in L^p(X)$ because

$$\int_X |f|^p \, d\mu \leq \|f\|_\infty^p \mu(X) \quad \text{or} \quad \left(\int_X |f|^p \, d\mu\right)^{\frac{1}{p}} \leq \|f\|_\infty (\mu(X))^{\frac{1}{p}}.$$

Since $(\mu(X))^{1/p} \to 1$ as $p \to \infty$, we have

$$\limsup \|f\|_p \leq \|f\|_\infty. \tag{8.19}$$

On the other hand, suppose that $|f(x)| \geq B$ on a set F of positive measure. Then

$$B(\mu(F))^{\frac{1}{p}} \leq \|f\|_p,$$

which implies

$$B \leq \liminf \|f\|_p.$$

Let $\varepsilon > 0$ be arbitrary. The set $\{x \in X: |f(x)| > \|f\|_\infty - \varepsilon\}$ has positive measure by definition of $\|f\|_\infty$. Therefore the preceding inequality yields

$$\|f\|_\infty - \varepsilon \leq \liminf \|f\|_p.$$

Since $\varepsilon > 0$ is arbitrary, we get $\|f\|_\infty \leq \liminf \|f\|_p$. Combining this with (8.19), we get the desired inequality.

Next, suppose that $f \notin L^\infty(X)$. We have to show that $\lim_{p\to\infty} \|f\|_p = \infty$. Consider an arbitrary $B > 0$. Since $\|f\|_\infty = \infty$, the inequality $|f(x)| \geq 2B$ must hold on a set F of (finite) positive measure $\mu(F)$. The inequality $(\mu(F))^{1/p} > \frac{1}{2}$ holds for all sufficiently large p. It follows (without any presumption of finiteness) that $\|f\|_p \geq 2B$ $(\mu(F))^{1/p} \geq B$. Since this has been shown to hold for all B and all sufficiently large p, we have $\lim_{p\to\infty} \|f\|_p = \infty$, as desired.

3.3.P12. **(a)** Let $0 < p < 1$ and let q be such that $\frac{1}{p} + \frac{1}{q} = 1$. Then $q < 0$. If $f \in L^p(X)$ and the function g on X is such that $g \neq 0$ a.e. and $\int_X |g|^q d\mu < \infty$, then $\int_X |g|^q d\mu \neq 0$ unless $\mu(X) = 0$, and

$$\int_X |fg| d\mu \geq \left(\int_X |f|^p \, d\mu\right)^{\frac{1}{p}} \left(\int_X |g|^q \, d\mu\right)^{\frac{1}{q}}.$$

(This is Hölder's Inequality for $0 < p < 1$.) If the hypothesis that $\int_X |g|^q d\mu < \infty$ is omitted, then the inequality is valid trivially, provided we interpret a negative power of ∞ to mean 0.

(b) Let $0 < p < 1$ and f, g be measurable functions on X such that $f \geq 0$, $g \geq 0$. Then

$$\|f + g\|_p \geq \|f\|_p + \|g\|_p.$$

(This is Minkowski's Inequality for $0 < p < 1$.)

(c) For a complex-valued measurable function ϕ on X, show that $|\int_X \phi d\mu| \leq \int_X |\phi| d\mu$.

Hint: **(a)** First note that $q < 0$ because $p < 1$ and $\frac{1}{p} + \frac{1}{q} = 1$. Set

$$P = \frac{1}{p} \quad \text{and} \quad Q = -\frac{q}{p}.$$

Then $P > 1$ and $Q > 1$ and $\frac{1}{P} + \frac{1}{Q} = p - \frac{p}{q} = p\left(1 - \frac{1}{q}\right) = 1$. Set

$$|f| = (uv)^{\frac{1}{p}}, |g| = v^{-\frac{1}{p}}.$$

These equations define nonnegative measurable functions u and v a.e., considering that $g \neq 0$ a.e., and

$$\int_X uv d\mu \leq \left(\int_X u^P d\mu\right)^{\frac{1}{P}} \left(\int_X v^Q d\mu\right)^{\frac{1}{Q}},$$

$$\int_X |f|^p d\mu \leq \left(\int_X |fg| d\mu\right)^p \left(\int_X |g|^q d\mu\right)^{\frac{1}{Q}} = \left(\int_X |fg| d\mu\right)^p \left(\int_X |g|^q d\mu\right)^{-\frac{p}{q}}.$$

Taking the pth root of both sides and multiplying by $(\int_X |g|^q d\mu)^{1/q}$, which is permissible because it has been assumed finite, we get

$$\int_X |fg| d\mu \geq \left(\int_X |f|^p d\mu\right)^{\frac{1}{p}} \left(\int_X |g|^q d\mu\right)^{\frac{1}{q}}.$$

(b) Since $f \geq 0$, $g \geq 0$, we have $f + g \geq f \geq 0$ and $f + g \geq g \geq 0$, so that $\|f + g\|_p \geq \|f\|_p$ as well as $\|f + g\|_p \geq \|g\|_p$. Therefore we may assume $f + g, f, g \in L^p(X)$. We may also assume that $\|f + g\|_p > 0$, because otherwise we would also have $\|f\|_p = 0 = \|g\|_p$, so that the inequality in question would be trivially true. Observe that

$$\int_X (f+g)^p d\mu = \int_X f(f+g)^{p-1} d\mu + \int_X g(f+g)^{p-1} d\mu.$$

We intend to apply (a) to each term on the right-hand side here. In order to do so, we need to know not only that $\int_X (f + g)^{(p-1)q} d\mu < \infty$, i.e., that $\int_X (f + g)^p d\mu < \infty$, but also that $f + g \neq 0$ a.e. It is legitimate to assume that $f + g > 0$ everywhere because, on the set where $f + g$ is 0, the functions f and g must also be 0 and integrals need be taken only over the complement of that set; in effect, we may take X to be the complement and $f + g > 0$ everywhere. Now, applying (a) and following the proof of Minkowski's Inequality 3.3.6, we arrive at the required conclusion.

(c) The case $p = \frac{1}{2}$ of (b) is that

$$\left[\int_X (f+g)^{\frac{1}{2}} d\mu\right]^2 \geq \left[\left(\int_X f^{\frac{1}{2}} d\mu\right)^2 + \left(\int_X g^{\frac{1}{2}} d\mu\right)^2\right],$$

which is equivalent to

$$\int_X (f^2+g^2)^{\frac{1}{2}} d\mu \geq \left[\left(\int_X |f| d\mu\right)^2 + \left(\int_X |g| d\mu\right)^2\right]^{\frac{1}{2}}$$

with real-valued measurable f and g, not presumed nonnegative. This in turn is equivalent to the inequality $\int_X |\phi| d\mu \geq |\int_X \phi d\mu|$ for complex-valued ϕ. In fact, for a measurable $\phi = f + ig$, where f and g are the real and imaginary parts respectively of ϕ, the integral $\int_X |\phi| d\mu$ is the left-hand side of the inequality displayed above and the absolute value $|\int_X \phi d\mu|$ is the right-hand side.

However, here is a quicker proof: Note that for complex ϕ, the integral $\int_X \phi d\mu$ is defined only when $\int_X |\phi| d\mu < \infty$. Since $\int_X \phi d\mu$ is a complex number, there exists a complex number α such that $|\alpha| = 1$ and $\alpha \int_X \phi d\mu = |\int_X \phi d\mu|$. Then $\Re(\alpha\phi) \leq |\alpha\phi| = |\phi|$. Therefore

$$\begin{aligned} \left|\int_X \phi\, d\mu\right| &= \Re\left(\left|\int_X \phi\, d\mu\right|\right) = \Re\left(\alpha \int_X \phi\, d\mu\right) = \Re\left(\int_X \alpha\phi\, d\mu\right) \\ &= \int_X \Re(\alpha\phi) d\mu \leq \int_X |\phi| d\mu. \end{aligned}$$

3.3.P13. Let $0 < p < 1$ and $X = [0, 1]$. Then there exist $f, g \in L^p(X)$ such that

$$\|f+g\|_p > \|f\|_p + \|g\|_p.$$

Hint: Let $f = \chi_{[0,\frac{1}{2}]}, g = \chi_{[\frac{1}{2},1]}$. Clearly, $f, g \in L^p(X)$. For any p, $0 < p < 1$, observe that $\|f+g\|_p = 1$, while $\|f\|_p + \|g\|_p = 2^{-1/p} + 2^{-1/p} = 2^{1-1/p} < 1$.

3.3.P14. [Cf. Problem 3.3.P11] Show that, if $\mu(X) < \infty$ and $0 < p < q \leq \infty$, then $L^q(X) \subseteq L^p(X)$. Also, show that when $\mu(X) = \infty$, the inclusion does not hold.
Hint: For $q < \infty$, let $f \in L^q(X)$; then $\int_X |f|^q d\mu < \infty$. It is evident that $|f|^q$ belongs to $L^r(X)$ where $r = q/p > 1$. Let s be such that $\frac{1}{r} + \frac{1}{s} = 1$. Then

$$\int_X |f|^p \, d\mu = \int_X |f|^p \cdot 1 d\mu \leq \left(\int_X |f|^{pr} \, d\mu\right)^{\frac{1}{r}} \left(\int_X 1^s \, d\mu\right)^{\frac{1}{s}}$$
$$= \left(\int_X |f|^q \, d\mu\right)^{\frac{p}{q}} (\mu(X))^{\frac{1}{s}} < \infty.$$

For $q = \infty$, $|f|^p \leq \|f\|_\infty^p$ a.e. and so, $|f|^p$ is integrable.

However, if $\mu(X) = \infty$, the inclusion $L^q(X) \subseteq L^p(X)$ need not hold because the function f: $[1, \infty] \to \mathbb{R}$ given by $f(x) = 1/x$ belongs to $L^q[1, \infty]$ for every $q > 1$ but not to $L^1[1, \infty]$.

3.3.P15. Let X be a measurable subset of $\mathbb{R}$ and $0 < p < q < \infty$. If $f \in L^p(X) \cap L^q(X)$, then $f \in L^r(X)$ for all r such that $p < r < q$.

Hint: Since $p < r < q$, there is a number t, $0 < t < 1$, such that $r = tp + (1 - t)q$. Observe that $|f|^{tp} \in L^{1/t}(X)$ and $|f|^{(1-t)q} \in L^{1/(1-t)}(X)$. Hence by Hölder's Inequality, we have

$$|f|^r = |f|^{tp} |f|^{(1-t)q} \in L^1(X).$$

3.3.P16. **(a)** Let $p \geq 1$ and let $\|f_n - f\|_p \to 0$ as $n \to \infty$. Show that $\|f_n\|_p \to \|f\|_p$ as $n \to \infty$.

(b) Suppose $\{f_n\}_{n \geq 1}$ is a sequence in $L^p(X)$, $f \in L^p(X)$, $f_n \to f$ a.e. and $\|f_n\|_p \to \|f\|_p$ as $n \to \infty$. Prove that $\|f_n - f\|_p \to 0$ as $n \to \infty$.

Hint: **(a)** By Minkowski's Inequality, we have

$$\left| \|f_n\|_p - \|f\|_p \right| \leq \|f_n - f\|_p.$$

The result follows on using the convergence $\|f_n - f\|_p \to 0$.

(b) Analogous to Problem 3.2.P20(b).

3.3.P17. Show that $\int_{[0,\pi]} x^{-1/4} \sin x \, dx \leq \pi^{3/4}$.

Hint: Apply the Cauchy–Schwarz Inequality to obtain

$$\int_{[0,\pi]} x^{-\frac{1}{4}} \sin x \, dx \leq \left(\int_{[0,\pi]} x^{-\frac{1}{2}} dx\right)^{\frac{1}{2}} \left(\int_{[0,\pi]} \sin^2 x \, dx\right)^{\frac{1}{2}} = \left(2\pi^{\frac{1}{2}}\right)^{\frac{1}{2}} \left(\frac{1}{2}\pi\right)^{\frac{1}{2}} = \pi^{\frac{3}{4}}.$$

3.3.P18. For each n, consider the functions f_n: $\mathbb{R} \to \mathbb{R}$ given by $f_n = \chi_{[n,n+1]}$, $n = 1, 2, \ldots$. Show that $f_n(x) \to 0$ as $n \to \infty$ for each $x \in \mathbb{R}$. Also, show that for all p such that $1 \leq p \leq \infty$, we have $f_n \in L^p(\mathbb{R})$, $\|f_n\|_p = 1$ for all n, so that $\|f_n\|_p \nrightarrow 0$ as $n \to \infty$. (This example shows that pointwise convergence does not imply convergence in any L^p norm, $1 \leq p \leq \infty$.)

Hint: Fix $x \in \mathbb{R}$. There exists an integer n_0 such that $n_0 > x$. So, $f_n(x) = 0$ for $n > n_0 + 1$ and hence $f_n(x) \to 0$ as $n \to \infty$. Also, $\int_{\mathbb{R}} |f_n|^p dx = \int_{\mathbb{R}} \chi^p_{[n,n+1]} dx = 1$, which implies $\|f_n\|_p = 1$ for all n and for all p such that $1 \leq p < \infty$. Being the characteristic function of a set of positive measure, f_n has essential supremum 1, i.e., $\|f_n\|_\infty = 1$ for all n. Consequently, $\|f_n\|_p \nrightarrow 0$ as $n \to \infty$ for $1 \leq p \leq \infty$.

3.3.P19. Give an example to show that convergence in the pth norm does not imply convergence a.e.

Hint: Consider the interval [0, 1]. For each n, divide [0, 1] into n subintervals:

$$\left[0, \frac{1}{n}\right], \left[\frac{1}{n}, \frac{2}{n}\right], \ldots, \left[\frac{n-1}{n}, 1\right].$$

Write

$$E_{n,k} = \left[\frac{k-1}{n}, \frac{k}{n}\right], \quad k = 1, 2, \ldots, n.$$

Enumerate all these intervals as follows:

$$E_{11}, E_{21}, E_{22}, E_{31}, E_{32}, E_{33}, \ldots.$$

Let $\{E_n\}_{n \geq 1}$ denote the above sequence of intervals. Define

$$f_n = \chi_{E_n}.$$

Since $m(E_n) \to 0$ as $n \to \infty$, $\int_{[0,1]} f_n^p = m(E_n) \to 0$ as $n \to \infty$, i.e. $\|f_n - 0\|_p \to 0$ as $n \to \infty$. However, $f_n(x)$ converges for no $x \in [0, 1]$. Indeed,

$$\left\{x \in [0, 1] : |f_n(x)| \geq \frac{1}{2}\right\} = E_n \quad \text{for all } n,$$

i.e., for each $x \in [0, 1]$, $f_n(x) = 1$ for infinitely many values of n. Hence $f_n(x) \nrightarrow 0$ as $n \to \infty$.

3.3.P20. Let p and q be conjugate indices and $\lim_{n\to\infty} \|f_n - f\|_p = 0 = \lim_{n\to\infty} \|g_n - g\|_q$, where $f_n, f \in L^p(X)$ and $g_n, g \in L^q(X)$, $n = 1, 2, \ldots$. Show that $\lim_{n\to\infty} \|f_n g_n - fg\|_1 = 0$.

Hint: We have

$$\|f_n g_n - fg\|_1 \le \|fg - f_n g\|_1 + \|f_n g - f_n g_n\|_1 \\ \le \|f - f_n\|_p \|g\|_q + \|f_n\|_p \|g - g_n\|_q,$$

using Hölder's Inequality. Since $\left|\|f_n\|_p - \|f\|_p\right| \le \|f_n - f\|_p$, using the Minkowski Inequality, for all large n, we have $\|f_n\|_p \le \|f\|_p + 1$. Hence, for large n,

$$\|f_n g_n - fg\|_1 \le \|f - f_n\|_p \|g\|_q + \left(\|f\|_p + 1\right) \|g - g_n\|_q.$$

The result now follows.

3.3.P21. For $1 \le p < \infty$, we denote by ℓ^p the space of all sequences $\{x_\nu\}_{\nu \ge 1}$ such that $\sum_{\nu=1}^{\infty} |x_\nu|^p < \infty$.

(a) Without interpreting sums as integrals with respect to the counting measure, show that, if $\{x_\nu\}_{\nu \ge 1} \in \ell^p$ and $\{y_\nu\}_{\nu \ge 1} \in \ell^q$, where $1 < p, q < \infty$ and $\frac{1}{p} + \frac{1}{q} = 1$, then

$$\sum_{\nu=1}^{\infty} |x_\nu y_\nu| \le \left(\sum_{\nu=1}^{\infty} |x_\nu|^p\right)^{\frac{1}{p}} \left(\sum_{\nu=1}^{\infty} |y_\nu|^q\right)^{\frac{1}{q}}.$$

For $p = 1$, $q = \infty$, show that

$$\sum_{\nu=1}^{\infty} |x_\nu y_\nu| \le \left(\sum_{\nu=1}^{\infty} |x_\nu|\right) \|\{y_\nu\}_{\nu \ge 1}\|_\infty.$$

(b) Using the inequality of part (a), show also that, if $\{x_\nu\}_{\nu \ge 1} \in \ell^p$ and $\{y_\nu\}_{\nu \ge 1} \in \ell^p$, then

$$\left(\sum_{\nu=1}^{\infty} |x_\nu + y_\nu|^p\right)^{\frac{1}{p}} \le \left(\sum_{\nu=1}^{\infty} |x_\nu|^p\right)^{\frac{1}{p}} + \left(\sum_{\nu=1}^{\infty} |y_\nu|^p\right)^{\frac{1}{p}}.$$

[The inequalities in (a) and (b) are known as the Hölder and Minkowski Inequalities respectively for sequences.]

Hint: (**a**) We need only consider the case when $\sum_{\nu=1}^{\infty}|x_\nu|^p \neq 0 \neq \sum_{\nu=1}^{\infty}|y_\nu|^q$. To begin with, we assume that

$$\sum_{\nu=1}^{\infty}|x_\nu|^p = 1 = \sum_{\nu=1}^{\infty}|y_\nu|^q.$$

In this case, the inequality in question reduces to

$$\sum_{\nu=1}^{\infty}|x_\nu y_\nu| \leq 1.$$

To obtain this, we put successively $a = |x_\nu|^p$ and $b = |y_\nu|^q$ for $\nu = 1, 2, \ldots, n$ in Lemma 3.3.3 and then add up the inequalities so obtained to arrive at

$$\sum_{\nu=1}^{n}|x_\nu y_\nu| \leq \frac{1}{p}\sum_{\nu=1}^{n}|x_\nu|^p + \frac{1}{q}\sum_{\nu=1}^{n}|y_\nu|^q \leq \frac{1}{p}\sum_{\nu=1}^{\infty}|x_\nu|^p + \frac{1}{q}\sum_{\nu=1}^{\infty}|y_\nu|^q = \frac{1}{p} + \frac{1}{q} = 1,$$

provided that $p > 1$ (so that $q \neq \infty$). This is true for each $n \in \mathbb{N}$. Letting $n \to \infty$, we obtain the inequality in question for the special case when $\sum_{\nu=1}^{\infty}|x_\nu|^p = 1 = \sum_{\nu=1}^{\infty}|y_\nu|^q$. The general case can be reduced to the foregoing special case if we take in place of x_ν and y_ν the numbers

$$x'_\nu = x_\nu \Big/ \left(\sum_{\nu=1}^{\infty}|x_\nu|^p\right)^{\frac{1}{p}}, \quad y'_\nu = y_\nu \Big/ \left(\sum_{\nu=1}^{\infty}|y_\nu|^q\right)^{\frac{1}{q}}$$

for which

$$\sum_{\nu=1}^{\infty}|x'_\nu|^p = 1 = \sum_{\nu=1}^{\infty}|y'_\nu|^q.$$

It follows by what has been proved in the paragraph above that

$$\sum_{\nu=1}^{\infty}|x_\nu y_\nu| \leq \left(\sum_{\nu=1}^{\infty}|x_\nu|^p\right)^{\frac{1}{p}}\left(\sum_{\nu=1}^{\infty}|y_\nu|^q\right)^{\frac{1}{q}},$$

provided that $p > 1$.The case $p = 1$, $q = \infty$ is trivial.

(b) For $p = 1$, observe that

$$|x_\nu + y_\nu| \le |x_\nu| + |y_\nu|, \quad \nu = 1, 2, \ldots.$$

Summing the inequalities for $\nu = 1, 2, \ldots, n$, we obtain

$$\sum_{\nu=1}^{n} |x_\nu + y_\nu| \le \sum_{\nu=1}^{n} |x_\nu| + \sum_{\nu=1}^{n} |y_\nu| \le \sum_{\nu=1}^{\infty} |x_\nu| + \sum_{\nu=1}^{\infty} |y_\nu|,$$

which implies the desired inequality on letting $n \to \infty$.

Now suppose $p > 1$. It is obvious that

$$\left[\sum_{\nu=1}^{n} |x_\nu + y_\nu|^p\right]^{\frac{1}{p}} \le \left[\sum_{\nu=1}^{n} (|x_\nu| + |y_\nu|)^p\right]^{\frac{1}{p}}. \tag{8.20}$$

Now,

$$(|x_\nu| + |y_\nu|)^p = (|x_\nu| + |y_\nu|)^{p-1}|x_\nu| + (|x_\nu| + |y_\nu|)^{p-1}|y_\nu|.$$

Summing these identities for $\nu = 1, 2, \ldots, n$, we obtain

$$\sum_{\nu=1}^{n} (|x_\nu| + |y_\nu|)^p = \sum_{\nu=1}^{n} (|x_\nu| + |y_\nu|)^{p-1}|x_\nu| + \sum_{\nu=1}^{n} (|x_\nu| + |y_\nu|)^{p-1}|y_\nu|.$$

The first term on the right may be estimated by the Hölder Inequality of part (a) (similarly for the second term), leading to

$$\begin{aligned}\sum_{\nu=1}^{n} (|x_\nu| + |y_\nu|)^{p-1}|x_\nu| &\le \left(\sum_{\nu=1}^{n} |x_\nu|^p\right)^{\frac{1}{p}} \left(\sum_{\nu=1}^{n} (|x_\nu| + |y_\nu|)^{(p-1)q}\right)^{\frac{1}{q}} \\ &= \left(\sum_{\nu=1}^{n} |x_\nu|^p\right)^{\frac{1}{p}} \left(\sum_{\nu=1}^{n} (|x_\nu| + |y_\nu|)^p\right)^{\frac{1}{q}}.\end{aligned}$$

Thus

$$\sum_{\nu=1}^{n} (|x_\nu| + |y_\nu|)^p \le \left(\left(\sum_{\nu=1}^{n} |x_\nu|^p\right)^{\frac{1}{p}} + \left(\sum_{\nu=1}^{n} |y_\nu|^p\right)^{\frac{1}{p}}\right)\left(\sum_{\nu=1}^{n} (|x_\nu| + |y_\nu|)^p\right)^{\frac{1}{q}}.$$

Dividing by $\left(\sum_{\nu=1}^{n} (|x_\nu| + |y_\nu|)^p\right)^{1/q}$, which we may suppose to be nonzero, we obtain

$$\left(\sum_{\nu=1}^{n}(|x_\nu|+|y_\nu|)^p\right)^{\frac{1}{p}} \leq \left(\sum_{\nu=1}^{n}|x_\nu|^p\right)^{\frac{1}{p}} + \left(\sum_{\nu=1}^{n}|y_\nu|^p\right)^{\frac{1}{p}} \leq \left(\sum_{\nu=1}^{\infty}|x_\nu|^p\right)^{\frac{1}{p}} + \left(\sum_{\nu=1}^{\infty}|y_\nu|^p\right)^{\frac{1}{p}}.$$

Thus $\left\{\left(\sum_{y=1}^{n}(|x_\nu|+|y_\nu|)^p\right)^{1/p}\right\}_{n\geq 1}$ is an increasing sequence of nonnegative numbers bounded above by the sum

$$\left(\sum_{\nu=1}^{\infty}|x_\nu|^p\right)^{\frac{1}{p}} + \left(\sum_{\nu=1}^{\infty}|y_\nu|^p\right)^{\frac{1}{p}}.$$

The desired inequality now follows by using (8.20).

3.3.P22. For the space ℓ^p $(1 \leq p < \infty)$ of Problem 3.3.P21, show that

$$d_p(x,y) = \left(\sum_{\nu=1}^{\infty}|x_\nu - y_\nu|^p\right)^{\frac{1}{p}}$$

defines a complete metric, without using Theorem 3.3.10
Hint: It is a consequence of Minkowski's Inequality [Problem 3.3.P21(b)] that $d_p(x, y) \in \mathbb{R}$. Evidently, $d_p(x, y) = 0$ if and only if $x = y$. The triangle inequality for d_p also follows from Minkowski's Inequality.
Let $\{x^{(m)}\}_{m\geq 1}, x^{(m)} = \left(x_1^{(m)}, x_2^{(m)}, \ldots\right)$ denote a Cauchy sequence in ℓ^p. Then for given $\varepsilon > 0$, there exists a positive integer $n_0(\varepsilon)$ such that

$$d_p\left(x^{(n)}, x^{(m)}\right) = \left(\sum_{\nu=1}^{\infty}|x_\nu^{(n)} - x_\nu^{(m)}|^p\right)^{\frac{1}{p}} < \varepsilon \tag{8.21}$$

for all $n, m \geq n_0(\varepsilon)$. This implies $\left|x_\nu^{(n)} - y_\nu^{(m)}\right| < \varepsilon$ for $n, m \geq n_0(\varepsilon)$, that is, for each ν, the sequence $\left\{x_\nu^{(m)}\right\}_{m\geq 1}$ is a Cauchy sequence of numbers. So by Cauchy's principle of convergence, $\lim_m x_\nu^{(m)} = x_\nu$, say. Let x be the sequence $(x_1, x_2, \ldots)$. It will be shown that $x \in \ell^p$ and $\lim_m x^{(m)} = x$. From (8.21), we have

$$\sum_{\nu=1}^{N}|x_\nu^{(n)} - x_\nu^{(m)}|^p < \varepsilon^p \tag{8.22}$$

for any positive integer N, provided $n, m \geq n_0(\varepsilon)$. Letting $m \to \infty$ in (8.22), we obtain

$$\sum_{\nu=1}^{N} \left|x_\nu^{(n)} - x_\nu^{(m)}\right|^p < \varepsilon^p$$

for any positive integer N and all $n \geq n_0(\varepsilon)$. The sequence $\left\{\sum_{\nu=1}^{N} \left|x_\nu^{(n)} - x_\nu\right|^p\right\}_{N\geq 1}$ is increasing and bounded above, and therefore has a finite limit $\sum_{\nu=1}^{\infty} \left|x_\nu^{(n)} - x_\nu\right|^p$, which is less than or equal to ε^p. Hence

$$\left(\sum_{\nu=1}^{\infty} \left|x_\nu^{(n)} - x_\nu\right|^p\right)^{\frac{1}{p}} \leq \varepsilon \quad \text{for all } n \geq n_0(\varepsilon). \tag{8.23}$$

Observe that

$$\left(\sum_{\nu=1}^{\infty} |x_\nu|^p\right)^{\frac{1}{p}} \leq \left(\sum_{\nu=1}^{\infty} \left|x_\nu^{(n)} - x_\nu\right|^p\right)^{\frac{1}{p}} + \left(\sum_{\nu=1}^{\infty} \left|x_\nu^{(n)}\right|^p\right)^{\frac{1}{p}}$$

by Minkowski's inequality, and consequently, $x \in \ell^p$ and $x^{(m)} \to x$ in view of (8.23).

3.3.P23. Show that, if $k_1, k_2, \ldots, k_n > 1, \sum_{i=1}^{n} \frac{1}{k_i} = 1$ and $f_i \in L^{k_i}(X)$ for each i, then

$$\int_X \left|\prod_{i=1}^{n} f_i\right| d\mu \leq \prod_{i=1}^{n} \left(\int_X |f_i|^{k_i} d\mu\right)^{\frac{1}{k_i}}. \tag{8.24}$$

Equality occurs if and only if either $f_i = 0$ a.e. for some i or there exist positive constants c_i, $1 \leq i \leq n$, such that

$$c_i|f_i|^{k_i} = c_j|f_j|^{k_j} \text{a.e.} 1 \leq i, j \leq n. \tag{8.25}$$

Hint: We omit writing "$d\mu$" in order to reduce clutter.
For $n = 1$, the inequality (8.24) is trivial. For $n = 2$, it is just Hölder's Inequality of Theorem 3.3.5. Suppose that it holds for some $n > 1$. Then if $\alpha = (1 - k_{n+1}^{-1})^{-1}$, we have $\sum_{i=1}^{n} \alpha k_i^{-1} = \alpha \sum_{i=1}^{n} k_i^{-1} = \alpha\left(1 - k_{n+1}^{-1}\right) = 1$. So,

$$\int_X \left(\prod_{i=1}^n |f_i|^\alpha\right) \le \prod_{i=1}^n \left(\int_X |f_i|^{k_i}\right)^{\alpha/k_i}.$$

But from Hölder's Inequality,

$$\int_X \left|\prod_{i=1}^{n+1} f_i\right| \le \left(\int_X \left(\left|\prod_{i=1}^n f_i\right|^\alpha\right)\right)^{\frac{1}{\alpha}} \left(\int_X |f_{n+1}|^{k_{n+1}}\right)^{1/k_{n+1}} \le \prod_{i=1}^{n+1} \left(\int_X |f_i|^{k_i}\right)^{1/k_i}.$$

So, (8.24) holds for $n + 1$. Thus by induction, the desired inequality follows for arbitrary $n \in \mathbb{N}$.

If $f_i = 0$ a.e. for some i, then equality clearly holds in (8.24). If on the other hand (8.25) holds, then a little computation shows that both sides of (8.24) are equal to

$$\frac{c_1}{c_1^{1/k_1} c_2^{1/k_2} \ldots c_n^{1/k_n}} \left(\int_X |f_1|^{k_1}\right).$$

This shows that for equality to hold in (8.24), it is sufficient that either $f_i = 0$ a.e. for some i or (8.25) holds for some positive $c_1, \ldots, c_n$. We go on to consider necessity.

When $n = 2$, the case of equality in (8.24) has been discussed in Theorem 3.3.5. Assume that the necessity in question is valid for n functions, where $n > 1$ (induction hypothesis). Consider the case of $n + 1$ functions, none of them being zero a.e., such that

$$\int_X \left|\prod_{i=1}^{n+1} f_i\right| = \prod_{i=1}^{n+1} \left(\int_X |f_i|^{k_i}\right)^{1/k_i}. \tag{8.26}$$

It is to be shown that there exist positive constants c_i, $1 \le i \le n + 1$, such that

$$c_i |f_i|^{k_i} = c_j |f_j|^{k_j} \quad \text{a.e.} \quad 1 \le i, j \le n+1.$$

Let α be as above. From (8.26) and Hölder's Inequality, we have

$$\prod_{i=1}^{n+1} \left(\int_X |f_i|^{k_i}\right)^{1/k_i} = \int_X \left|\prod_{i=1}^{n+1} f_i\right| \le \left(\int_X \left|\prod_{i=1}^n f_i\right|^\alpha\right)^{\frac{1}{\alpha}} \left(\int_X |f_{n+1}|^{k_{n+1}}\right)^{1/k_{n+1}},$$

which, upon cancelling the common last factor on each side and raising to power α, implies

$$\prod_{i=1}^{n}\left(\int_X |f_i|^{k_i}\right)^{\alpha/k_i} \leq \left(\int_X \left|\prod_{i=1}^{n} f_i\right|^{\alpha}\right).$$

Recalling from above that $\alpha \sum_{l=1}^{n} k_i^{-1} = 1$ and applying (8.24) to the right-hand side here, we obtain

$$\left(\int_X \left|\prod_{i=1}^{n} f_i\right|^{\alpha}\right) \leq \prod_{i=1}^{n}\left(\int_X |f_i|^{k_i}\right)^{\alpha/k_i}.$$

The preceding two inequalities lead to

$$\left(\int_X \left|\prod_{i=1}^{n} f_i\right|^{\alpha}\right) = \prod_{i=1}^{n}\left(\int_X |f_i|^{k_i}\right)^{\alpha/k_i}. \tag{8.27}$$

In particular, $\left|\prod_{i-1}^{n} f_i\right|$ cannot be 0 a.e. Besides, it follows by the induction hypothesis that there exist positive constants c_i, $1 \leq i \leq n$, such that

$$c_i |f_i|^{k_i} = c_j |f_j|^{k_j} \quad \text{a.e.} \quad 1 \leq i, j \leq n. \tag{8.28}$$

It remains to show that there exists a positive c_{n+1} such that $c_{n+1}|f_{n+1}|^{k_{n+1}} = c_1 |f_1|^{k_1}$. To this end, we substitute (8.27) into (8.26), thereby obtaining

$$\int_X \left|\prod_{i=1}^{n+1} f_i\right| = \left(\int_X \left|\prod_{i=1}^{n} f_i\right|^{\alpha}\right)^{\frac{1}{\alpha}} \left(\int_X |f_{n+1}|^{k_{n+1}}\right)^{1/k_{n+1}}.$$

Since $1/\alpha + 1/k_{n+1} = 1$ (by definition of α as being $(1 - k_{n+1}{}^{-1})^{-1}$) and neither $\left|\prod_{i=1}^{n} f_i\right|$ nor $|f_{n+1}|$ is 0 a.e., the second assertion of Theorem 3.3.5 yields a positive number γ such that $\gamma |f_{n+1}|^{k_{n+1}} = \left|\prod_{i=1}^{n} f_i\right|^{\alpha}$. We claim that $c_{n+1} = \left(\prod_{i=1}^{n} c_i^{\alpha/k_i}\right)\gamma$ serves our purpose. Indeed, on the basis of (8.28), almost everywhere we have

$$
\begin{aligned}
c_{n+1}|f_{n+1}|^{k_{n+1}} &= \left(\prod_{i=1}^{n} c_i^{\alpha/k_i}\right)\left(\gamma|f_{n+1}|^{k_{n+1}}\right) = \left(\prod_{i=1}^{n} c_i^{\alpha/k_i}\right)\left|\prod_{i=1}^{n} f_i\right|^{\alpha} \\
&= \left(\prod_{i=1}^{n}\left(c_i^{1/k_i}|f_i|\right)\right)^{\alpha} = \left(\prod_{i=1}^{n}\left(c_1^{1/k_1}|f_1|^{k_1/k_i}\right)\right)^{\alpha} \\
&= \left(c_1^{1/k_1+\cdots+1/k_n}|f_1|^{k_1(1/k_1+\cdots+1/k_n)}\right)^{\alpha} \\
&= c_1|f_1|^{k_1}, \text{once again using the equality } \alpha\sum_{i=1}^{n} k_i^{-1} = 1.
\end{aligned}
$$

3.3.P24. (a) Let $\infty \geq p \geq 1$ and $f_i \in L^p(X)$ for $i = 1, 2, \ldots, n$. Show that

$$\left\|\sum_{i=1}^{n} f_i\right\|_p \leq \sum_{i=1}^{n} \|f_i\|_p.$$

(b) Let $\infty > p > 1$. If equality holds in (a), show that there exist nonnegative constants c_i, $1 \leq i \leq n$, not all 0, such that $c_i f_i = c_j f_j$ a.e. for $1 \leq i, j \leq n$.
(c) If either $p = 1$ or $p = \infty$, show that the analogue of (b) does not hold.
Hint: **(a)** For $n = 2$, this is Theorem 3.3.8. Suppose the result holds for $n = j$, where $j > 1$. Then, using Theorem 3.3.8 once again,

$$
\begin{aligned}
\left\|\sum_{i=1}^{j+1} f_i\right\|_p &= \left\|\sum_{i=1}^{j} f_i + f_{j+1}\right\|_p \leq \left\|\sum_{i=1}^{j} f_i\right\|_p + \|f_{j+1}\|_p \leq \sum_{i=1}^{j} \|f_i\|_p + \|f_{j+1}\|_p \\
&= \sum_{i=1}^{j+1} \|f_i\|_p.
\end{aligned}
$$

So the result holds for $n = j + 1$. By induction, the inequality now follows.
(b) If $f_i = 0$ a.e. for some i, then we need only choose $c_i = 1$ and $c_j = 0$ for $j \neq i$. So, we consider only the case when none of the f_i is 0 a.e.

When $n = 2$, what is to be proved is immediate from the second part of Minkowski's Inequality in Theorem 3.3.6. Assume the assertion true for some $n > 1$ and consider $n + 1$ functions f_i such that $\left\|\sum_{l=1}^{n+1} f_l\right\|_p = \sum_{l=1}^{n+1} \|f_l\|_p$, none of them being 0 a.e. Then by Minkowski's Inequality,

$$\sum_{i=1}^{n+1} \|f_i\|_p = \left\|\sum_{i=1}^{n+1} f_i\right\|_p \leq \left\|\sum_{i=1}^{n} f_i\right\|_p + \|f_{n+1}\|_p,$$

so that $\left\|\sum_{l=1}^{n} f_i\right\|_p \leq \sum_{i=1}^{n} \|f_i\|_p$. But the reverse inequality must hold by (a) and therefore

$$\sum_{i=1}^{n} \|f_i\|_p = \left\|\sum_{i=1}^{n} f_i\right\|_p.$$

Now the induction hypothesis yields nonnegative constants c_i, $1 \leq i \leq n$, not all 0, such that $c_i f_i = c_j f_j$ a.e. for $1 \leq i, j \leq n$. It remains to show that there exists a nonnegative c_{n+1} such that $c_{n+1} f_{n+1} = c_1 f_1$ a.e. Since none of the f_i is 0 a.e., we know that none of the c_i is 0. It follows that $f_i = (c_1/c_i) f_1$ for every i and hence that $\sum_{l=1}^{n} f_i = \sum_{i=1}^{n} (c_1/c_i) f_1$. Now, it follows from the statements displayed above that

$$\left\|\sum_{i=1}^{n+1} f_i\right\|_p = \left\|\sum_{i=1}^{n} f_i\right\|_p + \|f_{n+1}\|_p.$$

Applying the second part of Minkowski's Inequality in Theorem 3.3.6, we find that there exist nonnegative constants α and β, not both zero, such that $\alpha \sum_{i=1}^{n} f_i = \beta f_{n+1}$ a.e. Since f_{n+1} is not 0 a.e., we know that $\alpha \neq 0$. Combining this with the equality $\sum_{i=1}^{n} f_i = \sum_{i=1}^{n} (c_1/c_i) f_1$ proved earlier, we get

$$(\beta/\alpha)\left(\sum_{i=1}^{n} (1/c_i)\right) f_{n+1} = c_1 f_1,$$

which means $c_{n+1} = (\beta/\alpha)\left(\sum_{i=1}^{n} (1/c_i)\right)$, obviously nonnegative, serves the purpose.

(c) Let X consist of two points called a and b, with the counting measure. Consider the functions f_1 and f_2 on X given by

$$f_1(a) = f_1(b) = f_2(a) = 1 \text{ and } f_2(b) = 2.$$

Then there can be no numbers c_1 and c_2 such that $c_1 f_1 = c_2 f_2$. But

$$\|f_1\|_1 = 2, \|f_2\|_1 = 3, \|f_1 + f_2\|_1 = 5 = \|f_1\|_1 + \|f_2\|_1;$$
$$\|f_1\|_\infty = 1, \|f_2\|_\infty = 2, \|f_1 + f_2\|_\infty = 3 = \|f_1\|_\infty + \|f_2\|_\infty.$$

3.3.P25. Show that, if for some p, where $0 < p < \infty$, $f \in L^p(X) \cap L^\infty(X)$, then for all q such that $p < q < \infty$, we have $f \in L^q(X)$ and

$$\|f\|_q \le \|f\|_p^{\frac{p}{q}} \|f\|_\infty^{\left(1-\frac{p}{q}\right)}.$$

Hint: Observe that

$$|f| = |f|^{\frac{p}{q}} |f|^{\left(1-\frac{p}{q}\right)} \le \|f\|_\infty^{\left(1-\frac{p}{q}\right)} |f|^{\frac{p}{q}},$$

which implies

$$\int_X |f|^q d\mu \le \|f\|_\infty^{\left(1-\frac{p}{q}\right)q} \int_X |f^p| d\mu,$$

i.e.

$$\|f\|_q^q \le \|f\|_\infty^{\left(1-\frac{p}{q}\right)q} \|f\|_p^p$$

and hence

$$\|f\|_q \le \|f\|_\infty^{\left(1-\frac{p}{q}\right)} \|f\|_p^{\frac{p}{q}}.$$

Problem Set 4.1

4.1.P1. **(a)** [Needed in **(b)**] Prove Abel's Lemma: If $b_1 \ge b_2 \ge \cdots \ge b_n \ge 0$ and $k \le \sum_{r=1}^{p} u_r \le K$ for $1 \le p \le n$, where $u_1, u_2, \ldots, u_n$ are any n real numbers, then

$$b_1 k \le \sum_{r=1}^{n} b_r u_r \le b_1 K.$$

The next two parts together constitute what is called *Bonnet's form of the Second Mean Value Theorem for Integrals* and will be needed in Theorem 4.2.8 below.

(b) If $\int_a^b f$ and $\int_a^b g$ both exist and f is decreasing on $[a, b]$ with $f(b) \ge 0$, then there exists a $\xi \in [a, b]$ such that

$$\int_a^b fg = f(a) \int_a^\xi g.$$

(c) If $\int_a^b f$ and $\int_a^b g$ both exist and f is increasing on $[a, b]$ with $f(a) \geq 0$, then there exists a $\xi \in [a, b]$ such that

$$\int_a^b fg = f(b) \int_\xi^b g.$$

Hint: **(a)** Let $S_p = \sum_{r=1}^p u_r, 1 \leq p \leq n$. Then $\sum_{r=1}^n b_r u_r = b_1 S_1 + b_2(S_2 - S_1) + \cdots + b_n(S_n - S_{n-1}) = (b_1 - b_2)S_1 + (b_2 - b_3)S_2 + \cdots + (b_{n-1} - b_n)S_{n-1} + b_n S_n$. [This is Abel's summation formula.] Therefore $\sum_{r=1}^n b_r u_r \geq (b_1 - b_2)k + (b_2 - b_3) k + \cdots + (b_{n-1} - b_n)k + b_n k = b_1 k$ and also $\sum_{r=1}^n b_r u_r \leq (b_1 - b_2)K + (b_2 - b_3) K + \cdots + (b_{n-1} - b_n)K + b_n K = b_1 K$.
(b) Let P: $a = x_0 < x_1 < \cdots < x_n = b$ be a partition of $[a, b]$, and m_r and M_r be such that $m_r \leq g(x) \leq M_r$ for all $x \in [x_{r-1}, x_r]$. Let t_r be any point in $[x_{r-1}, x_r]$. We then have

$$m_r(x_r - x_{r-1}) \leq \int_{x_{r-1}}^{x_r} g \leq M_r(x_r - x_{r-1})$$

and

$$m_r(x_r - x_{r-1}) \leq g(t_r)(x_r - x_{r-1}) \leq M_r(x_r - x_{r-1}).$$

Summing these over $1 \leq r \leq p$, we get

$$\sum_{r=1}^p m_r(x_r - x_{r-1}) \leq \int_a^{x_p} g \leq \sum_{r=1}^p M_r(x_r - x_{r-1})$$

and

$$\sum_{r=1}^p m_r(x_r - x_{r-1}) \leq \sum_{r=1}^p g(t_r)(x_r - x_{r-1}) \leq \sum_{r=1}^p M_r(x_r - x_{r-1}),$$

which give $\left| \int_a^{x_p} g - \sum_{r=1}^p g(t_r)(x_r - x_{r-1}) \right| \leq \sum_{r=1}^p (M_r - m_r)(x_r - x_{r-1}) \leq \omega(g, P)$, where $\omega(g, P)$ is the difference between the upper and lower sums of g over the partition P. It follows that

$$\int_a^{x_p} g - \omega(g, P) \leq \sum_{r=1}^p g(t_r)(x_r - x_{r-1}) \leq \int_a^{x_p} g + \omega(g, P).$$

Since $\int_a^t g$ is a continuous function of t, it has a minimum value A and a maximum value B. Using the preceding inequality, we get

$$A - \omega(g,P) \leq \sum_{r=1}^{p} g(t_r)(x_r - x_{r-1}) \leq B + \omega(g,P), \quad \text{for } 1 \leq p \leq n.$$

From this inequality, the hypothesis that f is decreasing with $f(b) \geq 0$ and from Abel's Lemma [see part **(a)**], it follows that

$$f(a)(A - \omega(g,P)) \leq \sum_{r=1}^{n} f(t_r)g(t_r)(x_r - x_{r-1}) \leq f(a)(B + \omega(g,P)),$$

which implies $f(a)A \leq \int_a^b fg \leq f(a)B$. So, $\int_a^b fg = f(a)\mu$ for some μ between A and B. But these are the minimum and maximum values of $\int_a^t g$ and the Intermediate Value Theorem 1.3.26 yields a point $\xi \in [a, b]$ such that $\mu = \int_a^\xi g$. Thus $\int_a^b fg = f(a) \int_a^\xi g$, as required.

(c) If we reset the value $f(a)$ to be 0, then f continues to be increasing and Riemann integrable over the interval. So we assume $f(a) = 0$. Since f is increasing, it follows that $\phi(x) = f(b) - f(x)$ is decreasing; also $\phi(b) = 0$. Applying **(b)** with ϕ in place of f and collecting terms suitably, we get $\int_a^b fg = f(b) \int_\xi^b g$.

4.1.P2. Find the Fourier coefficients of the periodic function f for which

$$f(x) = \begin{cases} -\alpha & -\pi \leq x < 0 \\ \alpha & 0 \leq x < \pi. \end{cases}$$

Hint: From Example 4.1.5(a), the Fourier series of the function g such that

$$g(x) = \begin{cases} 0 & -\pi \leq x < 0 \\ 2\alpha & 0 \leq x < \pi \end{cases}$$

is $g(x) \sim \alpha + \frac{4\alpha}{\pi} \sum_{n=0}^{\infty} \frac{\sin(2n+1)x}{2n+1}$. The given function f here can be obtained from g as $f = g - \alpha$ and therefore its Fourier series is

$$f(x) \sim \frac{4\alpha}{\pi} \sum_{n=0}^{\infty} \frac{\sin(2n+1)x}{2n+1}.$$

4.1.P3. Using Bonnet's form of the Second Mean Value Theorem for Integrals of 4.1.P1(c), prove that for a monotone function f on $[-\pi, \pi]$, the Fourier coefficients a_n and b_n ($n \neq 0$) satisfy the inequalities

$$|a_n| \le \frac{1}{n\pi}|f(\pi) - f(-\pi)|, \quad |b_n| \le \frac{1}{n\pi}|f(\pi) - f(-\pi)|.$$

Hint: When f is increasing, the function $f(x) - f(-\pi)$ is increasing as well as nonnegative, and since $\pi a_n = \int_{-\pi}^{\pi}(f(x) - f(-\pi))\cos nx\, dx$, by 4.1.P1(c), we have

$$\begin{aligned}\pi a_n = [f(\pi) - f(-\pi)]\int_{\xi}^{\pi}\cos nx\, dx &= [f(\pi) - f(-\pi)]\frac{\sin n\pi - \sin n\xi}{n}\\ &= [f(\pi) - f(-\pi)]\frac{-\sin n\xi}{n}.\end{aligned}$$

4.1.P4. Suppose f is Lebesgue integrable with period 2π and "modulus of continuity" ω, which means

$$\omega(\delta) = \sup\{|f(x) - f(y)| : |x - y| \le \delta\}.$$

Show that its Fourier coefficients satisfy $|a_n| \le \omega(\pi/n)$, $|b_n| \le \omega(\pi/n)$, $(n \neq 0)$. Hint: On substituting $t = x + \frac{\pi}{n}$ in $a_n = \frac{1}{\pi}\int_{-\pi}^{\pi} f(x)\cos nx\, dx$, we get another integral for a_n over $[-\pi, \pi]$:

$$a_n = \frac{1}{\pi}\int_{-\pi+\pi/n}^{\pi+\pi/n} f\left(x - \frac{\pi}{n}\right)\cos(nx - \pi)dx = -\frac{1}{\pi}\int_{-\pi}^{\pi} f\left(x - \frac{\pi}{n}\right)\cos nx\, dx.$$

Adding these two integrals over $[-\pi, \pi]$, we get

$$2a_n = \frac{1}{\pi}\int_{-\pi}^{\pi}\left[f(x) - f\left(x - \frac{\pi}{n}\right)\right]\cos nx\, dx.$$

Therefore

$$2|a_n| \le \frac{1}{\pi}\int_{-\pi}^{\pi}\left|f(x) - f\left(x - \frac{\pi}{n}\right)\right||\cos nx|dx \le \frac{1}{\pi}\omega\left(\frac{\pi}{n}\right)\cdot 2\pi = 2\omega\left(\frac{\pi}{n}\right).$$

Similarly for b_n.

4.1.P5. For the trigonometric series $\frac{1}{2}a_0 + \sum_{n=1}^{\infty}(a_n \cos nx + b_n \sin nx)$, suppose $\sum_{k=1}^{\infty}(|a_k| + |b_k|)$ converges. Show that the trigonometric series converges uniformly and that it is the Fourier series of its own sum.

Hint: Let $s_p(x) = \frac{1}{2}a_0 + \sum_{k=1}^{p}(a_k \cos kx + b_k \sin kx)$. Then for $n > m$, $|s_n(x) - s_m(x)|$ $= \left|\sum_{k=m+1}^{n}(a_k \cos kx + b_k \sin kx)\right| \leq \sum_{k=m+1}^{n}(|a_k| + |b_k|)$. Thus the sequence of partial sums $\{s_n(x)\}$ is uniformly Cauchy and hence converges (uniformly) to a continuous limit $f(x)$. Apply Proposition 4.1.3.

4.1.P6. A finite set of continuous functions $g_1, \ldots, g_n$ on $[a, b]$ is said to be an **orthonormal** system with weight function w (Lebesgue integrable and nonnegative) if

$$\langle g_i, g_j \rangle = \int_a^b g_i(x) g_j(x) w(x) dx = \begin{cases} 0 & i \neq j \\ 1 & i = j \end{cases}.$$

(Example: By Lemma 4.1.2, any finite number of functions among $\frac{1}{2}$, $\cos kx$, $\sin kx$ form an orthonormal set with domain $[-\pi, \pi]$ and weight $w(x) = \frac{1}{\pi}$.) For the function $\sum_{k=1}^{n} a_k g_k(x)$, show that

$$\sum_{k=1}^{n} |a_k| \leq \sqrt{n} \cdot \left(\int_a^b w(x) dx\right)^{\frac{1}{2}} \max\left\{\left|\sum_{k=1}^{n} a_k g_k(x)\right| : x \in [a, b]\right\}.$$

Hint: $\sum_{k=1}^{n} |a_k| \leq \sqrt{n} \cdot \left(\sum_{k=1}^{n} a_k^2\right)^{1/2}$ by the Cauchy–Schwarz Inequality. Also,

$$\sum_{k=1}^{n} a_k^2 = \left\langle \sum_{j=1}^{n} a_j g_j, \sum_{\ell=1}^{n} a_\ell g_\ell \right\rangle = \int_a^b \left(\sum_{k=1}^{n} a_k g_k(x)\right)^2 w(x) dx$$
$$\leq \left(\max\left\{\left|\sum_{k=1}^{n} a_k g_k(x)\right| : x \in [a, b]\right\}\right)^2 \int_a^b w(x) dx.$$

4.1.P7. Suppose the sequence of partial sums s_n of the trigonometric series a_0 + $(a_n \cos nx + b_n \sin nx)$ has a subsequence $s_{n(k)}$ that converges uniformly to a function f. Show that the trigonometric series is the Fourier series corresponding to f.

Hint: Consider any $p \in \mathbb{N}$. We shall show that $a_p = \frac{1}{\pi}\int_{-\pi}^{\pi} f(x) \cos px \, dx$. To this end, let ε be any positive number. Choose $k \in \mathbb{N}$ such that $|s_{n(k)} - f| < \varepsilon/2\pi$

everywhere on $[-\pi, \pi]$ and also $n(k) > p$. Then $|\frac{1}{\pi}\int_{-\pi}^{\pi}[s_{n(k)}(x) - f(x)]\cos px\, dx| < \varepsilon$. But $\frac{1}{\pi}\int_{\pi}^{-\pi} s_{n(k)}(x)\cos px\, dx = a_p$ in view of the choice $n(k) > p$. This shows that $|a_p - \frac{1}{\pi}\int_{\pi}^{-\pi} f(x)\cos px\, dx| < \varepsilon$.

Problem Set 4.2

4.2.P1. Suppose that f is a 2π-periodic function that is Lebesgue integrable on $[-\pi, \pi]$ and is differentiable at a point x_0. Prove that

$$\lim_{n\to\infty} s_n(x_0) = f(x_0),$$

where s_n denotes the partial sum of the Fourier series of f.
Hint: Using Remark 4.2.4 and Proposition 4.2.3(c), we may write

$$\begin{aligned} s_n(x_0) - f(x_0) &= \frac{1}{\pi}\int_{-\pi}^{\pi} [f(x_0 - t) - f(x_0)]D_n(t)dt \\ &= \frac{1}{\pi}\int_{-\pi}^{\pi} \frac{f(x_0 - t) - f(x_0)}{t}\frac{\frac{1}{2}t}{\sin\frac{1}{2}t}\sin\left(n + \frac{1}{2}t\right)dt. \end{aligned}$$

Since $\lim_{t\to 0}\frac{f(x_0-t)-f(x_0)}{t} = f'(x_0)$, it follows that, for $\varepsilon > 0$, there exists a $\delta > 0$ such that

$$\left|\frac{f(x_0 - t) - f(x_0)}{t} - f'(x_0)\right| < \varepsilon \quad \text{for } 0 < t < \delta$$

that is,

$$f'(x_0) - \varepsilon < \frac{f(x_0 - t) - f(x_0)}{t} < f'(x_0) + \varepsilon \quad \text{for } 0 < t < \delta.$$

Thus by the assumption on f, the difference quotient $\frac{f(x_0-t)-f(x_0)}{t}$ is bounded near zero and is Lebesgue integrable on $[-\pi, \pi]$. The factor $(t/2)/(\sin(t/2))$ is bounded on $[-\pi, \pi]$. Therefore

$$\frac{f(x_0 - t) - f(x_0)}{t}\frac{\frac{1}{2}t}{\sin\frac{1}{2}t}$$

is Lebesgue integrable over $[-\pi, \pi]$. The required conclusion now follows from the Riemann Lebesgue Theorem 4.2.5.

4.2.P2. **(Bessel's Inequality)** If f is any Riemann integrable periodic function, its Fourier coefficients satisfy the inequality

$$\frac{1}{2}a_0^2 + \sum_{n=1}^{\infty}\left(a_n^2 + b_n^2\right) \le \frac{1}{\pi}\int_{-\pi}^{\pi} f(x)^2 dx.$$

Note that it is part of the assertion that the series on the left is convergent. [It follows from this result that the Fourier coefficients a_n and b_n of a Riemann integrable function tend to zero as $n \to \infty$.]
Hint: For $n \ge 1$, let $s_n(x)$ be the partial sum of the Fourier series of f. By using Lemma 4.1.2 and Definition 4.1.4, it follows that

$$\frac{1}{\pi}\int_{-\pi}^{\pi} s_n(x)^2 dx = \frac{1}{2}a_0^2 + \sum_{k=1}^{n}\left(a_k^2 + b_k^2\right) = \frac{1}{\pi}\int_{-\pi}^{\pi} f(x)s_n(x)dx.$$

Now,

$$\begin{aligned}\frac{1}{\pi}\int_{-\pi}^{\pi} [f(x) - s_n(x)]^2 dx &= \frac{1}{\pi}\int_{-\pi}^{\pi} f(x)^2 dx - \frac{2}{\pi}\int_{-\pi}^{\pi} f(x)s_n(x)dx + \frac{1}{\pi}\int_{-\pi}^{\pi} s_n(x)^2 dx \\ &= \frac{1}{\pi}\int_{-\pi}^{\pi} f(x)^2 dx - \frac{1}{2}a_0^2 - \sum_{k=1}^{n}\left(a_k^2 + b_k^2\right).\end{aligned}$$

Since $[f(x) - s_n(x)]^2 \ge 0$ and the integral of a nonnegative function is nonnegative, the required inequality follows. Moreover, the square of a Riemann integrable function must be Riemann integrable, and therefore it also follows that the series involved is convergent.

4.2.P3. The series $\sum_{n=1}^{\infty} \frac{\cos nx}{\sqrt{n}}$ is uniformly convergent on every closed subinterval $[\alpha, 2\pi - \alpha]$, $0 < \alpha < \pi$ of $(0, 2\pi)$. Show that it is not the Fourier series of a Riemann integrable function.
Hint: The uniform convergence is proved as in Example 4.2.14(b) using Dirichlet's Test [28, Theorem 4.4.2] and the summation formula

$$\sum_{k=1}^{n-1} \cos kx = \frac{-\frac{1}{2}\left(\sin\left(n - \frac{1}{2}\right)x - \sin\frac{x}{2}\right)}{\sin\frac{x}{2}} \quad \text{for } \sin\frac{x}{2} \ne 0.$$

Since the series $\sum_{n=1}^{\infty} \frac{1}{n}$ is divergent, it follows from 4.2.P2 that the given series is not a Fourier series of a Riemann integrable function.

4.2.P4. Is the series $\sum_{n=1}^{\infty} \frac{\sin nx}{\ln(n+1)}$ the Fourier series of a continuous function?

Hint: No. If it were the Fourier series of any Riemann integrable function, then by 4.2.P2 the series $\sum_{n=1}^{\infty} \frac{1}{(\ln(n+1))^2}$ would be convergent, which it is not, as can be seen by comparison with $\sum \frac{1}{n}$.

4.2.P5. Show that $\frac{2}{\pi}\int_0^{\pi} \left|\frac{\sin\left(n+\frac{1}{2}t\right)}{2\sin\frac{1}{2}t}\right| dt \geq \frac{4}{\pi^2}\sum_{k=1}^{n}\frac{1}{k} \geq \frac{4}{\pi^2}\left(\frac{1}{n} + \ln n\right)$.

Hint: Since $|\sin x| \leq |x|$ for all x, we get

$$\frac{2}{\pi}\int_0^{\pi}\left|\frac{\sin\left(n+\frac{1}{2}t\right)}{2\sin\frac{1}{2}t}\right| dt \geq \frac{2}{\pi}\int_0^{\pi}\left|\sin\left(n+\frac{1}{2}t\right)\right|\frac{1}{t}dt = \frac{2}{\pi}\int_0^{\left(n+\frac{1}{2}\right)\pi} |\sin t|\frac{1}{t}dt$$
$$\geq \frac{2}{\pi}\sum_{k=1}^{n}\frac{1}{k\pi}\int_{(k-1)\pi}^{k\pi} |\sin t| dt = \frac{4}{\pi^2}\sum_{k=1}^{n}\frac{1}{k}.$$

It may be noted that since $\sum_{k=1}^{n}\frac{1}{k} \geq \frac{1}{n} + \int_1^n \frac{1}{x}dx = \frac{1}{n} + \ln n$, it follows that

$$\frac{2}{\pi}\int_0^{\pi}\left|\frac{\sin\left(n+\frac{1}{2}t\right)}{2\sin\frac{1}{2}t}\right| dt \geq \frac{4}{\pi^2}\left(\frac{1}{n} + \ln n\right).$$

4.2.P6. Show that $\frac{2}{\pi}\int_0^{\pi}\left|\frac{\sin\left(n+\frac{1}{2}t\right)}{2\sin\frac{1}{2}t}\right| dt \leq \ln n + \frac{1}{n\pi} + \ln \pi + \frac{2}{\pi}$.

Hint: Split the integral as $I + J$, where

$$I = \frac{2}{\pi}\int_0^{\frac{1}{n}}\left|\frac{\sin\left(n+\frac{1}{2}t\right)}{2\sin\frac{1}{2}t}\right| dt \quad \text{and} \quad J = \frac{2}{\pi}\int_{\frac{1}{n}}^{\pi}\left|\frac{\sin\left(n+\frac{1}{2}t\right)}{2\sin\frac{1}{2}t}\right| dt.$$

Since $\left|\frac{\sin\left(n+\frac{1}{2}\right)t}{2\sin\frac{1}{2}t}\right| = \left|\frac{1}{2} + \sum_{k=1}^{n}\cos kt\right| \leq n + \frac{1}{2}$, we have $I \leq \frac{2}{\pi}\left(1 + \frac{1}{2n}\right)$. Moreover, since $\sin t \geq 2t/\pi$ when $0 \leq t \leq \pi/2$, we also have $J \leq \frac{2}{\pi}\frac{\pi}{2}\int_{1/n}^{\pi}\frac{1}{t}dt = \ln n + \ln \pi$. Now add these two inequalities.

4.2.P7. Let the Fourier series corresponding to a continuous periodic function $f(x)$ be given by $\frac{1}{2}a_0 + \sum_{n=1}^{\infty}(a_n\cos nx + b_n\sin nx)$. Show that the series obtained by integrating this Fourier series term by term converges to $\int_0^x f(t)dt$. [The remarkable thing about this result is that the Fourier series of f is not assumed to converge to it. Note however that the integrated series is not a trigonometric series.]

Hint: Consider $g(x) = \int_0^x \left[f(t) - \frac{1}{2}a_0\right] dt$. By the Fundamental Theorem of Calculus, the function g is differentiable at each x and therefore satisfies a Lipschitz condition of order 1. By Theorem 4.2.13, it has a Fourier series which converges to it uniformly. Denote its Fourier coefficients by c_k, d_k. We evaluate them by integrating by parts and using the obvious fact that $g(0) = 0 = g(2\pi)$:

$$\begin{aligned}
c_k &= \frac{1}{\pi}\left[g(t)\frac{1}{k}\sin kt\Big|_{-\pi}^{\pi} - \int_{-\pi}^{\pi} g'(t)\frac{1}{k}\sin kt\,dt\right] \\
&= -\frac{1}{k\pi}\int_{-\pi}^{\pi}\left[f(t) - \frac{1}{2}a_0\right]\sin kt\,dt = -\frac{1}{k\pi}\int_{-\pi}^{\pi} f(t)\sin kt\,dt + \frac{1}{2}a_0\frac{1}{k\pi}\int_{-\pi}^{\pi}\sin kt\,dt \\
&= -\frac{1}{k}b_k,
\end{aligned}$$

and

$$\begin{aligned}
d_k &= \frac{1}{\pi}\left[-g(t)\frac{1}{k}\cos kt\Big|_{-\pi}^{\pi} + \int_{-\pi}^{\pi} g'(t)\frac{1}{k}\cos kt\,dt\right] \\
&= \frac{1}{k\pi}\int_{-\pi}^{\pi}\left[f(t) - \frac{1}{2}a_0\right]\cos kt\,dt = \frac{1}{k\pi}\int_{-\pi}^{\pi} f(t)\cos kt\,dt - \frac{1}{2}a_0\frac{1}{k\pi}\int_{-\pi}^{\pi}\cos kt\,dt \\
&= \frac{1}{k}a_k,
\end{aligned}$$

where we have used the fact that $g(\pi) = g(-\pi)$. Using these values of the Fourier coefficients of g and the fact that the series converges uniformly to g, we get

$$g(x) = \frac{1}{2}c_0 + \sum_{k=1}^{\infty}\frac{a_k\sin kx - b_k\cos kx}{k}.$$

Put $x = 0$ to obtain $0 = g(0) = \frac{1}{2}c_0 - \sum_{k=1}^{n}\frac{1}{k}b_k$ or $\frac{1}{2}c_0 = \sum_{k=1}^{n}\frac{1}{k}b_k$. From this we get

$$g(x) = \sum_{k=1}^{\infty}\frac{a_k\sin kx + b_k(1 - \cos kx)}{k}.$$

Together with the definition of g, this implies

$$\int_0^x f(t)dt = \frac{1}{2}a_0 x + \sum_{k=1}^{\infty} \frac{a_k \sin kx + b_k(1 - \cos kx)}{k}.$$

This is exactly what we would have obtained if we had integrated the Fourier series of f term by term.

4.2.P8. If f is continuously differentiable on $[a, b]$, use integration by parts (but *not* the Riemann–Lebesgue Theorem 4.2.5) to show that

$$\lim_{\lambda\to\infty} \int_a^b f(t) \sin \lambda t \, dt = 0 = \lim_{\lambda\to\infty} \int_a^b f(t) \cos \lambda t \, dt.$$

Hint: Integrating by parts, we get

$$\lim_{\lambda\to\infty} \int_a^b f(t) \sin \lambda t \, dt = f(t)\left(-\frac{\cos \lambda t}{\lambda}\right)\Bigg|_a^b + \int_a^b \frac{\cos \lambda t}{\lambda} f'(t) \, dt.$$

Now,

$$\left|\frac{f(a)\cos \lambda a - f(b)\cos \lambda b}{\lambda}\right| \le \frac{2M_1}{\lambda}, \quad \text{where } M_1 \ge \sup\{|f(x)| : x \in [a, b]\}$$

and

$$\left|\int_a^b \frac{\cos \lambda t}{\lambda} f'(t) dt\right| \le \frac{M_2}{\lambda}, \quad \text{where } M_2 \ge \int_a^b |f'(t)| \, dt.$$

Hence $\left| \int_a^b f(t) \sin \lambda t \, dt \right| \le (2M_1 + M_2)/\lambda$, which tends to 0 as $\lambda \to \infty$. The argument for the other integral is similar.

4.2.P9. Let f be continuous with period 2π and Fourier coefficients a_k, b_k. Show that

$$\sum_{k=0}^{n} (|a_k| + |b_k|) \le 2(2n+1)^{\frac{1}{2}} \sup\{|f(x)| : x \in [-\pi, \pi]\}.$$

Hint: By the Cauchy–Schwarz Inequality,

$$\sum_{k=0}^{n}(|a_k|+|b_k|)\leq(2n+1)^{\frac{1}{2}}\left[\sum_{k=0}^{n}(a_k^2+b_k^2)\right]^{\frac{1}{2}}.$$

In view of Bessel's Inequality [see Problem 4.2.P2]

$$\begin{aligned}\sum_{k=0}^{n}(a_k^2+b_k^2)&\leq\frac{1}{2}a_0^2+\frac{1}{\pi}\int_{-\pi}^{\pi}f(x)^2dx\\&\leq\frac{1}{2}\left(\frac{1}{\pi}\int_{-\pi}^{\pi}f(x)dx\right)^2+\frac{1}{\pi}\int_{-\pi}^{\pi}f(x)^2dx\\&\leq\frac{1}{2}\left(\frac{1}{\pi}2\pi(\sup\{|f(x)|:x\in[-\pi,\pi]\})\right)^2\\&\qquad+\frac{1}{\pi}2\pi(\sup\{|f(x)|:x\in[-\pi,\pi]\})^2\\&\leq 2(\sup\{|f(x)|:x\in[-\pi,\pi]\})^2+2(\sup\{|f(x)|:x\in[-\pi,\pi]\})^2\\&\leq 4(\sup\{|f(x)|:x\in[-\pi,\pi]\})^2.\end{aligned}$$

Therefore

$$\sum_{k=0}^{n}(|a_k|+|b_k|)\leq 2(2n+1)^{\frac{1}{2}}\sup\{|f(x)|:x\in[-\pi,\pi]\}.$$

4.2.P10. [Cf. Problem 3.2.P24] Show that the improper integral $\int_0^{\infty}\frac{\sin x}{x}dx$ converges and evaluate it by applying Dirichlet's Theorem 4.2.10. It is understood that $\frac{\sin x}{x}$ is to be replaced by 1 when $x = 0$.

Hint: Since $\frac{\sin x}{x}$ is continuous on [0, 1], $\int_0^1\frac{\sin x}{x}dx$ exists. Now, $\int_1^A\frac{\sin x}{x}\,dx=\cos 1-\frac{\cos A}{A}-\int_1^A\frac{\cos x}{x^2}dx$ and the improper integral $\int_1^{\infty}\frac{\cos x}{x^2}dx$ converges, as can be seen by comparing it with $\int_1^{\infty}\frac{1}{x^2}dx$. To evaluate, consider the periodic function $f:\mathbb{R}\to\mathbb{R}$ given on the interval $[-\pi, \pi]$ by $f(x)=\frac{\sin(x/2)}{x}$ with the understanding that $f(0)=\frac{1}{2}$. Then f is continuous everywhere and satisfies the hypotheses of Dirichlet's Theorem 4.2.10. Therefore

$$\frac{1}{2} = \lim_{n\to\infty} \frac{1}{\pi} \int_{-\pi}^{\pi} f(0-u)D_n(u)du = \lim_{n\to\infty} \frac{1}{\pi} \int_{-\pi}^{\pi} \frac{\sin\left(n+\frac{1}{2}\right)u}{2\sin\frac{1}{2}u} \cdot \frac{\sin\frac{1}{2}u}{u} du$$
$$= \lim_{n\to\infty} \frac{1}{\pi} \int_{-\pi}^{\pi} \frac{\sin\left(n+\frac{1}{2}\right)u}{2u} du$$
$$= \lim_{n\to\infty} \frac{1}{2\pi} \int_{-\left(n+\frac{1}{2}\pi\right)}^{\left(n+\frac{1}{2}\pi\right)} \frac{\sin x}{x} dx = \frac{1}{\pi} \int_0^{\infty} \frac{\sin x}{x} dx$$

since this integral converges.

Problem Set 4.3

4.3.P1. If f: $[a, b] \to \mathbb{R}$ is continuous and $\varepsilon > 0$, then prove by using Fejér's Theorem 4.3.4 that there exists a polynomial $P(x)$ such that

$$\sup\{|f(x) - P(x)| : x \in [a, b]\} < \varepsilon.$$

[This is known as the Weierstrass Polynomial Approximation Theorem.]
Hint: Step 1. Suppose f is a continuous even function on $\mathbb{R}$ which is of period 2π. By Fejér's Theorem 4.3.4, for $\varepsilon > 0$, there must exist an integer, $n + 1$ say, such that

$$\sup\{|f(x) - \sigma_{n+1}(x)| : x \in [-\pi, \pi]\} < \varepsilon.$$

where

$$\sigma_{n+1}(x) = \frac{s_0(x) + s_1(x) + \cdots + s_n(x)}{n+1},$$

s_k being the partial sums of the Fourier series of f. Observe that, in view of the evenness of f, the Fourier coefficients b_n are all zero and therefore

$$s_k(x) = \frac{1}{2}a_0 + a_1 \cos x + \cdots + a_k \cos kx.$$

Step 2. There exists a polynomial T_m such that

$$\cos mx = T_m(\cos x) \quad \text{for } x \in [-\pi, \pi].$$

[This is of independent interest apart from being a step in the present proof.] The result is true for $m = 0$ (take $T_0(t) = 1$) and for $m = 1$(take $T_1(t) = t$). Suppose it is true for all $n < m$, where $m \geq 2$. Then

$$\begin{aligned}\cos mx &= \cos mx + \cos(m-2)x - \cos(m-2)x \\ &= \cos((m-1)x + x) + \cos((m-1)x - x) - \cos(m-2)x \\ &= 2\cos(m-1)x\cos x - \cos(m-2)x \\ &= T_m(\cos x),\end{aligned}$$

where $T_m(t) = 2t{\bullet}T_{m-1}(t) - T_{m-2}(t)$.

The results of these two steps may be summarised as below.

Let f be a continuous even function defined on $\mathbb{R}$, having period 2π. Then, for $\varepsilon > 0$, there exists an integer n such that

$$\sup\left\{\left|f(x) - \sum_{k=0}^{n} c_k T_k(\cos x)\right| : x \in [-\pi, \pi]\right\} < \varepsilon.$$

Next we prove the existence of the polynomial P when the domain of f is $[0, 1]$.

Let f: $[0, 1] \to \mathbb{R}$ be a continuous function and let $\varepsilon > 0$ be given. Define $g:\mathbb{R} \to \mathbb{R}$ by

$$g(x) = f(|\cos x|).$$

Then g is continuous on $\mathbb{R}$, has period 2π and $g(-x) = g(x)$. It follows from what has been proved above that there exist polynomials $T_0, \ldots, T_n$ and constants $c_0, \ldots, c_n$ such that

$$\sup\left\{\left|g(x) - \sum_{k=0}^{n} c_k T_k(\cos x)\right| : x \in [-\pi, \pi]\right\} < \varepsilon.$$

Hence

$$\begin{aligned}&\sup\left\{\left|f(x) - \sum_{k=0}^{n} c_k T_k(x)\right| : x \in [0, 1]\right\} \\ &\leq \sup\left\{\left|f(|\cos x|) - \sum_{k=0}^{n} c_k T_k(|\cos x|)\right| : x \in [-\pi, \pi]\right\} < \varepsilon.\end{aligned}$$

Writing $P(x) = \sum_{k=0}^{n} c_k T_k(x)$, we obtain the required polynomial.

Finally we prove the existence of P when the domain is any closed bounded interval $[a, b]$. Consider $f(a + y(b - a))$. This function of y is defined and continuous on $[0, 1]$. Hence it follows from what has been proved so far that there exists a polynomial $Q(y)$ such that

$$\sup\{|f(a+y(b-a)) - Q(y)| : y \in [0, 1]\} < \varepsilon.$$

For $x \in [a, b]$, we have $y = \frac{x-a}{b-a} \in [0, 1]$ and $x = a + y(b - a)$. Thus

$$\sup\left\{\left|f(x) - Q\left(\frac{x-a}{b-a}\right)\right| : x \in [a, b]\right\} < \varepsilon.$$

Therefore $P(x) = Q\left(\frac{x-a}{b-a}\right)$ has the required property.

4.3.P2. Use the Weierstrass Polynomial Approximation Theorem to prove that if f and g are continuous functions on $[a, b]$ such that

$$\int_a^b x^n f(x)dx = \int_a^b x^n g(x)dx \quad \text{for all } n \geq 0,$$

then $f = g$.

Hint: We need only show that if $\int_a^b x^n f(x)dx = 0$ for all $n \geq 0$, then $f(x) = 0$ everywhere on $[a, b]$. If f satisfies this hypothesis, then $\int_a^b f(x)P(x)dx = 0$ for any polynomial function P. Let M be an upper bound for $|f|$ on $[a, b]$. For any $\varepsilon > 0$, there exists [by the Weierstrass Polynomial Approximation Theorem] a polynomial function P such that $|f - P| < \varepsilon/M$ on $[a, b]$. Then

$$0 \leq \int_a^b f(x)^2 dx \leq \left|\int_a^b f(x)(f(x) - P(x))dx\right| + \left|\int_a^b f(x)P(x)dx\right|$$
$$\leq M(\varepsilon/M) + 0 = \varepsilon.$$

Since $\varepsilon > 0$ is arbitrary, it follows that $\int_a^b f(x)^2 dx = 0$. It is now sufficient to show that if f is a continuous nonnegative-valued function on $[a, b]$ with $\int_a^b f = 0$, then $f(x) = 0$ everywhere on $[a, b]$. If not, then $f(x_0) > 0$, where $x_0 \in [a, b]$. As f is continuous, there exists a $\delta > 0$ for which $|f(x) - f(x_0)| < \frac{1}{2} f(x_0)$ whenever $|x - x_0| < \delta$. Then $f(x) > \frac{1}{2} f(x_0)$ whenever $|x - x_0| < \delta$. Consequently,

$$\int_a^b f(x)dx = \int_a^{x_0-\delta} f(x)dx + \int_{x_0-\delta}^{x_0+\delta} f(x)dx + \int_{x_0+\delta}^b f(x)dx$$

$$\geq \int_{x_0-\delta}^{x_0+\delta} f(x)dx \text{ by nonnegativity of } f$$

$$> \frac{1}{2}f(x_0)2\delta = f(x_0)\delta > 0, \text{ a contradiction.}$$

4.3.P3. If a_n and b_n are the Fourier coefficients of f, show that the Fejér sums σ_n are

$$\sigma_n(x) = \frac{1}{2}a_0 + \sum_{k=1}^{n-1}\left(1 - \frac{k}{n}\right)(a_k \cos kx + b_k \sin kx).$$

Use this fact to show that

$$\frac{1}{\pi}\int_{-\pi}^{\pi} \sigma_n(x)^2 dx = \frac{1}{2}a_0^2 + \sum_{k=1}^{n-1}\left(1 - \frac{k}{n}\right)^2 (a_k^2 + b_k^2).$$

Deduce **Parseval's Theorem** that if f has period 2π and is continuous with Fourier coefficients a_n, b_n, then

$$\frac{1}{\pi}\int_{-\pi}^{\pi} f(x)^2 dx = \frac{1}{2}a_0^2 + \sum_{k=1}^{\infty} (a_k^2 + b_k^2).$$

Hint: Since $s_p(x) = \frac{1}{2}a_0 + \sum_{k=1}^{p} (a_k \cos kx + b_k \sin kx)$, we have

$$\sigma_n(x) = \frac{1}{n}\sum_{p=0}^{n-1} s_p(x) = \frac{1}{2}a_0 + \frac{n-1}{n}(a_1 \cos x + b_1 \sin x) + \frac{n-2}{n}(a_2 \cos 2x + b_2 \sin 2x)$$

$$+ \cdots + \frac{n-k}{n}(a_k \cos kx + b_k \sin kx) + \cdots + \frac{1}{n}(a_{n-1} \cos(n-1)x + b_{n-1} \sin(n-1)x)$$

$$= \frac{1}{2}a_0 + \sum_{k=1}^{n-1}\left(1 - \frac{k}{n}\right)(a_k \cos kx + b_k \sin kx).$$

Now $\frac{1}{\pi}\int_{-\pi}^{\pi}\sigma_n(x)^2dx = \frac{1}{\pi}\int_{-\pi}^{\pi}\left[\frac{1}{2}a_0\sum_{k=1}^{n-1}\left(1-\frac{k}{n}\right)(a_k\cos kx + b_k\sin)kx\right]^2 dx = \frac{1}{2}a_0^2 + \sum_{k=1}^{n-1}\left(1-\frac{k}{n}\right)^2(a_k^2+b_k^2)$ by Lemma 4.1.2. Since $\sigma_n \to f$ uniformly on $[-\pi, \pi]$ by Fejér's Theorem 4.3.4, it follows that

$$\begin{aligned}\frac{1}{\pi}\int_{-\pi}^{\pi}f(x)^2dx &= \frac{1}{\pi}\int_{-\pi}^{\pi}\lim_{n\to\infty}\sigma_n(x)^2dx = \lim_{n\to\infty}\frac{1}{\pi}\int_{-\pi}^{\pi}\sigma_n(x)^2dx\\ &= \lim_{n\to\infty}\left[\frac{1}{2}a_0^2 + \sum_{k=1}^{n-1}\left(1-\frac{k}{n}\right)(a_k^2+b_k^2)\right]\\ &\le \frac{1}{2}a_0^2 + \sum_{k=1}^{\infty}(a_k^2+b_k^2).\end{aligned}$$

Now apply Bessel's Inequality (Problem 4.2.P2).

4.3.P4. Two different periodic functions can have the same Fourier series (e.g. if they differ only on a finite set of points). Can this happen if the functions are continuous?

Hint: No, in view of Corollary 4.3.6.

4.3.P5. If f and g are continuous functions of period 2π with Fourier coefficients a_n, b_n and α_n, β_n respectively, show that

$$\frac{1}{\pi}\int_{-\pi}^{\pi}f(x)g(x)dx = \frac{1}{2}a_0\alpha_0 + \sum_{n=1}^{\infty}(a_n\alpha_n + b_n\beta_n).$$

Hint: Observe that the Fourier coefficients of the function $f-g$ are $\frac{1}{2}(a_0-\alpha_0), a_n-\alpha_n$ and $b_n-\beta_n$, $n = 1, 2, 3, \ldots$. Applying Parseval's Theorem (see Problem 4.3.P3) to the function $f-g$ and also to f and g, we obtain

$$\frac{1}{\pi}\int_{-\pi}^{\pi}(f-g)^2 = \frac{1}{2}(a_0-\alpha_0)^2 + \sum_{n=1}^{\infty}(a_n-\alpha_n)^2 + \sum_{n=1}^{\infty}(b_n-\beta_n)^2,$$

from which the desired equality follows.

4.3.P6. Suppose $0 < \delta < \pi$ and $f(x) = 1$ if $|x| \le \delta$, $f(x) = 0$ if $\delta < |x| < \pi$, and $f(x + 2\pi) = f(x)$ for all x.

(a) Compute the Fourier coefficients of f.

(b) Conclude that $\sum_{n=1}^{\infty}\frac{\sin n\delta}{n} = \frac{\pi-\delta}{2}$ when $0 < \delta < \pi$.

(c) Deduce from Parseval's Theorem (see Problem 4.3.P3) that $\sum_{n=1}^{\infty} \frac{\sin^2 n\delta}{n^2\delta} = \frac{\pi-\delta}{2}$ when $0 < \delta < \pi$.

(d) Let $\delta \to 0$ and prove that $\int_0^{\infty} \left(\frac{\sin x}{x}\right)^2 dx = \frac{\pi}{2}$.

Hint: **(a)** $a_0 = \frac{1}{\pi}\int_{-\delta}^{\delta} dx = \frac{2\delta}{\pi}, a_n = \frac{1}{\pi}\int_{-\delta}^{\delta} \cos nx\, dx = \frac{2}{n\pi}\sin n\delta$ and every b_n is 0.
(b) Since the function is continuous at 0 and is monotone on $(-\delta, 0)$ and on $(0, \delta)$, the Fourier series converges to $f(0) = 1$ at 0 by Dirichlet's Theorem 4.2.10. So, $1 = \frac{1}{2}a_0 + \sum_{n=1}^{\infty} a_n = \frac{\delta}{\pi} + \sum_{n=1}^{\infty} \frac{2}{n\pi}\sin n\delta$ and hence $\frac{\pi-\delta}{2} = \sum_{n=1}^{\infty} \frac{\sin n\delta}{n}$.
(c) By Parseval's Theorem, $\frac{1}{2}a_0^2 + \sum_{n=1}^{\infty} a_n^2 = \frac{1}{2}\frac{4\delta^2}{\pi^2} + \sum_{n=1}^{\infty} \frac{4\sin^2 n\delta}{n^2\pi^2} = \frac{1}{\pi}\int_{-\delta}^{\delta} dx = \frac{2\delta}{\pi}$.

Therefore $\frac{\delta^2}{\pi^2} + \frac{2}{\pi^2}\sum_{n=1}^{\infty} \frac{\sin^2 n\delta}{n^2} = \frac{\delta}{\pi}$, so that $\frac{\delta}{\pi} + \frac{2}{\pi}\sum_{n=1}^{\infty} \frac{\sin^2 n\delta}{n^2\delta} = 1$, which implies for $0 < \delta < \pi$ that

$$\frac{\pi - \delta}{2} = \sum_{n=1}^{\infty} \frac{\sin^2 n\delta}{n^2\delta} = \sum_{n=1}^{\infty} \frac{\sin^2 n\delta}{n^2\delta^2}\delta. \tag{8.29}$$

(d) Now let $\varepsilon > 0$ be given. Since the improper integral $\int_0^{\infty} \left(\frac{\sin x}{x}\right)^2 dx$ is convergent (by comparison (see Proposition 1.7.6) with $1/x^2$) there exists an X such that

$$\left|\int_0^{\infty} \left(\frac{\sin x}{x}\right)^2 dx - \int_0^{X} \left(\frac{\sin x}{x}\right)^2 dx\right| < \varepsilon \tag{8.30}$$

and

$$X > \frac{1}{\varepsilon}. \tag{8.31}$$

Approximating the Riemann integral $\int_0^X \left(\frac{\sin x}{x}\right)^2 dx$ by Riemann sums, we find that there exists a positive integer m' such that $\delta = X/m'$ satisfies

$$\delta < 2\varepsilon \tag{8.32}$$

and

$$\left|\int_0^X \left(\frac{\sin x}{x}\right)^2 dx - \sum_{n=1}^{m'} \frac{\sin^2 n\delta}{n^2\delta^2}\delta\right| < \varepsilon. \tag{8.33}$$

In view of (8.29), there exists an $m > m'$ such that

$$\left|\frac{\pi-\delta}{2} - \sum_{n=1}^{m} \frac{\sin^2 n\delta}{n^2\delta^2}\delta\right| < \varepsilon,$$

i.e.

$$\left|\frac{\pi-\delta}{2} - \sum_{n=1}^{m'} \frac{\sin^2 n\delta}{n^2\delta^2}\delta - \sum_{n=m'+1}^{m} \frac{\sin^2 n\delta}{n^2\delta^2}\delta\right| < \varepsilon. \tag{8.34}$$

Now, $\sum\limits_{n=m'+1}^{m} \frac{\sin^2 n\delta}{n^2\delta^2}\delta \le \sum\limits_{n=m'+1}^{\infty} \frac{1}{n^2\delta} \le \frac{1}{\delta}\int_{m'}^{\infty} \frac{1}{x^2}dx$, because $\frac{1}{n^2} \le \frac{1}{x^2}$ on $[n - 1, n]$. Moreover, $\frac{1}{\delta}\int_{m'}^{\infty} \frac{1}{x^2}dx = \frac{1}{m'\delta} = \frac{1}{X}$ because $\delta = X/m'$. It follows by (8.31) that

$$\sum_{n=m'+1}^{\infty} \frac{\sin^2 n\delta}{n^2\delta^2}\delta < \varepsilon.$$

Taken with (8.34), (8.33) and (8.30), this implies

$$\left|\frac{\pi-\delta}{2} - \int_0^{\infty} \left(\frac{\sin x}{x}\right)^2 dx\right| < 4\varepsilon.$$

Using (8.32), we further obtain $\left|\frac{\pi}{2} - \int_0^{\infty} \left(\frac{\sin x}{x}\right)^2 dx\right| < 5\varepsilon$.

4.3.P7. Show that $\sum\limits_{n=1}^{\infty} \frac{\cos nx}{n^{3/2}}$ is the Fourier series of a continuous function and converges to it uniformly.

Hint: Since $\left|\frac{\cos nx}{n^{3/2}}\right| \le \frac{1}{n^{3/2}}$ for all x and $\sum\limits_{n=1}^{\infty} \frac{1}{n^{3/2}}$ is convergent, it follows by the Weierstrass M-test that the given series is uniformly convergent. The conclusion now follows from Proposition 4.1.3 and Corollary 4.3.6.

Problem Set 4.4

4.4.P1. Find the Fourier cosine series for the function given on $[0, \pi]$ by $\sin x$. Deduce from the series that $\frac{1}{2} = \sum\limits_{n=1}^{\infty} \frac{1}{4m^2-1}$.

Hint: The function $f(x) = \sin x$ may be extended to an even function on $[-\pi, \pi]$ by setting $f(x) = |\sin x|$ and then extended to a periodic function on the whole of $\mathbb{R}$. The (extended) function satisfies a Lipschitz condition on $[-\pi, \pi]$ with $M = 1$ and $\alpha = 1$. In fact by the Mean Value Theorem,

$$||\sin x| - |\sin y|| \leq |\sin x - \sin y| \leq |\cos \xi||x - y| \leq |x - y|,$$

where ξ lies between x and y. Thus its Fourier series converges uniformly on $[-\pi, \pi]$ by Theorem 4.2.13. We proceed to obtain the series.

$$a_0 = \frac{2}{\pi}\int_0^\pi \sin x\, dx = \frac{4}{\pi}$$

and

$$\begin{aligned} a_n &= \frac{2}{\pi}\int_0^\pi \sin x \cos nx\, dx = \frac{1}{\pi}\int_0^\pi [\sin(n+1)x - \sin(n-1)x]dx \\ &= \frac{1}{\pi}\left[-\frac{\cos(n+1)x}{n+1} + \frac{\cos(n-1)x}{n-1}\right]_0^\pi = -\frac{2}{\pi(n^2-1)}[(-1)^n + 1] \text{ for } n \neq 1, \end{aligned}$$

While $a_1 = \frac{2}{\pi}\int_0^\pi \sin x \cos x\, dx = 0$. So,

$$|\sin x| = \frac{2}{\pi} - \sum_{n=1}^{\infty} \frac{4\cos 2nx}{\pi(4n^2-1)}.$$

Set $x = 0$ to obtain the required equality.

4.4.P2. Find the Fourier series of the function

$$f(x) = \begin{cases} 0 & \text{if } -2 \leq x < -1 \\ k & \text{if } -1 \leq x \leq 1 \\ 0 & \text{if } 1 < x \leq 2, \end{cases}$$

extended periodically to the whole of $\mathbb{R}$ (period 4). Here k is some nonzero constant.

Hint: This is an even integrable function with period 4. Therefore it has a cosine series as in Proposition 4.4.1 with $l = 2$. Accordingly,

$$a_0 = \frac{2}{2}\int_0^2 f(x)dx = \int_0^1 k\,dx = k \quad \text{and} \quad a_n = \frac{2}{2}\int_0^2 f(x)\cos\frac{n\pi x}{2}dx$$
$$= \int_0^1 k\cos\frac{n\pi x}{2}dx = k\frac{2}{n\pi}\sin\frac{n\pi}{2}.$$

Thus $a_n = 0$ if n is even but not 0 while $a_n = \frac{2k}{n\pi}$ for $n = 1, 5, 9, \ldots$ and $-\frac{2k}{n\pi}$ for $n = 3, 7, 11, \ldots$. Hence

$$f(x) \sim \frac{k}{2} + \frac{2k}{\pi}\left(\cos\frac{\pi x}{2} - \frac{1}{3}\cos\frac{3\pi x}{2} + \frac{1}{5}\cos\frac{5\pi x}{2} - + \cdots\right).$$

Now f has jumps at all integral points except $4n$ ($n \in \mathbb{Z}$) and is monotone between consecutive jumps. Therefore by an obvious extension of Dirichlet's Theorem 4.2.10 to the case of an arbitrary period, the series converges to $f(x)$ at points of continuity; moreover it converges to $\frac{1}{2}[f(x+)+f(x-)] = \frac{k}{2}$ at points of jump discontinuity, as can also be verified directly.

Problem Set 4.5

4.5.P1. If $A \subseteq [-\pi, \pi]$ is measurable, prove that

$$\lim_{n\to\infty}\int_A \cos nx\, dm(x) = \lim_{n\to\infty}\int_A \sin nx\, dm(x) = 0.$$

Hint: $\int_A \cos nx\, dm(x) = \int_{[-\pi,\pi]} \chi_A(x)\cos nx\, dm(x) = \langle \chi_A, \cos nx\rangle$ in $L^2[-\pi, \pi]$. So, the first limit follows by Theorem 4.5.3. The second one is argued similarly.

4.5.P2. Let $n_1, n_2, \ldots$ be a sequence of positive integers and let

$$E = \{x \in [-\pi, \pi] : \lim_{k\to\infty} \sin n_k x \text{ exists}\}.$$

Prove that E has measure $m(E) = 0$.

Hint: Observe that E is measurable. Let $E_1 = \left\{x \in E : \lim_{k\to\infty} 2\sin^2 n_k x - 1 \geq 0\right\}$ and $E_2 = \left\{x \in E : \lim_{k\to\infty} 2\sin^2 n_k x - 1 < 0\right\}$. Then E_1 and E_2 are measurable and $E = E_1 \cup E_2$. By Problem 4.5.P1,

$$0 = \lim_{k\to\infty} \int_{E_1} (-\cos 2n_k x)dm(x) = \lim_{k\to\infty} \int_{E_1} (2\sin^2 n_k x - 1)dx$$
$$= \int_{E_1} \left[\lim_{k\to\infty} (2\sin^2 n_k x - 1)\right] dx,$$

where the final equality uses the Dominated Convergence Theorem 3.2.16, keeping in mind that $|2\sin^2 n_k x - 1| \leq 1$ and the constant function 1 on E is Lebesgue integrable. However, $\lim_{k\to\infty} (2\sin^2 n_k x - 1) \geq 0$ 1 on E_1 by definition of this set. It follows by 3.2.P1(a) that $\lim_{k\to\infty} (2\sin^2 n_k x - 1) = 0$ a.e. on E_1. This implies that $\lim_{k\to\infty} \sin n_k x = \pm 1/\sqrt{2}$ for $x \in E_1$. A similar argument shows that the same is true on E_2. Let E_3 [resp. E_4] be the subset of E_1 on which the aforementioned limit is $1/\sqrt{2}$ [resp. $-1/\sqrt{2}$]. Applying Theorem 4.5.3 to χ_{E_3}, we have

$$0 = \lim_{k\to\infty} \int_{E_3} \sin n_k x \, dm(x) = \frac{1}{\sqrt{2}} m(E_3), \text{whence } m(E_3) = 0.$$

Similarly, $m(E_4) = 0$ and hence $m(E_1) = 0$. Analogously, $m(E_2) = 0$.
4.5.P3. Show that each of the two sequences

$$1, \cos x, \cos 2x, \cos 3x, \ldots$$
$$\sin x, \sin 2x, \sin 3x, \ldots$$

is a complete orthogonal sequence in $L^2[0, \pi]$.
Hint: Orthogonality follows in the same manner as in Example 4.5.2(a). Suppose $\int_{[0,\pi]} f(x)\cos nx \, dm(x) = 0$ for $n = 0, 1, 2, \ldots$, where $f \in L^2[0, \pi]$. Extend f to $[-\pi, \pi]$ so that it becomes an even function. Then

$$\int_{[-\pi,\pi]} f(x)^2 dm(x) = 2\int_{[0,\pi]} f(x)^2 dm(x), \quad \text{so that } f \in L^2[-\pi, \pi].$$

Also,

$$\int_{[-\pi,\pi]} f(x) \cos nx \, dm(x) = 2\int_{[0,\pi]} f(x) \cos nx \, dm(x) \text{ for } n = 0, 1, 2, \ldots$$

and

$$\int_{[-\pi,\pi]} f(x) \sin nx \, dm(x) = 0 \text{ for } n = 1, 2, \ldots$$

because the integrand is odd. It follows by Theorem 4.5.8 that $f = 0$ a.e. on $[-\pi, \pi]$ and, in particular, on $[0, \pi]$.

For the other sequence, redefine $f(0)$ to be 0, so that f is unchanged a.e. Then extend it to be odd.

4.5.P4. Suppose E is a subset of $[-\pi, \pi]$ with positive measure and let $\delta > 0$ be given. Deduce from Bessel's Inequality that the number of positive integers k such that $\sin kx > \delta$ for all $x \in E$ is finite.

Hint: For any k of the kind in question, $\int_{[-\pi,\pi]} \chi_E \sin kx \, dm(x) \geq m(E)\delta$, i.e. the Fourier coefficient b_k of χ_E exceeds $m(E)\delta > 0$. If this happens for infinitely many k, then the series in Bessel's Inequality diverges, contrary to the assertion of the inequality that its sum cannot exceed $\int_{[-\pi,\pi]} \chi_E^2 = m(E) < \infty$.

Problem Set 5.2

5.2.P1. Give an example of a monotone function which is discontinuous at each rational number in [0, 1] (see Example 5.2.7).

Hint: Let $\{r_k\}_{k\geq 1}$ be an enumeration of the rational numbers in [0, 1]. Define f on [0, 1] by the rule

$$f(x) = \sum_{r_n \leq x} 2^{-n}$$

where the summation is extended over all indices n for which $r_n \leq x$. Clearly, the function f is increasing and, for each rational number r_i, $f(r_i) - f(r_i-) = 2^{-i}$. It may be noted that f is right continuous at each rational number r_i. Moreover, it is continuous at each irrational number.

5.2.P2. Let f be defined on an open interval (a, b) and assume that for each point x of the interval, there exists an open interval $\mathfrak{N}_x$ of x in which f is increasing. Prove that f is increasing throughout (a, b).

Hint: Let $x < y$. Consider the closed interval $[x, y]$. By hypothesis, there is an open interval about each of the points in $[x, y]$ in which f is increasing. By the Heine–Borel Theorem, there exists a finite cover $\{\mathfrak{N}_{x_i}\}_{i=1}^m$ of intervals which covers $[x, y]$ and in each of which f is increasing. Without loss of generality, we may assume $x \in \mathfrak{N}_{x_1}$. If $y \in \mathfrak{N}_{x_1}$, then $f(x) \leq f(y)$ because f is increasing on $\mathfrak{N}_{x_1}$. If y does not belong to $\mathfrak{N}_{x_1}$, then there exists an interval amongst $\{\mathfrak{N}_{x_i}\}_{i=2}^m$ which has a nonempty intersection with $\mathfrak{N}_{x_1}$. Call it $\mathfrak{N}_{x_2}$ and let $z_1 \in \mathfrak{N}_{x_1} \cap \mathfrak{N}_{x_2}$. Clearly, f is increasing in $\mathfrak{N}_{x_1} \cup \mathfrak{N}_{x_2}$. If $y \notin \mathfrak{N}_{x_2}$, then continue the process till $\cup_{i=1}^k \mathfrak{N}_{x_i}$ contains y, which must happen at some stage because there are only finitely many intervals and they cover $[x, y]$. Obviously, the function is increasing on $\cup_{i=1}^k \mathfrak{N}_{x_i}$. Consequently, $f(x) \leq f(y)$.

5.2.P3. Let f be continuous on a compact interval $[a, b]$ and assume that f does not have a local minimum or local maximum at any interior point. Prove that f must be monotonic on $[a, b]$.
Hint: Since f is continuous on the compact interval $[a, b]$, it assumes its minimum as well as maximum. It cannot have a minimum or maximum at an interior point, for that point will also be a local minimum or local maximum. If the function is constant, there is nothing to prove. Assume that the minimum is attained at a, in which case the maximum is attained at b. We shall show that f is increasing on $[a, b]$. Suppose not. Then there exist $x, y \in [a, b]$ such that $x < y$ but $f(x) > f(y)$. Set $\alpha = \frac{f(x)+f(y)}{2}$. Clearly, $f(a) \leq f(y) < \alpha < f(x)$. By the Intermediate Value Theorem 1.3.26, the value α is attained between a and x and again between x and y, say, at u and v respectively. Between u and v, it assumes its maximum, which cannot be at u or v, because $f(x) > \alpha = f(u) = f(v)$. Thus f has a maximum between u and v (by the Extreme Value Theorem 1.3.20). Hence f has a local maximum, which contradicts the hypothesis.

A similar argument shows that, when a is a maximum and b a minimum, the function is decreasing.

5.2.P4. A function f increases on a closed interval $[a, b]$ and it is true that, if $f(a) < \lambda < f(b)$, there exists an $x \in [a, b]$ such $f(x) = \lambda$. Prove that f is continuous on $[a, b]$. Does the same conclusion hold if the hypothesis that f increases is dropped but it is still required that $f(a) < f(b)$?

A function f increases on an open interval I and, for each $a, b \in I$, it is true that, if $f(a) < \lambda < f(b)$, then there exists an $x \in I$ such $f(x) = \lambda$. Prove that f is continuous on I. Does the same conclusion hold if the hypothesis that f increases is dropped?
Hint: In either case, the hypothesis implies that the range is an interval. Therefore continuity follows directly from Remark 5.2.3.

In the first case, the same conclusion does *not* hold if the hypothesis that f increases is dropped. Consider the function (see Fig. 8.2)

$$f(x) = \begin{cases} x & 0 \leq x < 1 \\ x - 1 & 1 \leq x \leq 2. \end{cases}$$

In the second case, just drop the endpoints.

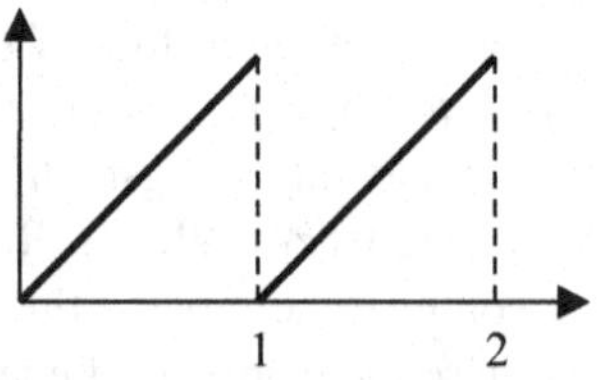

Fig. 8.2 Function in Problem 5.2.P4

Problem Set 5.3

5.3.P1. If the function f assumes its local maximum at an interior point c of its domain, then show that $D^+f(c) \leq 0$ and $D_-f(c) \geq 0$.
Hint: If $D^+f(c) > 0$, then there exist arbitrarily small positive values of h such that $\frac{f(c+h)-f(c)}{h} > 0$, which implies $f(c + h) - f(c) > 0$ for some $c + h$ in every neighbourhood of c, contradicting the fact that f assumes a local maximum at c.
If $D_-f(c) < 0$, then there exist arbitrarily small positive values of h such that $\frac{f(c-h)-f(c)}{-h} <0$, which implies $f(c - h) - f(c) > 0$ for some $c - h$ in every neighbourhood of c, contradicting the fact that f assumes its local maximum at c.
5.3.P2. Suppose f is continuous on $[a, b]$ and $D^+f(x) > 0$ for all x in $[a, b)$. Show that $f(b) \geq f(a)$. Also, give a counterexample to show that the continuity hypothesis cannot be dropped.
Hint: Let $E = \{x \in [a, b]: f(x) \geq f(a)\}$. The set E is a closed subset of $[a, b]$ because f is continuous on $[a, b]$. If $c = \sup E$, then $c \in E$ since E is closed. Therefore $f(c) \geq f(a)$. We need only show that $c = b$. If $c < b$, then $D^+f(c) > 0$ implies $\frac{f(x)-f(c)}{x-c} > 0$ for some x arbitrarily close to c and on the right of it. Thus $f(x) > f(c) \geq f(a)$ for some $x > c$. This is a contradiction since $c = \sup\{x \in [a, b]: f(x) \geq f(a)\}$. Hence $c = b$.
The function f on $[0, 2]$ given by $f(x) = x$ on $[0, 1)$ and $f(x) = x - 3$ on $[1, 2]$ is discontinuous at 1 and satisfies $D^+f(c) = 1$ everywhere, but $f(2) = -1 < 0 = f(0)$.
5.3.P3. Show that f may be discontinuous at x_0 when all four Dini derivatives are equal to ∞.
Hint: Let $0 < k < 1$ and consider the function

$$f(x) = \begin{cases} 1 & x > x_0 \\ 0 & x < x_0 \\ k & x = x_0. \end{cases}$$

Then

$$D^+f(x_0) = \limsup_{h\to 0+} \frac{f(x_0+h) - f(x_0)}{h} = \limsup_{h\to 0+} \frac{1-k}{h} = \infty,$$

$$D_+f(x_0) = \liminf_{h\to 0+} \frac{f(x_0+h) - f(x_0)}{h} = \liminf_{h\to 0+} \frac{1-k}{h} = \infty,$$

$$D^-f(x_0) = \limsup_{h\to 0-} \frac{f(x_0+h) - f(x_0)}{h} = \limsup_{h\to 0-} \frac{-k}{h} = \infty,$$

$$D_{-}f(x_0) = \liminf_{h\to 0-} \frac{f(x_0+h)-f(x_0)}{h} = \liminf_{h\to 0-} \frac{-k}{h} = \infty.$$

Thus all four Dini derivatives of f at x_0 are equal to ∞ and the function is discontinuous at x_0.

5.3.P4. Let f: $[0, \infty)\to\mathbb{R}$ be differentiable and suppose that $f(0) = 0$ and that f' is increasing. Prove that

$$g(x) = \begin{cases} \frac{f(x)}{x} & x > 0 \\ f'(0) & x = 0 \end{cases}$$

defines an increasing function of x.

Hint: Let $0 \le \alpha < \beta$. Applying the Lagrange Mean Value Theorem to g on $[\alpha, \beta]$, we obtain some $\gamma \in (\alpha, \beta)$ such that

$$\frac{g(\beta)-g(\alpha)}{\beta-\alpha} = g'(\gamma) = \frac{1}{\gamma}\left[f'(\gamma) - \frac{f(\gamma)}{\gamma}\right].$$

Applying the Lagrange Mean Value Theorem to f on $[0, \gamma]$, we obtain some $\delta \in (0, \gamma)$ such that

$$\frac{f(\gamma)-f(0)}{\gamma} = f'(\delta).$$

Therefore

$$\frac{g(\beta)-g(\alpha)}{\beta-\alpha} = \frac{1}{\gamma}\left[f'(\gamma) - \frac{f(\gamma)}{\gamma}\right] = \frac{1}{\gamma}[f'(\gamma) - f'(\delta)] \ge 0.$$

5.3.P5. (a) Show that if $f'(x)$ exists, then $D^{+}(f + g)(x) = f'(x) + D^{+}g(x)$ and similarly for other Dini derivatives.

(b) Give an example when $D^{+}(f + g)(x) \neq D^{+}f(x) + D^{+}g(x)$.

Hint: **(a)** Let $\eta > 0$ be given. There exists a δ' such that for all h satisfying $0 < h < \delta'$, we have

$$\frac{f(x+h)-f(x)}{h} > f'(x) - \eta$$

and

$$\frac{f(x+h)-f(x)}{h} + \frac{g(x+h)-g(x)}{h} < D^+(f+g)(x) + \eta.$$

Now,

$$\begin{aligned}\frac{g(x+h)-g(x)}{h} &= \frac{(f+g)(x+h)-(f+g)(x)}{h} - \frac{f(x+h)-f(x)}{h} \\ &< D^+(f+g)(x) + \eta - (f'(x) - \eta) \\ &= D^+(f+g)(x) - f'(x) + 2\eta.\end{aligned}$$

Therefore $D^+g(x) = \inf_{\delta>0} \sup_{0<h<\delta} \dfrac{g(x+h)-g(x)}{h} \leq D^+(f+g)(x) - f'(x) + 2\eta$.

Since η is arbitrarily chosen, it follows that $D^+g(x) \leq D^+(f+g)(x) - f'(x)$, which is equivalent to $D^+(f+g)(x) \geq D^+g(x) + f'(x)$. The reverse inequality is an immediate consequence of the definition of the lim sup.

(b) Let

$$f(x) = \begin{cases} 0 & x = 0 \\ x & x \in \mathbb{Q} \\ -x & x \notin \mathbb{Q} \end{cases}$$

and let $g(x) = -f(x)$. Then $D^+(f+g)(0) = 0$. However, $D^+f(0) = D^+g(0) = 1$. In fact, for $h > 0$,

$$\frac{f(h)-f(0)}{h} = \frac{f(h)}{h} = \begin{cases} 1 & x \in \mathbb{Q} \\ -1 & x \notin \mathbb{Q}. \end{cases}$$

Consequently, $\limsup_{h\to 0+} \frac{f(h)-f(0)}{h} = 1$. Similarly $D^+g(0) = 1$.

5.3.P6. **(a)** Let f be continuous on $[a, b]$. If $\frac{f(b)-f(a)}{b-a} < C$, then $D^+f(x) \leq C$ for uncountably many $x \in [a, b]$.

(b) Let f be continuous on $[a, b]$. If $\frac{f(b)-f(a)}{b-a} > C$, then $D^-f(x) \geq C$ for uncountably many $x \in [a, b]$.

Hint: **(a)** Let $k = C(b - a)$. Then $k > f(b) - f(a)$. Consider the function

$$g(x) = f(x) - f(a) - k\frac{x-a}{b-a}.$$

Then $g(a) = 0$ and $g(b) = f(b) - f(a) - k < 0$. Let s be such that $0 = g(a) > s > g(b)$. Let

$$E = \{x \in [a, b] : g(x) = s\};$$

E is closed, being the inverse image of a closed subset of $\mathbb{R}$ under the continuous map g. Since E is also bounded, it has a largest point x_s, say, with $g(x_s) = s$. Since $g(b) < s$, we have $x_s < b$. Since $g(x_s + h) < s$ for all $0 < h < b - x_s$ (Intermediate Value Theorem 1.3.26), we have $D^+g(x_s) \leq 0$, whence by Problem 5.3.P5,

$$D^+f(x_s) = D^+g(x_s) + \frac{k}{b-a} \leq \frac{k}{b-a} = C.$$

Different s's generate different x_s's and there are uncountably many s's between 0 and $g(b)$, i.e., there are uncountably many points x such that $D^+f(x) \leq C$.

(b) Similar.

Remark It is a consequence that if f is continuous and one Dini derivative is nonnegative except perhaps at a countable or finite number of points, then f is increasing. In fact, if $D^+f(x) \geq 0$ for $a \leq x \leq b$ except for a countable or finite number of points and f fails to be increasing, there must exist points x and y such that $y > x$ and $f(y) < f(x)$, then $\frac{f(y)-f(x)}{y-x} < C < 0$, which implies $D^+f(z) \leq C < 0$ for uncountably many points z between x and y, contradicting our hypothesis.

5.3.P7. Let f be a continuous function defined on $[a, b]$. Suppose that there exist real constants α, β such that $\alpha \leq D^+f(x) \leq \beta$ for all $x \in (a, b)$. Prove that $h\alpha \leq f(x + h) - f(x) \leq h\beta$ provided $a \leq x < x + h \leq b$.
Hint: Suppose $D^+f(x) \geq \alpha$. Let $g(x) = f(x) - \alpha x$. Since $D^+g(x) \geq 0$ [see Problem 5.3.P5], the function g is increasing [see Remark above]. If $h > 0$, we therefore have $g(x + h) \geq g(x)$, or in other words,

$$f(x+h) - f(x) - \alpha(x+h) + \alpha x \geq 0,$$

i.e.

$$\frac{f(x+h) - f(x)}{h} \geq \alpha. \tag{8.35}$$

This proves the first of the two inequalities in question. A similar argument proves the other.

Remark From (8.35), it follows that $D_+f(x) \geq \alpha$. Again, g is increasing. If $h > 0$, we therefore have $g(x) - g(x - h) \geq 0$,

$$f(x) - \alpha x - f(x-h) + \alpha(x-h) \geq 0,$$

$$\frac{f(x-h) - f(x)}{-h} \geq \alpha,$$

whence

$$D^- f(x) \geq D_- f(x) \geq \alpha.$$

If f is continuous, all four Dini derivatives have the same upper and lower bounds in any interval.

5.3.P8. Let f be a continuous real-valued function defined on $[a, b]$ and let $x \in (a, b)$ be such that D^+f is finite in a neighbourhood of x and continuous at x. Prove that $f'(x)$ exists.

Hint: Since D^+f is finite and continuous at x, its upper and lower bounds are arbitrarily close to $D^+f(x)$ in a sufficiently small neighbourhood of x. By the remark following Problem 5.3.P7, the same is true of the upper and lower bounds of the other three Dini derivatives. This means all the four Dini derivatives at x coincide with $f'(x)$.

5.3.P9. If one of the Dini derivatives of a continuous function is zero everywhere in an interval, the function is constant there.

Hint: $D^+f = 0 \Leftrightarrow D^+f \geq 0$ and $D^+f \leq 0$.

From the continuity of f and the fact that $D^+f \geq 0$, we conclude that

$$f \text{ is increasing.}$$

Similarly from the continuity of f and the fact that $D^+f \leq 0$, we conclude that

$$f \text{ is decreasing.}$$

Since f has been shown to be increasing as well as decreasing, it follows that f is constant.

5.3.P10. Construct monotonic jump functions f on $[0, 1)$ whose discontinuities have 0 as a limit point and such that $f'_+(0)$ is

(a) zero (b) ∞ (c) positive and finite.

Also, compute the quantum of jump at each of the discontinuities.

Hint: **(a)** Define a function on the domain $[0,1)$ as

$$f(x) = \begin{cases} 3^{-n} & 2^{-n} \le x < 2^{-n+1}, n = 1, 2, \ldots \\ 0 & x = 0. \end{cases}$$

Let $x_1 < x_2$. Then there exist m and n, $m \ge n$, such that $x_1 \in [2^{-m}, 2^{-m+1})$ and $x_2 \in [2^{-n}, 2^{-n+1})$. So, $f(x_1) = 3^{-m}$ and $f(x_2) = 3^{-n}$, i.e. f is increasing. At each of the points 2^{-n}, we have $f(2^{-n}+) - f(2^{-n}) = 3^{-n} - 3^{-n} = 0$ and $f(2^{-n}) - f(2^{-n}-) = 3^{-n} - 3^{-n-1} > 0$, i.e., at the point $x = 2^{-n}$, f is right continuous and has a jump of magnitude $3^{-n} - 3^{-n-1}$, which tends to 0 as $n \to \infty$. Moreover, $f'_+(0) = 0$. In fact, if $h > 0$, we have

$$\lim_{h \to 0} \frac{f(0+h) - f(0)}{h} = \lim_{h \to 0} \frac{f(h)}{h} = 0.$$

Indeed, for $2^{-m-1} \le h < 2^{-m}$

$$\frac{1}{3} \cdot \left(\frac{2}{3}\right)^m < \frac{f(h)}{h} \le \left(\frac{2}{3}\right)^{m+1}.$$

(b) Define a function on the domain [0, 1) as

$$f(x) = \begin{cases} 2^{-n} & 3^{-n} \le x < 3^{-n+1}, n = 1, 2, \ldots \\ 0 & x = 0. \end{cases}$$

As in (a) above, it is right continuous at 3^{-n}, with a jump of magnitude $2^{-n} - 2^{-n-1}$. Moreover, $f'_+(0) = \infty$. In fact, if $h > 0$, we have

$$\frac{f(0+h) - f(0)}{h} = \frac{f(h)}{h};$$

and if $3^{-m-1} \le h < 3^{-m}$, then $\frac{f(h)}{h} > \frac{3^m}{2^{m+1}}$, which tends to ∞ as $m \to \infty$.

(c) Define a function on the domain [0, 1) as

$$f(x) = \begin{cases} \frac{1}{n} & \frac{1}{n} \le x < \frac{1}{n-1}, n = 2, 3, \ldots \\ 0 & x = 0 \end{cases}$$

Then f has jumps of amount $\frac{1}{n(n+1)}$ at the points $\frac{1}{n}$. Also, for $h > 0$, we have

$$\frac{f(0+h)-f(0)}{h}=\frac{f(h)}{h}$$

and if $(m+1)^{-1} \le h < m^{-1}$, then

$$f(h)=\frac{1}{m+1},$$

so that

$$\frac{1}{m+1}m<\frac{f(h)}{h}\le 1,$$

which implies

$$\lim_{h\to 0+}\frac{f(h)}{h}=1.$$

5.3.P11. If all Dini derivatives of a function f satisfy $|Df(x)| \le K$ everywhere on an interval, then the function satisfies the condition $|f(x) - f(y)| \le K|x - y|$.
Hint: Suppose not. Then we have $|f(\alpha)-f(\beta)| > K'|\alpha - \beta|$ for some α and β in the interval and some $K' > K$. We may suppose $\alpha < \beta$. Then the inequality $|f(\alpha) - f(\beta)| > K'(\beta - \alpha)$ holds over the interval (α, β). For any $\gamma \in (\alpha, \beta)$, the same inequality holds either over the interval (α, γ) or over (γ, β), as can be deduced by applying the triangle inequality. By the usual bisection argument, we get a nested sequence of intervals (α_n, β_n), such that $|f(\alpha_n) - f(\beta_n)| > K'(\beta_n - \alpha_n)$ for each n and $0 < (\beta_n - \alpha_n) \to 0$. Let ξ be the unique common point of all these intervals. Then either $\alpha_n < \xi$ for all n or $\beta_n > \xi$ for all n. In the former case, $|D - f(\xi)| \ge K' > K$ and in the latter case, $|D^+f(\xi)| \ge K' > K$.

Problem Set 5.4

5.4.P1. Definition. Let E be a subset of $\mathbb{R}$ of finite outer measure. A family $\mathcal{I}$ of closed intervals, each of positive length, is called a **Vitali cover of *E*** if for each $x \in E$ and every $\varepsilon > 0$, there exists an $I \in \mathcal{I}$ such that $x \in I$ and $\ell(I) < \varepsilon$., i.e., each point of E belongs to an arbitrarily short interval of $\mathcal{I}$.

Example. Let $\{r_n\}_{n\ge 1}$ be an enumeration of the rationals in $[a, b]$. Then the collection $\{I_{n,i}\}$, where $I_{n,i} = \left[r_n - \frac{1}{i}, r_n + \frac{1}{i}\right], n, i \in \mathbb{N}$, forms a Vitali cover of $[a, b]$.
(Vitali Covering Theorem) Let E be a subset of $\mathbb{R}$ of finite outer measure and $\mathcal{I}$ be a Vitali cover of E. Then given any $\varepsilon > 0$, there is a finite disjoint collection $\{I_1, I_2, \ldots, I_N\}$ of intervals in $\mathcal{I}$ such that

$$m^*\left(E\backslash\bigcup_{i=1}^{N} I_i\right) < \varepsilon.$$

Hint: Let $\mathcal{O}$ be an open subset of $\mathbb{R}$ of finite measure containing E. Let $\mathcal{I}_0 = \{I \in \mathcal{I}: I \subseteq \mathcal{O}\}$. Obviously, $\mathcal{I}_0$ is still a Vitali cover of E. Let $h_1 = \sup\{\ \ell(I): I \in \mathcal{I}_0\}$. Choose an interval $I \in \mathcal{I}_0$ whose length is greater than $\frac{1}{2} h_1$, i.e., $\ell(I_n) > \frac{1}{2} h_1$. If $E \subseteq I_1$, the construction is complete. If not, suppose $I_1, I_2, \ldots, I_n \in \mathcal{I}_0$ have been selected such that they are disjoint. If $E \subseteq \cup_{k=1}^{n} I_k$, the construction is complete. Otherwise, write

$$A_n = \bigcup_{k=1}^{n} I_k, U_n = \mathcal{O} \cap A_n^c.$$

Clearly, A_n is closed, U_n is open and $U_n \cap E$ is nonempty. By definition of a Vitali covering, there must exist an $I \in \mathcal{I}_0$ such that $I \subseteq U_n$. Let

$$h_n = \sup\{\ell(I) : I \in \mathcal{I}_0, I \subseteq U_n\}.$$

Choose $I_{n+1} \in \mathcal{I}_0$ such that $I_{n+1} \subseteq U_n$ and $\ell(I_{n+1}) > \frac{1}{2} h_n$.

Thus we have a sequence $\{I_n\}$ of disjoint intervals of $\mathcal{I}_0$. Let $A = \cup_{n \geq 1} I_n$. Since $\cup_{n \geq 1} I_n \subseteq \mathcal{O}$, we have $\Sigma_n \ell(I_n) \leq m(\mathcal{O}) < \infty$. Hence we can find an integer N such that

$$\sum_{n=N+1}^{\infty} \ell(I_n) < \frac{\varepsilon}{3}. \tag{8.36}$$

Also, $\lim_{k\to\infty} h_k \leq 2 \cdot \lim_{k\to\infty} \ell(I_{k+1}) = 0$. Let

$$R = E\backslash\bigcup_{i=1}^{N} I_i.$$

We need to show that $m^*(R) < \varepsilon$. Let $x \in R$ be arbitrary. Since $\cup_{i=1}^{N} I_i$ is a closed set not containing x, we can find an interval $I \in \mathcal{I}_0$ which contains x and whose length is so small that I does not meet any of the intervals $I_1, I_2, \ldots, I_N$.

The interval I must have one point in common with some interval I_k, $k > N$. For, if I were disjoint from $I_{N+1}, I_{N+2}, \ldots$, then it would follow on noting that $\ell(I) \leq h_k$, $k = N+1, N+2, \ldots$ (by definition of h_k) and $\lim_{k\to\infty} h_k = 0$ (proved above) that $\ell(I) = 0$, so that I would not be an interval in a Vitali cover.

Let m be the smallest integer such that I meets I_m. Then we have $m > N$ and $\ell(I) \leq h_{m-1} < 2\,\ell(I_m)$.

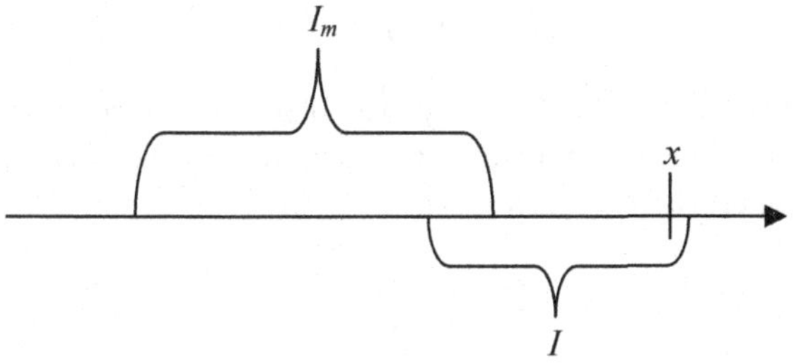

Fig. 8.3 For Problem 5.4.P1

Since x is in I and has a point in common with I_m, it follows that the distance of the midpoint of I_m from x is at most $\ell(I) + \frac{1}{2}\ell(I_m) \leq \frac{5}{2}\ell(I_m)$ (see Fig. 8.3 above and produce a formal proof!). Thus x belongs to the interval J_m having the same midpoint as I_m and five times its length. Thus we have shown

$$R \subseteq \bigcup_{m=N+1}^{\infty} J_m.$$

Hence $m^*(R) \leq \sum_{m=N+1}^{\infty} \ell(J_m) = 5\sum_{m=N+1}^{\infty} \ell(I_m) < \varepsilon$ by (8.36).

5.4.P2. (Lebesgue's Theorem) Let $[a, b]$ be a closed interval in $\mathbb{R}$ and let f be a real-valued monotone function on $[a, b]$. Using the Vitali Covering Theorem, prove that f is differentiable almost everywhere.
Hint: Without loss of generality, we may assume that f is increasing (otherwise consider $-f$). Let us show that the set where some two Dini derivatives are unequal is of measure zero. Let

$$E = \{x \in [a, b] : D_+f(x) < D^+f(x)\}.$$

For every pair of positive rational numbers u and v, such that $u < v$, let

$$E_{u,v} = \{x \in E : D_+f(x) < u < v < D^+f(x)\}.$$

Clearly, $E = \cup E_{u,v}$, where u and v are positive rational numbers such that $u < v$. Since E is a countable union of the sets $\cup E_{u,v}$, it suffices to show that $m(E_{u,v}) = 0$ for all $0 < u < v$ in $\mathbb{Q}$. Suppose not. Then there exist positive rational numbers u and v such that $m(E_{u,v}) = \alpha > 0$. For a given $\varepsilon > 0$, there exists an open set $O \supseteq E_{u,v}$ such that $m(O) < \alpha + \varepsilon$. For each $x \in E_{u,v}$, there exists an arbitrarily small positive h such that $[x, x + h] \subset [a, b) \cap O$ and

$$f(x+h)-f(x)<uh, \tag{8.37}$$

using the fact that $D_+f(x) < u$. This way, we obtain a Vitali covering $\mathcal{I}$ of $E_{u,v}$. By the Vitali Covering Theorem 5.4.P1, there is a finite subcollection $\{I_1, I_2, \ldots, I_N\}$ of disjoint closed intervals in $\mathcal{I}$ such that

$$m\left(E_{u,v}\backslash\bigcup_{i=1}^{N}I_i\right)<\varepsilon.$$

Write $I_i = [x_i, x_i + h_i]$, $i = 1, 2, \ldots, N$. Then the above inequality leads to

$$m\left(E_{u,v}\cap\left[\bigcup_{i=1}^{N}(x_i,x_i+h_i)\right]^c\right)<\varepsilon. \tag{8.38}$$

Then summing (8.37) over the n disjoint intervals, we have

$$\sum_{i=1}^{N}[f(x_i+h_i)-f(x_i)]<\sum_{i=1}^{N}uh_i<u\cdot m(O)<u\cdot(\alpha+\varepsilon).$$

Let y be an arbitrary point of the set

$$E_0=E_{u,v}\cap\left[\bigcup_{i=1}^{N}(x_i,x_i+h_i)\right].$$

Since $D^+f(y) > v$, there is a small interval of the form $[y, y + k] \subseteq I_i$ for some $i = 1, 2, \ldots, N$ such that

$$f(y+k)-f(y)>vk. \tag{8.39}$$

This way, we obtain a Vitali covering $\mathcal{I}^*$ of E_0. Using the Vitali Covering Theorem 5.4.P1, we obtain a subcollection $\{J_1, J_2, \ldots, J_M\}$ of disjoint intervals in $\mathcal{I}^*$ such that

$$m\left(E_0\backslash\bigcup_{j=1}^{M}J_j\right)<\varepsilon,$$

i.e.,

$$m\left(E_{u,v}\cap\left[\bigcup_{i=1}^{N}(x_i,x_i+h_i)\right]\setminus\bigcup_{j=1}^{M}J_j\right)<\varepsilon. \tag{8.40}$$

If we write $J_j = [y_j, y_j + k_j]$, $j = 1, 2, \ldots, M$, then

$$m\left(\bigcup_{i=1}^{N}(x_i,x_i+h_i)\right)=m\left(\bigcup_{j=1}^{M}J_j\right)\le\sum_{j=1}^{M}k_j.$$

Consequently,

$$\begin{aligned}\alpha=m(E_{u,v})&\le m\left(E_{u,v}\cap\left[\bigcup_{i=1}^{N}(x_i,x_i+h_i)\right]^c\right)+m\left(E_{u,v}\cap\left[\bigcup_{i=1}^{N}(x_i,x_i+h_i)\right]\right)\\&<\varepsilon+\left(\varepsilon+m\left(\bigcup_{j=1}^{M}J_j\right)\right)\text{by (8.38)and (8.40)}\\&<\varepsilon+\left(\varepsilon+\sum_{j=1}^{M}k_j\right),\end{aligned}$$

which implies $\sum_{j=1}^{M}k_j>\alpha-2\varepsilon$.

Summing (8.39) over the intervals $J_1, J_2, \ldots, J_M$, we obtain

$$\sum_{j=1}^{M}[f(y_j+k_j)-f(y_j)]>v\sum_{j=1}^{M}k_j>v\cdot(\alpha-2\varepsilon).$$

Since each J_j is contained in some I_i, we sum over those j for which $J_j \subset I_i$ to obtain

$$\sum[f(y_j+k_j)-f(y_j)]\le f(x_i+h_i)-f(x_i),$$

using the fact that f is an increasing function. Therefore

$$\sum_{i=1}^{N}[f(x_i+h_i)-f(x_i)]\ge\sum_{j=1}^{M}[f(y_j+k_j)-f(y_j)],$$

which implies

$$u \cdot (\alpha + \varepsilon) \geq v \cdot (\alpha - 2\varepsilon).$$

Since this is true for each $\varepsilon > 0$, we must have $u \geq v$. But $u < v$. This contradiction implies $\alpha = 0$. Hence $m(E) = 0$.

We next show that the set

$$F = \{x \in (a, b) : f'(x) = \infty\}$$

has measure zero. Let $\beta > 0$ be arbitrary. For each $x \in F$, there exists an arbitrarily small positive h such that $[x, x + h] \subset (a, b)$ and $f(x + h) - f(x) > \beta h$. Repeated application of the Vitali Covering Theorem 5.4.P1 followed by an application of Proposition 2.3.21(b) yields a countable disjoint family $\{[x_n, x_n + h_n]\}$ of these intervals such that

$$m\left(F \setminus \bigcup_{n=1}^{\infty} [x_n, x_n + h_n]\right) = 0.$$

Now,

$$\beta m(F) \leq \beta \sum h_n \leq \sum [f(x_n + h_n) - f(x_n)] \leq f(b) - f(a),$$

which implies $m(F) = 0$, since $\beta > 0$ is arbitrary.

5.4.P3. If three open intervals have a nonempty intersection, then at least one among them is contained in the union of the other two, or equivalently, the union of some two among them is the same as the union of all three.

Hint: The union of any family of intervals with a nonempty intersection is an interval. This is easy to prove by means of the characterisation of an interval as being a set that contains any number that lies between two of its numbers. So, the union of *any two* of the three intervals is an open interval and the union of all the three intervals is also an open interval, which we shall call (a, b). Then a must be the left endpoint of one of them, which we shall name as A. If the other endpoint of A is b, then both the other intervals are contained in A and the required conclusion follows. So, suppose b is the right endpoint of some other interval, which we shall call B. Now, $A \cup B$ is an interval. Its left endpoint must be a and its right endpoint must be b. Thus $A \cup B = (a, b)$, which is also the union $A \cup B \cup C$ of all three intervals. Therefore $C \subseteq A \cup B \cup C = A \cup B$. This shows that C is contained in the union of the other two.

5.4.P4. **(a)** Let μ be a finite measure on a set X and $A_1, \ldots, A_n$ be finitely many distinct measurable subsets of X, where $n \geq 3$. If no three sets among the A_i have a nonempty intersection and each $\mu(A_i)$ is finite, show that

$$\mu\left(\bigcup_{i=1}^{n} A_i\right) = \sum_{i=1}^{n} \mu(A_i) - \sum_{i \leq j \leq n} \mu(A_i \cap A_j).$$

(b) Let μ be a measure on a set X and $A_1, \ldots, A_n$ be finitely many distinct measurable subsets of X, where $n \geq 3$. If no three sets among the A_i have a nonempty intersection, show that

$$\sum_{i=1}^{n} \mu(A_i) \leq 2\mu\left(\bigcup_{i=1}^{n} A_i\right).$$

Hint: **(a)** We use induction. First consider three sets, i.e. $n = 3$. Since their intersection is empty, the sets $A_1 \cap A_3$ and $A_2 \cap A_3$ are disjoint. Therefore

$$\mu((A_1 \cup A_2) \cap A_3) = \mu((A_1 \cap A_3) \cup (A_2 \cap A_3)) = \mu(A_1 \cap A_3) + \mu(A_2 \cap A_3).$$

Using the equality $\mu(A \cup B) = \mu(A) + \mu(B) - \mu(A \cap B)$ twice and the above equality once, we obtain

$$\begin{aligned}\mu(A_1 \cup A_2 \cup A_3) &= \mu(A_1 \cup A_2) + \mu(A_3) - \mu((A_1 \cup A_2) \cap A_3) \\ &= \mu(A_1) + \mu(A_2) - \mu(A_1 \cap A_2) + \mu(A_3) - \mu(A_1 \cap A_3) - \mu(A_2 \cap A_3) \\ &= \mu(A_1) + \mu(A_2) + \mu(A_3) - \mu(A_1 \cap A_2) - \mu(A_1 \cap A_3) - \mu(A_2 \cap A_3).\end{aligned}$$

Now assume the equality for $n - 1$ sets $A_1, \ldots, A_{n-1}$ and consider n distinct sets $A_1, \ldots, A_n$ such that no three have a nonempty intersection. The same must be true of $A_1, \ldots, A_{n-1}$, and moreover, the $n-1$ sets $A_1 \cap A_n, A_2 \cap A_n, \ldots, A_{n-1} \cap A_n$ must be disjoint. So,

$$\mu\left(\left(\bigcup_{i=1}^{n-1} A_i\right) \cap A_n\right) = \mu\left(\bigcup_{i=1}^{n-1} (A_i \cap A_n)\right) = \sum_{i=1}^{n-1} \mu(A_i \cap A_n).$$

Using the equality $\mu(A \cup B) = \mu(A) + \mu(B) - \mu(A \cap B)$ once, then the induction hypothesis and finally the above equality, we obtain

$$\mu\left(\bigcup_{i=1}^{n} A_i\right) = \mu\left(\bigcup_{i=1}^{n-1} A_i\right) + \mu(A_n) - \mu\left(\left(\bigcup_{i=1}^{n-1} A_i\right) \cap A_n\right)$$
$$= \sum_{i=1}^{n-1} \mu(A_i) - \sum_{i<j\leq n-1} \mu(A_i \cap A_j) + \mu(A_n) - \mu\left(\left(\bigcup_{i=1}^{n-1} A_i\right) \cap A_n\right)$$
$$= \sum_{i=1}^{n} \mu(A_i) - \sum_{i<j\leq n-1} \mu(A_i \cap A_j) - \sum_{i=1}^{n-1} \mu(A_i \cap A_n)$$
$$= \sum_{i=1}^{n} \mu(A_i) - \sum_{i<j\leq n} \mu(A_i \cap A_j).$$

(b) If one of the measures $\mu(A_i)$ is ∞, the inequality holds trivially. So, assume each $\mu(A_i)$ is finite. By part (a), we have the equality

$$\sum_{i=1}^{n} \mu(A_i) = \mu\left(\bigcup_{i=1}^{n} A_i\right) + \sum_{i<j\leq n} \mu(A_i \cap A_j).$$

However, the sets $A_i \cap A_j$ are disjoint and their union is contained in $\cup_{i=1}^{n} A_i$; therefore

$$\sum_{i<j\leq n} \mu(A_i \cap A_j) \leq \mu\left(\bigcup_{i=1}^{n} A_i\right).$$

Problem Set 5.5

5.5.P1. Using the functions constructed in Examples 5.5.4(a) and (b), show that strict inequality can hold in Theorem 5.5.1.
Hint: Using Example 5.5.4(a): Since $f'(x) = 0$ a.e., we have $\int_{0,1} f' = 0$. However, $f(1) - f(0) = \sum_{0<r_k\leq 1} \frac{1}{2^k} > 0$.
(b) Using Example 5.5.4(b): Since $F'(x) = 0$ a.e., we have $\int_{[0,1]} F' = 0$. However, $F(1) - F(0) > 0$ because F is strictly increasing.
5.5.P2. Let f be Lebesgue's singular function. Compute all the four Dini derivatives of f at each point $x \in [0, 1]$. (Cf. Example 5.5.2).
Hint: Let x be a point in the complement of the Cantor set C. Since f is constant in a neighbourhood of x, the derivative $f'(x)$ exists and is zero.

We now investigate the differentiability at points of the Cantor set. At a left endpoint x of a complementary interval (the function being constant on the right of x),

$$f'_+(x) = \lim_{h\to 0+} \frac{f(x+h)-f(x)}{h} = 0 = D^+f(x) = D_+f(x);$$

and at the right endpoint x of a complementary interval (the function being constant on the left of x),

$$f'_-(x) = \lim_{h\to 0+} \frac{f(x-h)-f(x)}{-h} = 0 = D^-f(x) = D_-f(x).$$

Let x denote the right endpoint of a complementary interval. Then x written in the ternary representation without using 1's has the form $x = .a_1a_2\cdots a_n200\cdots$ and $f(x) = .b_1b_2\cdots b_n100\cdots$, where $b_i = a_i/2$, $i = 1, 2, \ldots, n$, interpreted as a number in the binary representation. If h is between 3^{-m-1} and 3^{-m}, and $m > n$, then $x + h$ would lie between $.a_1a_2\cdots a_n200\cdots + 3^{-m-1}$ and $.a_1a_2\cdots a_n200\cdots + 3^{-m}$. Using the definition of f, the difference $f(x + h) - f(x)$ would lie between $\sum_{k=m+2}^{\infty} \frac{1}{2^k} = \frac{1}{2^{m+1}}$ and $\sum_{k=m+1}^{\infty} \frac{1}{2^k} = \frac{1}{2^m}$; consequently,

$$\frac{2^{-m-1}}{3^{-m}} < \frac{f(x+h)-f(x)}{h} < \frac{2^{-m}}{3^{-m-1}}.$$

Hence, as $h \to 0$ (so that $m \to \infty$) this difference quotient becomes positively infinite. That is, at a right endpoint x, $f'_+(x) = \infty$. Similarly, $f'_-(x) = \infty$ at a left endpoint.

Let x be a point of the Cantor set but not the endpoint of a removed interval. Then $x = .a_1a_2\cdots a_n\cdots$, where $a_n = 2$ for infinitely many n. Let $x_n = .a_1a_2\cdots a_n00\cdots$. Then $\{x_n\}_{n\geq 1}$ is a sequence of right endpoints and is an increasing sequence converging to x. Therefore

$$\frac{f(x)-f(x_n)}{x-x_n} = \frac{.0\cdots 0b_{n+1}b_{n+2}\cdots(\text{ binary })}{.0\cdots 0a_{n+1}a_{n+2}\cdots(\text{ ternary })}$$

If b_N is the first nonzero b_k with $k \geq n + 1$, then the right-hand side is greater than or equal to

$$\frac{\frac{1}{2^N}}{\frac{1}{3^{N-1}}} = \frac{3^{N-1}}{2^N} \to \infty \text{ as } n \to \infty.$$

So, $D^-f(x) = \infty$. Similarly, it may be shown that $D^+f(x) = \infty$. The computation of $D_+f(x)$ and $D_-f(x)$ is left to the reader.

5.5.P3. Suppose f: $[a, b] \to \mathbb{R}$ has derivative 0 a.e. on $[a, \beta]$ whenever $\beta < b$. Show that f has derivative 0 a.e. on $[a, b]$.
Hint: We may assume for convenience that $b - a > 1$. For every n, there exists an $A_n \subseteq \left[a,\ b - \frac{1}{n}\right]$ with measure 0 such that f has derivative 0 everywhere on $[a, b] \backslash (A_n \cup B_n)$, where $B_n = \left(b - \frac{1}{n}, b\right]$. Consider any $x \in [a, b]$ such that $x \notin \cap_{n \geq 1}(A_n \cup B_n)$. Then $x \notin (A_n \cup B_n)$ for some n i.e., $x \in [a, b] \backslash (A_n \cup B_n)$ for some n and therefore f has derivative 0 at x. Therefore f has derivative 0 everywhere on the complement of $\cap_{n \geq 1}(A_n \cup B_n)$. Since $\cap_{n \geq 1}(A_n \cup B_n)$ has measure 0 by Problem 3.1.P9, this means f has derivative 0 a.e. on $[a, b]$.

Problem Set 5.6

5.6.P1. Show that

$$V([a,\ b], f+g) \leq V([a,\ b], f) + V([a,\ b], g) \text{ and } V([a,\ b], cf) = |c|V([a,\ b], f).$$

Hint: The first inequality has already been proved [see Remark 5.6.2(c)]. For a partition P: $a = x_0 < x_1 < \cdots < x_n = b$ of $[a, b]$,

$$T(P,\ cf) = \sum_{k=1}^{n} |cf(x_k) - cf(x_{k-1})| = \sum_{k=1}^{n} |c||f(x_k) - f(x_{k-1})| \leq |c|V([a,\ b],\ f),$$

which implies $V([a, b], cf) \leq |c|V([a, b], f)$.

Conversely, let $\varepsilon > 0$ be given. There exists a partition P: $a = x_0 < x_1 < \cdots < x_n = b$ of $[a, b]$ such that

$$\sum_{k=1}^{n} |f(x_k) - f(x_{k-1})| > V([a,\ b],\ f) - \frac{\varepsilon}{|c|}.$$

Now,

$$\sum_{k=1}^{n} |cf(x_k) - cf(x_{k-1})| = |c| \sum_{k=1}^{n} |f(x_k) - f(x_{k-1})| > |c|V([a,\ b],\ f) - \varepsilon,$$

which implies

$$V([a,\ b], cf) \leq |c|V([a,\ b], f) - \varepsilon.$$

5.6.P2. Let ϕ: $(a, b) \to \mathbb{R}$ have a (finite) left limit everywhere and $y \in [a, b)$ [resp. $y \in (a, b]$] be a point where the right limit $\phi(y+)$ [resp. left limit $\phi(y-)$] exists (and is finite). Then first,

$$\lim_{x \to y+} \phi(x-) = \phi(y+)$$

and second,

$$\lim_{x \to y-} \phi(x-) = \phi(y-).$$

In particular, $\lim_{x \to a+} \phi(x-) = \phi(a+)$ and $\lim_{x \to b-} \phi(x-) = \phi(b-)$.

Hint: Consider any $\varepsilon > 0$. By definition of the right-hand limit, there exists a $\delta > 0$ such that

$$\phi(y+) - \varepsilon < \phi(z) < \phi(y+) + \varepsilon \quad \text{for all } z \in (y, y+\delta). \tag{8.41}$$

Now let $x \in (y, y + \delta)$ be arbitrary. Then $(y, x) \subset (y, y + \delta)$, and therefore it follows in accordance with (8.41) that

$$\phi(y+) - \varepsilon < \phi(u) < \phi(y+) + \varepsilon \quad \text{for all } u \in (y, x).$$

Since this holds for all $u \in (y, x)$, we may take the limit as $u \to x$, which automatically means $u \to x-$. Upon taking the limit, we get

$$\phi(y+) - \varepsilon \leq \phi(x-) \leq \phi(y+) + \varepsilon.$$

As this has been shown to hold for an arbitrary $x \in (y, y + \delta)$, we may take the limit as $x \to y$, which automatically means $x \to y+$. Upon taking the limit, we obtain

$$\phi(y+) - \varepsilon \leq \lim_{x \to y+} \phi(x-) \leq \phi(y+) + \varepsilon.$$

As $\varepsilon > 0$ is arbitrary, we have

$$\lim_{x \to y+} \phi(x-) = \phi(y+).$$

For $y = a$, this means $\lim_{x \to a+} \phi(x-) = \phi(a+)$.

Consider any $\varepsilon > 0$. By definition of the left-hand limit, there exists a $\delta > 0$ such that

$$\phi(y-) - \varepsilon < \phi(z) < \phi(y-) + \varepsilon \quad \text{for all } z \in (y - \delta, y). \tag{8.42}$$

Now let $x \in (y - \delta, y)$ be arbitrary. Then $(y - \delta, x) \subset (y - \delta, y)$, and therefore it follows in accordance with (8.42) that

$$\phi(y-) - \varepsilon < \phi(u) < \phi(y-) + \varepsilon \quad \text{for all } u \in (y - \delta, x).$$

Since this holds for all $u \in (y - \delta, x)$, letting $u \to x$ ($u \to x-$), we get

$$\phi(y-) - \varepsilon \leq \phi(x-) \leq \phi(y-) + \varepsilon.$$

As this has been shown to hold for an arbitrary $x \in (y - \delta, y)$, we may take the limit as $x \to y$ ($x \to y-$), thereby obtaining

$$\phi(y-) - \varepsilon \leq \lim_{x \to y-} \phi(x-) \leq \phi(y-) + \varepsilon.$$

Therefore

$$\lim_{x \to y-} \phi(x-) = \phi(y-),$$

When $y = b$, the above equality becomes $\lim_{x \to b-} \phi(x-) = \phi(b-)$.
5.6.P3. The function f defined by $f(x) = \sin\frac{\pi}{x}, 0 < x \leq 1$ and $f(0) = 0$, is not of bounded variation on [0, 1].
Hint: Choose the partition of [0, 1] consisting of the points

$$0, 1 \text{ and } \frac{2}{2k+1}, k = 1, 2, \cdots, n,$$

i.e., the points of the partition (in increasing order) are

$$0 < \frac{2}{2n+1} < \frac{2}{2n-1} < \cdots < \frac{2}{5} < \frac{2}{3} < 1.$$

Then

$$\sum_{k=2}^{n}\left|f\left(\frac{2}{2k-1}\right)-f\left(\frac{2}{2k+1}\right)\right|+\left|f(1)-f\left(\frac{2}{3}\right)\right|+\left|f\left(\frac{2}{2n+1}\right)-f(0)\right|$$
$$= 2n+2 \to \infty \text{ as } n \to \infty.$$

5.6.P4. Define f: [0, 1]→ℝ by $f(0) = 0$ and $f(x) = x^2 \sin\frac{1}{x}$, $x \neq 0$. Show that f is of bounded variation on [0, 1].
Hint: $f'(x) = 2x\sin\frac{1}{x} - \cos\frac{1}{x}$ and so, $|f'(x)| \leq 3$. Using parts (e) and (f) of Remark 5.6.2, it follows that f is of bounded variation.
5.6.P5. Show that the function defined on [0, 1] by

$$f(x) = \begin{cases} x^2 \sin\frac{1}{x^2} & \text{for } x \neq 0 \\ 0 & \text{for } x = 0 \end{cases}$$

is not of bounded variation.
Hint: Modify the argument of Example 5.6.3(c).
5.6.P6. Suppose that f: $[a, b] \to [c, d]$ is monotone and that g is of bounded variation on $[c, d]$. Prove that $V([a, b], g \circ f) \leq V([c, d], g)$.
Hint: Assume that f is increasing. Let P: $a = x_0 < x_1 < \cdots < x_n = b$ be a partition of $[a, b]$. Then

$$c \leq f(x_0) \leq f(x_1) \leq \cdots \leq f(x_n) \leq d,$$

though not necessarily a partition, gives rise to a partition Q of $[c, d]$, and

$$\begin{aligned} T(P, g \circ f) &= \sum_{k=1}^{n} |g(f(x_k)) - g(f(x_{k-1})) + |g(f(x_0)) - g(c)| + |g(d) - g(f(x_0))| \\ &= T(Q, g) \\ &\leq V([c, d], g). \end{aligned}$$

5.6.P7. Let $\{f_n\}_{n\geq 1}$ be a sequence of real-valued functions defined on $[a, b]$ that converge pointwise to the function f. Prove that

$$V([a,\ b],\ f) \leq \liminf_{n\to\infty} V([a,\ b],\ f_n).$$

Hint: Let P: $a = x_0 < x_1 < \cdots < x_p = b$ be a partition of $[a, b]$. Since $f_n \to f$ pointwise,

1. for $x \in [a, b]$, and any $\varepsilon > 0$, there exists an m_x such that $n \geq m_x \Rightarrow |f_n(x) - f(x)| < \frac{\varepsilon}{2p}$;
2. $|f(x_j) - f(x_{j-1})| \leq |f(x_j) - f_n(x_j)| + |f_n(x_j) - f_n(x_{j-1})| + |f_n(x_{j-1}) - f(x_{j-1})|$
$\leq 2\frac{\varepsilon}{2p} + |f_n(x_j) - f_n(x_{j-1})|$ for $n \geq m_0 \geq \max\{m_{x_0}, m_{x_1}, \ldots, m_{x_p}\}$.
So,

$$\sum_{j=1}^{p} |f(x_j) - f(x_{j-1})| \leq \sum_{j=1}^{p} |f_n(x_j) - f_n(x_{j-1})| + \varepsilon \text{ for } n \geq m_0.$$

Thus,

$$T(P, f) \leq V([a, b], f_n) + \varepsilon \text{ for } n \geq m_0.$$

Consequently,

$$T(P, f) \leq \liminf_{n\to\infty} V([a, b], f_n) + \varepsilon.$$

Considering that P is an arbitrary partition of $[a, b]$, we obtain $V([a, b], f) \leq \liminf_{n\to\infty} V([a, b], f_n) + \varepsilon$ The rest is clear.

5.6.P8. Let $\{f_n\}_{n\geq 1}$ be a sequence of functions of bounded variation on $[a, b]$ such that $\Sigma_n f_n(a)$ converges absolutely and $\Sigma_n V([a, b], f_n) < \infty$. Prove that

(i) $\Sigma_n f_n(x)$ converges absolutely for each $x \in [a, b]$,
(ii) $V([a, b], \Sigma_n f_n) \leq \Sigma_n V([a, b], f_n)$.

Hint: (i) $|f_n(x)| \leq |f_n(x) - f_n(a)| + |f_n(a)| \leq V([a, b], f_n) + |f_n(a)|$. Therefore

$$\sum_n |f_n(x)| \leq \sum_n V([a, b], f_n) + \sum_n |f_n(a)| < \infty.$$

(ii) For $n = 2$, the result has already been established. Assume that it is true for $n = m$. Then

$$V\left([a, b], \sum_{k=1}^{m+1} f_k\right) = V\left([a, b], \sum_{k=1}^{m} f_k + f_{m+1}\right) \leq V\left([a, b], \sum_{k=1}^{m} f_k\right) + V([a, b], f_{m+1}) \leq \sum_{k=1}^{m+1} V([a, b], f_k).$$

This completes the induction proof that $V([a,\ b], \sum_{k=1}^{n} f_k) \leq \sum_{k=1}^{n} V([a,\ b], f_k)$. This implies $\liminf_{n\to\infty} V([a,\ b], \sum_{k=1}^{n} f_k) \leq \sum_{k=1}^{\infty} V([a,\ b], f_k)$. The required conclusion (ii) now follows upon applying Problem 5.6.P7.

5.6.P9. Suppose $g \in L^1[a,\ b]$ and set $f(x) = \int_{[a,x]} g, a \leq x \leq b$. Show that $V([a,\ b], f) = \int_{[a,\ b]} |g|$.
Hint: We have already shown just before Example 5.6.12 that f is of bounded variation on $[a,\ b]$ and

$$V([a,\ b],\ f) \leq \int_{[a,\ b]} |g|. \tag{8.43}$$

It is sufficient to prove the reverse inequality.

Suppose g is continuous. Let $\varepsilon > 0$ be given. By the uniform continuity of g, there is a $\delta > 0$ such that $|g(x) - g(y)| < \varepsilon/2(b-a)$ whenever $|x-y| < \delta$. Now let P: $a = x_0 < x_1 < \cdots < x_n = b$ be a partition of $[a,\ b]$ such that $x_{k+1} - x_k < \delta$, $k = 0,1, 2, \ldots, n-1$. Then

$$\begin{aligned}
|f(x_{k+1}) - f(x_k)| &= \left|\int_{[x_k,x_{k+1}]} g\right| = \left|\int_{[x_k,x_{k+1}]} \{g(x_k) - [g(x_k) - g]\}\right| \\
&\geq \left|\int_{[x_k,x_{k+1}]} g(x_k)\right| - \left|\int_{[x_k,x_{k+1}]} [g - g(x_k)]\right| \geq \left|\int_{[x_k,x_{k+1}]} g(x_k)\right| - (x_{k+1} - x_k)\frac{\varepsilon}{2(b-a)} \\
&= \int_{[x_k,x_{k+1}]} |g(x_k)| - (x_{k+1} - x_k)\frac{\varepsilon}{2(b-a)} \\
&\geq \int_{[x_k,x_{k+1}]} |g| - \int_{[x_k,x_{k+1}]} |g - g(x_k)| - (x_{k+1} - x_k)\frac{\varepsilon}{2(b-a)} \\
&\geq \int_{[x_k,x_{k+1}]} |g| - (x_{k+1} - x_k)\frac{\varepsilon}{(b-a)}.
\end{aligned}$$

Summing up both sides, we have

$$\sum_{k=1}^{n} |f(x_{k+1}) - f(x_k)| \geq \int_{[a,\ b]} |g| - \varepsilon,$$

which implies

$$V([a,\ b], f) \geq \int_{[a,\ b]} |g| - \varepsilon.$$

Since $\varepsilon > 0$ is arbitrary, we arrive at

$$V([a,\ b],f) \geq \int_{[a,\ b]} |g|, \tag{8.44}$$

where f is the function related to g by $f(x) = \int_{[a,\ x]} g, a \leq x \leq b$. This has been obtained by assuming that g is continuous.

For $g \in L^1[a, b]$ and arbitrary $\eta > 0$, there exists a continuous g_0 on $[a, b]$ such that

$$\int_{[a,b]} |g_0 - g| < \eta, \tag{8.45}$$

using Proposition 3.4.2. Now,

$$\int_{[a,\ b]} |g_0| = \int_{[a,\ b]} |g + (g_0 - g)| \geq \int_{[a,\ b]} |g| - \int_{[a,\ b]} |g_0 - g| \geq \int_{[a,\ b]} |g| - \eta. \tag{8.46}$$

Since g_0 is continuous, we may apply (8.44) to it, thereby obtaining

$$V([a,\ b],f_0) \geq \int_{[a,\ b]} |g_0|,$$

where f_0 is the function related to g_0 by $f_0(x) = \int_{[a,\ x]} g_0, a \leq x \leq b$. Combining this with (8.46), we obtain

$$V([a,\ b],\ f_0) \geq \int_{[a,\ b]} |g| - \eta. \tag{8.47}$$

Now, applying (8.43) to $g - g_0$ yields

$$V([a,\ b],\ f_0 - f) \leq \int_{[a,\ b]} |g_0 - g|. \tag{8.48}$$

However, by the first part of Problem 5.6.P1, we have

$$\begin{aligned} V([a,\ b],\ f) &\geq -V([a,\ b],\ f_0 - f) + V([a,\ b],\ f_0) \\ &\geq -\int_{[a,\ b]} |g_0 - g| + V([a,\ b],\ f_0) \quad \text{by(8.48)} \\ &> V([a,\ b],\ f_0) - \eta \qquad \text{by (8.45)} \\ &> \int_{[a,\ b]} |g| - 2\eta \qquad \text{by (8.47).} \end{aligned}$$

Since $\eta > 0$ is arbitrary, the result follows.

5.6.P10. Deduce from Problem 5.6.P9 that $\sin x$ is of bounded variation on $[0, 2\pi]$ and that its total variation is 4.
Hint: [We have already noted in Example 5.6.3(a) without using Problem 5.6.P9 that $\sin x$ is of bounded variation on any compact interval.] Since

$$\sin x = \int_{[0,x]} \cos t\, dt$$

and $\cos t \in L^1[0, 2\pi]$, it follows by Problem 5.6.P9 that $\sin x$ is of bounded variation on $[0, 2\pi]$ and that

$$\begin{aligned} V([0,2\pi], \sin x) &= \int_{[0,2\pi]} |\cos t| dt \\ &= \int_{[0,\frac{\pi}{2}]} \cos t\, dt - \int_{[\frac{\pi}{2},\frac{3\pi}{2}]} \cos t\, dt + \int_{[\frac{3\pi}{2},2\pi]} \cos t\, dt = 4. \end{aligned}$$

5.6.P11. Define f on $[0, 1]$ by

$$f(0) = 0 \quad \text{and} \quad f(x) = x \sin\frac{1}{x}, x \neq 0.$$

Show that f is not an indefinite integral of a Lebesgue integrable function on $[0, 1]$.
Hint: It is clear that f is continuous on $[0, 1]$. However, it is not of bounded variation on $[0, 1]$. The argument is essentially that of Example 5.6.3(c). Indeed, for the partition

$$0 < \frac{2}{(4n+1)\pi} < \frac{2}{4n\pi} < \frac{2}{(4n-1)\pi} < \cdots < \frac{2}{2\pi} < \frac{2}{\pi} < 1,$$

we have the inequality

$$\begin{aligned} &\left|f\left(\frac{2}{(4n+1)\pi}\right) - f(0)\right| + \left|f\left(\frac{2}{4n\pi}\right) - f\left(\frac{2}{(4n+1)\pi}\right)\right| + \cdots + \left|\sin 1 - \frac{2}{\pi}\right| \\ &\geq \frac{4}{\pi}\left(1 + \frac{1}{3} + \cdots + \frac{1}{4n+1}\right), \end{aligned}$$

the right-hand side of which tends to ∞ as $n \to \infty$.

Consequently, by Problem 5.6.P9 above, it is impossible to find an $L^1[0, 1]$ function g such that $x\sin\frac{1}{x} = \int_{[0,x]} g, 0<x\leq 1$, i.e. f is not an indefinite integral of a Lebesgue integrable function on [0, 1].

5.6.P12. Define

$$f(x) = \begin{cases} 1 & x \in \mathbb{Q} \\ 0 & x \notin \mathbb{Q}. \end{cases}$$

Show that f is not of bounded variation on any compact interval.
Hint: In Example 5.3.3(b), we have seen that for rational x,

$$D^+f(x) = 0, D_+f(x) = -\infty, D^-f(x) = \infty \text{ and } D_-f(x) = 0$$

and for irrational x,

$$D^+f(x) = \infty, D_+f(x) = 0, D^-f(x) = 0 \text{ and } D_-f(x) = -\infty.$$

Thus the function f does not possess even a one-sided derivative at any x in the compact interval. By Remark 5.6.7(c), it cannot be of bounded variation.

Alternative argument: Call the compact interval [a, b] and let P: $a = x_0 < x_1 < \cdots < x_{n+1} = b$ be a partition of it such that the interior points $x_1 < \cdots < x_n$ are alternately rational and irrational. Then $T(P, f) \geq \sum_{j=2}^{n} |f(x_j) - f(x_{i-1})| = n - 1$. Since n is arbitrary, it follows that f fails to be of bounded variation.

5.6.P13. Construct functions f_n, n = 0, 1, 2, ... on an interval [a, b] such that

(i) $f_n \to f$ uniformly,
(ii) each f_n is of bounded variation,
(iii) f is not of bounded variation.

Hint: Let $f_n(x) = \begin{cases} 0 & 0\leq x<\frac{1}{n} \\ f(x) & \frac{1}{n}\leq x\leq 1 \end{cases}$, where $f(x) = \begin{cases} 0 & x=0 \\ x\cos\frac{\pi}{2x} & 0<x\leq 1. \end{cases}$

(i) Let $\varepsilon > 0$ be given. Choose n_0 so large that $1/n_0 < \varepsilon$. For $n \geq n_0$, we have $|f(x) - f_n(x)| = \begin{cases} 0 & \frac{1}{n}\leq x\leq 1 \\ \left|x\cos\frac{\pi}{2x}\right| & 0<x<\frac{1}{n} \end{cases}$ and $\left|x\cos\frac{\pi}{2x}\right| < \frac{1}{n} < \varepsilon$ whenever $0<x<\frac{1}{n}$. So, $f_n \to f$ uniformly on [a, b].

(ii) The restriction of f_n to $\left[\frac{1}{n}, 1\right]$ satisfies $f_n'(x) = \cos\frac{\pi}{2x} - x\left(\sin\frac{\pi}{2x}\right)\left(-\frac{1}{2x^2}\right) = \cos\frac{\pi}{2x} + \frac{1}{2x}\sin\frac{\pi}{2x}$. Therefore $|f_n'(x)| \leq 1 + \frac{n}{2}$. So, by Remark 5.6.2(f), the restriction of f_n to $\left[\frac{1}{n}, 1\right]$ is of bounded variation; but the restriction to $\left[0, \frac{1}{n}\right]$ is a step function and is therefore also of bounded variation. It follows by Theorem 5.6.4 that f_n is of bounded variation on [0, 1].

(iii) It is not difficult to check that f is not of bounded variation. Indeed, for the partition $P: 0 < \frac{1}{2n} < \frac{1}{2n-1} < \frac{1}{2n-3} < \cdots < \frac{1}{3} < \frac{1}{2} < 1$, we have

$$T(P, f) = \frac{1}{2n} + \frac{1}{2n} + \frac{1}{2n-2} + \frac{1}{2n-2} + \ldots + \frac{1}{2} + \frac{1}{2}$$
$$= 1 + \frac{1}{2} + \cdots + \frac{1}{n-1} + \frac{1}{n},$$

which tends to ∞ as $n \to \infty$.

5.6.P14. Let f be defined on [0, 1] by the following formulae [see Fig. 8.4]:

$$f(x) = \begin{cases} x & x = \frac{1}{2n}, n \in \mathbb{N} \\ 0 & x = \frac{1}{2n+1}, n \in \mathbb{N} \\ \text{linear} & \frac{1}{n+1} \le x \le \frac{1}{n}, n \in \mathbb{N} \\ 0 & x = 0. \end{cases}$$

Show that f is not of bounded variation.
Hint: Clearly, the function is continuous. For the partition

$$P: 0, \frac{1}{2n+1}, \frac{1}{2n}, \frac{1}{2n-1}, \ldots, \frac{1}{2}, 1$$

of [0,1], we have

$$T(P,f) = \frac{1}{2n} + \frac{1}{2n} + \frac{1}{2n-2} + \frac{1}{2n-2} + \ldots + \frac{1}{2} + \frac{1}{2}$$
$$= 1 + \frac{1}{2} + \cdots + \frac{1}{n-1} + \frac{1}{n},$$

which tends to ∞ as $n \to \infty$.

5.6.P15. If $f(x) = \int_{[a,x]} g$, where $g \in L^1[a, b]$, prove that the positive and negative variations of f, namely, $\mathcal{P}([a, b], f)$ and $\mathcal{N}([a, b], f)$, are given by $\int_{[a,b]} g^+$ and $\int_{[a,b]} g^-$ respectively.

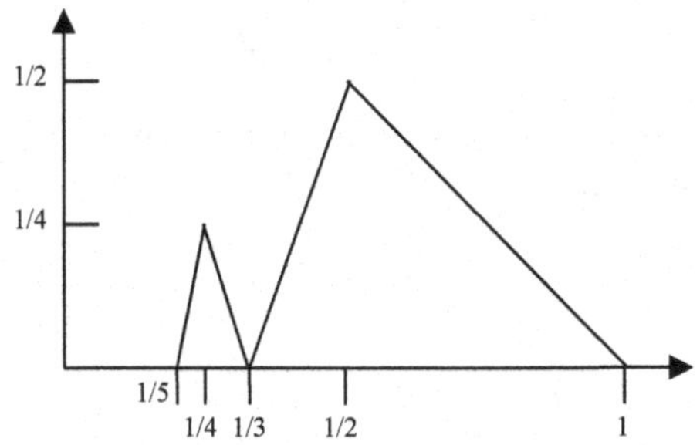

Fig. 8.4 Function in Problem 5.6.P14

Hint: By (5.64) of Proposition 5.6.9,

$$\begin{aligned}2\mathcal{P}([a,\ b],\ f) &= V([a,\ b],\ f) + f(b) - f(a)\\ &= \int_{[a,b]} |g| + f(b) - f(a) \text{by Problem 5.6.P9}\\ &= \int_{[a,b]} |g| + \int_{[a,b]} g + 0 \quad \text{by definition of } f\\ &= \int_{[a,b]} (g^+ + g^-) + \int_{[a,b]} (g^+ - g^-) = 2\int_{[a,b]} g^+.\end{aligned}$$

The argument for $\mathcal{N}([a, b], f)$ is analogous.

5.6.P16. Let $x_1, x_2, \ldots$ be the points of discontinuity in (a, b) of the function f of bounded variation with domain $[a, b]$. We define j: $[a, b] \to \mathbb{R}$ by the following formulae:

$$\begin{aligned}&j(a) = f(a) - f(a+)\\ \text{and } &j(x) = f(x) - f(x-) + \sum_{x_i < x} [f(x_i+) - f(x_i-)] \text{ if } a < x \leq b.\end{aligned} \tag{8.49}$$

By the "jump" of f at x, we mean $|f(x) - f(x-)| + |f(x+) - f(x)|$ if $a < x < b$ and $|f(a+) - f(a)|$ if $x = a$ and $|f(b) - f(b-)|$ if $x = b$. [Note that this agrees with Definition 5.2.4 for a monotone function.]

Show without using Theorem 5.3.1 that

(a) The function j defined in (8.49) is of bounded variation on $[a, b]$.
(b) $j(y-) = \sum_{x_i < y} [f(x_i+) - f(x_i-)]$ for $y \in (a, b]$,
 $j(y^+) = \sum_{x_i \leq y} [f(x_i+) - f(x_i-)]$ for $y \in (a, b)$
 and $j(a+) = 0$.
(c) If $F = f - j$, then F is continuous and of bounded variation.
(d) Show that $j' = 0$ a.e.
(e) Show that the representation of f as $F + j$, where F is continuous and $j' = 0$ a.e. is not unique (not even up to a constant) (cf. Problem 5.7.P4).

Hint: **(a)** We shall show first that

$$V([a,\ b],\ j) \leq J.$$

Let P: $a = y_0 < y_1 < \cdots < y_n = b$ be a partition of $[a, b]$. We shall estimate $|j(y_i) - j(y_{i-1})|$ for $i = 1, \ldots, n$.

For $i = 1$,

$$
\begin{aligned}
|j(y_1) - j(y_0)| &= |j(y_1) - j(a)| \\
&= \left| f(y_1) - f(y_1-) + \sum_{x_k < y_1} [f(x_k+) - f(x_k-)] - f(a) + f(a+) \right| \\
&\leq |f(y_1) - f(y_1-)| + |f(a) - f(a+)| \\
&\quad + \sum_{x_k < y_1} (|f(x_k) - f(x_k-)| + |f(x_k+) - f(x_k)|).
\end{aligned}
$$

For $i \geq 2$,

$$
\begin{aligned}
|j(y_i) - j(y_{i-1})| &= |f(y_i) - f(y_i-) + \sum_{x_k < y_i} [f(x_k+) - f(x_k-)] \\
&\quad - f(y_{i-1}) + f(y_{i-1}-) - \sum_{x_k < y_{i-1}} [f(x_k+) - f(x_k-)]| \\
&= \left| f(y_i) - f(y_i-) - f(y_{i-1}) + f(y_{i-1}-) + \sum_{y_{i-1} < x_k < y_i} [f(x_k+) - f(x_k-)] \right| \\
&= | f(y_i) - f(y_i-) - f(y_{i-1}) + f(y_{i-1}-) + f(y_{i-1}+) - f(y_{i-1}-) \\
&\quad \left. + \sum_{y_{i-1} < x_k < y_i} [f(x_k+) - f(x_k-)] \right| \\
&\leq |f(y_i) - f(y_i-)| + |f(y_{i-1}+) - f(y_{i-1})| \\
&\quad + \sum_{y_{i-1} < x_k < y_i} (|f(x_k) - f(x_k-)| + |f(x_k+) - f(x_k)|).
\end{aligned}
$$

Upon writing these estimates consecutively for $i = 2, \ldots, n$, we obtain

$$
\begin{aligned}
|j(y_2) - j(y_1)| &\leq |f(y_2) - f(y_2-)| + |f(y_1+) - f(y_1)| \\
&\quad \sum_{y_1 < x_k < y_2} (|f(x_k) - f(x_k-)| + |f(x_k+) - f(x_k)|) \\
|j(y_3) - j(y_2)| &\leq |f(y_3) - f(y_3-)| + |f(y_2+) - f(y_2)| \\
&\quad \sum_{y_2 < x_k < y_3} (|f(x_k) - f(x_k-)| + |f(x_k+) - f(x_k)|) \\
|j(y_4) - j(y_3)| &\leq |f(y_4) - f(y_4-)| + |f(y_3+) - f(y_3)| \\
&\quad \sum_{y_3 < x_k < y_4} (|f(x_k) - f(x_k-)| + |f(x_k+) - f(x_k)|) \\
&\vdots \qquad\qquad \vdots \qquad\qquad \vdots \\
|j(y_n) - j(y_{n-1})| &\leq |f(y_n) - f(y_n-)| + |f(y_{n-1}+) - f(y_{n-1})| \\
&\quad \sum_{y_{n-1} < x_k < y_n} (|f(x_k) - f(x_k-)| + |f(x_k+) - f(x_k)|).
\end{aligned}
$$

When we take the summation over $i = 1, \ldots, n$, the first term on the right-hand side of each estimate except the last one can be combined with the second term on the right-hand side of the next estimate. Upon doing so, we obtain

$$T(P, j) \leq J.$$

Consequently, $V([a, b], j) \leq J$, as was to be shown.

We proceed to prove that $J \leq V([a, b], f)$, which will complete the demonstration that j is of bounded variation. First we shall prove that if $y_1, y_2, \ldots, y_n$ are distinct points of (a, b), then

$$V(([a, b], f)) \geq s(a) + \sum_{i=1}^{n} s(y_i) + s(b),$$

where $s(x)$ means the jump at x.

Without loss of generality, we may suppose that $y_1 < y_2 < \cdots < y_n$. There exists a $\delta > 0$ such that the $n + 2$ intervals $(a, a + \delta)$, $(y_i - \delta, y_i + \delta)$, $(b - \delta, b)$ are all disjoint. Let $\eta > 0$ be arbitrary and put $\varepsilon = \frac{1}{3}\eta$. By definition of jump, for each i $(1 \leq i \leq n)$, we can select points $u_i \in (y_i - \delta, y_i)$ and $v_i \in (y_i, y_i + \delta)$ which satisfy $|f(y_i) - f(u_i)| + |f(y_i) - f(v_i)| > s(y_i) - \frac{\varepsilon}{n}$. We can also select points $v_0 \in (a, a + \delta)$ and $u_{n+1} \in (b - \delta, b)$ which satisfy $|f(a) - f(v_0)| > s(a) - \varepsilon$ and $|f(b) - f(v_{n+1})| > s(b) - \varepsilon$. On the basis of the disjointness noted earlier, we know that

$$a < v_0 < u_1 < y_1 < v_1 < \cdots < u_n < y_n < v_n < u_{n+1} < b.$$

Consequently, these points form a partition of $[a, b]$, which we shall denote by P. It follows from the manner in which the points u_i and v_i have been selected that

$$T(P, f) \geq s(a) + \sum_{i=1}^{n} s(y_i) + s(b) - \varepsilon - n\left(\frac{\varepsilon}{n}\right) - \varepsilon = s(a) + \sum_{i=1}^{n} s(y_i) + s(b) - \eta.$$

Since $\eta > 0$ is arbitrary and $V([a, b], f) \geq T(P, f)$, it follows that $V([a, b], f) \geq s(a) + \sum_{i=1}^{n} s(y_i) + s(b)$. On the other hand, J is the supremum of such sums, taken over all choices of finitely many distinct points $y_1, y_2, \ldots, y_n$ of (a, b). So, the desired inequality is now immediate.

(b) For $y \in (a, b]$,

$$j(y) = f(y) - f(y-) + \sum_{x_k < y} [f(x_k+) - f(x_k-)];$$

therefore

$$j(y-h)=f(y-h)-f((y-h)-)+\sum_{x_k<y-h}[f(x_k+)-f(x_k-)], \quad 0<h<y-a,$$

$$j(y)-j(y-h)=f(y)-f(y-)-f(y-h)+f((y-h)-) + \sum_{y>x_k\geq y-h}[f(x_k+)-f(x_k-)].$$

Letting $h\to 0$, we get

$$j(y)-j(y-)=f(y)-f(y-)-f(y-)+f(y-)+\sum_{y>x_k\geq y-h}[f(x_k+)-f(x_k-)]$$

by the second part of Problem 5-6.P2

$$=f(y)-f(y-).$$

Hence

$$j(y-)=j(y)-f(y)+f(y-)+\sum_{x_k<y}[f(x_k+)-f(x_k-)] = \sum_{x_k<y}[f(x_k+)-f(x_k-)].$$

For $y\in(a,b)$,

$$j(y)=f(y)-f(y-)+\sum_{x_k<y}[f(x_k+)-f(x_k-)];$$

therefore

$$j(y+h)=f(y+h)-f((y+h)-)+\sum_{x_k<y+h}[f(x_k+)-f(x_k-)], \quad 0<h<b-y,$$

$$j(y+h)-j(y)=f(y+h)-f((y+h)-)-f(y)+f(y-) + \sum_{y\leq x_k<y+h}[f(x_k+)-f(x_k-)].$$

Letting $h \to 0$, we get

$$j(y+)-j(y)=f(y+)-f(y+)-f(y)+f(y-)+f(y+)-f(y-)$$

↑ by the first part of Problem 5-6.P2

$$=[f(y+)-f(y)].$$

Hence

$$\begin{aligned} j(y+) &= j(y)+[f(y+)-f(y)] \\ &= f(y)-f(y-)+\sum_{x_k<y}[f(x_k+)-f(x_k-)]+f(y+)-f(y) \\ &= \sum_{x_k\le y}[f(x_k+)-f(x_k-)]. \end{aligned}$$

Finally, consider $y(a+)$. We have

$$j(a)=f(a)-f(a+)$$

and

$$j(a+h)=f(a+h)-f((a+h)-)+\sum_{x_k<a+h}[f(x_k+)-f(x_k-)], \quad 0<h<b-a$$

$$\begin{aligned} j(a+h)-j(a) &= f(a+h)-f((a+h)-)-f(a)+f(a+) \\ &\quad+\sum_{x_k<a+h}[f(x_k+)-f(x_k-)]. \end{aligned}$$

Letting $h \to 0$, we get

$$j(a+)-j(a)=f(a+)-f(a+)-f(a)+f(a+)+0$$

↑ by the first part of Problem 5-6.P2

$$=-f(a)+f(a+)=-j(a).$$

(c) By (8.49), we have $F(a) = f(a+)$. We shall first show that F is continuous and of bounded variation on $[a, b]$.

First we prove continuity at any $y \in (a, b)$. By Problem 5.6.P2,

$$j(y-) = \sum_{x_i < y} [f(x_i+) - f(x_i-)] \tag{8.50}$$

and

$$j(y+) = \sum_{x_i \leq y} [f(x_i+) - f(x_i-)]. \tag{8.51}$$

From (8.49) and (8.50),

$$j(y-) = j(y) - f(y) + f(y-),$$

whence

$$F(y-) = f(y-) - j(y-) = f(y-) - j(y) + f(y) - f(y-) = F(y). \tag{8.52}$$

From (8.50), (8.51) and (8.52), we obtain

$$\begin{aligned} F(y+) &= f(y+) - j(y+) \\ &= f(y+) - (f(y+) - f(y-)) - \sum_{x_i < y} [f(x_i+) - f(x_i-)], \quad \text{by (8.51)} \\ &= f(y-) - \sum_{x_i < y} [f(x_i+) - f(x_i-)] \\ &= f(y-) - j(y-) \text{by } \ (8.50) \\ &= F(y) \quad \text{by (8.52)}. \end{aligned}$$

Hence

$$F(y+) = F(y).$$

Thus F is continuous at every $y \in (a, b)$.

By Problem 5.6.P2, we know that

$$j(a+) = 0 \quad \text{and} \quad j(b-) = \sum_{x_i < b} [f(x_i+) - f(x_i-)].$$

Recall that $F(a) = f(a+)$. Now, $F(a+) = f(a+) - j(a+) = f(a+) = F(a)$. Therefore F is continuous at a.

$$\begin{aligned} F(b) &= f(b) - j(b) = f(b) - f(b) + f(b-) - \sum_{x_i < b} [f(x_i+) - f(x_i-)] \\ &= f(b-) - \sum_{x_i < b} [f(x_i+) - f(x_i-)] \\ &= f(b-) - j(b-) = F(b-). \end{aligned}$$

Thus the function F is continuous on $[a, b]$.

F is of bounded variation because f and j are.

(d) Since f is of bounded variation, we have $f = f_1 - f_2$, where f_1 and f_2 are nonnegative [nonnegativity will not be relevant in what follows] increasing functions on $[a, b]$. Let S_1 and S_2 be the saltus functions associated with f_1 and f_2 respectively.

Now,

$$\begin{aligned} j(x) &= f(x) - f(x-) + \sum_{x_i<x} [f(x_i+) - f(x_i-)] \quad \text{for } a<x\leq b \\ &= f_1(x) - f_2(x) - f_1(x-) + f_2(x-) \\ &\quad + \sum_{x_i<x} [f_1(x_i+) - f_2(x_i+) - f_1(x_i-) + f_2(x_i-)] \end{aligned}$$

$$\begin{aligned} &= f_1(x) - f_1(x-) + \sum_{x_i<x} [f_1(x_i+) - f_1(x_i-)] \\ &\quad - [f_2(x) - f_2(x-) + \sum_{x_i<x} [f_2(x_i+) - f_2(x_i-)]]. \end{aligned}$$

Recall that in the summation occurring in the definition of the associated saltus functions [see the beginning of Sect. 5.3], the left endpoint a is not excluded from being among the points x_k; but in the summation occurring in the definition of j, it is specifically excluded by taking $x_1, x_2, \ldots$ to be the points of discontinuity in (a, b). Taking this distinction into account, we find that

$$f_1(x) - f_1(x-) + \sum_{x_1<x} [f_1(x_i+) - f_1(x_i-)] = S_1(x) - [f_1(a+) - f_1(a)]$$

and

$$f_2(x) - f_2(x-) + \sum_{x_1<x} [f_2(x_i+) - f_2(x_i-)] = S_2(x) - [f_2(a+) - f_2(a)].$$

Substituting these two equalities into the one derived for $j(x)$ above, we arrive at

$$j(x) = S_1(x) - S_2(x) - [f(a+) - f(a)] \text{ for } a<x\leq b.$$

Since S_1 and S_2 are differentiable a.e. and $S_1'(x) = 0 = S_2'(x)$ a.e. on $[a, b]$ by Theorem 5.4.7, it now follows that j is differentiable a.e. and $j'(x) = 0$ a.e. on $[a, b]$.

(e) Suppose the domain $[a, b]$ is $[0, 1]$. Let g be the Cantor function as in Example 5.5.2. It was shown there that g is increasing (hence of bounded variation), continuous and has derivative 0 a.e. It follows on the one hand that $F + g$ is of bounded variation and continuous, and on the other hand that $j + g$ has derivative 0 a.e. However, their difference is f.

Problem Set 5.7

5.7.P1. If f: $[a, b] \to \mathbb{R}$ is absolutely continuous on $[a, c]$ as well as on $[c, b]$, then it is absolutely continuous on $[a, b]$.
Hint: For any interval $I = [\alpha, \beta] \subseteq [a, b]$, denote by $\Delta(I)$ the nonnegative number $|f(\beta) - f(\alpha)|$.
Let $\varepsilon > 0$ be given. There exists a $\delta > 0$ such that $\sum_{I \in \mathcal{K}} \Delta(I) < \frac{\varepsilon}{2}$ for an arbitrary finite system $\mathcal{K}$ of disjoint intervals contained in either $[a, c]$ or $[c, b]$ and having total length less than δ. Now consider an arbitrary finite system $\mathcal{J}$ of disjoint intervals contained in $[a, b]$ and having total length less than δ. If every interval of the system lies in either $[a, c]$ or $[c, b]$, it is clear that $\sum_{i \in \mathcal{J}} \Delta(I) < \frac{\varepsilon}{2} + \frac{\varepsilon}{2} = \varepsilon$. If not, then precisely one interval J fails to do so, in which case, c belongs to the interior of J. Moreover, all the other intervals are contained either in $[a, c]$ or $[c, b]$. Denote by $\mathcal{A}$ the subsystem of the former and by $\mathcal{B}$ the subsystem of the latter. If either of these systems is empty, then the sum relating to it is empty and therefore taken to be 0. Then

$$\sum_{i \in \mathcal{J}} \Delta(I) = \sum_{i \in \mathcal{A}} \Delta(I) + \sum_{i \in \mathcal{B}} \Delta(I) + \Delta(J). \tag{8.53}$$

Let J_A be the part of J that lies to the left of c and J_B be the part that lies to the right of c. Then

$$\Delta(J) \leq \Delta(J_A) + \Delta(J_B). \tag{8.54}$$

Also, the intervals of $\mathcal{A}$ and the interval J_A together form a finite system of disjoint intervals contained in $[a, c]$. Therefore $\sum_{I \in \mathcal{A}} \Delta(I) + \Delta(J_A) < \frac{\varepsilon}{2}$. Similarly, $\sum_{I \in \mathcal{B}} \Delta(I) + \Delta(J_B) < \frac{\varepsilon}{2}$. It follows that

$$\sum_{i \in \mathcal{A}} \Delta(I) + \Delta(J_A) + \sum_{i \in \mathcal{B}} \Delta(I) + \Delta(J_B) < \frac{\varepsilon}{2} + \frac{\varepsilon}{2} = \varepsilon.$$

Upon combining this with (8.53) and (8.54), we get $\sum_{I \in \mathcal{J}} \Delta(I) < \varepsilon$.

5.7.P2. Let f: $[a, b] \to \mathbb{R}$ be an absolutely continuous function and let $E \subset [a, b]$ be a set of measure zero. Then show that $m(f(E)) = 0$. Use this to show that the Cantor function is not absolutely continuous. Show that on a closed bounded interval, the class of absolutely continuous functions is a proper subclass of continuous functions of bounded variation. [This is trivial if the domain is unbounded, because $\sin x$ will do the trick.]

Hint: Let $\varepsilon > 0$ be given. We shall show that there exists a $\delta > 0$ such that, for an arbitrary countable system of disjoint intervals $\{(a_k, b_k)\}_{k \geq 1}$, the sum of whose lengths is less than δ, the inequality

$$\sum_{k=1}^{\infty} (M_k - m_k) < \varepsilon \tag{8.55}$$

holds, where

$m_k = \min\{f(x): a_k \leq x \leq b_k\}$ and $M_k = \max\{f(x): a_k \leq x \leq b_k\}$.

By absolute continuity of f and Remark 5.7.2(d), there exists a $\delta > 0$ such that the inequality (8.55) holds with $\frac{\varepsilon}{2}$ in place of ε for any finite system of disjoint intervals $\{(a_k, b_k)\}_{n \geq k \geq 1}$, the sum of whose lengths is less than δ, i.e.,

$$\sum_{k=1}^{n} (M_k - m_k) < \frac{\varepsilon}{2}.$$

Letting $n \to \infty$, we obtain (8.55).

Assume that the points a and b do not belong to E, so that $E \subset (a, b)$. Since $m(E) = 0$, there exists a bounded open set $G \supset E$ such that $m(G) < \delta$. Without loss of generality, we may assume that $G \subseteq (a, b)$, for otherwise we replace G by $G \cap (a, b)$ and call the intersection G. Now,

$$G = \bigcup_{k=1}^{\infty} (\alpha_k, \beta_k) \quad \text{and} \quad \sum_{k=1}^{\infty} (\beta_k - \alpha_k) < \delta.$$

This implies $f(E) \subseteq f(G) = \cup_{k \geq 1} f((\alpha_k, \beta_k)) \subseteq \cup_{k \geq 1} f([\alpha_k, \beta_k])$. Therefore

$$m^*(f(E)) \leq \sum_{k=1}^{\infty} m^*(f([\alpha_k, \beta_k])).$$

It is also clear that

$$f([\alpha_k, \beta_k]) = [m_k, M_k]$$

and consequently,

$$m^*(f(E)) \leq \sum_{k=1}^{\infty} (M_k - m_k) < \varepsilon,$$

which implies

$$m(f(E)) = 0,$$

since $\varepsilon > 0$ is arbitrary.
Note that the removal of the points a and b from the set E implies removal of at most two points, namely, $f(a)$ and $f(b)$, from $f(E)$. This has no effect on the measure of $f(E)$.

The Cantor function maps the Cantor set, which has measure 0, into the set [0, 1], which does not have measure 0. Therefore it is not absolutely continuous. Consequently, the Cantor function provides an example of a function that is of bounded variation on a closed bounded interval but is not absolutely continuous.

5.7.P3. Let f: $[a, b] \to \mathbb{R}$ be an absolutely continuous function. Then f maps measurable sets into measurable sets.
Hint: Let E be a measurable set in $[a, b]$. Then there exist an increasing sequence $\{F_n\}$ of closed sets contained in E such that

$$m(E) = \lim_{n\to\infty} m(F_n)$$

by Proposition 2.3.24. Therefore we can write

$$E = \left(\bigcup_{n=1}^{\infty} F_n\right) \cup N,$$

where N is a set of measure zero; hence

$$f(E) = f\left(\bigcup_{n=1}^{\infty} F_n\right) \cup f(N) = \bigcup_{n=1}^{\infty} f(F_n) \cup f(N).$$

By Problem 5.7.P2 above, $f(N)$ is of measure zero. Each $f(F_n)$, being the image of a compact subset F_n under the continuous map f, is compact and hence measurable. Consequently, $f(E)$ is measurable.

5.7.P4. Let f be defined and measurable on the closed interval $[a, b]$ and let E be any measurable subset on which f' exists. Then prove that

$$m^*(f(E)) \le \int_E |f'|.$$

Hint: First we suppose that $|f'(t)| < M$ on E, where M denotes a positive integer. Let

$$E_k^n = \left\{t \in E : \frac{k-1}{2^n} < |f'(t)| < \frac{k}{2^n}, k = 1, 2, \ldots, M2^n, n = 1, 2, \ldots\right\}.$$

Then for each n,

$$\begin{aligned} m^*(f(E)) &= m^*\left(f\left(\bigcup_k E_k^n\right)\right) = m^*\left(\bigcup_k f(E_k^n)\right) \\ &\leq \sum_k m^*(f(E_k^n)), \quad \text{using Proposition 2.2.9} \\ &= \sum_k \frac{k-1}{2^n} m(E_k^n) + \frac{1}{2^n}\sum_k m(E_k^n). \end{aligned}$$

Therefore

$$m^*(f(E)) \leq \lim_{n\to\infty}\left[\sum_k \frac{k-1}{2^n} m(E_k^n) + \frac{1}{2^n}\sum_k m(E_k^n)\right] = \int_E |f'|.$$

Now, assume that f' is not bounded. Let

$$A_k = \{t \in E : (k-1) \leq f'(t) < k\}, k = 1, 2, \ldots.$$

Then

$$\begin{aligned} m^*(f(E)) &= m^*\left(f\left(\bigcup_k A_k\right)\right) = m^*\left(\bigcup_k f(A_k)\right) \leq \sum_k m^*(f(A_k)) \leq \sum_k \int_{A_k} |f'| \\ &= \int_E |f'|. \end{aligned}$$

So far, we have proved (see Remark 5.7.2(b), Proposition 5.7.2 and Problem 5.7.P2) that an absolutely continuous function is continuous, of bounded variation and maps a set of measure zero into a set of measure zero. The following result shows that these three properties characterise an absolutely continuous function [Banach–Zarecki Theorem].

5.7.P5. If f is continuous and of bounded variation on $[a, b]$, and if f maps a set of measure zero into a set of measure zero, then f is absolutely continuous on $[a, b]$.
Hint: Let $\{(a_k, b_k)\}_{k\geq 1}$ be a finite system of disjoint open intervals contained in $[a, b]$ and let $E_k = \{x \in [a_k, b_k]: f' \text{ exists}\}$. Since $m([a_k, b_k]\backslash E_k) = 0$ and f maps sets of measure zero into sets of measure zero,

$$m(f([a_k, b_k])) = m(f(E_k)),$$

and therefore

$$\begin{aligned}\sum_{k=1}^{n} |f(b_k) - f(a_k)| &\le \sum_{k=1}^{n} m(f([a_k, b_k])) \\ &\le \sum_{k=1}^{n} m(f(E_k)) \\ &\le \sum_{k=1}^{n} \int_{E_k} |f'| \\ &\le \sum_{k=1}^{n} \int_{[a_k, b_k]} |f'|,\end{aligned}$$

which tends to 0 as $\sum_{k=1}^{n} (b_k - a_k) \to 0$, since f' is integrable in view of Theorem 5.5.1.

5.7.P6. Let f be absolutely continuous on [ε, 1] for each positive ε < 1. Give an example to show that continuity of f on [0, 1] does not imply absolute continuity on [0, 1]. What if f is also of bounded variation on [0, 1]? Show that x^p is absolutely continuous on [0, 1] when $0 \le p < 1$.

Hint: The function

$$f(x) = \begin{cases} x \sin\frac{1}{x} & x \neq 0 \\ 0 & x = 0 \end{cases}$$

is such that it is absolutely continuous on [ε, 1] for positive ε < 1, since

$$|f'(x)| = \left| \sin\frac{1}{x} - \frac{1}{x}\cos\frac{1}{x} \right| \le 1 + \frac{1}{\varepsilon}.$$

The function f is continuous on [0, 1]. However, it is not of bounded variation on [0, 1].

The answer is Yes if f is also of bounded variation on [0, 1]:

The function $v(x) = V([0, x], f)$, $0 \le x \le 1$, is continuous, since f is continuous on [0, 1]. Let $\varepsilon > 0$ be given. Then there exists a $\delta_1 > 0$ such that $v(x) < \frac{\varepsilon}{2}$ provided $x < \delta_1$. We assume, as we may, that $\delta_1 < 1$. On $[\delta_1, 1]$, the function is absolutely continuous. There exists a $\delta_2 > 0$ such that for any finite system $\{(\alpha_k, \beta_k)\}_{1 \le k \le n}$ of disjoint intervals contained in $[\delta_1, 1]$ and where the sum of lengths is less than δ_2, we have

$$\sum_{k=1}^{n} |f(\beta_k) - f(\alpha_k)| < \frac{\varepsilon}{2}.$$

Let $\delta = \min\{\delta_1, \delta_2\}$ and let $\{(a_i, b_i)\}_{1 \le i \le m}$ be a system of disjoint open intervals contained in [0,1] and having sum of lengths less than δ. If an (a_i, b_i) is such that $\delta_1 \in (a_i, b_i)$, we may replace (a_i, b_i) by $(a_i, \delta_1) \cup (\delta_1, b_i)$. This new system of intervals $\{(\alpha_i, \beta_i)\}_{1 \le i \le m'}$ are disjoint and the sum of their lengths is less than δ. We then have

$$\sum_{i=1}^{m} |f(b_i) - f(a_i)| \le \sum_{i=1}^{m} |f(\beta_i) - f(\alpha_i)|$$
$$= \sum_{i \in A} |f(\beta_i) - f(\alpha_i)| + \sum_{i \in B} |f(\beta_i) - f(\alpha_i)| < \frac{\varepsilon}{2} + \frac{\varepsilon}{2} = \varepsilon,$$

where A consists of those indices i in $\{1, 2, \ldots, m\}$ for which (α_i, β_i) lie in $[0, \delta_1]$ and B consists of the remaining indices.

For every positive $\varepsilon < 1$, the function x^p $(0 \le p < 1)$ is continuous on $[\varepsilon, 1]$ with bounded derivative and hence absolutely continuous by Remark 5.7.2(e). Since it is increasing on [0, 1], it is also of bounded variation on [0, 1]; besides, it is continuous on [0, 1]. It follows from what has been proved above that the function is absolutely continuous on [0, 1].

5.7.P7. Show that the function f given on $[0, \infty)$ by $f(x) = x^{\frac{1}{2}}$ for $0 \le x < 1$ and satisfying $f(x + k) = f(x) + k$ for $0 \le x < 1$ and $k \in \mathbb{N}$ is absolutely continuous on any bounded subinterval of its domain but not on the entire domain. [Note that the example of such a function given in Remark 5.7.2(a) satisfies a Lipschitz condition on every bounded subinterval of $[0, \infty)$ while the present one does not.]

Hint: From the second part of the definition of f, it follows that $f(1) = f(0+1) = f(0) + 1 = 1^{\frac{1}{2}}$, so that the equality $f(x) = x^{\frac{1}{2}}$ holds for $x = 1$ as well. Therefore the function is absolutely continuous on [0, 1] by Problem 5.7.P6. Now, for any $k \in \mathbb{N}$, we find that $x \in [k, k + 1]$ implies $x - k \in [0, 1]$, which further implies $f(x) = f((x-k)+k) = f(x-k) + k = (x-k)^{\frac{1}{2}} + k$. Therefore, by Problem 5.7.P6 and the fact that k is a constant, the function is absolutely continuous on the interval $[k, k + 1]$ for any $k \in \mathbb{N}$. It now follows by Problem 5.7.P1 that it is absolutely continuous on any bounded interval $[0, k + 1]$ and hence on any bounded subinterval of $[0, \infty)$.

To see why it is not absolutely continuous on $[0, \infty)$, consider some positive $\delta < 1$ and $A > \sum_{i=1}^{\infty} \frac{1}{i^2}$. For $x_i = i, y_i = i + \frac{\delta}{Ai^2}, i = 1, \ldots, n,$

$$\sum_{i=1}^{n} |y_i - x_i| = \sum_{i=1}^{n} \frac{\delta}{Ai^2} = \delta \sum_{i=1}^{n} \frac{1}{Ai^2} < \delta.$$

However,

$$\sum_{i=1}^{n}|f(y_i)-f(x_i)| = \left(\frac{\delta}{A}\right)^{\frac{1}{2}}\sum_{i=1}^{n}\frac{1}{i}.$$

As the right-hand side of the above equality tends to ∞ as $n \to \infty$, the function f cannot be absolutely continuous on $[0,\infty)$.

5.7.P8. Let f be an increasing function defined on $[a, b]$. Show that f can be decomposed into a sum of increasing functions $f = g + h$, where g is absolutely continuous and h is increasing with $h' = 0$ a.e.

Hint: For an increasing function f, the derivative f' exists a.e. (see Theorem 5.4.6) and is integrable (see Theorem 5.5.1). Set

$$g(x) = \int_{[a,x]} f'(t)dt.$$

Then g is absolutely continuous and

$$g(\beta) - g(\alpha) = \int_{[\alpha,\beta]} f'(t)dt \leq f(\beta) - f(\alpha). \tag{8.56}$$

If $h = f - g$, then

$$h(\beta) - h(\alpha) = [f(\beta) - f(\alpha)] - [g(\beta) - g(\alpha)] \geq 0,$$

using (8.56). Consequently, h is increasing. However,

$$h'(x) = f'(x) - g'(x) = 0 \text{ a.e.}$$

Problem Set 5.8

5.8.P1. Suppose f is absolutely continuous on $[a, b]$ and $f'(x) = 0$ a.e. on $[a, b]$. Then f is a constant function. Give a proof of this that is different from the one in Corollary 5.8.4.

Hint: Let $E = \{x \in (a, b): f'(x) = 0\}$. Then

$$m(f(E)) \leq \int_E |f'| = 0 \tag{8.57}$$

by Problem 5.7.P4. Also, f is absolutely continuous, and $[a, b]\backslash E$ is a set of measure 0; so it follows that

$$m(f([a,\ b]\backslash E)) = 0 \tag{8.58}$$

by Problem 5.7.P2. Now,

$$\begin{aligned} m(f([a,\ b])) &= m(f([a,\ b]\backslash E) \cup E) \\ &= m(f(([a,\ b]\backslash E)) \cup f(E)) \\ &\leq m(f(([a,\ b]\backslash E))) + m(f(E)) = 0, \end{aligned} \tag{8.59}$$

by (8.57) and (8.58). Since f is continuous, unless f is constant, $f([a, b])$ contains an interval of positive length, contradicting (8.59). Therefore f must be constant.

5.8.P2. If f and g are absolutely continuous on $[a, b]$, $f'(x) = g'(x)$ a.e. on $[a, b]$ and $f(x_0) = g(x_0)$ for some $x_0 \in [a, b]$, then show that $f(x) = g(x)$ for every $x \in [a, b]$.

Hint: If $h(x) = f(x) - g(x)$ for $x \in [a, b]$, then h is absolutely continuous and $h'(x) = 0$ a.e. on $[a, b]$. By Problem 5.8.P1, h is a constant. Since it vanishes at x_0, it follows that $f(x) = g(x)$ for every $x \in [a, b]$.

5.8.P3. Let f: $[a, b] \to \mathbb{R}$ be absolutely continuous and let

$$v(a) = 0,\ v(x) = V([a,\ x],\ f) \text{for } x \in [a,\ b].$$

Then v is absolutely continuous on $[a, b]$ by Proposition 5.7.9. Show that $v'(x) = |f'(x)|$ for almost all $x \in [a, b]$.
Hint: Since v is absolutely continuous, Theorem 5.8.2 leads to

$$v(x) = \int_{[a,\ x]} v',$$

which, in turn, implies

$$v(y) - v(x) = \int_{[x,y]} v' \quad \text{for all } y > x.$$

Let P be an arbitrary partition of $[a, b]$:

$$P : a = x_0 < x_1 < \cdots < x_n = b.$$

Then another application of Theorem 5.8.2 leads to

$$T(P,\ f) = \sum_{k=1}^{n} |f(x_k) - f(x_{k-1})| = \sum_{k=1}^{n} \left| \int_{[x_{k-1},x_k]} f' \right| \leq \sum_{k=1}^{n} \int_{[x_{k-1},x_k]} |f'| = \int_{[a,b]} |f'|.$$

So,

$$v(b) - v(a) = V([a, b], f) \leq \int_{[a,b]} |f'|.$$

Similarly,

$$v(y) - v(x) \leq \int_{[x,y]} |f'| \text{ for all } y > x.$$

Hence $v'(x) \leq |f'(x)|$ for almost all $x \in [a, b]$ by Theorem 5.8.5. Conversely, since

$$v(y) - v(x) \geq |f(y) - f(x)| \quad \text{for all } y > x,$$

we obtain

$$v'(x) \geq |f'(x)| \quad \text{for all } x \in [a, b].$$

5.8.P4. Show that, if g is an integrable function defined on $[a, b]$ with indefinite integral f, then $f'(x) = g(x)$ whenever x is a point of continuity of g.

Hint: For any x,

$$f(x) = \int_{[a,x]} g \quad \text{and} \quad f'(x) = \lim_{h\to 0}\frac{f(x+h) - f(x)}{h} = \lim_{h\to 0}\frac{1}{h}\int_{[x,x+h]} g = g(x)$$

since every point of continuity of g is a Lebesgue point of g.

5.8.P5. If $f(x) = \int_{[0,x]} g$, where g is an integrable function defined on [0, 1], then $f' = g$ need not hold even when f' exists.

Hint: Let $g(x) = \begin{cases} 1 & \text{if } x \text{ is rational} \\ 0 & \text{if } x \text{ is irrational.} \end{cases}$ Then for every x, $\int_{[0,x]} g = 0$ because the set over which g is not zero, being a subset of the set of all rationals, has measure zero. So, $f(x) = 0$ for all $x \in [0, 1]$. However, $f' = g$ at all irrational x and $f' \neq g$ on the set of rationals in [0, 1].

5.8.P6. (Cf. Theorem 5.5.1) Show that the inequality $\int_{[a,x]} f' \leq f(b) - f(a)$ need not hold when f is not monotone increasing.

Hint: Take any monotone increasing function for which strict inequality holds and multiply by -1. In particular, for the Cantor function (see Example 5.5.2), $f'(x) = 0$ a.e. on [0, 1] and $f(0) = 0$, $f(1) = 1$. So, $\int_{[0,1]} f' = 0 < f(1) - f(0) = 1$. On multiplying f by -1, we obtain $0 = \int_{[0,1]} -f' > -1$.

5.8.P7. (Lebesgue Decomposition Theorem for Functions) Suppose f is of bounded variation on $[a, b]$. Then there exists an absolutely continuous function g and a function h such that $f(x) = g(x) + h(x)$, $x \in [a, b]$, where $h'(x) = 0$ a.e. Up to constants, the decomposition is unique.

Hint: It is sufficient to prove this on the assumption that f is monotonic on $[a, b]$. Then the derivative f' exists a.e. by Lebesgue's Theorem 5.4.6 and is integrable by Theorem 5.5.1. Let

$$g(x) = \int_{[a,x]} f' \quad \text{for } a \leq x \leq b.$$

Then g is absolutely continuous by Proposition 5.8.1 and by Theorem 5.8.5, we have $f'(x) - g'(x) = 0$ a.e. on $[a, b]$. Now, all we need to do is to set

$$h(x) = f(x) - g(x), \quad x \in [a, b].$$

As for uniqueness, suppose $f = g_1 + h_1$, where g_1 is absolutely continuous and $h_1' = 0$ a.e. Then $g - g_1 = h_1 - h$ is absolutely continuous with derivative 0 a.e., which implies $g = g_1 + c$ by Corollary 5.8.4 and hence $h = h_1 - c$.

5.8.P8. [Application to Fourier Series; due to Lebesgue] Let f be a function in $L^1([-\pi, \pi])$. Then show that

$$\lim_{n\to\infty} \sigma_n(x) = f(x)$$

at every Lebesgue point x of f (and hence a.e. on $[-\pi, \pi]$ by Theorem 5.8.9). Here the notation is as in Fejér's Theorem 4.3.4.

Hint: All integrals in this solution are to be understood as Lebesgue integrals.

Write $\Phi_x(t) = \int_0^t |\phi_x(u)| du$, where $\phi_x(u) = f(x + u) + f(x - u) - 2f(x)$. Then $\Phi_x(u) = |\phi_x(u)|$ a.e. by Theorem 5.8.5. In view of Corollary 5.8.10, we have $\lim_{t\to 0} \Phi_x(t)/t = 0$ because x is a Lebesgue point of f.

Consider any $\varepsilon > 0$ and choose $\delta > 0$ such that

$$\left|\frac{1}{t}\Phi_x(t)\right| \leq \varepsilon \quad \text{whenever} \quad 0 < t \leq \delta. \tag{8.60}$$

We may suppose that $\delta < \pi$. If $n > \frac{1}{\delta}$, we consider

$$\pi[\sigma_n(x) - f(x)] = \int_0^\pi \phi_x(u) \frac{\sin^2 \frac{nu}{2}}{2n \sin^2 \frac{u}{2}} du = \int_0^{\frac{1}{n}} + \int_{\frac{1}{n}}^{\delta} + \int_\delta^\pi = I_1 + I_2 + I_3,$$

and estimate each of the quantities I_1, I_2, I_3.

Since $\sin t < t$ for $t > 0$, and $\frac{1}{\sin t} < \frac{\pi}{2t}$ for $0 < t < \frac{\pi}{2}$, therefore for $0 < t < \frac{1}{n}$, we have

$$\frac{\sin^2 \frac{nu}{2}}{2n \sin^2 \frac{u}{2}} \leq \frac{\left(\frac{nu}{2}\right)^2}{2n}\left(\frac{\pi}{u}\right)^2 = \frac{\pi^2 n}{8}.$$

Hence

$$|I_1| \leq \frac{\pi^2 n}{8}\int_0^{\frac{1}{n}} |\phi_x(u)|du = \frac{\pi^2}{8}\frac{1}{\left(\frac{1}{n}\right)}\int_0^{\frac{1}{n}} |\phi_x(u)|du < \frac{\pi^2}{8}\varepsilon, \tag{8.61}$$

using (8.60).

We next estimate I_2. For $\frac{1}{n} \leq u \leq \delta$, we have

$$|I_2| \leq \int_{\frac{1}{n}}^{\delta} |\phi_x(u)| \frac{\sin^2 \frac{nu}{2}}{2n \sin^2 \frac{u}{2}} du \leq \frac{\pi^2}{2n}\int_{\frac{1}{n}}^{\delta} \frac{|\phi_x(u)|}{u^2} du,$$

since $|\sin t| \leq 1$ for every t and $\frac{1}{\sin t} < \frac{\pi}{2t}$ when $0 < t < \frac{\pi}{2}$. Now, integration by parts for Lebesgue integrals can be justified by the usual argument, taking care to note that a product of absolutely continuous functions is absolutely continuous, whence the product rule for derivatives holds a.e. On integrating by parts, and using the fact that $\Phi_x(u) = |\phi_x(u)|$ a.e., we obtain

$$\begin{aligned}
|I_2| &\leq \frac{\pi^2}{2n}\left\{\left[\Phi_x(u)\cdot\frac{1}{u^2}\right]_{\frac{1}{n}}^{\delta} + \int_{\frac{1}{n}}^{\delta} \frac{2}{u^3}\Phi_x(u)du\right\} \\
&\leq \frac{\pi^2}{2n}\left\{\Phi_x(\delta)\cdot\frac{1}{\delta^2} - \Phi_x\left(\frac{1}{n}\right)\cdot n^2 + 2\int_{\frac{1}{n}}^{\delta}\frac{\varepsilon u}{u^3}du\right\} \quad \text{using (8.60)} \\
&= \frac{\pi^2}{2n}\left\{\Phi_x(\delta)\cdot\frac{1}{\delta^2} - \Phi_x\left(\frac{1}{n}\right)\cdot n^2 + 2\varepsilon\left(n - \frac{1}{\delta}\right)\right\} \\
&\leq \frac{\pi^2}{2n}\left\{\frac{\varepsilon}{\delta} - \Phi_x\left(\frac{1}{n}\right)\cdot n^2 + 2\varepsilon\left(n - \frac{1}{\delta}\right)\right\} \quad \text{using (8.60) again} \\
&\leq \frac{\pi^2}{2n}\{2\varepsilon n\} = \pi^2\varepsilon,
\end{aligned} \tag{8.62}$$

neglecting the terms which are negative.

Finally, we estimate I_3.

$$\begin{aligned}
|I_3| &\leq \int_{\delta}^{\pi} |\phi_x(u)| \frac{\sin^2 \frac{nu}{2}}{2n \sin^2 \frac{u}{2}} du \\
&\leq M_n(\delta)\int_{\delta}^{\pi} |\phi_x(u)|du,
\end{aligned} \tag{8.63}$$

where $M_n(\delta) = \max_{\delta \le u \le \pi} \frac{\sin^2 \frac{mu}{2}}{2n \sin^2 \frac{u}{2}} = \left[2n \sin^2 \frac{\delta}{2}\right]^{-1}$, which tends to 0 as $n \to \infty$.

On using (8.61), (8.62) and (8.63), we obtain

$$|\sigma_n(x) - f(x)| \to 0 \text{ as } n \to \infty.$$

Definition Let $f(x) = g(x) + ih(x)$ be a complex function on the closed bounded interval $[a, b]$, where $g(x)$ and $h(x)$ are real. For each partition P: $a = x_0 < x_1 < \cdots < x_n = b$ of $[a, b]$, we set

$$T(P, f) = \sum_{k=1}^{n} \|f(x_k) - f(x_{k-1})\|,$$

where $\|f(x) - f(y)\|^2 = (g(x) - g(y))^2 + (h(x) - h(y))^2$. The number

$$V([a, b], f) = \sup_P T(P, f)$$

is called the **total variation** of f over $[a, b]$.

Additivity over adjacent intervals as in Theorem 5.6.4 can be established by making obvious modifications in the proof of that theorem. It will be used in Problems 5.8.P9 and 5.8.P10.

5.8.P9. Show that the complex function $f = g + ih$ satisfies

$$V([a, b], f) \le V([a, b], g) + V([a, b], h),$$

is of finite variation if and only if g and h are so, and is absolutely continuous if and only if g and h are so.
Hint: This is an immediate consequence of the inequalities

$$\max\{T(P, g), T(P, h)\} \le T(P, f) \le T(P, g) + T(P, h)$$

and additivity over adjacent intervals.

Definition Let g and h be continuous real-valued functions on the closed bounded interval $[a, b]$. The set of points $(g(x), h(x))$, $a \le x \le b$, in $\mathbb{R}^2$ is called a **continuous curve** and total variation $V([a, b], f)$ of $f(x) = g(x) + ih(x)$ is called the **length of the curve**. Similarly, if

$$v(x) = V([a, x], f), \quad a \le x \le b$$

then $v(x)$ is called the **length of the arc** of the curve **between the points $(g(a), h(a))$ and $(g(x), h(x))$**. If the curve has finite length, it is called a **rectifiable** curve.

5.8.P10. Show that the curve given by g and h is rectifiable if and only if the functions are of bounded variation on $[a, b]$. Show that in this case,

$$v(x) \geq \int_{[a,x]} [(g')^2 + (h')^2]^{\frac{1}{2}} \quad \text{for all } x$$

and equality holds for all x if and only if g and h are absolutely continuous.

In particular, if g is a function of bounded variation and $v(x)$ is its total variation over $[a, x]$, then $v(x) \geq \int_{[a,x]}|g'|$ for all x. Equality holds for all x if and only if g is absolutely continuous.

Hint: The curve is rectifiable if and only if the complex function $f = g + ih$ is of finite variation and, by Problem 5.8.P9, this is so if and only if g and h are of finite variation. When this so, each of v, g and h is differentiable a.e. and hence they are simultaneously differentiable a.e. Let $x \in (a, b)$ be an arbitrary point where all three are differentiable. Then for sufficiently small $\Delta > 0$, by additivity of V over adjacent intervals, we have

$$v(x+\Delta) - v(x) = V([x, x+\Delta], f) \geq \left[(g(x+\Delta) - g(x))^2 + (h(x+\Delta) - h(x))^2\right]^{\frac{1}{2}}.$$

Upon dividing by Δ and letting $\Delta \to 0$, we obtain $v'(x) \geq \left[g'(x)^2 + h'(x)^2\right]^{\frac{1}{2}}$. This holds for almost all $x \in [a, b]$. It follows by Theorem 5.5.1 that

$$v(x) \geq \int_{[a,x]} v'(u)du \geq \int_{[a,x]} \left[(g')^2 + (h')^2\right]^{\frac{1}{2}}.$$

Now suppose g and h are absolutely continuous. We wish to show that

$$v(x) = \int_{[a,x]} \left[(g')^2 + (h')^2\right]^{\frac{1}{2}} \quad \text{for almost all } x.$$

Since g and h are absolutely continuous, we have

$$g(\beta) - g(\alpha) = \int_{[\alpha,\beta]} g' \quad \text{and} \quad h(\beta) - h(\alpha) = \int_{[\alpha,\beta]} h',$$

where $a \leq \alpha < \beta \leq b$. Hence

$$\begin{aligned} \|f(\beta) - f(\alpha)\| &= \left\{[g(\beta) - g(\alpha)]^2 + [h(\beta) - h(\alpha)]^2\right\}^{\frac{1}{2}} \\ &= \left\{\left|\int_{[\alpha,\beta]} g'\right|^2 + \left|\int_{[\alpha,\beta]} h'\right|^2\right\}^{\frac{1}{2}}. \end{aligned} \tag{8.64}$$

Let us write $A = \|f(\beta) - f(\alpha)\|$. Assume that $A > 0$. Choose c_1 and c_2 defined by

$$c_1 A = \int_{[\alpha,\beta]} g', \quad c_2 A = \int_{[\alpha,\beta]} h'.$$

It follows from (8.64) that $c_1^2 + c_2^2 = 1$. Moreover,

$$\int_{[\alpha,\beta]} [c_1 g' + c_2 h'] = c_1^2 A + c_2^2 A = A.$$

But

$$\begin{aligned}|c_1 g' + c_2 h'| &\le (c_1^2 + c_2^2)^{\frac{1}{2}}((g')^2 + (h')^2)^{\frac{1}{2}}\\ &= ((g')^2 + (h')^2)^{\frac{1}{2}}.\end{aligned}$$

So,

$$A = \int_{[\alpha,\beta]} [c_1 g' + c_2 h'] \le \int_{[\alpha,\beta]} |c_1 g' + c_2 h'| \le \int_{[\alpha,\beta]} [(g')^2 + (h')^2]^{\frac{1}{2}}.$$

As A was defined to mean $\|f(\beta) - f(\alpha)\|$, the above inequality says that

$$\|f(\beta) - f(\alpha)\| \le \int_{[\alpha,\beta]} [(g')^2 + (h')^2]^{\frac{1}{2}}. \tag{8.65}$$

Now, let $a = x_0 < x_1 < \cdots < x_n = x$ be any partition of $[a, x]$.

$$\begin{aligned}T(P,f) &= \sum_{k=1}^{n} \|f(x_k) - f(x_{k-1})\|\\ &\le \sum_{k=1}^{n} \int_{[x_{k-1},x_k]} [(g')^2 + (h')^2]^{\frac{1}{2}} \text{ by (8.65).}\end{aligned}$$

So $v(x)$ does not exceed $\int_{[a,x]}[(g')^2 + (h')^2]^{1/2}$.

Conversely, suppose that $v(x) = \int_{[a,x]}[(g')^2 + (h')^2]^{1/2}$. Appealing to Proposition 5.8.1, we obtain absolute continuity of v. The inequalities $|g(x + \Delta) - g(x)| \le \|f(x + \Delta) - f(x)\| \le v(x + \Delta) - v(x)$ and $|h(x + \Delta) - h(x)| \le \|f(x + \Delta) - f(x)\|$ $v(x + \Delta) - v(x)$ now lead to the conclusion that g and h are absolutely continuous, as was to be proved. [By appealing to Theorem 5.8.5, we further obtain $v'(x) = [g'(x)^2 + h'(x)^2]^{1/2}$, which the problem does not ask for.]

5.8.P11. Let g be a real-valued function on [0, 1] and $\gamma(t) = t + ig(t)$. The length of the graph of g is, by definition, the total variation of γ on [0, 1]. Show that the length is finite if and only if g is of bounded variation.
Hint: Immediate from Problem 5.8.P9 above.

5.8.P12. Assume that g is continuous and increasing on [0, 1], $g(0) = 0$, $g(1) = \alpha$. By L_g we denote the length of the graph of g. Show that $L_g = 1 + \alpha$ if and only if $g'(x) = 0$ a.e. In particular, the length of the graph of the Cantor function constructed in Example 5.5.2 is 2.

Hint: If $\alpha = 0$, then g, being increasing, is identically zero. Moreover, the length of the graph of g equals the length of the interval between 0 and 1, which is 1.

Next, assume that $\alpha > 0$. Let $g'(x) = 0$ almost everywhere and write $f(x) = x + ig(x)$. Since $L_g = V([0, 1], f)$ by definition, we have

$$L_g = V([0,\ 1],\ x+ig(x)) \leq 1+V([0,\ 1],\ g) \text{ by Problem 5.8.P9}$$
$$= 1+\alpha.$$

It remains to show that $L_g \geq 1 + \alpha$. Select a sequence $\{h_n\}_{n\geq 1}$ of nonnegative real numbers satisfying $h_n \to 0$ and let $\varepsilon > 0$. If

$$A = \{x : g'(x) = 0\}$$

and

$$A_n = \{x : g(x+h_n) - g(x) \leq \varepsilon h_n\},$$

then $m(A) = 1$. Moreover, $x \in A$ implies $x \in A_n$ for all $n \geq n_x$, so $A \subseteq \liminf A_n$. Then $m(\liminf A_n) = 1$ and since $m(\liminf A_n) \leq \liminf m(A_n)$ by Proposition 2.3.21(c), we have also $\liminf m(A_n) = 1$. Hence $\lim m(A_n) = 1$. Now select the index p and the integer N such that $m(A_p) > 1 - \varepsilon$ and $h_p N > 1$.

Let $\alpha = \inf A_p$ and $\beta = \sup A_p$. Then $\beta - \alpha > 1 - \varepsilon$ since $[\alpha, \beta] \supseteq A_p$ and $m(A_p) > 1 - \varepsilon$. Take $x_1 \in A_p$ such that $x_1 - \alpha < \varepsilon_1 = \varepsilon/N$ and $y_1 = x_1 + h_p$. Then

$$y_1 = x_1 + h_p < \alpha + \frac{\varepsilon}{N} + h_p.$$

Claim. $\alpha + \varepsilon/N + h_p < \beta$ i.e. $\beta - \alpha > \varepsilon/N + h_p$.
It is enough to show that

$$1-\varepsilon > \tfrac{\varepsilon}{N} + h_p$$
$$\Leftrightarrow 1 - h_p > \varepsilon(1+\tfrac{1}{N})$$
$$\Leftrightarrow \varepsilon(1+\tfrac{1}{N}) < 1 - h_p < (1-\tfrac{1}{N}) \text{ because } h_p > \tfrac{1}{N}$$
$$\Leftrightarrow \varepsilon(N+1) < N-1$$
$$\Leftrightarrow \varepsilon < \tfrac{N-1}{N+1}, \text{ which is true.}$$

Let $\varepsilon_1 = \varepsilon/N$. Take $x_1 \in A_p$ such that

$$x_1 - \inf\{x : x \in A_p\} < \varepsilon_1$$

and let $y_1 = x_1 + h_p$. Next, take $x_2 \in A_p$ such that

$$0 \le x_2 - \inf\{x : x \in A_p \cap [y_1, 1]\} < \varepsilon_1$$

and let $y_2 = x_2 + h_p$. Continuing with this procedure, we obtain at most N closed intervals $[x_i, y_i]$ with disjoint interiors, and

$$m([0, 1] \backslash [x_i, y_i]) \le m([0, 1] \backslash A_p) + N\varepsilon_1 < \varepsilon + \varepsilon = 2\varepsilon.$$

So, $m(\cup_i [x_i, y_i]) > 1 - 2\varepsilon$. Now consider the partition P of [0, 1] having any partition points, besides the endpoints of [0, 1], all x_i and all y_i. The part

$$\sum |f(y_i) - f(x_i)|$$

of $T(P, f)$ is at least $1 - 2\varepsilon$ and since

$$\sum |g(y_i) - g(x_i)| \le \varepsilon \sum (y_i - x_i) \le \varepsilon,$$

where $g(1) - g(0) = \alpha$, the remaining terms of $T(P, f)$ yield at least $\alpha - \varepsilon$. Hence $T(P, f) \ge 1 - 2\varepsilon + \alpha - \varepsilon = 1 + \alpha - 3\varepsilon$. Therefore

$$L_g = V([0, 1], f) \ge 1 + \alpha.$$

Conversely, let $L_g = 1 + \alpha$ and let $g = g_1 + g_2$, where $g_1(x) = \int_{[0,x]} g'$ and $g_2'(x) = 0$ a.e,. be the Lebesgue decomposition of g. Since g_1 and g_2 are increasing, $V([0, 1], g) = V([0, 1], g_1) + V([0, 1], g_2)$. Then

$$\begin{aligned} 1 + \alpha &= V([0, 1], x + ig_1 + ig_2) \le V([0, 1], x + ig_1) + V([0, 1], g_2) \\ &\le 1 + V([0, 1], g_1) + V([0, 1], g_2) = 1 + V([0, 1], g) = 1 + \alpha, \end{aligned}$$

so that $V([0, 1], x + ig_1) = 1 + V([0, 1], g_1)$. It follows by Problem 5.8.P10 that

$$\int_{[0,1]} \left\{1 + (g_1')^2\right\}^{\frac{1}{2}} = \int_{[0,1]} \{1 + g_1'\}.$$

But $\{1 + (g_1')^2\}^{1/2} \le 1 + g_1'$ for almost all x. Hence the functions are equal a.e., and so, $g_1' = 0$ a.e. Since $g'(x) = g_1'(x)$ holds a.e., we arrive at the desired result.

5.8.P13. Let $f \in L^1[-\pi, \pi]$ and $\int_{-\pi}^{\pi} f(x)dx = 0$, $\int_{-\pi}^{\pi} f(x)\cos nx\, dx = 0$, $\int_{-\pi}^{\pi} f(x)\sin nx\, dx = 0$, $n = 1, 2, \ldots$. Then $f(x) = 0$ a.e. on $[-\pi, \pi]$.

Hint: Let $F(y) = \int_{-\pi}^{y} f(x)dx$. Observe that F is continuous and $F(\pi) = \int_{\pi}^{-\pi} f(x)\, dx = 0 = F(-\pi)$. Also, $F'(y) = f(y)$ by Theorem 5.8.5. Using integration by parts (see Theorem 5.8.12), we have

$$\int_{-\pi}^{\pi} F(y)\cos ny\,dy = \int_{-\pi}^{\pi}\Big(\int_{-\pi}^{y} f(x)dx\Big)\cos ny\,dy$$
$$= \left[\Big(\int_{-\pi}^{y} f(x)dx\Big)\frac{\sin ny}{n}\right]_{-\pi}^{\pi} - \int_{-\pi}^{\pi} f(y)\frac{\sin ny}{n}dy = 0.$$

Similarly, $\int_{-\pi}^{\pi} F(y)\sin ny\, dy = 0$. By Corollary 4.3.6, it follows that $F = 0$ everywhere on $[-\pi, \pi]$. Now, $\int_{\pi}^{-y} f(x)dx = 0$ for every $y \in [-\pi, \pi]$, which implies $f = 0$ a.e. on $[-\pi, \pi]$ by Problem 3.2.P14(b).

Problem Set 5.9

5.9.P1. Let μ be a signed measure on $(X, \mathcal{F})$. Show that

$$|\mu|(E) = \sup \sum_{j=1}^{n} |\mu(E_j)|,$$

where the sets $E_j \in \mathcal{F}$ are disjoint and $E = \cup_{1\le j\le n} E_j$.
Hint: Fix $E \in \mathcal{F}$ and suppose that $E = \cup_{1\le j\le n} E_j$, where the sets $E_j \in \mathcal{F}$ are disjoint. Let β denote the right-hand side of the equality to be proved. Then we have

$$\begin{aligned}\sum_{j=1}^{n} |\mu(E_j)| &= \sum_{j=1}^{n} |\mu^+(E_j) - \mu^-(E_j)| \\ &\le \sum_{j=1}^{n} |\mu^+(E_j)| + \sum_{j=1}^{n} |\mu^-(E_j)| \\ &= \sum_{j=1}^{n} \mu^+(E_j) + \sum_{j=1}^{n} \mu^-(E_j) \\ &= \sum_{j=1}^{n} (\mu^+(E_j) + \mu^-(E_j)) \\ &= \sum_{j=1}^{n} |\mu|(E_j) = |\mu|(E).\end{aligned}$$

Consequently,

$$\beta \le |\mu|(E).$$

Consider $\{E \cap A, E \cap B\}$, where A, B is a Hahn decomposition of X for μ.

$\beta \ge |\mu(E \cap A)| + |\mu(E \cap B)| = |\mu^+(E)| + |\mu^-(E)|$ as seen in the proof of the Jordan Decomposition Theorem 5.9.16
$= \mu^+(E) + \mu^-(E) = |\mu|(E)$.

5.9.P2. If $\mu = \mu_1 - \mu_2$, where μ_1 and μ_2 are nonnegative measures and either μ_1 or μ_2 is finite, then $\mu_1 \geq \mu^+$ and $\mu_2 \geq \mu^-$, where $\mu = \mu^+ - \mu^-$ is the Jordan decomposition of μ.

Hint: Let E be a measurable subset of X and $E_1 = E \cap A$, $E_2 = E \cap B$, where A, B is a Hahn decomposition of X for μ. Then $\mu^-(E_1) = 0$ and therefore $\mu_2(E_1) \geq \mu^-(E_1)$. But $\mu_1 - \mu^+ = \mu_2 - \mu^-$ and hence $\mu_1(E_1) \geq \mu^+(E_1)$.

Now, $\mu^+(E) = \mu^+(E_1) + \mu^+(E_2) \leq \mu_1(E_1) \leq \mu_1(E)$ since $\mu^+(E_2) = 0$. This shows that $\mu_1 \geq \mu^+$, which immediately leads to $\mu_2 \geq \mu^-$ when combined with the fact that $\mu_1 - \mu^+ = \mu_2 - \mu^-$.

5.9.P3. Let $(X, \mathcal{F}, \mu)$ be a finite signed measure space. Show that there is an $M > 0$ such that $|\mu(E)| < M$ for every $E \in \mathcal{F}$.

Hint: Let $\mu = \mu^+ - \mu^-$ be the Jordan decomposition of μ. Since $\mu(X) = \mu^+(X) - \mu^-(X)$ is given to be finite, both $\mu^+(X)$ and $\mu^-(X)$ must be finite. It follows that $\mu^+(X) + \mu^-(X)$ is finite. Now, μ^+ and μ^- are nonnegative measures and therefore $\mu^+(E) \leq \mu^+(X)$ and $\mu^-(E) \leq \mu^-(X)$ for every $E \in \mathcal{F}$. Hence

$$|\mu(E)| = |\mu^+(E) - \mu^-(E)| \leq \mu^+(E) + \mu^-(E) \leq \mu^+(X) + \mu^-(X) \quad \text{for every } E \in \mathcal{F}.$$

Thus $M = \mu^+(X) + \mu^-(X)$ has the required property.

5.9.P4. Show that a signed measure μ on $(X, \mathcal{F})$ is finite [resp. σ-finite] if and only if $|\mu|$ is finite [resp. σ-finite] if and only if μ^+ and μ^- are both finite [resp. σ-finite].

Hint: If μ is finite, it follows as in Problem 5.9.P3 that μ^+ and μ^- are both finite and hence that $|\mu| = \mu^+ + \mu^-$ is finite. Conversely, if $|\mu| = \mu^+ + \mu^-$ is finite, it is immediate that μ^+ and μ^- are both finite and hence that $\mu = \mu^+ - \mu^-$ is finite.

The corresponding results on σ-finiteness are straighforward consequences.

5.9.P5. Let μ be a signed measure on $(X, \mathcal{F})$. Then, for every $E \in \mathcal{F}$, the following hold:

(a) $\mu^+(E) = \sup\{\mu(F): F \subseteq E, F \in \mathcal{F}\}$;
(b) $\mu^-(E) = \sup\{-\mu(F): F \subseteq E, F \in \mathcal{F}\}$.

Hint: **(a)** Let A, B be a Hahn decomposition of X for μ. For $E \in \mathcal{F}$,

$$\mu^+(E) = \mu(E \cap A) \leq \sup\{\mu(F) : F \subseteq E, F \in \mathcal{F}\}.$$

Since B is a negative set for μ, if $F \subseteq E$ with $F \in \mathcal{F}$, then

$$\mu(F) = \mu(F \cap A) + \mu(F \cap B) \leq \mu(F \cap A) = \mu^+(F) \leq \mu^+(E).$$

Hence

$$\sup\{\mu(F) : F \subseteq E, E \in \mathcal{F}\} = \mu^+(E).$$

(b) Since $\mu^- = (-\mu)^+$, assertion (b) follows from (a).

5.9.P6. Let μ_1 and μ_2 be nonnegative measures on $(X, \mathcal{F})$ such that for all $\alpha > 0$ and $\beta > 0$, there exist sets $A_{\alpha,\beta}$, $B_{\alpha,\beta} \in \mathcal{F}$ satisfying $A_{\alpha,\beta} \cup B_{\alpha,\beta} = X$, $\mu_1(A_{\alpha,\beta}) < \alpha$, $\mu_2(B_{\alpha,\beta}) < \beta$. Show that $\mu_1 \perp \mu_2$.

Hint: We prove this in two steps.
Step 1: *Given* $\varepsilon > 0$, *there exist* A_ε, $B_\varepsilon \in \mathcal{F}$ *such that* $A_\varepsilon \cup B_\varepsilon = X$, $\mu_1(A_\varepsilon) = 0$ *and* $\mu_2(B_\varepsilon) < \varepsilon$.

Proof Consider an arbitrary $p \in \mathbb{N}$. Choose $\alpha = \frac{1}{p}$ and $\beta = \frac{\varepsilon}{2^p}$. In accordance with the hypothesis, we obtain sets $A_{\alpha,\beta}$, $B_{\alpha,\beta} \in \mathcal{F}$ satisfying $A_{\alpha,\beta} \cup B_{\alpha,\beta} = X, \mu_1(A_{\alpha,\beta}) < \alpha = \frac{1}{p}, \mu_2(B_{\alpha,\beta}) < \beta = \frac{\varepsilon}{2^p}$. In order to emphasise the special choice of α and β, we shall denote the sets $A_{\alpha,\beta}$, $B_{\alpha,\beta}$ respectively by $A_{p,\varepsilon}$, $B_{p,\varepsilon}$. Now let $A_\varepsilon = \bigcap_{p \geq 1} A_{p,\varepsilon}$ and $B_\varepsilon = \bigcup_{p \geq 1} B_{p,\varepsilon}$. Since $A_{p,\varepsilon}$, $B_{p,\varepsilon}$ have union equal to X, the same is true of A_ε, B_ε. Moreover, $\mu_1(A_\varepsilon) = 0$ and $\mu_2(B_\varepsilon) < \sum_{p=1}^{\infty} \left(\frac{\varepsilon}{2^p}\right) = \varepsilon$. This completes the proof of Step 1.

Step 2: $\mu_1 \perp \mu_2$.
Proof Consider an arbitrary $p \in \mathbb{N}$ and choose $\varepsilon = \frac{1}{p}$. From Step 1, we obtain disjoint A_ε, $B_\varepsilon \in \mathcal{F}$ such that $A_\varepsilon \cup B_\varepsilon = X$, $\mu_1(A_\varepsilon) = 0$ and $\mu_2(B_\varepsilon) < \varepsilon = \frac{1}{p}$. In order to emphasise the special choice of ε, we shall denote the sets A_ε, B_ε respectively by A_p and B_p. Now let $A = \bigcup_{p \geq 1} A_p$ and $B = \bigcap_{p \geq 1} B_p$. Since A_p, B_p have union equal to X, the same is true of A, B. Moreover, $\mu_1(A) = 0$ and $\mu_2(B) = 0$. Using Remark 5.9.14(a), we arrive at $\mu_1 \perp \mu_2$.

Problem Set 5.10

5.10.P1. Let $\mu = m + \delta$, where m is Lebesgue measure on [0, 1] and δ is the "Dirac" measure, which is defined by setting $\delta(E) = 1$ if $0 \in E$ and 0 if $0 \notin E$. Determine $\left[\frac{dm}{du}\right]$.
Hint: The Radon–Nikodým derivative $\left[\frac{dm}{du}\right]$ is defined [see Remark 5.10.9(a)] as being characterized a.e.[μ] by the equality

$$m(E) = \int_E \left[\frac{dm}{d\mu}\right] d\mu, \quad \text{for every measurable } E. \tag{8.66}$$

Note that δ satisfies

$$\int_X f\,d\delta = f(0) \text{for any Lebesgue measurable } f. \tag{8.67}$$

Since $\chi_{(0,1]} = 1$ a.e.[m], for arbitrary Lebesgue measurable E, we have

$$\begin{aligned} m(E) &= \int_E 1dm = \int_E \chi_{[0,1]}dm = \int_X \chi_{[0,1]}\chi_E\,dm = \int_X \chi_{[0,1]}\chi_E\,dm + 0 \\ &= \int_X \chi_{[0,1]}\chi_E\,dm + \left(\chi_{[0,1]}\chi_E\right)(0) \\ &= \int_X \chi_{[0,1]}\chi_E\,dm + \int_X \chi_{[0,1]}\chi_E d\delta \text{ by (8.67)} \\ &= \int_X \chi_{[0,1]}\chi_E\,d\mu \,\text{because}\ \mu = m + \delta \\ &= \int_E \chi_{[0,1]}\,d\mu. \end{aligned}$$

Taking into account (8.66), we conclude that $\left[\frac{dm}{d\mu}\right] = \chi(0,1]$ a.e.$[\mu]$.

5.10.P2. Let

$$f(x) = \begin{cases} \sqrt{1-x} & \text{if } x \le 1 \\ 0 & \text{if } x > 1, \end{cases} \quad g(x) = \begin{cases} x^2 & \text{if } x \ge 0 \\ 0 & \text{if } x < 0 \end{cases}$$

$$\nu(E) = \int_E f\,dx \quad \text{and} \quad \mu(E) = \int_E g\,dx, \quad E \in \mathfrak{M}.$$

Find the Lebesgue decomposition of ν with respect to μ.
Hint: Observe that μ and ν are nonnegative measures. Now,

$$\nu(E) = \int_{E\cap(-\infty,0)} f\,dx + \int_{E\cap(0,\infty)} f\,dx.$$

Claim: $\nu_0(E) = \int_{E\cap(-\infty,0)} f\,dx$ and $\nu_1(E) = \int_{E\cap(0,\,\infty)} f\,dx$ are such that $\nu_0 \perp \mu$ and $\nu_1 \ll \mu$.

(i) Since $\nu_0[0, \infty) = 0$ and $\mu(-\infty, 0) = 0$, it follows that $\nu_0 \perp \mu$.
(ii) In order to show that $\nu_1 \ll \mu$, one needs to consider only those $E \in$ which satisfy $E \subseteq (0, \infty)$, because $\nu_1(-\infty, 0] = 0$. Let E be one such set with $\mu(E) = 0$. From the definition of μ, it follows that $m(E) = 0$ since $g > 0$ on $(0, \infty)$. Consequently, $\nu_1(E) = 0$, which implies $\nu_1 \ll \mu$.

This proves the claim.

It now follows from the obvious equality $\nu = \nu_0 + \nu_1$ and Definition 5.10.11 that $\nu = \nu_0 + \nu_1$ is a Lebesgue decomposition of ν with respect to μ. By Theorem 5.10.13, it is unique.

Comment: The solution depends only on the fact that f is nonnegative while g is 0 on $(-\infty, 0)$ but > 0 on $(0, \infty)$.

5.10.P3. Let $\mu = m + \delta$, where m is the usual Lebesgue measure on $\mathbb{R}$ and δ is the "Dirac" measure defined by

$$\delta(E) = \begin{cases} 1 & \text{if } 0 \in E \\ 0 & \text{if } 0 \notin E, \end{cases} \qquad E \in \mathfrak{M}.$$

Determine the Lebesgue decomposition of μ with respect to m.
Hint: For $E \in \mathfrak{M}$,

$$\mu(E) = \mu(E \cap \mathbb{R} \backslash \{0\}) + \mu(E \cap \{0\}) = \nu_1(E) + \nu_0(E), \text{say},$$

where $\nu_0 \perp m$ since ν_0 is concentrated at 0 and $m\{0\} = 0$. Also, $\nu_1 \ll \mu$. Indeed, if $E \in \mathfrak{M}$ and $\mu(E) = 0$, then $\mu(E \cap \mathbb{R}\backslash\{0\}) \leq \mu(E) = 0$. Consequently, $\nu_1(E) = 0$.

5.10.P4. Let μ_1, μ_2 and ν be σ-finite measures on $(X, \mathcal{F})$. Prove the following:
(a) If $\mu_i \ll \nu$, $i = 1, 2$, then $\mu_1 + \mu_2 \ll \nu$ and

$$\left[\frac{d(\mu_1 + \mu_2)}{d\nu}\right] = \left[\frac{d\mu_1}{d\nu}\right] + \left[\frac{d\mu_2}{d\nu}\right] \quad \text{a.e.}[\nu].$$

(b) If $\mu_1 \ll \mu_2$ and $\mu_2 \ll \mu_1$, then $\left[\frac{d\mu_1}{d\mu_2}\right]\left[\frac{d\mu_2}{d\mu_1}\right] = 1$ a.e.$[\mu_1]$ and a.e.$[\mu_2]$.
Hint: **(a)** Clearly, $\mu_1 + \mu_2$ is a σ-finite measure and $\mu_1 + \mu_2 \ll \nu$. For $E \in \mathcal{F}$,

$$(\mu_1 + \mu_2)(E) = \mu_1(E) + \mu_2(E) = \int_E \left[\frac{d\mu_1}{d\nu}\right] d\nu + \int_E \left[\frac{d\mu_2}{d\nu}\right] d\nu.$$

So, the uniqueness of the Radon–Nikodým derivative gives the result.
(b) Write $\left[\frac{d\mu_1}{d\mu_2}\right] = f$ and $\left[\frac{d\mu_2}{d\mu_1}\right] = g$. Since μ_1 is nonnegative, $\int_E f\, d\mu_2 = \mu_1(E) \geq 0$ for every $E \in \mathcal{F}$. Consequently, $f \geq 0$ a.e.$[\mu_2]$.

Let $\{f_n\}_{n \geq 1}$ be an increasing sequence of simple nonnegative functions converging everywhere to f. It follows by using the Monotone Convergence Theorem 3.2.4 that

$$\lim_{n\to\infty}\int_E f_n\,d\mu_2 = \int_E f\,d\mu_2 \quad\text{and}\quad \lim_{n\to\infty}\int_E f_n g\,d\mu_1 = \int_E fg\,d\mu_1 \quad \text{for every } E \in \mathcal{F}$$

Let $F \in \mathcal{F}$. Then

$$\int_E \chi_F\,d\mu_2 = \mu_2(E\cap F) = \int_{E\cap F} g\,d\mu_1 = \int_E \chi_F g\,d\mu_1.$$

So,

$$\int_E f_n\,d\mu_2 = \int_E f_n g\,d\mu_1, \quad n = 1, 2, \ldots$$

and therefore

$$\mu_1(E) = \int_E f\,d\mu_2 = \int_E fg\,d\mu_1,$$

which implies

$$\left[\frac{d\mu_1}{d\mu_2}\right]\left[\frac{d\mu_2}{d\mu_1}\right] = 1 \quad \text{a.e.}[\mu_1] \text{ and a.e.}[\mu_2].$$

5.10.P5. $(X, \mathcal{F}, \nu)$ be a σ-finite measure space and μ_1, μ_2 be σ-finite signed measures such that $\mu_1 + \mu_2$ is also a signed measure on $(X, \mathcal{F})$. Prove the following:
(a) If $\mu_i \ll \nu$, $i = 1, 2$, then $\mu_1 + \mu_2 \ll \nu$ and

$$\left[\frac{d(\mu_1+\mu_2)}{d\nu}\right] = \left[\frac{d\mu_1}{d\nu}\right] + \left[\frac{d\mu_2}{d\nu}\right] \quad \text{a.e.}[\nu].$$

(b) If μ is a σ-finite signed measure on $(X, \mathcal{F})$ such that $\mu \ll \nu$, then

$$\left[\frac{d|\mu|}{d\nu}\right] = \left|\left[\frac{d\mu}{d\nu}\right]\right| \quad \text{a.e.}[\nu].$$

Hint: **(a)** Let $\mu_i = \mu_i^+ + \mu_i^-$ be the Jordan decomposition of μ_i. Since $\mu_i^+ + \ll \nu$ and $\mu_i^- + \ll \nu$ (see Proposition 5.10.4), it follows that $(\mu_1^+ + \mu_2^+) \ll \nu$ and $(\mu_1^- + \mu_2^-) \ll \nu$. Consequently, it follows from Problem 5.10.P4 that

$$\left[\frac{d(\mu_1^+ + \mu_2^+)}{d\nu}\right] = \left[\frac{d\mu_1^+}{d\nu}\right] + \left[\frac{d\mu_2^+}{d\nu}\right] \text{ a.e. } [\nu]$$

and

$$\left[\frac{d(\mu_1^- + \mu_2^-)}{d\nu}\right] = \left[\frac{d\mu_1^-}{d\nu}\right] + \left[\frac{d\mu_2^-}{d\nu}\right] \text{ a.e. } [\nu].$$

Since $\mu_1 + \mu_2$ is a signed measure, it follows first of all that $(\mu_1^+ + \mu_2^+)(E)$ and $(\mu_1^- + \mu_2^-)(E)$ cannot both be ∞ for any $E \in \mathcal{F}$. For, such an E would satisfy either $\mu_1^+(E) = \infty$ or $\mu_2^+(E) = \infty$ and we need handle only the possibility that $\mu_1^+(E) = \infty$. When this occurs, we have $\mu_1(E) = \infty$ together with $\mu_1^-(E) = 0$, which leads to $\mu_2^-(E) = \infty$ and hence to $\mu_2(E) = -\infty$, contradicting the hypothesis that $(\mu_1 + \mu_2)(E)$ is well defined.

Next, since $(\mu_1^+ + \mu_2^+)(E)$ and $(\mu_1^- + \mu_2^-)(E)$ cannot both be ∞ for any $E \in \mathcal{F}$, it follows from the equalities displayed above that, for every $E \in \mathcal{F}$ with $|(\mu_1 + \mu_2)(E)| < \infty$, we have

$$\begin{aligned}(\mu_1 + \mu_2)(E) &= (\mu_1^+ + \mu_2^+)(E) - (\mu_1^- + \mu_2^-)(E) \\ &= \int_E \left[\frac{d(\mu_1^+ + \mu_2^+)}{d\nu}\right] d\nu - \int_E \left[\frac{d(\mu_1^- + \mu_2^-)}{d\nu}\right] d\nu \\ &= \int_E \left[\frac{d\mu_1^+}{d\nu}\right] - \left[\frac{d\mu_1^-}{d\nu}\right] d\nu + \int_E \left[\frac{d\mu_2^+}{d\nu}\right] - \left[\frac{d\mu_2^-}{d\nu}\right] d\nu \\ &= \int_E \left[\frac{d\mu_1}{d\nu}\right] d\nu + \int_E \left[\frac{d\mu_2}{d\nu}\right] d\nu.\end{aligned} \tag{8.68}$$

In the last step, we have used the easily proven equality [employing the Radon–Nikodým Theorem and Example 5.9.18]

$$\int_E \left[\frac{d\mu^+}{d\nu}\right] - \left[\frac{d\mu^-}{d\nu}\right] d\nu = \int_E \left[\frac{d\mu}{d\nu}\right] d\nu,$$

where ν is a σ-finite measure and μ is a σ-finite signed measure, $\nu \ll \mu$. Now, from (8.68) and the uniqueness of the Radon–Nikodým derivative,

$$\left[\frac{d(\mu_1 + \mu_2)}{d\nu}\right] = \left[\frac{d\mu_1}{d\nu}\right] + \left[\frac{d\mu_2}{d\nu}\right] \text{ a.e. } [\nu]$$

(b) $\mu \ll \nu$ if and only if $|\mu| \ll \nu$, using Proposition 5.10.4. So, for $E \in \mathcal{F}$, we have

$$\mu(E) = \int_E \left[\frac{d\mu}{d\nu}\right] d\nu \quad \text{and} \quad |\mu|(E) = \int_E \left|\frac{d|\mu|}{d\nu}\right| d\nu.$$

Let A, B be a Hahn decomposition of X for μ. Then $\mu^+(E) = \mu(E) = |\mu|(E)$ for every $E \subseteq A$ and $\mu-(E) = -\mu(E) = |\mu|(E)$ for every $E \subseteq B$. Thus,

$$\left[\frac{d\mu}{d\nu}\right](x) = \left[\frac{d|\mu|}{d\nu}\right](x) \quad \text{a.e. } [\nu] \text{ for } x \in A$$

and

$$-\left[\frac{d\mu}{d\nu}\right](x) = \left[\frac{d|\mu|}{d\nu}\right](x) \quad \text{a.e. } [\nu] \text{ for } x \in B.$$

Hence

$$\left|\left[\frac{d\mu}{d\nu}\right]\right|(x)| = \left[\frac{d|\mu|}{d\nu}\right](x) \text{ a.e. } [\nu].$$

5.10.P6. Let $(X, \mathcal{F}, \mu)$ be a σ-finite measure space. Let ν be a measure on $(X, \mathcal{F})$ for which the conclusion of the Radon–Nikodým Theorem holds. Prove that ν is σ-finite.
Hint: It follows from the hypothesis that

$$\nu(X) = \int_X f \, d\mu,$$

where f is a real-valued nonnegative measurable function on X. Since μ is σ-finite, we have $X = \cup_{n \geq 1} X_n$, $\mu(X_n) < \infty$. Let $Y_k = \{x \in X: 0 \leq f(x) \leq k\}$. Then X can be written as $X = \cup_{k \geq 0} Y_k$, considering that f is real-valued. Consequently, $X = \cup_{n \geq 1, k \geq 0}(X_n \cap Y_k)$ and

$$\nu(X_n \cap Y_k) = \int_{X_n \cap Y_k} f \, d\mu \leq k\mu(X_n) < \infty.$$

This implies that ν is σ-finite.

Problem Set 5.11

5.11.P1. Suppose μ is a measure on $\tilde{I}$ and μ^* is the outer measure induced by μ. Let I_σ denote the family of countable unions of sets of $\tilde{I}$. Given any set E and any $\varepsilon > 0$, show that there is a set $A \in I_\sigma$ such that $E \subseteq A$ and

$$\mu^*(A) \le \mu^*(E) + \varepsilon.$$

Hint: By the definition of μ^*, there exists a sequence $\{E_i\}_{i\geq 1}$ of sets $E_i \in \tilde{I}$ such that $E \subseteq \cup_{i\geq 1}E_i$ and $\sum_{i=1}^{\infty} \mu(E_i) < \mu^*(E) + \varepsilon$. Set $A = \cup_{i\geq 1}E_i$. Then $A \in I_\sigma$ and by countable subadditivity of μ^* [Remark 5.11.16(c)] and the fact that μ^* coincides with μ on $\tilde{I}$ [Theorem 5.11.21], we have

$$\mu^*(A) \le \sum_{i=1}^{\infty} \mu^*(E_i) = \sum_{i=1}^{\infty} \mu(E_i) < \mu^*(E) + \varepsilon.$$

5.11.P2. Suppose μ is a measure on $\tilde{I}$ and μ^* is the outer measure induced by μ. Let I_σ denote the family of countable unions of sets of $\tilde{I}$ and $I_{\sigma\delta}$ denote the family of countable intersections of sets of I_σ. Given any set E, show that there is a set $A \in I_{\sigma\delta}$ such that $E \subseteq A$ and $\mu^*(E) = \mu^*(A)$.
Hint: For every positive integer n, there exists a set $A_n \in I_\sigma$ with $E \subseteq A_n$ and $\mu^*(A_n) \le \mu^*(E) + \frac{1}{n}$. Let $A = \cap_{n\geq 1}A_n$. Then $A \in I_{\sigma\delta}$ and $E \subseteq A$. Since $A \subseteq A_n$, we have $\mu^*(A) \le \mu^*(A_n) \le \mu^*(E) + \frac{1}{n}$ for each n. Therefore $\mu^*(A) \le \mu^*(E)$. But $E \subseteq A$ and so, $\mu^*(E) \le \mu^*(A)$ by monotonicity. Hence $\mu^*(E) = \mu^*(A)$.
5.11.P3. Let F be a real-valued bounded increasing right continuous function on $\mathbb{R}$ and for any right closed interval $(a, b]$, define $\mu((a, b]) = F(b) - F(a)$. Show that:

(i) Let $\{E_i\}_{1\leq i\leq n}$ be disjoint right closed intervals ($E_i = (a_i, b_i]$, $1 \le i \le n$) such that $\cup_{1\leq i\leq n}E_i \subseteq I$, where I is a right closed interval. Then $\sum_{i=1}^{n} \mu(E_i) \le \mu(I)$.
(ii) If the right closed interval $(a_0, b_0]$ is contained in the union of a sequence of right closed intervals $(a_i, b_i]$, $i = 1, 2, \ldots$, then

$$\mu((a_0, b_0]) \le \sum_{i=1}^{\infty} \mu((a_i, b_i]);$$

moreover, if the intervals in the sequence $(a_i, b_i]$, $i = 1, 2, \ldots$, are disjoint and $(a_0, b_0] = \cup_{i\geq 1}(a_i, b_i]$, then

$$\mu((a_0, b_0]) = \sum_{i=1}^{\infty} \mu((a_i, b_i]).$$

Hint: Modify the proofs of Lemmas 5.11.3 and 5.11.5.

5.11.P4. Let h be a real-valued bounded increasing left continuous function on $\mathbb{R}$ such that $h' = 0$ a.e. and $\mu = \mu_h$ be the Lebesgue–Stieltjes measure induced by h. Suppose also that there is a closed bounded interval I such that h is constant on each of the unbounded open intervals that constitute the complement I^c. Show that μ and Lebesgue measure m (on Borel sets) are mutually singular (see Definition 5.9.13).

Hint: Since $h' = 0$ a.e., there exist disjoint Borel sets A_0 and B_0 with union $\mathbb{R}$ [i.e., mutually complementary Borel subsets of $\mathbb{R}$] such that

$$m(A_0) = 0$$

and $h'(x) = 0$ for all $x \in B_0$. We may assume that B_0 contains the complement $I^c = \mathbb{R}\backslash I$ of the interval I. Note that

$$\mu(I^c) = 0,$$

because h is constant on each of the unbounded open intervals that constitute the complement I^c.

Let $\alpha > 0$ and $\beta > 0$. Each $x \in B_0$ is contained in a closed bounded interval $[s, t]$ of arbitrarily small length such that $h(t) - h(s) \leq (\beta/m(I))\bullet(t - s)$. If $[s, t]$ lies outside I, then $h(t) - h(s) = 0$. Such intervals provide a Vitali cover of $B_0 \cap I$ in the sense defined in Problem 5.4.P1 and therefore by the Vitali Covering Theorem of that problem, there exists a finite disjoint collection of intervals in the Vitali cover, whose union B_1 satisfies

$$m((B_0 \cap I)\backslash B_1) < \alpha.$$

If we replace each closed interval $[\alpha,\beta]$ by the corresponding left closed interval $[\alpha, \beta)$, their union satisfies the same inequality. We may thus assume that B_1 is the union of the left closed intervals. Then we can assert on the basis of the disjointness assured by the Vitali Covering Theorem that

$$\mu(B_1) \leq \left(\frac{\beta}{m(I)}\right) \cdot m(B_1) < \beta.$$

Since $\mu(I^c) = 0$, it follows that $\mu(B_1 \cup I^c) < \beta$. Now take

$$A = A_0 \cup ((B_0 \cap I)\backslash B) \quad \text{and} \quad B = B_1 \cup I^c.$$

Then we have

$$m(A) \leq m(A_0) + m((B_0 \cap I)\backslash B_1) = 0 + m((B_0 \cap I)\backslash B_1) < \alpha$$

and

$$\mu(B) \le \mu(B_1) + \mu(I^c) = \mu(B_1) + 0 < \beta.$$

In view of Problem 5.9.P6, all we need prove is that $A \cup B = \mathbb{R}$, which we do in the next paragraph.

Consider any $x \in \mathbb{R}$ such that $x \notin A$; we must show that $x \in B$. Since $x \notin A$, we must have $x \notin A_0$, which implies $x \in B_0$, considering that A_0,B_0 are mutually complementary. By the definition of A, we must also have $x \notin (B_0 \cap I) \cap B^c = B_0 \cap (I \cap B^c)$. However, as already noted, $x \in B_0$. It follows that $x \notin (I \cap B^c)$. But the definition of B implies $I \cap B^c = I \cap (B_1 \cup I^c)^c = I \cap B_1^c = B^c$. Since we have shown that $x \notin (I \cap B^c)$, it is now immediate that $x \in B$, as desired. This completes the argument that $A \cup B = \mathbb{R}$, which is all that we needed to prove.

5.11.P5. Suppose f is an increasing function on $[a, b]$ that is left continuous and let $f = g + h$ be a Lebesgue decomposition in accordance with Problem 5.8.P7. Extend each of the three functions to $\mathbb{R}$ so as to take the same value on $(-\infty, a)$ as at a and to take any constant value on (b, ∞) not less than the value at b, but so chosen that the equality $f = g + h$ holds on (b, ∞). Prove the following:

(a) There is some constant c such that $g(x) = c + \int_{[a,x]} f'$ for all $x \in \mathbb{R}$.
(b) The functions f, g and h on $\mathbb{R}$ are also bounded, increasing and left continuous; moreover, $f = g + h$ on $\mathbb{R}$.
(c) If μ_f, μ_g and μ_h are the Lebesgue–Stieltjes measures induced by f, g and h respectively, then $\mu_f = \mu_g + \mu_h$ and this equality constitutes the unique Lebesgue decomposition of μ_f with respect to Lebesgue measure m.

Hint: **(a)** As in the solution of Problem 5.8.P7, g is given on $[a, b]$ by $g(x) = c + \int_{[a,x]} f'$, where c is some constant. The manner of extending the functions to all of $\mathbb{R}$ makes f' vanish outside $[a, b]$ and therefore g is given by the same equality on all of $\mathbb{R}$.
(b) Since f is increasing on $[a, b]$, the derivative f' must be nonnegative wherever it exists, and therefore g is increasing on $[a, b]$. By the inequality (5.35) of Theorem 5.5.1, the same is true of h. Moreover h is left continuous on $(a, b]$ because f and g both are. The manner of extending the functions to all of $\mathbb{R}$ ensures that the functions f, g and h, when extended to all of $\mathbb{R}$, continue to be increasing and left continuous, and to satisfy $f = g + h$.
(c) The equality $f = g + h$ shows by an easy computation that when E is a left closed interval, $\mu_f(E) = \mu_g(E) + \mu_h(E)$. By the uniqueness part of Theorem 5.11.22, we obtain $\mu_f = \mu_g + \mu_h$. Since g is absolutely continuous, it follows from Remark 5.11.24(c) that $\mu_g \ll m$. By Problem 5.8.P7, in accordance with which the functions g and h are formed, we have $h' = 0$ a.e. on $[a, b]$ and hence also on $\mathbb{R}$. It follows by Problem 5.11.P4 that μ_h and m are mutually singular. By the uniqueness part of Theorem 5.10.13, $\mu_f = \mu_g + \mu_h$ is the unique Lebesgue decomposition of μ_f with respect to Lebesgue measure.

5.11.P6. Let μ be a finite nonnegative measure on $(\mathbb{R}, \mathfrak{B})$ and $f = f_\mu$. Suppose $[a, b]$ is an arbitrary interval and ϕ is the function on $\mathbb{R}$ obtained by extending the restriction of f to $[a, b]$ so as to be equal to $f(a)$ on $(-\infty, a)$ and equal to $f(b+)$ on (b, ∞). Then obviously, ϕ is bounded, increasing and left continuous. Show that the Lebesgue–Stieltjes measure μ_ϕ induced by ϕ is given by $\mu_\phi(E) = \mu(E \cap [a, b])$ for every Borel set E.

Hint: Since $\mu(E \cap [a, b])$ defines a finite measure on Borel sets E, it is sufficient in view of the uniqueness part of Theorem 5.11.22 to show that $\mu_\phi(E) = \mu(E \cap [a, b])$ for every left closed interval $E = [\alpha, \beta)$. That is, $\mu([\alpha, \beta) \cap [a, b]) = \phi(\beta) - \phi(\alpha)$ whenever $\alpha \leq \beta$. Since this is trivial when $\alpha = \beta$, we may assume $\alpha < \beta$.

Case 1: $\alpha < a$. Here, $f(a) = \phi(\alpha)$.

Subcase $\beta \leq a$. This makes

$$[\alpha, \beta) \cap [a, b] = \varnothing \quad \text{and} \quad \phi(\beta) = f(a).$$

So, $\mu([\alpha, \beta) \cap [a, b]) = 0 = f(a) - f(a) = \phi(\beta) - \phi(\alpha)$, as desired.

Subcase $a < \beta \leq b$. Here,

$$[\alpha, \beta) \cap [a, b] = [a, \beta) \quad \text{and} \quad f(\beta) = \phi(\beta).$$

So, $\mu([\alpha, \beta) \cap [a, b]) = f(\beta) - f(a) = \phi(\beta) - \phi(\alpha)$.

Subcase $b < \beta$. In this subcase,

$$[\alpha, \beta) \cap [a, b] = [\alpha, b] \quad \text{and} \quad f(b+) = \phi(\beta)$$

So, on the basis of Remark 5.11.24(d),

$$\begin{aligned}\mu([\alpha, \beta) \cap [a, b]) &= \mu([a, b]) = \mu([a, b)) + \mu(\{b\}) \\ &= f(b) - f(a) + \mu(\{b\}) = f(b) - f(a) + f(b+) - f(b) \\ &= f(b+) - f(a) = \phi(\beta) - \phi(\alpha).\end{aligned}$$

Case 2: $a \leq \alpha \leq b$. Here, $f(\alpha) = \phi(\alpha)$.

Subcase $\beta \leq b$. This makes

$$[\alpha, \beta) \cap [a, b] = [\alpha, \beta) \quad \text{and} \quad f(\beta) = \phi(\beta).$$

So, $\mu([\alpha, \beta) \cap [a, b]) = \mu([\alpha, \beta)) = f(\beta) - f(\alpha) = \phi(\beta) - \phi(\alpha)$.

Subcase $b < \beta$. Here,

$$[\alpha, \beta) \cap [a, b] = [\alpha, b] \quad \text{and} \quad f(b+) = \phi(\beta).$$

So, on the basis of Remark 5.11.24(d),

$$\begin{aligned}\mu([\alpha,\beta)\cap[a,\ b]) &= \mu([a,\ b]) = \mu([a,b)) + \mu(\{b\})\\ &= f(b) - f(\alpha) + \mu(\{b\}) = f(b) - f(\alpha) + f(b+) - f(b)\\ &= f(b+) - f(\alpha) = \phi(\beta) - \phi(\alpha).\end{aligned}$$

Case 3: $b < \alpha$. Here, $\phi(\alpha) = \phi(\beta) = f(b+)$ and $[\alpha,\ \beta) \cap [a,\ b] = \varnothing$.
So, $\mu([\alpha,\ \beta) \cap [a,\ b]) = 0 = f(b+) - f(b+) = \phi(\beta) - \phi(\alpha)$.

5.11.P7. Let μ be a finite nonnegative measure on $(\mathbb{R}, \mathfrak{B})$ and $f = f_\mu$. If μ and Lebesgue measure m (on $\mathfrak{B}$) are mutually singular, show that $f' = 0$ a.e.

Hint: Let $[a,\ b]$ be an arbitrary interval and ϕ be the function on $\mathbb{R}$ obtained as in Problem 5.11.P6 by extending the restriction of f to $[a,\ b]$ so as to be equal to $f(a)$ on $(-\infty,\ a)$ and equal to $f(b+)$ on $(b,\ \infty)$; it is sufficient to show that $\phi' = 0$ a.e. Note that $f(b+) \geq f(b)$, so that the constant value assigned to ϕ on (b,∞) is not less than $\phi(b)$ and thus ϕ is an extension that fulfils the requirement of Problem 5.11.P5 as well.

Let $f = g + h$ be a Lebesgue decomposition of the restriction of f to $[a,\ b]$ in accordance with Problem 5.8.P7. Then $h' =$ a.e. on $[a,\ b]$. The function f already has an extension ϕ to $\mathbb{R}$ of the kind required in Problem 5.11.P5. Now extend g and h also to $\mathbb{R}$ in a corresponding manner while ensuring the equality $\phi = g + h$. Then $h' = 0$ a.e. on $\mathbb{R}$. Moreover, the Lebesgue–Stieltjes measures μ_ϕ, μ_g and μ_h induced by ϕ, g and h respectively satisfy $\mu_\phi = \mu_g + \mu_h$ and this equality constitutes the unique Lebesgue decomposition of μ_ϕ with respect to Lebesgue measure m. Consider any Borel set E. By Problem 5.11.P6,

$$\mu_\phi(E) = \mu(E \cap [a, b]).$$

Since $\mu(E \cap [a,\ b]) \leq \mu(E)$ for every Borel set E, the measures μ_ϕ and m must be mutually singular. Since $\mu_\phi = 0 + \mu_\phi$, this equality constitutes the unique Lebesgue decomposition of μ_ϕ, and hence $\mu_g = 0$. By Remark 5.11.24(b), the function g must be a constant. It is now immediate that $f' = g' + h' = 0$ a.e.

5.11.P8. Let μ be a finite nonnegative measure on $(\mathbb{R}, \mathfrak{B})$ and $\mu = \nu + \lambda$ be its Lebesgue decomposition with respect to Lebesgue measure m. Then f_ν is absolutely continuous on any interval $[a,\ b]$ and $f_\lambda' = 0$ a.e. Also, $f_\mu = f_\nu + f_\lambda$. [*Remark* This means the equality $f_\mu = f_\nu + f_\lambda$ is a Lebesgue decomposition of f_μ on any interval $[a,\ b]$ in accordance with Problem 5.8.P7.]

Hint: The equality $f_\mu = f_\nu + f_\lambda$ is trivial from the definition of the function f_μ associated with a measure μ. Since $\nu \ll m$, the function f_ν is absolutely continuous by Remark 5.11.24(c). And since λ and m are mutually singular, $f_\lambda' = 0$ a.e. by Problem 5.11.P7.

Problem Set 6.1

6.1.P1. If $f \in L^p(X)$, $\infty > p > 1$, then $\|f\|_p = \sup\{\int_X |fg|d\mu\colon g \in L^q(X),\ \|g\|_q = 1\}$, where q is given by $\frac{1}{p} + \frac{1}{q} = 1$.

If $f \in L^1(X)$, then $\|f\|_1 = \sup\{\int_X |fg| d\mu: g \in L^\infty(X), \|g\|_\infty = 1\}$.
Note that this differs from Proposition 6.1.3 in that the absolute value is inside the integral rather than outside it.
Hint: We assume $\|f\|_p \neq 0$, because otherwise there is nothing to prove.

By Hölder's Inequality,

$$\int_X |fg| d\mu \leq \|f\|_p \|g\|_q \leq \|f\|_p \text{ if } \|g\|_q = 1.$$

Hence

$$\|f\|_p \geq \sup\left\{\int_X |fg|\, d\mu : g \in L^q(X), \|g\|_q = 1\right\}.$$

Taking into account that $\|f\|_p \neq 0$, let

$$g = |f|^{p-1} \|f\|_p^{-\frac{p}{q}}.$$

Then

$$\int_X |g|^q\, d\mu = \int_X |f|^{(p-1)q} \|f\|_p^{-p}\, d\mu = 1,$$

so that $\|g\|_q = 1$. But

$$\int_X |fg| d\mu = \int_X |f|^p\, d\mu \|f\|_p^{-\frac{p}{q}} = \|f\|_p^{p-\frac{p}{q}} = \|f\|_p.$$

For the assertion about $f \in L^1(X)$, proceed as above but take $g = 1$ everywhere.

6.1.P2. If $f \in L^\infty(X)$, then $\|f\|_\infty = \sup\{\int_X |fg| d\mu: g \in L^1(X), \|g\|_1 = 1\}$, provided that every set of positive measure contains a subset of finite positive measure. Show also that this is false if the proviso about the measure is omitted. Note that this differs from Proposition 6.1.2(v) in that the absolute value is inside the integral rather than outside it.
Hint: The case when $\|f\|_\infty = 0$ is trivial. Given any positive $\varepsilon < \|f\|_\infty$, let $E \subseteq X$ be a measurable subset of positive measure on which $|f| > \|f\|_\infty - \varepsilon > 0$. Since every set of positive measure contains a subset of finite positive measure, we may take E to have finite positive measure, so that $\chi_E \in L^1(X)$. Consequently, the function $g = (\text{sgn} f)\chi_E \in L^1(X)$ and is nonzero. Besides,

$$\int_X |g|d\mu = \int_E |(\text{sgn}f)f|d\mu = \int_E |f|d\mu \geq (\|f\|_\infty - \varepsilon)\|g\|_1.$$

Replacing g by $g/\|g\|_1$, we get $\int_X |fg|d\mu \geq \|f\|_\infty - \varepsilon$ and $\|g\|_1 = 1$. Since the positive number $\varepsilon < \|f\|_\infty$ is arbitrary, we arrive at

$$\sup\left\{\int_X |fg|d\mu : g \in L^1(X), \|g\|_1 = 1\right\} \geq \|f\|_\infty.$$

The reverse inequality is trivial.
For the second part, let X consist of two points called a and b. Consider the measure which is 1 on $\{a\}$ and ∞ on $\{b\}$. Then L^1 consists of functions that are 0 at b. The function f defined to be 0 at a and 1 at b belongs to L^∞ with $\|f\|_\infty = 1$. However, $\int_X |fg|d\mu = 0$ for every $g \in L^1(X)$.
6.1.P3. Let $\{f_n\}_{n\geq 1}$ be a sequence of functions in $L^\infty(X)$. Prove that $\{f_n\}_{n\geq 1}$ converges to $f \in L^\infty(X)$ if and only if there is a set E of measure zero such that $f_n \to f$ uniformly on E^c.
Hint: Let $E_n = \{x \in X: |f_n(x) - f(x)| > \|f_n - f\|_\infty\}$ and $E = \cup_{n\geq 1} E_n$. Since $m(E_n) = 0$, it follows that $m(E) \leq \sum_{n=1}^\infty m(E_n) = 0$. Then $|f_n(x) - f(x)| \leq \|f_n - f\|_\infty$ for all n and all $x \in E^c$. The hypothesis that $\|f_n - f\|_\infty < \varepsilon$ for all $n \geq n_0$ implies $|f_n(x) - f(x)| < \varepsilon$ for all $n \geq n_0$ and all $x \in E^c$, that is, $\{f_n\}_{n\geq 1}$ converges to f uniformly on E^c.
On the other hand, suppose $\{f_n\}_{n\geq 1}$ converges to f uniformly a.e., that is, for all $\varepsilon > 0$, there exists an n_0 such that $n \geq n_0$ implies

$$|f_n(x) - f(x)| < \varepsilon \text{ for all } x \in E^c,$$

where $m(E) = 0$; that is, $n \geq n_0$ implies

$$m(\{x \in X : |f_n(x) - f(x)| \geq \varepsilon\}) = 0.$$

Then, for $n \geq n_0$,

$$\|f_n - f\|_\infty \leq \varepsilon.$$

6.1.P4. Suppose f is integrable on $\mathbb{R}$ and for fixed $h \in \mathbb{R}$, let $f_h(x) = f(x + h)$ be a translate of f. Show that f_h is also integrable and that

$$\int_{\mathbb{R}} f_h dm = \int_{\mathbb{R}} f\, dm.$$

Hint: Clearly, $(f_h)^+ = (f^+)_h$ and $(f_h)^- = (f^-)_h$; so it is sufficient to prove the result for $f \geq 0$.

Let $s = \sum_{k=1}^{m} c_k \chi_{E_k}$, where $c_k > 0$ and E_k is a measurable set of finite measure, $1 \leq k \leq m$. Then

$$s_h = \sum_{k=1}^{m} c_k (\chi_{E_k})_h = \sum_{k=1}^{m} c_k \chi_{E_k - h}$$

and

$$\int_{\mathbb{R}} s_h dm = \sum_{k=1}^{m} c_k m(E_k - h) = \sum_{k=1}^{m} c_k m(E_k) = \int_{\mathbb{R}} s\, dm,$$

since Lebesgue measure is translation invariant (Proposition 2.3.23).

By Theorem 2.5.9, there exists an increasing sequence $\{s_n\}$ of measurable simple functions such that $s_n \to f$. But then the sequence $\{(s_n)_h\}$ is also increasing and $(s_n)_h \to f_h$. So, by the Monotone Convergence Theorem, and the result of the paragraph above, we have

$$\int_{\mathbb{R}} f_h dm = \lim_{n\to\infty} \int_{\mathbb{R}} (s_n)_h dm = \lim_{n\to\infty} \int_{\mathbb{R}} s_n dm = \int_{\mathbb{R}} f\, dm.$$

6.1.P5. Show that if $|f_n| \leq M$ a.e. and $f_n \to f$ in $L^p(X)$, where $\mu(X) < \infty$ and $p \geq 1$, then $f_n \to f$ in $L^{p'}(X)$ for $1 \leq p' < \infty$.

Hint: Suppose $p' < p$ and $p_1 = p/p'$; let q_1 be such that $\frac{1}{p_1} + \frac{1}{q_1} = 1$. Then

$$\int_X |f_n - f|^{p'} d\mu \leq \left\| |f_n - f|^{p'} \right\|_{p_1} (\mu(X))^{\frac{1}{q_1}} = (\|f_n - f\|_p)^{p'} (\mu(X))^{\frac{1}{q_1}} \to 0$$

as $n \to \infty$, that is, $f_n \to f$ in $L^{p'}(X)$.

Assume $p' > p$. Since $|f_n| \leq M$ a.e., we have $|f_n - f| \leq 2M$ a.e. for all n. Now,

$$(\|f_n - f\|_{p'})^{p'} \leq (2M)^{p'-p} \|f_n - f\|_p^p \to 0$$

as $n \to \infty$.

Remark The reader will note that the hypothesis $|f_n| \le M$ a.e. is used only for $p' > p$ and the hypothesis $m(X) < \infty$ is used only for $p' < p$.

6.1.P6. Let $I = [a, b] \subset \mathbb{R}$ and $1 < p < \infty$. If $f \in L^p(I)$ and $F(x) = C + \int_{[a,x]} f\,dm$, then show that

$$\sup_{\mathcal{P}} \sum_{k=0}^{n-1} \frac{|F(x_{k+1}) - F(x_k)|^p}{(x_{k+1} - x_k)^{p-1}} \le \int_I |f|^p dm, \tag{8.69}$$

where $\mathcal{P} : a = x_0 < x_1 < \cdots < x_n = b$ is a partition of $[a, b]$.

Conversely, if

$$K = \sup_{\mathcal{P}} \sum_{k=0}^{n-1} \frac{|F(x_{k+1}) - F(x_k)|^p}{(x_{k+1} - x_k)^{p-1}} < \infty, \tag{8.70}$$

where $\mathcal{P}$ is a partition of I, then there exists a $\phi \in L^p(I)$ such that $F(x) = C + \int_{[a,x]} \phi\,dm$; show also that $\int_I |\phi|^p dm \le K$.

Use this to show that equality holds in (8.69).

Hint: By Hölder's Inequality,

$$|F(x_{k+1}) - F(x_k)| = \left| \int_{[x_k, x_{k+1}]} f\,dm \right| \le (x_{k+1} - x_k)^{\frac{1}{q}} \left(\int_{[x_k, x_{k+1}]} |f|^p dm \right)^{\frac{1}{p}},$$

where $\frac{1}{p} + \frac{1}{q} = 1$. Hence

$$\frac{|F(x_{k+1}) - F(x_k)|^p}{(x_{k+1} - x_k)^{p-1}} \le \int_{[x_k, x_{k+1}]} |f|^p dm,$$

which is to say, inequality (8.69) holds.

For the converse, note that the given condition (8.70) is strengthened if some of the terms in it are omitted. So, for arbitrary disjoint open intervals (a_k, b_k), $k = 1, 2, 3, \ldots, n$, contained in $[a, b]$, we have

$$\sum_{k=1}^{n} \frac{|F(b_k) - F(a_k)|^p}{(b_k - a_k)^{p-1}} \le K.$$

By Hölder's Inequality,

$$\sum_{k=1}^{n}|F(b_k)-F(a_k)| = \sum_{k=1}^{n}\frac{|F(b_k)-F(a_k)|}{(b_k-a_k)^{1-\frac{1}{p}}}(b_k-a_k)^{\frac{1}{q}}$$
$$\leq \left\{\sum_{k=1}^{n}\frac{|F(b_k)-F(a_k)|^p}{(b_k-a_k)^{p-1}}\right\}^{\frac{1}{p}}\left\{\sum_{k=1}^{n}(b_k-a_k)\right\}^{\frac{1}{q}}.$$

Therefore

$$\sum_{k=1}^{n}|F(b_k)-F(a_k)| \leq K^{\frac{1}{p}}\left\{\sum_{k=1}^{n}(b_k-a_k)\right\}^{\frac{1}{q}},$$

which implies F is absolutely continuous and is, therefore, representable in the form

$$F(x) = C + \int_{[a,x]}\phi\, dm,$$

where $\phi \in L^1(I)$ by Corollary 5.8.3. The reader will note that $\phi = F'$ a.e. by Theorem 5.8.5 (although this will play no role in the argument). It remains to show that $\phi \in L^p(I)$. For this purpose, we divide $[a, b]$ into n equal parts by the points

$$x_k^{(n)} = a + \frac{k}{n}(b-a), \quad k = 0, 1, 2, \ldots, n$$

and introduce the function $f_n(t)$ by setting

$$f_n(t) = \begin{cases} \frac{F(x_{k+1}^{(n)})-F(x_k^{(n)})}{x_{k+1}^{(n)}-x_k^{(n)}} & x_k^{(n)} < t < x_{k+1}^{(n)} \\ 0 & t = x_k^{(n)}. \end{cases}$$

We shall show that $\lim_{n\to\infty} f_n = \phi$ almost everywhere. Let x be a point which is not a point of subdivision and for which $F'(x)$ exists and is finite. Thus x lies in some open interval $(x_{k_n}^{(n)}, x_{k_n+1}^{(n)})$ for all natural numbers n. Since $x_{k_{n+1}}^{(n)} - x^{(n)}k_n = \frac{b-a}{n} \to 0$ as $n \to \infty$, it follows that each of the expressions

$$\frac{F(x_{k_{n+1}}^{(n)}) - F(x)}{x_{k_{n+1}}^{(n)} - x} \quad \text{and} \quad \frac{F(x_{k_n}^{(n)}) - F(x)}{x_{k_n}^{(n)} - x} \tag{8.71}$$

tends to $F'(x)$ as $n \to \infty$. However,

$$f_n(x) = \frac{F(x_{k+1}^{(n)}) - F(x_k^{(n)})}{x_{k+1}^{(n)} - x_k^{(n)}},$$

and accordingly, the number $f_n(x)$ lies between the numbers (8.71). Therefore $\lim_{n\to\infty} f_n(x) = F'(x)$. By Fatou's Lemma,

$$\int_I |\phi|^p dm \le \sup \int_I |f_n^p| dm$$

and since

$$\int_I |f_n|^p dm = \sum_{k=0}^{n-1} \int_{[x_k^{(n)}, x_{k+1}^{(n)}]} |f_n|^p dm = \sum_{k=0}^{n-1} \frac{\left|F(x_{k+1}^{(n)}) - F(x_k^{(n)})\right|^p}{\left(x_{k+1}^{(n)} - x_k^{(n)}\right)^{p-1}} \le K,$$

it follows that

$$\int_I |\phi|^p dm \le K,$$

completing the proof of the converse.

We now proceed to prove that equality holds in (8.69). Since $f \in L^p(I)$, inequality (8.69) implies that condition (8.70) holds. It follows that there exists a $\phi \in L^p(I)$ such that $F(x) = C + \int_{[a,x]} \phi dm$ and $\int_I |\phi|^p dm \le K$. However, it is given that $F(x) = C + \int_{[a,x]} f\, dm$ and therefore $\phi = f$ a.e. Hence $\int_I |f|^p dm = \int_I |\phi|^p dm \le K$, which is precisely the reverse of inequality (8.69). Thus, equality must hold in (8.69).

6.1.P7. (Chebychev's Inequality) Let $f \in L^p(X)$, where $1 \le p < \infty$. Show that, for every $\lambda > 0$,

$$\lambda^p \mu(\{x : |f(x)| > \lambda\}) \le \int_X |f|^p \, d\mu.$$

Moreover,

$$\lim_{\lambda\to\infty} \lambda^p \mu(\{x : |f(x)| > \lambda\}) = 0.$$

Hint: Let $A_\lambda = \{x: |f(x)| > \lambda\}$. Then A_λ is measurable and $\lambda^p \chi_{A_\lambda} \le |f|^p$ a.e. So,

$$\int_X \lambda^p \chi_{A_\lambda}\, d\mu = \lambda^p \mu(\{x : |f(x)| > \lambda\}) \le \int_X |f|^p\, d\mu.$$

Also,

$$\lambda^p \mu(\{x : |f(x)| > \lambda\}) \le \int_{A_\lambda} |f|^p\, d\mu \to 0 \text{ as } \lambda \to \infty$$

since $f \in L^p(X)$.

6.1.P8. Let $\{f_n\}_{n\geq 1}$ be a sequence in $L^p(X)$, $1 \le p \le \infty$, which converges to a function f in $L^p(X)$. Then show that, for each g in $L^q(X)$, where $\frac{1}{p} + \frac{1}{q} = 1$, we have

$$\int_X fg\, dm = \lim_{n\to\infty} \int_X f_n g\, dm.$$

Hint: The mapping that carries $f \in L^p(X)$ into $\int_X fg\ dm$ is continuous by Proposition 6.1.3 if $1 \le p < \infty$ and by Proposition 6.1.2(iv) if $p = \infty$.

Problem Set 6.2

6.2.P1. For the sequence $\{f_k\}_{k\geq 1}$of Example 6.2.17, determine whether it converges (a) almost uniformly (b) in measure.
Hint: (a) Yes (b) Yes.

6.2.P2. Suppose that $\{f_k\}_{k\geq 1}$is a sequence of nonnegative measurable functions defined on X that converges in measure to f. Prove that

$$\int_X f\, d\mu \le \liminf \int_X f_k\, d\mu.$$

Hint: Suppose that $\liminf \int_X f_k\, d\mu < \int_X f\ d\mu < \infty$. Then there exists a suitable $\delta > 0$ and a subsequence $\{f_i\}_{i\geq 1}$of $\{f_k\}_{k\geq 1}$such that $\int_X f_i\, d\mu < \int_X f\ d\mu - \delta$ for all i. By Corollary 6.2.7, we can find a further subsequence $\{f_j\}_{j\geq 1}$such that $f_j \overset{ae}{\to} f$. But then by Fatou's Lemma 3.2.8, we have

$$\int_X f\, d\mu \le \liminf \int_X f_j\, d\mu \le \int_X f\, d\mu - \delta.$$

This is a contradiction.

On the other hand, suppose that $\liminf \int_X f_k\, d\mu < \int_X f\, d\mu = \infty$. Then there exists a constant $M > 0$ and a subsequence $\{f_i\}_{i\geq 1}$of $\{f_k\}_{k\geq 1}$such that $\int_X f_i\, d\mu < M$ for all i. By Corollary 6.2.7, we can find a further subsequence $\{f_j\}_{j\geq 1}$ such that $f_j \overset{ae}{\to} f$. But then by Fatou's Lemma 3.2.8, we have

$$\int_X f\,d\mu \leq \liminf \int_X f_j\,d\mu \leq M.$$

This is again a contradiction.

6.2.P3. If a sequence $\{f_k\}_{k\geq 1}$ of measurable functions is Cauchy in measure and there exists a measurable function f to which a subsequence $\{f_j\}_{j\geq 1}$ converges in measure, then $f_k \overset{meas}{\longrightarrow} f$.

Hint: The following is the key to the result:

$$X(|f_k - f| > a) \subseteq X(|f_k - f_j| > \frac{a}{2}) \cup X(|f_j - f| > \frac{a}{2}).$$

6.2.P4. Let $\mu(X) < \infty$ and $\{f_k\}_{k\geq 1}$ be a sequence of measurable functions such that, for every $\varepsilon > 0$, $\sum_{n=1}^{\infty} \mu(X(|f_k - f| \geq \varepsilon)) < \infty$. Show that $f_k \overset{meas}{\longrightarrow} f$.

Hint: Since the series of nonnegative terms is convergent, it follows that, for large k, $\mu(X(|f_k - f| \geq \varepsilon)) < \varepsilon$, i.e. $f_k \overset{meas}{\longrightarrow} f$.

6.2.P5. Let $\mu(X) < \infty$. Define $\rho(f, g) = \int_X \frac{|f-g|}{1+|f-g|}\,d\mu$ for every pair of measurable functions f and g. Show that
(a) $\infty > \rho(f, g) \geq 0$, $\rho(f, g) = \rho(g, f)$ and $\rho(f, g) = 0$ if and only if $f = g$ a.e.;
(b) $\rho(f, h) \leq \rho(f, g) + \rho(g, h)$ [triangle inequality];

Remark This means ρ is a peusdometric in the sense of Definition 1.3.3.

(c) $f_k \overset{meas}{\longrightarrow} f$ if and only if $\rho(f_k, f) \to 0$ as $k \to \infty$;
(d) a sequence $\{f_k\}_{k\geq 1}$ of measurable functions is Cauchy in measure if and only if it is Cauchy in the sense of the pseudometric ρ;
(e) setting $\|f\| = \rho\,(f, 0)$ does not provide a norm on the space of equivalence classes of measurable functions that agree a.e., except in the trivial case when $\mu(X) = 0$.

Hint: **(a)** Clearly, $\rho(f, g) \geq 0$ and $\rho(f, g) = \rho(g, f)$. Since $\mu(X) < \infty$, we have $\rho(f, g) < \infty$. Also, $\rho(f, g) = 0$ if and only if $\frac{|f-g|}{1+|f-g|} = 0$ a.e., which is equivalent to $f = g$ a.e.

(b) For all real a and b, we have $\frac{|a+b|}{1+|a+b|} \leq \frac{|a|}{1+|a|} + \frac{|b|}{1+|b|}$. [See Theorem 1.1.2 of [26, p. 24]]. Put $a = f - g$ and $b = h - g$ and integrate to obtain the required triangle inequality.

(c) Let f_k and f all be measurable functions and set $X_k = X(|f_k - f| > a)$, where $a > 0$. On using the fact that the function given by $\frac{t}{1+t}$, $t > 0$, is increasing, we obtain

$$\rho(f_k,f) = \int_X \frac{|f_k - f|}{1+|f_k - f|}d\mu \geq \int_{X_k} \frac{|f_k - f|}{1+|f_k - f|}d\mu$$
$$> \int_{X_k} \frac{a}{1+a}d\mu = \frac{a}{1+a}\mu(X_k).$$

So, if $\rho(f_k, f) \to 0$, then $\mu(X_k) \to 0$, that is $f_k \overset{meas}{\to} f$.

On the other hand, suppose $f_k \overset{meas}{\to} f$. Now we have

$$\rho(f_k, f) = \int_{X_k} \frac{|f_k - f|}{1+|f_k - f|}d\mu \geq \int_{X\backslash X_k} \frac{|f_k - f|}{1+|f_k - f|}d\mu$$
$$< \mu(X_k) + \varepsilon\mu(X\backslash X_k) < \mu(X_k) + \varepsilon\mu(X).$$

Since $f_k \overset{meas}{\to} f$, we have $\mu(X_k) \to 0$, and therefore

$$\rho(f_k,f) < \varepsilon(1+\mu(X)) \text{ for all large enough } k,$$

that is, $\rho(f_k, f) \to 0$.

(d) The argument is similar to (c).
(e) Let f be the constant function 1. Then $\rho(f, 0) = \frac{1}{2}\mu(X)$ but $\rho(2f, 0) = \frac{2}{3}\,\mu(X) \neq 2\rho(f, 0)$ when $\mu(X) \neq 0$.

6.2.P6. For any two measurable functions f and g on X, define

$$\rho(f,g) = \inf\{c + \mu(X(|f - g| > c)) : c > 0\}.$$

Show that, if either $\mu(X) < \infty$ or $f, g, f_k \in L^\infty$, then

(a) $\infty > \rho(f, g) \geq 0$, and $\rho(f, g) = 0$ if and only if $f = g$ a.e.; also, $\rho(f, g) = \rho(g, f)$;
(b) $\rho(f, h) \leq \rho(f, g) + \rho(g, h)$ [triangle inequality];

Remark This means ρ is a pseudometric in the sense of Definition 1.3.3.
(c) $f_k \overset{meas}{\to} f$ if and only if $\rho(f_k, f) \to 0$ as $k \to \infty$;
(d) a sequence $\{f_k\}_{k\geq 1}$ of measurable functions is Cauchy in measure if and only if it is Cauchy in the sense of the pseudometric ρ;
(e) if $A \subseteq X$ is measurable and $a > 0$, then $\rho(a\chi_A, 0) = \min\{a, \mu(A)\}$;
(f) setting $\|f\| = \rho(f, 0)$ does not provide a norm on the space of equivalence classes of measurable functions that agree a.e., except in the trivial case that no subset $A \subseteq X$ satisfies $0 < \mu(A) < \infty$.
Hint: **(a)** Clearly, $\rho(f, g) \geq 0$ and $\rho(f, g) = \rho(g, f)$. Now, $\rho(f, g) < \infty$ if either $\mu(X) < \infty$ or $f, g \in L^\infty$; in fact, $\rho(f, g) \leq \|f - g\|_\infty$ if $f, g \in L^\infty$. The rest of the arguments require only that $\rho(f, g)$ always be finite and therefore remain valid when

either $\mu(X) < \infty$ or all functions are in L^∞. Assume $\rho(f, g) = 0$ and consider any $\varepsilon > 0$. There is a sequence $\{c_k\}_{k \geq 1}$ of positive numbers such that

$$c_k < \frac{\varepsilon}{2^k} \text{ and } \mu(X(|f - g| > c_k)) < \frac{\varepsilon}{2^k} \text{ for each } k.$$

The inequality $c_k < \frac{\varepsilon}{2^k}$ for each k implies that $c_k \to 0$. Therefore $X(f \neq g) = X(|f - g| > 0) = \cup_{k \geq 1} X(|f - g| > c_k)$ and hence $\mu(X(f \neq g)) \leq \sum_{k=1}^{\infty} \mu(X(|f - g| > c_k)) \leq \sum_{k=1}^{\infty} \frac{\varepsilon}{2^k} = \varepsilon$. Consequently, $\mu(X(f \neq g)) = 0$.

Conversely, assume $\mu(X(f \neq g)) = 0$, i.e. $\mu(X(|f - g| > 0)) = 0$, and consider any $\varepsilon > 0$. Then $\mu(X(|f - g| > 0)) < \frac{\varepsilon}{2}$ and therefore $\frac{\varepsilon}{2} + \mu(X(|f - g| > 0)) < \varepsilon$. But this implies $c + \mu(X(|f - g| > c)) < \varepsilon$ when $c = \frac{\varepsilon}{2}$. Therefore $\inf\{c + \mu(X(|f - g| > c)): c > 0\} < \varepsilon$, which means $\rho(f, g) < \varepsilon$, by definition of $\rho(f, g)$. Since this holds for any $\varepsilon > 0$, it follows that $\rho(f, g) = 0$.

(b) It is obvious from the definition that $\rho(f, g) = \rho(f - g, 0)$. Therefore the triangle inequality will follow if we can show that $\rho(f + g, 0) \leq \rho(f, 0) + \rho(g, 0)$. To establish this, consider any $\varepsilon > 0$. Then there exist $c, c' > 0$ such that

$$c + \mu(X(|f| > c)) < \rho(f, 0) + \frac{\varepsilon}{2} \text{ and } c' + \mu(X(|g| > c')) < \rho(g, 0) + \frac{\varepsilon}{2}.$$

This implies

$$(c + c') + \mu(X(|f| > c)) + \mu(X(|g| > c')) < \rho(f, 0) + \rho(g, 0) + \varepsilon.$$

Now, $|f + g|(x) > c + c' \Rightarrow [|f|(x) > c$ or $|g|(x) > c']$ and therefore

$$\mu(X(|f + g| > c + c')) \leq \mu(X(|f| > c)) + \mu(X(|g| > c')),$$

so that

$$(c + c') + \mu(X(|f + g| > c + c')) < \rho(f, 0) + \rho(g, 0) + \varepsilon.$$

It follows that

$$\rho(f + g, 0) < \rho(f, 0) + \rho(g, 0) + \varepsilon.$$

Since this holds for any $\varepsilon > 0$, we have $\rho(f + g, 0) \leq \rho(f, 0) + \rho(g, 0)$, as required.

(c) To begin with, observe that by definition of $\rho(f, g)$, the inequality $\rho(f_k, f) < \varepsilon$ means $\inf\{c + \mu(X(|f_k - f| > c)): c > 0\} < \varepsilon$, and by definition of infimum, this means

$$c' + \mu(X(|f_k - f| > c')) < \varepsilon \text{ for some } c' > 0. \tag{8.72}$$

Assume $f_k \xrightarrow{meas} f$. We must show that, for any $\varepsilon > 0$, there exists an N such that $k \geq N$ implies that (8.72) holds. With this in view, consider any $\varepsilon > 0$. The assumption that $f_k \xrightarrow{meas} f$ means $\mu(X(|f_k - f| > c) \to 0$, no matter which positive number c may be. In particular, this is so when $c = \frac{\varepsilon}{2}$. Hence there exists an N such that $k \geq N$ implies $\mu(X(|f_k - f| > \frac{\varepsilon}{2}) < \frac{\varepsilon}{2}$, which in turn implies

$$\frac{\varepsilon}{2} + \mu(X(|f_k - f| > \frac{\varepsilon}{2})) < \varepsilon,$$

that is, (8.72) holds with $c' = \frac{\varepsilon}{2}$.

Conversely, assume $\rho(f_k, f) \to 0$. We must show that, for any $c > 0$ and any $\eta > 0$, there exists an N such that $k \geq N$ implies $\mu(X(|f_k - f| > c) < \eta$. Accordingly, consider any $c > 0$ and any $\eta > 0$. The assumption that $\rho(f_k, f) \to 0$ means that for $\varepsilon = \min\{c, \eta\}$, there exists an N such that $k \geq N$ implies that (8.72) holds, that is,

$$c' + \mu(X(|f_k - f| > c')) < \min\{c, \eta\} \text{ for some } c' > 0.$$

The positive number c' must satisfy $c' < \min\{c, \eta\} \leq c$ and also $\mu(X(|f_k - f| > c')) < \min\{c, \eta\} \leq \eta$. However, $\mu(X(|f_k - f| > c) \leq \mu(X(|f_k - f| > c')$ because $c' < c$. Therefore $\mu(X(|f_k - f| > c) < \eta$.

Remark Since $\rho(f, g) \leq \|f - g\|_\infty$, we have another proof that L^∞ convergence implies convergence in measure.

(d) The argument is similar to (c).

(e) Denote $a\chi_A$ by f. Let $c \geq a$. Then $X(|f| > c)$ is empty and $c + \mu(X(|f| > c)) = c$. Therefore

$$\inf\{c + \mu(X(|f| > c)) : c \geq a\} = a.$$

But if $c < a$, then $X(|f| > c) = A$ and $c + \mu(X(|f| > c)) = c + \mu(A)$. Therefore

$$\inf\{c + \mu(X(|f| > c)) : c < a\} = \mu(A).$$

It follows that $\rho(f, 0) = \inf\{c + \mu(X(|f| > c)) : c > 0\} = \min\{a, \mu(A)\}$.

(f) Consider any subset $A \subseteq X$ that satisfies $0 < \mu(A) < \infty$ and let $f(x) = \frac{1}{2}\mu(A)\chi_A$. Then $\rho(f, 0) = \frac{1}{2}\mu(A)$ and $\rho(4f, 0) = \mu(A)$ in view of part (e), which means $\rho(4f, 0) = 2\rho(f, 0) \neq 4\rho(f, 0)$, considering that $\mu(A) > 0$.

6.2.P7. If $f_k \overset{meas}{\to} f$ and $g \in L^\infty(X)$, then show that $f_k g \overset{meas}{\to} fg$.
Hint: Let M denote the essential sup of g. Then the set $X_M = X(|g| > M)$ has measure 0.

$$X(|f_n g - fg| > a) \subseteq \{x \in X \backslash X_M : |f_n g - fg| > a\} \cup \{x \in X_M : |f_n g - fg| > a\}$$
$$\subseteq \{x \in X \backslash X_M : |f_n - f| > \frac{a}{M}\} \cup \{x \in X_M : |f_n g - fg| > a\}.$$

6.2.P8. If f and $\{f_k\}_{k \geq 1}$ are measurable functions defined on X with $\mu(X) < \infty$, show that the following are equivalent:

(i) $f_k \overset{meas}{\to} f$,
(ii) every subsequence of $\{f_k\}_{k \geq 1}$ has a subsequence converging to f a.e.

Hint: (i) ⇒ (ii). Since $f_k \overset{meas}{\to} f$, the same is true of every subsequence. By Corollary 6.2.7, any subsequence has a subsequence converging to f a.e.
(ii) ⇒ (i). Let $a > 0$ be given. Let $E_k = X(|f_k - f| \geq a\}$; we then have to show that $\mu(E_k) \to 0$. If this fails to hold, then $\mu(E_{k_j}) > \delta$ for some $\delta > 0$ and some subsequence $\{f_{k_j}\}_{j \geq 1}$. In view of the hypothesis, by selecting a subsequence of $\{k_j\}$ if necessary, we may assume that $f_{k_j} \overset{ae}{\to} f$, so that $\mu(\limsup E_{k_j}) = 0$. Hence by Proposition 2.3.21(d), which can be shown to be valid for a general measure,

$$0 = \mu\left(\limsup E_{k_j}\right) \geq \limsup \mu\left(E_{k_j}\right) \geq \delta > 0,$$

and this is a contradiction.

6.2.P9. For Lebesgue measure on [0, 1], show that there cannot exist a ρ satisfying (a) and (b) of Problems 6.2.P5 and 6.2.P6, but satisfying

$$f_k \overset{ae}{\longrightarrow} f \quad \text{if and only if} \quad \rho(f_k, f) \to 0 \text{ as } k \to \infty$$

instead of (c).

Hint: Suppose such a ρ exists. The sequence $\{f_k\}_{k \geq 1}$ of Example 6.2.20 converges to 0 in measure and by Problem 6.2.P8, we conclude that every subsequence of it has a subsequence converging to 0 a.e. We can use ρ to show that the sequence converges to 0 a.e., which was observed not to be the case in Example 6.2.20. If the sequence were not to converge to 0 a.e., then there would exist some $\eta > 0$ and some subsequence $\{f_{k_j}\}_{j \geq 1}$ of $\{f_k\}_{k \geq 1}$ satisfying $\rho(f_{k_j}, 0) > \eta$ for every j. It follows that every subsequence of the subsequence $\{f_{k_j}\}_{j \geq 1}$ also satisfies the same and therefore fails to converge to 0 a.e., contradicting the conclusion obtained above by applying Problem 6.2.P8.

6.2.P10. (a) If $f_k \xrightarrow{meas} f$, and $f_k \xrightarrow{ae} g$, then $g = f$ a.e.
(b) If $f_k \xrightarrow{meas} f$, and $f_k \le f_{k+1}$ a.e. for each k, then $f_k \xrightarrow{ae} f$.
Hint: **(a)** Since $f_k \xrightarrow{meas} f$, by Corollary 6.2.7, there exists a subsequence converging to f a.e. Since $f_k \xrightarrow{ae} g$, the subsequence converges to g as well. Therefore $g = f$ a.e. Note that this works even if g is extended real-valued.
(b) The hypothesis $f_k \le f_{k+1}$ a.e. for each k means that for each k, there exists a set E_k of measure 0 such that $f_k \le f_{k+1}$ on the complement E_k^c. The union $E = \cup_{k \ge 1} E_k$ then has measure 0 and, every $x \in E^c$ satisfies the inequality $f_k \le f_{k+1}$ for every k. Thus $\{f_k(x)\}_{k \ge 1}$ is an increasing sequence for every $x \in E^c$ and therefore has an extended real limit for every $x \in E^c$. Let g be the function defined to be this limit for $x \in E^c$ and 0 on E. Then $f_k \xrightarrow{ae} g$ and by part (a), it follows that $g = f$ a.e. Consequently, $f_k \xrightarrow{ae} f$.
6.2.P11. Show that if $X = \mathbb{Z}$ with the counting measure, then convergence in measure is equivalent to uniform convergence. Does this equivalence hold for the counting measure on an arbitrary set?

Hint: $\mu(X(|f_k - f| \ge a)) \to 0$. There exists a positive integer K such that

$$k \ge K \Rightarrow \mu(X(|f_k - f| \ge a)) < \frac{1}{2}.$$

Since each nonempty subset of $\mathbb{Z}$ has measure at least 1, it follows that $X(|f_k - f| \ge a) = \varnothing$ for $k \ge K$. That is to say,

$$k \ge K \Rightarrow |f_k(x) - f(x)| < a \text{ for all } x \in \mathbb{Z}.$$

Thus $f_k \xrightarrow{unif} g$.

Yes, the equivalence holds for the counting measure on an arbitrary set.

6.2.P12. Let $(X, \mathcal{F}, \mu)$ be a measure space with $\mu(X) < \infty$ and suppose f and $\{f_k\}_{k \ge 1}$ are measurable functions on X. For $\varepsilon > 0$ and integer $k \ge 1$, put

$$E_k^\varepsilon = X(|f_k - f| \ge \varepsilon).$$

Prove that $f_k \to f$ (pointwise) if and only if, for all $\varepsilon > 0$, $\left\{ \bigcup_{k \ge 1} E_k^\varepsilon \right\}_{n \ge 1}$ decreases to $\varnothing$.

Hint: First suppose $f_k \to f$ and consider any $\varepsilon > 0$. Observe that $\cup_{k \ge n+1} E_k^\varepsilon \subseteq \cup_{k \ge n} E_k^\varepsilon$. Let $f^\varepsilon = \cap_{n \ge 1}(\cup_{k \ge n} E_k^\varepsilon)$. We have to show that $f^\varepsilon = \varnothing$. Consider any $x \in f^\varepsilon$. By supposition, there exists an n_0 such that $|f_k(x) - f(x)| < \varepsilon$ for all $k \ge n_0$. Since $x \in f^\varepsilon$, we have $x \in \cup_{k \ge n} E_k^\varepsilon$ for all n. In particular, $x \in \cup_{k \ge n_0} E_k^\varepsilon$. So there exists a $k \ge n_0$ such that $x \in E_k^\varepsilon$, which implies $|f_k(x) - f(x)| \ge \varepsilon$. This contradiction shows that $f^\varepsilon = \varnothing$, proving the "only if" part.

Now, suppose that for any $\varepsilon > 0$, $\{\cup_{k \ge n} E_k^\varepsilon\}_{n \ge 1}$ decreases to $\varnothing$ and consider any $x \in X$. We must show that $f_k(x) \to f(x)$. Consider any $\eta > 0$. There exists an n_0

such that $x \notin \cup_{k \geq n_0} E_k^{\eta}$, which implies $x \notin E_k^{\varepsilon}$ for every $k \geq n_0$. Thus for every $k \geq n_0$, we have $|f_n(x) - f(x)| < \varepsilon$. So, $f_k(x) \to f(x)$.

6.2.P13. Let $(X,\mathcal{F}, \mu)$ be a measure space with $\mu(X) < \infty$ and suppose f and $\{f_k\}_{k \geq 1}$ are measurable functions on X. For $\varepsilon > 0$ and integer $k \geq 1$, put

$$E_k^{\varepsilon} = X(|f_k - f| \geq \varepsilon).$$

Prove that $f_k \xrightarrow{ae} f$ if and only if $\lim_{n\to\infty} \mu\left(\cup_{k \geq n} E_k^{\varepsilon}\right) = 0$ for all $\varepsilon > 0$.

Hint: First suppose $f_k \xrightarrow{ae} f$.

By definition, this means there exists a measurable set G with $\mu(G^c) = 0$ and $f_k(x) \to f(x)$ for every $x \in G$. Consider any $\varepsilon > 0$. Observe that $\cup_{k \geq n+1} E_k^{\varepsilon} \subseteq \cup_{k \geq n} E_k^{\varepsilon}$. Therefore by Proposition 3.1.8(b), we need only prove that

$$\mu\Big(\bigcap_{n \geq 1}\Big(\bigcup_{k \geq n} E_k^{\varepsilon}\Big)\Big) = 0.$$

Let $f^{\varepsilon} = \cap_{n \geq 1}(\cup_{k \geq n} E_k^{\varepsilon})$. We have to show that $\mu(f^{\varepsilon}) = 0$. Consider any $x \in G \cap f^{\varepsilon}$. By supposition, there exists an n_0 such that $|f_k(x) - f(x)| < \varepsilon$ for all $k \geq n_0$. Since $x \in f^{\varepsilon}$, we have $x \in \cup_{k \geq n} E_k^{\varepsilon}$ for all n. In particular, $x \in \cup_{k \geq n_0} E_k^{\varepsilon}$. So there exists a $k \geq n_0$ such that $x \in E_k^{\varepsilon}$, which implies $|f_k(x) - f(x)| \geq \varepsilon$. This contradiction shows that $G \cap f^{\varepsilon} = \emptyset$. Therefore $f^{\varepsilon} = (G \cap f^{\varepsilon}) \cup (G^c \cap f^{\varepsilon}) = G^c \cap f^{\varepsilon} \subseteq G^c$. Since $\mu(G^c) = 0$, it follows that $\mu(f^{\varepsilon}) = 0$.

Now, suppose that $\lim_{n\to\infty} \mu\left(\cup_{k \geq n} E_k^{\varepsilon}\right) = 0$ for any $\varepsilon > 0$.

Since $\cup_{k \geq n+1} E_k^{\varepsilon} \subseteq \cup_{k \geq n} E_k^{\varepsilon}$, we have $\mu(\cap_{n \geq 1}(\cup_{k \geq n} E_k^{\varepsilon})) = 0$ for any $\varepsilon > 0$ in view of Proposition 3.1.8(b). Upon considering $\varepsilon = 1/p$ for $p = 1, 2, \ldots$, we find that $\mu(\cup_{p \geq 1}(\cap_{n \geq 1}(\cup_{k \geq n} E_k^{1/p}))) = 0$. So, the set $G = (\cup_{p \geq 1}(\cap_{n \geq 1}(\cup_{k \geq n} E_k^{1/p})))^c$ satisfies $\mu(G^c) = 0$. Consider any $x \in G$. We claim that $f_k(x) \to f(x)$. Consider an arbitrary integer $p > 0$. There exists an n_0 such that $x \in \cup_{k \geq n_0} E_k^{1/p}$, which implies $x \notin E_k^{1/p}$ for every $k \geq n_0$. Thus for every $k \geq n_0$, we have $|f_k(x) - f(x)| < 1/p$. Note that we have shown such an n_0 to exist for an arbitrary integer $p > 0$. Consequently, $f_k(x) \to f(x)$.

6.2.P14. Let $(X, \mathcal{F}, \mu)$ be a measure space with $\mu(X) < \infty$ and suppose f and $\{f_k\}_{k \geq 1}$ are measurable functions on X. For $\varepsilon > 0$ and integer $m \geq 1$, let

$$E_m^{\varepsilon} = X(|f_{m \geq n} - f| \geq \varepsilon).$$

Prove that $f_k \xrightarrow{ae} f$ if and only if $\lim_{n\to\infty} \mu\left(\bigcup_{m \geq n} E_m^{\varepsilon}\right) = 0$ for all $\varepsilon > 0$.

Hint: This is a direct consequence of Problem 6.2.P13 and Egorov's Theorem 6.2.21.

Problem Set 7.1

7.1.P1. Let $X = Y = [0, 1]$, $\mathcal{F} = \mathcal{G}$ be the σ-algebra of measurable subsets of $[0, 1]$ and $\mu = \nu$ be Lebesgue measure on $[0, 1]$. Let

$$f(x, y) = \frac{x^2 - y^2}{(x^2 + y^2)^2}, \quad (x, y) \in (0, 1) \times (0, 1).$$

Prove that each of the iterated integrals of f exists. The function f is, however, not in $L^1([0, 1] \times [0, 1])$.

Hint: For each fixed x, the function $f(x, y)$, though undefined at $y = 0$ and $y = 1$, becomes Riemann (and hence Lebesgue) integrable on $[0, 1]$, when defined in any manner at $y = 0$ and $y = 1$. It is easy to verify that the Riemann integral is $\int_0^1 f(x, y)dy = \frac{1}{1+x^2}$. It follows that the Lebesgue integral $\int_{(0,1)} f(x, y)d\nu(y)$ is also equal to $\frac{1}{1+x^2}$. Hence $\int_{(0,1)} \left(\int_{(0,1)} f(x,y)d\nu(y)\right) d\mu(x) = \frac{\pi}{4}$. Similarly, it can be checked that $\int_{(0,1)} \left(\int_{(0,1)} f(x, y)d\mu(x)\right)d\nu(y) = -\frac{\pi}{4}$. Since the iterated integrals are unequal, it follows by Fubini's Theorem that $f \notin L^1([0, 1] \times [0, 1])$. Alternatively, by Tonelli's Theorem 7.1.23,

$$\begin{aligned}\int_{(0,1)\times(0,1)} |f(x, y)|d(\mu \times \nu) &= \int_{(0,1)} (\int_{(0,1)} |f(x,y)|d\nu(x))d\mu(y) \\ &= \int_{(0,1)} (\int_{(0,1)} \left|\frac{x^2 - y^2}{(x^2 + y^2)^2}\right| d\nu(x))d\mu(y) \\ &\geq \int_{(0,1)} (\int_{(0,y)} \left|\frac{x^2 - y^2}{(x^2 + y^2)^2}\right| d\nu(x))d\mu(y)\end{aligned}$$

$$\begin{aligned}&= \int_{(0,1)} (\int_{(0,y)} \frac{y^2 - x^2}{(x^2 + y^2)^2} d\nu(x))d\mu(y) \\ &= \int_{(0,1)} \left(\frac{x}{x^2 + y^2}\bigg|_{x=y} - \frac{x}{x^2 + y^2}\bigg|_{x=0}\right) d\mu(y)\end{aligned}$$

$$= \int_{(0,1)} \frac{1}{2y} d\mu(y) = \infty.$$

Thus $f \notin L^1((0, 1) \times (0, 1))$. Computing the last integral above as an improper integral is justified by the Monotone Convergence Theorem, keeping in mind that the integrand is nonnegative.

7.1.P2. Let $X = Y = [-1, 1]$ and $\mathcal{F} = \mathcal{G}$ be the σ-algebra of measurable subsets of $[-1, 1]$ and $\mu = \nu$ be Lebesgue measure on $[-1, 1]$. Let

$$f(x,\ y) = \begin{cases} \frac{xy}{(x^2+y^2)^2} & \text{if } (x,y) \neq (0,\ 0) \\ 0 & \text{otherwise.} \end{cases}$$

Show that

$$\int_{[-1,1]} (\int_{[-1,1]} f(x,y)d\nu(y))d\mu(x) = 0 = \int_{[-1,1]} (\int_{[-1,1]} f(x,y)d\mu(x))d\nu(y),$$

but the function is not Lebesgue integrable over [-1, 1] × [-1, 1].
Hint: For $y \neq 0$, $\int_{[-1,1]} f(x,\ y)d\mu(x) = 0$ and for $x \neq 0$, $\int_{[-1,1]} f(x,\ y)d\nu(y) = 0$; so each of the iterated integrals is zero. However, f is not integrable on $[-1,\ 1] \times [-1,\ 1]$, as the following argument shows.

Since

$$\int_{[0,1]\times[0,1]} |f(x,y)|d(\mu \times \nu) \leq \int_{[-1,1]\times[-1,1]} |f(x,\ y)|d(\mu \times \nu) < \infty,$$

it will follow that f is integrable on $[0,\ 1] \times [0,\ 1]$ if it is integrable on $[-1,\ 1] \times [-1,\ 1]$. By Tonelli's Theorem 7.1.23

$$\int_{[0,1]\times[0,1]} |f(x,\ y)|d(\mu \times \nu) = \int_{(0,1)} \left(\int_{(0,1)} \frac{xy}{(x^2+y^2)^2} d\nu(y)\right) d\mu(x).$$

Simple calculations show that

$$\int_{(0,1)} \frac{xy}{(x^2+y^2)^2} d\nu(y) = \frac{1}{2x} - \frac{x}{2(1+x^2)}.$$

However, $\int_{(0,1)} \frac{1}{2x} - \frac{x}{2(1+x^2)} d\mu(x) = \infty$. Computing this integral as an improper integral is justified by the Monotone Convergence Theorem, keeping in mind that the integrand is nonnegative.

7.1.P3. Let $f \in L^1(X,\ \mathcal{F},\ \mu)$ and $g \in L^1(Y,\ \mathcal{G},\ \nu)$, and suppose $\phi(x,\ y) = f(x)g(y)$, $x \in X$, $y \in Y$. Show that $\phi \in L^1(X \times Y,\ \mathcal{F} \times \mathcal{G},\ \mu \times \nu)$ and

$$\int_{X\times Y} \phi(x,\ y)d(\mu \times \nu) = (\int_X f(x)d\mu(x))(\int_Y g(y)d\nu(y)).$$

Hint: The function $f^*(x,\ y) = f(x)$ is $\mathcal{F} \times \mathcal{G}$-measurable. Indeed,

$$\{(x, y) \in X \times Y : f^*(x, y) > \alpha\} = \{x \in X : f(x) > \alpha\} \times Y.$$

Similarly, the function $g^*(x, y) = g(y)$ is $\mathcal{F}\times\mathcal{G}$-measurable. The function $f(x)g(y) = f^*(x, y)g^*(x, y)$, being the product of two $\mathcal{F}\times\mathcal{G}$-measurable functions, is $\mathcal{F}\times\mathcal{G}$-measurable. Since the iterated integral

$$\int_X \left(\int_Y |f(x)g(y)|dv(y)\right) d\mu(x) = \left(\int_Y |g(y)|dv(y)\right)\left(\int_X |f(x)|d\mu(x)\right) < \infty,$$

Fubini's Theorem 7.1.25 is applicable and the desired result follows.

7.1.P4. Let $X = Y = [0, 1]$, $\mathcal{F} = \mathcal{G} =$ the σ-algebra of Lebesgue measurable subsets of [0, 1] and $\mu = \nu =$ Lebesgue measure. Suppose that either $f \in L^1([0, 1] \times [0, 1])$ or f is a measurable nonnegative function on $[0, 1] \times [0, 1]$. Prove that

$$\int_{[0,1]} \left(\int_{[0,x]} f(x,y)dy\right) dx = \int_{[0,1]} \left(\int_{[y,1]} f(x,y)dx\right) dy.$$

Hint: Depending on whether $f \in L^1([0, 1] \times [0, 1])$ or f is a measurable nonnegative function, apply either Fubini's Theorem 7.1.25 or Tonelli's Theorem 7.1.23 to the function g given on the domain $[0, 1] \times [0, 1]$ by $g = f\chi_A$, where $A = \{(x, y) \in [0, 1] \times [0, 1]: y \in [0, x]\}$ after arguing that A is (i) measurable and (ii) the same set as $\{(x, y) \in [0, 1] \times [0, 1]: x \in [y, 1]\}$. Here, (ii) is obvious because the conjunction of inequalities

$$0 \le x \le 1 \text{ and } 0 \le y \le x$$

is plainly equivalent to the conjunction

$$0 \le y \le 1 \text{ and } y \le x \le 1.$$

Regarding (i), for any positive integer n, set $I_j = \left[\frac{j-1}{n}, \frac{j}{n}\right], J_j = \left[0, \frac{j}{n}\right]$ and $V_n = \cup_{1 \le j \le n}(I_j \times J_j)$. Clearly, each V_n, being a union of measurable rectangles, is measurable, and hence so is $\cap_{n \ge 1} V_n$. However, this intersection is A.

7.1.P5. Let $f \in L^1((0, a))$ and $g(x) = \int_{[x,a]} \frac{f(t)}{t} dt, 0 < x \le a$. Show that $g \in L^1((0, a))$ and that $\int_{[0,a]} g(x)dx = \int_{[0,a]} f(t)dt$.

Hint: We wish to show that

$$\int_{[0,a]} |g|dm = \int_{[0,a]} \left| \int_{[x,a]} \frac{f(t)}{t} dt \right| dx < \infty.$$

It is enough to show that $\int_{[0,a]} \int_{[x,a]} \frac{|f(t)|}{t} dtdx < \infty$. The functions given by $\frac{f(t)}{t}$ and $\frac{|f(t)|}{t}$ on (0, a) × (0, a) are both measurable. Interchanging the order of integration and using the obvious analogue of the result of 7.1.P4 above, we transform the last integral as

$$\int_{[0,a]} \int_{[0,t]} \frac{|f(t)|}{t} dxdt = \int_{[0,a]} |f(t)|dt < \infty.$$

Thus $g \in L^1((0, a))$. Now, by the definition of g,

$$\int_{[0,a]} g(x)dx = \int_{[0,a]} \int_{[x,a]} \frac{f(t)}{t} dtdx.$$

Using the obvious analogue of the result of 7.1.P4 above once again,

$$\int_{[0,a]} g(x)dx = \int_{[0,a]} \int_{[0,t]} \frac{f(t)}{t} dxdt$$
$$= \int_{[0,a]} f(t)dt.$$

7.1.P6. Let $X = Y = [0, 1]$, $\mathcal{F} = \mathcal{G}$ = the σ-algebra of Lebesgue measurable subsets of [0, 1] and $\mu = \nu$ = Lebesgue measure. Define for $x, y \in [0, 1]$,

$$f(x, y) = \begin{cases} 1 & x \in \mathbb{Q} \\ 2y & x \notin \mathbb{Q}. \end{cases}$$

Compute

$$\int_{[0,1]} \left(\int_{[0,1]} f(x,y)d\nu(y) \right) d\mu(x) \text{ and } \int_{[0,1]} \left(\int_{[0,1]} f(x,y)d\mu(x) \right) d\nu(y).$$

Does $f \in L^1([0, 1] \times [0, 1])$?

Hint: The function can be expressed as $f(x, y) = \chi_{I \cap \mathbb{Q}}(x)\cdot 1 + \chi_{I/\mathbb{Q}}(x)\cdot 2y$. Since each of the two terms is a measurable function on $[0, 1] \times [0, 1]$, the sum f is a measurable function too. Its values lie between 0 and 2 and it is therefore integrable. Moreover, by Tonelli's Theorem 7.1.23, its integral equals each of its repeated integrals. It is sufficient therefore to compute either one of the two repeated integrals asked for.

For each $x \in [0, 1]$, we have $f(x, y) = 1$ or $2y$ according as $x \in \mathbb{Q}$ or $x \notin \mathbb{Q}$, so that $f(x, y) = 2y$ for almost all $x \in [0, 1]$. Therefore

$$\int_{[0,1]} f(x,y)d\nu(y) = 1 \text{ for almost all } x \in [0,1].$$

Hence

$$\int_{[0,1]} \left(\int_{[0,1]} f(x,y)d\nu(y) \right) d\mu(x) = 1.$$

Alternatively, we may compute the other repeated integral: For each $y \in [0, 1]$, we have $f(x, y) = 1$ or $2y$ according as $x \in \mathbb{Q}$ or $x \notin \mathbb{Q}$, so that $f(x, y) = 2y$ for almost all $x \in [0, 1]$. Therefore

$$\int_{[0,1]} f(x,y)d\mu(x) = 2y \text{ for all } y \in [0,1].$$

Hence

$$\int_{[0,1]} \left(\int_{[0,1]} f(x,y)d\mu(x) \right) d\nu(y) = 1, \text{ as expected.}$$

7.1.P7. Suppose that $\{a_{m,n}\}_{m,n \geq 1}$ is a double sequence of nonnegative real numbers. Then show that

$$\sum_{m=1}^{\infty}\sum_{n=1}^{\infty} a_{m,n} = \sum_{n=1}^{\infty}\sum_{m=1}^{\infty} a_{m,n}.$$

Hint: Let $(X, \mathcal{F}, \mu) = (Y, \mathcal{G}, \nu) = (\mathbb{N}, \mathcal{P}(\mathbb{N}), \gamma)$, where γ is the counting measure on $\mathbb{N}$, obviously σ-finite. The product σ-algebra $\mathcal{P}(\mathbb{N}) \times \mathcal{P}(\mathbb{N})$ is easily verified to be $\mathcal{P}(\mathbb{N} \times \mathbb{N})$, so that every function on $\mathbb{N} \times \mathbb{N}$ is measurable. Define $f: \mathbb{N} \times \mathbb{N} \to \mathbb{R}$ by $f(m, n) = a_{m,n}$. Then by Tonelli's Theorem,

$$\sum_{m=1}^{\infty}\sum_{n=1}^{\infty} a_{m,n} = \sum_{m=1}^{\infty}\left(\int_{\mathbb{N}} f(m,n)d\gamma(n)\right) = \int_{\mathbb{N}}\int_{\mathbb{N}} f(m,n)d\gamma(n)d\gamma(m)$$
$$= \int_{\mathbb{N}}\int_{\mathbb{N}} f(m,n)d\gamma(m)d\gamma(n) = \sum_{n=1}^{\infty}\sum_{m=1}^{\infty} a_{m,n}.$$

7.1.P8. Let $X = Y = [0, 1]$, $\mathcal{F} = \mathcal{G}$ = the σ-algebra of Borel subsets of [0, 1]; suppose μ is Lebesgue measure on Borel subsets of X and let γ be the counting measure on Y. Show that $V = \{(x, y) \in [0, 1] \times [0, 1]: x = y\}$ is $\mathcal{F}\times \mathcal{G}$-measurable and

$$\int_Y d\gamma \int_X \chi_V \, d\mu = 0 \text{ but } \int_Y d\mu \int_X \chi_V d\gamma = 1.$$

Hint: For any positive integer n, set $I_j = \left[\frac{j-1}{n}, \frac{j}{n}\right]$ and $V_n = \cup_{1\leq j\leq n}(I_j\times I_j)$. Clearly, each V_n, being a union of measurable rectangles, is measurable, and so is $V = \cap_{n\geq 1}V_n$. Now,

$$\int_Y d\gamma \int_X \chi_V \, d\mu = \int_Y \mu(V^y)d\gamma = 0$$

and

$$\int_X d\mu \int_Y \chi_V d\gamma = \int_X \gamma(V_x)d\mu = 1.$$

Fubini's and Tonelli's Theorem both fail because γ is not σ-finite.

7.1.P9. Let $X = Y = \mathbb{N}$ and μ = ν = counting measure on $\mathbb{N}$. Let

$$f(x, y) = \begin{cases} 2 - 2^{-x} & \text{if } x = y \\ -2 + 2^{-x} & \text{if } x = y + 1 \\ 0 & \text{otherwise.} \end{cases}$$

Show that the iterated integrals of f are not equal and $f \notin L^1(\mu \times \nu)$. [This shows that neither the integrability condition in Fubini's Theorem nor the nonnegativity condition in Tonelli's Theorem can be omitted.]
Hint: For each y, we have $\int_X f^y d\mu = (2 - 2^{-y})\cdot 1 + (-2 + 2^{-y-1})\cdot 1 = -2^{-y-1}$. So,

$$\int_Y d\nu \int_X f \, d\mu = \sum_{y=1}^{\infty} -2^{-y-1} = -\frac{1}{2}.$$

However,

$$\int_X d\mu \int_Y f\, d\nu = \int_X [(2-2^{-x})\cdot 1 + (-2+2^{-x+1})\cdot 1]d\mu = \sum_{x=1}^{\infty} 2^{-x} = 1$$

but

$$\int_{X\times Y} f^+ d(\mu\times\nu) = \infty \text{ and } \int_{X\times Y} f^- d(\mu\times\nu) = \infty.$$

Indeed, $f^+(x,y) = \begin{cases} 2-2^{-x} & \text{if } x=y \\ 0 & \text{otherwise} \end{cases}$, $f^-(x,y) = \begin{cases} 2-2^{-x} & \text{if } x=y+1 \\ 0 & \text{otherwise} \end{cases}$

7.1.P10. Let $(X, \mathcal{F}, \mu)$ and $(Y, \mathcal{G}, \nu)$ be measure spaces with $\mu(X) = 1$ and ν σ-finite. If E is a measurable subset of $X \times Y$ and if $\nu(E_x) \le \frac{1}{2}$ for almost all $x \in X$, then

$$\nu(\{y \in X : \mu(E^y) = 1\}) \le \frac{1}{2}.$$

Hint: $\int_X d\mu(x) \int_Y \chi_{E_x}\, d\nu(y) = \int_Y d\nu(y) \int_X \chi_{E^y} d\mu(x)$. Since

$$\int_X d\mu(x) \int_Y \chi_{E_x} d\nu(y) = \int_X \nu(E_x) d\mu(x) \le \frac{1}{2},$$

it follows that

$$\int_Y d\nu(y) \int_X \chi_{E^y}\, d\mu(x) = \int_Y \mu(E^y) d\nu(y) \le \frac{1}{2}.$$

Now denote $\{y \in Y: \mu(E^y) = 1\}$ by A. (It must be measurable by Lemma7.1.19(b).) Then

$$\int_A \mu(E^y) d\nu(y) + \int_{A^c} \mu(E^y) d\nu(y) \le \frac{1}{2},$$

the second integral above being nonnegative. This implies the required inequality.

7.1.P11. Let $X = Y = [0, 1]$, equipped with the σ-algebra of Lebesgue measurable subsets and Lebesgue measure. Consider a real sequence $\{\alpha_n\}_{n\ge 1}$, $0 < \alpha_1 < \alpha_2 < \cdots$, satisfying $\lim_{n\to\infty} \alpha_n = 1$. For each n, choose a continuous function g_n such that $\{t: g_n(t) \ne 0\} \subseteq (\alpha_n, \alpha_{n+1})$ and also $\int_{[0,1]} g_n(t) dt = 1$. Define

$$f(x, y) = \sum_{n=1}^{\infty} [g_n(x) - g_{n+1}(x)]g_n(y).$$

Note that for each (x, y), at most two terms in the sum can be nonzero. Thus no convergence problem arises in the definition of f. Show that f is not integrable and that its repeated integrals do not agree.

Hint: Denote the interval $[\alpha_n, \alpha_{n+1}]$ by I_n. Then $\int_{I_n} g_n(t)dt = \int_{[0,1]} g_n(t)dt = 1$ for each n. Also, the function f is measurable. For each fixed y,

$$\begin{aligned}\int_{[0,1]} f(x,y)dx &= \int_{[0,1]} \sum_{n=1}^{\infty} [g_n(x) - g_{n+1}(x)]g_n(y)dx \\ &= g_n(y)\left(\int_{I_n} g_n(x)dx - \int_{I_{n+1}} g_{n+1}(x)dx\right) \\ &= 0.\end{aligned}$$

So, $\int_{[0,1]} dy \int_{[0,1]} f(x, y)dx = 0$. On the other hand,

$$\int_{[0,1]} f(x,y)dy = \sum_{n=1}^{\infty} [g_n(x) - g_{n+1}(x)] \int_{[0,1]} g_n(y)dy = g_1(x),$$

and hence

$$\int_{[0,1]} dx \int_{[0,1]} f(x, y)dy = 1.$$

The iterated integrals are unequal. But Fubini's Theorem is not contradicted, since

$$\begin{aligned}\int_{[0,1]\times[0,1]} |f(x, y)|d(x, y) &= \sum_{i,j=1}^{\infty} \int_{I_i \times I_j} \left|\sum_{n=1}^{\infty} [g_n(x) - g_{n+1}(x)]g_n(y)\right| d(x, y) \\ &= \sum_{i,j=1}^{\infty} \int_{I_i \times I_j} \left|\left[g_j(x) - g_{j+1}(x)\right]g_j(y)\right| d(x, y) \\ &= \sum_{j=1}^{\infty} \int_{I_j \times I_j} \left|\left[g_j(x) - g_{j+1}(x)\right]g_j(y)\right| d(x, y)\end{aligned}$$

$$+\sum_{j=1}^{\infty}\int_{I_{j+1}\times I_j}\left|\left[g_j(x)-g_{j+1}(x)\right]g_j(y)\right|d(x,\ y)$$

$$=\sum_{j=1}^{\infty}\int_{I_j\times I_j}\left|g_j(x)g_j(y)\right|d(x,\ y)$$

$$+\sum_{j=1}^{\infty}\int_{I_{j+1}\times I_j}\left|g_{j+1}(x)g_j(y)\right|d(x,\ y)$$

$$=\sum_{j=1}^{\infty}\int_{I_j}\left(\int_{I_j}\left|g_j(x)g_j(y)\right|dx\right)dy$$

$$+\sum_{j=1}^{\infty}\int_{I_{j+1}}\left(\int_{I_j}\left|g_{j+1}(x)g_j(y)\right|dx)\right)dy$$

by Tonelli's Theorem 7.1.23

$$\geq\sum_{j=1}^{\infty}1+\sum_{j=1}^{\infty}1=\infty.$$

7.1.P12. This problem deals with a measurable function on $[0, 1] \times [0, 1] \subseteq \mathbb{R}^2$ for which the iterated integrals exist and are equal though the function fails to be integrable. Let $I = [0, 1] \times [0, 1]$; divide I into four equal squares. Let $I_1 = \left[0, \frac{1}{2}\right] \times \left[0, \frac{1}{2}\right]$. Next, divide the square $\left[\frac{1}{2}, 1\right] \times \left[\frac{1}{2}, 1\right]$ into four equal squares and set $I_2 = \left[\frac{1}{2}, \frac{3}{2^2}\right] \times \left[\frac{1}{2}, \frac{3}{2^2}\right]$, and so on. On each square I_k, $k = 1, 2, \ldots$, define a function φ_k as follows; divide I_k into four equal squares and let φ_k be -1 on the interiors of the left bottom and right upper squares, 1 on the interiors of the right bottom and left upper squares and zero elsewhere. Set

$$f(x,\ y)=\sum_{k=1}^{\infty}\frac{1}{|I_k|}\varphi_k(x,\ y),$$

where $|I_k|$ is the area of the square I_k. The function $f(x, y)$ is well defined since each point $(x, y) \in I$ belongs to the interior of at most one of the I_k. Show that the repeated integrals exist and are equal but the function is not integrable, though it is measurable.

Hint: For each $x \in [0, 1]$, we have $\int_{[0,1]} f(x, y)dy = 0$; so, $\int_{[0,1]}\int_{[0,1]} f(x, y)dydx = 0$. A similar argument shows that the other iterated integral is also zero. Thus both iterated integrals exist and are equal. Since $|\varphi_k(x, y)| = 1$ for $(x, y) \in I_k$, it follows that

$$\int_{[0,1]\times[0,1]} |f(x, y)|d(x, y) = \sum_{k=1}^{\infty} \frac{1}{|I_k|} \int_{I_k} |\varphi_k(x, y)|d(x, y) = \sum_{k=1}^{\infty} 1 = \infty.$$

7.1.P13. This is another example (besides 7.1.P2 and 7.1.P12) which shows that the integrability condition cannot be omitted from Fubini's Theorem. Let $X = Y = \mathbb{Z}$, the set of integers, $\mathcal{P}(\mathbb{Z})$ the σ-algebra of all subsets of $\mathbb{Z}$ and γ be the counting measure on $\mathcal{P}(\mathbb{Z})$. Let

$$f(x, y) = \begin{cases} x & \text{if } y = x \\ -x & \text{if } y = x + 1 \\ 0 & \text{otherwise .} \end{cases}$$

Show that the repeated integrals exist and are unequal. Also, show by a direct computation that the function is not integrable.
Hint: $\int_{\mathbb{Z}} f(x, y)d\gamma(y) = x + (-x) = 0$ for each $x \in \mathbb{Z}$; it follows from here that $\int_{\mathbb{Z}} \int_{\mathbb{Z}} f(x, y)d\gamma(y)d\gamma(x) = 0$. On the other hand, $\int_{\mathbb{Z}} f(x, y)d\gamma(x) = -(y - 1) + y = 1$ for each $y \in \mathbb{Z}$. Consequently,

$$\int_{\mathbb{Z}} \int_{\mathbb{Z}} f(x, y)d\gamma(x)d\gamma(y) = \sum_{y=1}^{\infty} 1 = \infty.$$

If $x \geq 0$, then

$$f^+(x, y) = \begin{cases} x & \text{if } x = y \\ 0 & \text{otherwise .} \end{cases}$$

If $x < 0$, then

$$f^+(x, y) = \begin{cases} -x & \text{if } x = y + 1 \\ 0 & \text{otherwise .} \end{cases}$$

Hence

$$\int_{\mathbb{Z}\times\mathbb{Z}} f^+ d(\gamma \times \gamma) = \int_A x d(\gamma \times \gamma) + \int_B -x d(\gamma \times \gamma) = \infty,$$

where $A = \{(x, y): x \geq 0 \text{ and } x = y\}$ and $B = \{(x, y): x < 0 \text{ and } x = y + 1\}$.

It would be instructive for the reader to compute $\int_{\mathbb{Z}\times\mathbb{Z}} f^- \, d(\gamma \times \gamma)$. No prizes for guessing the answer.

7.1.P14. Let $(X, \mathcal{F}, \mu)$ and $(Y, \mathcal{G}, \nu)$ be measure spaces, $A \subseteq X$ and $B \subseteq Y$. If $A \times B \in \mathcal{F} \times \mathcal{G}$ show that, if $A \neq \varnothing$, then B is measurable and if $B \neq \varnothing$, then A is measurable. Hence show that $A \times B \notin \mathcal{F} \times \mathcal{G}$ if and only if $A \neq \varnothing \neq B$ and either A or B is nonmeasurable.

Hint: Assume $A \times B \in \mathcal{F} \times \mathcal{G}$. If $A \neq \varnothing$, choose any $x \in A$. Then $(A \times B)_x = B$ and must be measurable by Proposition 7.1.12. Similarly when $B \neq \varnothing$.

Suppose $A \times B \notin \mathcal{F} \times \mathcal{G}$. If either A or B is $\varnothing$, then $A \times B = \varnothing$, which is ruled out by the supposition. So, $A \neq \varnothing \neq B$. If A and B were both to be measurable, then by definition of $\mathcal{F} \times \mathcal{G}$ we would have $A \times B \in \mathcal{F} \times \mathcal{G}$, contrary to the supposition. So, either A or B is nonmeasurable.

Now suppose conversely that $A \neq \varnothing \neq B$ and either A or B is nonmeasurable. By the contrapositive of what has been proved in the first paragraph, $A \times B \notin \mathcal{F} \times \mathcal{G}$.

7.1.P15. [Cf. Problem 3.2.P13(d)] Let $(X, \mathcal{F}, \mu)$ be a measure space. Verify for every real-valued measurable function f that

$$\int_X |f|^p \, d\mu = \int_{(0,\,\infty)} pt^{p-1} \mu(\{|f| > t\}) dt, \quad 0<p<\infty.$$

Hint: We apply Tonelli's Theorem 7.1.23 to the function g given on $X \times (0, \infty)$ by $g(x, t) = pt^{p-1}\chi_A(x, t)$, where $A = \{(x, t) \in X \times (0, \infty): 0 < t < |f(x)|\}$ after noting that A is measurable (considering that $|f(x)| - t$ defines a measurable function on $X \times (0, \infty)$). By doing so, we get

$$\int_X d\mu \int_{(0,|f(x)|)} pt^{p-1} dt = \int_{(0,\infty)} pt^{p-1} dt \int_{\{|f(x)| > t\}} d\mu.$$

Observe that

$$|f(x)|^p = \int_{(0,|f(x)|)} tp^{p-1} dt.$$

Integrating, we get

$$\begin{aligned}\int_X |f|^p\, d\mu &= \int_X d\mu \int_{(0,|f(x)|)} tp^{p-1} dt \\ &= \int_{(0,\infty)} pt^{p-1} dt \int_{\{|f(x)|>t\}} d\mu \\ &= \int_{(0,\infty)} pt^{p-1} \mu(\{|f(x)|>t\}) dt.\end{aligned}$$

Problem Set 7.2

7.2.P1. Show that if μ is not complete, then f measurable and $f = g$ a.e. do not imply that g is measurable.
Hint: Let $X = \{a, b, c\}$, $\mathcal{F} = \{\varnothing, \{a\}, \{b, c\}, X\}$ and $\mu(\{a\}) = 1 = \mu(X)$, $\mu(\varnothing) = \mu(\{b, c\}) = 0$. Define f on X by the rule $f(a) = f(b) = f(c) = 3$. Then f is measurable. The function g defined by $g(a) = 3$, $g(b) = 2$, $g(c) = 1$ agrees with f except on the set $\{b, c\}$ and $\mu(\{b, c\}) = 0$. However, $X(g < 2) = \{c\}$, which is not measurable. Thus g is not measurable.
Alternative Hint: Let μ denote the Borel measure on $[0, 1]$ and f a Borel measurable function on $[0, 1]$. Let G denote the Cantor set and $E \subseteq G$ be a measurable set of measure zero, which is not a Borel set. As noted in the Problem 2.3.P20 such a set E exists. Define $g = \chi_E + \chi_G + f$. Then $f = g$ a.e. In fact, $\{x \in X: f(x) \neq g(x)\} = G$. But g is not Borel measurable. For otherwise, $\chi_E = g - f - \chi_G$ would be Borel measurable, which is false.
7.2.P2. Suppose $\mathcal{F}$ and $\mathcal{G}$ are σ-algebras of subsets of X with measures μ and ν respectively. If $\mathcal{F} \subseteq \mathcal{G}$ and $\mu = \nu|_{\mathcal{F}}$, show that $\overline{\mathcal{F}} \subseteq \overline{\mathcal{G}}$.
Hint: Any set in the completion $\overline{\mathcal{F}}$ is of the form $E \cup F$, where $E \in \mathcal{F}$ and $F \subseteq N \in \mathcal{F}$, $\mu(N) = 0$. Since $\mathcal{F} \subseteq \mathcal{G}$, this implies $E \in \mathcal{G}$ and $F \subseteq N \in \mathcal{G}$, $\nu(N) = \nu|_{\mathcal{F}}(N) = \mu(N) = 0$. So, the set belongs to $\overline{\mathcal{G}}$.
7.2.P3. Suppose $\mathcal{F}$ and $\mathcal{G}$ are σ-algebras of subsets of X with measures μ and ν respectively. If $\mathcal{F} \subseteq \mathcal{G} \subseteq \overline{\mathcal{F}}$ and $\nu = \bar{\mu}|_{\mathcal{G}}$, show that $\overline{\mathcal{G}} = \overline{\mathcal{F}}$ and $\bar{\nu} = \bar{\mu}$.
Hint: By Problem 7.2.P2, $\overline{\mathcal{G}} \subseteq \overline{\overline{\mathcal{F}}}$, whereas $\overline{(\overline{\mathcal{F}})} = \overline{\mathcal{F}}$ because $\overline{\mathcal{F}}$ is already complete. Thus, $\overline{\mathcal{G}} \subseteq \overline{\mathcal{F}}$. We shall argue that $\overline{\mathcal{F}} \subseteq \overline{\mathcal{G}}$. In the light of Problem 7.2.P2, it is sufficient to know that $\mu = \nu|_{\mathcal{F}}$. For any $E \in \mathcal{F}$, we have $\mu(E) = \bar{\mu}(E)$ by Theorem 7.2.2; but we also have $E \in \mathcal{G}$ because $\mathcal{F} \subseteq \mathcal{G}$ and hence $\bar{\mu}(E) = \bar{\mu}|_{\mathcal{G}}(E) = \nu(E)$ because $\nu = \bar{\mu}|_{\mathcal{G}}$. Thus, any $E \in \mathcal{F}$ satisfies $\mu(E) = \nu(E)$, which means $\mu = \nu|I_{\mathcal{F}}$. This completes the proof that $\overline{\mathcal{G}} = \overline{\mathcal{F}}$.
It now follows that $\bar{\mu}$ is a measure on $\overline{\mathcal{G}}$ while the equality $\nu = \bar{\mu}|_{\mathcal{G}}$ holds by hypothesis. In other words, $\bar{\mu}$ is an extension of ν to the completion $\overline{\mathcal{G}}$. The uniqueness part of Theorem 7.2.2 therefore leads to the equality $\bar{\nu} = \bar{\mu}$.
7.2.P4. [Application to Fourier Series] Prove the following: Let f be a function in $L^p[-\pi, \pi]$, where $1 \leq p < \infty$. Then the Cesàro means of the Fourier series for f converge to f in the L^p norm.

Hint: All integrals are intended to be Lebesgue integrals even though they have been written like Riemann integrals.
We extend f to the real line so as to be periodic. The symbols σ_n and F_n will have the meanings as in Remark 4.3.2.
Since the Fourier series of any measurable function that is a.e. equal to f is the same as that of f, we may replace f by any such function. In effect, we may assume f to have any property that some measurable function that is a.e. equal to f has. In particular, we may assume that f is real-valued and, by Problem 2.5.P4, that it is Borel measurable. Now, the map $(x, t) \to x - t$ from $\mathbb{R}\times \mathbb{R}$ to $\mathbb{R}$ is continuous and hence Borel measurable. It follows from Problem 3.1.P11 that the composition $(x, t) \to f(x - t)$ from $\mathbb{R}\times \mathbb{R}$ to $\mathbb{R}$ is Borel measurable. Using Theorem 7.2.16, we find that the composition in question is also measurable with respect to the product of the σ-algebras of Lebesgue measurable subsets of $\mathbb{R}$. This justifies applying Tonelli's Theorem 7.1.23 to the function on $\mathbb{R}\times \mathbb{R}$ given by $(x, t) \to |[f(x - t) - f(x)]g(x)|$, where $g:\mathbb{R}\to\mathbb{R}$ is Lebesgue measurable. (It is trivial that $(x, t) \to g(x)$ from $\mathbb{R}\times \mathbb{R}$ to $\mathbb{R}$ is measurable with respect to the product of the σ-algebras of Lebesgue measurable subsets of $\mathbb{R}$.)
For $f \in L^p$, we wish to estimate

$$\|\sigma_n - f\|_p.$$

Then by a simple change of variables,

$$\sigma_n(x) = \frac{1}{\pi}\int_{-\pi}^{\pi} f(x-t)F_n(t)dt,$$

where

$$F_n(x) = \frac{\sin^2 \frac{nx}{2}}{2n\sin^2 \frac{x}{2}}.$$

Let g be any function in L^q, where p and q are conjugate exponents, such that $\|g\|_q = 1$ and which, for $p = 1$, is the function identically 1. We may assume that g is real-valued. By applying Tonelli's Theorem 7.1.23 to the function $(x, t) \to |[f(x - t) - f(x)]g(x)|$, we obtain

$$\int_{-\pi}^{\pi}\int_{-\pi}^{\pi}|[f(x-t)-f(x)]g(x)|d(x,t) = \int_{-\pi}^{\pi}\int_{-\pi}^{\pi}|[f(x-t)-f(x)]g(x)|dx\, F_n(t)dt$$
$$= \int_{-\pi}^{\pi}\int_{-\pi}^{\pi}|[f_t(x)-f(x)]g(x)|dxF_n(t)dt, \text{ where } f_t(x)=f(x-t)$$
$$\leq \int_{-\pi}^{\pi}\|f_t-f\|_p\|g\|_q F_n(t)dt \text{ by Hölder's Inequality}$$
$$\leq \int_{-\pi}^{\pi}\left(\|f_t\|_p+\|f\|_p\right)\|g\|_q F_n(t)dt$$
$$\leq \int_{-\pi}^{\pi} 2\|f\|_p\|g\|_q F_n(t)dt$$
$$= 2\|f\|_p\|g\|_q\int_{-\pi}^{\pi}F_n(t)dt$$
$$= 2\pi\|f\|_p \quad \text{by Prop.4.3.3(d)}$$
$$<\infty.$$

In view of the finiteness just proved, Fubini's Theorem 7.1.25 can be applied to the function $(x, t) \to [f(x-t) - f(x)]g(x)$ on $[-\pi, \pi] \times [-\pi, \pi]$. This leads, in conjunction with Proposition 4.3.3(d), to

$$\frac{1}{2\pi}\int_{-\pi}^{\pi}[\sigma_n(x)-f(x)]g(x)\ dx = \frac{1}{2\pi^2}\int_{-\pi}^{\pi}\int_{-\pi}^{\pi}[f(x-t)-f(x)]g(x)dx\ F_n(t)dt$$

and thus

$$\left|\frac{1}{2\pi}\int_{-\pi}^{\pi}[\sigma_n(x)-f(x)]g(x)dx\right| \leq \frac{1}{\pi}\int_{-\pi}^{\pi}\left|\frac{1}{2\pi}\int_{-\pi}^{\pi}[f(x-t)-f(x)]g(x)dx\right|F_n(t)dt.$$

Using Hölder's Inequality, we find that the inside integral is no larger than

$$\|g\|_q\|f_t-f\|_p.$$

Consequently,

$$\left|\frac{1}{2\pi}\int_{-\pi}^{\pi}[\sigma_n(x)-f(x)]g(x)dx\right| \leq \frac{1}{2\pi}\|g\|_q\,\frac{1}{\pi}\int_{-\pi}^{\pi}\|f_t-f\|_p F_n(t)dt$$

for every $g \in L^q$ with $\|g\|_q = 1$. By Proposition 6.1.3, it follows that

$$\begin{aligned} &\|\sigma_n - f\|_p \leq \tfrac{1}{\pi}\int_{-\pi}^{\pi} \|f_t - f\|_p F_n(t)dt \\ &= \tfrac{1}{\pi}\int_{|t|\leq\delta} \|f_t - f\|_p F_n(t)dt + \tfrac{1}{\pi}\int_{|t|>\delta} \|f_t - f\|_p F_n(t)dt \\ &\leq \sup\left\{\|f_t - f\|_p : |t| \leq \delta\right\} \tfrac{1}{\pi}\int_{|t|\leq\delta} F_n(t)dt + 2\|f\|_p \tfrac{1}{\pi}\int_{|t|>\delta} F_n(t)dt. \end{aligned} \tag{8.73}$$

By Proposition 6.1.6, the sup in (8.73) is arbitrarily small if δ is sufficiently small. Using part (f) of Proposition 4.3.3, we find that the limit of the last expression as $n \to \infty$ is zero, no matter how small δ is. Therefore $\lim\limits_{n\to\infty} \|\sigma_n - f\|_p = 0$.

References

1. Apostol, T.M.: Mathematical Analysis. Narosa Publishing House, New Delhi (1974)
2. Bartle, R.G.: The Elements of Integration. Wiley, New York (1966)
3. Berberian, S.K.: Measure and Integration. Macmillan & Co, New York (1965)
4. Berberian, S.K.: A First Course in Real Analysis. Springer, New York (1974)
5. Boas, R.P.: Differentiability of jump functions. Colloq. Math. **8**, 81–82 (1961)
6. Depree, J., Swartz, C.W.: Introduction to Real Analysis. Wiley, New York (1988)
7. Davis, M., Insall, M.: Mathematics and design, yes, but will it fly? Nexus Netw. J. **4**(2), 9–13. https://www.nexusjournal.com/volume-4/number-2-november-2002.html (2002). Accessed 06 June 2018
8. Goldberg, R.: Methods of Real Analysis. Blaisdell Pub Co, Waltham, Mass (1964)
9. Grabisch, M., Murofushi, T., Sugeno, M., Kacprzyk, J.: Fuzzy Measures and Integrals: Theory and Applications. Physica Verlag, Berlin (2000)
10. Halmos, P.R.: Naïve Set Theory. Affiliated East West Press Pvt Ltd, New Delhi (1972)
11. Halmos, P.R.: Measure Theory. D van Nostrand Co, Princeton, NJ (1950)
12. Hewitt, E., Stromberg, K.: Real and Abstract Analysis. Springer, New York (1965)
13. Modave, F., Grabisch, M.: Preference representation by the Choquet integral: The commensurability hypothesis. https://www.researchgate.net/profile/Francois_Modave/publication/228544097_Preference_representation_by_the_Choquet_integral_The_commensurability_hypothesis/links/544103760cf2a6a049a3f425.pdf. Accessed 06 June 2018
14. Modave, F., Klir, G.J.: Fuzzy Measure Theory. Plenum Press, New York (1992)
15. Pugh, C.C.: Real Mathematical Analysis, 2nd ed. Springer, New York (1991)
16. Rana, I.K.: Introduction to Measure and Integration. Narosa Publishing House, New Delhi (1997)
17. Riesz, F., Sz-Nagy, B.: Functional Analysis. Frederick Ungar Publishing Co, New York (1955)
18. Rubel, L.A.: Differentiability of monotonic functions. Colloq. Math. **10**, 277–279 (1963)
19. Rudin, W.: Principles of Mathematical Analysis, International edn. McGraw-Hill Book Co, Singapore (1987)
20. Rudin, W.: Real and Complex Analysis. McGraw-Hill Book Company, New York (1966)
21. Sz-Nagy, B.: Introduction to Real Functions and Orthogonal Expansions. Oxford University Press, New York (1965)
22. Schechter, E.: An Introduction to the Gauge Integral. https://math.vanderbilt.edu/schectex/ccc/gauge (2009). Accessed 06 June 2018
23. Schmeidler, D.: Integral representations without additivity. Proc. Amer. Math. Soc. **97**, 255–261 (1986)

S. Shirali and H. L. Vasudeva, *Measure and Integration*,
Springer Undergraduate Mathematics Series,
https://doi.org/10.1007/978-3-030-18747-7

24. Serrin, J., Varberg, D.E.: A general chain rule for derivatives and the change of variables formula for the Lebesgue integral. Amer. Math. Mon. **76**, 514–520 (1969)
25. McShane, E.J.: Integration. Princeton University Press, Princeton (1947)
26. Shirali, S., Vadsudeva, H.L.: Metric Spaces. Springer, London (2006)
27. Shirali, S., Vadsudeva, H.L.: Mathematical Analysis. Narosa Publishing House, New Delhi (2006)
28. Shirali, S., Vadsudeva, H.L.: An Introduction to Mathematical Analysis. Narosa Publishing House, New Delhi (2014)
29. Taylor, A.E.: General Theory of Functions and Integration. Blaisdell Publishing Co, New York (1965)
30. Ulam, S.H.: Zur Masstheorie in allgemeinen Megenlehre. Fund. Math. **16**, 141–150 (1930)
31. Wang, Z.: Convergence theorems for sequences of Choquet integrals. Int. J. Gen. Syst. **26**, 133–143 (1997)
32. Wang, Z., Klir, G.J.: Fuzzy Measure Theory. Plenum Press, New York (1992)
33. Wang, Z., Leung, K.S., Wang, M.L., Fang, J., Xu, K.: Nonlinear nonnegative multiregression based on Choquet integrals. Int. J. Approx. Reason. **25**, 71–87 (2000)
34. Zygmund, A.: Trigonometric Series. Cambridge University Press (1959)

Index

S. Shirali and H. L. Vasudeva, *Measure and Integration*,
Springer Undergraduate Mathematics Series,
https://doi.org/10.1007/978-3-030-18747-7

Printed in the United States
By Bookmasters